Introductory Algebra
with Basic Mathematics

Second Edition

Introductory Algebra

with Basic Mathematics

Second Edition

Richard N. Aufmann
Palomar College, California

Vernon C. Barker
Palomar College, California

Joanne S. Lockwood
Plymouth State College, New Hampshire

HOUGHTON MIFFLIN COMPANY Boston Toronto
Geneva, Illinois Palo Alto Princeton, New Jersey

Cover Design: Harold Burch, Harold Burch Design, New York City
Cover Image: Eiji Yanagi/PHOTONICA

Senior Sponsoring Editor: *Maureen O'Connor*
Associate Editor: *Dawn Nuttall*
Senior Project Editor: *Maria A. Morelli*
Editorial Assistant: *Christina M. Lillios*
Senior Production/Design Coordinator: *Sarah Ambrose*
Senior Manufacturing Coordinator: *Priscilla Bailey*

Printed in the U.S.A.

ISBN: 0-395-74621-3

6789-BBS-03 02

CONTENTS

4 *Decimals and Real Numbers* 125

5 *Variable Expressions* 179

PREFACE

The second edition of *Introductory Algebra with Basic Mathematics* is designed for students who need to review basic concepts in mathematics before studying beginning algebra. The first four chapters of the text review integer and rational number concepts. Variables are introduced in Chapter 1 and used in conjunction with the review of number concepts. In this new edition of *Introductory Algebra with Basic Mathematics,* careful attention has been given to implementing the standards suggested by NCTM and AMATYC. Each chapter begins with a Focus on Problem Solving that introduces students to various successful problem-solving strategies. At the end of each section are Critical Thinking Exercises that include writing, synthesis, and challenge problems. Each chapter ends with Projects in Mathematics. The Projects in Mathematics feature is an extension or application of a concept covered in the chapter. These projects can be used for cooperative learning activities or extra credit.

Instructional Features

Interactive Approach

Introductory Algebra with Basic Mathematics uses an interactive style that provides a student with the opportunity to try a skill as it is presented. Each section is divided into objectives, and every objective contains one or more sets of matched-pair examples. The first example in each pair is worked out; the second example, labeled You Try It, is for the student to work. By solving this problem, the student practices concepts as they are presented in the text. There are *complete* worked-out solutions to these problems at the end of the book. By comparing their solutions to model solutions in the appendix, students are able to obtain immediate feedback on and reinforcement of the concepts.

Emphasis on Problem-Solving Strategies

Introductory Algebra with Basic Mathematics features a carefully sequenced approach to application problems that emphasizes proven strategies to solve problems. Students are encouraged to develop their own strategies, draw diagrams, and to write their strategies as part of the solution to each application problem. In each case, model strategies are presented as guides for students to follow as they attempt the matched-pair You Try It. Having students provide strategies is a natural way to incorporate writing into the math curriculum.

Emphasis on Applications

The traditional approach to teaching algebra covers only the straightforward manipulation of numbers and variables and thereby fails to teach students the practical value of algebra. By contrast, *Introductory Algebra with Basic Mathematics* contains an extensive collection of contemporary application problems. Wherever appropriate, the last objective of a section presents applications that require the student to use the skills covered in that section to solve practical problems. This carefully integrated applied approach generates the student's awareness of algebra as a real-life tool.

Integrated Learning System Organized by Objectives

Each chapter begins with a list of the learning objectives included within that chapter. Each objective is then restated in the chapter to remind the student of the current topic of discussion. The same objectives that organize the text are used as the structure for exercises, the testing programs, and the Computer Tutor. For every objective in the text there is a corresponding computer tutorial and a corresponding set of test questions associated with that objective.

The Interactive Approach

Instructors have long realized the need for a text that requires the student to use a skill as it is being taught. *Introductory Algebra with Basic Mathematics* uses an interactive technique that meets this need. Every objective, including the one shown below, contains at least one pair of examples. One of the examples is worked. The second example in the pair (You Try It) is not worked so that the student may interact with the text by solving a similar problem. To provide immediate feedback, a complete solution to this problem is provided at the back of the book. The benefit of this interactive style is that the student can check that a new skill has been learned before attempting a homework assignment.

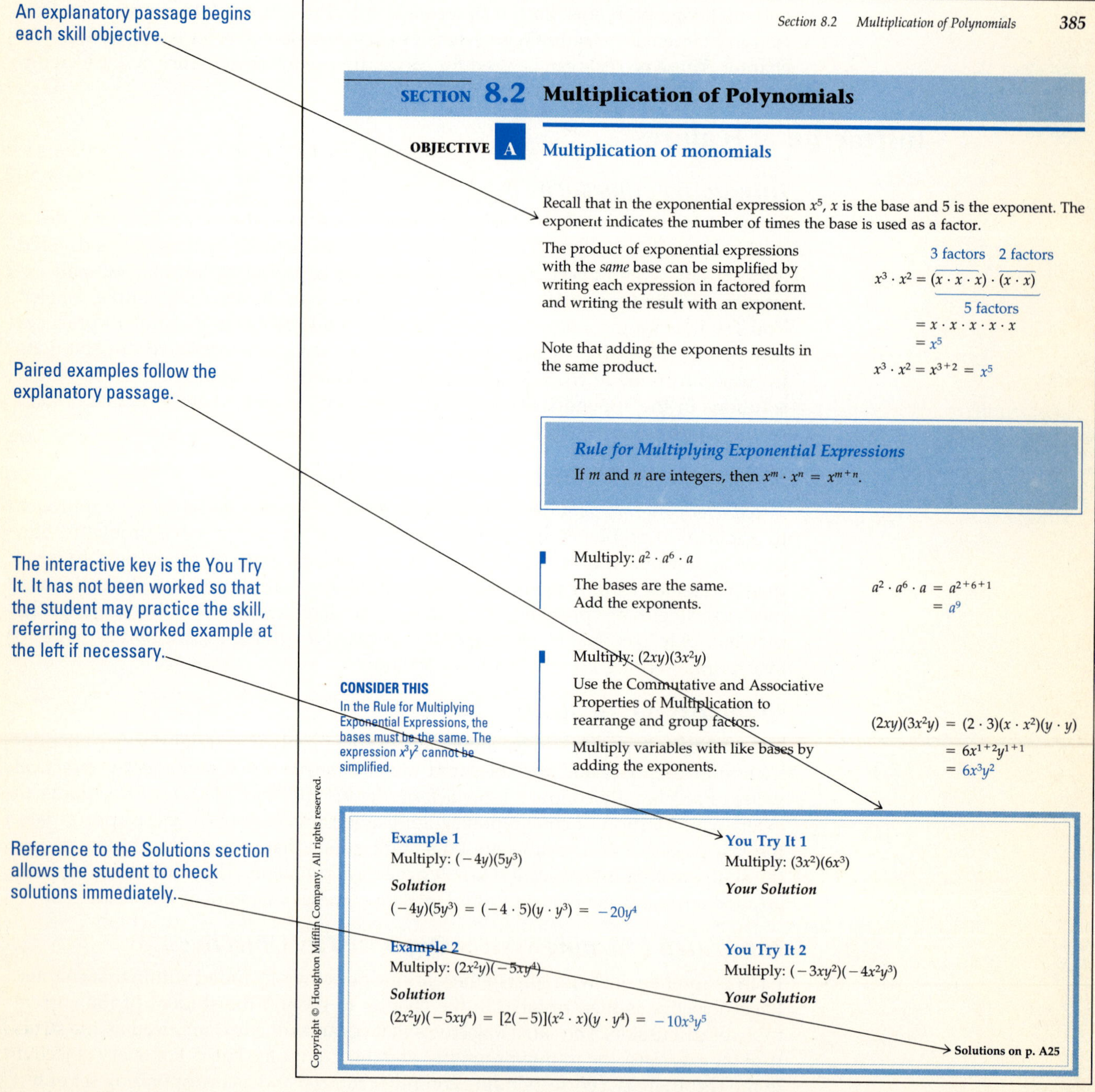

Section 8.2 Multiplication of Polynomials 385

SECTION 8.2 Multiplication of Polynomials

OBJECTIVE A Multiplication of monomials

Recall that in the exponential expression x^5, x is the base and 5 is the exponent. The exponent indicates the number of times the base is used as a factor.

The product of exponential expressions with the *same* base can be simplified by writing each expression in factored form and writing the result with an exponent.

$$x^3 \cdot x^2 = \underbrace{\overbrace{(x \cdot x \cdot x)}^{3\text{ factors}} \cdot \overbrace{(x \cdot x)}^{2\text{ factors}}}_{5\text{ factors}} = x \cdot x \cdot x \cdot x \cdot x = x^5$$

Note that adding the exponents results in the same product.

$$x^3 \cdot x^2 = x^{3+2} = x^5$$

Rule for Multiplying Exponential Expressions
If m and n are integers, then $x^m \cdot x^n = x^{m+n}$.

Multiply: $a^2 \cdot a^6 \cdot a$

The bases are the same. Add the exponents.

$$a^2 \cdot a^6 \cdot a = a^{2+6+1} = a^9$$

Multiply: $(2xy)(3x^2y)$

Use the Commutative and Associative Properties of Multiplication to rearrange and group factors.

Multiply variables with like bases by adding the exponents.

$$(2xy)(3x^2y) = (2 \cdot 3)(x \cdot x^2)(y \cdot y) = 6x^{1+2}y^{1+1} = 6x^3y^2$$

CONSIDER THIS
In the Rule for Multiplying Exponential Expressions, the bases must be the same. The expression x^3y^2 cannot be simplified.

Example 1
Multiply: $(-4y)(5y^3)$

Solution
$(-4y)(5y^3) = (-4 \cdot 5)(y \cdot y^3) = -20y^4$

You Try It 1
Multiply: $(3x^2)(6x^3)$

Your Solution

Example 2
Multiply: $(2x^2y)(-5xy^4)$

Solution
$(2x^2y)(-5xy^4) = [2(-5)](x^2 \cdot x)(y \cdot y^4) = -10x^3y^5$

You Try It 2
Multiply: $(-3xy^2)(-4x^2y^3)$

Your Solution

Solutions on p. A25

Emphasis on Applications

The solution of an application problem in *Introductory Algebra with Basic Mathematics* is presented in two parts: Strategy and Solution. The strategy is a written description of the steps that are necessary to solve the problem; the solution is the implementation of the strategy. Using this format provides students with a structure for problem solving. It also encourages students to write strategies for solving problems which, in turn, fosters organizing problem-solving strategies in a logical way. Having students write strategies is a natural way to incorporate writing into the math curriculum.

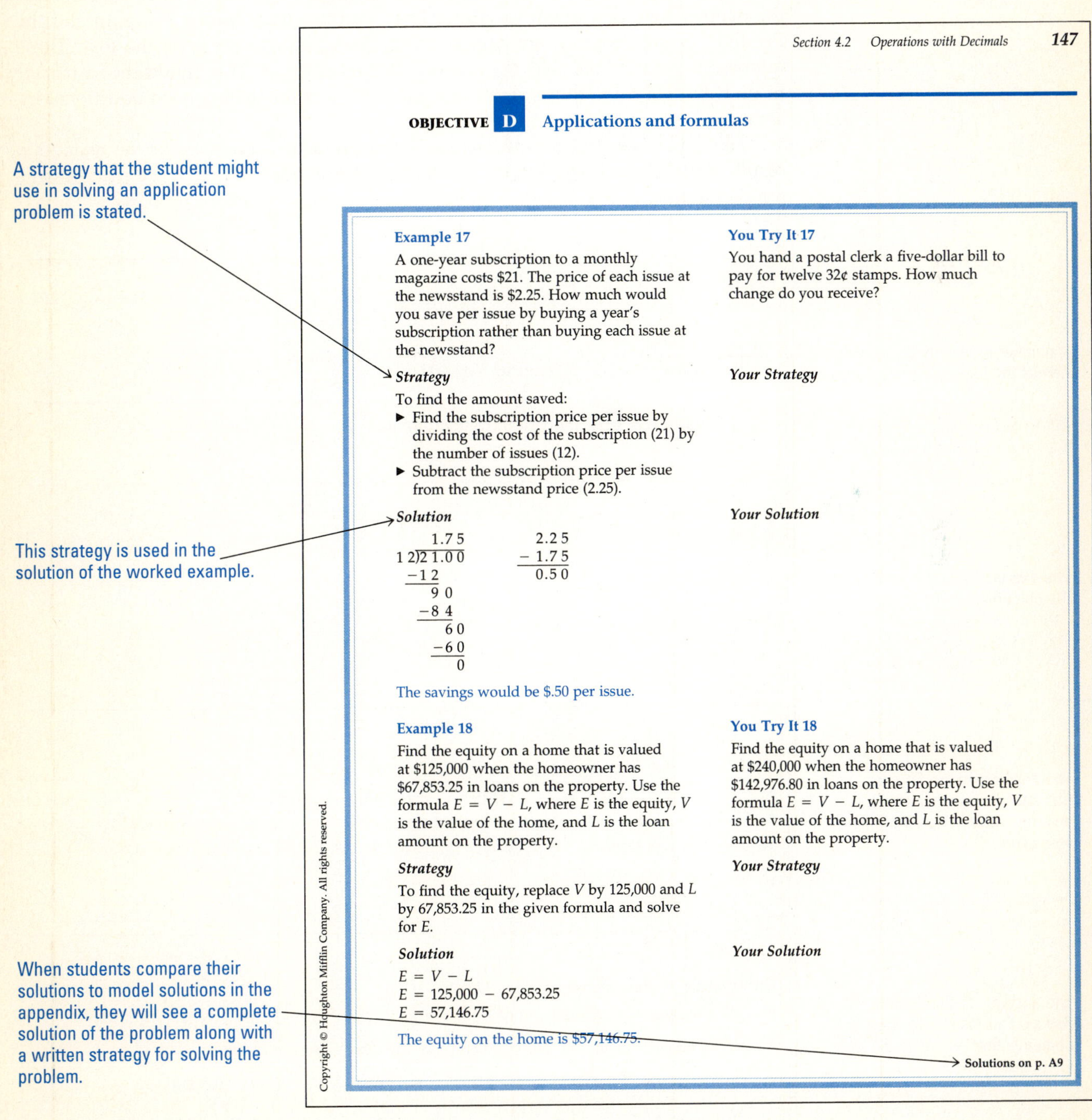

Section 4.2 Operations with Decimals 147

OBJECTIVE D Applications and formulas

Example 17

A one-year subscription to a monthly magazine costs \$21. The price of each issue at the newsstand is \$2.25. How much would you save per issue by buying a year's subscription rather than buying each issue at the newsstand?

Strategy

To find the amount saved:

- Find the subscription price per issue by dividing the cost of the subscription (21) by the number of issues (12).
- Subtract the subscription price per issue from the newsstand price (2.25).

Solution

$$12\overline{)21.00} = 1.75;\quad -12,\ 90,\ -84,\ 60,\ -60,\ 0$$

$$2.25 - 1.75 = 0.50$$

The savings would be \$.50 per issue.

You Try It 17

You hand a postal clerk a five-dollar bill to pay for twelve 32¢ stamps. How much change do you receive?

Your Strategy

Your Solution

Example 18

Find the equity on a home that is valued at \$125,000 when the homeowner has \$67,853.25 in loans on the property. Use the formula $E = V - L$, where E is the equity, V is the value of the home, and L is the loan amount on the property.

Strategy

To find the equity, replace V by 125,000 and L by 67,853.25 in the given formula and solve for E.

Solution

$E = V - L$
$E = 125{,}000 - 67{,}853.25$
$E = 57{,}146.75$

The equity on the home is \$57,146.75.

You Try It 18

Find the equity on a home that is valued at \$240,000 when the homeowner has \$142,976.80 in loans on the property. Use the formula $E = V - L$, where E is the equity, V is the value of the home, and L is the loan amount on the property.

Your Strategy

Your Solution

Solutions on p. A9

Objective-Specific Approach

Many mathematics textbooks are not organized in a manner that facilitates management of learning. Typically, students are left to wander through a maze of apparently unrelated lessons, exercise sets, and tests. *Introductory Algebra with Basic Mathematics* solves this problem by organizing all lessons, exercise sets, computer tutorials, and tests around a carefully constructed hierarchy of objectives. The advantage of this objective-by-objective organization is that it enables the student who is uncertain at any step in the learning process to refer easily to the original presentation and review that material.

The objective-specific approach also allows the instructor greater control over the management of student progress. The Computerized Testing Program and the Printed Testing Program are organized by the same objectives as the text. These references are provided with the answers to the test items. This allows the instructor to quickly determine those objectives on which a student may need additional instruction.

The Computer Tutor is also organized around the objectives of the text. As a result, supplemental instruction is available for any objectives that are troublesome for a student.

A numbered objective statement names the topic of each lesson.

Section 6.3 General Equations 263

SECTION 6.3 General Equations

OBJECTIVE A Equations of the form $ax + b = c$

The exercise sets correspond to the objectives in the text.

6.3 EXERCISES

Objective A

Solve.

1. $5y + 1 = 11$ 2. $3x + 5 = 26$ 3. $2z - 9 = 11$

The answers to the Chapter Review Exercises show the objective from which a problem was taken.

Chapter Review Exercises *page 317*

1. -4 [6.1C] 2. $\frac{3}{2}$ [6.3A] 3. $x > 2$ (number line −5 to 5) [6.6A] 4. 44 cm [6.4A] 5. 3 [6.3C] 6. 5.625% [6.2A] 7. $x \geq -4$ [6.6B] 8. -3 [6.1B] 9. $\angle x = 22°$, $\angle y = 158°$ [6.4C] 10. $\angle a = 138°$, $\angle b = 42°$ [6.4B] 11. $\frac{1}{2}$ [6.3B] 12. Yes [6.1A] 13. 562.50 [6.2A] 14. $x \geq -3$ [6.6A]

The answers to the Cumulative Review Exercises also show the objective that relates to the exercise.

Cumulative Review Exercises *page 378*

1. $-17x + 28$ [5.1D] 2. $\frac{1}{15}$ [4.3B] 3. $\frac{3}{2}$ [6.3A] 4. -12 [2.4A] 5. $-\frac{5}{8}$ [2.4A] 6. $x < -20$ [6.6A] 7. $\angle a = 43°$; $\angle b = 137°$ [6.4B] 8. $\angle x = 29°$ [6.4A] 9. $6y^4 + 8y^3 - 16y^2$ [5.1C] 10. $2\frac{3}{11}$ [3.3C] 11. -6.8 [2.1C] 12. $\frac{5}{9} > 0.5$ [4.1B] 13. 7 [2.2B] 14. $\frac{7}{4}$ [6.3B] 15. $\frac{2}{3}$ [6.1C] 16. $15x^2$ [3.2C]

Additional Learning Aids

Focus on Problem Solving

The Focus on Problem Solving feature introduces students to various proven problem-solving strategies. Some of the topics discussed include inductive reasoning, deductive reasoning, finding counterexamples, and the calculator as a problem-solving tool.

Projects in Mathematics

The Projects in Mathematics feature occurs at the end of each chapter. These projects can be used as extra credit or cooperative learning activities. The projects cover various aspects of mathematics including the field of geometry, statistics, science, and business.

Chapter Summaries

The Chapter Summaries are a useful guide to students as they review for a test. The Chapter Summary includes the Key Words and Essential Rules that were covered in the chapter.

Study Skills

To the Student provides suggestions for using this text and approaches to creating good study habits.

Computer Tutor

The Computer Tutor is an interactive computer tutorial that covers *every* objective in the text.

Glossary

A glossary at the back of the book includes definitions of terms used in the text.

Exercises

End-of-Section Exercises

Introductory Algebra with Basic Mathematics contains more than 5000 exercises. At the end of each section are exercise sets keyed to the corresponding learning objectives. The exercises are carefully developed to ensure that students can apply the concepts in the section to a variety of problem situations.

Critical Thinking Exercises

The end-of-section exercises are followed by Critical Thinking Exercises, which feature exercise types such as:

- challenge problems,
- writing exercises,
- problems that ask students to determine incorrect procedures, and
- problems that require a more in-depth analysis.

Writing Exercises

Within the Critical Thinking Exercises are writing exercises denoted by **[W]**. These exercises ask students to write about a topic in the section or to research and report on a related topic. There are also writing exercises in some of the application problems. These exercises ask students to write a sentence that describes the meaning of their answer in the context of the problem.

Chapter Review Exercises

Review Exercises are found at the end of each chapter, and are selected to help the student integrate all of the topics presented in the chapter. The answers to all review exercises are given in Answers to Selected Exercises at the back of the book. Along with the answer is a reference to the objective that pertains to each exercise.

Cumulative Review Exercises

Cumulative Review Exercises, which appear at the end of each chapter beginning with Chapter 2, help the student maintain skills learned in previous chapters. The answers to all cumulative review exercises are given in the Answers to Selected Exercises. Along with the answers is a reference to the objective that pertains to each exercise.

New to this Edition

Topical Coverage

Introductory Algebra with Basic Mathematics retains its strong commitment to applications of mathematics. We have added many new and contemporary applications to this edition. The new organization also gives more prominence to problems from the field of geometry, although we have endeavored to provide a balanced approach to application problems that reoccurs consistently throughout the text.

We have added many new application problems to the text. Within some of these problems, we have incorporated short writing exercises. These exercises ask students to explain, in the context of the application, the significance of their answer.

Since the first edition of this text, we have emphasized to students that they should write the strategy they will use to solve a word problem. As we have used this in our classes, we have found that students will frequently not include a diagram to aid them in their attempt to solve a problem. With this edition, we have included "Draw a diagram" in those applications where it is appropriate.

Chapter 1 has been completely rewritten to integrate variables with arithmetic operations. As each arithmetic operation is introduced, verbal phrases for those operations are also discussed. This allows students to practice translation problems from the beginning of the text. Application problems that use formulas are given as applications of variables.

Integers and operations with integers are now presented in Chapter 2. Operations with rational numbers begins in Chapter 3, which is devoted entirely to fractions. Chapter 4 contains operations with decimals.

Although variables are used throughout the first four chapters, we have added a new chapter devoted to variable expressions and their properties. Geometry formulas are used in this chapter to relate variable expressions to solving real applications.

The chapter on proportions in the first edition has been integrated into Chapter 10, Algebraic Fractions. Solving percent application problems is integrated into Chapter 6, Solving Equations and Inequalities.

Graphing linear equations and inequalities is now covered earlier in the text (Chapter 7). By placing graphing in an earlier chapter, it is possible to explore application problems from not only an algebraic setting, but a graphical setting as well.

Chapter 12 of the first edition, Applied Geometry, is now incorporated throughout the text. This integrated approach to algebra and geometry is one of the ways in which we have attempted to implement the suggestions of the American Mathematical Association of Two-Year Colleges.

Margin notes are interspersed throughout the text. These notes are called *Con-*

sider This or *Point of Interest.* The *Consider This* note alerts students that a procedure may be particularly involved or reminds students that certain checks of their work should be performed. The *Point of Interest* note is an interesting sidelight of the topic being discussed.

Focus on Problem Solving

Although successful problem solvers use a variety of techniques, it has been well established that basic recurring strategies exist. Among these are trying to solve a related problem, trying to solve an easier problem, working backward, trial and error, and other techniques. The Focus on Problem Solving feature presents some of these methods in the context of a problem to be solved. Students are encouraged to apply these strategies on similar problems.

Projects in Mathematics

Through the Projects in Mathematics feature, some of the standards suggested by the NCTM can be implemented. These Projects offer an opportunity for students to explore several topics relating to geometry, statistics, science, and business.

Computer Tutor

The Computer Tutor has been completely revised. It is now an algorithmically based tutor that includes color and animation. The algorithmic feature provides an infinite number of practice problems for students to attempt. The algorithms have been carefully crafted to present a variety of problem types from easy to difficult. A complete solution to each problem is available.

An interactive feature of the Computer Tutor requires students to respond to questions about the topic in the current lesson. In this way, students can assess their understanding of concepts as they are presented. A Glossary can be accessed at any time so that students can look up words whose definitions they may have forgotten.

When the student completes a lesson, a printed report is available. This optional report gives the student's name, the objectives studied, the number of problems attempted, the number of problems answered correctly, the percent correct, and the time spent working on the exercises in that objective.

Supplements for the Instructor

Instructor's Resource Manual with Solutions Manual

The Instructor's Resource Manual includes suggestions for course sequencing and outlines for the answers to the writing exercises. The Solutions Manual contains worked-out solutions for all end-of-section exercises, Critical Thinking exercises, Chapter Review Exercises, and Cumulative Review Exercises.

Computerized Test Generator

The Computerized Test Generator is the first of three testing materials. The data base contains more than 2000 test items. These questions are unique to the Test Generator. The Test Generator is designed to provide an unlimited number of tests for each chapter, cumulative chapter tests, and a final exam. It is available for the IBM PC and compatible computers and the Macintosh. The IBM version provides new on-line testing and also contains algorithms, which produce an unlimited number of certain types of test questions.

Printed Test Bank with Chapter Tests

The Printed Test Bank, the second component of the testing material, is a printout of all items in the Computerized Test Generator. Instructors who do not have access to a computer can use the test bank to select items to include on a test being prepared by hand. Chapter Tests contain the printed testing program, which is the third source of testing material. Four printed tests, two free response and two multiple choice, are provided for each chapter.

Supplements for the Student

Student Solutions Manual

The Student Solutions Manual contains the complete solutions to all odd-numbered exercises in the text.

Computer Tutor

The Computer Tutor is an interactive instructional computer program for student use. Each objective of the text is supported by a tutorial in the Computer Tutor. These tutorials contain an interactive lesson which covers the material in the objective. Following the lesson are randomly generated exercises for the student to attempt. A record of the student's progress is available.

The Computer Tutor can be used in several ways: (1) to cover material the student missed because of an absence; (2) to reinforce instruction on a concept that the student has not yet mastered; (3) to review material in preparation for exams. This tutorial is available for the Macintosh and IBM PC and compatible computers running Windows.

Videotapes

The videotape series contains lessons that accompany *Introductory Algebra with Basic Mathematics.* These lessons follow the format and style of the text and are closely tied to specific sections of the text. Each videotape begins with an application and then the necessary mathematics needed to solve that application are presented. The tape closes with the solution of the application problem.

Acknowledgments

The authors would like to thank these reviewers for their valuable suggestions:

Dr. Frank Cerreto, *Richard Stockton College of New Jersey;* **Jay Hollowell,** *Commonwealth College, VA;* **Cindy A. Noftle; Jim Price,** *Tulsa Junior College, OK;* **Deana J. Richmond**

TO THE STUDENT

Many students feel that they will never understand math while others appear to do very well with little effort. Oftentimes what makes the difference is that successful students take an active role in the learning process.

Learning mathematics requires your *active* participation. Although doing homework is one way you can actively participate, it is not the only way. First, you must attend class regularly and become an active participant. Secondly, you must become actively involved with the textbook.

Introductory Algebra with Basic Mathematics was written and designed with you in mind as a participant. Here are some suggestions on how to use the features of this textbook.

There are 12 chapters in this text. Each chapter is divided into sections and each section is subdivided into learning objectives. Each learning objective is labeled with a letter from A–E.

First, read each objective statement carefully so you will understand the learning goal that is being presented. Next, read the objective material carefully, being sure to note each bold word. These words indicate important concepts that you should familiarize yourself with. Study each in-text example carefully, noting the techniques and strategies used to solve the example.

You will then come to the key learning feature of this text, the *boxed examples.* These examples have been designed to aid you in a very specific way. Notice that in each example box, the example on the left is completely worked out and the "You Try It" example on the right is not. *You* are expected to work the right-hand example (in the space provided) in order to immediately test your understanding of the material you have just studied.

You should study the worked-out example carefully by working through each step presented. This allows you to focus on each step and reinforces the technique for solving that type of problem. You can then use the worked-out example as a model for solving similar problems.

Solve the "You Try It" example using the problem-solving techniques that you have just studied. When you have completed your solution, check your work by turning to the page in the appendix where the complete solution can be found. The page number on which the solution appears is printed at the bottom of the example box in the right-hand corner. By checking your solution, you will know immediately whether or not you fully understand the skill just studied.

When you have completed studying an objective, do all of the exercises in the exercise set that correspond with that objective. The exercises will be labeled with the same letter as the objective. Algebra is a subject that needs to be learned in small sections and practiced continually in order to be mastered. Doing all of the exercises in each exercise set will help you master the problem-solving techniques necessary for success.

Once you have completed the exercises to an objective, you should check your answers to the odd-numbered exercises with those found in the back of the book.

After completing a chapter, read the Chapter Summary. This summary highlights the important topics covered in the chapter. Following the Chapter Summary are Chapter Review Exercises, and a Cumulative Review (beginning with Chapter 2). Doing the review exercises is an important way of testing your understanding of the chapter. The answer to each review exercise is in an appendix at the back of the book. Each answer is followed by a reference that tells which objective that exercise was taken from. For example, (4.2B) means Section 4.2, Objective B. After checking your answers, restudy any objective that you missed. It may be very helpful to retry some of the exercises for that objective to reinforce your problem-solving techniques.

The Cumulative Review allows you to refresh the skills you have learned in previous chapters. This is very important in mathematics. By consistently reviewing previous materials, you will retain the previous skills as you build new ones.

Remember, to be successful, attend class regularly; read the textbook carefully; actively participate in class; work with your textbook using the "You Try It" examples for immediate feedback and reinforcement of each skill; do all the homework assignments; review constantly; and work carefully.

INDEX OF APPLICATIONS

CHAPTER

1

Whole Numbers and Variable Expressions

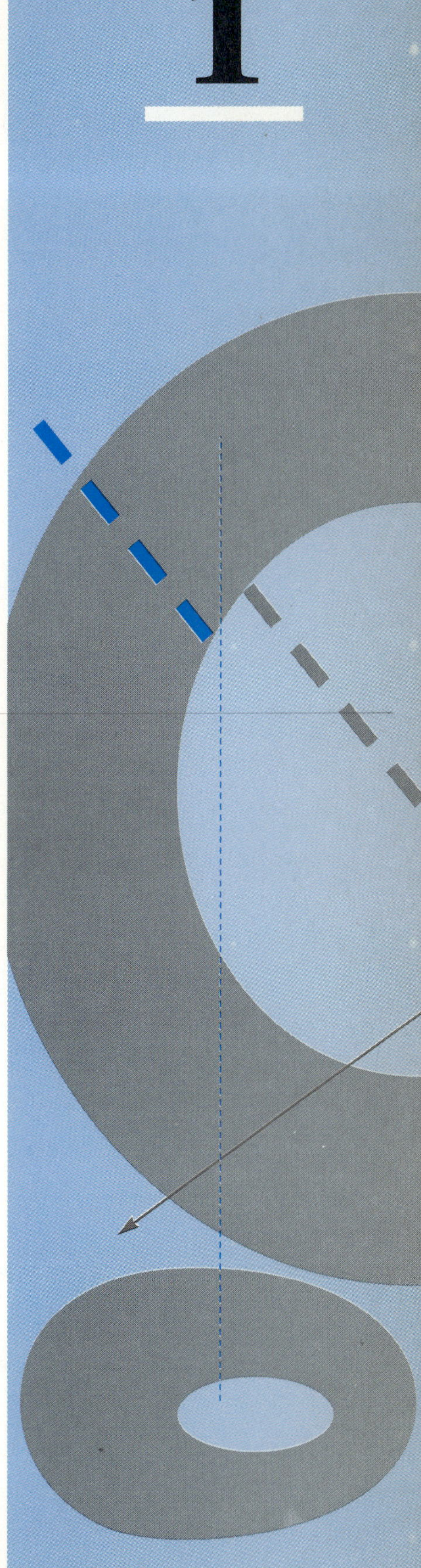

1.1 *Addition and Subtraction of Whole Numbers*

A Place value
B Addition of whole numbers
C Subtraction of whole numbers
D Applications and formulas

1.2 *Multiplication and Division of Whole Numbers*

A Multiplication of whole numbers
B Exponents
C Division of whole numbers
D Factors
E Applications and formulas

1.3 *Rounding and Estimation*

A Rounding
B Estimation

1.4 *Prime Factorization and the Order of Operations Agreement*

A Prime factorization
B The least common multiple (LCM)
C The greatest common factor (GCF)
D The Order of Operations Agreement

Focus on Problem Solving

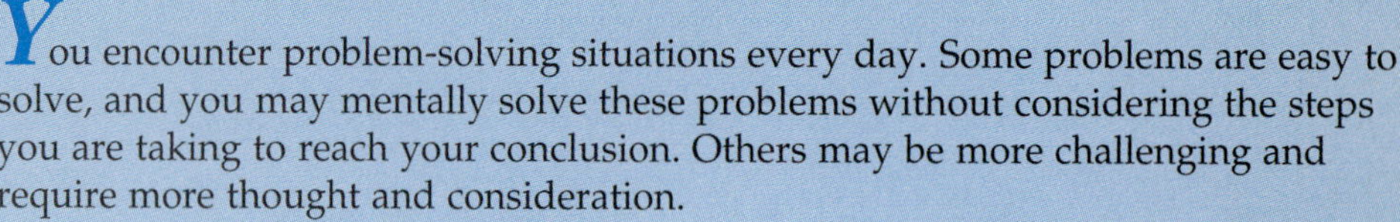

You encounter problem-solving situations every day. Some problems are easy to solve, and you may mentally solve these problems without considering the steps you are taking to reach your conclusion. Others may be more challenging and require more thought and consideration.

Suppose a friend suggests you both take a trip over spring break. You'd like to go. What questions go through your mind? You might ask yourself some of the following questions:

> How much will the trip cost? What will be the cost for travel, hotel rooms, meals, etc.?
> Are some costs going to be shared by both me and my friend?
> Can I afford it?
> How much money do I have in the bank?
> How much more money than I have now do I need?
> How much time is there to earn that much money?
> How much can I earn in that amount of time?
> How much money must I keep in the bank in order to pay the next tuition bill (or some other expense)?

These questions require different mathematical skills. Determining the cost of the trip requires *estimation*; for example, you must use your knowledge of air fares or the cost of gasoline to arrive at an estimate of these costs. If some of the costs are going to be shared, you need to *divide* those costs by 2 in order to determine your share of the expense. The question of how much more money you need requires *subtraction*: the amount you need minus the amount you have in the bank now. To determine how much money you can earn in the given amount of time requires *multiplication*—for example, the amount you earn per week times the number of weeks you will work. To determine if the amount you can earn in the given amount of time is sufficient, you need to use your knowledge of *order relations* to compare the amount you can earn with the amount you need.

Facing the problem-solving situation described above may not seem difficult to you. The reason may be that you have faced similar situations before and, therefore, you know how to work through this one. You may feel better prepared to deal with a circumstance such as this one because you know what questions to ask. An important aspect of learning to solve problems is learning what questions to ask. As you work through application problems in this text, try to become more conscious of the mental process you are going through. You might begin by asking yourself the following questions whenever you are solving an application problem:

1. Have I read the problem enough times to be able to understand the situation being described?
2. Will restating the problem in different words help me to understand the problem situation better?
3. What facts are given? (You might make a list of the information contained in the problem.)
4. What information is being asked for?
5. What relationship exists between the given facts? What relationship exists between the given facts and the solution?
6. What mathematical operations are needed in order to solve the problem?

Try to focus on the problem-solving situation and not on the computation or on getting the answer quickly. And remember, the more problems you solve, the better able you are to solve other problems in the future, partly because you are learning what questions to ask.

SECTION 1.1 Addition and Subtraction of Whole Numbers

OBJECTIVE A Place value

POINT OF INTEREST

The numerals that we use have not always appeared as they do today. In the second century, the symbols for 1 through 9 were

– = ≡ ¥ ⊢ б

1 2 3 4 5 6

5 6 7

7 8 9

By the eighth century, the symbols 2 and 3 were used instead of horizontal lines, but five looked like 4 and 9 was shown as ℇ.

The **natural numbers** are 1, 2, 3, 4, 5, 6, 7, 8, 9, 10, 11, . . .

The three dots mean that the list continues on and on and that there is no largest natural number. The natural numbers are also called the **counting numbers.**

The **whole numbers** are 0, 1, 2, 3, 4, 5, 6, 7, 8, 9, 10, 11, . . . Note that the whole numbers include the natural numbers and zero.

When a whole number is written using the digits 0, 1, 2, 3, 4, 5, 6, 7, 8, and 9, it is said to be in **standard form.** The position of each digit in the number determines the digit's **place value.** The diagram below shows a **place-value chart** naming the first twelve place values. The number 64,273 is in **standard form** and has been entered in the chart.

In the number 64,273, the position of the digit 6 determines that its place value is ten-thousands.

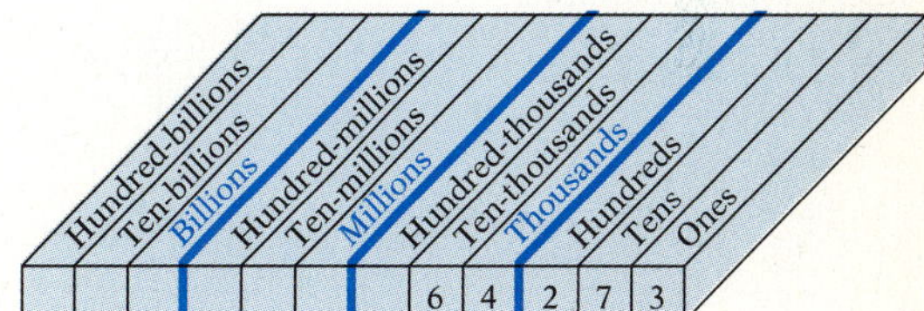

When a number is written in standard form, each group of digits separated by a comma is called a **period.** The number 5,316,709,842 has four periods. The period names are shown in color in the place-value chart above.

The whole number 37,286 can be written in **expanded form** as

30,000 + 7000 + 200 + 80 + 6

POINT OF INTEREST

The Babylonians had a place-value system based on 60. Its influence is still with us in angle measurement and time: 60 seconds in 1 minute, 60 minutes in 1 hour. It appears that the earliest record of a base 10 place-value system for natural numbers was developed in the eighth century.

The place-value chart can be used to find the expanded form of a number.

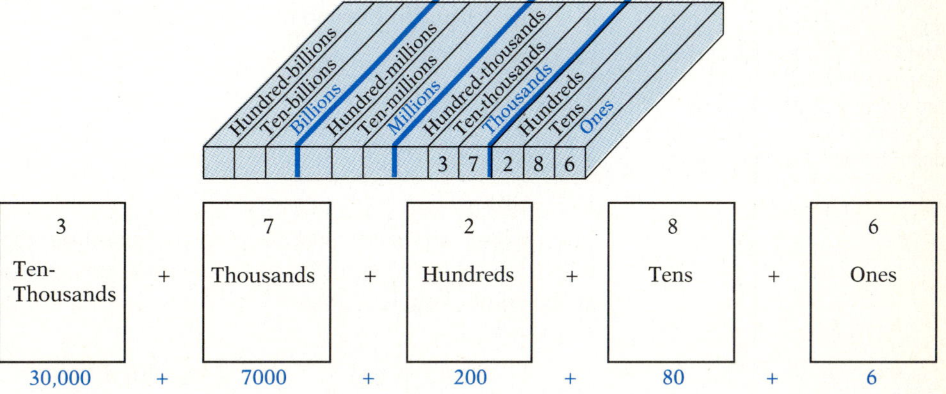

Write the number 510,409 in expanded form.

500,000 + 10,000 + 0 + 400 + 0 + 9
= 500,000 + 10,000 + 400 + 9 • Note the effect of having zeros in the number.

Example 1
Write 103,452 in expanded form.

Solution
100,000 + 3000 + 400 + 50 + 2

You Try It 1
Write 507,204 in expanded form.

Your Solution

Solution on p. A1

OBJECTIVE B Addition of whole numbers

POINT OF INTEREST
The first use of the plus sign appeared in 1489 in *Mercantile Arithmetic*. It was used to indicate a surplus and not to symbolize the addition operation. That use did not appear until around 1515.

Addition is the process of finding the total of two or more numbers. The numbers being added are called **addends** and the answer is called the **sum.**

5	+	9	=	14
Addend		**Addend**		**Sum**

To add large numbers, begin by arranging the numbers vertically, keeping the digits of the same place value in the same column.

Find the sum of 359 and 478.

Remember that a *sum* is the answer to an addition problem.
Arrange the numbers vertically.
Add the ones column.
9 + 8 = 17 (1 ten + 7 ones).
Write the 7 in the ones column and carry the 1 ten to the tens column.

Hundreds	Tens	Ones
	1	
3	5	9
4	7	8
		7

Add the tens column.
1 + 5 + 7 = 13 (1 hundred + 3 tens).
Write the 3 in the tens column and carry the 1 hundred to the hundreds column.

$$\begin{array}{r} {}^{1}\,{}^{1} \\ 359 \\ +\,478 \\ \hline 37 \end{array}$$

Add the hundreds column.
1 + 3 + 4 = 8 (8 hundreds).
Write the 8 in the hundreds column.

$$\begin{array}{r} {}^{1}\,{}^{1} \\ 359 \\ +\,478 \\ \hline 837 \end{array}$$

CONSIDER THIS
Carrying can be modeled with money. For instance, to add \$87 + \$45, think \$7 + \$5 is \$12, which can be exchanged for 1 ten dollar bill and 2 one dollar bills. Add the 1 ten dollar bill to the 8 tens and 4 tens. The result is 13 ten dollar bills, which can be exchanged for 1 one hundred dollar bill and 3 ten dollar bills.

The phrase *the sum of* was used in the example above to indicate the operation of addition. All the phrases listed below indicate addition. An example of each is shown at the right of each phrase.

added to	6 added to 9	9 + 6
more than	3 more than 8	8 + 3
the sum of	the sum of 7 and 4	7 + 4
increased by	2 increased by 5	2 + 5
the total of	the total of 1 and 6	1 + 6
plus	8 plus 10	8 + 10

Just as the word *it* is used in language to stand for an object, a letter of the alphabet can be used in mathematics to stand for a number. Such a number is called a **variable.**

A mathematical expression that contains one or more variables is a **variable expression.** Replacing the variables in a variable expression with numbers and then simplifying the numerical expression is called **evaluating the variable expression.**

Evaluate $a + b$ when $a = 678$ and $b = 294$.

Replace a with 678 and b with 294. $a + b$; $678 + 294$

Arrange the numbers vertically.

Add.

$$\begin{array}{r} \scriptstyle 1\,1 \\ 678 \\ +\,294 \\ \hline 972 \end{array}$$

Variables are often used in algebra to describe mathematical relationships. Variables are used below to describe three properties, or rules, of addition. Examples of each property are shown at the right.

The Addition Property of Zero

$a + 0 = a$ or $0 + a = a$

$5 + 0 = 5$
$0 + 9 = 9$

The Addition Property of Zero states that the sum of a number and zero equals the number. Here the variable a is used to represent any whole number. It can even represent the number zero because $0 + 0 = 0$.

The Commutative Property of Addition

$a + b = b + a$

$4 + 2 = 2 + 4$
$3 + 6 = 6 + 3$

The Commutative Property of Addition states that two numbers can be added in either order; the sum will be the same. Here the variables a and b represent any whole numbers. Therefore, for example, if you know that the sum of 5 and 7 is 12, then you also know that the sum of 7 and 5 is 12 because $5 + 7 = 7 + 5$.

The Associative Property of Addition

$(a + b) + c = a + (b + c)$

$(2 + 3) + 4 = 2 + (3 + 4)$
$5 + 4 = 2 + 7$
$9 = 9$

The Associative Property of Addition states that when adding three or more numbers, the numbers can be grouped in any order; the sum will be the same. Note in the example at the right above that we can add the sum of 2 and 3 to 4, or we can add 2 to the sum of 3 and 4. In either case, the sum of the three numbers is 9.

Example 2 Find the total of 439 and 57.

Solution

$$\begin{array}{r} \scriptstyle 1 \\ 439 \\ +\ \ 57 \\ \hline 496 \end{array}$$

You Try It 2 What is 94 more than 47?

Your Solution

Example 3 Evaluate $x + y + z$ when $x = 8427$, $y = 3659$, and $z = 6281$.

Solution $x + y + z$

$8427 + 3659 + 6281$

$$\begin{array}{r} \scriptstyle 1\,1\,1\ \ \\ 8427 \\ 3659 \\ +\ \ 6281 \\ \hline 18{,}367 \end{array}$$

You Try It 3 Evaluate $x + y + z$ when $x = 1692$, $y = 4783$, and $z = 5046$.

Your Solution

Solutions on p. A1

OBJECTIVE C Subtraction of whole numbers

POINT OF INTEREST

The use of the minus sign dates from the same period as the plus sign, about 1515.

Subtraction is the process of finding the difference between two numbers.

18	−	11	=	7
Minuend	**−**	**Subtrahend**	**=**	**Difference**

Note that addition and subtraction are related.

11	+	7	=	18
Subtrahend	+	Difference	=	Minuend

Because the sum of the subtrahend and the difference equals the minuend, you can use addition to check your subtraction.

To subtract large numbers, begin by arranging the numbers vertically, keeping the digits of the same place value in the same column. Note that when the lower digit is larger than the upper digit in any column, you must borrow.

CONSIDER THIS

Borrowing can be related to money. For instance, if Kelly has $27 as 2 ten dollar bills and 7 one dollar bills and Chris wants to borrow $9, then Kelly can exchange a ten dollar bill for 10 one dollar bills. Kelly then has 1 ten dollar bill and 17 one dollar bills. Kelly now can give Chris 9 one dollar bills. This leaves Kelly with 1 ten and 8 one dollar bills.

Find the difference between 692 and 378.

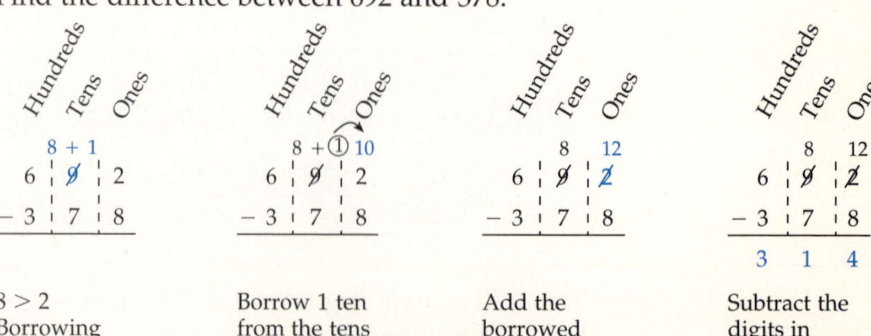

Subtraction may involve repeated borrowing.

Subtract: 7325 − 4698

$$\begin{array}{r} 7\;3\;\overset{1}{\cancel{2}}\;\overset{15}{\cancel{5}} \\ -\;4\;6\;9\;8 \\ \hline 7 \end{array}$$

Borrow 1 ten (10 ones) from the tens column and add 10 to the 5 in the ones column. Subtract 15 − 8.

$$\begin{array}{r} 7\;\overset{2}{\cancel{3}}\;\overset{11}{\overset{1}{\cancel{2}}}\;\overset{15}{\cancel{5}} \\ -\;4\;6\;9\;8 \\ \hline 2\;7 \end{array}$$

Borrow 1 hundred (10 tens) from the hundreds column and add 10 to the 1 in the tens column. Subtract 11 − 9.

$$\begin{array}{r} \overset{6}{\cancel{7}}\;\overset{12}{\overset{2}{\cancel{3}}}\;\overset{11}{\overset{1}{\cancel{2}}}\;\overset{15}{\cancel{5}} \\ -\;4\;6\;9\;8 \\ \hline 2\;6\;2\;7 \end{array}$$

Borrow 1 thousand (10 hundreds) from the thousands column and add 10 to the 2 in the hundreds column. Subtract 12 − 6 and 6 − 4.

When there is a zero in the minuend, borrowing can be performed as shown below.

Subtract: 3904 − 1775

Borrow 1 from 90. (90 − 1 = 89. The 8 is in the hundreds column. The 9 is in the tens column.) Add 10 to the 4 in the ones column. Then subtract the numbers in each column.

$$\begin{array}{r} 3\;\overset{8}{\cancel{9}}\;\overset{9}{\cancel{0}}\;\overset{14}{\cancel{4}} \\ -\;1\;7\;7\;5 \\ \hline 2\;1\;2\;9 \end{array}$$

Evaluate $c - d$ when $c = 6183$ and $d = 2759$.

$c - d$

Replace c with 6183 and d with 2759. $6183 - 2759$

Arrange the numbers vertically and then subtract.

$$\begin{array}{r} \overset{5}{\cancel{6}}\;\overset{11}{\cancel{1}}\;\overset{7}{\cancel{8}}\;\overset{13}{\cancel{3}} \\ -\;2\;7\;5\;9 \\ \hline 3\;4\;2\;4 \end{array}$$

The phrase *the difference between* was used in this objective to indicate the operation of subtraction. All the phrases listed below indicate subtraction. An example of each is shown at the right of each phrase.

minus	10 minus 3	10 − 3
less	8 less 4	8 − 4
less than	2 less than 9	9 − 2
the difference between	the difference between 6 and 1	6 − 1
decreased by	7 decreased by 5	7 − 5

Caution: Note the order in which the numbers are subtracted when the phrase "less than" is used. Suppose that you have $10 and I have $6 *less than* you do; then I have $6 less than $10, or $10 − $6 = $4.

Example 4 Find 5283 minus 764.

Solution

$$\begin{array}{r} \overset{4}{\cancel{5}}\,\overset{12}{\cancel{2}}\,\overset{7}{\cancel{8}}\,\overset{13}{\cancel{3}} \\ -\quad 7\;6\;4 \\ \hline 4\;5\;1\;9 \end{array}$$

You Try It 4 What is 2673 decreased by 814?

Your Solution

Example 5 Evaluate $x - y$ when $x = 3506$ and $y = 2477$.

Solution $x - y$

$3506 - 2477$

$$\begin{array}{r} 3\,\overset{4}{\cancel{5}}\,\overset{9}{\cancel{0}}\,\overset{16}{\cancel{6}} \\ -\;2\;4\;7\;7 \\ \hline 1\;0\;2\;9 \end{array}$$

You Try It 5 Evaluate $x - y$ when $x = 7061$ and $y = 3229$.

Your Solution

Solutions on p. A1

OBJECTIVE D Applications and formulas

To solve an application problem, first read the problem carefully. The **Strategy** involves identifying the quantity to be found and planning the steps that are necessary to find that quantity. The **Solution** involves performing each operation stated in the Strategy and writing the answer.

Example 6

What is the price of a pair of skates that cost a business \$109 and has a markup of \$49? Use the formula $P = C + M$, where P is the price of a product to the consumer, C is the cost paid by the store for the product, and M is the markup.

Strategy

To find the price, replace C by 109 and M by 49 in the given formula and solve for P.

Solution

$P = C + M$

$P = 109 + 49$

$P = 158$

The price of the skates is \$158.

You Try It 6

What is the price of a leather jacket that cost a business \$148 and has a markup of \$74? Use the formula $P = C + M$, where P is the price of a product to the consumer, C is the cost paid by the store for the product, and M is the markup.

Your Strategy

Your Solution

Solution on p. A1

1.1 EXERCISES

Objective A

Write the number in expanded form.

1. 6398 **2.** 7245 **3.** 46,182 **4.** 532,791

5. 328,476 **6.** 5064 **7.** 90,834 **8.** 20,397

9. 400,635 **10.** 402,708 **11.** 504,603 **12.** 8,000,316

Objective B

Solve.

13. What is 88,123 increased by 80,451?

14. What is 44,765 more than 82,003?

15. What is 654 added to 7293?

16. Find the sum of 658, 2709, and 10,935.

17. Find the total of 216, 8707, and 90,714.

18. Write the sum of x and y.

Evaluate the variable expression $x + y$ for the given values of x and y.

19. $x = 328; y = 471$ **20.** $x = 652; y = 137$ **21.** $x = 4752; y = 7398$

22. $x = 6047; y = 9283$ **23.** $x = 38{,}229; y = 51{,}671$ **24.** $x = 74{,}376; y = 19{,}528$

Evaluate the variable expression $a + b + c$ for the given values of a, b, and c.

25. $a = 693$; $b = 508$; $c = 371$

26. $a = 177$; $b = 892$; $c = 405$

27. $a = 4938$; $b = 2615$; $c = 7038$

28. $a = 6059$; $b = 3774$; $c = 5136$

29. $a = 12{,}897$; $b = 36{,}075$; $c = 48{,}441$

30. $a = 52{,}847$; $b = 49{,}036$; $c = 24{,}717$

Identify the property that justifies the statement.

31. $9 + 12 = 12 + 9$

32. $8 + 0 = 8$

33. $11 + (13 + 5) = (11 + 13) + 5$

34. $0 + 16 = 16 + 0$

Use the given property of addition to complete the statement.

35. The Addition Property of Zero
$28 + 0 = ?$

36. The Commutative Property of Addition
$16 + ? = 7 + 16$

37. The Associative Property of Addition
$9 + (? + 17) = (9 + 4) + 17$

38. The Addition Property of Zero
$0 + ? = 51$

Objective C

Solve.

39. Find the difference between 2536 and 918.

40. What is 1623 minus 287?

41. What is 5426 less than 12,804?

Solve.

42. Find 14,801 less 3522.

43. Find 85,423 decreased by 67,875.

44. Write the difference between x and y.

Evaluate the variable expression $x - y$ for the given values of x and y.

45. $x = 50; y = 37$

46. $x = 80; y = 33$

47. $x = 914; y = 271$

48. $x = 623; y = 197$

49. $x = 740; y = 385$

50. $x = 870; y = 243$

51. $x = 8672; y = 3461$

52. $x = 7814; y = 3512$

53. $x = 1605; y = 839$

54. $x = 1406; y = 968$

55. $x = 23{,}409; y = 5178$

56. $x = 56{,}397; y = 8249$

Objective D

Solve.

57. Find the difference between the smallest four-digit number and the largest two-digit number.

58. You eat an apple and one cup of cornflakes with one tablespoon of sugar and one cup of milk for breakfast. Find the total number of calories consumed if one apple contains 80 calories, one cup of cornflakes has 95 calories, one tablespoon of sugar has 45 calories, and one cup of milk has 150 calories.

59. According to the graph at the right, **(a)** how much more is spent per year on fast food than on airline tickets? **(b)** How much less is spent on cable TV and video than on prescription drugs? (Source: Ernst & Young)

60. The repair bill on your car includes \$179 for parts, \$78 for labor, and \$15 for sales tax. What is the total amount owed?

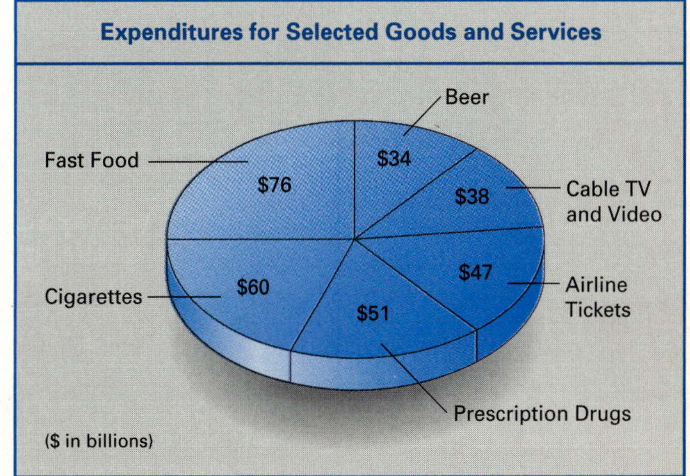

Solve.

61. The computer system you would like to purchase includes an operating system priced at \$830, a monitor that costs \$245, an extended keyboard priced at \$175, and a printer that sells for \$395. What is the total cost of the computer system?

62. Use the formula $A = P + I$, where A is the value of an investment, P is the original investment, and I is the interest earned, to find the value of an investment that earned \$775 in interest on an original investment of \$12,500.

63. Use the formula $A = P + I$, where A is the value of an investment, P is the original investment, and I is the interest earned, to find the value of an investment that earned \$484 in interest on an original investment of \$8800.

64. What is the mortgage loan amount on a home that sells for \$145,000 with a down payment of \$14,500? Use the formula $M = S - D$, where M is the mortgage loan amount, S is the selling price, and D is the down payment.

65. What is the mortgage loan amount on a home that sells for \$118,000 with a down payment of \$23,600? Use the formula $M = S - D$, where M is the mortgage loan amount, S is the selling price, and D is the down payment.

66. What is the ground speed of an airplane flying at an airspeed of 375 mph into a 25-mph head wind? Use the formula $g = a - h$, where g is the ground speed, a is the airspeed, and h is the speed of the head wind.

67. Find the ground speed of an airplane traveling into a 15-mph head wind with an airspeed of 425 mph. Use the formula $g = a - h$, where g is the ground speed, a is the airspeed, and h is the speed of the head wind.

Critical Thinking

68. If you roll two ordinary six-sided dice and add the two numbers that appear on top, how many different sums are possible?

69. How many two-digit numbers are there? How many three-digit numbers are there?

70. In how many different ways can a panel of four on-off switches be set if no two adjacent switches can be off?

71. Determine whether the statement is always true, sometimes true, or never true.

a. If a is any whole number, then $a - 0 = a$.
b. If a is any whole number, then $a - a = 0$.

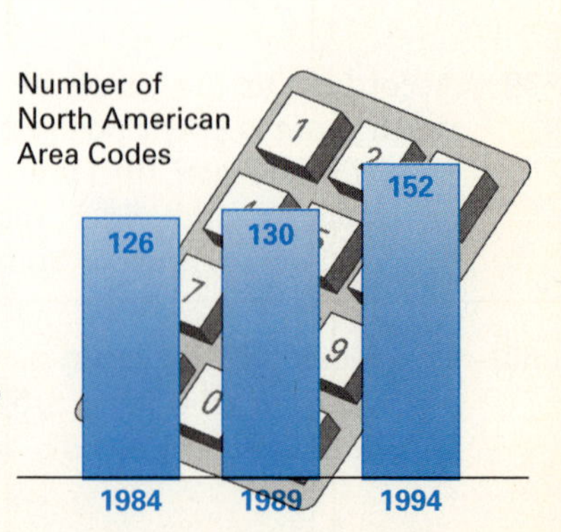

72. [W] Write a paragraph explaining the data presented in the graph at the right.

73. [W] What is the national debt of the United States? What does this figure mean?

SECTION 1.2 Multiplication and Division of Whole Numbers

OBJECTIVE A Multiplication of whole numbers

A store manager orders six cases of telephone answering machines. Each case contains eight answering machines. The total number of answering machines ordered can be calculated by adding 6 eights.

$$8 + 8 + 8 + 8 + 8 + 8 = 48$$

This problem involves repeated addition of the same number; therefore, the answer can be calculated by the shorter process of multiplication. **Multiplication** is the repeated addition of the same number.

The numbers that are multiplied are called **factors.** The answer is called the **product.**

$$8 + 8 + 8 + 8 + 8 + 8 = 48$$

or

6	$\times$	8	=	48
Factor		Factor		Product

POINT OF INTEREST

The cross $\times$ was first used as a symbol for multiplication in 1631 in a book titled *The Key to Mathematics.* Also in that year, another book, *Practice of the Analytical Art,* advocated the use of a dot to indicate multiplication.

The times sign $\times$ is one symbol that is used to mean multiplication. Each of the expressions below represents multiplication.

$6 \cdot 8$ $\quad$ $6(8)$ $\quad$ $(6)(8)$ $\quad$ $6a$ $\quad$ $6(a)$ $\quad$ ab

The expression $6a$ means "6 times a." The expression ab means "a times b."

Multiply: 37(4)

Multiply $4 \cdot 7$.
$4 \cdot 7 = 28$ (2 tens + 8 ones).
Write the 8 in the ones column and carry the 2 to the tens column.

$$\begin{array}{r} {}^{2}\\ 37 \\ \times\ \ 4 \\ \hline 8 \end{array}$$

The 3 in 37 is 3 tens.
Multiply $4 \cdot 3$ tens. $\quad 4 \cdot 3$ tens = 12 tens
Add the carry digit. $\quad$ + 2 tens
$\quad$ 14 tens

Write the 14.

$$\begin{array}{r} {}^{2}\\ 37 \\ \times\ \ 4 \\ \hline 148 \end{array}$$

Multiply: (47)(23)

Multiply by the ones digit. $3 \cdot 47 = 141$.

$$\begin{array}{r} 47 \\ \times\ 23 \\ \hline 141 \end{array}$$

The last digit is written in the ones column.

Multiply by the tens digit. $2 \cdot 47 = 94$.

$$\begin{array}{r} 47 \\ \times\ 23 \\ \hline 141 \\ 94\ \end{array}$$

The last digit is written in the tens column.

Add.

$$\begin{array}{r} 47 \\ \times\ 23 \\ \hline 141 \\ 94\ \\ \hline 1081 \end{array}$$

	Thousands	Hundreds	Tens	Ones	
			4	7	
×			2	3	
		1	4	1	← 3 × 47
		9	4	0	← 20 × 47
	1	0	8	1	← 141 + 940

The place-value chart illustrates the placement of the products.

Find the product of 600 and 70.

Remember that a *product* is the answer to a multiplication problem. Multiply the nonzero parts of the factors ($6 \cdot 7 = 42$). Write the same number of zeros in the product as the total number of zeros in the factors.

$600 \cdot 70 = 42{,}000$

3 zeros — 3 zeros

Evaluate *abcd* when $a = 3$, $b = 20$, $c = 10$, and $d = 4$.

abcd means *a* times *b* times *c* times *d*. Replace each variable with its value.

$abcd$
$3(20)(10)4$

Multiply the first two numbers. $= 60(10)4$

Multiply the product by the third number. $= (600)4$

Continue multiplying until all the numbers have been multiplied. $= 2400$

Just as there are properties of addition, there are properties of multiplication.

The Multiplication Property of Zero

$a \cdot 0 = 0$ or $0 \cdot a = 0$

$8 \cdot 0 = 0$
$0 \cdot 5 = 0$

The Multiplication Property of Zero states that the product of a number and zero equals zero. Here the variable *a* is used to represent any whole number. It can even represent the number zero because $0 \cdot 0 = 0$.

The Multiplication Property of One

$a \cdot 1 = a$ or $1 \cdot a = a$

$6 \cdot 1 = 6$
$1 \cdot 9 = 9$

The Multiplication Property of One states that the product of a number and one equals the number. Multiplying a number by 1 does not change the number.

The Commutative Property of Multiplication

$a \cdot b = b \cdot a$

$5 \cdot 3 = 3 \cdot 5$
$2 \cdot 6 = 6 \cdot 2$

The Commutative Property of Multiplication states that two numbers can be multiplied in either order; the product will be the same. Here the variables *a* and *b* represent any whole numbers. Therefore, for example, if you know that the product of 4 and 9 is 36, then you also know that the product of 9 and 4 is 36 because $4 \cdot 9 = 9 \cdot 4$

The Associative Property of Multiplication

$(a \cdot b) \cdot c = a \cdot (b \cdot c)$

$$\begin{aligned}(2 \cdot 3) \cdot 4 &= 2 \cdot (3 \cdot 4)\\ 6 \cdot 4 &= 2 \cdot 12\\ 24 &= 24\end{aligned}$$

The Associative Property of Multiplication states that when multiplying three numbers, the numbers can be grouped in any order; the product will be the same. Note in the example at the right above that we can multiply the product of 2 and 3 by 4, or we can multiply 2 by the product of 3 and 4. In either case, the product is 24.

The phrase *the product of* was used in this objective to indicate the operation of multiplication. All the phrases below indicate multiplication. An example of each is shown at the right of each phrase.

times	8 times 4	$8 \cdot 4$
the product of	the product of 9 and 5	$9 \cdot 5$
multiplied by	7 multiplied by 3	$3 \cdot 7$
twice	twice 6	$2 \cdot 6$

Example 1 What is 528 times 703?

Solution

$$\begin{array}{r} 528 \\ \times \quad 703 \\ \hline 1584 \\ 36960 \\ \hline 371{,}184 \end{array}$$

You Try It 1 What is 657 multiplied by 408?

Your Solution

Example 2 Evaluate $3ab$ when $a = 10$ and $b = 40$.

Solution $3ab$

$$\begin{aligned}3(10)(40) &= 30(40)\\ &= 1200\end{aligned}$$

You Try It 2 Evaluate $5xy$ when $x = 20$ and $y = 60$.

Your Solution

Solutions on p. A1

OBJECTIVE B Exponents

Repeated multiplication of the same factor can be written in two ways:

$4 \cdot 4 \cdot 4 \cdot 4 \cdot 4$ or 4^5 ← exponent; 4 ← base

The expression 4^5 is in **exponential form.** The **exponent** 5 indicates how many times the base 4 occurs as a factor in the multiplication. The display below explains how to read numbers written in exponential form.

$2 = 2^1$	read "two to the first power" or just "two." Usually the 1 is not written.
$2 \cdot 2 = 2^2$	read "two squared" or "two to the second power."
$2 \cdot 2 \cdot 2 = 2^3$	read "two cubed" or "two to the third power."
$2 \cdot 2 \cdot 2 \cdot 2 = 2^4$	read "two to the fourth power."
$2 \cdot 2 \cdot 2 \cdot 2 \cdot 2 = 2^5$	read "two to the fifth power."

Variable expressions can contain exponents.

$x^1 = x$	x to the first power is usually written simply as x.
$x^2 = x \cdot x$	x^2 means x times x.
$x^3 = x \cdot x \cdot x$	x^3 means x occurs as a factor 3 times.
$x^4 = x \cdot x \cdot x \cdot x$	x^4 means x occurs as a factor 4 times.

Each place value in the place-value chart can be expressed as a power of 10.

Ten =	10 =	$10 = 10^1$
Hundred =	100 =	$10 \cdot 10 = 10^2$
Thousand =	1000 =	$10 \cdot 10 \cdot 10 = 10^3$
Ten-thousand =	10,000 =	$10 \cdot 10 \cdot 10 \cdot 10 = 10^4$
Hundred-thousand =	100,000 =	$10 \cdot 10 \cdot 10 \cdot 10 \cdot 10 = 10^5$
Million =	1,000,000 =	$10 \cdot 10 \cdot 10 \cdot 10 \cdot 10 \cdot 10 = 10^6$

Note that the exponent on 10 when the number is written in exponential form is the same as the number of zeros in the number written in standard form. For example, $10^5 = 100{,}000$; the exponent on 10 is 5, and the number 100,000 has 5 zeros.

To evaluate a numerical expression containing exponents, write each factor as many times as indicated by the exponent and then multiply.

Evaluate the variable expression c^4d^3 when $c = 2$ and $d = 5$.

Replace c with 2 and d with 5. Then evaluate the exponential expression.

$$c^4d^3$$
$$2^4 \cdot 5^3 = (2 \cdot 2 \cdot 2 \cdot 2) \cdot (5 \cdot 5 \cdot 5)$$
$$= 16 \cdot 125 = 2000$$

A calculator can be used to evaluate an exponential expression. The y^x key is used to enter the exponent.

Example 3 Write $7 \cdot 7 \cdot 7 \cdot 4 \cdot 4$ in exponential notation.

Solution $7 \cdot 7 \cdot 7 \cdot 4 \cdot 4 = 7^3 \cdot 4^2$

You Try It 3 Write $2 \cdot 2 \cdot 2 \cdot 3 \cdot 3 \cdot 3 \cdot 3$ in exponential notation.

Your Solution

Example 4 Evaluate x^2y^3 when $x = 4$ and $y = 2$.

Solution x^2y^3 (x^2y^3 means x^2 times y^3.)

$$4^2 \cdot 2^3 = (4 \cdot 4) \cdot (2 \cdot 2 \cdot 2)$$
$$= 16 \cdot 8$$
$$= 128$$

You Try It 4 Evaluate x^4y^2 when $x = 1$ and $y = 3$.

Your Solution

Solutions on p. A1

OBJECTIVE C Division of whole numbers

Division is used to separate objects into equal groups.

A grocer wants to equally distribute 24 new products on 4 shelves. How many products should the grocer place on each shelf? The grocer's problem can be written:

$$\begin{array}{rcl} & 6 \swarrow & \text{Number on each shelf (Quotient)} \\ \text{Number of shelves (Divisor)} \rightarrow & 4\overline{)24} & \leftarrow \text{Number of products (Dividend)} \end{array}$$

Notice that the quotient multiplied by the divisor equals the dividend.

$4\overline{)24}$ with quotient 6 because [6 Quotient] × [4 Divisor] = [24 Dividend]

Division is also represented by the symbol ÷ or by a fraction bar. Both are read "divided by."

$9\overline{)54}$ with quotient 6 $\qquad 54 \div 9 = 6 \qquad \frac{54}{9} = 6$

The fact that the quotient times the divisor equals the dividend can be used to illustrate properties of division.

$0 \div 4 = 0$ because $0 \cdot 4 = 0$.

$4 \div 4 = 1$ because $1 \cdot 4 = 4$.

$4 \div 1 = 4$ because $4 \cdot 1 = 4$.

$4 \div 0 = ?$ What number can be multiplied by 0 to get 4? There is no number whose product with 0 is 4 because the product of a number and zero is 0. Division by zero is undefined. $\quad ? \cdot 0 = 4$

The properties of division are stated below. In these statements, the symbol ≠ is read "is not equal to." Recall that the variable a represents any whole number. Therefore, for the first two properties, we must state that $a \neq 0$ in order to ensure that we are not dividing by zero.

Division Properties of Zero and One

If $a \neq 0$, $0 \div a = 0$.	Zero divided by any number other than zero is zero.
If $a \neq 0$, $a \div a = 1$.	Any number other than zero divided by itself is one.
$a \div 1 = a$	A number divided by one is the number.
$a \div 0$ is undefined.	Division by zero is undefined.

Divide and check: $3192 \div 4$

```
    7
4)3 1 9 2      Think 31 ÷ 4.
 -2 8          Subtract 7 × 4.
    3 9        Bring down the 9.
```

```
    7 9
4)3 1 9 2
 -2 8
    3 9        Think 39 ÷ 4.
   -3 6        Subtract 9 × 4.
      3 2      Bring down the 2.
```

```
    7 9 8
4)3 1 9 2
 -2 8
    3 9
   -3 6
      3 2      Think 32 ÷ 4.      Check.    7 9 8
     -3 2      Subtract 8 × 4.            ×     4
        0                                 3 1 9 2
```

Sometimes it is not possible to separate objects into a whole number of equal groups.

A packer at a bakery has 14 muffins to pack into 3 boxes. Each box will hold 4 muffins. From the diagram, we see that after the packer places 4 muffins in each box, there are 2 muffins left over. The 2 is called the **remainder.**

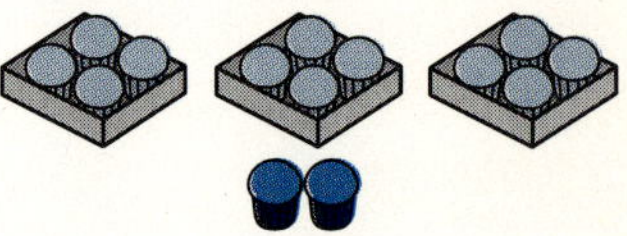

The packer's division problem could be written:

```
                               ↙ Number in each box
                            4    Quotient                    or      4 r2
Number of boxes  →       3)1 4 ← Total number of muffins           3)14
     Divisor              -1 2     Dividend
                             2 ← Number left over
                                 Remainder
```

For any division problem, **(quotient · divisor) + remainder = dividend.**

Find the quotient of 389 and 24.

```
      1 6 r5
2 4)3 8 9
   -2 4
    1 4 9
   -1 4 4
        5        Check: (16 · 24) + 5 = 384 + 5 = 389
```

The phrase *the quotient of* was used in the example above to indicate the operation of division. The phrase *divided by* also indicates division.

the quotient of	the quotient of 8 and 4	$8 \div 4$
divided by	9 divided by 3	$9 \div 3$

Example 5 Evaluate $\frac{x}{y}$ when $x = 342$ and $y = 9$.

Solution $\frac{x}{y}$

$\frac{342}{9}$

$$\begin{array}{r} 38 \\ 9\overline{)342} \\ -27 \\ \hline 72 \\ -72 \\ \hline 0 \end{array}$$

You Try It 5 Evaluate $\frac{x}{y}$ when $x = 672$ and $y = 8$.

Your Solution

Solution on p. A1

OBJECTIVE D Factors

Natural number factors of a number divide that number evenly (there is no remainder).

1, 2, 3, and 6 are natural number factors of 6 because they divide 6 evenly.

$1\overline{)6}$ = 6 $\quad 2\overline{)6}$ = 3 $\quad 3\overline{)6}$ = 2 $\quad 6\overline{)6}$ = 1

Notice that both the divisor and the quotient are factors of the dividend.

To find the factors of a number, try dividing the number by 1, 2, 3, 4, 5, . . . Those numbers that divide the number evenly are its factors. Continue this process until the factors start to repeat.

Find all the factors of 42.

$42 \div 1 = 42$	1 and 42 are factors of 42.
$42 \div 2 = 21$	2 and 21 are factors of 42.
$42 \div 3 = 14$	3 and 14 are factors of 42.
$42 \div 4$	4 will not divide 42 evenly.
$42 \div 5$	5 will not divide 42 evenly.
$42 \div 6 = 7$	6 and 7 are factors of 42.
$42 \div 7 = 6$	7 and 6 are factors of 42.

The factors are repeating.
All the factors of 42 have been found.

The factors of 42 are 1, 2, 3, 6, 7, 14, 21, and 42.

The following rules are helpful in finding the factors of a number.

2 is a factor of a number if the ones digit is 0, 2, 4, 6, or 8.

436 ends in 6.
Therefore, 2 is a factor of 436.
($436 \div 2 = 218$)

3 is a factor of a number if the sum of its digits is divisible by 3.

The sum of the digits of 489 is $4 + 8 + 9 = 21$. 21 is divisible by 3. Therefore, 3 is a factor of 489. $(489 \div 3 = 163)$

4 is a factor of a number if its last two digits are divisible by 4.

556 ends in 56. 56 is divisible by 4. $(56 \div 4 = 14)$ Therefore, 4 is a factor of 556. $(556 \div 4 = 139)$

5 is a factor of a number if the ones digit is 0 or 5.

520 ends in 0. Therefore, 5 is a factor of 520. $(520 \div 5 = 104)$

Example 6 Find all the factors of 40.

Solution

$40 \div 1 = 40$
$40 \div 2 = 20$
$40 \div 3$ Does not divide evenly.
$40 \div 4 = 10$
$40 \div 5 = 8$
$40 \div 6$ Does not divide evenly.
$40 \div 7$ Does not divide evenly.
$40 \div 8 = 5$ The factors are repeating.

The factors of 40 are 1, 2, 4, 5, 8, 10, 20, and 40.

You Try It 6 Find all the factors of 30.

Your Solution

Solution on p. A1

OBJECTIVE  Applications and formulas

Example 7

At what speed would you need to travel in order to drive a distance of 294 mi in 6 h? Use the formula $r = \frac{d}{t}$, where r is the average speed, d is the distance, and t is the time.

Strategy

To find the speed, replace d by 294 and t by 6 in the given formula and solve for r.

Solution

$$r = \frac{d}{t}$$

$$r = \frac{294}{6} = 49$$

You would need to travel at 49 mph.

You Try It 7

At what speed would you need to travel to drive 486 mi in 9 h? Use the formula $r = \frac{d}{t}$, where r is the average speed, d is the distance, and t is the time.

Your Strategy

Your Solution

Solution on p. A1

1.2 EXERCISES

Objective A

Solve.

1. Find the product of 500 and 3.

2. Find 30 multiplied by 80.

3. What is 40 times 50?

4. What is the product of 400, 3, 20, and 0?

5. Write the product of f and g.

Evaluate the expression for the given values of the variables.

6. $7a$, when $a = 465$

7. $6n$, when $n = 382$

8. xyz, when $x = 5$, $y = 12$, and $z = 30$

9. abc, when $a = 4$, $b = 20$, and $c = 50$

10. $2xy$, when $x = 67$ and $y = 23$

11. $4ab$, when $a = 95$ and $b = 33$

12. $5cd$, when $c = 312$ and $d = 134$

13. $4st$, when $s = 523$ and $t = 127$

Identify the property that justifies the statement.

14. $1 \cdot 29 = 29$

15. $6(4 \cdot 20) = (6 \cdot 4)20$

16. $37 \cdot 0 = 0$

17. $43 \cdot 1 = 1 \cdot 43$

Use the given property of multiplication to complete the statement.

18. The Commutative Property of Multiplication
$20(44) = (?)20$

19. The Associative Property of Multiplication
$(? \cdot 6)100 = 5(6 \cdot 100)$

Use the given property of multiplication to complete the statement.

20. The Multiplication Property of Zero
$45 \cdot 0 = ?$

21. The Multiplication Property of One
$? \cdot 77 = 77$

Objective B

Write in exponential form.

22. $2 \cdot 2 \cdot 2 \cdot 7 \cdot 7 \cdot 7 \cdot 7 \cdot 7$

23. $3 \cdot 3 \cdot 3 \cdot 3 \cdot 3 \cdot 3 \cdot 5 \cdot 5 \cdot 5$

24. $2 \cdot 2 \cdot 3 \cdot 3 \cdot 3 \cdot 5 \cdot 5 \cdot 5 \cdot 5$

25. $7 \cdot 7 \cdot 11 \cdot 11 \cdot 11 \cdot 19 \cdot 19 \cdot 19 \cdot 19$

26. $x \cdot x \cdot x \cdot y \cdot y \cdot y$

27. $a \cdot a \cdot b \cdot b \cdot b \cdot b$

Evaluate.

28. $3^2 \cdot 10^3$

29. $5^2 \cdot 3^3$

30. $4^3 \cdot 0^3$

31. $2 \cdot 5^2 \cdot 7^3$

32. $5^2 \cdot 2 \cdot 3^4$

Evaluate the expression for the given values of the variables.

33. x^2y, when $x = 3$ and $y = 4$

34. ab^6, when $a = 5$ and $b = 2$

35. ab^3, when $a = 7$ and $b = 4$

36. c^2d^2, when $c = 3$ and $d = 5$

37. p^2q^4, when $p = 3$ and $q = 10$

38. m^3n^3, when $m = 5$ and $n = 10$

Objective C

Solve.

39. Find the quotient of 7256 and 8.

40. What is the quotient of 8172 and 9?

41. What is 6168 divided by 7?

42. Write the quotient of c and d.

Evaluate the variable expression $\frac{x}{y}$ for the given values of x and y.

43. $x = 48; y = 1$

44. $x = 56; y = 56$

45. $x = 79; y = 0$

46. $x = 0; y = 23$

47. $x = 208; y = 7$

48. $x = 309; y = 6$

49. $x = 39{,}200; y = 4$

50. $x = 16{,}200; y = 3$

51. $x = 7128; y = 24$

Objective D

Find all the factors of the number.

52. 20

53. 12

54. 16

55. 18

56. 13

57. 17

58. 24

59. 36

60. 56

61. 45

62. 28

63. 32

64. 26

65. 52

66. 57

67. 48

68. 64

69. 54

70. 75

71. 80

Objective E

Solve.

72. The table at the right shows the number of breakfast items prepared and served at Denny's restaurants in a year. **(a)** About how many pounds of sausage are cooked each month? **(b)** About how many cups of coffee are served each month?

Breakfast Item	*Numbers Served*
Eggs	231 million
Six-inch pancakes	74 million
Sausage	12 million pounds
Bacon	15 million pounds
Coffee	384 million cups

73. How many of the first one hundred whole numbers are divisible by all of the numbers 2, 3, 4, and 5?

74. Each side of a CAV-format laserdisc can hold 30 min of video. If the laserdisc can store 30 still frames per second, how many still frames of video can be stored on each side of a laserdisc?

Solve.

75. Refer to the nutrition facts shown at the right. **(a)** How many times more grams of sugars than of protein are in each serving? **(b)** How many times more grams of total carbohydrate than of dietary fiber are in each serving?

Nutrition Facts

Serving Size 1 1/4 Cups (55g/2.0 oz.)
Servings per Container

Amount Per Serving	Cereal	Cereal with 1/2 Cup Vitamins A & D Skim Milk
Calories	130	170
Calories from Fat	15	20
	% Daily Value **	
Total Fat 2.0g*	3 %	3 %
Saturated Fat 1.0g	2 %	3 %
Cholesterol 0mg	0 %	0 %
Sodium 60mg	3 %	5 %
Potassium 120mg	3 %	7 %
Total Carbohydrate 38g	13 %	15 %
Dietary Fiber 10g	40 %	40 %
Sugars 12g		
Other Carbohydrate 18g		
Protein 7g		

76. Find the total amount paid on a loan when the monthly payment is \$285 and the loan is paid off in 24 months. Use the formula $A = MN$, where A is the total amount paid, M is the monthly payment, and N is the number of payments.

77. Find the total amount paid on a loan when the monthly payment is \$187 and the loan is paid off in 36 months. Use the formula $A = MN$, where A is the total amount paid, M is the monthly payment, and N is the number of payments.

78. Use the formula $t = \frac{d}{r}$, where t is the time, d is the distance, and r is the average speed, to find the time it would take to drive 513 mi at an average speed of 57 mph.

79. Use the formula $t = \frac{d}{r}$, where t is the time, d is the distance, and r is the average speed, to find the time it would take to drive 432 mi at an average speed of 54 mph.

80. The current value of the stocks in a mutual fund is \$10,500,000. The number of shares outstanding is 500,000. Find the value per share of the fund. Use the formula $V = \frac{C}{S}$, where V is the value per share, C is the current value of the stocks in the fund, and S is the number of shares outstanding.

81. The current value of the stocks in a mutual fund is \$4,500,000. The number of shares outstanding is 250,000. Find the value per share of the fund. Use the formula $V = \frac{C}{S}$, where V is the value per share, C is the current value of the stocks in the fund, and S is the number of shares outstanding.

Critical Thinking

82. 13,827 is not divisible by 4. By rearranging the digits, find the largest possible number that is divisible by 4.

83. Look at the columns of numbers at the right. In what column is the number 1 million?

A	B	C
1	8	27
64	125	216
.	.	.
.	.	.
.	.	.

84. Determine whether the statement is always true, sometimes true, or never true.
a. If a is any whole number, then $a \cdot 0 = a$.
b. If a is any whole number, then $a \cdot 1 = 1$.

85. [W] According to the National Safety Council, in a recent year, one death resulting from an accident occurred every 5 min. At this rate, how many accidental deaths occurred each hour? Each day? Throughout the year? Explain how you arrived at your answers.

SECTION 1.3 Rounding and Estimation

OBJECTIVE  Rounding

When the distance to the sun is given as 93,000,000 mi, the number represents an approximation to the true distance. Giving an approximate value for an exact number is called **rounding.** A number is rounded to a given place value.

48 is closer to 50 than it is to 40. 48 rounded to the nearest ten is 50.

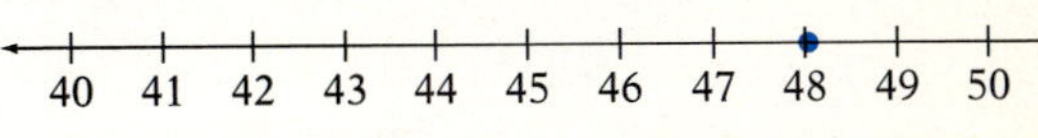

4872 rounded to the nearest ten is 4870.

4872 rounded to the nearest hundred is 4900.

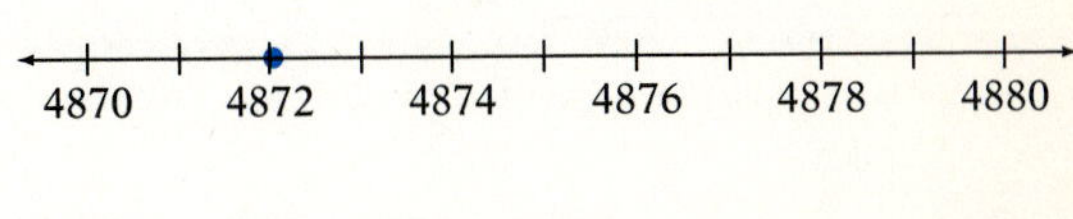

A number is rounded to a given place value without using the number line by looking at the first digit to the right of that place value.

If the digit to the right of the given place value is less than 5, that digit and all digits to the right are replaced by zeros.

Round 12,743 to the nearest hundred.

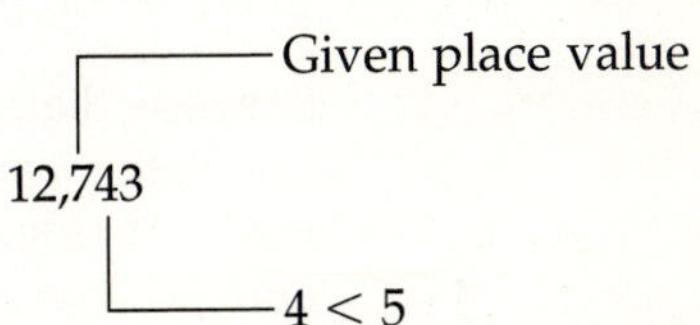

12,743 rounded to the nearest hundred is 12,700.

If the digit to the right of the given place value is greater than or equal to 5, increase the digit in the given place value by 1, and replace all other digits to the right by zeros.

Round 46,738 to the nearest thousand.

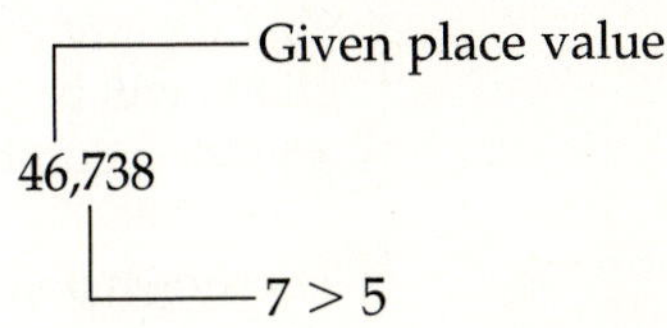

46,738 rounded to the nearest thousand is 47,000.

Round 29,873 to the nearest thousand.

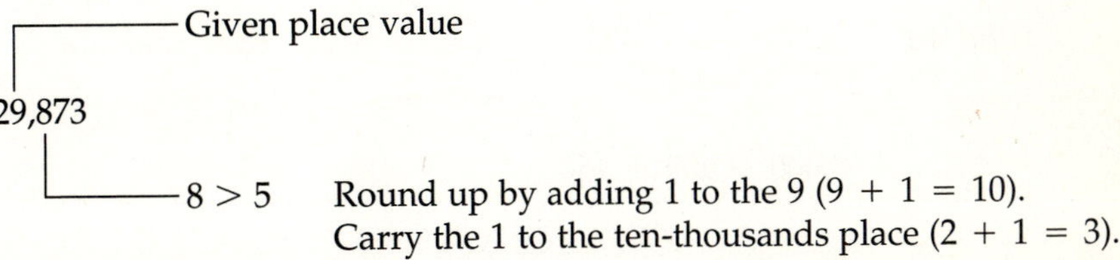

Round up by adding 1 to the 9 $(9 + 1 = 10)$.
Carry the 1 to the ten-thousands place $(2 + 1 = 3)$.

29,873 rounded to the nearest thousand is 30,000.

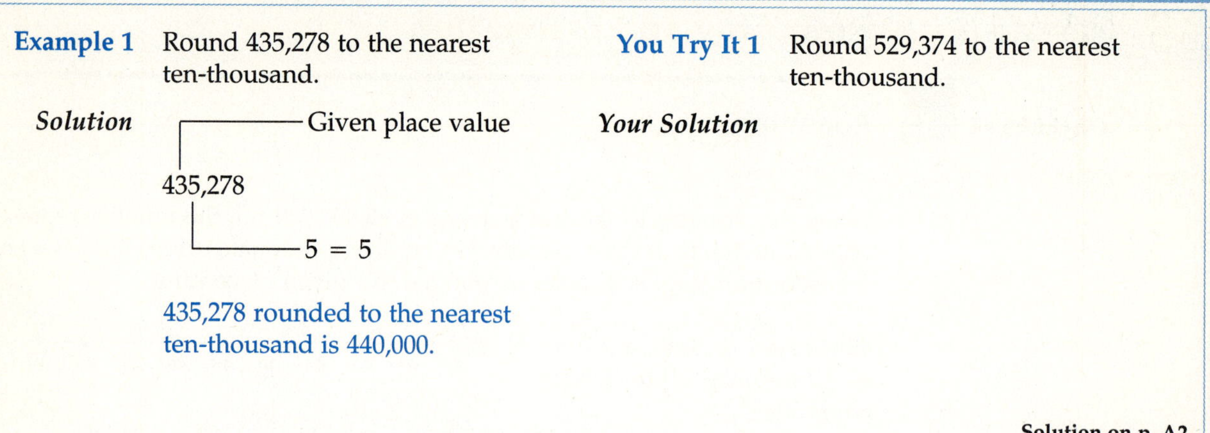

Example 1 Round 435,278 to the nearest ten-thousand.

Solution

435,278 rounded to the nearest ten-thousand is 440,000.

You Try It 1 Round 529,374 to the nearest ten-thousand.

Your Solution

Solution on p. A2

OBJECTIVE B Estimation

An important skill in mathematics is determining whether or not an answer to a problem is reasonable. One method of determining if an answer is reasonable is to use estimation. An **estimate** is an approximation.

Estimation is especially valuable when using a calculator. Suppose that you are adding 1497 and 2568 on a calculator. You enter the number 1497 correctly, but you inadvertently enter 256 instead of 2568 for the second addend. The sum reads 1753. If you quickly make an estimate of the answer, you will recognize that the sum 1753 is not reasonable and that an error has been made.

$$\begin{array}{r} 1497 \\ +\ 2568 \\ \hline 4065 \end{array} \qquad \begin{array}{r} 1497 \\ +\ 256 \\ \hline 1753 \end{array}$$

To estimate the answer to a calculation, round each number to the highest place value of the number; the first digit of each number will be nonzero and all other digits will be zero. Perform the calculation using the rounded numbers.

$$\begin{array}{rr} 1497 \longrightarrow & 1000 \\ 2568 \longrightarrow & +\ 3000 \\ \hline & 4000 \end{array}$$

As shown above, the sum 4000 is an estimate of the sum of 1497 and 2568; it is very close to the actual sum 4065. 4000 is not close to the incorrectly calculated sum 1753.

Example 2 Estimate the quotient of 55,272 and 392.

Solution 55,272 ⟶ 60,000
392 ⟶ 400

$60{,}000 \div 400 = 150$

You Try It 2 Estimate the quotient of 216,936 and 207.

Your Solution

Solution on p. A2

1.3 EXERCISES

Objective A

Round the number to the given place value.

1. 837 Tens
2. 925 Tens
3. 3049 Tens
4. 7108 Tens
5. 1638 Hundreds
6. 4962 Hundreds
7. 17,639 Hundreds
8. 28,551 Hundreds
9. 5326 Thousands
10. 6809 Thousands
11. 84,608 Thousands
12. 93,825 Thousands
13. 389,702 Thousands
14. 629,513 Thousands
15. 746,898 Ten-thousands
16. 352,876 Ten-thousands
17. 36,702,599 Millions
18. 71,834,250 Millions

Objective B

Estimate by rounding. Then find the exact answer.

19. 6742 + 8298
20. 5426 + 1732
21. 7355 − 5219
22. 8953 − 2217
23. 3467 · 359
24. 8745(63)
25. 36,472 ÷ 47
26. 62,176 ÷ 58
27. 972,085 + 416,832

Estimate by rounding. Then find the exact answer.

28. 23,774 + 38,026

29. 59,126 − 20,843

30. 63,051 − 29,478

31. (39,246)(29)

32. 64,409 · 67

33. 389,804 ÷ 76

34. 637,072 ÷ 29

35. $\begin{array}{r} 8941 \\ \times\ 726 \\ \hline \end{array}$

36. $\begin{array}{r} 2837 \\ \times\ 216 \\ \hline \end{array}$

37. $219\overline{)332{,}004}$

38. $219\overline{)632{,}034}$

39. $\begin{array}{r} 387 \\ 295 \\ 614 \\ +\ 702 \\ \hline \end{array}$

40. $\begin{array}{r} 528 \\ 163 \\ 947 \\ +\ 275 \\ \hline \end{array}$

41. $\begin{array}{r} 224{,}196 \\ -\ 98{,}531 \\ \hline \end{array}$

42. $\begin{array}{r} 873{,}925 \\ -\ 28{,}744 \\ \hline \end{array}$

Critical Thinking

43. There are 52 weeks in a year. Is this an exact figure or an approximation?

44. If 3846 is rounded to the nearest ten and then that number is rounded to the nearest hundred, is the result the same as when you round 3846 to the nearest hundred? If not, which of the two methods is correct for rounding to the nearest hundred?

45. [W] List at least three situations in which it is helpful to round numbers. List at least three situations in which estimates are helpful.

46. [W] What is the enrollment of your school? To what place value would it be reasonable to round this number? Why? To what place value is the population of your town or city rounded? Why? To what place value is the population of your state rounded? To what place value is the population of the United States rounded?

47. [W] What estimate is given for the size of the population in your state by the year 2000? What is the estimate of the size of the population in the United States by the year 2000? Estimates differ. On what basis was the estimate you recorded derived?

SECTION 1.4 Prime Factorization and the Order of Operations Agreement

OBJECTIVE A Prime factorization

POINT OF INTEREST

Prime numbers are an important part of cryptology, the study of secret codes. To ensure that codes cannot be broken, a cryptologist uses prime numbers that have hundreds of digits.

A whole number is a **prime** number if its only natural number factors are 1 and itself. 7 is prime because its only factors are 1 and 7. If a whole number is not prime, it is a **composite** number. Because 6 has factors of 2 and 3, 6 is a composite number. 1 is not considered a prime number; therefore, it is not included in the following list of prime numbers less than 50: 2, 3, 5, 7, 11, 13, 17, 19, 23, 29, 31, 37, 41, 43, 47

The **prime factorization** of a number is the expression of the number as a product of its prime factors. To find the prime factors of 90, begin with the smallest prime number as a trial divisor and continue with prime numbers as trial divisors until the final quotient is prime.

Find the prime factorization of 90.

$\begin{array}{r} 45 \\ 2\overline{)90} \end{array}$	$\begin{array}{r} 15 \\ 3\overline{)45} \\ 2\overline{)90} \end{array}$	$\begin{array}{r} 5 \\ 3\overline{)15} \\ 3\overline{)45} \\ 2\overline{)90} \end{array}$
Divide 90 by 2.	45 is not divisible by 2. Divide 45 by 3.	Divide 15 by 3. 5 is prime.

The prime factorization of 90 is $2 \cdot 3 \cdot 3 \cdot 5$, or $2 \cdot 3^2 \cdot 5$.

Finding the prime factorization of larger numbers can be more difficult. Try each prime number as a trial divisor. Stop when the square of the trial divisor is greater than the number being factored.

Find the prime factorization of 201.

$\begin{array}{r} 67 \\ 3\overline{)201} \end{array}$ 67 cannot be divided evenly by 2, 3, 5, 7, or 11. Prime numbers greater than 11 need not be tried because $11^2 = 121$ and $121 > 67$.

The prime factorization of 201 is $3 \cdot 67$.

Example 1 Find the prime factorization of 84.

Solution

$\begin{array}{r} 7 \\ 3\overline{)21} \\ 2\overline{)42} \\ 2\overline{)84} \end{array}$

$84 = 2 \cdot 2 \cdot 3 \cdot 7 = 2^2 \cdot 3 \cdot 7$

You Try It 1 Find the prime factorization of 88.

Your Solution

Solution on p. A2

OBJECTIVE B The least common multiple (LCM)

The **multiples** of a number are the products of that number and the numbers 1, 2, 3, 4, 5, . . .

$4 \cdot 1 = 4$
$4 \cdot 2 = 8$
$4 \cdot 3 = 12$
$4 \cdot 4 = 16$
$4 \cdot 5 = 20$ The multiples of 4 are 4, 8, 12, 16, 20, . . .
⋮

A number that is a multiple of two or more numbers is a **common multiple** of those numbers.

The multiples of 6 are 6, 12, 18, 24, 30, 36, 42, 48, 54, 60, 66, 72, . . .
The multiples of 8 are 8, 16, 24, 32, 40, 48, 56, 64, 72, 80, 88, 96, . . .
Some common multiples of 6 and 8 are 24, 48, and 72.

The **least common multiple (LCM)** is the smallest common multiple of two or more numbers.

The least common multiple of 6 and 8 is 24.

Listing the multiples of each number is one way to find the LCM. Another way to find the LCM uses the prime factorization of each number.

Find the LCM of 32 and 36.

Write the prime factorization of each number and circle the highest power of each prime factor.

$32 = \textcircled{2^5}$
$36 = 2^2 \cdot \textcircled{3^2}$

The LCM is the product of the circled factors. $2^5 \cdot 3^2 = 32 \cdot 9 = 288$

The LCM of 32 and 36 is 288.

Example 2 Find the LCM of 12, 18, and 40.

Solution
$12 = 2^2 \cdot 3$
$18 = 2 \cdot \textcircled{3^2}$
$40 = \textcircled{2^3} \cdot \textcircled{5}$
The LCM $= 2^3 \cdot 3^2 \cdot 5$
$= 8 \cdot 9 \cdot 5 = 360$

You Try It 2 Find the LCM of 16, 24, and 28.

Your Solution

Solution on p. A2

OBJECTIVE C The greatest common factor (GCF)

Recall that a number that divides another number evenly is a **factor** of the number.

18 can be evenly divided by 1, 2, 3, 6, 9, and 18.
1, 2, 3, 6, 9, and 18 are factors of 18.

A number that is a factor of two or more numbers is a **common factor** of those numbers.

The factors of 24 are 1, 2, 3, 4, 6, 8, 12, and 24.

The factors of 36 are 1, 2, 3, 4, 6, 9, 12, 18, and 36.

The common factors of 24 and 36 are 1, 2, 3, 4, 6, and 12.

CONSIDER THIS

The following model, in which *a* and *b* are any whole numbers, may help you with the LCM and GCF.

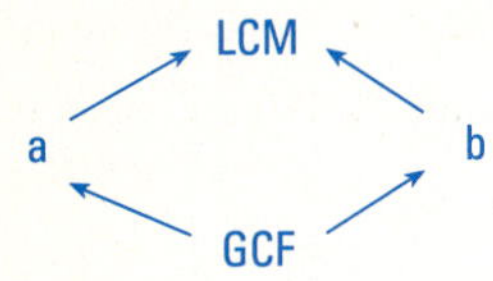

The arrows indicate "divides into."

The **greatest common factor (GCF)** is the largest common factor of two or more numbers.

The greatest common factor of 24 and 36 is 12.

Listing the factors of each number is one way to find the GCF. Another way to find the GCF uses the prime factorization of each number.

Find the GCF of 12 and 30.

Write the prime factorization of each number and circle the lowest power of each prime factor that occurs in both factorizations. The prime factor 5 occurs in the prime factorization of 30 but not in the prime factorization of 12. Since 5 is not a factor in both factorizations, do not circle 5.

$12 = 2^2 \cdot \textcircled{3}$

$30 = \textcircled{2} \cdot 3 \cdot 5$

The GCF is the product of the circled factors. $2 \cdot 3 = 6$

The GCF of 12 and 30 is 6.

Example 3 Find the GCF of 14 and 27.

Solution

$14 = 2 \cdot 7$

$27 = 3^3$

No common prime factor occurs in the factorizations.

The GCF is 1.

You Try It 3 Find the GCF of 25 and 52.

Your Solution

Example 4 Find the GCF of 16, 20, and 28.

Solution

$16 = 2^4$

$20 = \textcircled{2^2} \cdot 5$

$28 = 2^2 \cdot 7$

The GCF $= 2^2 = 4$.

You Try It 4 Find the GCF of 32, 40, and 56.

Your Solution

Solutions on p. A2

OBJECTIVE D The Order of Operations Agreement

In order to prevent more than one answer to the same problem, an Order of Operations Agreement is followed. The Order of Operations Agreement is shown on the next page.

The Order of Operations Agreement

Step 1 Do all operations inside parentheses.
Step 2 Simplify any numerical expressions containing exponents.
Step 3 Do multiplication and division as they occur from left to right.
Step 4 Do addition and subtraction as they occur from left to right.

Simplify: $2(4 + 1) - 2^3 + 6 \div 2$

	$2(4 + 1) - 2^3 + 6 \div 2$
Perform operations in parentheses.	$= 2(5) - 2^3 + 6 \div 2$
Simplify expressions with exponents.	$= 2(5) - 8 + 6 \div 2$
Do multiplication and division as they occur from left to right.	$= 10 - 8 + 6 \div 2$ $= 10 - 8 + 3$
Do addition and subtraction as they occur from left to right.	$= 2 + 3$ $= 5$

Evaluate $5a - (b + c)^2$ when $a = 6$, $b = 1$, and $c = 3$.

	$5a - (b + c)^2$
Replace a with 6, b with 1, and c with 3.	$5(6) - (1 + 3)^2$
Use the Order of Operations Agreement to simplify the resulting numerical expression. Perform operations inside parentheses.	$= 5(6) - (4)^2$
Simplify expressions with exponents.	$= 5(6) - 16$
Do the multiplication.	$= 30 - 16$
Do the subtraction.	$= 14$

One or more of the above steps may not be needed to simplify an expression. In that case, proceed to the next step in the Order of Operations Agreement.

Example 5

Evaluate $(a - b)^2 + 3c$ when $a = 6$, $b = 4$, and $c = 1$.

Solution

$(a - b)^2 + 3c$

$$\begin{aligned}(6 - 4)^2 + 3(1) &= (2)^2 + 3(1)\\ &= 4 + 3(1)\\ &= 4 + 3\\ &= 7\end{aligned}$$

You Try It 5

Evaluate $(a - b)^2 + 5c$ when $a = 7$, $b = 2$, and $c = 4$.

Your Solution

Solution on p. A2

1.4 EXERCISES

Objective A

Find the prime factorization of the number.

1. 16 **2.** 24 **3.** 12 **4.** 27 **5.** 15

6. 36 **7.** 40 **8.** 50 **9.** 37 **10.** 83

11. 65 **12.** 87 **13.** 80 **14.** 90 **15.** 28

16. 49 **17.** 42 **18.** 62 **19.** 81 **20.** 51

21. 89 **22.** 101 **23.** 66 **24.** 46 **25.** 120

Objective B

Find the LCM of the numbers.

26. 4 and 8 **27.** 3 and 9 **28.** 2 and 7 **29.** 5 and 11

30. 6 and 10 **31.** 8 and 12 **32.** 9 and 15 **33.** 14 and 21

34. 12 and 16 **35.** 8 and 14 **36.** 4 and 10 **37.** 9 and 30

38. 14 and 42 **39.** 16 and 48 **40.** 24 and 36 **41.** 16 and 28

42. 30 and 40 **43.** 20 and 28 **44.** 45 and 60 **45.** 72 and 108

46. 2, 5, and 8 **47.** 3, 5, and 10 **48.** 5, 10, and 20 **49.** 4, 8, and 12

Find the LCM of the numbers.

50. 3, 12, and 18 **51.** 9, 36, and 45 **52.** 9, 36, and 72 **53.** 14, 42, and 70

Objective C

Find the GCF of the numbers.

54. 9 and 12 **55.** 6 and 15 **56.** 18 and 30 **57.** 15 and 35

58. 14 and 42 **59.** 25 and 50 **60.** 16 and 80 **61.** 17 and 51

62. 21 and 55 **63.** 32 and 35 **64.** 8 and 36 **65.** 12 and 80

66. 16 and 20 **67.** 24 and 30 **68.** 16 and 28 **69.** 24 and 36

70. 30 and 40 **71.** 45 and 75 **72.** 12 and 54 **73.** 84 and 120

74. 6, 10, and 12 **75.** 8, 12, and 20 **76.** 6, 15, and 36 **77.** 15, 20, and 30

78. 3, 17, and 51 **79.** 21, 63, and 84 **80.** 24, 36, and 60 **81.** 32, 56, and 72

Objective D

Simplify.

82. $8 \div 4 + 2$ **83.** $12 - 9 \div 3$ **84.** $6 \cdot 4 + 5$

Simplify.

85. $5 \cdot 7 + 3$

86. $4^2 - 3$

87. $6^2 - 14$

88. $5 \cdot (6 - 3) + 4$

89. $8 + (6 + 2) \div 4$

90. $9 + (7 + 5) \div 6$

91. $14 \cdot (3 + 2) \div 10$

92. $13 \cdot (1 + 5) \div 13$

93. $14 - 2^3 + 9$

94. $6 \cdot 3^2 + 7$

95. $18 + 5 \cdot 3^2$

96. $14 + 5 \cdot 2^3$

97. $20 + (9 - 4) \cdot 2$

98. $10 + (8 - 5) \cdot 3$

99. $3^2 + 5 \cdot (6 - 2)$

100. $2^3 + 4(10 - 6)$

101. $3^2 \cdot 2^2 + 3 \cdot 2$

102. $6(7) + 4^2 \cdot 3^2$

103. $14 - 2(6)$

104. $18 + 3(7)$

105. $2(9 - 2) + 5$

106. $6(8 - 3) - 12$

107. $15 - (7 - 1) \div 3$

108. $16 - (13 - 5) \div 4$

109. $11 + 2 - 3 \cdot 4 \div 3$

110. $17 + 1 - 8 \cdot 2 \div 4$

111. $3(5 + 3) \div 8$

Evaluate the expressions for the given values of the variable.

112. $x - 2y$, when $x = 8$ and $y = 3$

113. $x + 6y$, when $x = 5$ and $y = 4$

114. $x^2 + 3y$, when $x = 6$ and $y = 7$

115. $3x^2 + y$, when $x = 2$ and $y = 9$

116. $x^2 + y \div x$, when $x = 2$ and $y = 8$

117. $x + y^2 \div x$, when $x = 4$ and $y = 8$

118. $4x + (x - y)^2$, when $x = 8$ and $y = 2$

119. $(x + y)^2 - 2y$, when $x = 3$ and $y = 6$

120. $x^2 + 3(x - y) + z^2$, when $x = 2$, $y = 1$, and $z = 3$

121. $x^2 + 4(x - y) \div z^2$, when $x = 8$, $y = 6$, and $z = 2$

Critical Thinking

122. Twin primes are two prime numbers that differ by 2. For instance, 17 and 19 are twin primes. Find three other sets of twin primes.

123. Find the LCM of x and $2x$. Find the GCF of x and $2x$.

124. If x is a prime number and y is a prime number, find the LCM of x and y. Find the GCF of x and y.

125. Write an expression that can be used to determine whether or not your calculator uses the Order of Operations Agreement. Use the expression to determine if your calculator uses the Order of Operations Agreement.

126. [W] In your own words, define the least common multiple of two numbers and the greatest common factor of two numbers.

127. [W] Explain why 2 is the only even prime number.

128. [W] Is the LCM of two numbers always divisible by the GCF of the two numbers? If so, explain why. If not, give an example.

Projects in Mathematics

Garbology

Save your garbage for one week. Weigh it. Estimate how much trash you throw away per year.

How much of the garbage you collected is recyclable? If these materials were recycled, how much refuse would this save over a year's time?

Multiply the amount of garbage you collected by the number of people living in your community. Determine the amount of trash discarded by your community during a year.

Investigate a recycling program at your school or in your community.

How many tons of garbage are recycled each year?

What materials are accepted as recyclable in the program? Classify the products as glass, aluminum, paper, cardboard, plastic, yard waste, and so on.

What is the cost of the recycling program? Include the cost of labor in your figure.

How much money does your school or community receive in payment for the recyclables?

What is the cost of trash removal? What would be the cost of discarding the recyclables rather than recycling them?

Does your school or community realize a profit as a result of recycling materials rather than discarding them? If so, how large a profit?

Can the benefits of recycling be measured only by the profit or loss associated with a recycling program? Why or why not?

Applications of Patterns in Mathematics

1. For the first circle shown below, use a straight line to connect each dot on the circle with every other dot on the circle. How many different straight lines are there?
2. Follow the same procedure described above for each of the other circles. How many different straight lines are there in each?

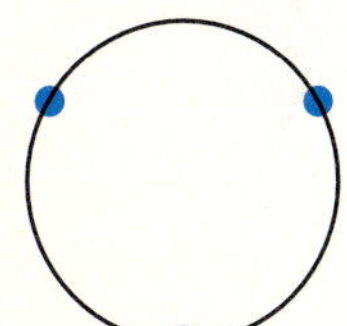

3. Find a pattern to describe the number of dots on a circle and the corresponding number of different lines drawn. Use the pattern to determine the number of different lines that would be drawn in a circle with 7 dots and in a circle with 8 dots.
4. You are arranging a tennis tournament with nine players. How many singles matches will be played among the nine players if each player plays each of the other players once?

Chapter Summary

Key Words The *natural numbers* or *counting numbers* are 1, 2, 3, 4, 5, 6, 7, 8, 9, 10, . . . The *whole numbers* are 0, 1, 2, 3, 4, 5, 6, 7, 8, 9, 10, . . . When a whole number is written using the digits 0, 1, 2, 3, 4, 5, 6, 7, 8, and 9, it is said to be in *standard form.* The position of each digit in the number determines the digit's *place value.*

Addition is the process of finding the total of two or more numbers. The numbers being added are called *addends.* The answer is the *sum. Subtraction* is the process of finding the difference between two numbers. The *minuend* minus the *subtrahend* equals the *difference.*

Multiplication is the repeated addition of the same number. The numbers that are multiplied are called *factors.* The answer is the *product. Division* is used to separate objects into equal groups. The *dividend* divided by the *divisor* equals the *quotient.*

The expression 3^5 is in *exponential form.* The *exponent* 5 indicates how many times the *base* 3 occurs as a factor in the multiplication.

Natural number *factors* of a number divide that number evenly. A number is a *prime* number if its only whole number factors are 1 and itself. The *prime factorization* of a number is the expression of the number as a product of its prime factors.

A *variable* is a letter that is used to stand for a number. A mathematical expression that contains one or more variables is a *variable expression.*

Essential Rules **To round a number to a given place value:** If the digit to the right of the given place value is less than 5, that digit and all digits to the right are replaced by zeros. If the digit to the right of the given place value is greater than or equal to 5, increase the digit in the given place value by 1, and replace all other digits to the right by zeros.

To estimate the answer to a calculation: Round each number to the highest place value of that number. Perform the calculation using the rounded numbers.

To find the LCM of two or more numbers, write the prime factorization of each number and circle the highest power of each prime factor. The LCM is the product of the circled factors.

To find the GCF of two or more numbers, write the prime factorization of each number and circle the lowest power of each prime factor that occurs in each factorization. The GCF is the product of the circled factors.

Addition Property of Zero	$a + 0 = a$ or $0 + a = a$
Commutative Property of Addition	$a + b = b + a$
Associative Property of Addition	$(a + b) + c = a + (b + c)$
Multiplication Property of Zero	$a \cdot 0 = 0$ or $0 \cdot a = 0$
Multiplication Property of One	$a \cdot 1 = a$ or $1 \cdot a = a$
Commutative Property of Multiplication	$a \cdot b = b \cdot a$
Associative Property of Multiplication	$(a \cdot b) \cdot c = a \cdot (b \cdot c)$
Division Properties of Zero and One	If $a \neq 0$, $0 \div a = 0$. $a \div 1 = a$ If $a \neq 0$, $a \div a = 1$. $a \div 0$ is undefined.

The Order of Operations Agreement

Step 1 Do all operations inside parentheses.
Step 2 Simplify expressions containing exponents.
Step 3 Do multiplication and division as they occur from left to right.
Step 4 Do addition and subtraction as they occur from left to right.

Chapter Review Exercises

1. Find the prime factorization of 90.

2. Evaluate $10^4 \cdot 3^2$.

3. Find the difference between 4207 and 1624.

4. Write $3 \cdot 3 \cdot 5 \cdot 5 \cdot 5 \cdot 5$ in exponential notation.

5. Evaluate $\frac{x}{y}$ when $x = 480$ and $y = 6$.

6. Round 38,729 to the nearest hundred.

7. Complete the statement by using the Multiplication Property of One

$$? \cdot 82 = 82$$

8. Identify the property that justifies the statement.

$$10 + 33 = 33 + 10$$

9. Evaluate $2xy$ when $x = 50$ and $y = 7$.

10. Find the quotient of 15,642 and 6.

11. Evaluate x^3y^2 when $x = 3$ and $y = 5$.

12. Estimate the sum of 482, 319, 570, and 146.

13. Find all the factors of 50.

14. Evaluate $x + y$ when $x = 683$ and $y = 249$.

15. Simplify: $16 + 4(7 - 5)^2 \div 8$

16. Find the GCF of 60 and 80.

17. Find the product of 4 and 659.

18. Evaluate $x - y$ when $x = 270$ and $y = 133$.

19. Find the LCM of 30 and 42.

20. Write 906,378 in expanded form.

21. Find the prime factorization of 102.

22. Evaluate $10^5 \cdot 7^2$.

23. Find the difference between 906 and 478.

24. Write $x \cdot x \cdot x \cdot x \cdot y \cdot y$ in exponential notation.

25. Evaluate $\frac{x}{y}$ when $x = 252$ and $y = 12$.

26. Round 74,965 to the nearest thousand.

27. Complete the statement by using the Associative Property of Addition.

$$? + (3 + 7) = (8 + 3) + 7$$

28. Identify the property that justifies the statement.

$$12(2) = 2(12)$$

29. Evaluate $8xy$ when $x = 20$ and $y = 5$.

30. Find the quotient of 174,298 and 25.

31. Evaluate a^4b^2 when $a = 10$ and $b = 3$.

32. Estimate the sum of 572, 431, 809, and 675.

33. Find all the factors of 104.

34. Evaluate $x + y$ when $x = 1394$ and $y = 5827$.

35. Simplify: $20 + 24(8 - 5) \div 2^2$

36. Find the GCF of 30 and 36.

37. Find the product of 8 and 473.

38. Evaluate $x - y$ when $x = 1024$ and $y = 736$.

39. Find the LCM of 18 and 45.

40. Write 60,705 is expanded form.

41. Evaluate $a^2b - c$ when $a = 4$, $b = 8$, and $c = 12$.

42. Evaluate $(a + b)^2 - 2c$ when $a = 5$, $b = 3$, and $c = 4$.

43. During his professional basketball career, Kareem Abdul-Jabbar had 17,440 rebounds. Elvin Hayes had 16,279 rebounds during his professional basketball career. How many more rebounds than Hayes did Abdul-Jabbar have during his career?

44. The profits of a firm are shared equally by its four partners. If this year's profits were $128,000, what is each partner's share of the profits?

45. The amount financed, A, on an installment purchase is the price, P, of the product minus the down payment, D. Use the formula $A = P - D$ to find the amount financed to buy an air conditioner if the price of the air conditioner is $478 and the down payment is $99.

46. The annual dividend, A, paid by a company to a stockholder is the product of the annual dividend per share, S, and the number of shares, N, owned by the stockholder. Use the formula $A = SN$ to find the annual dividend to be paid to a stockholder who owns 425 shares when the annual dividend per share is $3.

47. Before your workout, the odometer on the exercise bike reads 382. You plan on cycling 25 mi. What should the odometer read after your workout?

48. A contractor quotes the cost of work on a new house, which is to have 2800 ft^2 of floor space, at $65 per square foot. Estimate the total cost of the contractor's work on the house.

49. Use the formula $d = rt$, where d is distance, r is speed, and t is time, to find the distance traveled in 3 h by a cyclist traveling at a speed of 14 mph.

50. Find the markup on a word processor that cost a business $1775 and that was sold for $2224. Use the formula $M = S - C$, where M is the markup on a product, S is the selling price of the product, and C is the cost of the product to the business.

CHAPTER

2

Integers and Variable Expressions

Focus on Problem Solving

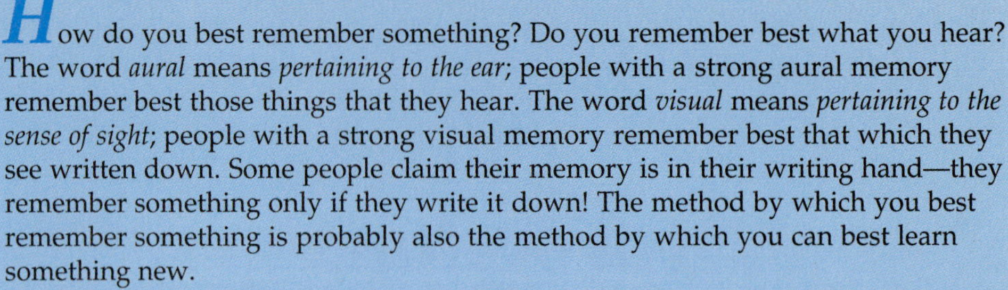

How do you best remember something? Do you remember best what you hear? The word *aural* means *pertaining to the ear*; people with a strong aural memory remember best those things that they hear. The word *visual* means *pertaining to the sense of sight*; people with a strong visual memory remember best that which they see written down. Some people claim their memory is in their writing hand—they remember something only if they write it down! The method by which you best remember something is probably also the method by which you can best learn something new.

In problem-solving situations, try to capitalize on your strengths. If you understand material better when you hear it spoken, read application problems aloud or have someone else read them to you. If writing helps you organize ideas, rewrite application problems in your own words.

No matter what your main strength, visualizing a problem can be a valuable aid in problem solving. A drawing, sketch, diagram, or chart can be a useful tool in problem solving, just as calculators and computers are tools. A diagram can be helpful in gaining an understanding of the relationships inherent in a problem-solving situation. A sketch will help you organize the given information, and can help you focus on the method by which the solution can be determined.

A tour bus drives 5 mi south, then 4 mi west, then 3 mi north, then 4 mi east. How far is the tour bus from the starting point?

Draw a diagram of the given information.

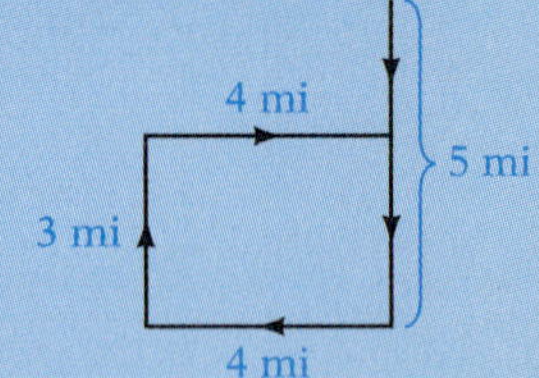

From the diagram, we can see that the solution can be determined by subtracting 3 from 5: $5 - 3 = 2$.

The bus is 2 mi from the starting point.

If you roll two ordinary six-sided dice and multiply the two numbers that appear on top, how many different products are there?

Make a chart of the possible products. In the chart below, repeated products are marked with an asterisk.

$1 \cdot 1 = 1$	$2 \cdot 1 = 2$ (*)	$3 \cdot 1 = 3$ (*)	$4 \cdot 1 = 4$ (*)	$5 \cdot 1 = 5$ (*)	$6 \cdot 1 = 6$ (*)
$1 \cdot 2 = 2$	$2 \cdot 2 = 4$ (*)	$3 \cdot 2 = 6$ (*)	$4 \cdot 2 = 8$ (*)	$5 \cdot 2 = 10$ (*)	$6 \cdot 2 = 12$ (*)
$1 \cdot 3 = 3$	$2 \cdot 3 = 6$ (*)	$3 \cdot 3 = 9$	$4 \cdot 3 = 12$ (*)	$5 \cdot 3 = 15$ (*)	$6 \cdot 3 = 18$ (*)
$1 \cdot 4 = 4$	$2 \cdot 4 = 8$	$3 \cdot 4 = 12$ (*)	$4 \cdot 4 = 16$	$5 \cdot 4 = 20$ (*)	$6 \cdot 4 = 24$ (*)
$1 \cdot 5 = 5$	$2 \cdot 5 = 10$	$3 \cdot 5 = 15$	$4 \cdot 5 = 20$	$5 \cdot 5 = 25$	$6 \cdot 5 = 30$ (*)
$1 \cdot 6 = 6$	$2 \cdot 6 = 12$	$3 \cdot 6 = 18$	$4 \cdot 6 = 24$	$5 \cdot 6 = 30$	$6 \cdot 6 = 36$

By counting the products that are not repeats, we can see that there are 18 different products.

In this chapter, you will notice that frequently a number line is used to help you to visualize the integers or order the integers, to help you to understand the concepts of opposite and absolute value, and to add integers. As you begin your work with integers, you may find that sketching a number line helps you understand a problem or work through a calculation.

SECTION 2.1 Introduction to Integers

OBJECTIVE  Order relations between integers

POINT OF INTEREST

Chinese manuscripts dating from about 250 B.C. contain the first recorded use of negative numbers. However, it was not until late in the 14th century that most mathematicians generally accepted these numbers.

In Chapter 1, only zero and numbers greater than zero were discussed. In this chapter, numbers less than zero are introduced. Phrases such as "7 degrees below zero," "$50 in debt," and "20 feet below sea level" refer to numbers less than zero.

The numbers greater than zero are called **positive numbers.** Numbers less than zero are called **negative numbers.**

Positive and Negative Numbers

A number n is positive if n is greater than 0.
A number n is negative if n is less than 0.

A positive number can be indicated by placing the sign + in front of the number. For example, we can write +4 instead of 4. Both +4 and 4 represent positive 4. Usually, however, the plus sign is omitted and it is understood that the number is a positive number.

A negative number is indicated by placing a negative sign (−) in front of the number. The number −1 is read "negative one," −2 is read "negative two," and so on.

Just as distances are associated with markings on the edge of a ruler, positive and negative numbers can be associated with points on a line. This line is called the **number line** and is shown below. The arrowheads at both ends indicate that the number line continues in both directions.

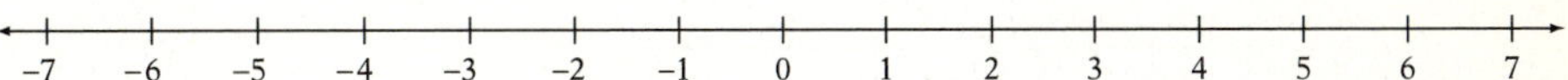

The **integers** are . . . −4, −3, −2, −1, 0, 1, 2, 3, 4, . . . The integers to the right of zero are the **positive integers.** The integers to the left of zero are the **negative integers.** Zero is an integer, but it is neither positive nor negative. The point corresponding to 0 on the number line is called the **origin.**

The **graph** of an integer is shown by placing a heavy dot on the number line directly above the number. Shown below is the graph of −6 on the number line.

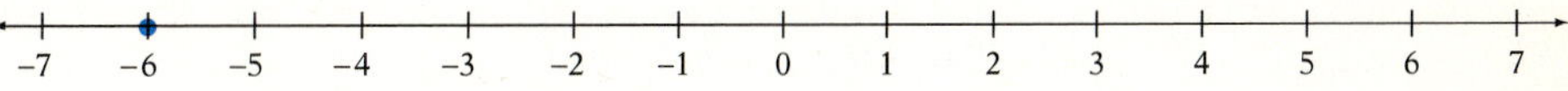

On a number line, the numbers get larger as we move from left to right. The numbers get smaller as we move from right to left. Therefore, a number line can be used to visualize the order relation between two integers.

POINT OF INTEREST

The symbols for "is less than" and "is greater than" were introduced by Thomas Harriot about 1630. Before that, ⊏ and ⊐ were used for > and <, respectively.

A number that appears to the right of a given number is **greater than** the given number. The symbol for "is greater than" is >.

2 is to the right of -3 on the number line.
2 is greater than -3.
$2 > -3$

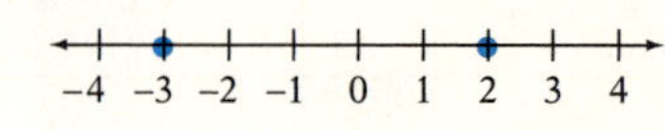

A number that appears to the left of a given number is **less than** the given number. The symbol for "is less than" is <.

-4 is to the left of 1 on the number line.
-4 is less than 1.
$-4 < 1$

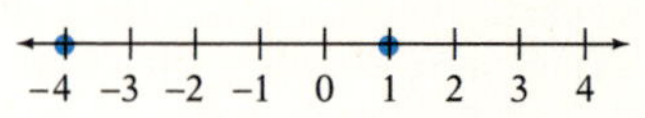

An **inequality** expresses the relative order of two mathematical expressions. $2 > -3$ and $-4 < 1$ are inequalities.

> ***Order Relations***
>
> $a > b$ if a is to the right of b on the number line.
> $a < b$ if a is to the left of b on the number line.

Example 1 On the number line, what number is 5 units to the right of -2?

Solution

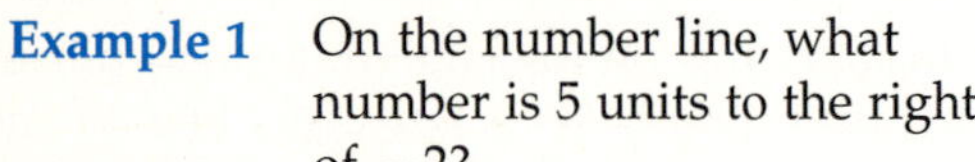

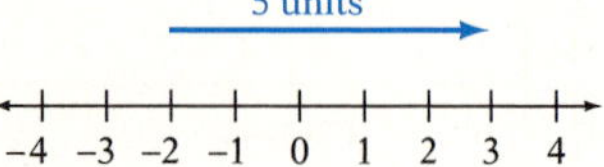

3 is 5 units to the right of -2.

You Try It 1 On the number line, what number is 4 units to the left of 1?

Your Solution

Example 2 If G is 2 and I is 4, what numbers are B and D?

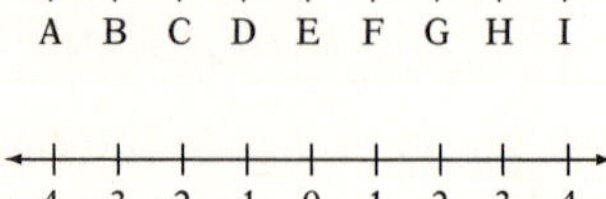

Solution

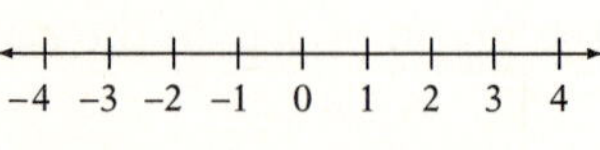

B is -3, and D is -1.

You Try It 2 If G is 1 and H is 2, what numbers are A and C?

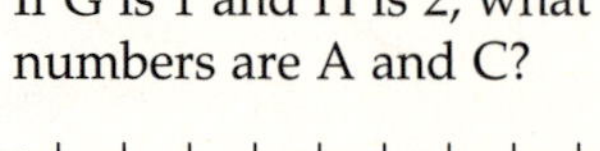

Your Solution

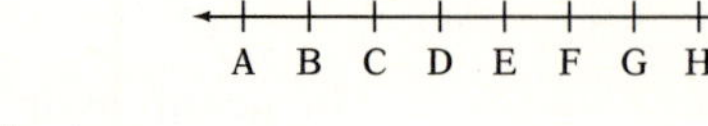

Solutions on p. A2

Example 3 Place the correct symbol, < or >, between the two numbers.

a. $-3 \quad -1$ b. $1 \quad -2$

Solution a. -3 is to the left of -1 on the number line.

$-3 < -1$

b. 1 is to the right of -2 on the number line.

$1 > -2$

You Try It 3 Place the correct symbol, < or >, between the two numbers.

a. $2 \quad -5$ b. $-4 \quad 3$

Your Solution

Example 4 Write the given numbers in order from smallest to largest.

5, − 2, 3, 0, −6

Solution −6, −2, 0, 3, 5

You Try It 4 Write the given numbers in order from smallest to largest.

−7, 4, −1, 0, 8

Your Solution

Solutions on p. A2

OBJECTIVE B Opposites

The distance from 0 to 3 on the number line is 3 units. The distance from 0 to -3 on the number line is 3 units. 3 and -3 are the same distance from 0 on the number line, but 3 is to the right of 0 and -3 is to the left of 0.

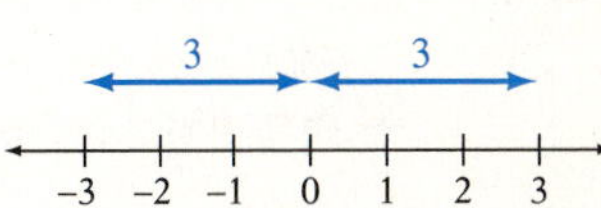

CONSIDER THIS
The [+/−] key on a calculator illustrates the idea of opposite. Entering 4 [+/−] gives the opposite of 4. Pressing the [+/−] key again gives 4, the opposite of −4.

Two numbers that are the same distance from zero on the number line but on opposite sides of zero are called **opposites.**

-3 is the opposite of 3
and
3 is the opposite of -3.

For any number n, the opposite of n is $-n$ and the opposite of $-n$ is n.

We can now define the **integers** as the whole numbers and their opposites.

A negative sign can be read as "the opposite of."

$-(3) = -3$ The opposite of positive 3 is negative 3.

$-(-3) = 3$ The opposite of negative 3 is positive 3.

Therefore, $-(a) = -a$ and $-(-a) = a$.

Note that with the introduction of negative integers and opposites, the symbols + and − can be read in different ways.

$6 + 2$	"six plus two"	+ is read "plus"
$+2$	"positive two"	+ is read "positive"
$6 - 2$	"six minus two"	− is read "minus"
-2	"negative two"	− is read "negative"
$-(-6)$	"the opposite of negative six"	− is read first as "the opposite of" and then as "negative"

When the symbols + and − indicate the operations of addition and subtraction, spaces are inserted before and after the symbol. When the symbols + and − indicate the sign of a number (positive or negative), there is no space between the symbol and the number.

Example 5 Find the opposite number.

a. -8 b. 15 c. a

Solution a. 8 b. -15 c. $-a$

You Try It 5 Find the opposite number.

a. 24 b. -13 c. $-b$

Your Solution

Example 6 Write the expression in words.

a. $7 - (-9)$ b. $-4 + 10$

Solution a. seven minus negative nine
b. negative four plus ten

You Try It 6 Write the expression in words.

a. $-3 - 12$ b. $8 + (-5)$

Your Solution

Example 7 Simplify.

a. $-(-27)$ b. $-(-c)$

Solution a. $-(-27) = 27$
b. $-(-c) = c$

You Try It 7 Simplify.

a. $-(-59)$ b. $-(y)$

Your Solution

Solutions on p. A2

OBJECTIVE C Absolute value

The **absolute value** of a number is the distance from zero to the number on the number line. Distance is never a negative number. Therefore, the absolute value of a number is a positive number or zero. The symbol for absolute value is "| |."

CONSIDER THIS

The important point to understand about absolute value is magnitude. If you run 5 miles west or 5 miles east, the distance is the same. Only the direction is different.

The distance from 0 to 3 is 3 units. Thus $|3| = 3$ (the absolute value of 3 is 3).

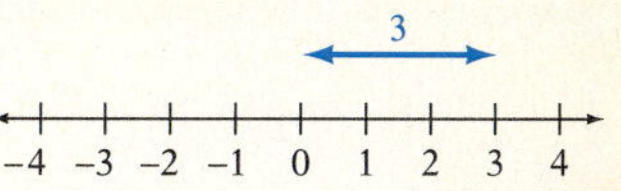

The distance from 0 to -3 is 3 units. Thus $|-3| = 3$ (the absolute value of -3 is 3).

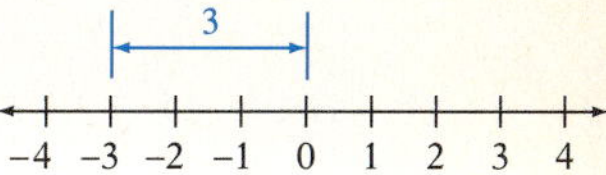

Therefore, $|3| = |-3| = 3$.
In general, $|a| = |-a|$.

The absolute value of a positive number is the number itself. $|5| = 5$

The absolute value of a negative number is the opposite of the negative number. $|-5| = 5$

The absolute value of zero is zero. $|0| = 0$

Evaluate $-|7|$.

The negative sign is *in front of* the absolute value symbol. Recall that a negative sign can be read as "the opposite of." $-|7|$ can be read "the opposite of the absolute value of 7."

$-|7| = -7$

Although $-4 < 1$, $|-4| > |1|$ because the distance from 0 to -4 on the number line is greater than the distance from 0 to 1.

Example 8 Find the absolute value of **a.** 6 and **b.** -9.

Solution **a.** $|6| = 6$

b. $|-9| = 9$

You Try It 8 Find the absolute value of **a.** -8 and **b.** 12.

Your Solution

Example 9 Evaluate **a.** $|-27|$ and **b.** $-|-14|$.

Solution **a.** $|-27| = 27$

b. $-|-14| = -14$

You Try It 9 Evaluate **a.** $|0|$ and **b.** $-|35|$.

Your Solution

Solutions on p. A2

Example 10 Evaluate $|-x|$, where $x = -4$.

Solution $|-x| = |-(-4)| = |4| = 4$

You Try It 10 Evaluate $|-y|$, where $y = 2$.

Your Solution

Example 11 Place the correct symbol, < or >, between the two numbers.

$|-6| \quad |-2|$

Solution $|-6| = 6, |-2| = 2$

$6 > 2$

$|-6| > |-2|$

You Try It 11 Place the correct symbol, < or >, between the two numbers.

$|-3| \quad |5|$

Your Solution

Example 12 Write the given numbers in order from smallest to largest.

$|-7|, -5, |0|, -(-4), -|-3|$

Solution $|-7| = 7, |0| = 0,$
$-(-4) = 4, -|-3| = -3$

$-5, -|-3|, |0|, -(-4), |-7|$

You Try It 12 Write the given numbers in order from smallest to largest.

$|6|, |-2|, -(-1), -4, -|-8|$

Your Solution

Solutions on p. A2

OBJECTIVE D Applications

Example 13

Which is the colder temperature, $-18°F$ or $-15°F$?

Strategy

To determine which is the colder temperature, compare the numbers -18 and -15. The lower number corresponds to the colder temperature.

Solution

$-18 < -15$

The colder temperature is $-18°F$.

You Try It 13

Which is closer to blast-off, -9 s and counting or -7 s and counting?

Your Strategy

Your Solution

Solution on p. A3

2.1 EXERCISES

Objective A

On the number line, which number is:

1. 3 units to the right of -2

2. 5 units to the right of -3

3. 4 units to the left of 3

4. 2 units to the left of -1

5. 6 units to the right of -3

6. 5 units to the left of 1

For Exercises 7–10, use the following number line.

A B C D E F G H I

7. If F is 1 and G is 2, what numbers are A and C?

8. If G is 3 and H is 4, what numbers are A and B?

9. If H is 0 and I is 1, what numbers are A and D?

10. If G is 2 and I is 4, what numbers are B and E?

Place the correct symbol, $<$ or $>$, between the two numbers.

11. -36 49

12. 21 -34

13. 53 -46

14. -27 -39

15. -51 -20

16. -136 0

17. -131 101

18. 127 -150

Write the given numbers in order from smallest to largest.

19. $-8, 21, -16, -13$

20. $17, -5, -19, 13$

21. $8, -1, -3, -6, 2$

22. $-10, 4, 12, -5, -7$

23. $11, -8, -1, 7, -6$

24. $10, -11, -2, 5, -7$

Objective B

Find the opposite of the number.

25. 4

26. 16

27. -2

28. -9

29. -31

30. -88

31. c

32. n

33. $-w$

34. $-d$

Write the expression in words.

35. $-(-11)$ 36. $-(-13)$ 37. $-(-d)$ 38. $-(-p)$

39. $-2 + (-5)$ 40. $5 + (-10)$ 41. $6 - (-7)$ 42. $-14 - (-3)$

43. $9 - 12$ 44. $-13 - 8$ 45. $-a - b$ 46. $m + (-n)$

Simplify.

47. $-(-38)$ 48. $-(-61)$ 49. $-(29)$ 50. $-(46)$ 51. $-(-52)$

52. $-(-73)$ 53. $-(-m)$ 54. $-(-z)$ 55. $-(b)$ 56. $-(p)$

Objective C

Find the absolute value of the number.

57. 4 58. -4 59. -7 60. 9 61. -11

62. 10 63. 12 64. -38 65. -65 66. 98

Evaluate.

67. $|-15|$ 68. $|-23|$ 69. $-|33|$ 70. $-|27|$

71. $|32|$ 72. $|25|$ 73. $-|-36|$ 74. $-|-41|$

75. $-|-81|$ 76. $-|-93|$ 77. $|x|$, where $x = 7$ 78. $|x|$, where $x = -10$

79. $|-x|$, where $x = 2$ 80. $|-x|$, where $x = 8$ 81. $|-y|$, where $y = -3$ 82. $|-y|$, where $y = -6$

Place the correct symbol, <, =, or >, between the two numbers.

83. $|7| \quad |-9|$

84. $|-12| \quad |8|$

85. $|-5| \quad |-2|$

86. $|6| \quad |13|$

87. $|-8| \quad |3|$

88. $|-1| \quad |-17|$

89. $|-14| \quad |14|$

90. $|x| \quad |-x|$

Write the given numbers in order from smallest to largest.

91. $|-8|, -(-3), |2|, -|-5|$

92. $-|6|, -(4), |-7|, -(-9)$

93. $-(-1), |-6|, |0|, -|3|$

94. $-|-7|, -9, -(5), |4|$

95. $-|2|, -(-8), 6, |1|, -7$

96. $-(-3), -|-8|, |5|, -|10|, -(-2)$

Objective D

The table below gives equivalent temperatures for combinations of temperature and wind speed. For example, a temperature of 15°F with a wind blowing at 10 mph has a cooling power equal to −3°F. Use the table for Exercises 97–100.

Windchill Factors

Wind Speed (mph)	*Temperature (degrees Fahrenheit)*															
	35	30	25	20	15	10	5	0	−5	−10	−15	−20	−25	−30	−35	−40
5	33	27	21	19	12	7	0	−5	−10	−15	−21	−26	−31	−36	−42	−47
10	22	16	10	3	−3	−9	−15	−22	−27	−34	−40	−46	−52	−58	−64	−71
15	16	9	2	−5	−11	−18	−25	−31	−38	−45	−51	−58	−65	−72	−78	−85
20	12	4	−3	−10	−17	−24	−31	−39	−46	−53	−60	−67	−74	−81	−88	−95
25	8	1	−7	−15	−22	−29	−36	−44	−51	−59	−66	−74	−81	−88	−96	−103
30	6	−2	−10	−18	−25	−33	−41	−49	−56	−64	−71	−79	−86	−93	−101	−109
35	4	−4	−12	−20	−27	−35	−43	−52	−58	−67	−74	−82	−89	−97	−105	−113
40	3	−5	−13	−21	−29	−37	−45	−53	−60	−69	−76	−84	−92	−100	−107	−115

97. Find the windchill factor when the temperature is 5°F and the wind speed is 15 mph.

98. Find the windchill factor when the temperature is 10°F and the wind speed is 20 mph.

99. Which feels colder, a temperature of 0°F with a 15 mph wind or a temperature of 10°F with a 25 mph wind?

100. Which feels colder, a temperature of −30°F with a 5 mph wind or a temperature of −20°F with a 10 mph wind?

Solve.

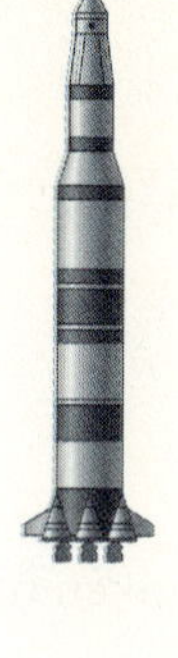

101. Which is closer to blast-off, -12 min and counting or -17 min and counting?

102. Which is the colder temperature, -19°C or -14°C?

103. In the stock market, the net change in the price of a share of stock is recorded as a positive or a negative number. If the price rises, the net change is positive. If the price falls, the net change is negative. If the net change for a share of Stock A is -2 and the net change for a share of Stock B is -1, which stock showed the least net change?

104. Some businesses record a profit as a positive number and a loss as a negative number. During the first quarter of this year, the loss experienced by a company was recorded as $-12{,}575$. During the second quarter of this year, the loss experienced by the company was $-11{,}350$. During which quarter was the loss greater?

105. Some businesses record a profit as a positive number and a loss as a negative number. During the third quarter of last year, the loss experienced by a company was recorded as $-26{,}800$. During the fourth quarter of last year, the loss experienced by the company was $-24{,}900$. During which quarter was the loss greater?

Critical Thinking

106. A is a point on the number line halfway between -9 and 3. B is a point halfway between A and the graph of 1 on the number line. B is the graph of what number?

107. Given x is an integer, find all values of x for which $|x| < 7$.

108. **a.** Name two numbers that are 4 units from 2 on the number line.
b. Name two numbers that are 5 units from 3 on the number line.

109. Determine whether the statement is always true, sometimes true, or never true.
a. The number $-n$ is a negative number.
b. A number and its opposite are different numbers.
c. $|x| > x$
d. $|x| > -x$
e. If n is a negative number, $-n$ is a positive number.
f. If n is a positive number, $-n$ is a negative number.

110. [W] In your own words, describe **(a)** the opposite of a number, **(b)** the absolute value of a number, and **(c)** the difference between the words *negative* and *minus.*

111. [W] Students A, B, C, and D were being questioned by their teacher. The teacher knew that one of the students had left an apple on the teacher's desk but did not know which one. Student A said it was either Student B or Student D. Student D said it was neither Student B nor Student C. If both statements were false, who left the apple on the teacher's desk? Explain how you arrived at your solution.

SECTION 2.2 Addition and Subtraction of Integers

OBJECTIVE A Addition of integers

Not only can an integer be graphed on a number line, an integer can be represented anywhere along a number line by an arrow. A positive number is represented by an arrow pointing to the right. A negative number is represented by an arrow pointing to the left. The absolute value of the number is represented by the length of the arrow. The integers 5 and -4 are shown on the number line in the figure below.

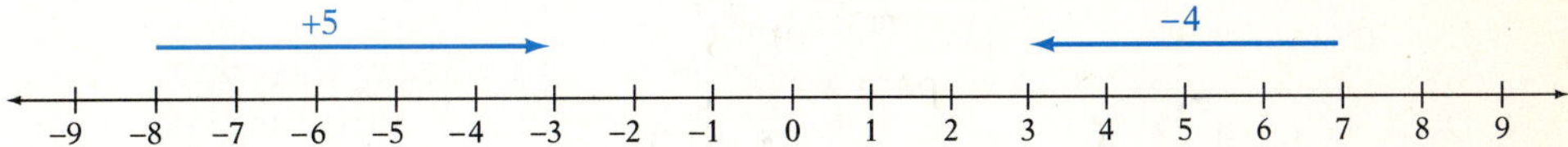

The sum of two integers can be shown on a number line. To add two integers, find the point on the number line corresponding to the first addend. At that point, draw an arrow representing the second addend. The sum is the number directly below the tip of the arrow.

CONSIDER THIS

There are several ways to model the addition of integers. The model at the right uses arrows. Another model uses money. For instance, if you are \$8 in debt (-8) and you repay \$5, then you are only \$3 in debt (-3). This can also be related to credit card debt. If you owe \$100 (-100) and charge \$25 more (-25), then you owe \$125 (-125).

$4 + 2 = 6$

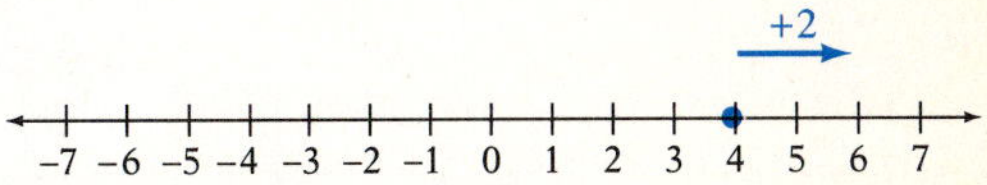

$-4 + (-2) = -6$

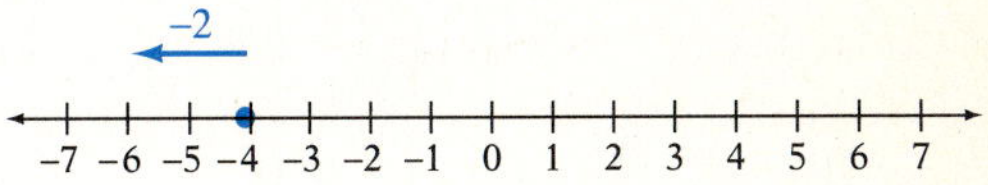

$-4 + 2 = -2$

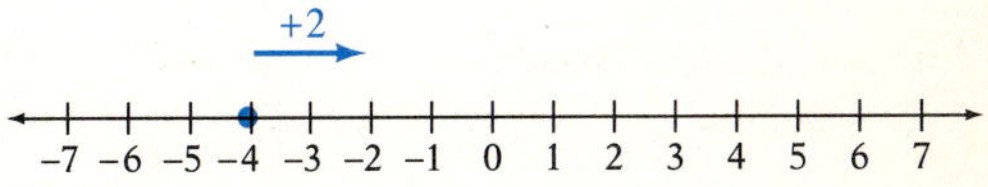

$4 + (-2) = 2$

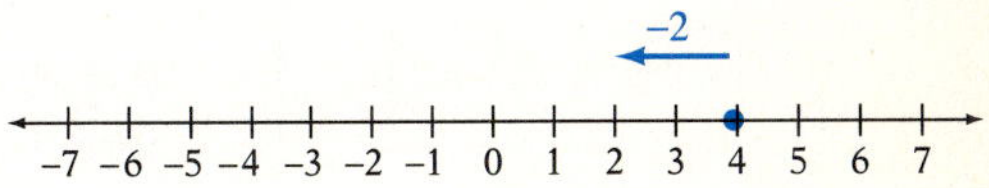

The sums shown above can be categorized by the signs of the addends.

The addends have the same sign.

$4 + 2$	positive 4 plus positive 2
$-4 + (-2)$	negative 4 plus negative 2

The addends have different signs.

$-4 + 2$	negative 4 plus positive 2
$4 + (-2)$	positive 4 plus negative 2

The rule for adding two integers depends on whether the signs of the addends are the same or different.

Rule for Adding Two Integers

TO ADD INTEGERS WITH THE SAME SIGN, add the absolute values of the numbers. Then attach the sign of the addends.

TO ADD INTEGERS WITH DIFFERENT SIGNS, find the absolute values of the numbers. Subtract the smaller absolute value from the larger absolute value. Then attach the sign of the addend with the larger absolute value.

Add: $(-4) + (-9)$

The signs of the addends are the same.
Add the absolute values of the numbers.
$|-4| = 4, |-9| = 9, 4 + 9 = 13$
Attach the sign of the addends.
(Both addends are negative. The sum is negative.) $(-4) + (-9) = -13$

Add: $6 + (-13)$

The signs of the addends are different.
Find the absolute values of the numbers.
$|6| = 6, |-13| = 13$
Subtract the smaller absolute value from the larger absolute value.
$13 - 6 = 7$
Attach the sign of the number with the larger absolute value.
$|-13| > |6|$. Attach the negative sign. $6 + (-13) = -7$

Add: $162 + (-247)$

The signs are different. Find the difference between the absolute values of the numbers.
$247 - 162 = 85$
Attach the sign of the number with the larger absolute value. $162 + (-247) = -85$

Add: $-14 + (-47)$

The signs are the same.
Add the absolute values of the numbers.
Attach the sign of the addends. $-14 + (-47) = -61$

Evaluate $-x + y$ when $x = -15$ and $y = -5$.

$-x + y$

Replace x with -15 and y with -5. $-(-15) + (-5)$

Simplify $-(-15)$. $= 15 + (-5)$

Add. $= 10$

POINT OF INTEREST

The Alexandrian astronomer Ptolemy began using *omicron, o,* the first letter of the Greek word that means "nothing," as the symbol for zero in A.D. 150. It was not until the 13th century, however, that Fibonacci introduced 0 to the Western world as a placeholder so that we could distinguish, for example, 45 from 405.

Add: $-8 + 8$

The signs are different. Find the difference between the absolute values of the numbers.

$8 - 8 = 0$ $\qquad -8 + 8 = 0$

Note in this example that we are adding a number and its opposite (-8 and 8), and the sum is 0. The opposite of a number is called its **additive inverse.** The opposite or additive inverse of -8 is 8, and the opposite or additive inverse of 8 is -8. **The sum of a number and its additive inverse is always zero.** This is known as the Inverse Property of Addition.

The properties of addition presented in Chapter 1 hold true for integers as well as whole numbers. These properties are repeated below, along with the Inverse Property of Addition.

The Addition Property of Zero	$a + 0 = a$ or $0 + a = a$
The Commutative Property of Addition	$a + b = b + a$
The Associative Property of Addition	$(a + b) + c = a + (b + c)$
The Inverse Property of Addition	$a + (-a) = 0$ or $-a + a = 0$

Add: $(-4) + (-6) + (-8) + 9$

	$(-4) + (-6) + (-8) + 9$
Add the first two numbers.	$= (-10) + (-8) + 9$
Add the sum to the third number.	$= (-18) + 9$
Continue until all the numbers have been added.	$= -9$

Check that, for the above example, the sum is the same if the numbers are added in a different order.

Example 1 Add: $42 + (-12) + (-30)$

Solution $42 + (-12) + (-30)$
$= 30 + (-30) = 0$

You Try It 1 Add: $-36 + 17 + (-21)$

Your Solution

Example 2 What is -162 increased by 98?

Solution $-162 + 98 = -64$

You Try It 2 Find the sum of -154 and -37.

Your Solution

Example 3 Evaluate $-x + y$ when $x = -11$ and $y = -2$.

Solution $-x + y$
$-(-11) + (-2) = 11 + (-2)$
$= 9$

You Try It 3 Evaluate $-x + y$ when $x = -3$ and $y = -10$.

Your Solution

Solutions on p. A3

OBJECTIVE B Subtraction of integers

Recall that the sign − can indicate the sign of a number, as in −3 (negative 3), or can indicate the operation of subtraction, as in 9 − 3 (nine minus three).

Look at each of the four subtraction expressions below and state whether the second number in each expression is a positive number or a negative number.

1. $(-10) - 8$
2. $(-10) - (-8)$
3. $10 - (-8)$
4. $10 - 8$

In **1** and **4**, the second number is positive 8. In **2** and **3**, the second number is negative 8.

Opposites are used to rewrite subtraction problems as related addition problems. Notice below that the subtraction of whole numbers is the same as the addition of the opposite number.

Subtraction		Addition of the Opposite	
$8 - 4$	=	$8 + (-4)$	= 4
$7 - 5$	=	$7 + (-5)$	= 2
$9 - 2$	=	$9 + (-2)$	= 7

Subtraction of integers can be written as the addition of the opposite number. To subtract two integers, rewrite the subtraction expression as the first number plus the opposite of the second number. Some examples are shown below.

First number	−	second number	=	First number	+	opposite of the second number
8	−	15	=	8	+	(−15) = −7
8	−	(−15)	=	8	+	15 = 23
−8	−	15	=	−8	+	(−15) = −23
−8	−	(−15)	=	−8	+	15 = 7

Rule for Subtracting Two Integers

To subtract two integers, add the opposite of the second integer to the first integer.

Subtract: $(-15) - 75$

Rewrite the subtraction operation as the sum of the first number and the opposite of the second number. The opposite of 75 is −75.

$(-15) - 75$

$= (-15) + (-75)$

Add.

$= -90$

When subtraction occurs several times in an expression, rewrite each subtraction as addition of the opposite and then add.

Subtract: $-13 - 5 - (-8)$

$-13 - 5 - (-8)$

Rewrite each subtraction as addition of the opposite. $= -13 + (-5) + 8$

Add. $= -18 + 8$
$= -10$

Simplify: $-14 + 6 - (-7)$

This problem involves both addition and subtraction. Rewrite the subtraction as addition of the opposite.

$-14 + 6 - (-7)$
$= -14 + 6 + 7$

Add. $= -8 + 7$
$= -1$

Evaluate $a - b$ when $a = -2$ and $b = -9$.

$a - b$

Replace a with -2 and b with -9. $-2 - (-9)$

Rewrite the subtraction as addition of the opposite. $= -2 + 9$

Add. $= 7$

Example 4 What is -12 minus 8?

Solution $-12 - 8 = -12 + (-8)$
$= -20$

You Try It 4 What is 14 less than -8?

Your Solution

Example 5 Simplify:
$-8 - 30 - (-12) - 7 - (-14)$

Solution $-8 - 30 - (-12) - 7 - (-14)$
$= -8 + (-30) + 12 + (-7) + 14$
$= -38 + 12 + (-7) + 14$
$= -26 + (-7) + 14$
$= -33 + 14$
$= -19$

You Try It 5 Simplify:
$-4 - (-3) + 12 - (-7) - 20$

Your Solution

Example 6 Evaluate $-x - y$ when $x = -4$ and $y = -3$.

Solution $-x - y$
$-(-4) - (-3) = 4 - (-3)$
$= 4 + 3$
$= 7$

You Try It 6 Evaluate $x - y$ when $x = -9$ and $y = 7$.

Your Solution

Solutions on p. A3

OBJECTIVE C Applications and formulas

Example 7

Find the temperature after an increase of 8°C from −5°C.

Strategy

To find the temperature, add the increase (8) to the previous temperature (−5).

Solution

$-5 + 8 = 3$

The temperature is 3°C.

You Try It 7

Find the temperature after an increase of 10°C from −3°C.

Your Strategy

Your Solution

Example 8

The average temperature on the sunlit side of the moon is approximately 215°F. The average temperature on the dark side is approximately −250°F. Find the difference between these average temperatures.

Strategy

To find the difference, subtract the average temperature on the dark side of the moon (−250) from the average temperature on the sunlit side (215).

Solution

$$\begin{aligned} 215 - (-250) &= 215 + 250 \\ &= 465 \end{aligned}$$

The difference is 465°F.

You Try It 8

The average temperature on the earth's surface is 57°F. The average temperature throughout the earth's stratosphere is −70°F. Find the difference between these average temperatures.

Your Strategy

Your Solution

Example 9

The distance, d, between point a and point b on the number line is given by the formula $d = |a - b|$. Use the formula to find d when $a = 7$ and $b = -8$.

Strategy

To find d, replace a by 7 and b by −8 in the given formula and solve for d.

Solution

$d = |a - b|$
$d = |7 - (-8)|$
$d = |7 + 8|$
$d = |15|$
$d = 15$

The distance between the two points is 15 units.

You Try It 9

The distance, d, between point a and point b on the number line is given by the formula $d = |a - b|$. Use the formula to find d when $a = -6$ and $b = 5$.

Your Strategy

Your Solution

Solutions on p. A3

2.2 EXERCISES

Objective A

Add.

1. $3 + (-5)$
2. $6 + (-7)$
3. $-3 + (-8)$
4. $-12 + (-1)$
5. $-4 + (-5)$
6. $-12 + (-12)$
7. $6 + (-9)$
8. $4 + (-6)$
9. $-6 + 7$
10. $-9 + 8$
11. $(-5) + (-10)$
12. $(-3) + (-17)$
13. $8 + 12$
14. $16 + 23$
15. $-7 + 7$
16. $-11 + 11$
17. $9 + (-6)$
18. $7 + (-8)$
19. $(-2) + (-15)$
20. $(-17) + (-5)$
21. $(-3) + 14$
22. $(-16) + 12$
23. $-8 + (-11)$
24. $-7 + (-10)$
25. $(-15) + (-6)$
26. $(-18) + (-3)$
27. $0 + (-14)$
28. $-19 + 0$
29. $73 + (-54)$
30. $-89 + 62$
31. $-46 + 93$
32. $-31 + 80$
33. $2 + (-3) + (-4)$
34. $7 + (-2) + (-8)$
35. $-3 + (-12) + (-15)$
36. $9 + (-6) + (-16)$
37. $-17 + (-3) + 29$
38. $13 + 62 + (-38)$
39. $11 + (-22) + 4 + (-5)$
40. $-14 + (-3) + 7 + (-6)$
41. $-22 + 10 + 2 + (-18)$
42. $-6 + (-8) + 13 + (-4)$
43. $-25 + (-31) + 24 + 19$
44. $10 + (-14) + (-21) + 8$

Solve.

45. What is 3 increased by -21?

46. Find 12 plus -9.

47. What is 16 more than -5?

48. What is 17 added to -7?

49. Find the total of -3, -8, and 12.

50. Find the sum of 5, -16, and -13.

Evaluate the expression for the given values of the variables.

51. $x + y$, when $x = -5$ and $y = -7$

52. $-x + y$, when $x = -5$ and $y = -7$

53. $x - y$, when $x = -5$ and $y = -7$

54. $-x - y$, when $x = -5$ and $y = -7$

55. $-a + b$, when $a = -8$ and $b = -3$

56. $a - b$, when $a = -8$ and $b = -3$

57. $a + b + c$, when $a = -4$, $b = 6$, and $c = -9$

58. $a + b + c$, when $a = -10$, $b = -6$, and $c = 5$

59. $x + y + (-z)$, when $x = -3$, $y = 6$, and $z = -17$

60. $-x + (-y) + z$, when $x = -2$, $y = 8$, and $z = -11$

Identify the property that justifies the statement.

61. $-12 + 5 = 5 + (-12)$

62. $-33 + 0 = -33$

63. $-46 + 46 = 0$

64. $-7 + (3 + 2) = (-7 + 3) + 2$

Use the given property of addition to complete the statement.

65. The Inverse Property of Addition
$16 + ? = 0$

66. The Commutative Property of Addition
$-5 + 17 = 17 + ?$

67. The Associative Property of Addition
$-11 + (6 + 9) = (? + 6) + 9$

68. The Addition Property of Zero
$-13 + ? = -13$

69. The Commutative Property of Addition
$-2 + ? = -4 + (-2)$

70. The Inverse Property of Addition
$? + (-18) = 0$

Objective B

Subtract.

71. $7 - 14$

72. $6 - 9$

73. $-7 - 2$

74. $-9 - 4$

75. $7 - (-2)$

76. $3 - (-4)$

77. $-6 - (-6)$

78. $-4 - (-4)$

79. $16 - 8$

80. $12 - 3$

81. $6 - (-12)$

82. $8 - (-5)$

83. $-12 - 16$

84. $-10 - 7$

85. $(-9) - (-3)$

86. $(-7) - (-4)$

87. $3 - 18$

88. $9 - 12$

89. $(-6) - 9$

90. $(-8) - 7$

91. $4 - (-14)$

92. $6 - (-15)$

93. $-4 - (-16)$

94. $-6 - (-15)$

95. $(-8) - (-12)$

96. $(-14) - (-7)$

97. $3 - (-24)$

98. $9 - (-9)$

99. $(-41) - 65$

100. $(-39) - 78$

101. $57 - 86$

102. $-95 - (-28)$

Solve.

103. Find the difference between 2 and -16.

104. How much larger is 5 than -11?

105. What is -10 decreased by -4?

106. Find -13 minus -8.

107. What is 6 less than -9?

108. What is the difference between -3 and 12?

Simplify.

109. $-4 - 3 - 2$

110. $4 - 5 - 12$

111. $12 - (-7) - 8$

112. $-12 - (-3) - (-15)$

113. $4 - 12 - (-8)$

114. $13 - 7 - 15$

115. $-30 - (-65) - 29 - 4$

116. $-16 - 47 - 63 - 12$

117. $42 - (-30) - 65 - (-11)$

118. $3 - 10 - 3$

119. $12 - (-6) + 8$

120. $-7 + 9 - (-3)$

121. $-5 + 12 - (-4)$

122. $-8 - (-14) + 7$

123. $-4 + 6 - 8 - 2$

124. $9 - 12 + 0 - 5$

125. $11 - (-2) - 6 + 10$

126. $5 + 4 - (-3) - 7$

127. $-1 - 8 + 6 - (-2)$

128. $-13 + 9 - (-10) - 4$

129. $6 - (-13) - 14 + 7$

Evaluate the expression for the given values of the variables.

130. $x - y$, when $x = -3$ and $y = 9$

131. $-x - y$, when $x = -3$ and $y = 9$

132. $x - (-y)$, when $x = -3$ and $y = 9$

133. $-x - (-y)$, when $x = -3$ and $y = 9$

134. $-a - b$, when $a = -6$ and $b = 10$

135. $a - (-b)$, when $a = -6$ and $b = 10$

136. $a - b - c$, when $a = 4$, $b = -2$, and $c = 9$

137. $a - b - c$, when $a = -1$, $b = 7$, and $c = -15$

138. $x - y - (-z)$, when $x = -9$, $y = 3$, and $z = 30$

139. $-x - (-y) - z$, when $x = 8$, $y = 1$, and $z = -14$

Objective C

Solve.

The elevation, or height, of places on the earth is measured in relation to sea level, or the average level of the ocean's surface. The table below shows height above sea level as a positive number and depth below sea level as a negative number. Use the table for Exercises 140–143.

Continent	*Highest Elevation (in meters)*		*Lowest Elevation (in meters)*	
Africa	Mt. Kilimanjaro	5895	Qattara Depression	−133
Asia	Mt. Everest	8848	Dead Sea	−400
Europe	Mt. Elbrus	5634	Caspian Sea	−28
America	Mt. Aconcagua	6960	Death Valley	−86

140. What is the difference in elevation between Mt. Elbrus and the Caspian Sea?

141. What is the difference in elevation between Mt. Aconcagua and Death Valley?

142. Find the difference in elevation between Mt. Kilimanjaro and the Qattara Depression.

143. Find the difference in elevation between Mt. Everest and the Dead Sea.

144. Find the temperature after a rise of 9°C from −6°C.

Solve.

145. Find the temperature afer a rise of 7°C from -18°C.

146. The high temperature for the day was 10°C. The low temperature was -4°C. Find the difference between the high and low temperatures for the day.

147. The low temperature for the day was -2°C. The high temperature was 11°C. Find the difference between the high and low temperatures for the day.

148. Use the equation $S = N - P$, where S is a golfer's score above or below par, N is the number of strokes made by the golfer, and P is par, to find a golfer's score when the golfer made 204 strokes and par is 216.

149. Use the equation $S = N - P$, where S is a golfer's score above or below par, N is the number of strokes made by the golfer, and P is par, to find a golfer's score when the golfer made 69 strokes and par is 72.

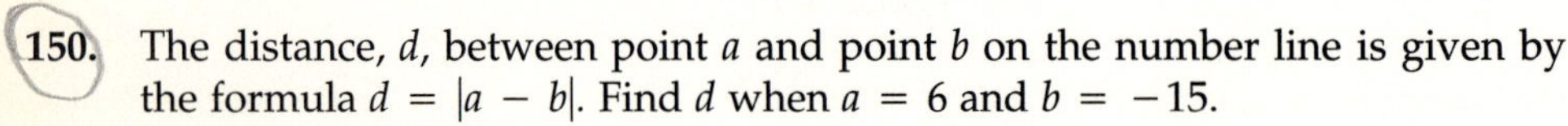

150. The distance, d, between point a and point b on the number line is given by the formula $d = |a - b|$. Find d when $a = 6$ and $b = -15$.

151. The distance, d, between point a and point b on the number line is given by the formula $d = |a - b|$. Find d when $a = 7$ and $b = -12$.

Critical Thinking

152. Given the list of numbers at the right, find the largest difference that can be obtained by subtracting one number in the list from a different number in the list. What is the smallest difference?

5, -2, -9, 11, 14

153. Use a reference at the library to find the record high temperature and the record low temperature for your state. Calculate the difference between these extremes. What are the highest recorded temperature and the lowest recorded temperature for North America? What is the difference between the two? What are the highest recorded temperature and the lowest recorded temperature worldwide? Find the difference between these extreme temperatures. Provide answers on both Fahrenheit and Celsius scales.

154. Determine whether the statement is always true, sometimes true, or never true.

a. The difference between a number and its additive inverse is zero.

b. The sum of a negative number and a negative number is a negative number.

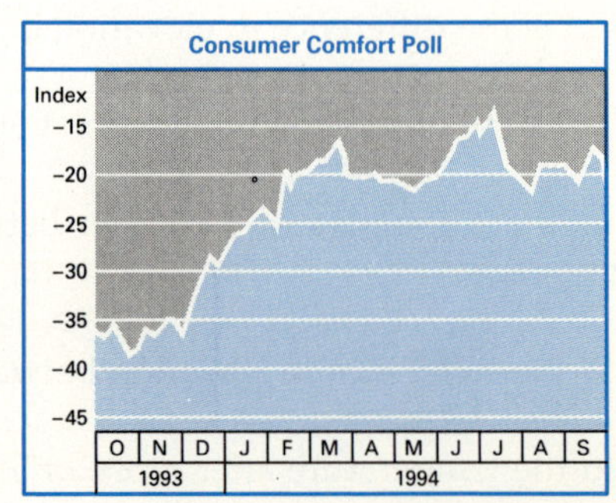

155. [W] The graph at the right indicates consumer confidence in the economy. Approximate the difference between the highest and lowest points of confidence on the graph. Write a paragraph describing your interpretation of the data presented in the graph.

156. [W] Describe the steps involved in using a calculator to simplify the expression $-17 - (-8) + (-5)$.

SECTION 2.3 Multiplication and Division of Integers

OBJECTIVE  Multiplication of integers

When 5 is multiplied by a sequence of decreasing integers, each product decreases by 5.

$5(3) = 15$
$5(2) = 10$
$5(1) = 5$
$5(0) = 0$

CONSIDER THIS

Another way to illustrate that positive times negative equals negative is to use repeated addition. For instance $(-5)(3)$ is $(-5) + (-5) + (-5) = (-15)$.

The pattern developed can be continued so that 5 is multiplied by a sequence of negative numbers. The resulting products must be negative in order to maintain the pattern of decreasing by 5.

$5(-1) = -5$
$5(-2) = -10$
$5(-3) = -15$
$5(-4) = -20$

This example illustrates that the product of a positive number and a negative number is negative.

When -5 is multiplied by a sequence of decreasing integers, each product increases by 5.

$-5(3) = -15$
$-5(2) = -10$
$-5(1) = -5$
$-5(0) = 0$

POINT OF INTEREST

Operations with negative numbers were not accepted until the late 13th century. One of the first attempts to prove that the product of two negative numbers is positive was made in the book *Ars Magna* by Gerolamo Cardano in 1545.

The pattern developed can be continued so that -5 is multiplied by a sequence of negative numbers. The resulting products must be positive in order to maintain the pattern of increasing by 5.

$-5(-1) = 5$
$-5(-2) = 10$
$-5(-3) = 15$
$-5(-4) = 20$

This example illustrates that the product of two negative numbers is positive.

The pattern for multiplication shown above is summarized in the following rule for multiplying integers.

Rule for Multiplying Two Integers

To multiply integers with the same sign, multiply the absolute values of the factors. The product is positive.

$4(8) = 32$

$-4(-8) = 32$

To multiply integers with different signs, multiply the absolute values of the factors. The product is negative.

$-4(8) = -32$

$4(-8) = -32$

Evaluate $-ab$ when $a = -2$ and $b = -9$.

$-ab$

Replace a with -2 and b with -9. $\quad -(-2)(-9)$

Multiply -2 times -9. $\quad = -(18)$

Simplify. $\quad = -18$

The properties of multiplication presented in Chapter 1 hold true for integers as well as whole numbers. These properties are repeated below.

The Multiplication Property of Zero	$a \cdot 0 = 0$	or	$0 \cdot a = 0$
The Multiplication Property of One	$a \cdot 1 = a$	or	$1 \cdot a = a$
The Commutative Property of Multiplication	$a \cdot b = b \cdot a$		
The Associative Property of Multiplication	$(a \cdot b) \cdot c = a \cdot (b \cdot c)$		

Multiply: $2(-3)(-5)(-7)$

$2(-3)(-5)(-7)$

Multiply the first two numbers. $= -6(-5)(-7)$

Then multiply the product by the third number. $= 30(-7)$

Continue until all the numbers have been multiplied. $= -210$

Check that, for the above example, the product is the same if the numbers are multiplied in a different order.

Example 1 Multiply: $-15 \cdot 4$

Solution $-15(4) = -60$

You Try It 1 Multiply: $-3 \cdot 12$

Your Solution

Example 2 Find -42 times 62.

Solution $-42 \cdot 62 = -2604$

You Try It 2 What is -38 multiplied by 51?

Your Solution

Example 3 Multiply: $-5(-4)(6)(-3)$

Solution

$$-5(-4)(6)(-3) = 20(6)(-3) = 120(-3) = -360$$

You Try It 3 Multiply: $-7(-8)(9)(-2)$

Your Solution

Example 4 Evaluate $-5x$ when $x = -11$.

Solution $-5x$

$-5(-11) = 55$

You Try It 4 Evaluate $-9y$ when $y = 20$.

Your Solution

Solutions on pp. A3–4

OBJECTIVE B Division of integers

Recall that the fraction bar can be read "divided by." Therefore, $\frac{8}{2}$ can be read "8 divided by 2."

For every division problem, there is a related multiplication problem.

Division: $\frac{8}{2} = 4$ Related multiplication: $4 \cdot 2 = 8$

This fact can be used to illustrate the rule for dividing integers.

Rule for Dividing Two Integers

To divide two numbers with the same sign, divide the absolute values of the numbers. The quotient is positive.

To divide two numbers with different signs, divide the absolute values of the numbers. The quotient is negative.

$\frac{12}{3} = 4$ because $4(3) = 12$.

$\frac{-12}{-3} = 4$ because $4(-3) = -12$.

$\frac{12}{-3} = -4$ because $-4(-3) = 12$.

$\frac{-12}{3} = -4$ because $-4(3) = -12$.

Note that $\frac{12}{-3}$, $\frac{-12}{3}$, and $-\frac{12}{3}$ are all equal to -4.

If a and b are two nonzero integers, then $\frac{a}{-b} = \frac{-a}{b} = -\frac{a}{b}$.

Evaluate $a \div (-b)$ when $a = -28$ and $b = -4$.

	$a \div (-b)$
Replace a with -28 and b with -4.	$-28 \div (-(-4))$
Simplify $-(-4)$.	$= -28 \div (4)$
Divide.	$= -7$

The division properties of zero and one, which were presented in Chapter 1, hold true for integers as well as whole numbers. These properties are repeated below.

Division Properties of Zero and One

If $a \neq 0$, $0 \div a = 0$.

If $a \neq 0$, $a \div a = 1$.

$a \div 1 = a$

$a \div 0$ is undefined.

Example 5 Divide: $(-120) \div (-8)$

Solution $(-120) \div (-8) = 15$

You Try It 5 Divide: $(-135) \div (-9)$

Your Solution

Example 6 Find the quotient of -23 and -23.

Solution $-23 \div (-23) = 1$

You Try It 6 What is 0 divided by -17?

Your Solution

Example 7 Divide: $\frac{95}{-5}$

Solution $\frac{95}{-5} = -19$

You Try It 7 Divide: $\frac{84}{-6}$

Your Solution

Example 8 Divide: $x \div 0$

Solution Division by zero is not defined.

$x \div 0$ is undefined.

You Try It 8 Divide: $x \div 1$

Your Solution

Example 9 Evaluate $\frac{-a}{b}$ when $a = -6$ and $b = -3$.

Solution $\frac{-a}{b}$

$$\frac{-(-6)}{-3} = \frac{6}{-3} = -2$$

You Try It 9 Evaluate $\frac{a}{-b}$ when $a = -14$ and $b = -7$.

Your Solution

Solutions on p. A4

OBJECTIVE C Applications

Example 10

The combined scores of the top five golfers in a tournament equaled -10 (10 under par). What was the average score of the five golfers?

Strategy

To find the average score, divide the combined scores (-10) by the number of golfers (5).

Solution

$-10 \div 5 = -2$

The average score was -2.

You Try It 10

The melting point of mercury is -38°C. The melting point of argon is five times the melting point of mercury. Find the melting point of argon.

Your Strategy

Your Solution

Solution on p. A4

2.3 EXERCISES

Objective A

Multiply.

1. $-4 \cdot 6$
2. $-7 \cdot 3$
3. $-2(-3)$
4. $-5(-1)$
5. $(9)(2)$
6. $(3)(8)$
7. $5(-4)$
8. $4(-7)$
9. $-8(2)$
10. $-9(3)$
11. $(-5)(-5)$
12. $(-3)(-6)$
13. $(-7)(0)$
14. $-11(1)$
15. $14(3)$
16. $62(9)$
17. $-32(4)$
18. $-24(3)$
19. $(-8)(-26)$
20. $(-4)(-35)$
21. $9(-27)$
22. $8(-40)$
23. $-5 \cdot (23)$
24. $-6 \cdot (38)$
25. $-7(-34)$
26. $-4(-51)$
27. $4 \cdot (-8) \cdot 3$
28. $5 \cdot 7 \cdot (-2)$
29. $(-6)(5)(7)$
30. $(-9)(-9)(2)$
31. $-8(-7)(-4)$
32. $-1(4)(-9)$

Solve.

33. What is twice -20?
34. Find the product of 100 and -7.
35. What is -30 multiplied by -6?
36. What is -9 times -40?
37. Write the product of $-q$ and r.
38. Write the product of $-f$, g, and h.

Evaluate the expression for the given values of the variables.

39. xy, when $x = -3$ and $y = -8$

40. $-xy$, when $x = -3$ and $y = -8$

41. $x(-y)$, when $x = -3$ and $y = -8$

42. $-xyz$, when $x = -6$, $y = 2$, and $z = -5$

43. $-8a$, when $a = -24$

44. $-7n$, when $n = -51$

45. $5xy$, when $x = -9$ and $y = -2$

46. $8ab$, when $a = 7$ and $b = -1$

47. $-4cd$, when $c = 25$ and $d = -8$

48. $-5st$, when $s = -40$ and $t = -8$

Identify the property that justifies the statement.

49. $0(-7) = 0$

50. $1(-22) = -22$

51. $-8(-5) = -5(-8)$

52. $-3(9 \cdot 4) = (-3 \cdot 9)4$

Use the given property of multiplication to complete the statement.

53. The Commutative Property of Multiplication
$-3(-9) = -9(?)$

54. The Associative Property of Multiplication
$?(5 \cdot 10) = (-6 \cdot 5)(10)$

55. The Multiplication Property of Zero
$-81 \cdot ? = 0$

56. The Multiplication Property of One
$?(-14) = -14$

Objective B

Divide.

57. $12 \div (-6)$

58. $18 \div (-3)$

59. $(-72) \div (-9)$

60. $(-64) \div (-8)$

Divide.

61. $0 \div (-6)$

62. $-49 \div 1$

63. $81 \div (-9)$

64. $-40 \div (-5)$

65. $\frac{72}{-3}$

66. $\frac{44}{-4}$

67. $\frac{-93}{-3}$

68. $\frac{-98}{-7}$

69. $-114 \div (-6)$

70. $-91 \div (-7)$

71. $-53 \div 0$

72. $(-162) \div (-162)$

73. $-128 \div 4$

74. $-130 \div (-5)$

75. $(-200) \div 8$

76. $(-92) \div (-4)$

Solve.

77. Find the quotient of -700 and 70.

78. Find 550 divided by -5.

79. What is -670 divided by -10?

80. What is the quotient of -333 and -3?

Evaluate the expression for the given values of the variables.

81. $a \div b$, when $a = -36$ and $b = -4$

82. $-a \div b$, when $a = -36$ and $b = -4$

83. $a \div (-b)$, when $a = -36$ and $b = -4$

84. $(-a) \div (-b)$, when $a = -36$ and $b = -4$

85. $\frac{x}{y}$, when $x = -42$ and $y = -7$

86. $\frac{-x}{y}$, when $x = -42$ and $y = -7$

87. $\frac{x}{-y}$, when $x = -42$ and $y = -7$

88. $\frac{-x}{-y}$, when $x = -42$ and $y = -7$

Objective C

Solve.

89. The boiling point of radon is −62°C. The boiling point of argon is three times the boiling point of radon. Find the boiling point of argon.

90. The boiling point of chlorine is −35°C. The boiling point of neon is seven times the boiling point of chlorine. Find the boiling point of neon.

91. The combined scores of the top four golfers in a tournament equaled −12 (12 under par). What was the average score of the four golfers?

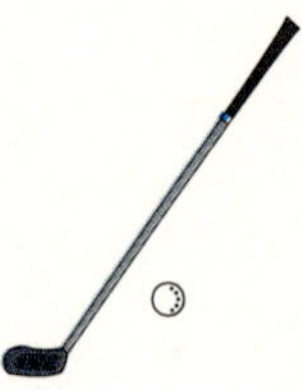

92. The windchill factor when the temperature is −15°F and the wind is blowing at 20 mph is five times the windchill factor when the temperature is 25°F and the wind is blowing at 35 mph. If the windchill factor at 25°F with a 35 mph wind is −12°F, what is the windchill factor at −15°F with a 20 mph wind?

A **geometric sequence** is a list of numbers in which each number after the first is found by multiplying the preceding number in the list by the same number. For example, in the sequence 1, 3, 9, 27, 81, . . . , each number after the first is found by multiplying the preceding number in the list by 3. To find the multiplier in a geometric sequence, divide the second number in the sequence by the first number; for the example above, $3 \div 1 = 3$.

93. Find the next three numbers in the geometric sequence −5, 15, −45, . . .

94. Find the next three numbers in the geometric sequence 2, −4, 8, . . .

95. Find the next three numbers in the geometric sequence −3, −12, −48, . . .

96. Find the next three numbers in the geometric sequence −1, −5, −25, . . .

Critical Thinking

97. Use repeated addition to show that the product of two integers with different signs is a negative number.

98. **a.** Find the largest possible product of two negative integers whose sum is −18. $x + y = -18$

b. Find the smallest possible sum of two negative integers whose product is 16. $xy = 16$

99. Determine whether the statement is always true, sometimes true, or never true.

a. The product of a number and its additive inverse is a negative number.
b. The product of an odd number of negative numbers is a negative number.
c. The square of a negative number is a positive number.

100. Find all negative integers x such that $1 - 3x < 12$.

101. [W] Describe the steps involved in using a calculator to simplify $(-2491) \div (-47)$.

102. [W] In your own words, describe the rules for multiplying and dividing integers.

SECTION 2.4 The Order of Operations Agreement

OBJECTIVE A The Order of Operations Agreement

The Order of Operations Agreement used in Chapter 1 is repeated here for your reference.

> *The Order of Operations Agreement*
>
> **Step 1** Do all operations inside parentheses.
>
> **Step 2** Simplify any numerical expressions containing exponents.
>
> **Step 3** Do multiplication and division as they occur from left to right.
>
> **Step 4** Do addition and subtraction as they occur from left to right.

Note how the following expressions containing exponents are simplified.

$(-3)^2 = (-3)(-3) = 9$ The (-3) is squared. Multiply -3 by -3.

$-(3^2) = -(3 \cdot 3) = -9$ Read $-(3^2)$ as "the opposite of three squared." 3^2 is 9. The opposite of 9 is -9.

$-3^2 = -(3^2) = -9$ The expression -3^2 is the same as $-(3^2)$.

Simplify: $8 - 4 \div (-2)$

There are no operations inside parentheses (Step 1).
There are no exponents (Step 2).
Do the division (Step 3). $8 - 4 \div (-2) = 8 - (-2)$

Do the subtraction (Step 4). $= 8 + 2 = 10$

Simplify: $(-3)^2 - 2(8 - 3) + (-5)$

$(-3)^2 - 2(8 - 3) + (-5)$

Perform operations inside parentheses. $= (-3)^2 - 2(5) + (-5)$

Simplify expressions with exponents. $= 9 - 2(5) + (-5)$

Do multiplication and division as they occur from left to right. $= 9 - 10 + (-5)$

Do addition and subtraction as they occur from left to right. $= 9 + (-10) + (-5)$
$= -1 + (-5)$
$= -6$

Evaluate $ab - b^2$ when $a = 2$ and $b = -6$.

	$ab - b^2$
Replace a with 2 and each b with -6.	$2(-6) - (-6)^2$
Use the Order of Operations Agreement to simplify the resulting numerical expression. Simplify the exponential expression.	$= 2(-6) - 36$
Do the multiplication.	$= -12 - 36$
Do the subtraction.	$= -12 + (-36)$ $= -48$

Example 1 Simplify $(-4)^2$ and -4^2.

Solution $(-4)^2 = (-4)(-4) = 16$

$-4^2 = -(4 \cdot 4) = -16$

You Try It 1 Simplify $(-5)^2$ and -5^2.

Your Solution

Example 2 Simplify: $12 \div (-2)^2 - 5$

Solution

$$\begin{aligned} 12 \div (-2)^2 - 5 &= 12 \div 4 - 5 \\ &= 3 - 5 \\ &= 3 + (-5) \\ &= -2 \end{aligned}$$

You Try It 2 Simplify: $8 \div 4 \cdot 4 - (-2)^2$

Your Solution

Example 3 Simplify: $(-3)^2(5 - 7)^2 - (-9) \div 3$

Solution

$$\begin{aligned} &(-3)^2(5 - 7)^2 - (-9) \div 3 \\ &= (-3)^2(-2)^2 - (-9) \div 3 \\ &= (9)(4) - (-9) \div 3 \\ &= 36 - (-9) \div 3 \\ &= 36 - (-3) \\ &= 36 + 3 \\ &= 39 \end{aligned}$$

You Try It 3 Simplify: $(-2)^2(3 - 7)^2 - (-16) \div (-4)$

Your Solution

Example 4 Evaluate $6a \div (-b)$ when $a = -2$ and $b = -3$.

Solution

$$\begin{aligned} &6a \div (-b) \\ &6(-2) \div (-(-3)) \\ &= 6(-2) \div (3) \\ &= -12 \div 3 \\ &= -4 \end{aligned}$$

You Try It 4 Evaluate $3a - 4b$ when $a = -2$ and $b = 5$.

Your Solution

Solutions on p. A4

2.4 EXERCISES

Objective A

Simplify.

1. $3 - 12 \div 2$

2. $-16 \div 2 + 8$

3. $2(3 - 5) - 2$

4. $2 - (8 - 10) \div 2$

5. $4 - (-3)^2$

6. $(-2)^2 - 6$

7. $4 \cdot (2 - 4) - 4$

8. $6 - 2 \cdot (1 - 3)$

9. $4 - (-2)^2 + (-3)$

10. $-3 + (-6)^2 - 1$

11. $3^3 - 4(2)$

12. $9 \div 3 - (-3)^2$

13. $3 \cdot (6 - 2) \div 6$

14. $4 \cdot (2 - 7) \div 5$

15. $2^3 - (-3)^2 + 2$

16. $6(8 - 2) \div 4$

17. $6 - 2(1 - 5)$

18. $(-2)^2 - (-3)^2 + 1$

19. $6 - (-4)(-3)^2$

20. $4 - (-5)(-2)^2$

21. $4 \cdot 2 - 3 \cdot 7$

22. $16 \div 2 - 9 \div 3$

23. $(-2)^2 - 5(3) - 1$

24. $4 - 2 \cdot 7 - 3^2$

25. $3 \cdot 2^3 + 5 \cdot (3 + 2) - 17$

26. $3 \cdot 4^2 - 16 - 4 + 3 - (1 - 2)^2$

27. $-12(6 - 8) + 1^3 \cdot 3^2 \cdot 2 - 6(2)$

28. $-3 \cdot (-2)^2 \cdot 4 \div 8 - (-12)$

29. $10(9) - (8 + 7) \div 5 + 6 - 7 + 8$

30. $-27 - (-3)^2 - 2 - 7 + 6 \cdot 3$

31. $(-1) \cdot (4 - 7)^2 \div 9 + 6 - 3 - 4(2)$

32. $16 - 4 \cdot 8 + 4^2 - (-18) - (-9)$

Evaluate the variable expression given $a = -2$, $b = 4$, $c = -1$, and $d = 3$.

33. $3a + 2b$

34. $a - 2c$

35. $16 \div (ac)$

36. $6b \div (-a)$

37. $bc \div (2a)$

38. $a^2 - b^2$

39. $b^2 - c^2$

40. $2a - (c + a)^2$

41. $(b - a)^2 + 4c$

42. $(b + c) \div d$

43. $(d - b) \div c$

44. $(2d + b) \div (-a)$

45. $(b - d) \div (c - a)$

46. $bd \div a + c$

47. $(d - a)^2 \div 5$

48. $(b + c)^2 + (a + d)^2$

49. $(d - a)^2 - 3c$

50. $(b + d)^2 - 4a$

Critical Thinking

51. What is the smallest integer greater than $-2^2 - (-3)^2 + 5(4) \div 10 - (-6)$?

52. Evaluate.
 a. $1^3 + 2^3 + 3^3 + 4^3$
 b. $(-1)^3 + (-2)^3 + (-3)^3 + (-4)^3$
 c. $1^3 + 2^3 + 3^3 + 4^3 + 5^3$
 Based on your answers to parts **a, b,** and **c,** evaluate $(-1)^3 + (-2)^3 + (-3)^3 + (-4)^3 + (-5)^3$.

53. The variables w, x, y, and z represent four different numbers such that:
$$z + w = z$$
$$w \cdot z = w$$
$$x + y = w$$
$$x(w + z) = z$$
$$x - y = z$$
Find the values of w, x, y, and z.

54. [W] Using the Order of Operations Agreement, explain how to evaluate Exercise 33.

55. [W] Evaluate $a \div bc$ and $a \div (bc)$ when $a = 16$, $b = 2$, and $c = -4$. Explain why the answers are not the same.

Projects in Mathematics

Morse Code/Binary Code

A	.–	S	...
B	–...	T	–
C	–.–.	U	..–
D	–..	V	...–
E	.	W	.––
F	..–.	X	–..–
G	––.	Y	–.––
H	....	Z	––..
I	..	0	–––––
J	.–––	1	.––––
K	–.–	2	..–––
L	.–..	3	...––
M	––	4	....–
N	–.	5	.....
O	–––	6	–....
P	.––.	7	––...
Q	––.–	8	–––..
R	.–.	9	––––.

In this chapter, the symbol "–" was explained as meaning *minus, negative,* or *opposite.* You have also seen this symbol as a hyphen or a dash.

Morse code is a system of communication in which letters and numbers are represented using dots and dashes. At the left is the representation of the letters of the alphabet and the numbers 0 through 9 in Morse code. Describe the pattern of dots and dashes used to represent the numbers 0 to 9. Why might the letters E and T have been chosen to have the shortest representations (only one symbol)?

Note that in Morse code, only two symbols are used. In our number system, the **decimal system,** we use 10 digits to represent numbers: 0, 1, 2, 3, 4, 5, 6, 7, 8, and 9. Because 10 digits are used, the decimal system is a **base ten** system. Another number system, the **binary number system,** uses only two digits to represent numbers: 0 and 1. Because two digits are used, the binary number system is a **base two** system.

The first five places of the digits in the decimal system are

Ten-thousands   Thousands   Hundreds   Tens   Ones

Each place represents a power of 10 ($10{,}000 = 10^4$, $1000 = 10^3$, $100 = 10^2, \ldots$).

The first five places in the binary number system are

Sixteens   Eights   Fours   Twos   Ones

Each place represents a power of 2 ($16 = 2^4$, $8 = 2^3$, $4 = 2^2, \ldots$).

The whole numbers can be represented in base two.

0
1
10 (1 two, zero ones)
11 (1 two, 1 one)
100 (1 four)
101 (1 four, 1 one)
110 (1 four, 1 two)
111 (1 four, 1 two, 1 one)
1000 (1 eight)
1001 (1 eight, 1 one) and so on

Write the next ten whole numbers using base two. What would be the values of the sixth and seventh places in base two?

One of the most important applications of the binary number system is to computers. Just as information in Morse code is relayed using only two symbols, computers use only two symbols, 0 and 1. Because all data, or information, in a computer is represented using only 0's and 1's, the data are said to be in **binary code.** A computer works on electric power, and an electric switch has only two positions, on and off. A 0 in binary code indicates that an electric switch is to be turned off; a 1 indicates that the switch is to be turned on.

A **bit** is the smallest unit of code that computers read; it is a **bi**nary digi**t**, either 0 or 1. Usually bits are grouped into **bytes** of 8 bits. Each byte stands for a letter, number, or any other symbol we might use in communicating information. For example, the letter W can be represented as 01010111.

Find the definition of each of the following words: gigabit, kilobyte, megabyte, terabit. Find out how the speed of a computer is specified. Write a description of machine language.

Chapter Summary

Key Words

Positive numbers are numbers greater than zero. *Negative numbers* are numbers less than zero.

The *integers* are . . . −4, −3, −2, −1, 0, 1, 2, 3, 4, . . . *Positive integers* are to the right of zero on the number line. *Negative integers* are to the left of zero on the number line.

Opposite numbers are two numbers that are the same distance from zero on the number line but on opposite sides of zero. The opposite of a number is called its *additive inverse.*

The *absolute value* of a number is the distance from zero to the number on the number line. The absolute value of a number is a positive number or zero. The symbol for absolute value is $|\ |$.

Essential Rules

To add integers with the same sign, add the absolute values of the numbers. Then attach the sign of the addends.

To add integers with different signs, find the absolute values of the numbers. Subtract the smaller absolute value from the larger absolute value. Then attach the sign of the addend with the larger absolute value.

To subtract two integers, add the opposite of the second integer to the first integer.

To multiply integers with the same sign, multiply the absolute values of the factors. The product is positive.

To multiply integers with different signs, multiply the absolute values of the factors. The product is negative.

To divide two numbers with the same sign, divide the absolute values of the numbers. The quotient is positive.

To divide two numbers with different signs, divide the absolute values of the numbers. The quotient is negative.

Order Relations $a > b$ if a is to the right of b on the number line.
$a < b$ if a is to the left of b on the number line.

Addition Property of Zero	$a + 0 = a$ or $0 + a = a$
Commutative Property of Addition	$a + b = b + a$
Associative Property of Addition	$(a + b) + c = a + (b + c)$
Inverse Property of Addition	$a + (-a) = 0$ or $-a + a = 0$
Multiplication Property of Zero	$a \cdot 0 = 0$ or $0 \cdot a = 0$
Multiplication Property of One	$a \cdot 1 = a$ or $1 \cdot a = a$
Commutative Property of Multiplication	$a \cdot b = b \cdot a$
Associative Property of Multiplication	$(a \cdot b) \cdot c = a \cdot (b \cdot c)$
Division Properties of Zero and One	If $a \neq 0$, $0 \div a = 0$. If $a \neq 0$, $a \div a = 1$. $a \div 1 = a$ $a \div 0$ is undefined.

The Order of Operations Agreement

Step 1 Do all operations inside parentheses.
Step 2 Simplify any numerical expressions containing exponents.
Step 3 Do multiplication and division as they occur from left to right.
Step 4 Do addition and subtraction as they occur from left to right.

Chapter Review Exercises

1. Write the expression $8 - (-1)$ in words.

2. Evaluate $-|-36|$.

3. Find the difference between -15 and -28.

4. Simplify: $-9 + 16 - (-7)$

5. Multiply: $-5(2)(-6)(-1)$

6. Evaluate $a + b$ when $a = -5$ and $b = 8$.

7. Evaluate $-a \div b$ when $a = -27$ and $b = -3$.

8. Find the quotient of 840 and -4.

9. Subtract: $-6 - (-7) - 15 - (-12)$

10. Evaluate $-ab$ when $a = -2$ and $b = -9$.

11. Find the sum of 18, -13, and -6.

12. Evaluate $-x - y$ when $x = -1$ and $y = 3$.

13. Write the numbers in order from smallest to largest.

    $|5|, -7, -(-6), 4, |-2|$

14. Identify the property that justifies the statement.

    $-11(-50) = -50(-11)$

15. Simplify: $(-2)^2 - (-3)^2 \div (1 - 4)^2 \cdot 2 - 6$

16. Find the product of -40 and -5.

17. Divide: $\dfrac{0}{-17}$

18. Find the opposite of -15.

19. Add: $3 + (-9) + 4 + (-10)$

20. Evaluate $(a - b)^2 - 2a$ when $a = -2$ and $b = -3$.

21. Place the correct symbol, $<$ or $>$, between the two numbers.

    $-8 \quad -10$

22. Complete the statement by using the Inverse Property of Addition.

    $-21 + ? = 0$

23. The melting point of radon is -71°C. The melting point of oxygen is three times as high as the melting point of radon. Find the melting point of oxygen.

24. Find the temperature after an increase of 5°C from -8°C.

25. The distance, d, between point a and point b on the number line is given by the formula $d = |a - b|$. Find d when $a = 7$ and $b = -5$.

Cumulative Review Exercises

1. Find the difference between -27 and -32.

2. Estimate the product of 439 and 28.

3. Divide: $19{,}254 \div 6$

4. Simplify: $16 \div (3 + 5) \cdot 9 - 2^4$

5. Evaluate $-|-82|$.

6. Round 629,874 to the nearest thousand.

7. Evaluate $5xy$ when $x = 80$ and $y = 6$.

8. What is -294 divided by -14?

9. What is -18 multiplied by -7?

10. Find the sum of -24, 16, and -32.

11. Find all the factors of 44.

12. Evaluate x^4y^2 when $x = 2$ and $y = 11$.

13. What is 5 less than -21?

14. Estimate the sum of 356, 481, 294, and 117.

15. Evaluate $-a - b$ when $a = -4$ and $b = -5$.

16. Find the product of -100 and 25.

17. Find the prime factorization of 69.

18. Write 3,047,953 in expanded form.

19. Simplify: $(1 - 5)^2 \div (-6 + 4) + 8(-3)$

20. Evaluate $-c \div d$ when $c = -32$ and $d = -8$.

21. Evaluate $\frac{a}{b}$ when $a = 39$ and $b = -13$.

22. Evaluate $4a + (a - b)^3$ when $a = 5$ and $b = 2$.

23. Albert Einstein was born on March 14, 1879. He died on April 18, 1955. How old was Albert Einstein when he died?

24. Find the temperature after an increase of 7°C from -12°C.

25. As a sales representative, your goal is to sell \$120,000 in merchandise during the year. You sold \$28,550 in merchandise during the first quarter of the year, \$34,850 during the second quarter, and \$31,700 during the third quarter. What must your sales for the fourth quarter be if you are to meet your goal for the year?

26. A golfer made 198 strokes and par is 206. Use the equation $S = N - P$, where S is a golfer's score in a tournament, N is the number of strokes made by the golfer, and P is par, to find the golfer's score.

CHAPTER

3

Fractions

Focus on Problem Solving

An application problem may not provide all the information that is necessary to solve the problem. Sometimes, however, the necessary information is common knowledge.

You are traveling by bus from Boston to New York. The trip is 4 h long. If the bus leaves Boston at 10 A.M., what time should you arrive in New York?

What other information do you need to solve this problem?

You need to know that, using a 12-hour clock, the hours run

10 A.M.
11 A.M.
12 P.M.
1 P.M.
2 P.M.

Four hours after 10 A.M. is 2 P.M.

You should arrive in New York at 2 P.M.

You purchase a 32¢ stamp at the post office and hand the clerk a one-dollar bill. How much change do you receive?

What other information do you need to solve this problem?

You need to know that there are 100¢ in one dollar.

Your change is 100¢ − 32¢.

$100 - 32 = 68$

You receive 68¢ in change.

What information do you need to know to solve each of the following problems?

1. You sell a dozen tickets to a fundraiser. Each ticket costs $10. How much money do you collect?
2. The weekly lab period for your science course is one hour and twenty minutes long. Find the length of the science lab period in minutes.
3. An employee's monthly salary is $1750. Find the employee's annual salary.
4. A survey revealed that eighth graders spend an average of 3 h each day watching television. Find the total time an eighth grader spends watching TV each week.

Answers:

1. You need to know that there are 12 in one dozen.
2. You need to know that there are 60 min in one hour.
3. You need to know that there are 12 months in one year.
4. You need to know that there are 7 days in one week.

SECTION 3.1 Introduction to Fractions

OBJECTIVE 

Proper fractions, improper fractions, and mixed numbers

A recipe calls for $\frac{1}{2}$ cup butter and $\frac{1}{4}$ teaspoon vanilla. The numbers $\frac{1}{2}$ and $\frac{1}{4}$ are fractions.

A **fraction** can represent the number of equal parts of a whole.

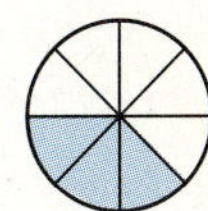

The circle at the right is divided into 8 equal parts. 3 of the 8 parts are shaded. The shaded portion of the circle is represented by the fraction $\frac{3}{8}$.

POINT OF INTEREST

As early as A.D. 630 the Hindu mathematician Brahmagupta wrote a fraction as one number over another separated by a space. The Arab mathematician al Hassar (about A.D. 1050) was the first to show a fraction with the horizontal bar separating the numerator and denominator.

Each part of a fraction has a name.

Fraction bar → $\frac{3}{8}$ ← **Numerator** (3), ← **Denominator** (8)

In a **proper fraction,** the numerator is smaller than the denominator. A proper fraction is a number less than 1.

$\frac{3}{8}$ is a proper fraction.

In an **improper fraction,** the numerator is greater than or equal to the denominator. An improper fraction is a number greater than or equal to 1.

The shaded portion of the circles at the right is represented by the improper fraction $\frac{7}{3}$.

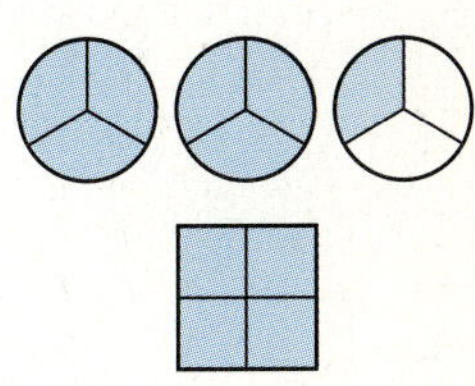

The shaded portion of the square at the right is represented by the improper fraction $\frac{4}{4}$.

A fraction bar can be read "divided by." Therefore, the fraction $\frac{4}{4}$ can be read "$4 \div 4$." Because a number divided by itself is equal to 1, $4 \div 4 = 1$ and $\frac{4}{4} = 1$. The shaded portion of the square shown above is equal to $\frac{4}{4}$ or 1 square.

Since the fraction bar can be read "divided by" and any number divided by one equals itself, any whole number can be written as an improper fraction. For example, $5 = \frac{5}{1}$ and $7 = \frac{7}{1}$.

Because zero divided by any number other than zero equals zero, the numerator of a fraction can be zero. For example, $\frac{0}{6} = 0$ because $0 \div 6 = 0$.

Recall that division by zero is not defined. Therefore, the denominator of a fraction cannot be zero. For example, $\frac{9}{0}$ is not defined because $\frac{9}{0} = 9 \div 0$, and division by zero is not defined.

POINT OF INTEREST

Archimedes (c. 287–212 B.C.) is the person who calculated that $\pi \approx 3\frac{1}{7}$. He actually showed that $3\frac{10}{71} < \pi < 3\frac{1}{7}$. The approximation $3\frac{10}{71}$ is more accurate but more difficult to use.

A **mixed number** is a number greater than 1 with a whole number part and a fractional part.

The shaded portion of the circles at the right is represented by the mixed number $2\frac{1}{2}$. Note from the diagram that the improper fraction $\frac{5}{2}$ is equal to the mixed number $2\frac{1}{2}$.

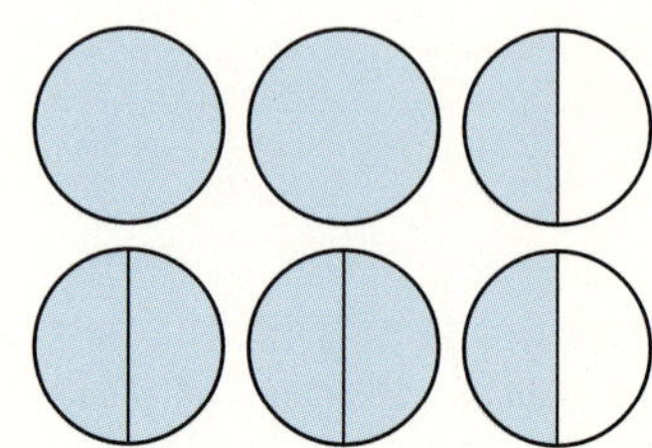

An improper fraction can be written as a mixed number.
To write $\frac{5}{2}$ as a mixed number, read $\frac{5}{2}$ as $5 \div 2$.

Divide the numerator by the denominator.	To write the fractional part of the mixed number, write the remainder over the divisor.	Write the answer.
$\begin{array}{r} 2 \\ 2\overline{)5} \\ -4 \\ \hline 1 \end{array}$	$\begin{array}{r} 2\frac{1}{2} \\ 2\overline{)5} \\ -4 \\ \hline 1 \end{array}$	$\frac{5}{2} = 2\frac{1}{2}$

To write a mixed number as an improper fraction, multiply the denominator of the fractional part of the mixed number by the whole number part. The sum of this product and the numerator of the fractional part is the numerator of the improper fraction. The denominator remains the same.

Write $4\frac{5}{6}$ as an improper fraction.

$$4\frac{5}{6} = \frac{(6 \cdot 4) + 5}{6} = \frac{24 + 5}{6} = \frac{29}{6}$$

In general, mixed numbers are not used in computations in algebra. Instead, calculations are made using improper fractions. If an application problem includes a mixed number, you will rewrite the mixed number as an improper fraction before working the problem. Also, if the answer to an application problem is an improper fraction, the improper fraction is written as a mixed number when writing the answer; for example, $\frac{5}{2}$ min would be written as $2\frac{1}{2}$ min.

Example 1 Write $\frac{14}{5}$ as a mixed number.

Solution
$$\begin{array}{r} 2 \\ 5\overline{)14} \\ -10 \\ \hline 4 \end{array} \qquad \frac{14}{5} = 2\frac{4}{5}$$

You Try It 1 Write $\frac{26}{3}$ as a mixed number.

Your Solution

Example 2 Write $12\frac{5}{8}$ as an improper fraction.

Solution
$$12\frac{5}{8} = \frac{(8 \cdot 12) + 5}{8} = \frac{96 + 5}{8}$$
$$= \frac{101}{8}$$

You Try It 2 Write $9\frac{4}{7}$ as an improper fraction.

Your Solution

Solutions on p. A4

OBJECTIVE  **B**

Equivalent fractions

Fractions can be graphed as points on the number line.

The number lines at the right show thirds and sixths graphed from 0 to 1.

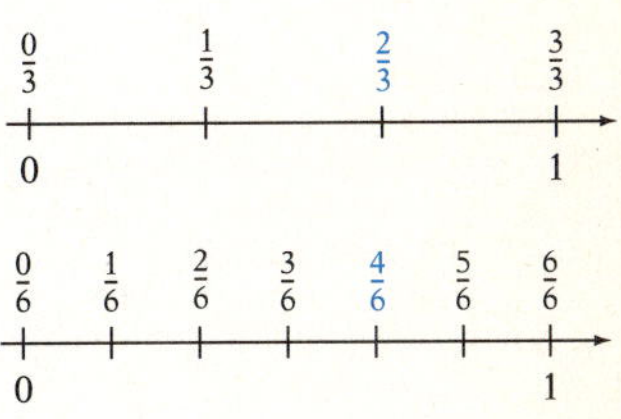

A particular point on the number line may be represented by different fractions, all of which are equal.

For example, $\frac{0}{3} = \frac{0}{6}$, $\frac{1}{3} = \frac{2}{6}$, $\frac{2}{3} = \frac{4}{6}$, and $\frac{3}{3} = \frac{6}{6}$.

Equal fractions with different denominators are called **equivalent fractions.**

$\frac{2}{3}$ and $\frac{4}{6}$ are equivalent fractions.

Note that we could rewrite $\frac{2}{3}$ as $\frac{4}{6}$ by multiplying both the numerator and denominator of $\frac{2}{3}$ by 2.

$$\frac{2}{3} = \frac{2 \cdot 2}{3 \cdot 2} = \frac{4}{6}$$

Also, we could rewrite $\frac{4}{6}$ as $\frac{2}{3}$ by dividing both the numerator and denominator of $\frac{4}{6}$ by 2.

$$\frac{4}{6} = \frac{4 \div 2}{6 \div 2} = \frac{2}{3}$$

This suggests the following property of fractions.

Equivalent Fractions

The numerator and denominator of a fraction can be multiplied by or divided by the same nonzero number. The resulting fraction is equivalent to the original fraction.

$$\frac{a}{b} = \frac{a \cdot c}{b \cdot c}, \text{ and } \frac{a}{b} = \frac{a \div c}{b \div c}, \text{ where } b \neq 0 \text{ and } c \neq 0$$

Write a fraction that is equivalent to $\frac{3}{8}$ and has a denominator of 40.

Divide the larger denominator by the smaller. $40 \div 8 = 5$

Multiply the numerator and denominator of the given fraction by the quotient (5). $\frac{3}{8} = \frac{3 \cdot 5}{8 \cdot 5} = \frac{15}{40}$

$\frac{15}{40}$ is equivalent to $\frac{3}{8}$

A fraction is in **simplest form** when the numerator and denominator have no common factors other than 1. The fraction $\frac{3}{8}$ is in simplest form because 3 and 8 have no common factors other than 1. The fraction $\frac{15}{40}$ is not in simplest form because the numerator and denominator have a common factor of 5.

To write a fraction in simplest form, divide its numerator and denominator by their common factors.

Write $\frac{12}{15}$ in simplest form.

12 and 15 have a common factor of 3. Divide the numerator and denominator by 3. $\frac{12}{15} = \frac{12 \div 3}{15 \div 3} = \frac{4}{5}$

The fraction $\frac{4}{5}$ is in simplest form because 4 and 5 have no common factors other than 1.

To simplify fractions by this method you must find any factors common to the numerator and denominator. Common factors can be found by writing the prime factorizations of the numerator and denominator. Then divide the numerator and denominator by the common prime factors.

$$\frac{12}{15} = \frac{2 \cdot 2 \cdot \overset{1}{\cancel{3}}}{\underset{1}{\cancel{3}} \cdot 5} = \frac{4}{5}$$

Write $\frac{30}{42}$ in simplest form.

Write the prime factorization of the numerator and denominator. Divide by the common factors.

$$\frac{30}{42} = \frac{\overset{1}{\cancel{2}} \cdot \overset{1}{\cancel{3}} \cdot 5}{\underset{1}{\cancel{2}} \cdot \underset{1}{\cancel{3}} \cdot 7} = \frac{5}{7}$$

Example 3 Write a fraction that is equivalent to $\frac{2}{5}$ and has a denominator of 25.

Solution $25 \div 5 = 5$

$$\frac{2}{5} = \frac{2 \cdot 5}{5 \cdot 5} = \frac{10}{25}$$

$\frac{10}{25}$ is equivalent to $\frac{2}{5}$.

You Try It 3 Write a fraction that is equivalent to $\frac{5}{8}$ and has a denominator of 48.

Your Solution

Example 4 Write 3 as an equivalent fraction that has a denominator of 15.

Solution $3 = \frac{3}{1}$ $\quad 15 \div 1 = 15$

$$3 = \frac{3}{1} = \frac{3 \cdot 15}{1 \cdot 15} = \frac{45}{15}$$

$\frac{45}{15}$ is equivalent to 3.

You Try It 4 Write 8 as an equivalent fraction that has a denominator of 12.

Your Solution

Example 5 Write $\frac{18}{54}$ in simplest form.

Solution

$$\frac{18}{54} = \frac{\overset{1}{\cancel{2}} \cdot \overset{1}{\cancel{3}} \cdot \overset{1}{\cancel{3}}}{\underset{1}{\cancel{2}} \cdot \underset{1}{\cancel{3}} \cdot \underset{1}{\cancel{3}} \cdot 3} = \frac{1}{3}$$

You Try It 5 Write $\frac{21}{84}$ in simplest form.

Your Solution

Solutions on p. A4

OBJECTIVE C Order relations between two fractions

The number line can be used to determine the order relation between two fractions.

A fraction that appears to the left of a given fraction is less than the given fraction.

$\frac{3}{8}$ is to the left of $\frac{5}{8}$.

$$\frac{3}{8} < \frac{5}{8}$$

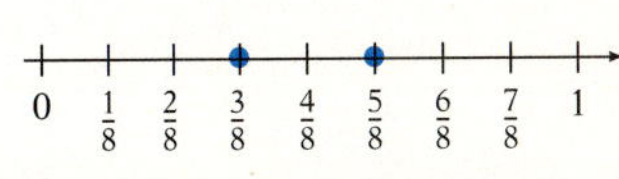

A fraction that appears to the right of a given fraction is greater than the given fraction.

$\frac{7}{8}$ is to the right of $\frac{3}{8}$.

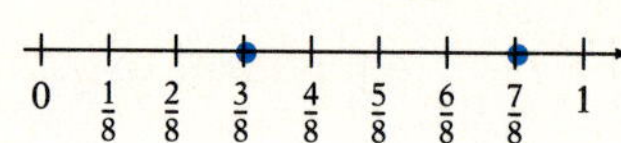

$\frac{7}{8} > \frac{3}{8}$

To find the order relation between two fractions with the same denominator, compare the numerators. The fraction that has the smaller numerator is the smaller fraction. The fraction that has the larger numerator is the larger fraction.

$\frac{3}{8}$ and $\frac{5}{8}$ have the same denominator. $\frac{3}{8} < \frac{5}{8}$ because $3 < 5$.

$\frac{7}{8}$ and $\frac{3}{8}$ have the same denominator. $\frac{7}{8} > \frac{3}{8}$ because $7 > 3$.

When the denominators of the fractions are different, begin by writing equivalent fractions with a common denominator. Then compare the numerators.

Find the order relation between $\frac{5}{12}$ and $\frac{7}{18}$.

Find the LCM of the denominators. The LCM of 12 and 18 is 36.

Write each fraction as an equivalent fraction with the LCM as the denominator.

$$\frac{5}{12} = \frac{5 \cdot 3}{12 \cdot 3} = \frac{15}{36}$$

$$\frac{7}{18} = \frac{7 \cdot 2}{18 \cdot 2} = \frac{14}{36}$$

Compare the fractions. $\frac{15}{36} > \frac{14}{36}$ because $15 > 14$.

$\frac{5}{12} > \frac{7}{18}$

Example 6 Place the correct symbol, < or >, between the two numbers.

$\frac{2}{3} \quad \frac{4}{7}$

Solution $\frac{2}{3} = \frac{14}{21} \qquad \frac{4}{7} = \frac{12}{21}$

$\frac{14}{21} > \frac{12}{21}$

$\frac{2}{3} > \frac{4}{7}$

You Try It 6 Place the correct symbol, < or >, between the two numbers.

$\frac{4}{9} \quad \frac{8}{21}$

Your Solution

Solution on p. A4

3.1 EXERCISES

Objective A

Write the improper fraction as a mixed number or a whole number.

1. $\frac{31}{3}$ **2.** $\frac{56}{8}$ **3.** $\frac{27}{9}$ **4.** $\frac{17}{9}$ **5.** $\frac{8}{3}$

6. $\frac{12}{5}$ **7.** $\frac{19}{8}$ **8.** $\frac{18}{1}$ **9.** $\frac{21}{1}$ **10.** $\frac{32}{15}$

11. $\frac{39}{14}$ **12.** $\frac{8}{8}$ **13.** $\frac{12}{12}$ **14.** $\frac{28}{3}$ **15.** $\frac{43}{5}$

Write the mixed number or whole number as an improper fraction.

16. $2\frac{1}{4}$ **17.** $4\frac{2}{5}$ **18.** $5\frac{1}{2}$ **19.** $3\frac{2}{3}$ **20.** $2\frac{4}{5}$

21. $6\frac{3}{8}$ **22.** $7\frac{5}{6}$ **23.** $9\frac{1}{5}$ **24.** 7 **25.** 4

26. $8\frac{1}{4}$ **27.** $1\frac{7}{9}$ **28.** $10\frac{1}{3}$ **29.** $6\frac{3}{7}$ **30.** $4\frac{7}{12}$

Objective B

Write an equivalent fraction with the given denominator.

31. $\frac{1}{2} = \frac{}{12}$ **32.** $\frac{1}{4} = \frac{}{20}$ **33.** $\frac{3}{8} = \frac{}{24}$ **34.** $\frac{9}{11} = \frac{}{44}$ **35.** $\frac{2}{17} = \frac{}{51}$

36. $\frac{9}{10} = \frac{}{80}$ **37.** $\frac{3}{4} = \frac{}{32}$ **38.** $\frac{5}{8} = \frac{}{32}$ **39.** $6 = \frac{}{18}$ **40.** $5 = \frac{}{35}$

41. $\frac{1}{3} = \frac{}{90}$ **42.** $\frac{3}{16} = \frac{}{48}$ **43.** $\frac{2}{3} = \frac{}{21}$ **44.** $\frac{4}{9} = \frac{}{36}$ **45.** $\frac{6}{7} = \frac{}{49}$

Write the fraction in simplest form.

46. $\frac{3}{12}$ **47.** $\frac{10}{22}$ **48.** $\frac{33}{44}$ **49.** $\frac{6}{14}$ **50.** $\frac{4}{24}$

51. $\frac{25}{75}$ **52.** $\frac{8}{33}$ **53.** $\frac{9}{25}$ **54.** $\frac{0}{8}$ **55.** $\frac{0}{11}$

56. $\frac{42}{36}$ **57.** $\frac{30}{18}$ **58.** $\frac{16}{16}$ **59.** $\frac{24}{24}$ **60.** $\frac{21}{35}$

61. $\frac{11}{55}$ **62.** $\frac{16}{60} = \frac{4}{15}$ **63.** $\frac{8}{84}$ **64.** $\frac{12}{20} = \frac{3}{5}$ **65.** $\frac{24}{36}$

Objective C

Place the correct symbol, < or >, between the two numbers.

66. $\frac{3}{8}$ $\frac{2}{5}$ **67.** $\frac{5}{7}$ $\frac{2}{3}$ **68.** $\frac{3}{4} < \frac{7}{9}$ **69.** $\frac{7}{12}$ $\frac{5}{8}$

70. $\frac{2}{3}$ $\frac{7}{11}$ **71.** $\frac{11}{14}$ $\frac{3}{4}$ **72.** $\frac{17}{24} > \frac{11}{16}$ **73.** $\frac{11}{12}$ $\frac{7}{9}$

74. $\frac{7}{15}$ $\frac{5}{12}$ **75.** $\frac{5}{8}$ $\frac{4}{7}$ **76.** $\frac{4}{5} > \frac{7}{9}$ **77.** $\frac{11}{30}$ $\frac{7}{24}$

Critical Thinking

78. Is the expression $x < \frac{4}{9}$ true when $x = \frac{3}{8}$? Is it true when $x = \frac{5}{12}$?

79. $\frac{2}{3} < \frac{3}{4}$. Is $\frac{2+3}{3+4}$ less than $\frac{2}{3}$, greater than $\frac{3}{4}$, or between $\frac{2}{3}$ and $\frac{3}{4}$?

80. [W] Find the business section of a newspaper. Choose a stock and record the fluctuations in the stock price for one month. Explain the part that fractions play in reporting the price and change in price of the stock.

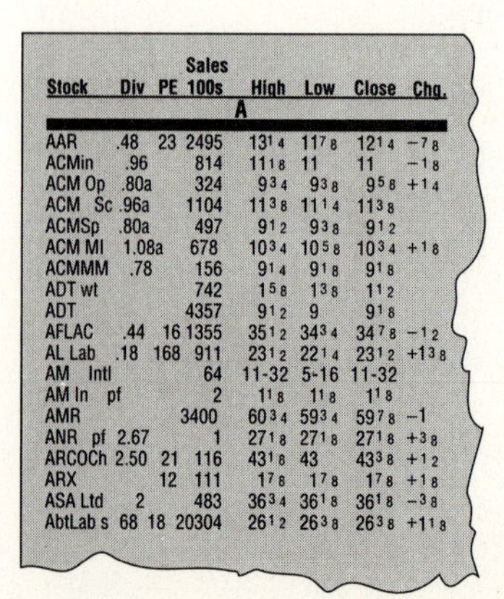

Stock	Div	PE	Sales 100s	High	Low	Close	Chg.
A							
AAR	.48	23	2495	$13\frac{1}{4}$	$11\frac{7}{8}$	$12\frac{1}{4}$	$-\frac{7}{8}$
ACMin	.96		814	$11\frac{1}{8}$	11	11	$-\frac{1}{8}$
ACM Op	.80a		324	$9\frac{3}{4}$	$9\frac{3}{8}$	$9\frac{5}{8}$	$+\frac{1}{4}$
ACM Sc	.96a		1104	$11\frac{3}{8}$	$11\frac{1}{4}$	$11\frac{3}{8}$	
ACMSp	.80a		497	$9\frac{1}{2}$	$9\frac{3}{8}$	$9\frac{1}{2}$	
ACM MI	1.08a		678	$10\frac{3}{4}$	$10\frac{5}{8}$	$10\frac{3}{4}$	$+\frac{1}{8}$
ACMMM	.78		156	$9\frac{1}{4}$	$9\frac{1}{8}$	$9\frac{1}{8}$	
ADT wt			742	$1\frac{5}{8}$	$1\frac{3}{8}$	$1\frac{1}{2}$	
ADT			4357	$9\frac{1}{2}$	9	$9\frac{1}{8}$	
AFLAC	.44	16	1355	$35\frac{1}{2}$	$34\frac{3}{4}$	$34\frac{7}{8}$	$-\frac{1}{2}$
AL Lab	.18	168	911	$23\frac{1}{2}$	$22\frac{1}{4}$	$23\frac{1}{2}$	$+1\frac{3}{8}$
AM Intl			64	11-32	5-16	11-32	
AM In pf			2	$1\frac{1}{8}$	$1\frac{1}{8}$	$1\frac{1}{8}$	
AMR			3400	$60\frac{3}{4}$	$59\frac{3}{4}$	$59\frac{7}{8}$	−1
ANR pf	2.67		1	$27\frac{1}{8}$	$27\frac{1}{8}$	$27\frac{1}{8}$	$+\frac{3}{8}$
ARCOCh	2.50	21	116	$43\frac{1}{8}$	43	$43\frac{3}{8}$	$+\frac{1}{2}$
ARX		12	111	$1\frac{7}{8}$	$1\frac{7}{8}$	$1\frac{7}{8}$	$+\frac{1}{8}$
ASA Ltd	2		483	$36\frac{3}{4}$	$36\frac{1}{8}$	$36\frac{1}{8}$	$-\frac{3}{8}$
AbtLab s	68	18	20304	$26\frac{1}{2}$	$26\frac{3}{8}$	$26\frac{3}{8}$	$+1\frac{1}{8}$

SECTION 3.2 Operations on Fractions

OBJECTIVE 

Addition of fractions

POINT OF INTEREST

Leonardo of Pisa, who was also called Fibonacci (c. 1175–1250), is credited with bringing the Hindu-Arabic number system to the western world and promoting its use instead of the cumbersome Roman numeral system.

He was also influential in promoting the idea of the fraction bar. His notation, however, was very different from what we have today. For instance, he wrote $\frac{3\ \ 5}{4\ \ 7}$ to mean $\frac{5}{7} + \frac{3}{7 \cdot 4}$.

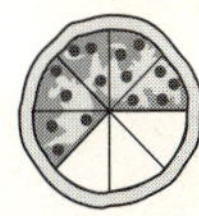

Suppose you and a friend order a pizza. The pizza has been cut into 8 equal pieces. If you eat 3 pieces of the pizza and your friend eats 2 pieces, then together you have eaten $\frac{5}{8}$ of the pizza.

Note that in adding the fractions $\frac{3}{8}$ and $\frac{2}{8}$, the numerators are added and the denominator remains the same.

$$\frac{3}{8} + \frac{2}{8} = \frac{3+2}{8} = \frac{5}{8}$$

Addition of Fractions

To add fractions with the same denominator, add the numerators and place the sum over the common denominator.

$$\frac{a}{b} + \frac{c}{b} = \frac{a+c}{b}, \text{ where } b \neq 0$$

Add: $\frac{5}{16} + \frac{7}{16}$

The denominators are the same. Add the numerators and place the sum over the common denominator.

$$\frac{5}{16} + \frac{7}{16} = \frac{5+7}{16}$$

Write the answer in simplest form.

$$= \frac{12}{16} = \frac{3}{4}$$

Add: $\frac{4}{x} + \frac{8}{x}$

The denominators are the same. Add the numerators and place the sum over the common denominator.

$$\frac{4}{x} + \frac{8}{x} = \frac{4+8}{x} = \frac{12}{x}$$

Before two fractions can be added, the fractions must have the same denominator. To add fractions with different denominators, first rewrite the fractions as equivalent fractions with a common denominator. The common denominator is the least common multiple (LCM) of the denominators of the fractions. The LCM of denominators is sometimes called the least common denominator (LCD).

CONSIDER THIS

You can find the LCD by multiplying the denominators and then dividing by the *common factor* of the two denominators. For 6 and 8, $6 \cdot 8 = 48$. Now divide by 2, the common factor of 6 and 8. $48 \div 2 = 24$

Find the sum of $\frac{5}{6}$ and $\frac{3}{8}$.

The common denominator is the LCM of 6 and 8. — The LCM of 6 and 8 is 24.

Write the fractions as equivalent fractions with the common denominator.

$$\frac{5}{6} + \frac{3}{8} = \frac{20}{24} + \frac{9}{24}$$

Add the fractions.

$$= \frac{20 + 9}{24}$$

$$= \frac{29}{24} = 1\frac{5}{24}$$

Recall that the $-$ sign indicates the opposite of a number. $-\frac{1}{2}$ indicates the opposite of $\frac{1}{2}$.

The negative sign in a fraction can be written in front, in the numerator, or in the denominator of the fraction.

$$-\frac{2}{3} = \frac{-2}{3} = \frac{2}{-3}$$

To add fractions with negative signs, rewrite the fractions with the negative signs in the numerators. Then add the numerators and place the sum over the common denominator.

Add: $-\frac{3}{8} + \frac{1}{8}$

To add these fractions, recall that $-\frac{3}{8} = \frac{-3}{8}$.

$$-\frac{3}{8} + \frac{1}{8} = \frac{-3}{8} + \frac{1}{8}$$

Add the fractions by adding the numerators and placing the sum over the common denominator.

$$= \frac{-3 + 1}{8}$$

Simplify the numerator.

$$= \frac{-2}{8}$$

Reduce the sum to simplest form. Although the sum could be left as $\frac{-1}{4}$, all answers in this text that are negative fractions are written with the negative sign in front of the fraction.

$$= \frac{-1}{4} = -\frac{1}{4}$$

Add: $-\frac{y}{2} + \frac{z}{2}$

The denominators are the same.

$$-\frac{y}{2} + \frac{z}{2} = \frac{-y}{2} + \frac{z}{2}$$

Add the numerators.

$$= \frac{-y + z}{2}$$

Add: $-\frac{5}{6} + \frac{3}{4}$

The common denominator is the LCM of 4 and 6.

The LCM of 4 and 6 is 12.

Rewrite with the negative sign in the numerator.

$$-\frac{5}{6} + \frac{3}{4} = \frac{-5}{6} + \frac{3}{4}$$

Rewrite each fraction as an equivalent fraction with the LCM as the denominator.

$$= \frac{-10}{12} + \frac{9}{12}$$

Add the fractions by adding the numerators and placing the sum over the common denominator.

$$= \frac{-10 + 9}{12}$$

Simplify the numerator and write the negative sign in front of the fraction.

$$= \frac{-1}{12} = -\frac{1}{12}$$

Add: $-\frac{2}{3} + \left(-\frac{4}{5}\right)$

Rewrite the addition problem with the sign in the numerator of each fraction.

$$-\frac{2}{3} + \left(-\frac{4}{5}\right) = \frac{-2}{3} + \frac{-4}{5}$$

Rewrite each fraction as an equivalent fraction using the LCM as the denominator.

$$= \frac{-10}{15} + \frac{-12}{15}$$

Add the fractions.

$$= \frac{-10 + (-12)}{15}$$

$$= \frac{-22}{15} = -1\frac{7}{15}$$

The mixed number $2\frac{1}{2}$ is the sum of 2 and $\frac{1}{2}$.

$$2\frac{1}{2} = 2 + \frac{1}{2}$$

Therefore, the sum of a whole number and a fraction is a mixed number.

$$2 + \frac{1}{2} = 2\frac{1}{2}$$

$$3 + \frac{4}{5} = 3\frac{4}{5}$$

$$8 + \frac{7}{9} = 8\frac{7}{9}$$

The sum of a whole number and a mixed number is a mixed number.

Evaluate $x + y$ when $x = 5$ and $y = 4\frac{2}{7}$.

$$x + y$$

Replace x with 5 and y with $4\frac{2}{7}$.

$$5 + 4\frac{2}{7}$$

Add the whole numbers (4 + 5). Write the fraction.

$$= 9\frac{2}{7}$$

Example 1

Find the sum of $\frac{4}{5}$, $\frac{3}{4}$, and $\frac{5}{8}$.

Solution

$$\frac{4}{5} + \frac{3}{4} + \frac{5}{8} = \frac{32}{40} + \frac{30}{40} + \frac{25}{40} = \frac{87}{40} = 2\frac{7}{40}$$

You Try It 1

Find the sum of $\frac{3}{5}$, $\frac{2}{3}$, and $\frac{5}{6}$.

Your Solution

Example 2

Evaluate $x + y + z$ when $x = -\frac{3}{8}$, $y = \frac{3}{4}$, and $z = -\frac{5}{6}$.

Solution

$x + y + z$

$$-\frac{3}{8} + \frac{3}{4} + \left(-\frac{5}{6}\right) = \frac{-3}{8} + \frac{3}{4} + \frac{-5}{6}$$

$$= \frac{-9}{24} + \frac{18}{24} + \frac{-20}{24}$$

$$= \frac{-9 + 18 + (-20)}{24}$$

$$= \frac{-11}{24} = -\frac{11}{24}$$

You Try It 2

Evaluate $x + y + z$ when $x = -\frac{5}{12}$, $y = \frac{5}{8}$, and $z = -\frac{1}{6}$.

Your Solution

Solutions on p. A5

OBJECTIVE B Subtraction of fractions

In the last objective, it was stated that to add fractions, the fractions must have the same denominator. The same is true for subtracting fractions: the two fractions must have the same denominator.

Subtraction of Fractions

To subtract fractions with the same denominator, subtract the numerators and place the difference over the common denominator.

$$\frac{a}{b} - \frac{c}{b} = \frac{a - c}{b}, \text{ where } b \neq 0$$

Subtract: $\frac{5}{8} - \frac{3}{8}$

The denominators are the same. Subtract the numerators and place the difference over the common denominator.

$$\frac{5}{8} - \frac{3}{8} = \frac{5-3}{8}$$

Write the answer in simplest form.

$$= \frac{2}{8} = \frac{1}{4}$$

To subtract fractions with different denominators, first rewrite the fractions as equivalent fractions with a common denominator. As in the addition of fractions, the common denominator is the least common multiple (LCM) of the denominators of the fractions.

Subtract: $\frac{5}{12} - \frac{3}{8}$

The common denominator is the LCM of 12 and 8.

The LCM of 12 and 8 is 24.

Write the fractions as equivalent fractions with the common denominator.

$$\frac{5}{12} - \frac{3}{8} = \frac{10}{24} - \frac{9}{24}$$

Subtract the fractions.

$$= \frac{10-9}{24} = \frac{1}{24}$$

To subtract fractions with negative signs, rewrite the fractions with the negative signs in the numerators. Then subtract the numerators and place the difference over the common denominator.

Subtract: $-\frac{2}{9} - \frac{5}{12}$

Rewrite with the negative sign in the numerator.

$$-\frac{2}{9} - \frac{5}{12} = \frac{-2}{9} - \frac{5}{12}$$

Write the fractions as equivalent fractions with a common denominator.

$$= \frac{-8}{36} - \frac{15}{36}$$

Subtract the numerators and place the difference over the common denominator.

$$= \frac{-8-15}{36} = \frac{-23}{36}$$

Write the negative sign in front of the fraction.

$$= -\frac{23}{36}$$

Subtract: $\frac{2}{3} - \left(-\frac{4}{5}\right)$

Rewrite the fraction $-\frac{4}{5}$ with the negative sign in the numerator.

$$\frac{2}{3} - \left(-\frac{4}{5}\right) = \frac{2}{3} - \left(\frac{-4}{5}\right)$$

Write the fractions as equivalent fractions with a common denominator.

$$= \frac{10}{15} - \left(\frac{-12}{15}\right)$$

Subtract the numerators.

$$= \frac{10-(-12)}{15}$$

$$= \frac{10+12}{15}$$

Simplify the fraction.

$$= \frac{22}{15} = 1\frac{7}{15}$$

Find the difference between 1 and $\frac{2}{3}$.

Rewrite 1 as a fraction with the same denominator as is in the fraction $\frac{2}{3}$.

$$1 - \frac{2}{3} = \frac{3}{3} - \frac{2}{3}$$

Subtract the fractions.

$$= \frac{3 - 2}{3} = \frac{1}{3}$$

Example 3

Find the difference between $\frac{13}{16}$ and $\frac{5}{24}$.

Solution

$$\frac{13}{16} - \frac{5}{24} = \frac{39}{48} - \frac{10}{48} = \frac{39 - 10}{48} = \frac{29}{48}$$

You Try It 3

Find the difference between $\frac{7}{12}$ and $\frac{4}{9}$.

Your Solution

Example 4

Evaluate $x - y$ when $x = -\frac{5}{6}$ and $y = -\frac{3}{8}$.

Solution

$x - y$

$$-\frac{5}{6} - \left(-\frac{3}{8}\right) = \frac{-5}{6} - \frac{-3}{8} = \frac{-20}{24} - \frac{-9}{24}$$

$$= \frac{-20 - (-9)}{24} = \frac{-20 + 9}{24}$$

$$= \frac{-11}{24} = -\frac{11}{24}$$

You Try It 4

Evaluate $x - y$ when $x = -\frac{5}{6}$ and $y = -\frac{7}{9}$.

Your Solution

Solutions on p. A5

OBJECTIVE C Multiplication of fractions

To multiply two fractions, multiply the numerators and multiply the denominators.

Multiplication of Fractions

The product of two fractions is the product of the numerators over the product of the denominators.

$$\frac{a}{b} \cdot \frac{c}{d} = \frac{ac}{bd}, \text{ where } b \neq 0 \text{ and } d \neq 0$$

Note that fractions do not need to have the same denominator in order to be multiplied.

Multiply: $\frac{2}{5} \cdot \frac{1}{3}$

Multiply the numerators.
Multiply the denominators.

$$\frac{2}{5} \cdot \frac{1}{3} = \frac{2 \cdot 1}{5 \cdot 3} = \frac{2}{15}$$

The product $\frac{2}{5} \cdot \frac{1}{3}$ can be read "$\frac{2}{5}$ times $\frac{1}{3}$" or "$\frac{2}{5}$ of $\frac{1}{3}$."

Reading the times sign as "of" is useful in diagraming the product of two fractions.

$\frac{1}{3}$ of the bar at the right is shaded.

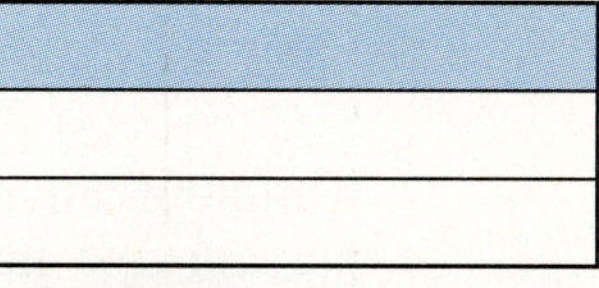

Shade $\frac{2}{5}$ of the $\frac{1}{3}$ already shaded.

$\frac{2}{15}$ of the bar is now shaded.

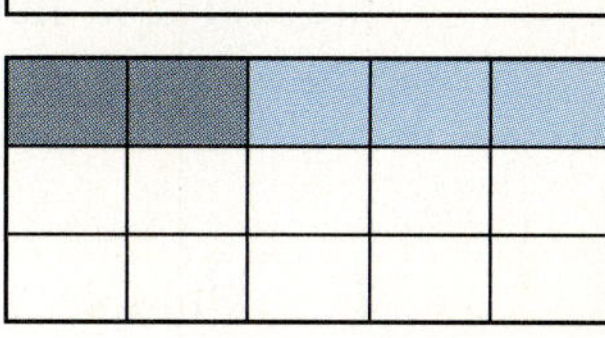

$$\frac{2}{5} \text{ of } \frac{1}{3} = \frac{2}{5} \cdot \frac{1}{3} = \frac{2}{15}$$

Multiply: $\frac{x}{7} \cdot \frac{y}{5}$

Multiply the numerators.
Multiply the denominators.

$$\frac{x}{7} \cdot \frac{y}{5} = \frac{x \cdot y}{7 \cdot 5}$$

Write the product in simplest form.

$$= \frac{xy}{35}$$

Multiply: $\frac{3}{8} \cdot \frac{4}{9}$

Multiply the numerators.
Multiply the denominators.

$$\frac{3}{8} \cdot \frac{4}{9} = \frac{3 \cdot 4}{8 \cdot 9}$$

Express the fraction in simplest form by first writing the prime factorization of each number.

$$= \frac{3 \cdot 2 \cdot 2}{2 \cdot 2 \cdot 2 \cdot 3 \cdot 3}$$

Divide by the common factors and write the product in simplest form.

$$= \frac{1}{6}$$

The sign rules for multiplying positive and negative fractions are the same rules used to multiply integers.

The product of two numbers with the same sign is positive.
The product of two numbers with different signs is negative.

Multiply: $-\frac{3}{4} \cdot \frac{8}{15}$

The signs are different. The product is negative. $-\frac{3}{4} \cdot \frac{8}{15} = -\left(\frac{3}{4} \cdot \frac{8}{15}\right)$

Multiply the numerators. Multiply the denominators. $= -\frac{3 \cdot 8}{4 \cdot 15}$

Write the product in simplest form. $= -\frac{3 \cdot 2 \cdot 2 \cdot 2}{2 \cdot 2 \cdot 3 \cdot 5}$

$= -\frac{2}{5}$

If multiplication with fractions involves a whole number, first rewrite the whole number as a fraction with a denominator of 1. When a multiplication involves a mixed number, first rewrite the mixed number as an improper fraction. Then multiply the fractions.

Multiply: $3 \cdot \frac{5}{8}$

Write the whole number 3 as the fraction $\frac{3}{1}$. $3 \cdot \frac{5}{8} = \frac{3}{1} \cdot \frac{5}{8}$

Multiply the fractions. There are no common factors in the numerator and denominator. $= \frac{3 \cdot 5}{1 \cdot 8}$

Write the improper fraction as a mixed number. $= \frac{15}{8} = 1\frac{7}{8}$

Evaluate xy when $x = 5\frac{2}{3}$ and $y = 6$.

xy

Replace x with $5\frac{2}{3}$ and y with 6. $5\frac{2}{3} \cdot 6$

Write the mixed number as an improper fraction and the whole number as a fraction with a denominator of 1. $= \frac{17}{3} \cdot \frac{6}{1}$

Multiply the fractions. $= \frac{17 \cdot 6}{3 \cdot 1}$

Write the product in simplest form. $= \frac{17 \cdot 2 \cdot 3}{3 \cdot 1} = \frac{34}{1} = 34$

If a is a natural number, then $\frac{1}{a}$ is called the **reciprocal** or **multiplicative inverse** of a. Note that $a \cdot \frac{1}{a} = \frac{a}{1} \cdot \frac{1}{a} = \frac{a}{a} = 1$.

The product of a number and its multiplicative inverse is 1. $\frac{1}{8} \cdot 8 = 8 \cdot \frac{1}{8} = 1$

Example 5 Multiply: $-\frac{3}{4}\left(\frac{1}{2}\right)\left(-\frac{8}{9}\right)$

Solution
$$-\frac{3}{4}\left(\frac{1}{2}\right)\left(-\frac{8}{9}\right)$$
The product of two negative fractions is positive.
$$= \frac{3}{4}\cdot\frac{1}{2}\cdot\frac{8}{9}$$
$$= \frac{3\cdot 1\cdot 8}{4\cdot 2\cdot 9}$$
$$= \frac{3\cdot 1\cdot 2\cdot 2\cdot 2}{2\cdot 2\cdot 2\cdot 3\cdot 3} = \frac{1}{3}$$

You Try It 5 Multiply: $-\frac{1}{3}\left(-\frac{5}{12}\right)\left(\frac{8}{15}\right)$

Your Solution

Example 6 What is the product of $\frac{7}{12}$ and 4?

Solution
$$\frac{7}{12}\cdot 4 = \frac{7}{12}\cdot\frac{4}{1}$$
$$= \frac{7\cdot 4}{12\cdot 1}$$
$$= \frac{7\cdot 2\cdot 2}{2\cdot 2\cdot 3\cdot 1}$$
$$= \frac{7}{3}$$
$$= 2\frac{1}{3}$$

You Try It 6 Find the product of $\frac{8}{9}$ and 6.

Your Solution

Example 7 Evaluate the variable expression xy when $x = 1\frac{4}{5}$ and $y = -\frac{5}{6}$.

Solution xy
$$1\frac{4}{5}\cdot\left(-\frac{5}{6}\right) = -\left(\frac{9}{5}\cdot\frac{5}{6}\right)$$
$$= -\frac{9\cdot 5}{5\cdot 6}$$
$$= -\frac{3\cdot 3\cdot 5}{5\cdot 2\cdot 3}$$
$$= -\frac{3}{2} = -1\frac{1}{2}$$

You Try It 7 Evaluate the variable expression xy when $x = 5\frac{1}{8}$ and $y = \frac{2}{3}$.

Your Solution

Solutions on p. A5

OBJECTIVE D Division of fractions

Two numbers whose product is 1 are called **reciprocals** of each other. Every number except 0 has a reciprocal.

The reciprocal of $\frac{3}{4}$ is $\frac{4}{3}$ because $\frac{3}{4} \cdot \frac{4}{3} = 1$.

The reciprocal of $\frac{a}{b}$ is $\frac{b}{a}$ $(a \neq 0, b \neq 0)$ because $\frac{a}{b} \cdot \frac{b}{a} = 1$.

The **reciprocal** of a fraction is the fraction with the numerator and denominator interchanged. The process of interchanging the numerator and denominator of a fraction is called **inverting** the fraction.

To find the reciprocal of a whole number, first rewrite the whole number as a fraction with a denominator of 1. Then invert the fraction.

$6 = \frac{6}{1}$

The reciprocal of 6 is $\frac{1}{6}$.

Reciprocals are used to rewrite division problems as related multiplication problems. Look at the following two problems:

$6 \div 2 = 3$ — 6 divided by 2 equals 3.

$6 \cdot \frac{1}{2} = 3$ — 6 times the reciprocal of 2 equals 3.

Division is defined as multiplication by the reciprocal. Therefore, "divided by 2" is the same as "times $\frac{1}{2}$." Fractions are divided by making this substitution.

Division of Fractions

To divide two fractions, multiply by the reciprocal of the divisor.

$$\frac{a}{b} \div \frac{c}{d} = \frac{a}{b} \cdot \frac{d}{c}, \text{ where } b \neq 0, c \neq 0, \text{ and } d \neq 0$$

Divide: $\frac{2}{5} \div \frac{3}{4}$

Rewrite the division as multiplication by the reciprocal of the divisor.

$$\frac{2}{5} \div \frac{3}{4} = \frac{2}{5} \cdot \frac{4}{3}$$

Multiply the fractions.

$$= \frac{2 \cdot 4}{5 \cdot 3}$$

$$= \frac{2 \cdot 2 \cdot 2}{5 \cdot 3} = \frac{8}{15}$$

The sign rules for dividing positive and negative fractions are the same rules used to divide integers.

The quotient of two numbers with the same sign is positive.
The quotient of two numbers with different signs is negative.

Divide: $-\frac{7}{10} \div \left(-\frac{14}{15}\right)$

The signs are the same. The quotient is positive. $\quad -\frac{7}{10} \div \left(-\frac{14}{15}\right) = \frac{7}{10} \div \frac{14}{15}$

Rewrite the division as multiplication by the reciprocal. $\quad = \frac{7}{10} \cdot \frac{15}{14}$

Multiply the fractions. $\quad = \frac{7 \cdot 15}{10 \cdot 14}$

$$= \frac{7 \cdot 3 \cdot 5}{2 \cdot 5 \cdot 2 \cdot 7}$$

$$= \frac{3}{4}$$

Evaluate $-x \div y$ when $x = -\frac{3}{8}$ and $y = -\frac{5}{12}$.

$$-x \div y$$

Replace x with $-\frac{3}{8}$ and y with $-\frac{5}{12}$. $\quad -\left(-\frac{3}{8}\right) \div \left(-\frac{5}{12}\right)$

Simplify $-\left(-\frac{3}{8}\right)$. $\quad = \frac{3}{8} \div \left(-\frac{5}{12}\right)$

The signs are different. The quotient is negative. $\quad = -\left(\frac{3}{8} \div \frac{5}{12}\right)$

Rewrite the division as multiplication by the reciprocal. $\quad = -\left(\frac{3}{8} \cdot \frac{12}{5}\right)$

Multiply the fractions. $\quad = -\frac{3 \cdot 12}{8 \cdot 5}$

$$= -\frac{3 \cdot 2 \cdot 2 \cdot 3}{2 \cdot 2 \cdot 2 \cdot 5}$$

$$= -\frac{9}{10}$$

As with multiplication of fractions, if division of fractions involves a whole number, first rewrite the whole number as a fraction with a denominator of 1. When the division involves a mixed number, first rewrite the mixed number as an improper fraction. Then divide the fractions.

Example 8 What is the quotient of 6 and $-\frac{3}{5}$?

Solution

$$6 \div \left(-\frac{3}{5}\right) = -\left(\frac{6}{1} \div \frac{3}{5}\right)$$
$$= -\left(\frac{6}{1} \cdot \frac{5}{3}\right)$$
$$= -\frac{6 \cdot 5}{1 \cdot 3}$$
$$= -\frac{2 \cdot 3 \cdot 5}{1 \cdot 3}$$
$$= -\frac{10}{1} = -10$$

You Try It 8 Find the quotient of 4 and $-\frac{6}{7}$.

Your Solution

Example 9 Divide: $\frac{x}{2} \div \frac{y}{4}$

Solution

$$\frac{x}{2} \div \frac{y}{4} = \frac{x}{2} \cdot \frac{4}{y}$$
$$= \frac{x \cdot 4}{2 \cdot y}$$
$$= \frac{x \cdot 2 \cdot 2}{2 \cdot y} = \frac{2x}{y}$$

You Try It 9 Divide: $\frac{x}{8} \div \frac{y}{6}$

Your Solution

Example 10 Evaluate $x \div y$ when $x = 3\frac{1}{8}$ and $y = 5$.

Solution $x \div y$

$$3\frac{1}{8} \div 5 = \frac{25}{8} \div \frac{5}{1}$$
$$= \frac{25}{8} \cdot \frac{1}{5}$$
$$= \frac{25 \cdot 1}{8 \cdot 5}$$
$$= \frac{5 \cdot 5 \cdot 1}{2 \cdot 2 \cdot 2 \cdot 5} = \frac{5}{8}$$

You Try It 10 Evaluate $x \div y$ when $x = 2\frac{1}{4}$ and $y = 9$.

Your Solution

Solutions on p. A5

OBJECTIVE E Applications and formulas

Example 11

A 12-foot board is cut into pieces $2\frac{1}{2}$ ft long for use as bookshelves. What is the length remaining after as many shelves as possible are cut?

Strategy

To find the length remaining:

- Divide the total length (12) by the length of each shelf $\left(2\frac{1}{2}\right)$. The quotient is the number of shelves cut, with a certain fraction of a shelf left over.
- Multiply the fraction left over by the length of each shelf.

Solution

$$12 \div 2\frac{1}{2} = \frac{12}{1} \div \frac{5}{2} = \frac{12}{1} \cdot \frac{2}{5} = \frac{24}{5} = 4\frac{4}{5}$$

4 shelves, each $2\frac{1}{2}$ ft long, can be cut from the board. The piece remaining is $\frac{4}{5}$ of $2\frac{1}{2}$ ft long.

$$\frac{4}{5} \cdot 2\frac{1}{2} = \frac{4}{5} \cdot \frac{5}{2} = \frac{4 \cdot 5}{5 \cdot 2} = 2$$

The length of the remaining piece is 2 ft.

You Try It 11

The Booster Club is making 22 sashes for the high school band members. Each sash requires $1\frac{3}{8}$ yd of material at a cost of \$8 per yard. Find the total cost of the material.

Your Strategy

Your Solution

Example 12

The formula $C = \frac{5}{9}(F - 32)$, where C is Celsius and F is Fahrenheit, is used to convert Fahrenheit to Celsius. Use this formula to find the temperature in degrees Celsius when the Fahrenheit temperature is 86°.

Strategy

To find the Celsius temperature, replace F by 86 in the given formula and solve for C.

Solution

$$C = \frac{5}{9}(F - 32)$$

$$C = \frac{5}{9}(86 - 32) = \frac{5}{9}(54) = \frac{5}{9} \cdot \frac{54}{1} = 30$$

The Celsius temperature is 30°.

You Try It 12

The formula $C = \frac{5}{9}(F - 32)$, where C is Celsius and F is Fahrenheit, is used to convert Fahrenheit to Celsius. Use this formula to find the temperature in degrees Celsius when the Fahrenheit temperature is 68°.

Your Strategy

Your Solution

Solutions on p. A6

Example 13

A unit fraction is a fraction with a numerator of 1. The following formula can be used to express one unit fraction as the sum of two other unit fractions.

$$\frac{1}{n} = \frac{1}{n+1} + \frac{1}{n(n+1)}$$

In this formula, n is a whole number greater than 1. Use this formula to express the unit fraction $\frac{1}{2}$ as the sum of two other unit fractions. Verify the results.

Strategy

- ▶ To express $\frac{1}{2}$ as the sum of two other unit fractions, replace n with 2 in the given formula and simplify.
- ▶ To verify the results, find the sum of the two unit fractions found.

Solution

$$\frac{1}{n} = \frac{1}{n+1} + \frac{1}{n(n+1)}$$

$$\frac{1}{2} = \frac{1}{2+1} + \frac{1}{2(2+1)}$$

$$\frac{1}{2} = \frac{1}{3} + \frac{1}{6}$$

$$\frac{1}{3} + \frac{1}{6} = \frac{2}{6} + \frac{1}{6} = \frac{3}{6} = \frac{1}{2}$$

The unit fraction $\frac{1}{2}$ is equal to the sum of the unit fractions $\frac{1}{3}$ and $\frac{1}{6}$.

You Try It 13

A unit fraction is a fraction with a numerator of 1. The following formula can be used to express one unit fraction as the sum of two other unit fractions.

$$\frac{1}{n} = \frac{1}{n+1} + \frac{1}{n(n+1)}$$

In this formula, n is a whole number greater than 1. Use this formula to express the unit fraction $\frac{1}{3}$ as the sum of two other unit fractions. Verify the results.

Your Strategy

Your Solution

Solution on p. A6

3.2 EXERCISES

Objective A

Add.

1. $\frac{2}{3} + \frac{1}{3}$

2. $\frac{1}{2} + \frac{1}{2}$

3. $\frac{7}{b} + \frac{9}{b}$

4. $\frac{3}{y} + \frac{6}{y}$

5. $\frac{a}{9} + \frac{b}{9}$

6. $\frac{y}{12} + \frac{z}{12}$

7. $\frac{7}{18} + \frac{13}{18} + \frac{1}{18}$

8. $\frac{8}{15} + \frac{2}{15} + \frac{11}{15}$

9. $\frac{1}{x} + \frac{4}{x} + \frac{6}{x}$

10. $\frac{8}{n} + \frac{5}{n} + \frac{3}{n}$

11. $-\frac{a}{2} + \frac{b}{2}$

12. $\frac{x}{4} + \left(-\frac{y}{4}\right)$

13. $\frac{1}{4} + \frac{2}{3}$

14. $\frac{2}{3} + \frac{1}{2}$

15. $\frac{7}{15} + \frac{9}{20}$

16. $\frac{4}{9} + \frac{1}{6}$

17. $\frac{2}{3} + \frac{1}{12} + \frac{5}{6}$

18. $\frac{3}{8} + \frac{1}{2} + \frac{5}{12}$

19. $-\frac{3}{4} + \frac{2}{3}$

20. $-\frac{7}{12} + \frac{5}{8}$

21. $\frac{2}{5} + \left(-\frac{11}{15}\right)$

22. $\frac{1}{4} + \left(-\frac{1}{7}\right)$

23. $\frac{3}{8} + \left(-\frac{1}{2}\right) + \frac{7}{12}$

24. $-\frac{7}{12} + \frac{2}{3} + \left(-\frac{4}{5}\right)$

25. $\frac{2}{3} + \left(-\frac{5}{6}\right) + \frac{1}{4}$

26. $-\frac{5}{8} + \frac{3}{4} + \frac{1}{2}$

27. $8 + 7\frac{2}{3}$

28. $6 + 9\frac{3}{5}$

Solve.

29. What is $-\frac{5}{6}$ added to $\frac{4}{9}$?

30. Find the total of $\frac{2}{7}$, $\frac{3}{14}$, and $\frac{1}{4}$.

31. What is $-\frac{2}{3}$ more than $-\frac{5}{6}$?

32. Find $-\frac{4}{9}$ plus $-\frac{5}{6}$.

33. Find $\frac{7}{8}$ increased by $-\frac{1}{3}$.

34. Find the sum of $\frac{11}{15}$, $-\frac{7}{10}$, and $\frac{2}{5}$.

Evaluate the variable expression $x + y$ for the given values of x and y.

35. $x = \frac{3}{5}; y = \frac{4}{5}$

36. $x = \frac{5}{8}; y = \frac{3}{8}$

37. $x = \frac{1}{3}; y = -\frac{3}{4}$

38. $x = -\frac{3}{8}; y = \frac{2}{9}$

39. $x = \frac{5}{6}; y = \frac{8}{9}$

40. $x = \frac{3}{5}; y = \frac{4}{7}$

41. $x = \frac{4}{15}; y = \frac{7}{10}$

42. $x = \frac{8}{21}; y = \frac{5}{14}$

43. $x = -\frac{5}{6}; y = \frac{4}{9}$

44. $x = \frac{3}{10}; y = -\frac{7}{15}$

45. $x = -\frac{5}{8}; y = -\frac{1}{6}$

46. $x = -\frac{3}{8}; y = -\frac{5}{6}$

Evaluate the variable expression $x + y + z$ for the given values of x, y, and z.

47. $x = \frac{3}{8}; y = \frac{1}{4}; z = \frac{7}{12}$

48. $x = \frac{5}{6}; y = \frac{2}{3}; z = \frac{7}{24}$

49. $x = -\frac{1}{2}; y = \frac{3}{4}; z = -\frac{5}{12}$

Objective B

Subtract.

50. $\frac{17}{20} - \frac{9}{20}$

51. $\frac{11}{24} - \frac{7}{24}$

52. $\frac{8}{d} - \frac{3}{d}$

53. $\frac{12}{y} - \frac{7}{y}$

54. $\frac{a}{9} - \frac{b}{9}$

55. $\frac{y}{4} - \frac{z}{4}$

56. $\frac{c}{12} - \left(-\frac{d}{12}\right)$

57. $\frac{m}{30} - \left(-\frac{n}{30}\right)$

58. $\frac{3}{4} - \frac{1}{8}$

59. $\frac{2}{3} - \frac{1}{6}$

60. $\frac{3}{7} - \frac{5}{14}$

61. $\frac{7}{8} - \frac{5}{16}$

62. $\frac{7}{10} - \frac{2}{5}$

63. $\frac{5}{9} - \frac{4}{15}$

64. $2 - \frac{4}{5}$

65. $1 - \frac{8}{9}$

66. $\frac{5}{21} - \frac{1}{6}$

67. $\frac{11}{12} - \frac{2}{3}$

68. $\frac{9}{20} - \frac{1}{30}$

69. $\frac{7}{9} - \frac{5}{18}$

Subtract.

70. $-\frac{1}{2} - \frac{3}{8}$

71. $-\frac{5}{6} - \frac{1}{9}$

72. $-\frac{3}{10} - \frac{4}{5}$

73. $-\frac{7}{15} - \frac{3}{10}$

74. $-\frac{5}{12} - \left(-\frac{2}{3}\right)$

75. $-\frac{3}{10} - \left(-\frac{5}{6}\right)$

76. $-\frac{5}{9} - \left(-\frac{11}{12}\right)$

77. $-\frac{5}{8} - \left(-\frac{7}{12}\right)$

Solve.

78. What is $-\frac{7}{12}$ minus $\frac{7}{9}$?

79. What is $\frac{3}{5}$ minus $-\frac{7}{10}$?

80. Find $\frac{1}{2}$ decreased by $-\frac{3}{7}$.

81. Find $-\frac{5}{6}$ decreased by $-\frac{2}{3}$.

82. What is $-\frac{2}{3}$ less than $-\frac{7}{8}$?

83. What is $-\frac{8}{9}$ less than $-\frac{1}{6}$?

84. Find 1 less $\frac{7}{12}$.

85. Find 1 less $-\frac{3}{20}$.

86. What is the difference between $-\frac{5}{8}$ and $-\frac{11}{12}$?

87. What is the difference between $-\frac{2}{15}$ and $\frac{7}{9}$?

Evaluate the variable expression $x - y$ for the given values of x and y.

88. $x = \frac{8}{9}; y = \frac{5}{9}$

89. $x = \frac{5}{6}; y = \frac{1}{6}$

90. $x = -\frac{11}{12}; y = \frac{5}{12}$

91. $x = -\frac{15}{16}; y = \frac{5}{16}$

92. $x = -\frac{2}{3}; y = -\frac{3}{4}$

93. $x = -\frac{5}{12}; y = -\frac{5}{9}$

94. $x = -\frac{3}{10}; y = -\frac{7}{15}$

95. $x = -\frac{5}{6}; y = -\frac{2}{15}$

96. $x = \frac{7}{9}; y = -\frac{2}{3}$

97. $x = \frac{5}{8}; y = -\frac{3}{16}$

98. $x = -\frac{9}{10}; y = \frac{1}{2}$

99. $x = -\frac{4}{9}; y = \frac{1}{6}$

Objective C

Multiply.

100. $\frac{2}{3} \cdot \frac{9}{10}$

101. $\frac{3}{8} \cdot \frac{4}{5}$

102. $-\frac{6}{7} \cdot \frac{11}{12}$

103. $\frac{5}{6} \cdot \left(-\frac{2}{5}\right)$

104. $-\frac{6}{7} \cdot \frac{0}{10}$

105. $\frac{5}{12} \cdot \frac{3}{0}$

106. $\left(-\frac{4}{15}\right) \cdot \left(-\frac{3}{8}\right)$

107. $\left(-\frac{3}{4}\right) \cdot \left(-\frac{2}{9}\right)$

108. $-\frac{3}{4} \cdot \frac{1}{2}$

109. $-\frac{8}{15} \cdot \frac{5}{12}$

110. $\frac{9}{x} \cdot \frac{7}{y}$

111. $\frac{4}{c} \cdot \frac{8}{d}$

112. $-\frac{y}{5} \cdot \frac{z}{6}$

113. $-\frac{a}{10} \cdot \left(-\frac{b}{6}\right)$

114. $-\frac{7}{12} \cdot \frac{5}{8} \cdot \frac{16}{25}$

115. $\frac{5}{12} \cdot \left(-\frac{1}{3}\right) \cdot \left(-\frac{8}{15}\right)$

116. $\left(-\frac{3}{5}\right) \cdot \frac{1}{2} \cdot \left(-\frac{5}{8}\right)$

117. $\frac{5}{6} \cdot \left(-\frac{2}{3}\right) \cdot \frac{3}{25}$

118. $6 \cdot \frac{1}{6}$

119. $\frac{1}{10} \cdot 10$

120. $12 \cdot \left(-\frac{5}{8}\right)$

121. $24 \cdot \left(-\frac{3}{8}\right)$

122. $-16 \cdot \frac{7}{30}$

123. $-9 \cdot \frac{7}{15}$

124. $3\frac{1}{3} \cdot (-9)$

125. $-2\frac{1}{2} \cdot 4$

126. $8 \cdot 5\frac{1}{4}$

127. $3 \cdot 2\frac{1}{9}$

Solve.

128. Find the product of $\frac{3}{4}$ and $\frac{14}{15}$.

129. Find the product of $\frac{12}{25}$ and $\frac{5}{16}$.

130. Find $-\frac{9}{16}$ multiplied by $\frac{4}{27}$.

131. Find $\frac{3}{7}$ multiplied by $-\frac{14}{15}$.

132. What is the product of $\frac{4}{21}$, $\frac{9}{10}$, and $-\frac{7}{8}$?

133. What is the product of $-\frac{5}{13}$, $-\frac{26}{75}$, and $\frac{5}{8}$?

Evaluate the variable expression xy for the given values of x and y.

134. $x = \frac{2}{3}; y = \frac{7}{8}$

135. $x = \frac{3}{8}; y = \frac{6}{7}$

136. $x = -\frac{5}{16}; y = \frac{7}{15}$

137. $x = -\frac{2}{5}; y = -\frac{5}{6}$

138. $x = -49; y = \frac{5}{14}$

139. $x = -\frac{3}{10}; y = -35$

140. $x = \frac{3}{13}; y = -6\frac{1}{2}$

141. $x = -3\frac{1}{2}; y = \frac{2}{7}$

Evaluate the variable expression xyz for the given values of x, y, and z.

142. $x = \frac{3}{8}; y = \frac{2}{3}; z = \frac{4}{5}$

143. $x = \frac{5}{9}; y = \frac{6}{7}; z = \frac{3}{5}$

144. $x = 4; y = \frac{0}{8}; z = -\frac{5}{9}$

145. $x = \frac{3}{8}; y = -\frac{3}{19}; z = -\frac{4}{9}$

146. $x = \frac{3}{5}; y = -\frac{1}{2}; z = \frac{2}{3}$

147. $x = \frac{4}{5}; y = -15; z = \frac{7}{8}$

Objective D

Divide.

148. $\frac{5}{7} \div \frac{2}{5}$

149. $\frac{3}{8} \div \frac{2}{3}$

150. $\frac{4}{7} \div \left(-\frac{4}{7}\right)$

151. $\left(-\frac{5}{7}\right) \div \left(-\frac{5}{6}\right)$

152. $\left(-\frac{1}{3}\right) \div \frac{1}{2}$

153. $\left(-\frac{3}{8}\right) \div \frac{7}{8}$

154. $-\frac{5}{16} \div \left(-\frac{3}{8}\right)$

155. $\left(-\frac{3}{4}\right) \div \left(-\frac{5}{6}\right)$

156. $\frac{4}{15} \div \frac{2}{5}$

157. $\frac{7}{15} \div \frac{14}{5}$

158. $6 \div \frac{3}{4}$

159. $8 \div \frac{2}{3}$

160. $\frac{3}{4} \div (-6)$

161. $-\frac{2}{3} \div 8$

162. $\frac{9}{10} \div 0$

163. $0 \div \frac{4}{5}$

164. $\frac{5}{12} \div \left(-\frac{15}{32}\right)$

165. $\frac{3}{8} \div \left(-\frac{5}{12}\right)$

166. $\left(-\frac{2}{3}\right) \div (-4)$

167. $\left(-\frac{4}{9}\right) \div (-6)$

Divide.

168. $\frac{8}{x} \div \left(-\frac{y}{4}\right)$

169. $-\frac{9}{m} \div \frac{n}{7}$

170. $\frac{b}{6} \div \frac{5}{d}$

171. $\frac{y}{10} \div \frac{4}{z}$

172. $5\frac{3}{5} \div \left(-\frac{7}{10}\right)$

173. $6\frac{8}{9} \div \left(-\frac{31}{36}\right)$

174. $5\frac{2}{7} \div 1$

175. $9\frac{5}{6} \div 1$

Solve.

176. Find the quotient of $\frac{9}{10}$ and $\frac{3}{4}$.

177. Find the quotient of $\frac{3}{5}$ and $\frac{12}{25}$.

178. What is $-\frac{15}{24}$ divided by $\frac{3}{5}$?

179. What is $\frac{5}{6}$ divided by $-\frac{10}{21}$?

180. Find $\frac{7}{8}$ divided by $-\frac{1}{4}$.

181. What is the quotient of $-\frac{5}{11}$ and $\frac{4}{5}$?

Evaluate the variable expression $x \div y$ for the given values of x and y.

182. $x = \frac{2}{5}; y = \frac{4}{7}$

183. $x = \frac{3}{8}; y = \frac{5}{12}$

184. $x = -\frac{5}{8}; y = -\frac{15}{2}$

185. $x = -\frac{14}{3}; y = -\frac{7}{9}$

186. $x = \frac{5}{6}; y = \frac{1}{9}$

187. $x = \frac{5}{7}; y = \frac{2}{7}$

188. $x = -18; y = \frac{3}{8}$

189. $x = 20; y = -\frac{5}{6}$

190. $x = \frac{1}{7}; y = 0$

191. $x = \frac{4}{0}; y = 12$

192. $x = \frac{1}{2}; y = \frac{3}{5}$

193. $x = \frac{2}{3}; y = \frac{7}{9}$

Objective E

Solve.

194. Two inlet pipes are being used to fill a tank. After one hour, the larger pipe has filled $\frac{2}{5}$ of the tank and the smaller pipe has filled $\frac{1}{4}$ of the tank. How much of the tank remains to be filled? Can the two pipes complete the job within another hour?

Solve.

195. A roofer and an apprentice are roofing a newly constructed house. In one day, the roofer completes $\frac{1}{3}$ of the job and the apprentice completes $\frac{1}{4}$ of the job. How much of the job remains to be done? Working at the same rate, can the roofer and the apprentice complete the job in one more day?

196. A chukker is one period of play in a polo match. A chukker lasts $7\frac{1}{2}$ min. Find the length of time in four chukkers.

197. The Assyrian calendar was based on the phases of the moon. One lunation was $29\frac{1}{2}$ days long. There were 12 lunations in one year. Find the number of days in one year in the Assyrian calendar.

198. Find the length of the shaft shown at the right.

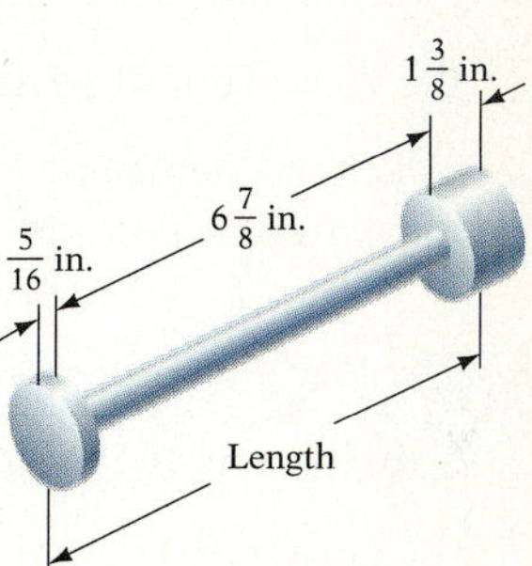

199. One rod is equal to $5\frac{1}{2}$ yd. How many feet are in one rod? How many inches are in one rod?

200. A recipe for chocolate chip cookies calls for $1\frac{3}{4}$ c flour. If you are halving the recipe, how much flour do you need?

201. A car used $12\frac{1}{2}$ gal of gasoline on a 275-mile trip. How many miles can this car travel on one gallon of gasoline?

202. In the 1980 presidential election, approximately $\frac{1}{2}$ of the voters voted for Ronald Reagan and $\frac{2}{5}$ of the voters voted for Jimmy Carter. What fraction of the voters did not vote for either the Republican or Democratic candidate?

203. A wooden travel game board has hinges that allow the board to be folded in half. If the dimensions of the open board are 14 in. by 14 in. by $\frac{7}{8}$ in., what are the dimensions of the board when it is closed?

204. According to a national survey, the average couple spends $4\frac{1}{2}$ h cleaning house each weekend. How many hours does the average couple spend cleaning house each year?

205. A factory worker can assemble a product in $7\frac{1}{2}$ min. How many products can the worker assemble in one hour?

206. Find the total wages of an employee who worked $26\frac{1}{2}$ h this week and who earns an hourly wage of $12.

Solve.

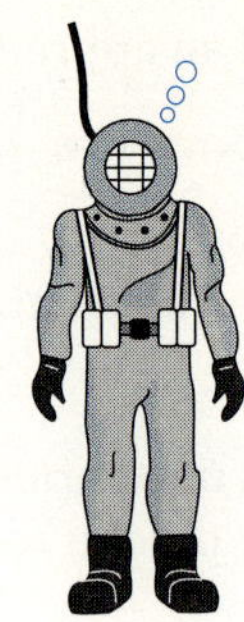

207. The pressure on a submerged object is given by $P = 15 + \frac{1}{2}D$, where D is the depth in feet and P is the pressure measured in pounds per square inch. Find the pressure on a diver who is at a depth of $12\frac{1}{2}$ ft.

208. A hiker covers a distance of $7\frac{1}{2}$ mi in 3 h. Use the equation $r = \frac{d}{t}$, where r is the rate in miles per hour, d is the distance, and t is the time, to find the rate at which the hiker walked.

209. Find the amount of force necessary to push a 75-pound crate across a floor if the coefficient of friction is $\frac{3}{8}$. Use the equation $F = \mu N$, where F is the force, μ is the coefficient of friction, and N is the weight of the crate. Force is measured in pounds.

210. The formula $\frac{2}{3n} = \frac{1}{2n} + \frac{1}{6n}$, where n is a natural number, can be used to express a fraction as the sum of two unit fractions. Use this formula to express $\frac{2}{9}$ as the sum of two unit fractions. Verify the results.

211. The formula $\frac{2}{3n} = \frac{1}{2n} + \frac{1}{6n}$, where n is a natural number, can be used to express a fraction as the sum of two unit fractions. Use this formula to express $\frac{2}{15}$ as the sum of two unit fractions. Verify the results.

Critical Thinking

212. Draw two diagrams, one to illustrate the addition of two fractions with the same denominator and another to illustrate the addition of two fractions with different denominators.

213. Determine whether the statement is always true, sometimes true, or never true.

a. If n is an even number, then $\frac{1}{2}n$ is a whole number.

b. If n is an odd number, then $\frac{1}{2}n$ is an improper fraction.

214. The figure at the right is divided into 5 parts.

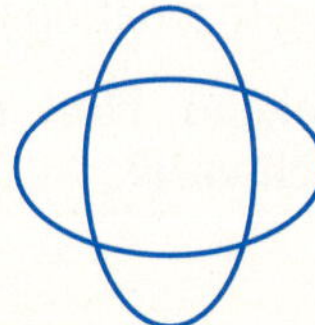

a. Is each part of the figure $\frac{1}{5}$ of the figure? Why or why not?

b. Can you trace over all the lines of the figure without lifting the pencil from the paper and without retracing over a line?

215. Find the sum of **a.** the 3 largest unit fractions, and **b.** the 4 largest unit fractions.

216. [W] On page 111, Exercise 197 describes the Assyrian calendar. Our calendar is based on the solar year. One solar year is $365\frac{1}{4}$ days. Use this fact to explain leap years.

SECTION 3.3 Exponents, Complex Fractions, and the Order of Operations Agreement

OBJECTIVE  Exponents

POINT OF INTEREST

Rene Descartes (1596–1650) was the first mathematician to extensively use exponential notation as it is used today. However, for some unknown reason, he always used xx for x^2.

Recall that an exponent indicates the repeated multiplication of the same factor. For example,

$$3^5 = 3 \cdot 3 \cdot 3 \cdot 3 \cdot 3$$

The exponent, 5, indicates how many times the base, 3, occurs as a factor in the multiplication.

The base of an exponential expression can be a fraction, for example, $\left(\frac{2}{3}\right)^4$. To evaluate this expression, write the factor as many times as indicated by the exponent and then multiply.

$$\left(\frac{2}{3}\right)^4 = \frac{2}{3} \cdot \frac{2}{3} \cdot \frac{2}{3} \cdot \frac{2}{3} = \frac{2 \cdot 2 \cdot 2 \cdot 2}{3 \cdot 3 \cdot 3 \cdot 3} = \frac{16}{81}$$

Evaluate $\left(-\frac{3}{5}\right)^2 \cdot \left(\frac{5}{6}\right)^3$.

$$\left(-\frac{3}{5}\right)^2 \cdot \left(\frac{5}{6}\right)^3$$

Write each factor as many times as indicated by the exponent.

$$= \left(-\frac{3}{5}\right) \cdot \left(-\frac{3}{5}\right) \cdot \frac{5}{6} \cdot \frac{5}{6} \cdot \frac{5}{6}$$

Multiply. The product of two negative numbers is positive.

$$= \frac{3}{5} \cdot \frac{3}{5} \cdot \frac{5}{6} \cdot \frac{5}{6} \cdot \frac{5}{6}$$

$$= \frac{3 \cdot 3 \cdot 5 \cdot 5 \cdot 5}{5 \cdot 5 \cdot 6 \cdot 6 \cdot 6}$$

Write the product in simplest form.

$$= \frac{5}{24}$$

Example 1 Evaluate x^2y^2 when $x = 1\frac{1}{2}$ and $y = \frac{2}{3}$.

Solution x^2y^2

$$\left(1\frac{1}{2}\right)^2 \cdot \left(\frac{2}{3}\right)^2 = \left(\frac{3}{2}\right)^2 \cdot \left(\frac{2}{3}\right)^2$$

$$= \frac{3}{2} \cdot \frac{3}{2} \cdot \frac{2}{3} \cdot \frac{2}{3}$$

$$= \frac{3 \cdot 3 \cdot 2 \cdot 2}{2 \cdot 2 \cdot 3 \cdot 3} = 1$$

You Try It 1 Evaluate x^4y^3 when $x = 2\frac{1}{3}$ and $y = \frac{3}{7}$.

Your Solution

Solution on p. A6

OBJECTIVE B Complex fractions

A **complex fraction** is a fraction whose numerator or denominator contains one or more fractions. Examples of complex fractions are shown below.

Main fraction bar ⟶ $\dfrac{\frac{3}{4}}{\frac{7}{8}}$ $\qquad \dfrac{4}{3-\frac{1}{2}}$ $\qquad \dfrac{\frac{9}{10}+\frac{3}{5}}{\frac{5}{6}}$ $\qquad \dfrac{3\frac{1}{2}\cdot 2\frac{5}{8}}{\left(4\frac{2}{3}\right)\div\left(3\frac{1}{5}\right)}$

Look at the first example above; recall that the fraction bar can be read "divided by."

Therefore, $\dfrac{\frac{3}{4}}{\frac{7}{8}}$ can be read "$\frac{3}{4}$ divided by $\frac{7}{8}$" and can be written $\frac{3}{4} \div \frac{7}{8}$, which is the division of two fractions and can be simplified by multiplying by the reciprocal.

$$\frac{\frac{3}{4}}{\frac{7}{8}} = \frac{3}{4} \div \frac{7}{8} = \frac{3}{4} \cdot \frac{8}{7} = \frac{3 \cdot 8}{4 \cdot 7} = \frac{6}{7}$$

Note that in this complex fraction, the numerator and denominator each consist of only one number. In the other examples above, this is not the case.

To simplify a complex fraction, first simplify the expression above the main fraction bar and the expression below the main fraction bar; the result is one number in the numerator and one number in the denominator. Then rewrite the complex fraction as a division problem by reading the main fraction bar as "divided by."

Simplify: $\dfrac{-\frac{9}{10}+\frac{3}{5}}{1\frac{1}{4}}$

Simplify the expression in the numerator.

Note: $-\frac{9}{10} + \frac{3}{5} = \frac{-9}{10} + \frac{6}{10} = \frac{-3}{10} = -\frac{3}{10}$

$$\frac{-\frac{9}{10}+\frac{3}{5}}{1\frac{1}{4}} = \frac{-\frac{3}{10}}{\frac{5}{4}}$$

Write the mixed number in the denominator as an improper fraction.

Rewrite the complex fraction as division. The quotient will be negative.

$$= -\left(\frac{3}{10} \div \frac{5}{4}\right)$$

Divide by multiplying by the reciprocal.

$$= -\left(\frac{3}{10} \cdot \frac{4}{5}\right)$$

$$= -\frac{6}{25}$$

Example 2

Evaluate the variable expression $\frac{x-y}{z}$ when $x = \frac{1}{8}$, $y = \frac{5}{8}$, and $z = -\frac{3}{4}$.

Solution

$$\frac{x-y}{z}$$

$$\frac{\frac{1}{8}-\frac{5}{8}}{-\frac{3}{4}} = \frac{-\frac{1}{2}}{-\frac{3}{4}} = -\frac{1}{2} \div \left(-\frac{3}{4}\right) = \frac{1}{2} \cdot \frac{4}{3} = \frac{2}{3}$$

You Try It 2

Evaluate the variable expression $\frac{x}{y-z}$ when $x = -\frac{4}{9}$, $y = 3$, and $z = \frac{1}{3}$.

Your Solution

Solution on p. A6

OBJECTIVE C The Order of Operations Agreement

The Order of Operations Agreement applies to simplifying expressions containing fractions.

The Order of Operations Agreement

Step 1 Do all operations inside parentheses.
Step 2 Simplify any numerical expressions containing exponents.
Step 3 Do multiplication and division as they occur from left to right.
Step 4 Do addition and subtraction as they occur from left to right.

Simplify: $\left(\frac{1}{2}\right)^2 + \left(\frac{2}{3} \div \frac{5}{9}\right) \cdot \frac{5}{6}$

$$\left(\frac{1}{2}\right)^2 + \left(\frac{2}{3} \div \frac{5}{9}\right) \cdot \frac{5}{6}$$

Do the operation inside the parentheses (Step 1). $= \left(\frac{1}{2}\right)^2 + \left(\frac{6}{5}\right) \cdot \frac{5}{6}$

Simplify the exponential expression (Step 2). $= \frac{1}{4} + \left(\frac{6}{5}\right) \cdot \frac{5}{6}$

Do the multiplication (Step 3). $= \frac{1}{4} + 1$

Do the addition (Step 4). $= 1\frac{1}{4}$

A fraction bar acts like parentheses. Therefore, simplify the numerator and denominator of a fraction as part of Step 1 in the Order of Operations Agreement.

Simplify: $6 - \frac{2+1}{15-8} \div \frac{3}{14}$

$$6 - \frac{2+1}{15-8} \div \frac{3}{14}$$

Perform operations above and below the fraction bar.

$$= 6 - \frac{3}{7} \div \frac{3}{14}$$

Do the division.

$$= 6 - \left(\frac{3}{7} \cdot \frac{14}{3}\right)$$

$$= 6 - 2$$

Do the subtraction.

$$= 4$$

Evaluate $\frac{w+x}{y} - z$ when $w = \frac{3}{4}$, $x = \frac{1}{4}$, $y = 2$, and $z = \frac{1}{3}$.

$$\frac{w+x}{y} - z$$

Replace each variable with its given value.

$$\frac{\frac{3}{4} + \frac{1}{4}}{2} - \frac{1}{3}$$

Simplify the numerator of the complex fraction.

$$= \frac{1}{2} - \frac{1}{3}$$

Do the subtraction.

$$= \frac{1}{6}$$

Example 3 Simplify: $\left(-\frac{2}{3}\right)^2 \div \frac{7-2}{13-4} - \frac{1}{3}$

Solution

$$\left(-\frac{2}{3}\right)^2 \div \frac{7-2}{13-4} - \frac{1}{3}$$

$$= \left(-\frac{2}{3}\right)^2 \div \frac{5}{9} - \frac{1}{3}$$

$$= \frac{4}{9} \div \frac{5}{9} - \frac{1}{3}$$

$$= \frac{4}{9} \cdot \frac{9}{5} - \frac{1}{3}$$

$$= \frac{4}{5} - \frac{1}{3} = \frac{7}{15}$$

You Try It 3 Simplify: $\left(-\frac{1}{2}\right)^3 \cdot \frac{7-3}{4-9} + \frac{4}{5}$

Your Solution

Solution on p. A7

3.3 EXERCISES

Objective A

Evaluate.

1. $\left(\frac{3}{4}\right)^2$

2. $\left(\frac{5}{8}\right)^2$

3. $\left(-\frac{1}{6}\right)^3$

4. $\left(-\frac{2}{7}\right)^3$

5. $\left(2\frac{1}{4}\right)^2$

6. $\left(3\frac{1}{2}\right)^2$

7. $\left(\frac{5}{8}\right)^3 \cdot \left(\frac{2}{5}\right)^2$

8. $\left(\frac{3}{5}\right)^3 \cdot \left(\frac{1}{3}\right)^2$

9. $\left(\frac{18}{25}\right)^2 \cdot \left(\frac{5}{9}\right)^3$

10. $\left(\frac{2}{3}\right)^3 \cdot \left(\frac{5}{6}\right)^2$

11. $\left(\frac{4}{5}\right)^4 \cdot \left(-\frac{5}{8}\right)^3$

12. $\left(-\frac{9}{11}\right)^2 \cdot \left(\frac{1}{3}\right)^4$

13. $7^2 \cdot \left(\frac{2}{7}\right)^3$

14. $4^3 \cdot \left(\frac{5}{12}\right)^2$

15. $4 \cdot \left(\frac{4}{7}\right)^2 \cdot \left(-\frac{3}{4}\right)^3$

16. $3 \cdot \left(\frac{2}{5}\right)^2 \cdot \left(-\frac{1}{6}\right)^2$

Evaluate the variable expression for the given values of x and y.

17. x^4, when $x = \frac{2}{3}$

18. y^3, when $y = -\frac{3}{4}$

19. x^3, when $x = 3\frac{1}{3}$

20. y^2, when $y = 6\frac{2}{3}$

21. x^4y^2, when $x = \frac{5}{6}$ and $y = -\frac{3}{5}$

22. x^5y^3, when $x = -\frac{5}{8}$ and $y = \frac{4}{5}$

23. x^3y^2, when $x = \frac{2}{3}$ and $y = 1\frac{1}{2}$

24. x^2y^4, when $x = 2\frac{1}{3}$ and $y = \frac{3}{7}$

25. x^3y^4, when $x = 2\frac{1}{4}$ and $y = 1\frac{1}{3}$

26. x^4y^3, when $x = 1\frac{2}{3}$ and $y = 1\frac{1}{5}$

Objective B

Simplify.

27. $\dfrac{\frac{9}{16}}{\frac{3}{4}}$

28. $\dfrac{\frac{7}{24}}{\frac{3}{8}}$

29. $\dfrac{-\frac{5}{6}}{\frac{15}{16}}$

30. $\dfrac{\frac{7}{12}}{-\frac{5}{18}}$

31. $\dfrac{\frac{3}{14}}{6}$

32. $\dfrac{9}{\frac{6}{7}}$

33. $\dfrac{\frac{2}{3}+\frac{1}{2}}{7}$

34. $\dfrac{\frac{5}{6}-\frac{1}{3}}{4}$

35. $\dfrac{-5}{\frac{3}{8}-\frac{1}{4}}$

36. $\dfrac{7}{\frac{3}{5}+\frac{1}{3}}$

37. $\dfrac{2+\frac{1}{4}}{\frac{3}{8}}$

38. $\dfrac{1-\frac{3}{4}}{\frac{5}{12}}$

39. $\dfrac{\frac{9}{25}}{\frac{4}{5}-\frac{1}{10}}$

40. $\dfrac{-\frac{5}{7}}{\frac{4}{7}-\frac{3}{14}}$

41. $\dfrac{\frac{1}{3}-\frac{3}{4}}{\frac{1}{6}+\frac{2}{3}}$

42. $\dfrac{\frac{9}{14}-\frac{1}{7}}{\frac{9}{14}+\frac{1}{7}}$

43. $\dfrac{3+\frac{1}{3}}{\frac{1}{6}-1}$

44. $\dfrac{4-\frac{5}{8}}{-\frac{1}{2}-\frac{3}{4}}$

45. $\dfrac{\frac{2}{3}-\frac{1}{6}}{-\frac{5}{8}-\frac{1}{4}}$

46. $\dfrac{\frac{1}{4}-\frac{1}{2}}{-\frac{3}{4}+\frac{1}{2}}$

Evaluate the expression for the given values of the variables.

47. $\frac{x+y}{z}$, when $x = \frac{2}{3}$, $y = \frac{3}{4}$, and $z = \frac{1}{12}$

48. $\frac{x}{y+z}$, when $x = \frac{8}{15}$, $y = \frac{3}{5}$, and $z = \frac{2}{3}$

49. $\frac{xy}{z}$, when $x = \frac{3}{4}$, $y = -\frac{2}{3}$, and $z = \frac{5}{8}$

50. $\frac{x}{yz}$, when $x = -\frac{5}{12}$, $y = \frac{8}{9}$, and $z = -\frac{3}{4}$

Evaluate the expression for the given values of the variables.

51. $\frac{x-y}{z}$, when $x = \frac{5}{8}$, $y = \frac{1}{4}$, and $z = \frac{3}{8}$

52. $\frac{x}{y-z}$, when $x = \frac{3}{10}$, $y = \frac{2}{5}$, and $z = \frac{4}{5}$

53. $\frac{wx}{y+z}$, when $w = \frac{1}{2}$, $x = \frac{1}{3}$, $y = \frac{1}{6}$, and $z = \frac{2}{3}$

54. $\frac{w+x}{yz}$, when $w = \frac{5}{6}$, $x = \frac{2}{3}$, $y = \frac{1}{3}$, and $z = \frac{1}{2}$

Objective C

Simplify.

55. $\frac{3}{7} \cdot \frac{14}{15} + \frac{4}{5}$

56. $\frac{3}{5} \div \frac{6}{7} + \frac{4}{5}$

57. $\left(\frac{5}{6}\right)^2 - \frac{5}{9}$

58. $\left(\frac{3}{5}\right)^2 - \frac{3}{10}$

59. $\frac{3}{4} \cdot \left(\frac{11}{12} - \frac{7}{8}\right) + \frac{5}{16}$

60. $\frac{7}{18} + \frac{5}{6} \cdot \left(\frac{2}{3} - \frac{1}{6}\right)$

61. $\frac{11}{16} - \left(\frac{3}{4}\right)^2 + \frac{7}{8}$

62. $\left(-\frac{2}{3}\right)^2 - \frac{7}{18} + \frac{5}{6}$

63. $\left(\frac{1}{3} - \frac{5}{6}\right) + \frac{7}{8} \div \left(-\frac{1}{2}\right)^2$

64. $\left(\frac{1}{4}\right)^2 \div \left(\frac{1}{2} - \frac{3}{4}\right) + \frac{5}{7}$

65. $\left(\frac{2}{3}\right)^2 + \frac{8-7}{3-9} \div \frac{3}{8}$

66. $\left(\frac{1}{3}\right)^3 \cdot \frac{14-5}{6-10} + \frac{3}{4}$

67. $\frac{1}{2} + \frac{\frac{13}{25}}{4 - \frac{3}{4}} \div \frac{1}{5}$

68. $\frac{4}{5} + \frac{3 - \frac{7}{9}}{\frac{5}{6}} \cdot \frac{3}{8}$

69. $\left(\frac{2}{3}\right)^2 + \frac{\frac{5}{8} - \frac{1}{4}}{\frac{2}{3} - \frac{1}{6}} \cdot \frac{8}{9}$

Evaluate the expression for the given values of the variables.

70. $x^2 + \frac{y}{z}$, when $x = -\frac{2}{3}$, $y = \frac{5}{8}$, and $z = \frac{3}{4}$

71. $\frac{x}{y} - z^2$, when $x = \frac{5}{6}$, $y = \frac{1}{3}$, and $z = -\frac{3}{4}$

72. $x - y^3z$, when $x = \frac{5}{6}$, $y = \frac{1}{2}$, and $z = \frac{8}{9}$

73. $xy^3 + z$, when $x = \frac{9}{10}$, $y = \frac{1}{3}$, and $z = \frac{7}{15}$

74. $\frac{wx}{y} + z$, when $w = \frac{4}{5}$, $x = \frac{5}{8}$, $y = \frac{3}{4}$, and $z = \frac{2}{3}$

75. $\frac{w}{xy} - z$, when $w = \frac{1}{2}$, $x = 4$, $y = \frac{3}{8}$, and $z = \frac{2}{3}$

76. $\frac{w}{x + y} - z$, when $w = \frac{3}{5}$, $x = \frac{1}{3}$, $y = \frac{1}{6}$, and $z = \frac{3}{10}$

77. $\frac{w - x}{y} + z$, when $w = \frac{3}{8}$, $x = \frac{3}{8}$, $y = \frac{1}{4}$, and $z = \frac{5}{13}$

Critical Thinking

78. Simplify: $\dfrac{\frac{3}{x} + \frac{2}{x}}{\frac{5}{6}}$

79. A computer can perform 600,000 operations in one second. To the nearest minute, how many minutes will it take for the computer to perform 10^8 operations?

80. Given that x is a whole number, for what value of x will the expression $\left(\frac{3}{4}\right)^2 + x^5 \div \frac{7}{8}$ have a minimum value? What is the minimum value?

81. When 6 gal of gasoline are put into a car's tank, the indicator goes from $\frac{1}{4}$ tank to $\frac{5}{8}$ tank. What is the capacity of the gasoline tank?

82. A videotape can record 2 h on short play, 4 h on long play, or 6 h on extended play. After recording 32 min on short play and 44 min on long play, how many minutes remain that can be recorded on extended play?

83. [W] A rancher died and left 17 horses to be divided among 3 children. The first child was to receive $\frac{1}{2}$ of the horses, the second child $\frac{1}{3}$ of the horses, and the third child $\frac{1}{9}$ of the horses. The executor for the family's estate realized that 17 horses could not be divided by halves, thirds, or ninths and so added a neighbor's horse to the farmer's. With 18 horses, the executor gave 9 horses to the first child, 6 horses to the second child, and 2 horses to the third child. This accounted for the 17 horses, so the executor returned the borrowed horse to the neighbor. Explain why this worked.

Projects in Mathematics

Music

In musical notation, notes are printed on a staff, which is a set of five horizontal lines and the spaces between them. The notes of a musical composition are grouped into measures, or bars. Vertical lines separate measures on a staff. The shape of a note indicates how long it should be held. The whole note has the longest time value of any note. Each time value is divided by two in order to find the next smallest note value.

The time signature is a fraction that appears at the beginning of a piece of music. The numerator of the fraction indicates the number of beats in a measure. The denominator indicates what kind of note receives one beat. For example, music written in $\frac{2}{4}$ time has 2 beats to a measure, and a quarter note receives one beat. One measure in $\frac{2}{4}$ time may have 1 half note, 2 quarter notes, 4 eighth notes, or any other combination of notes totaling 2 beats. Other common time signatures include $\frac{4}{4}$, $\frac{3}{4}$, and $\frac{6}{8}$.

Explain the meaning of the 6 and the 8 in the time signature $\frac{6}{8}$. Give some possible combinations of notes in one measure of a piece written in $\frac{4}{4}$ time.

What does a dot at the right of a note indicate? What is the effect of a dot at the right of a half note? A quarter note? An eighth note?

Symbols called rests are used to indicate periods of silence in a piece of music. What symbols are used to indicate the different time values of rests?

Find some examples of musical compositions written in different time signatures. Use a few measures from each to show that the sum of the time values of the notes and rests in each measure equals the numerator of the time signature.

Construction

Suppose you are building your own home. Design a stairway from the first floor of the house to the second floor. Some of the questions you will need to answer include:

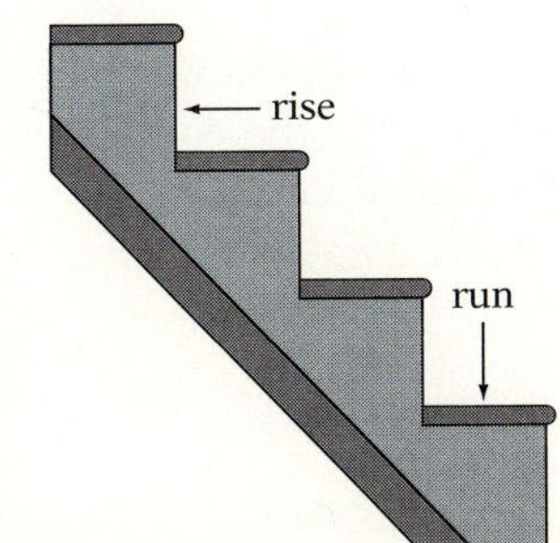

What is the distance from the floor of the first story to the floor of the second story?

Typically, what is the number of steps in a stairway?

What is a reasonable length for the run of each step?

What width lumber will be used to build the staircase?

In designing the stairway, remember that each riser should be the same height and each run should be the same length and that the width of the lumber used for the steps must be considered.

Chapter Summary

Key Words

A *fraction* can represent the number of equal parts of a whole. In a fraction, the *fraction bar* separates the *numerator* and the *denominator.*

In a *proper fraction,* the numerator is smaller than the denominator; a proper fraction is a number less than 1. In an *improper fraction,* the numerator is greater than or equal to the denominator; an improper fraction is a number greater than or equal to 1. A *mixed number* is a number greater than 1 with a whole number part and a fractional part.

Equal fractions with different denominators are called *equivalent fractions.*

A fraction is in *simplest form* when the numerator and denominator have no common factors other than 1.

The *reciprocal* of a fraction is the fraction with the numerator and denominator interchanged.

A *complex fraction* is a fraction whose numerator or denominator contains one or more fractions.

Essential Rules

To write an improper fraction as a mixed number, divide the numerator by the denominator.

To write a mixed number as an improper fraction, multiply the denominator of the fractional part by the whole number part. Add this product and the numerator of the fractional part. The sum is the numerator of the improper fraction. The denominator remains the same.

To write a fraction in simplest form, divide the numerator and denominator of the fraction by their common factors.

To add fractions with the same denominators, add the numerators and place the sum over the common denominator.

To subtract fractions with the same denominators, subtract the numerators and place the difference over the common denominator.

To add or subtract fractions with different denominators, first rewrite the fractions as equivalent fractions with a common denominator. The common denominator is the least common multiple (LCM) of the denominators of the fractions. Then add or subtract the fractions.

To multiply two fractions, multiply the numerators and place the product over the product of the denominators.

To divide two fractions, multiply by the reciprocal of the divisor.

To simplify a complex fraction, simplify the expression above the main fraction bar and simply the expression below the main fraction bar. Then rewrite the complex fraction as a division problem by reading the main fraction bar as "divided by."

The Order of Operations Agreement

Step 1 Do all operations inside parentheses.
Step 2 Simplify any numerical expressions containing exponents.
Step 3 Do multiplication and division as they occur from left to right.
Step 4 Do addition and subtraction as they occur from left to right.

Chapter Review Exercises

1. Write $\frac{19}{2}$ as a mixed number.

2. Subtract: $\frac{2}{9} - \frac{7}{18}$

3. Evaluate $x \div y$ when $x = \frac{5}{8}$ and $y = \frac{3}{4}$.

4. Evaluate $ab^2 - c$ when $a = 4$, $b = -\frac{1}{2}$, and $c = \frac{5}{7}$.

5. Divide: $3 \div \frac{7}{8}$

6. Find the product of 3 and $-\frac{8}{9}$.

7. Evaluate $\frac{x}{y + z}$ when $x = \frac{7}{8}$, $y = \frac{4}{5}$, and $z = -\frac{1}{2}$.

8. Multiply: $\frac{1}{4} \cdot \frac{8}{9} \cdot (-3)$

9. Evaluate xy when $x = -8$ and $y = \frac{5}{12}$.

10. Add: $-\frac{11}{15} + \frac{7}{10}$

11. Place the correct symbol, < or >, between the two numbers.
$\frac{7}{8} \quad \frac{17}{20}$

12. Write a fraction that is equivalent to $\frac{4}{9}$ and has a denominator of 72.

13. Find the difference between $\frac{2}{3}$ and $\frac{11}{18}$.

14. Evaluate x^2y^3 when $x = \frac{8}{9}$ and $y = -\frac{3}{4}$.

15. Write $2\frac{5}{14}$ as an improper fraction.

16. Evaluate $x + y + z$ when $x = \frac{5}{8}$, $y = -\frac{3}{4}$, and $z = \frac{1}{2}$.

17. Find the quotient of $\frac{5}{9}$ and $-\frac{2}{3}$.

18. Simplify: $\frac{2}{5} \div \frac{4}{7} + \frac{3}{8}$

19. Evaluate $a - b$ when $a = -\frac{3}{5}$ and $b = \frac{3}{10}$.

20. Evaluate $\left(-\frac{3}{8}\right)^2 \cdot 4^2$.

21. Find the sum of $\frac{7}{12}$ and $-\frac{1}{2}$.

22. Write $\frac{30}{105}$ in simplest form.

Solve.

23. Two inlet pipes are filling a tank. After one hour, the larger pipe has filled $\frac{3}{5}$ of the tank and the smaller pipe has filled $\frac{1}{4}$ of the tank. How much of the tank remains to be filled?

24. An employee hired for piecework can assemble a unit in $2\frac{1}{2}$ min. How many units can this employee assemble during an 8-hour day?

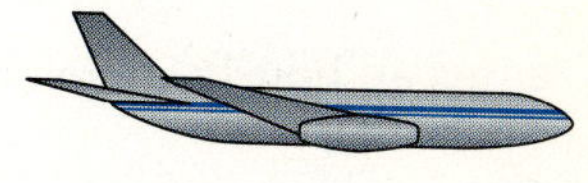

25. What is the final velocity, in feet per second, of an object dropped from a plane with a starting velocity of 0 ft/s and a fall of $15\frac{1}{2}$ s? Use the equation $V = S + 32t$, where V is the final velocity of a falling object, S is its starting velocity, and t is the time of the fall.

Cumulative Review Exercises

1. Add: $-\frac{7}{9} + \frac{5}{6}$

2. Find the product of -4 and $\frac{7}{8}$.

3. Find the GCF of 72 and 108.

4. Evaluate $2a - (b - a)^2$ when $a = 2$ and $b = -3$.

5. Evaluate abc when $a = \frac{4}{7}$, $b = -\frac{1}{6}$, and $c = 3$.

6. Subtract: $\frac{3}{4} - \left(-\frac{5}{7}\right)$

7. Subtract: $-8 - (-12) - (-15) - 32$

8. Simplify: $\frac{2}{5} \div \frac{9 - 6}{3 + 7} + \left(-\frac{1}{2}\right)^2$

9. Find the quotient of $\frac{8}{9}$ and $-\frac{4}{5}$.

10. Find the sum of $\frac{9}{16}$ and $-\frac{5}{8}$.

11. Divide: $-\frac{1}{3} \div \frac{5}{9}$

12. Add: $6847 + 3501 + 924$

13. Find the difference between $\frac{5}{14}$ and $\frac{9}{42}$.

14. Evaluate x^3y^4 when $x = \frac{7}{12}$ and $y = \frac{6}{7}$.

15. Evaluate $(x - y)^3 + 5x$ when $x = 8$ and $y = 6$.

16. Find the LCM of 15 and 25.

17. Estimate the difference between 84,357 and 66,042.

18. Evaluate $a - b$ when $a = \frac{3}{4}$ and $b = -\frac{7}{8}$.

19. Write $7\frac{3}{4}$ as an improper fraction.

20. Find the prime factorization of 140.

Solve.

21. It is projected that the population along the eastern seaboard from Boston, Massachusetts, to Brunswick, Maine, will increase to 700,000 in 2010 from 370,000 in 1980. Find the projected increase in the population in that area during the 30-year period.

22. In a recent year, the Fish and Game Department in a New England state raised 573,000 lb of trout and salmon and released all the fish into the public waters of the state. It is estimated that $\frac{4}{5}$ of these fish were eventually caught by anglers. How many pounds of fish released by the department that year were caught by anglers?

23. Find the cost of purchasing 50 shares of a stock selling for $\$2\frac{1}{4}$ per share.

24. Find the total wages of an employee who worked $18\frac{3}{4}$ h this week and who earns an hourly wage of $8.

25. The pressure on a submerged object is given by $P = 15 + \frac{1}{2}D$, where D is the depth in feet and P is the pressure in pounds per square inch. Find the pressure on a diver who is at a depth of $14\frac{3}{4}$ ft.

CHAPTER

4

Decimals and Real Numbers

4.1 ***Introduction to Decimals***

A Place value
B Order relations between decimals
C Rounding
D Applications

4.2 ***Operations with Decimals***

A Addition and subtraction of decimals
B Estimation
C Multiplication and division of decimals
D Applications and formulas

4.3 ***Percents, Fractions, and Decimals***

A Fractions and decimals
B Percents as decimals or fractions
C Fractions and decimals as percents

4.4 ***Real Numbers***

A Real numbers and the real number line
B Inequalities in one variable
C Applications

Focus on Problem Solving

As you progress in your study of algebra, you will find that the problems become less concrete and more abstract. Problems that are concrete provide information pertaining to a specific instance. Abstract problems are theoretical; they are stated without reference to a specific instance. Let's look at an example of an abstract problem.

How many cents are in d dollars?

How can you solve this problem? Are you able to solve the same problem if the information given is concrete?

How many cents are in 5 dollars?

You know that there are 100 cents in 1 dollar. To find the number of cents in 5 dollars, multiply 5 by 100.

$100 \cdot 5 = 500$ There are 500 cents in 5 dollars.

Use the same procedure to find the number of cents in d dollars: multiply d by 100.

$100 \cdot d = 100d$ There are $100d$ cents in d dollars.

This problem might be taken a step further.

If one pen costs c cents, how many pens can be purchased with d dollars?

Consider the same problem using numbers in place of variables.

If one pen costs 25 cents, how many pens can be purchased with 2 dollars?

To solve this problem, you need to calculate the number of cents in 2 dollars (multiply 2 by 100), and divide the result by the cost per pen (25 cents).

$\frac{100 \cdot 2}{25} = \frac{200}{25} = 8$ If one pen costs 25 cents, 8 pens can be purchased with 2 dollars.

Use the same procedure to solve the related abstract problem. Calculate the number of cents in d dollars (multiply d by 100), and divide the result by the cost per pen (c cents).

$\frac{100 \cdot d}{c} = \frac{100d}{c}$ If one pen costs c cents, $\frac{100d}{c}$ pens can be purchased with d dollars.

At the heart of the study of algebra is the use of variables. The variables in these problems make them abstract. But the variables also allow us to generalize situations and state rules about mathematics.

Try each of the following problems.

1. If you travel m miles on one gallon of gasoline, how far can you travel on g gallons of gasoline?
2. If you walk a mile in x minutes, how far can you walk in h hours?
3. If one photocopy costs n nickels, how many photocopies can you make for q quarters?

Answers:

1. If you travel m miles on one gallon of gasoline, you can travel gm miles on g gallons of gasoline.
2. If you walk a mile in x minutes, you can walk $\frac{60h}{x}$ miles in h hours.
3. If one photocopy costs n nickels, you can make $\frac{5q}{n}$ photocopies for q quarters.

SECTION 4.1 Introduction to Decimals

OBJECTIVE A Place value

POINT OF INTEREST

Leonardo of Pisa (Fibonacci) anticipated the decimal system. He wrote  $\frac{2}{10}\frac{7}{10}5$ to mean $5 + \frac{7}{10} + \frac{2}{10 \cdot 10}$.

The price tag on a sweater reads \$31.88. The number 31.88 is in **decimal notation.** A number written in decimal notation is often called simply a **decimal.**

A number written in decimal notation has three parts.

31	.	88
Whole number part	**Decimal point**	**Decimal part**

The decimal part of the number represents a number less than one. For example, \$.88 is less than one dollar. The decimal point (.) separates the whole number part from the decimal part.

The position of a digit in a decimal determines the digit's place value. The place-value chart is extended to the right to show the place value of digits to the right of a decimal point.

In the decimal 458.302719, the position of the digit 7 determines that its place value is ten-thousandths.

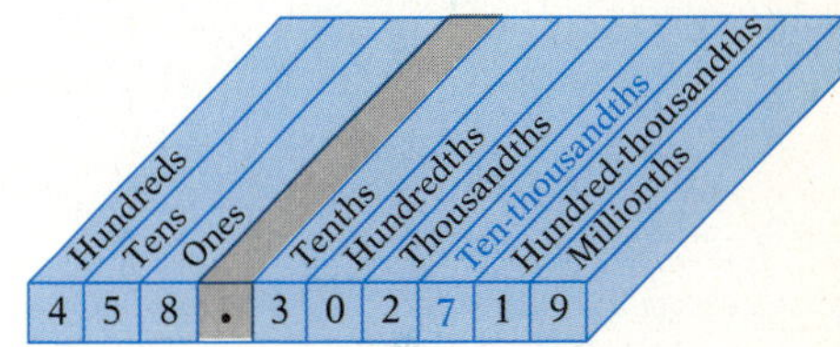

Note the relationship between fractions and numbers written in decimal notation.

seven tenths	seven hundredths	seven thousandths
$\frac{7}{10} = 0.7$	$\frac{7}{100} = 0.07$	$\frac{7}{1000} = 0.007$
1 zero in 10	2 zeros in 100	3 zeros in 1000
1 decimal place in 0.7	2 decimal places in 0.07	3 decimal places in 0.007

To write a decimal in words, write the decimal part of the number as if it were a whole number, then name the place value of the last digit.

0.9684 nine thousand six hundred eighty-four ten-thousandths

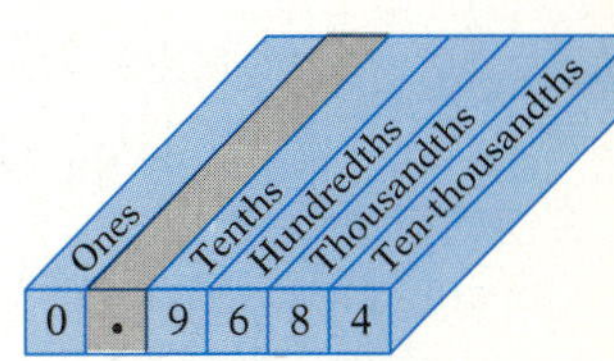

The decimal point in a decimal is read as "and."

372.516 three hundred seventy-two and five hundred sixteen thousandths

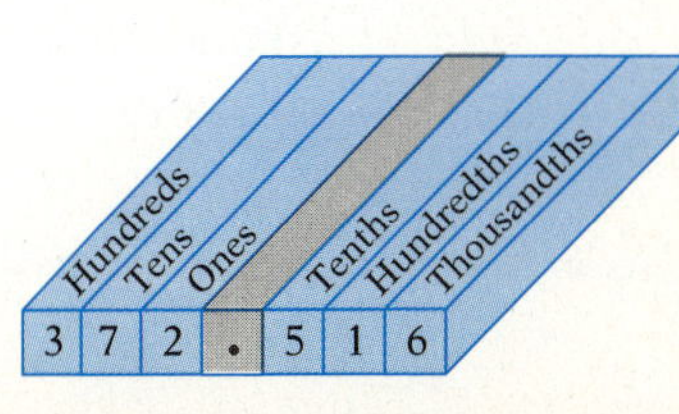

To write a decimal in standard form when it is written in words, write the whole number part, replace the word "and" with a decimal point, and write the decimal part so that the last digit is in the given place-value position.

four and twenty-three hundredths

3 is in the hundredths place. 4.23

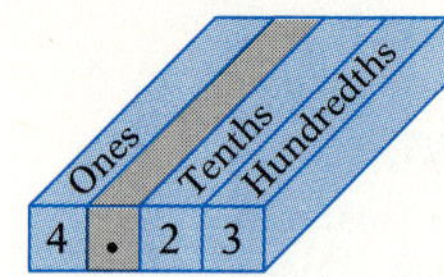

When writing a decimal in standard form, you may need to insert zeros after the decimal point so that the last digit is in the given place-value position.

ninety-one and eight thousandths

8 is in the thousandths place. Insert 2 zeros so that the 8 is in the thousandths place. 91.008

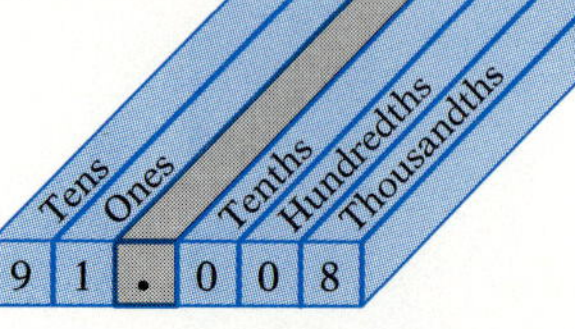

sixty-five ten-thousandths

5 is in the ten-thousandths place. Insert 2 zeros so that the 5 is in the ten-thousandths place. 0.0065

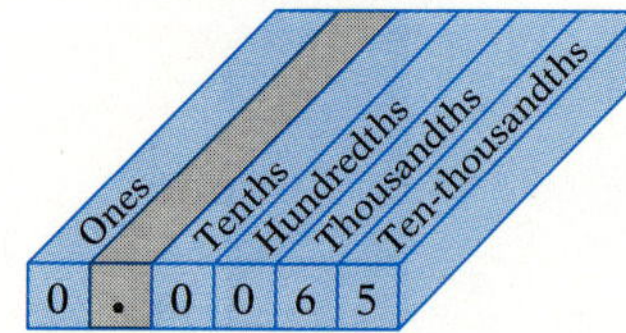

Example 1 Name the place value of the digit 8 in the number 45.687.

Solution The digit 8 is in the hundredths place.

You Try It 1 Name the place value of the digit 4 in the number 907.1342.

Your Solution

Example 2 Write $\frac{43}{100}$ as a decimal.

Solution $\frac{43}{100} = 0.43$

[forty-three hundredths]

You Try It 2 Write $\frac{501}{1000}$ as a decimal.

Your Solution

Example 3 Write 0.289 as a fraction.

Solution $0.289 = \frac{289}{1000}$

[289 thousandths]

You Try It 3 Write 0.67 as a fraction.

Your Solution

Example 4 Write 293.50816 in words.

Solution two hundred ninety-three and fifty thousand eight hundred sixteen hundred-thousandths

You Try It 4 Write 55.6083 in words.

Your Solution

Solutions on p. A7

Example 5 Write twenty-three and two hundred forty-seven millionths in standard form.

Solution 23.000247

You Try It 5 Write eight hundred six and four hundred ninety-one hundred-thousandths in standard form.

Your Solution

Solution on p. A7

OBJECTIVE B Order relations between decimals

A whole number can be written as a decimal by writing a decimal point to the right of the last digit. For example:

62 = 62. 497 = 497.

You know that $62 and $62.00 both represent sixty-two dollars. Any number of zeros may be written to the right of the decimal point in a whole number without changing the value of the number.

62 = 62.00 = 62.0000 497 = 497.0 = 497.000

Also, any number of zeros may be written to the right of the last digit in a decimal without changing its value.

0.8 = 0.80 = 0.800 1.35 = 1.350 = 1.3500 = 1.35000 = 1.350000

This fact is used to find the order relation between two decimals.

To compare two decimals, write the decimal part of each number so that each has the same number of decimal places. Then compare the two numbers.

Place the correct symbol, < or >, between the two numbers 0.693 and 0.71.

0.693 has 3 decimal places.
0.71 has 2 decimal places.
Write 0.71 with 3 decimal places. 0.71 = 0.710

Compare 0.693 and 0.710.
693 thousandths < 710 thousandths 0.693 < 0.710

Remove the zero written in 0.710. 0.693 < 0.71

Place the correct symbol, < or >, between the two numbers 5.8 and 5.493.

Write 5.8 with 3 decimal places. 5.8 = 5.800

Compare 5.800 and 5.493.
The whole number part (5) is the same.
800 thousandths > 493 thousandths 5.800 > 5.493

Remove the extra zeros written in 5.800. 5.8 > 5.493

Example 6 Place the correct symbol, $<$ or $>$, between the two numbers.

0.039 0.1001

Solution $0.039 = 0.0390$

$0.0390 < 0.1001$

$0.039 < 0.1001$

You Try It 6 Place the correct symbol, $<$ or $>$, between the two numbers.

0.065 0.0802

Your Solution

Example 7 Write the given numbers in order from smallest to largest.

1.01, 1.2, 1.002, 1.1, 1.12

Solution 1.002, 1.010, 1.100, 1.120, 1.200

1.002, 1.01, 1.1, 1.12, 1.2

You Try It 7 Write the given numbers in order from smallest to largest.

3.03, 0.33, 0.3, 3.3, 0.03

Your Solution

Solutions on p. A7

OBJECTIVE  Rounding

POINT OF INTEREST

Not all rounding is done as shown here. For example, when sales tax is computed, the decimal is always rounded up to the nearest cent. Thus, a sales tax of \$.132 would be \$.14.

In general, rounding decimals is similar to rounding whole numbers except that the digits to the right of the given place value are dropped instead of being replaced by zeros.

If the digit to the right of the given place value is less than 5, that digit and all digits to the right are dropped.

Round 6.9237 to the nearest hundredth.

6.9237 — Given place value (hundredths)

$3 < 5$ Drop the digits 3 and 7.

6.9237 rounded to the nearest hundredth is 6.92.

If the digit to the right of the given place value is greater than or equal to 5, increase the digit in the given place value by 1, and drop all digits to its right.

Round 12.385 to the nearest tenth.

12.385 — Given place value (tenths)

$8 > 5$ Increase 3 by 1 and drop all digits to the right of 3.

12.385 rounded to the nearest tenth is 12.4.

Round 0.46972 to the nearest thousandth.

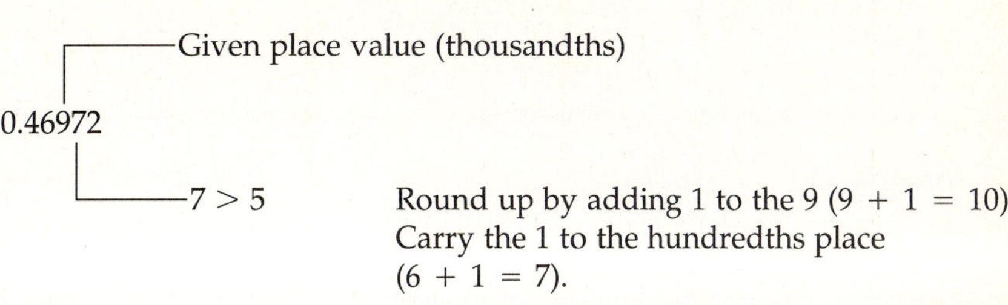

Round up by adding 1 to the 9 (9 + 1 = 10). Carry the 1 to the hundredths place (6 + 1 = 7).

0.46972 rounded to the nearest thousandth is 0.470.

Note that in this example, the zero in the given place value is not dropped. This indicates that the number is rounded to the nearest thousandth. If we dropped the zero and wrote 0.47, it would indicate that the number was rounded to the nearest hundredth.

Example 8 Round 0.9375 to the nearest thousandth.

Solution

Given place value
0.9375
5 = 5

0.9375 rounded to the nearest thousandth is 0.938.

You Try It 8 Round 3.675849 to the nearest ten-thousandth.

Your Solution

Example 9 Round 2.5963 to the nearest hundredth.

Solution

Given place value
2.5963
6 > 5

2.5963 rounded to the nearest hundredth is 2.60.

You Try It 9 Round 48.907 to the nearest tenth.

Your Solution

Example 10 Round 72.416 to the nearest whole number.

Solution

Given place value
72.416
4 < 5

72.416 rounded to the nearest whole number is 72.

You Try It 10 Round 31.8652 to the nearest whole number.

Your Solution

Solutions on p. A7

OBJECTIVE D Applications

Example 11

On Wednesday of a recent week, a British pound would have been exchanged for 1.6340 American dollars. On Thursday of the same week, a British pound would have been exchanged for 1.6372 American dollars. On which day was a British pound worth more money in American dollars, Wednesday or Thursday?

Strategy

To determine on which day the British pound was worth more money in American dollars, compare the numbers 1.6340 and 1.6372.

Solution

$1.6372 > 1.6340$

The British pound was worth more money in American dollars on Thursday of that week.

You Try It 11

In a recent year, in the parallel bars event of the N.C.A.A. gymnastics championships, Mark Warburton scored 9.675 points and Patrick Kirkey scored 9.725 points. Who had the higher score, Warburton or Kirkey?

Your Strategy

Your Solution

Example 12

On average, an American goes to the movies 4.56 times per year. To the nearest whole number, how many times per year does an American go to the movies?

Strategy

To find the number, round 4.56 to the nearest whole number.

Solution

4.56 rounded to the nearest whole number is 5.

An American goes to the movies about 5 times per year.

You Try It 12

In Bremen, Germany, is a special tower that enables scientists to study objects in a gravity-free environment. The tower is 479 ft tall and contains a pipe 361 ft high. Objects placed in a capsule inside the pipe take 4.74 s to descend the length of the pipe. To the nearest tenth of a second, how long does it take an object to descend the length of the pipe?

Your Strategy

Your Solution

Solutions on p. A7

4.1 EXERCISES

Objective A

Name the place value of the digit 5.

1. 76.31587 **2.** 291.508 **3.** 432.09157

4. 0.0006512 **5.** 38.2591 **6.** 0.0000853

Write the fraction as a decimal.

7. $\frac{3}{10}$ **8.** $\frac{9}{10}$ **9.** $\frac{21}{100}$ **10.** $\frac{87}{100}$

11. $\frac{461}{1000}$ **12.** $\frac{853}{1000}$ **13.** $\frac{93}{1000}$ **14.** $\frac{61}{1000}$

Write the decimal as a fraction.

15. 0.1 **16.** 0.3 **17.** 0.47 **18.** 0.59

19. 0.289 **20.** 0.601 **21.** 0.09 **22.** 0.013

Write the number in words.

23. 0.37 **24.** 25.6 **25.** 9.4

26. 1.004 **27.** 0.0053 **28.** 41.108

29. 16.3152 **30.** 0.045 **31.** 3.157

Write the number in words.

32. 26.04

33. 9.37

34. 2.0001

Write the number in standard form.

35. six hundred seventy-two thousandths

36. three and eight hundred six ten-thousandths

37. nine and four hundred seven ten-thousandths

38. four hundred seven and three hundredths

39. six hundred twelve and seven hundred four thousandths

40. two hundred forty-six and twenty-four thousandths

41. eight thousand thirty-four and three thousand three ten-thousandths

42. two thousand sixty-seven and nine thousand two ten-thousandths

43. seventy-three and two thousand six hundred eighty-four hundred-thousandths

44. ninety-eight and seven thousand six hundred fourteen hundred-thousandths

Objective B

Place the correct symbol, < or >, between the two numbers.

45. 0.16 0.6

46. 0.7 0.56

47. 5.54 5.45

48. 3.605 3.065

49. 0.047 0.407

50. 9.004 9.04

51. 1.0008 1.008

52. 9.31 9.031

53. 7.6005 7.605

54. 4.6 40.6

55. 0.31502 0.3152

56. 0.07046 0.07036

Write the numbers in order from smallest to largest.

57. 0.39, 0.309, 0.399

58. 0.66, 0.699, 0.696, 0.609

59. 0.24, 0.024, 0.204, 0.0024

60. 1.327, 1.237, 1.732, 1.372

61. 0.06, 0.059, 0.061, 0.0061

62. 21.87, 21.875, 21.805, 21.78

Objective C

Round the number to the given place value.

63. 6.249 Tenths

64. 5.398 Tenths

65. 21.007 Tenths

66. 30.0092 Tenths

67. 18.40937 Hundredths

68. 413.5972 Hundredths

69. 72.4983 Hundredths

70. 6.061745 Thousandths

71. 936.2905 Thousandths

72. 96.8027 Whole number

73. 47.3192 Whole number

74. 5439.83 Whole number

75. 7014.96 Whole number

76. 0.023591 Ten-thousandths

77. 2.975268 Hundred-thousandths

Objective D

Solve.

78. A nickel weighs about 0.1763668 oz. Find the weight of a nickel to the nearest hundredth of an ounce.

79. The total cost of a parka, including sales tax, is $83.7188. Round the total cost to the nearest cent to find the amount a customer pays for the parka.

80. Runners in the Boston Marathon run a distance of 26.21875 mi. To the nearest tenth of a mile, find the distance an entrant who finishes the Boston Marathon runs.

81. The average life expectancy in Great Britain is 75.3 years. The average life expectancy in Italy is 75.5 years. In which country is the average life expectancy longer, Great Britain or Italy?

Solve.

82. In a recent year, in the vault event of the N.C.A.A. gymnastics championships, Michelle Bryant scored 9.85 points and Kristi Pinnick scored 9.8375 points. Who had the higher score, Bryant or Pinnick?

83. Charge accounts generally require a minimum payment on the balance in the account each month. Use the minimum payment schedule shown below to determine the minimum payment due on the account balances in **a** through **g.**

a. \$187.93
b. \$342.55
c. \$261.48
d. \$16.99
e. \$310.00
f. \$158.32
g. \$200.10

If the New Balance is:	*The Minimum Required Payment is:*
Up to \$20.00	The new balance
\$20.01 to \$200.00	\$20.00
\$200.01 to \$250.00	\$25.00
\$250.01 to \$300.00	\$30.00
\$300.01 to \$350.00	\$35.00
\$350.01 to \$400	\$40.00

84. Shipping and handling charges on catalog mail orders generally are based on the dollar amount of the order. Use the table shown below to determine the cost of shipping the orders in **a** through **g.**

a. \$12.42
b. \$23.56
c. \$47.80
d. \$66.91
e. \$35.75
f. \$20.00
g. \$18.25

If the Amount Ordered is:	*The Shipping and Handling Charge is:*
\$10.00 and under	\$1.60
\$10.01 to \$20.00	\$2.40
\$20.01 to \$30.00	\$3.60
\$30.01 to \$40.00	\$4.70
\$40.01 to \$50.00	\$6.00
\$50.01 and up	\$7.00

Critical Thinking

85. The size of an automobile engine is generally given in liters. For each of the following categories, find an example and give its engine size: a compact car, a sports car, a mid-sized car, a pick-up truck, and a luxury car. Compare the relative sizes of the engines in your examples.

86. Find a number between **a.** 0.1 and 0.2, **b.** 1 and 1.1, and **c.** 0 and 0.005.

87. [W] To what decimal place are timed events in the Olympics recorded? Provide some examples of events and the winning times in each.

88. [W] Provide an example of a situation in which a decimal is always rounded up, even if the digit to the right is less than 5. Provide an example of a situation in which a decimal is always rounded down, even if the digit to the right is 5 or greater than 5. (*Hint:* Think about situations in which money changes hands.)

89. [W] Prepare a report on the Richter scale. Include in your report the magnitudes that classify an earthquake as strong or moderate, the magnitudes that classify an earthquake as a microearthquake, and the largest known recorded shocks.

SECTION 4.2 Operations with Decimals

OBJECTIVE  Addition and subtraction of decimals

POINT OF INTEREST

Simon Stevin (1548–1620) was the first to name decimal numbers. He wrote the number 2.345 as 2 0 3 1 4 2 5 3 .
He called the whole number part the *commencement;* the tenths digit, *prime;* the hundredths digit, *second;* the thousandths digit, *thirds,* and so on.

To add decimals, write the numbers so that the decimal points are on a vertical line. Add as you would with whole numbers. Then write the decimal point in the sum directly below the decimal points in the addends.

Add: 0.326 + 4.8 + 57.23

Note that by placing the decimal points on a vertical line, digits of the same place value are added.

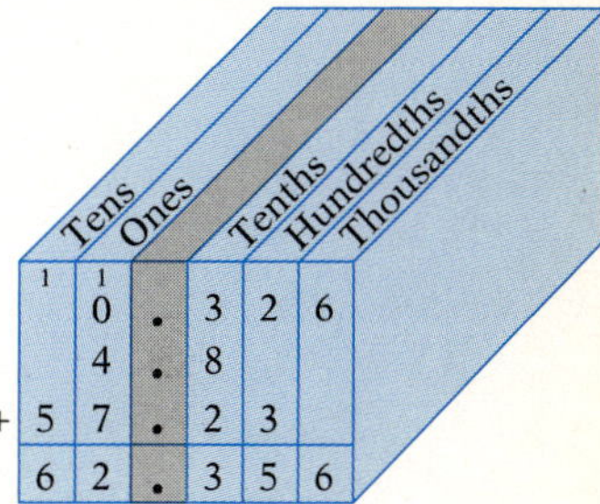

Find the sum of 0.64, 8.731, 12, and 5.9.

Arrange the numbers vertically, placing the decimal points on a vertical line.

Add the numbers in each column. Write the decimal point in the sum directly below the decimal points in the addends.

```
 1 2
   0.6 4
   8.7 3 1
 1 2.
+  5.9
 2 7.2 7 1
```

To subtract decimals, write the numbers so that the decimal points are on a vertical line. Subtract as you would with whole numbers. Then write the decimal point in the difference directly below the decimal point in the subtrahend.

Subtract and check: 31.642 − 8.759

Note that by placing the decimal points on a vertical line, digits of the same place value are subtracted.

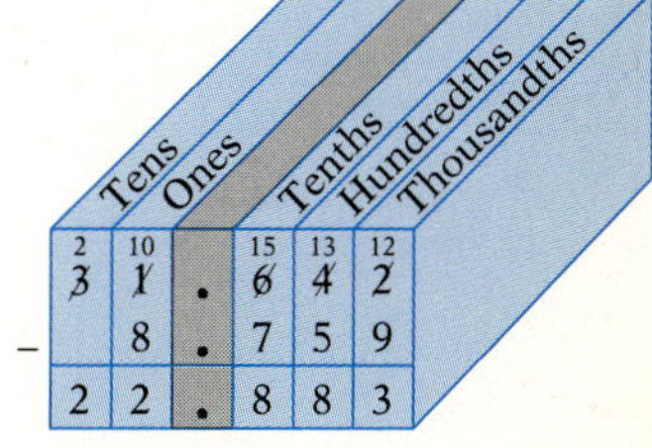

Check:	Subtrahend	8.7 5 9
	+ Difference	+ 2 2.8 8 3
	= Minuend	3 1.6 4 2

Subtract and check: 5.4 − 1.6832

Insert zeros in the minuend so that it has the same number of decimal places as the subtrahend.

$$\begin{array}{r} 5.4000 \\ -\ 1.6832 \\ \hline \end{array}$$

Subtract and then check.

$$\begin{array}{r} \overset{4\ 13\ 9\ 9\ 10}{5.4000} \\ -\ 1.6832 \\ \hline 3.7168 \end{array}$$

Check:
$$\begin{array}{r} 1.6832 \\ +\ 3.7168 \\ \hline 5.4000 \end{array}$$

Find the difference between 8 and 3.97. Check the answer.

Write 8 with two decimal places.

$$\begin{array}{r} \overset{7\ 9\ 10}{8.00} \\ -\ 3.97 \\ \hline 4.03 \end{array}$$

Check:
$$\begin{array}{r} 3.97 \\ +\ 4.03 \\ \hline 8.00 \end{array}$$

Recall that the absolute value of a number is the distance from zero to the number on the number line. The absolute value of a number is a positive number or zero.

$|54.29| = 54.29$

$|-36.087| = 36.087$

The sign rules for adding and subtracting decimals are the same rules used to add and subtract integers.

Add: $-36.087 + 54.29$

The signs of the addends are different. Subtract the smaller absolute value from the larger absolute value.

$54.29 - 36.087 = 18.203$

Attach the sign of the number with the larger absolute value.

$|54.29| > |-36.087|$

The sum is positive.

$-36.087 + 54.29 = 18.203$

Recall that the opposite or additive inverse of n is $-n$ and the opposite of $-n$ is n. To find the opposite of a number, change the sign of the number.

Subtract: $-2.86 - 10.3$

Rewrite subtraction as addition of the opposite. The opposite of 10.3 is −10.3.

$-2.86 - 10.3$

$= -2.86 + (-10.3)$

The signs of the addends are the same. Add the absolute values of the numbers. Attach the sign of the addends.

$= -13.16$

Evaluate $c - d$ when $c = 6.731$ and $d = -2.48$.

	$c - d$
Replace c with 6.731 and d with -2.48.	$6.731 - (-2.48)$
Rewrite subtraction as addition of the opposite.	$= 6.731 + 2.48$
Add.	$= 9.211$

Example 1 Add:
$35.8 + 182.406 + 71.0934$

Solution

$$\begin{array}{r} \scriptstyle 1\ \ 1\quad\quad\quad \\ 35.8 \\ 182.406 \\ +\ \ 71.0934 \\ \hline 289.2994 \end{array}$$

You Try It 1 Add:
$8.64 + 52.7 + 0.39105$

Your Solution

Example 2 Subtract and check:
$73 - 8.16$

Solution

$$\begin{array}{r} \scriptstyle 6\ 12\ \ 9\ 10 \\ \not{7}\not{3}.\not{0}\not{0} \\ -\ \ 8.16 \\ \hline 64.84 \end{array} \qquad \text{Check:} \quad \begin{array}{r} 8.16 \\ +\ 64.84 \\ \hline 73.00 \end{array}$$

You Try It 2 Subtract and check:
$25 - 4.91$

Your Solution

Example 3 Subtract: $8.94 - (-13.7)$

Solution $8.94 - (-13.7) = 8.94 + 13.7$
$= 22.64$

You Try It 3 Add: $-41.65 + 29.303$

Your Solution

Example 4 What is -251.49 more than -638.7?

Solution $-638.7 + (-251.49)$
$= -890.19$

You Try It 4 What is 4.002 minus 9.378?

Your Solution

Example 5 Evaluate $x + y + z$ when $x = -1.6$, $y = 7.9$, and $z = -4.8$.

Solution $x + y + z$
$-1.6 + 7.9 + (-4.8)$
$= 6.3 + (-4.8) = 1.5$

You Try It 5 Evaluate $x + y + z$ when $x = -7.84$, $y = -3.05$, and $z = 2.19$.

Your Solution

Solutions on p. A8

OBJECTIVE B Estimation

Recall that to estimate the answer to a calculation, round each number to the highest place value of the number; the first digit of each number will be nonzero and all other digits will be zero. Perform the calculation using the rounded numbers.

Estimate the sum of 23.037 and 16.7892.

Round each number to the nearest ten.

23.037 ⟶ 20
16.7892 ⟶ + 20

Add the rounded numbers.

40

40 is an estimate of the sum of 23.037 and 16.7892. Note that 40 is very close to the actual sum, 39.8262.

23.037
+ 16.7892
39.8262

When a number in an estimation is a decimal less than one, round the decimal so that there is one nonzero digit.

Estimate the difference between 4.895 and 0.6193.

Round 4.895 to the nearest one.
Round 0.6193 to the nearest tenth.
Subtract the rounded numbers.

4.895 ⟶ 5.0
0.6193 ⟶ − 0.6
4.4

4.4 is an estimate of the difference between 4.895 and 0.6193. It is close to the actual difference, 4.2757.

4.8950
− 0.6193
4.2757

Example 6 Estimate the sum of 0.3927, 0.4856, and 0.2104.

Solution
0.3927 ⟶ 0.4
0.4856 ⟶ 0.5
0.2104 ⟶ + 0.2
1.1

You Try It 6 Estimate the sum of 6.514, 8.903, and 2.275.

Your Solution

Example 7 Estimate the difference between 6.908 and 0.4751.

Solution
6.908 ⟶ 7.0
0.4751 ⟶ − 0.5
6.5

You Try It 7 Estimate the difference between 487.52 and 61.903.

Your Solution

Solutions on p. A8

OBJECTIVE C Multiplication and division of decimals

POINT OF INTEREST
Benjamin Banneker (1731–1806) was the first African American to earn distinction as a mathematician and scientist. He was on the survey team that determined the boundaries of Washington, D.C. Surveying requires extensive use of decimals.

Decimals are multiplied as if they were whole numbers; then the decimal point is placed in the product. Writing the decimals as fractions shows where to write the decimal point in the product.

$$0.4 \cdot 2 = \frac{4}{10} \cdot \frac{2}{1} = \frac{8}{10} = 0.8$$

1 decimal place in 0.4 — 1 decimal place in 0.8

$$0.4 \cdot 0.2 = \frac{4}{10} \cdot \frac{2}{10} = \frac{8}{100} = 0.08$$

1 decimal place in 0.4
1 decimal place in 0.2 — 2 decimal places in 0.08

$$0.4 \cdot 0.02 = \frac{4}{10} \cdot \frac{2}{100} = \frac{8}{1000} = 0.008$$

1 decimal place in 0.4
2 decimal places in 0.02 — 3 decimal places in 0.008

To multiply decimals, multiply the numbers as you would whole numbers. Then write the decimal point in the product so that the number of decimal places in the product is the sum of the numbers of decimal places in the factors.

Multiply: (32.41)(7.6)

32.41	2 decimal places
× 7.6	1 decimal place
19446	
22687	
246.316	3 decimal places

Estimating the product of 32.41 and 7.6 shows that the decimal point has been correctly placed.

Round 32.41 to the nearest ten. 32.41 ⟶ 30
Round 7.6 to the nearest one. 7.6 ⟶ × 8
Multiply the two numbers. 240

240 is an estimate of (32.41)(7.6). It is close to the actual product 246.316.

Multiply: 0.061(0.08)

0.061	3 decimal places
× 0.08	2 decimal places
0.00488	5 decimal places

Insert two zeros between the 4 and the decimal point so that there are 5 decimal places in the product.

To multiply a decimal by a power of 10 (10, 100, 1000, . . .), move the decimal point to the right the same number of places as there are zeros in the power of 10.

$2.7935 \cdot \underline{10} = 27.935$

1 zero — 1 decimal place

$2.7935 \cdot \underline{100} = 279.35$

2 zeros — 2 decimal places

$2.7935 \cdot \underline{1000} = 2793.5$

3 zeros — 3 decimal places

$2.7935 \cdot \underline{10{,}000} = 27{,}935.$

4 zeros — 4 decimal places

$2.7935 \cdot \underline{100{,}000} = 279{,}350.$

5 zeros — 5 decimal places

A zero must be inserted before the decimal point.

Note that if the power of 10 is written in exponential notation, the exponent indicates how many places to move the decimal point.

$2.7935 \cdot 10^1 = 27.935$

1 decimal place

$2.7935 \cdot 10^2 = 279.35$

2 decimal places

$2.7935 \cdot 10^3 = 2793.5$

3 decimal places

$2.7935 \cdot 10^4 = 27{,}935.$

4 decimal places

$2.7935 \cdot 10^5 = 279{,}350.$

5 decimal places

Find the product of 64.18 and 10^3.

The exponent on 10 is 3. Move the decimal point in 64.18 three places to the right. $64.18 \cdot 10^3 = 64{,}180$

Evaluate $100x$ when $x = 5.714$.

$100x$

Replace x with 5.714. $100(5.714)$

Multiply. There are two zeros in 100. Move the decimal point two places to the right. $= 571.4$

To divide decimals, move the decimal point in the divisor to the right so that the divisor is a whole number. Move the decimal point in the dividend the same number of places to the right. Place the decimal point in the quotient directly above the decimal point in the dividend. Then divide as you would with whole numbers.

Divide: 29.585 ÷ 4.85

4.85.)29.58.5

Move the decimal point 2 places to the right in the divisor. Move the decimal point 2 places to the right in the dividend. Place the decimal point in the quotient. Then divide as shown at the right.

```
         6.1
485.)2958.5
    -2910
       485
      -485
         0
```

Moving the decimal point the same number of places in the divisor and the dividend does not change the quotient because the process is the same as multiplying the numerator and denominator of a fraction by the same number. For the last example,

$$4.85\overline{)29.585} = \frac{29.585}{4.85} = \frac{29.585 \cdot 100}{4.85 \cdot 100} = \frac{2958.5}{485} = 485\overline{)2958.5}$$

Estimating the quotient of 29.585 and 4.85 shows that the decimal point has been correctly placed in the example above.

Round 29.585 to the nearest ten. 29.585 ⟶ 30
Round 4.85 to the nearest one. 4.85 ⟶ 5

Divide the two numbers. 30 ÷ 5 = 6

6 is an estimate of 29.585 ÷ 4.85.
It is very close to the actual quotient 6.1.

CONSIDER THIS

An alternate method for rounding quotients is

```
  1.2
7)8.6
 -7
  1 6
 -1 4
    2
```

Since 2 is less than one half of the divisor, the digit in the hundredths place will be less than 5. The quotient is rounded to 1.2.

In division of decimals, rather than writing the quotient with a remainder, the quotient is usually rounded to a specified place value. The symbol ≈, which is read "is approximately equal to," indicates that the quotient has been rounded.

Divide and round to the nearest tenth: 0.86 ÷ 0.7.

```
      1.22 ≈ 1.2
0.7.)0.8.60
    - 7
      16
    - 14
       20
     - 14
        6
```

⟵ In order to round the quotient to the nearest tenth, the division must be carried to the hundredths place. Therefore, a zero must be inserted in the dividend so that the quotient has a digit in the hundredths place.

To divide a decimal by a power of 10 (10, 100, 1000, 10,000, . . .), move the decimal point to the left the same number of places as there are zeros in the power of 10.

462.81 ÷ 10 (1 zero)	= 46.281 (1 decimal place)	
462.81 ÷ 100 (2 zeros)	= 4.6281 (2 decimal places)	
462.81 ÷ 1000 (3 zeros)	= 0.46281 (3 decimal places)	
462.81 ÷ 10,000 (4 zeros)	= 0.046281 (4 decimal places)	A zero must be inserted between the decimal point and the 4.
462.81 ÷ 100,000 (5 zeros)	= 0.0046281 (5 decimal places)	Two zeros must be inserted between the decimal point and the 4.

If the power of 10 is written in exponential notation, the exponent indicates how many places to move the decimal point.

$462.81 \div 10^1 = 46.281$
1 decimal place

$462.81 \div 10^2 = 4.6281$
2 decimal places

$462.81 \div 10^3 = 0.46281$
3 decimal places

$462.81 \div 10^4 = 0.046281$
4 decimal places

$462.81 \div 10^5 = 0.0046281$
5 decimal places

Find the quotient of 3.59 and 100.

There are two zeros in 100. Move the decimal point in 3.59 two places to the left.

$3.59 \div 100 = 0.0359$

What is the quotient of 64.79 and 10^4?

The exponent on 10 is 4. Move the decimal point in 64.79 four places to the left.

$64.79 \div 10^4 = 0.006479$

The sign rules for multiplying and dividing integers are the same rules used to multiply and divide decimals.

The product or quotient of two numbers with the same sign is positive.
The product or quotient of two numbers with different signs is negative.

Multiply: $(-3.2)(-0.008)$

The signs are the same.
The product is positive.
Multiply the absolute values of the numbers. $(-3.2)(-0.008) = 0.0256$

Divide: $-1.16 \div 2.9$

The signs are different.
The quotient is negative.
Divide the absolute values of the numbers. $-1.16 \div 2.9 = -0.4$

Evaluate ab when $a = -4.3$ and $b = 0.6$.

Replace a with -4.3 and b with 0.6. ab $\quad -4.3(0.6)$

The signs are different.
The product is negative.
Multiply the absolute values of the numbers. $= -2.58$

Example 8 Multiply: 36.084(0.057)

Solution

$$\begin{array}{r} 36.084 \\ \times \quad 0.057 \\ \hline 252588 \\ 180420 \\ \hline 2.056788 \end{array}$$

You Try It 8 Multiply: 44.356(0.72)

Your Solution

Example 9 Estimate the product of 0.7639 and 0.2188.

Solution $0.7639 \longrightarrow 0.8$
$0.2188 \longrightarrow \times\ 0.2$
0.16

You Try It 9 Estimate the product of 6.407 and 0.959.

Your Solution

Example 10 What is the product of 1.756 and 10^4?

Solution $1.756 \cdot 10^4 = 17{,}560$

You Try It 10 What is 835.294 multiplied by 1000?

Your Solution

Solutions on p. A8

Example 11 Divide and round to the nearest hundredth: $448.2 \div 53$

Solution

$$
\begin{array}{r}
8.456 \approx 8.46 \\
53\overline{)448.200} \\
\underline{-424} \\
24\,2 \\
\underline{-21\,2} \\
3\,00 \\
\underline{-2\,65} \\
350 \\
\underline{-318} \\
32
\end{array}
$$

You Try It 11 Divide and round to the nearest thousandth: $519.37 \div 86$

Your Solution

Example 12 What is 63.7 divided by 100?

Solution $63.7 \div 100 = 0.637$

You Try It 12 Find the quotient of 592.4 and 10^4.

Your Solution

Example 13 Divide and round to the nearest tenth: $-6.94 \div (-1.5)$

Solution The quotient is positive.

$-6.94 \div (-1.5) \approx 4.6$

You Try It 13 Divide and round to the nearest tenth: $-25.7 \div 0.31$

Your Solution

Example 14 Multiply: $-3.42(6.1)$

Solution $-3.42(6.1) = -20.862$

You Try It 14 Multiply: $(-0.7)(-5.8)$

Your Solution

Example 15 Evaluate $50ab$ when $a = -0.9$ and $b = -0.2$.

Solution $50ab$

$50(-0.9)(-0.2) = -45(-0.2)$
$= 9$

You Try It 15 Evaluate $25xy$ when $x = -0.8$ and $y = 0.6$.

Your Solution

Example 16 Evaluate $\frac{x}{y}$ when $x = -76.8$ and $y = 0.8$.

Solution $\frac{x}{y}$

$\frac{-76.8}{0.8} = -96$

You Try It 16 Evaluate $\frac{x}{y}$ when $x = -40.6$ and $y = -0.7$.

Your Solution

Solutions on p. A8

OBJECTIVE D Applications and formulas

Example 17

A one-year subscription to a monthly magazine costs \$21. The price of each issue at the newsstand is \$2.25. How much would you save per issue by buying a year's subscription rather than buying each issue at the newsstand?

Strategy

To find the amount saved:

- ▶ Find the subscription price per issue by dividing the cost of the subscription (21) by the number of issues (12).
- ▶ Subtract the subscription price per issue from the newsstand price (2.25).

Solution

$$\begin{array}{r} 1.75 \\ 12\overline{)21.00} \\ -12 \\ \hline 9\,0 \\ -8\,4 \\ \hline 6\,0 \\ -6\,0 \\ \hline 0 \end{array} \qquad \begin{array}{r} 2.25 \\ -\,1.75 \\ \hline 0.50 \end{array}$$

The savings would be \$.50 per issue.

You Try It 17

You hand a postal clerk a five-dollar bill to pay for twelve 32¢ stamps. How much change do you receive?

Your Strategy

Your Solution

Example 18

Find the equity on a home that is valued at \$125,000 when the homeowner has \$67,853.25 in loans on the property. Use the formula $E = V - L$, where E is the equity, V is the value of the home, and L is the loan amount on the property.

Strategy

To find the equity, replace V by 125,000 and L by 67,853.25 in the given formula and solve for E.

Solution

$E = V - L$
$E = 125{,}000 - 67{,}853.25$
$E = 57{,}146.75$

The equity on the home is \$57,146.75.

You Try It 18

Find the equity on a home that is valued at \$240,000 when the homeowner has \$142,976.80 in loans on the property. Use the formula $E = V - L$, where E is the equity, V is the value of the home, and L is the loan amount on the property.

Your Strategy

Your Solution

Solutions on p. A9

Example 19

On an overseas flight, the airline charges \$6.40 for each kilogram or part of a kilogram over 50 kg of luggage weight. How much extra must be paid for three pieces of luggage weighing 21.4 kg, 19.3 kg, and 16.8 kg?

Strategy

To find the extra charge:

- Add the three weights (21.4, 19.3, and 16.8) to find the total weight of the luggage.
- Subtract 50 kg from the total weight of the luggage to find the excess weight.
- Round the difference up to the nearest whole number.
- Multiply the charge per kilogram of excess weight (6.40) by the excess weight.

Solution

$21.4 + 19.3 + 16.8 = 57.5$

$57.5 - 50 = 7.5$

7.5 rounded to the nearest whole number is 8.

$6.40(8) = 51.20$

The extra charge for the luggage is \$51.20.

You Try It 19

A health food store buys nuts in 100-pound containers and repackages the nuts for resale. The store packages the nuts in 2-pound bags and sells them for \$8.50 per bag. Find the profit on a 100-pound container of nuts costing \$325.

Your Strategy

Your Solution

Example 20

Use the formula $P = BF$, where P is the insurance premium, B is the base rate, and F is the rating factor, to find the insurance premium due on an insurance policy with a base rate of \$342.50 and a rating factor of 2.2.

Strategy

To find the insurance premium due, replace B by 342.50 and F by 2.2 and solve for P.

Solution

$P = BF$
$P = 342.50(2.2)$
$P = 753.50$

The insurance premium due is \$753.50.

You Try It 20

Use the formula $P = BF$, where P is the insurance premium, B is the base rate, and F is the rating factor, to find the insurance premium due on an insurance policy with a base rate of \$276.25 and a rating factor of 1.8.

Your Strategy

Your Solution

Solutions on p. A9

4.2 EXERCISES

Objective A

Add or subtract.

1. 56.4 + 35.97
2. 4.291 + 6.78
3. 14.93 + 1.0625 + 18.7
4. 1.864 + 39 + 25.0781
5. 2.04 + 35.6 + 4.918
6. 12 + 73.59 + 6.482
7. 85.69 − 2.13
8. 62.039 − 14.81
9. 51.3702 − 26.1049
10. 53.24 − 9.376
11. 28 − 6.74
12. 5 − 1.386
13. 6.02 − 3.252
14. 0.92 − 0.0037
15. −42.1 − 8.6
16. −6.57 − 8.933
17. 5.73 − 9.042
18. −31.894 + 7.5
19. −9.37 + 3.465
20. 1.09 − (−8.3)
21. 257.8 − (−11.4)
22. −19 − (−2.65)
23. −26 − (−13.75)
24. 3.18 − 5.72 − 6.4
25. −12.3 − 4.07 + 6.82
26. −8.9 + 7.36 − 14.2
27. −5.6 − (−3.82) − 17.409

Solve.

28. Find 9.044 more than 73.812.
29. What is 4.872 added to 16.48?
30. Find the sum of 2.536, 14.97, 8.014, and 21.67.
31. Find the total of 6.24, 8.573, 19.06, and 22.488.
32. What is the difference between 7.3509 and 0.1628?
33. What is 6.9217 decreased by 3.4501?
34. How much larger is 5 than 1.63?
35. What is the sum of −65.47 and −32.91?
36. Find 382.9 plus −430.6.
37. What is 1.093 increased by −2.807?

Solve.

38. What is -16.44 more than -73.5?

39. Find -138.72 minus 510.64.

40. What is 4.793 less than -6.82?

41. Find the difference between -31 and -62.09.

Evaluate the variable expression $x + y$ for the given values of x and y.

42. $x = -125.41; y = 361.55$

43. $x = -28.33; y = 46.72$

44. $x = -8.729; y = -23.46$

45. $x = -6.175; y = -19.49$

Evaluate the variable expression $x + y + z$ for the given values of x, y, and z.

46. $x = 41.33; y = -26.095; z = 70.08$

47. $x = -6.059; y = 3.884; z = 15.71$

48. $x = 81.72; y = 36.067; z = -48.93$

49. $x = -16.219; y = 47; z = -2.3885$

Evaluate the variable expression $x - y$ for the given values of x and y.

50. $x = 43.29; y = 18.76$

51. $x = 58.27; y = 34.91$

52. $x = 6.029; y = -4.708$

53. $x = 9.552; y = -1.764$

54. $x = -16.329; y = 4.54$

55. $x = -21.073; y = 6.48$

56. $x = -3.69; y = -1.527$

57. $x = -8.21; y = -6.798$

58. $x = -2.6; y = 0.354$

Objective B

Estimate by rounding. Then find the exact answer.

59. $37.92 + 81.63$

60. $3.582 + 4.193$

61. $5.37 + 26.49$

62. $184.27 + 65.092$

63. $0.24 + 0.38 + 0.96$

64. $0.52 + 0.455 + 0.93$

Estimate by rounding. Then find the exact answer.

65. 6.408 + 5.917

66. 87.65 − 49.032

67. 48.6 − 19.753

68. 6.272 − 1.848

69. 91.02 − 18.53

70. 85.3 − 67.011

71. 0.931 − 0.628

72. 0.894 − 0.456

73. 387.6 − 54.92

Objective C

Multiply or divide.

74. 0.9(0.3)

75. (0.54)(0.6)

76. (8.2)(1.8)

77. 8.29(0.004)

78. (0.75)(0.32)

79. 56.4(0.0097)

80. −5.2(0.8)

81. (−6.3)(−2.4)

82. (1.9)(−3.7)

83. −1.3(4.2)

84. −8.1(−7.5)

85. 1.31(−0.006)

86. −10(0.59)

87. (−100)(4.73)

88. −9(−7.6)

89. (−0.47)(−0.3)

90. 6.93 ÷ 0.9

91. 16.15 ÷ 0.5

92. 8.721 ÷ 0.9

93. 7.02 ÷ 3.6

94. 27.08 ÷ (−0.4)

95. −8.919 ÷ 0.9

96. (−0.396) ÷ (−3.6)

97. 84.66 ÷ (−1.7)

98. −2.501 ÷ 0.41

99. (−3.312) ÷ (−0.8)

100. 1.003 ÷ (−0.59)

101. −26.22 ÷ 6.9

Divide. Round to the nearest tenth.

102. (−52.8) ÷ (−9.1)

103. −6.824 ÷ 0.053

104. 0.0416 ÷ (−0.53)

105. (−31.792) ÷ (−0.86)

Solve.

106. Find the product of 0.48 and 10.

107. What is the product of 5.92 and 100?

108. Find the quotient of 52.78 and 10.

109. What is 37,942 divided by 1000?

110. What is 1000 times 4.25?

111. What is 3.587 multiplied by 1000?

112. What is the quotient of 498.3 and 100?

113. Find 248.1 divided by 1000.

114. What is the product of 0.82 and 10^2?

115. Find the product of 71.92 and 10^3.

116. What is the quotient of 48.05 and 10^2?

117. Find the quotient of 382.55 and 10^3.

118. Find the product of 6.71 and 10^4.

119. What is the product of 0.0354 and 10^5?

120. What is 0.13 divided by 10^4?

121. Find 9.407 divided by 10^3.

122. Find the product of 2.7, -16, and 3.04.

123. What is the product of -36.2, 9.1, and -10?

Estimate by rounding. Then find the exact answer.

124. 86.4(4.2)

125. (9.81)(0.77)

126. (0.764)(5.3)

127. 0.238(8.2)

128. 4.35(2.58)

129. (6.88)(9.97)

130. (8.432)(0.043)

131. 28.45(1.13)

Estimate by rounding. Then find the exact answer and round to the nearest hundredth.

132. $42.43 \div 3.8$

133. $678 \div 0.71$

134. $6.398 \div 5.5$

135. $0.994 \div 0.456$

136. $1.237 \div 0.021$

137. $421.0935 \div 4.0827$

138. $33.14 \div 4.6$

139. $129.38 \div 4.47$

Evaluate the expression for the given values of the variables.

140. xy, when $x = 5.68$ and $y = 0.2$

141. ab, when $a = 6.27$ and $b = 8$

142. $40c$, when $c = 2.5$

143. $20d$, when $d = 8.7$

144. $10t$, when $t = -4.8$

145. $50p$, when $p = -6.3$

Evaluate the expression for the given values of the variables.

146. $30x$, when $x = 5.9$

147. $100z$, when $z = -8.7$

148. $40q$, when $q = -20.6$

149. xy, when $x = -3.71$ and $y = 2.9$

150. xy, when $x = -1.95$ and $y = 6.8$

151. ab, when $a = 0.379$ and $b = -0.22$

152. ab, when $a = 452$ and $b = -0.86$

153. cd, when $c = -2.537$ and $d = -9.1$

154. cd, when $c = -4.259$ and $d = -6.3$

Evaluate the variable expression $\frac{x}{y}$ for the given values of x and y.

155. $x = 52.8$; $y = 0.4$

156. $x = 3.542$; $y = 0.7$

157. $x = -2.436$; $y = 0.6$

158. $x = -0.396$; $y = 3.6$

159. $x = 0.648$; $y = -2.7$

160. $x = 26.22$; $y = -6.9$

161. $x = -8.034$; $y = -3.9$

162. $x = -64.05$; $y = -6.1$

163. $x = 1.003$; $y = -0.59$

Objective D

Solve.

164. According to the chart at the right, the Dow Jones Industrial Average fell 2.69 points to 3932.68. **(a)** What did the Dow Jones Industrial Average close at the day before? **(b)** What did the New York Stock Exchange (NYSE) close at the day before?

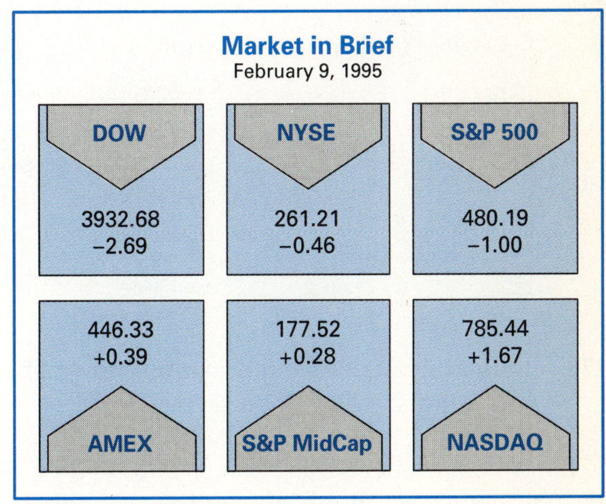

165. A survey revealed that eighth graders spend on average 21.7 h each week watching television and 5.6 h doing homework. On average, how much more time each week does an eighth grader spend watching TV than doing homework?

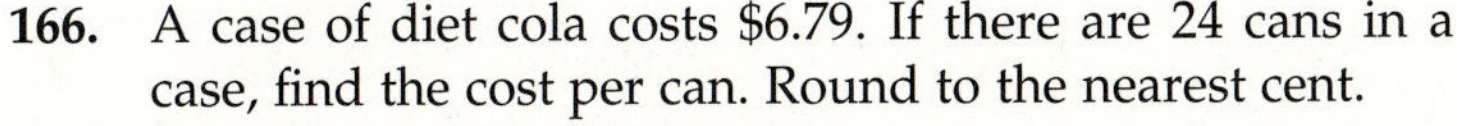

166. A case of diet cola costs \$6.79. If there are 24 cans in a case, find the cost per can. Round to the nearest cent.

167. You travel 295 mi on 12.5 gal of gasoline. How many miles can you travel on one gallon of gasoline?

168. A customer purchases a product that sells for \$39.88. The sales tax is \$2.39. How much change should the customer receive from a \$50 bill?

169. You make a down payment of \$225 on a camcorder and agree to make payments of \$34.17 a month for the next 18 months. Find the total cost of the camcorder.

Solve.

170. Using the menu shown below, estimate the bill for the following order: 1 soup, 1 cheese sticks, 1 blackened swordfish, 1 chicken divan, and 1 carrot cake.

Appetizers

Soup of the Day	$2.75
Cheese Sticks	$3.25
Potato Skins	$3.50

Entrees

Roast Prime Rib	$18.95
Blackened Swordfish	$16.95
Chicken Divan	$14.95

Desserts

Carrot Cake	$4.25
Ice Cream Pie	$5.50
Cheese Cake	$6.75

171. Using the menu shown above, estimate the bill for the following order: 1 potato skins, 1 cheese sticks, 1 roast prime rib, 1 chicken divan, 1 ice cream pie, and 1 cheese cake.

172. Using the catalogue information shown below, estimate the cost of ordering 4 boxes of 3.5″ DS/DD diskettes.

Catalog Number	Description	Cost
DC-4392	5.25" DS/DD Diskettes, Box of 10	$3.90
DC-4397	5.25" DS/HD Diskettes, Box of 50	$19.50
DC-5108	3.5" DS/DD Diskettes, Box of 10	$5.90
DC-5260	3.5" SS/DD Diskettes, Box of 25	$8.75
DC-5914	3.5" DS/HD Diskettes, Box of 25	$14.75

173. Using the catalogue information shown above, estimate the cost of ordering 6 boxes of 5.25″ DS/HD diskettes.

174. A bookkeeper earns a salary of $340 for a 40-hour week. This week the bookkeeper worked 6 h of overtime at a rate of $12.75 for each hour of overtime worked. Find the bookkeeper's total income for the week.

175. In the United States today, on average each person throws away 3.6 lb of garbage per day. On average, how many pounds per year does a family of four discard?

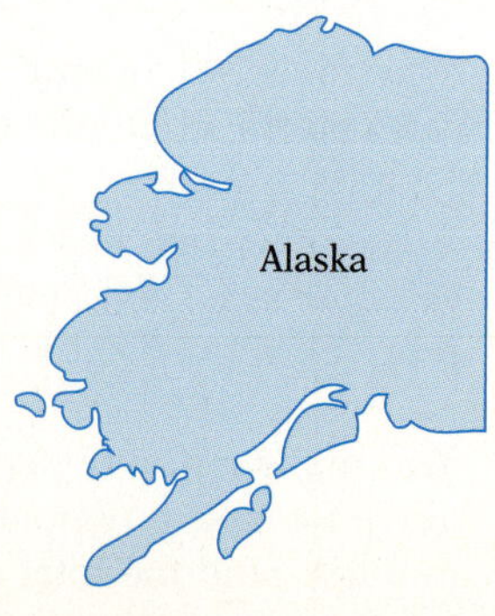

176. In Fairbanks, Alaska, the average temperature during the month of July is 61.5°F, while the average temperature during January is −12.7°F. What is the difference between the average temperatures in Fairbanks during July and January?

Solve.

177. The boiling point of oxygen is -182.962°C. The melting point of oxygen is -218.4°C. What is the difference between the boiling point and the melting point of oxygen?

178. In the chart at the right, a positive number indicates money placed in funds, and a negative number indicates money withdrawn from funds. For the month of October, what is the difference between the amount invested in stock funds and the amount withdrawn from bond funds? (Source: AMG Data Services)

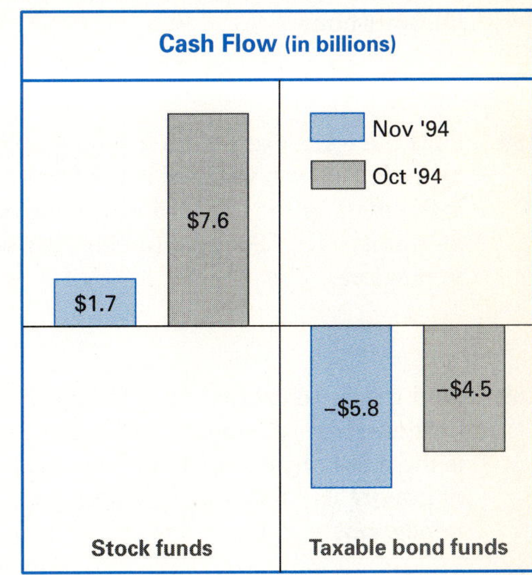

179. On January 22, 1943, in Spearfish, South Dakota, the temperature fell from 12.22°C at 9:00 A.M. to -20°C at 9:27 A.M. How many degrees did the temperature fall during the 27-minute period?

180. Use the formula $M = S - C$, where M is the markup on a consumer product, S is the selling price, and C is the cost of the product to the business, to find the markup on a product that cost a business \$1653.19 and has a selling price of \$2231.81.

181. Use the formula $M = S - C$, where M is the markup on a consumer product, S is the selling price, and C is the cost of the product to the business, to find the markup on a product that cost a business \$30.73 and has a selling price of \$87.80.

182. The amount of an employee's earnings that is subject to federal withholding is called federal earnings. Find the federal earnings for an employee who earns \$694.89 and has a withholding allowance of \$132.69. Use the formula $F = E - W$, where F is the federal earnings, E is the employee's earnings, and W is the withholding allowance.

183. The amount of an employee's earnings that is subject to federal withholding is called federal earnings. Find the federal earnings for an employee who earns \$572.45 and has a withholding allowance of \$88.46. Use the formula $F = E - W$, where F is the federal earnings, E is the employee's earnings, and W is the withholding allowance.

184. Use the formula $M = \frac{C}{N}$, where M is the cost per mile for a rental car, C is the total cost, and N is the number of miles driven, to find the cost per mile when the total cost of renting a car is \$260.16 and you drive the car 542 mi.

185. Use the formula $M = \frac{C}{N}$, where M is the cost per mile for a rental car, C is the total cost, and N is the number of miles driven, to find the cost per mile when the total cost of renting a car is \$311.88 and you drive the car 678 mi.

186. Find the cost of operating an 1800-watt TV set for 5 h at a cost of \$.06 per kilowatt-hour. Use the formula $c = \frac{1}{1000}wtk$, where c is the cost of operating an appliance, w is the number of watts, t is the time in hours, and k is the cost per kilowatt-hour.

Solve.

187. Find the cost of operating a 200-watt stereo for 3 h at a cost of \$.10 per kilowatt-hour. Use the formula $c = \frac{1}{1000}wtk$, where c is the cost of operating an appliance, w is the number of watts, t is the time in hours, and k is the cost per kilowatt-hour.

188. Find the force exerted on a falling object that has a mass of 4.25 kg. Use the formula $F = ma$, where F is the force exerted by gravity on a falling object, m is the mass of the object, and a is the acceleration of gravity. The acceleration of gravity is -9.80 m/s² (meters per second squared). The force is measured in newtons.

189. Find the force exerted on a falling object that has a mass of 6.75 kg. Use the formula $F = ma$, where F is the force exerted by gravity on a falling object, m is the mass of the object, and a is the acceleration of gravity. The acceleration of gravity is -9.80 m/s² (meters per second squared). The force is measured in newtons.

Critical Thinking

190. Write the number in standard form.
a. 3.8 million **b.** 5.7 billion **c.** 294.6 million **d.** 1.2 trillion

191. Indicate which digits of the number, if any, need not be entered on a calculator.
a. 1.500 **b.** 0.908 **c.** 60.07 **d.** 0.0032

192. Using the method, presented in this section, of estimating the sum of two decimals, what is the largest amount by which the estimate of the sum of two decimals with tenths, hundredths, and thousandths places could differ from the exact sum? Assume the number in the thousandths place is not zero.

193. Determine whether the statement is always true, sometimes true, or never true.
a. The product of an even number of negative factors is a negative number.
b. The sum of an odd number of negative addends is a negative number.
c. If $a \geq 0$, then $|a| = a$.
d. If $a \leq 0$, then $|a| = -a$.

194. A ball point pen priced at 50¢ was not selling. When the price was reduced to a different whole number of cents, the entire stock sold for \$31.93. How many cents were charged per pen when the price was reduced?

195. [W] Explain how baseball batting averages are determined.

196. [W] What does the term "population density" mean? How is population density determined? What is the population density of the state you live in? How does this compare with the population density of the country as a whole?

197. [W] What is the national debt of the United States of America? Divide the national debt by the number of United States citizens. What do these numbers mean?

SECTION 4.3 Percents, Fractions, and Decimals

OBJECTIVE A Fractions and decimals

Since the fraction bar can be read "divided by," any fraction can be written as a decimal. To write a fraction as a decimal, divide the numerator of the fraction by the denominator.

CONSIDER THIS
Think of the fraction bar as "divided by." $\frac{3}{4}$ is 3 divided by 4.

Convert $\frac{3}{4}$ to a decimal.

$$\begin{array}{r} 0.75 \\ 4\overline{)3.00} \\ -2\,8 \\ \hline 20 \\ -20 \\ \hline 0 \end{array}$$

0.75 ⟵ This is a **terminating decimal.**

0 ⟵ The remainder is zero.

$\frac{3}{4} = 0.75$

Convert $\frac{5}{11}$ to a decimal.

$$\begin{array}{r} 0.4545 \\ 11\overline{)5.0000} \\ -4\,4 \\ \hline 60 \\ -55 \\ \hline 50 \\ -44 \\ \hline 60 \\ -55 \\ \hline 5 \end{array}$$

0.4545 ⟵ This is a **repeating decimal.**

5 ⟵ The remainder is never zero.

$\frac{5}{11} = 0.\overline{45}$ The bar over the digits 45 is used to show that these digits repeat.

To convert a decimal to a fraction, remove the decimal point and place the decimal part over a denominator equal to the place value of the last digit in the decimal.

hundredths: $0.57 = \frac{57}{100}$

hundredths: $7.65 = 7\frac{65}{100} = 7\frac{13}{20}$

tenths: $8.6 = 8\frac{6}{10} = 8\frac{3}{5}$

Convert 4.375 to a fraction.

The 5 in 4.375 is in the thousandths place. Write 0.375 as a fraction with a denominator of 1000. $4.375 = 4\frac{375}{1000}$

Simplify the fraction. $= 4\frac{3}{8}$

To find the order relation between a fraction and a decimal, first rewrite the fraction as a decimal. Then compare the two decimals.

Find the order relation between $\frac{6}{7}$ and 0.855.

Write the fraction as a decimal. Round to one more place value than the given decimal. (0.855 has 3 decimal places; round to 4 decimal places.) $\frac{6}{7} \approx 0.8571$

Compare the two decimals. $0.8571 > 0.8550$

Replace the decimal approximation of $\frac{6}{7}$ with $\frac{6}{7}$. $\frac{6}{7} > 0.855$

Example 1 Convert $\frac{5}{8}$ to a decimal.

Solution

$$8\overline{)5.000} = 0.625 \qquad \frac{5}{8} = 0.625$$

You Try It 1 Convert $\frac{4}{5}$ to a decimal.

Your Solution

Example 2 Convert $3\frac{1}{3}$ to a decimal.

Solution Write $\frac{1}{3}$ as a decimal.

$$3\overline{)1.000} = 0.333 = 0.\overline{3}$$

$$3\frac{1}{3} = 3.\overline{3}$$

You Try It 2 Convert $1\frac{2}{3}$ to a decimal.

Your Solution

Example 3 Convert 7.25 to a fraction.

Solution $7.25 = 7\frac{25}{100} = 7\frac{1}{4}$

You Try It 3 Convert 6.2 to a fraction.

Your Solution

Example 4 Place the correct symbol, < or >, between the two numbers.

$0.845 \quad \frac{5}{6}$

Solution $\frac{5}{6} \approx 0.8333$

$0.8450 > 0.8333$

$0.845 > \frac{5}{6}$

You Try It 4 Place the correct symbol, < or >, between the two numbers.

$0.588 \quad \frac{7}{12}$

Your Solution

Solutions on p. A9

OBJECTIVE B Percents as decimals or fractions

Percent means "parts of 100." In the figure at the right, there are 100 parts. Because 19 of the 100 parts are shaded, 19% of the figure is shaded.

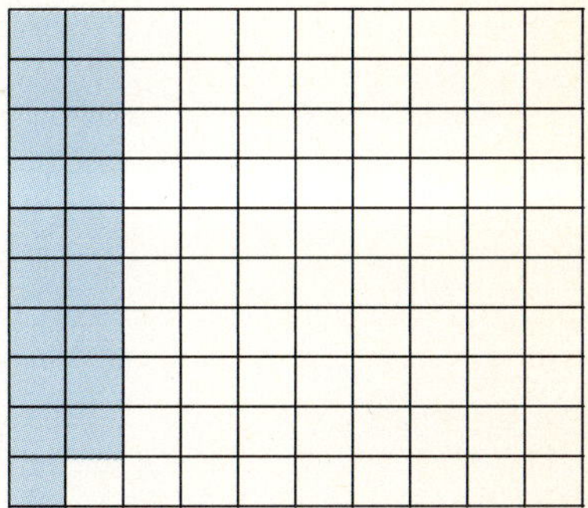

19 of 100 equal parts can be expressed as the fraction $\frac{19}{100}$. One percent can be expressed as 1 of 100 equal parts or $\frac{1}{100}$. Thus 1% is $\frac{1}{100}$ or 0.01.

"A population growth rate of 5%," "a manufacturer's discount of 40%," and "an 8% increase in pay" are typical examples of the many ways in which percent is used in applied problems. When solving problems involving a percent, it is usually necessary either to rewrite the percent as a fraction or a decimal, or to rewrite a fraction or a decimal as a percent.

To write a percent as a fraction, remove the percent sign and multiply by $\frac{1}{100}$.

Write 67% as a fraction.

Remove the percent sign and multiply by $\frac{1}{100}$.

$$67\% = 67\left(\frac{1}{100}\right) = \frac{67}{100}$$

To write a percent as a decimal, remove the percent sign and multiply by 0.01.

Write 19% as a decimal.

Remove the percent sign and multiply by 0.01. This is the same as moving the decimal point two places to the left.

$$19\% = 19(0.01) = 0.19$$

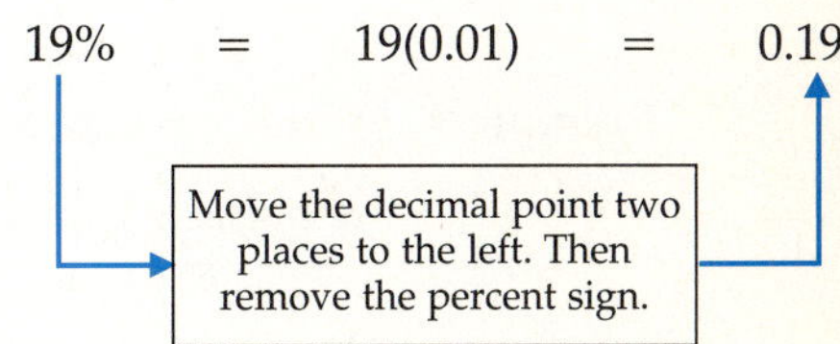

Example 5 Write 150% as a fraction and as a decimal.

Solution $150\% = 150\left(\frac{1}{100}\right) = \frac{150}{100} = 1\frac{1}{2}$

$150\% = 150(0.01) = 1.50$

You Try It 5 Write 110% as a fraction and as a decimal.

Your Solution

Example 6 Write 0.35% as a decimal.

Solution $0.35\% = 0.35(0.01) = 0.0035$

You Try It 6 Write 0.8% as a decimal.

Your Solution

Solutions on p. A9

OBJECTIVE C Fractions and decimals as percents

A fraction or decimal can be written as a percent by multiplying by 100%. Since 100% is $\frac{100}{100} = 1$, **multiplying by 100% is the same as multiplying by 1.**

Write $\frac{7}{8}$ as a percent.

Multiply $\frac{7}{8}$ by 100%. $\frac{7}{8} = \frac{7}{8}(100\%) = \frac{700}{8}\% = 87.5\%$

Write 0.64 as a percent.

Multiply by 100%. This is the same as moving the decimal point two places to the right. $0.64 = 0.64(100\%) = 64\%$

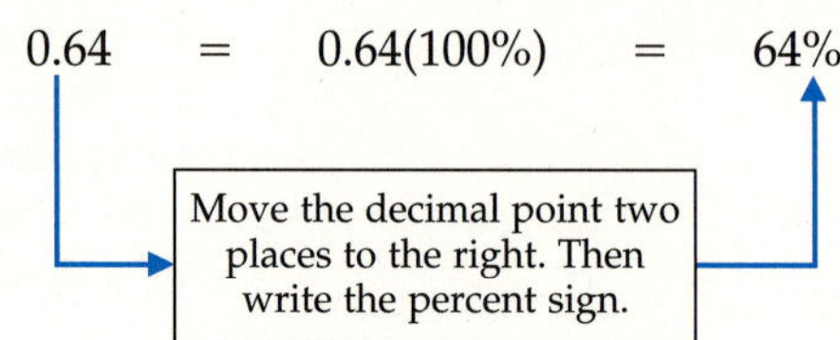

Example 7 Write 0.055 as a percent.

Solution $0.055 = 0.055(100\%) = 5.5\%$

You Try It 7 Write 0.038 as a percent.

Your Solution

Example 8 Write 1.78 as a percent.

Solution $1.78 = 1.78(100\%) = 178\%$

You Try It 8 Write 1.05 as a percent.

Your Solution

Example 9 Write $\frac{3}{8}$ as a percent. Write the remainder as a fraction.

Solution $\frac{3}{8} = \frac{3}{8}(100\%) = \frac{300}{8}\%$

$= 37\frac{1}{2}\%$

You Try It 9 Write $\frac{9}{7}$ as a percent. Write the remainder as a fraction.

Your Solution

Example 10 Write $1\frac{1}{7}$ as a percent. Round to the nearest tenth of a percent.

Solution $1\frac{1}{7} = \frac{8}{7} = \frac{8}{7}(100\%)$

$= \frac{800}{7}\% \approx 114.3\%$

You Try It 10 Write $1\frac{5}{9}$ as a percent. Round to the nearest tenth of a percent.

Your Solution

Solutions on p. A9

4.3 EXERCISES

Objective A

Convert the fraction to a decimal.
Place a bar over repeating digits of a repeating decimal.

1. $\frac{3}{8}$ **2.** $\frac{7}{10}$ **3.** $\frac{7}{15}$ **4.** $\frac{8}{11}$ **5.** $\frac{9}{16}$

6. $\frac{7}{12}$ **7.** $\frac{5}{3}$ **8.** $\frac{7}{4}$ **9.** $\frac{3}{25}$ **10.** $\frac{9}{1000}$

11. $2\frac{3}{4}$ **12.** $1\frac{1}{2}$ **13.** $3\frac{2}{9}$ **14.** $4\frac{1}{6}$ **15.** $2\frac{1}{4}$

16. $6\frac{3}{5}$ **17.** $4\frac{5}{24}$ **18.** $7\frac{8}{9}$ **19.** $4\frac{11}{20}$ **20.** $5\frac{21}{25}$

Convert the decimal to a fraction.

21. 0.6 **22.** 0.2 **23.** 0.25 **24.** 0.75 **25.** 0.48

26. 0.32 **27.** 0.125 **28.** 0.325 **29.** 0.045 **30.** 0.085

31. 0.028 **32.** 0.096 **33.** 2.5 **34.** 3.4 **35.** 4.55

36. 9.95 **37.** 1.72 **38.** 5.68 **39.** 7.431 **40.** 2.297

Place the correct symbol, < or >, between the two numbers.

41. $\frac{9}{10}$ 0.89 **42.** $\frac{7}{20}$ 0.34 **43.** $\frac{4}{5}$ 0.803 **44.** $\frac{3}{4}$ 0.706

Place the correct symbol, < or >, between the two numbers.

45. $0.444 \quad \frac{4}{9}$ **46.** $0.72 \quad \frac{5}{7}$ **47.** $0.13 \quad \frac{3}{25}$ **48.** $0.25 \quad \frac{13}{50}$

49. $\frac{5}{16} \quad 0.312$ **50.** $\frac{7}{18} \quad 0.39$ **51.** $\frac{10}{11} \quad 0.909$ **52.** $\frac{8}{15} \quad 0.543$

Solve.

53. According to the chart at the right, how many inches of rain fell in West Palm Beach during Tropical Storm Gordon? Write your answer in fractional form.

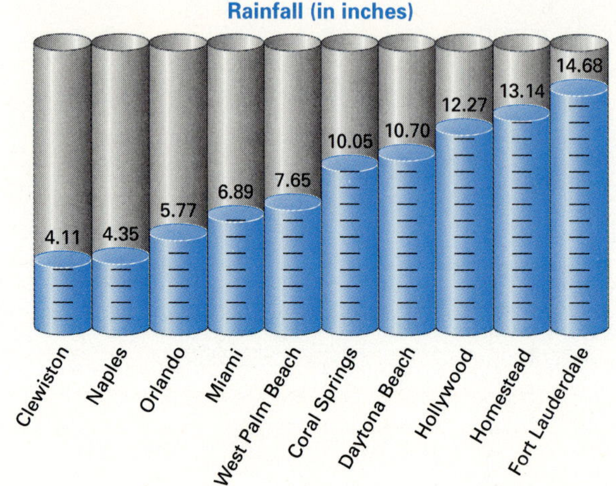

Objective B

Write as a fraction and as a decimal.

54. 5% **55.** 60% **56.** 30% **57.** 90% **58.** 250%

59. 140% **60.** 28% **61.** 66% **62.** 35% **63.** 85%

64. 6% **65.** 8% **66.** 122% **67.** 165% **68.** 3%

69. 9% **70.** 29% **71.** 83% **72.** 36% **73.** 64%

Write as a fraction.

74. $11\frac{1}{9}\%$ **75.** $4\frac{2}{7}\%$ **76.** $12\frac{1}{2}\%$ **77.** $37\frac{1}{2}\%$ **78.** $31\frac{1}{4}\%$

79. $3\frac{1}{8}\%$ **80.** $66\frac{2}{3}\%$ **81.** $45\frac{5}{11}\%$ **82.** $15\frac{3}{8}\%$ **83.** $6\frac{2}{3}\%$

84. $\frac{3}{8}\%$ **85.** $\frac{1}{4}\%$ **86.** $\frac{1}{2}\%$ **87.** $5\frac{3}{4}\%$ **88.** $68\frac{3}{4}\%$

89. $83\frac{1}{3}\%$ **90.** $6\frac{1}{4}\%$ **91.** $8\frac{2}{3}\%$ **92.** $87\frac{1}{2}\%$ **93.** $3\frac{1}{3}\%$

Write as a decimal.

94. 7.3% **95.** 9.1% **96.** 15.8% **97.** 16.7% **98.** 0.3%

99. 0.9% **100.** 121.2% **101.** 18.23% **102.** 62.14% **103.** 0.15%

104. 8.25% **105.** 5.05% **106.** 80.4% **107.** 0.06% **108.** 0.08%

Solve.

109. According to the graph at the right, what fraction of the baby boomers surveyed would use a $50,000 inheritance for retirement?

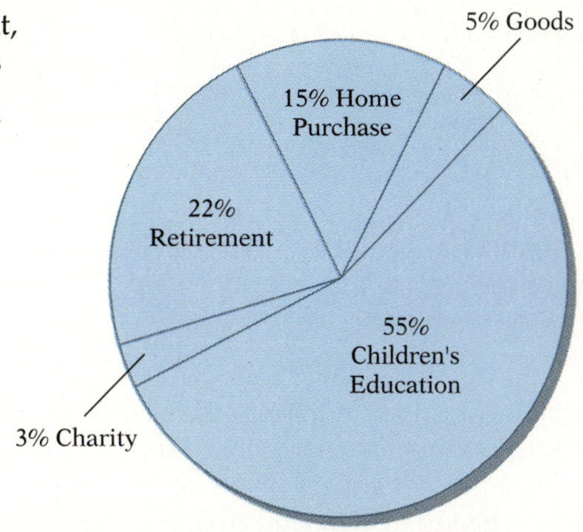

How baby boomers would use a $50,000 inheritance

Objective C

Write as a percent.

110. 0.15 **111.** 0.37 **112.** 0.05 **113.** 0.02 **114.** 0.175

115. 0.125 **116.** 1.15 **117.** 1.36 **118.** 0.62 **119.** 0.96

120. 1.012 **121.** 2.09 **122.** 1.03 **123.** 0.2 **124.** 0.7

Write as a percent. Round to the nearest tenth of a percent.

125. $\frac{27}{50}$ **126.** $\frac{4}{20}$ **127.** $\frac{37}{200}$ **128.** $\frac{1}{3}$ **129.** $\frac{3}{8}$

130. $\frac{5}{11}$ **131.** $\frac{4}{9}$ **132.** $\frac{7}{8}$ **133.** $2\frac{1}{2}$ **134.** $1\frac{2}{7}$

135. $1\frac{11}{12}$ **136.** $\frac{37}{40}$ **137.** $\frac{2}{5}$ **138.** $\frac{1}{6}$ **139.** $1\frac{7}{9}$

Write as a percent. Write the remainder in fractional form.

140. $\frac{225}{200}$ **141.** $\frac{17}{50}$ **142.** $\frac{17}{25}$ **143.** $\frac{3}{8}$ **144.** $\frac{9}{16}$

145. $\frac{5}{14}$ **146.** $\frac{3}{19}$ **147.** $\frac{4}{7}$ **148.** $1\frac{1}{4}$ **149.** $1\frac{5}{9}$

150. $2\frac{5}{6}$ **151.** $\frac{7}{30}$ **152.** $\frac{4}{15}$ **153.** $\frac{3}{11}$ **154.** $\frac{2}{9}$

Critical Thinking

155. Write the part of the square that is shaded as a fraction, as a decimal, and as a percent. Write the part of the square that is not shaded as a fraction, as a decimal, and as a percent.

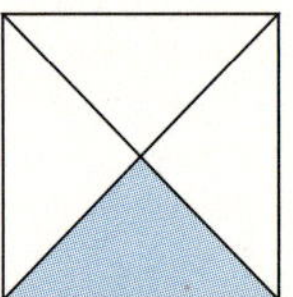

156. Determine whether the statement is true or false. If the statement is false, give an example to show that the statement is false.
a. Multiplying a number by a percent always decreases the number.
b. A percent sign can be attached to or removed from a number as needed.
c. The word percent means "per hundred."
d. A percent is always less than one.

157. **(a)** A sale on computers advertised $\frac{1}{3}$ off the regular price. What percent of the regular price does this represent? **(b)** A suit was priced at 50% off the regular price. What fraction of the regular price does this represent?

158. If $\frac{2}{5}$ of the population voted in an election, what percent of the population did not vote?

159. A medication loses one-half its potency each week. At the end of how many weeks will the medication have less than 10% of its original potency?

160. According to the chart at the right, what fraction of the passengers on American Airlines filed complaints?

Airline Customers' Complaints (per 100,000 fliers)

Airline	Complaints
Southwest	0.28
Delta	0.44
Northwest	0.53
United	0.61
American	0.68
USAir	0.81
America West	1.01
Continental	1.82
TWA	2.25

Source: Transportation Department

161. [W] Convert $\frac{1}{9}$, $\frac{2}{9}$, $\frac{3}{9}$, and $\frac{4}{9}$ to decimals. Describe the pattern. Use the pattern to convert $\frac{5}{9}$, $\frac{7}{9}$, and $\frac{8}{9}$ to decimals.

162. [W] Explain in your own words how to change a percent to a decimal and a decimal to a percent.

SECTION 4.4 Real Numbers

OBJECTIVE A Real numbers and the real number line

A **rational number** is the quotient of two integers.

> *Rational Numbers*
>
> A rational number is a number that can be written in the form $\frac{a}{b}$, where a and b are integers and $b \neq 0$.

Each of the three numbers shown at the right is a rational number. $\frac{3}{4}$ $\frac{-2}{9}$ $\frac{13}{-5}$

An integer can be written as the quotient of the integer and 1. Therefore, **every integer is a rational number.** $6 = \frac{6}{1}$ $-8 = \frac{-8}{1}$

A mixed number can be written as the quotient of two integers. Therefore, **every mixed number is a rational number.** $1\frac{4}{7} = \frac{11}{7}$ $3\frac{2}{5} = \frac{17}{5}$

Recall that a fraction can be written as a decimal by dividing the numerator of the fraction by the denominator. The result is either a terminating decimal or a repeating decimal.

To convert $\frac{3}{8}$ to a decimal, read the fraction bar as "divided by."

$\frac{3}{8} = 3 \div 8 = 0.375$. This is an example of a terminating decimal.

To convert $\frac{6}{11}$ to a decimal, divide 6 by 11.

$\frac{6}{11} = 6 \div 11 = 0.\overline{54}$. This is an example of a repeating decimal.

Every rational number can be written either as a terminating decimal or as a repeating decimal. All terminating and repeating decimals are rational numbers.

Some numbers have decimal representations that never terminate or repeat, for example,

0.12122122212222 . . .

The pattern in this number is one more 2 following each successive 1 in the number. There is no repeating block of digits. This number is an **irrational number.** Other examples of irrational numbers include π (which is presented in Chapter 5) and some square roots (the topic of Chapter 11).

LOOK CLOSELY
Rational numbers are fractions such as $-\frac{6}{7}$ or $\frac{10}{3}$ where the numerator and denominator are integers. Rational numbers are also represented by repeating decimals, for instance 0.25767676 . . . or terminating decimals such as 1.73. An irrational number is neither a repeating decimal nor a terminating decimal. For instance 2.45445444544445 . . . is an irrational number.

Irrational Numbers

An **irrational number** is a number whose decimal representation never terminates or repeats.

The rational numbers and the irrational numbers taken together are called the **real numbers.**

Real Numbers

The **real numbers** are all the rational numbers together with all the irrational numbers.

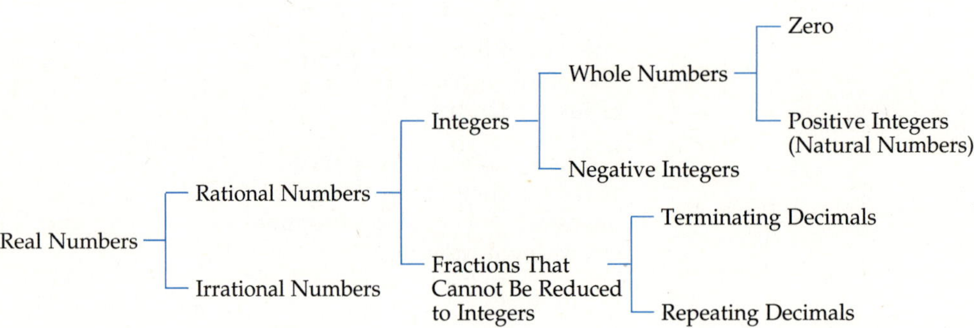

The number line is also called the **real number line.** Every real number corresponds to a point on the real number line, and every point on the real number line corresponds to a real number.

Graph $3\frac{1}{2}$ on the real number line.

$3\frac{1}{2}$ is a positive number and is, therefore, to the right of zero on the number line. Draw a solid dot three and one-half units to the right of zero on the number line.

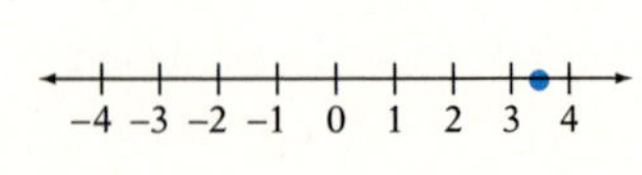

Graph -2.5 on the real number line.

-2.5 is a negative number and is, therefore, to the left of zero on the number line. Draw a solid dot two and one-half units to the left of zero on the number line.

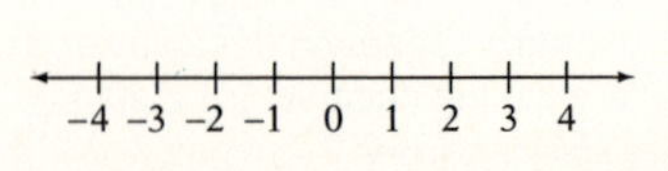

Graph the real numbers greater than 2.

To graph the real numbers greater than 2, a solid dot should be placed above every number to the right of 2 on the number line. It is not possible to list all the real numbers greater than 2. It is not even possible to list all the real numbers between 2 and 3, or even to give the smallest real number greater than 2. The number 2.0000000001 is greater than 2 and is certainly very close to 2, but even smaller numbers greater than 2 can be written by inserting more and more zeros after the decimal point. Therefore, the graph of the real numbers greater than 2 is shown by drawing a heavy line to the right of 2.

The arrow indicates that the heavy line continues without end. The real numbers greater than 2 do not include the number 2. The circle on the graph indicates that 2 is not included in the graph.

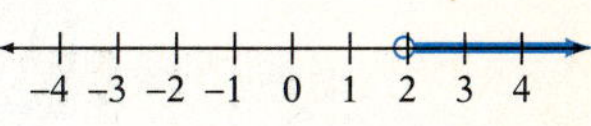

Graph the real numbers between 1 and 3.

The real numbers between 1 and 3 do not include the number 1 or the number 3; thus circles are drawn at 1 and 3. Draw a heavy line between 1 and 3 to indicate all the real numbers between these two numbers.

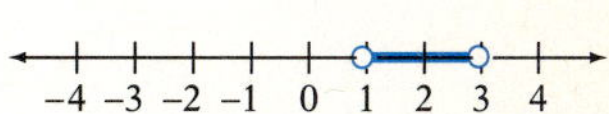

Example 1 Graph 0.5 on the real number line.

Solution Draw a solid dot one-half unit to the right of zero on the number line.

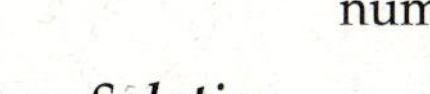

You Try It 1 Graph $-1\frac{1}{2}$ on the real number line.

Your Solution

–4 –3 –2 –1 0 1 2 3 4

Example 2 Graph the real numbers less than -1.

Solution The real numbers less than -1 are to the left of -1 on the number line.
Draw a circle at -1.
Draw a heavy line to the left of -1.
Draw an arrow at the left of the line.

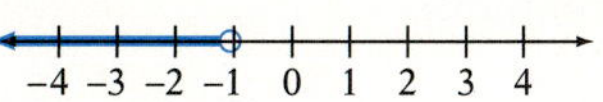

You Try It 2 Graph the real numbers greater than -2.

Your Solution

–4 –3 –2 –1 0 1 2 3 4

Solutions on p. A10

Example 3 Graph the real numbers between -3 and 0.

Solution Draw a circle at -3 and a circle at 0. Draw a heavy line between -3 and 0.

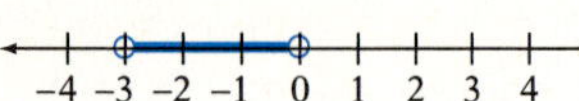

You Try It 3 Graph the real numbers between -1 and 4.

Your Solution

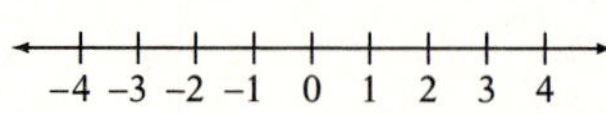

Solution on p. A10

OBJECTIVE B Inequalities in one variable

Recall that the symbol for "is greater than" is $>$, and the symbol for "is less than" is $<$. The symbol $\geq$ means "is greater than or equal to." The symbol $\leq$ means "is less than or equal to."

The statement $5 < 5$ is a false statement because 5 is not less than 5. $5 < 5$ False

The statement $5 \leq 5$ is a true statement because 5 is "less than or equal to" 5; 5 is equal to 5. $5 \leq 5$ True

An **inequality** contains the symbol $>$, $<$, $\geq$, or $\leq$, and expresses the relative order of two mathematical expressions.

$4 > -3$
$-9.7 < 0$
$6 + 2 \geq 1$
$x \leq 5$

Inequalities

The inequality $x \leq 5$ is read "x is less than or equal to 5."

For the inequality $x > -3$, which values of the variable listed below make the inequality true?

a. -6 **b.** -3.9 **c.** 0 **d.** 0.5

Replace x in $x > -3$ with each number, and determine whether or not each inequality is true.

a.	**b.**	**c.**	**d.**
$x > -3$	$x > -3$	$x > -3$	$x > -3$
$-6 > -3$	$-3.9 > -3$	$0 > -3$	$0.5 > -3$
False	False	True	True

The numbers 0 and 0.5 make the inequality true.

There are many values of the variable x that will make the inequality $x > -3$ true; any number greater than -3 makes the inequality true. Replacing x with any number less than -3 will result in a false statement.

What values of the variable x make the inequality $x \leq 4$ true?

All real numbers less than or equal to 4 make the inequality true.

The numbers that make an inequality true can be graphed on the real number line.

Graph $x > 1$.

The numbers that, when substituted for x, make this inequality true are all the real numbers greater than 1. The numbers greater than 1 are all the numbers to the right of 1 on the number line. The circle on the graph indicates that 1 is not included in the numbers greater than 1.

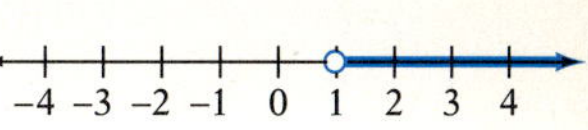

Graph $x \geq 1$.

The numbers that make this inequality true are all the real numbers greater than or equal to 1. The solid dot at 1 indicates that 1 is included in the numbers greater than or equal to 1.

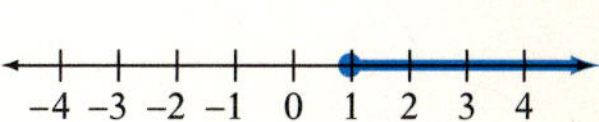

Remember, for $<$ or $>$, draw a circle on the graph. For $\leq$ or $\geq$, draw a solid dot.

Example 4 For the inequality $x \leq -6$, which values of the variable listed below make the inequality true?

a. -12 **b.** -6 **c.** 0 **d.** 5

Solution

a. $x \leq -6$
$-12 \leq -6$ True

b. $x \leq -6$
$-6 \leq -6$ True

c. $x \leq -6$
$0 \leq -6$ False

d. $x \leq -6$
$5 \leq -6$ False

The numbers -12 and -6 make the inequality true.

You Try It 4 For the inequality $x \geq 4$, which values of the variable listed below make the inequality true?

a. -1 **b.** 0 **c.** 4 **d.** 6

Your Solution

Example 5 What values of the variable x make the inequality $x < 8$ true?

Solution All real numbers less than 8 make the inequality true.

You Try It 5 What values of the variable x make the inequality $x > -7$ true?

Your Solution

Solutions on p. A10

Example 6 Graph $x \leq 3$.

Solution Draw a solid dot at 3.
Draw an arrow to the left of 3.

You Try It 6 Graph $x \geq -4$.

Your Solution

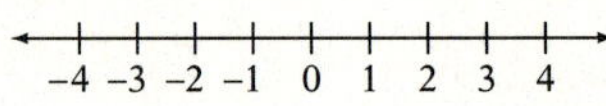

Solution on p. A10

OBJECTIVE C Applications

Solving application problems requires recognition of the verbal phrases that translate into mathematical symbols. Below is a partial list of the phrases used to indicate each of the four inequality symbols.

$<$	is less than	$>$	is greater than is more than exceeds
$\leq$	is less than or equal to maximum at most or less	$\geq$	is greater than or equal to minimum at least or more

Example 7

The minimum wage at the company you work for is \$5.25 an hour. Write an inequality for the wages at the company. Is it possible for an employee to earn \$5.15 an hour?

Strategy

- ▶ To write the inequality, let w represent the minimum wage. Since it is a minimum wage, all wages are greater than or equal to \$5.25.
- ▶ To determine if a wage of \$5.15 is possible, replace w in the inequality by 5.15. If the inequality is true, it is possible. If the inequality is false, it is not possible.

Solution

$w \geq 5.25$

$5.15 \geq 5.25$ False

It is not possible for an employee to earn \$5.15 an hour.

You Try It 7

On the highway near your home, motorists who exceed a speed of 55 mph are ticketed. Write an inequality for the speeds at which a motorist is ticketed. Will a motorist traveling at 58 mph be ticketed?

Your Strategy

Your Solution

Solution on p. A10

4.4 EXERCISES

Objective A

Graph the number on the real number line.

1. $2\frac{1}{2}$

−6 −5 −4 −3 −2 −1 0 1 2 3 4 5 6

2. $-2\frac{1}{2}$

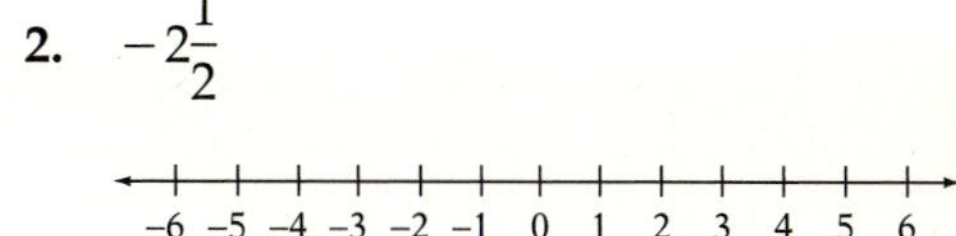

3. -3.5

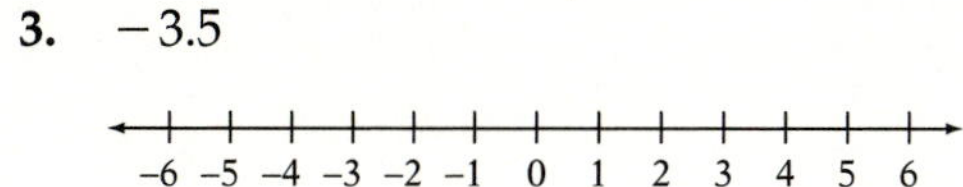

4. -0.5

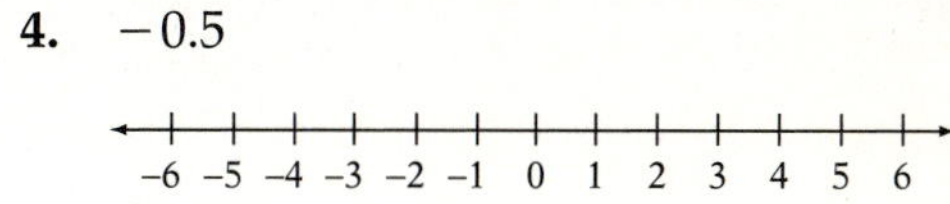

5. $-4\frac{1}{2}$

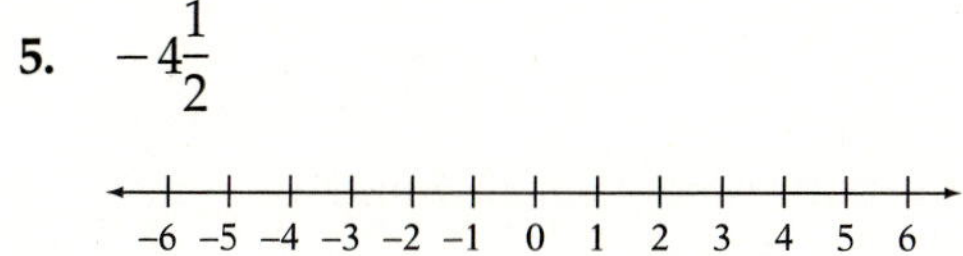

6. $\frac{1}{2}$

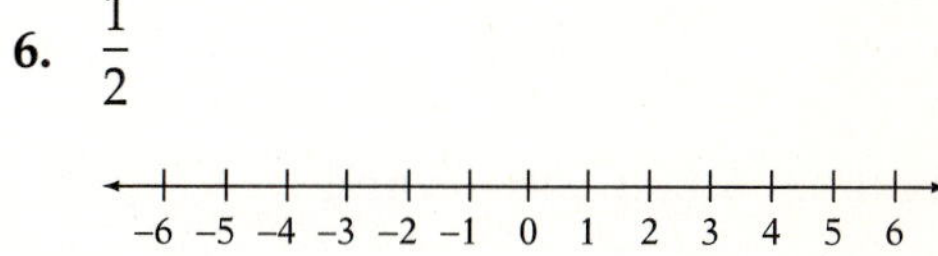

7. 1.5

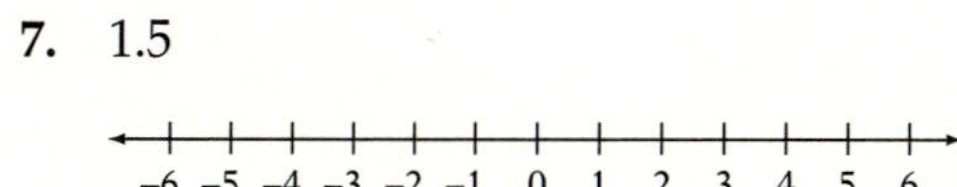

8. 5.5

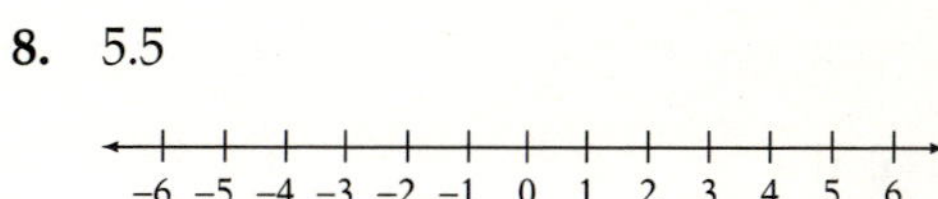

Graph.

9. the real numbers greater than 6

−6 −5 −4 −3 −2 −1 0 1 2 3 4 5 6

10. the real numbers greater than 1

−6 −5 −4 −3 −2 −1 0 1 2 3 4 5 6

11. the real numbers less than 0

−6 −5 −4 −3 −2 −1 0 1 2 3 4 5 6

12. the real numbers less than 2

−6 −5 −4 −3 −2 −1 0 1 2 3 4 5 6

13. the real numbers greater than -1

−6 −5 −4 −3 −2 −1 0 1 2 3 4 5 6

14. the real numbers greater than -4

−6 −5 −4 −3 −2 −1 0 1 2 3 4 5 6

Graph.

15. the real numbers less than -5

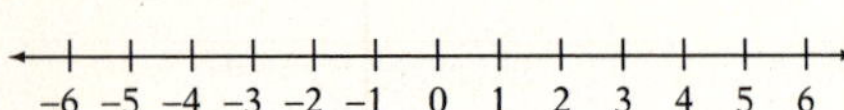

16. the real numbers less than -3

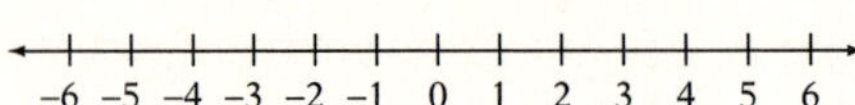

17. the real numbers between 2 and 5

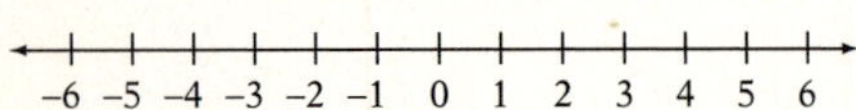

18. the real numbers between 4 and 6

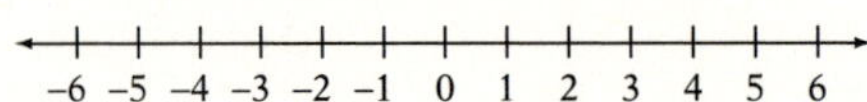

19. the real numbers between -4 and 0

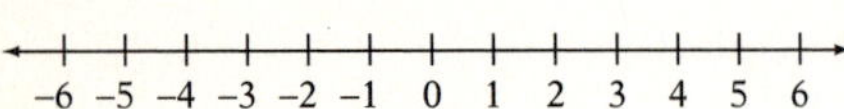

20. the real numbers between 0 and 3

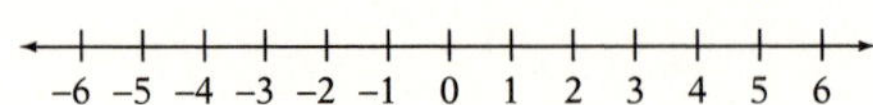

21. the real numbers between -2 and 6

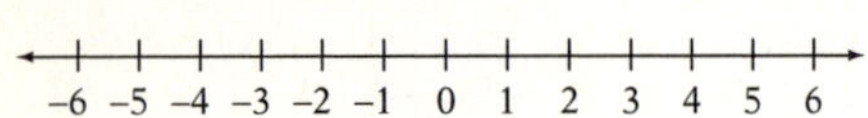

22. the real numbers between -1 and 5

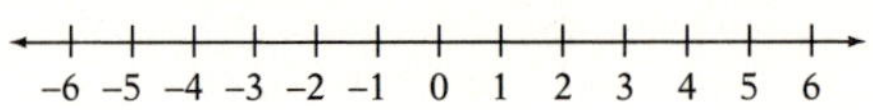

23. the real numbers between -6 and 1

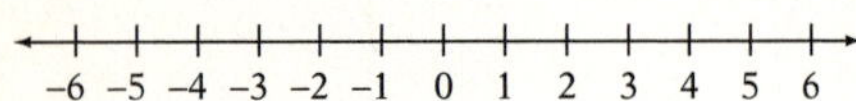

24. the real numbers between -5 and 0

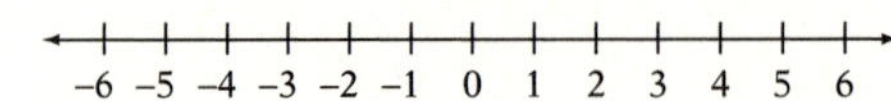

Objective B

Solve.

25. For the inequality $x > 9$, which numbers listed below make the inequality true?
 a. -3.8 b. 0 c. 9 d. 10.1

26. For the inequality $x \leq 5$, which numbers listed below make the inequality true?
 a. -1.11 b. 0 c. 5 d. 5.01

27. For the inequality $x \geq -2$, which numbers listed below make the inequality true?
 a. -6 b. -2 c. 0.4 d. 2.1

28. For the inequality $x \leq -7$, which numbers listed below make the inequality true?
 a. -14 b. -7 c. -1.3 d. -5.2

What values of the variable x make the inequality true?

29. $x < 3$

30. $x > -6$

31. $x \geq -1$

32. $x \leq 5$

Graph the inequality on the real number line.

33. $x < -2$

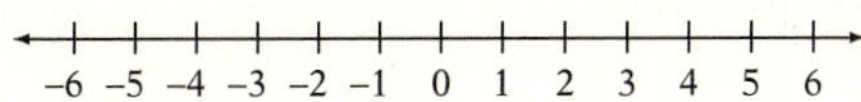

34. $x > 4$

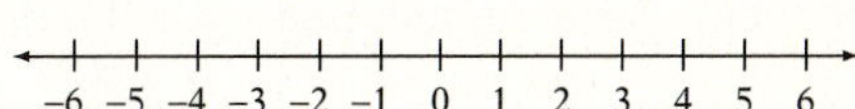

35. $x \geq 0$

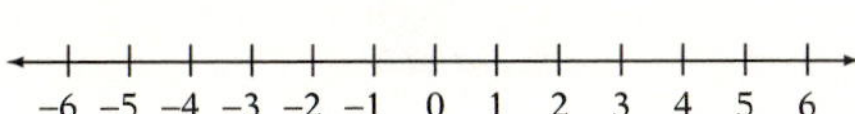

36. $x \leq -3$

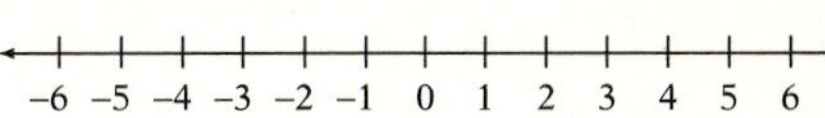

37. $x > -5$

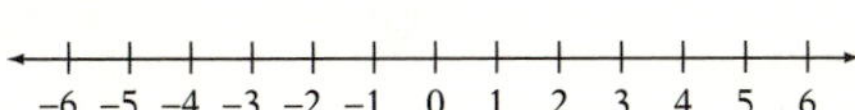

38. $x < -1$

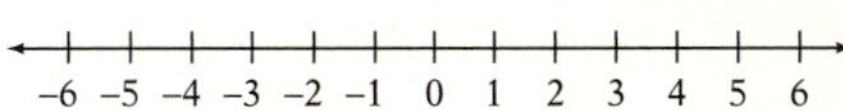

39. $x \leq 2$

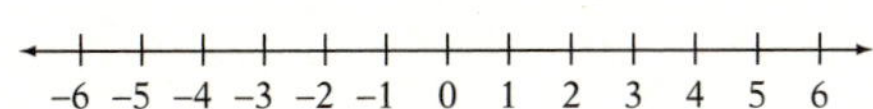

40. $x \geq 6$

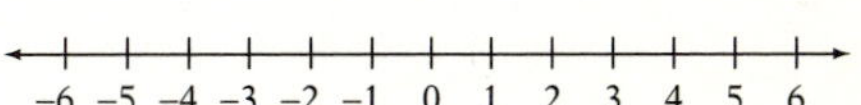

Objective C

Solve.

41. Each sales representative for a company must sell at least 50,000 units per year. Write an inequality for the number of units a sales representative must sell. Has a representative who sold 49,000 units this past year met the sales goal?

42. A health official recommends a cholesterol level of less than 220 units. Write an inequality for the acceptable cholesterol level. Is a cholesterol level of 238 within the recommended levels?

43. A part-time student can take a maximum of 9 credit hours per semester. Write an inequality for the number of credit hours a part-time student can take. Does a student taking 8.5 credit hours fulfill the requirement for being a part-time student?

44. A service organization will receive a bonus of \$200 for collecting more than 1750 lb of aluminum cans during a collection drive. Write an inequality for the number of pounds of cans that must be collected in order to earn the bonus. If 1705.5 lb of aluminum cans are collected, will the organization receive the bonus?

45. Your monthly budget allows you to spend at most \$1200 per month. Write an inequality for the amount of money you can spend per month. Have you kept within your budget during a month you spent \$1190.50?

Solve.

46. In order to get a B in a history course, you must earn more than 80 points on the final exam. Write an inequality for the number of points you need to score on the final exam. Will a score of $80\frac{1}{2}$ earn you a B in the course?

47. Computer disks should be stored at temperatures greater than 50°F. Write an inequality for the temperatures at which computer disks should be stored. Is it safe to store a computer disk at a temperature of 47.5°F?

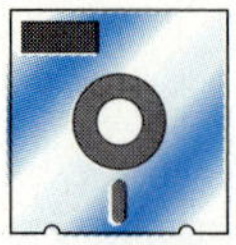

48. According to NCAA rules, the diameter of metal which forms the rim on a basketball hoop is to be $\frac{5}{8}$ in. or less. Write an inequality for the diameter of the ring on a basketball hoop. Does a ring with a diameter of $\frac{9}{16}$ in. meet the NCAA regulations?

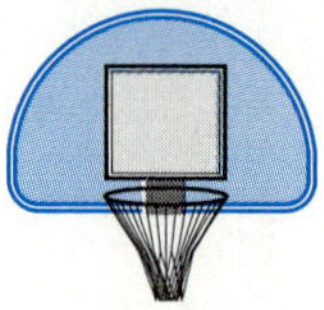

Critical Thinking

49. Classify each number as a whole number, an integer, a positive integer, a negative integer, a rational number, an irrational number, and/or a real number.

a. -2 **b.** 18 **c.** $-\frac{9}{37}$ **d.** -6.606 **e.** $4.5\overline{6}$ **f.** 3.050050005 . . .

50. Using the variable x, write an inequality to represent the graph.

a. 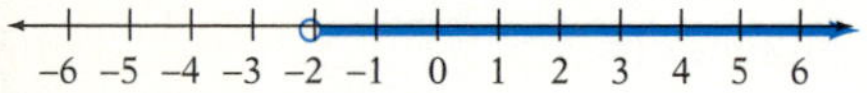

b. 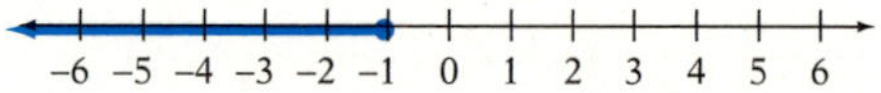

51. For the given inequality, which of the numbers in parentheses make the inequality true?

a. $|x| < 9$ $(-2.5, 0, 9, 15.8)$
b. $|x| > -3$ $(-6.3, -3, 0, 6.7)$
c. $|x| \geq 4$ $(-1.5, 0, 4, 13.6)$
d. $|x| \leq 5$ $(-4.9, 0, 2.1, 5)$

52. Given that a, b, c, and d are real numbers, which will ensure that $a + c < b + d$?

a. $a < b$ and $c < d$
b. $a > b$ and $c > d$
c. $a < b$ and $c > d$
d. $a > b$ and $c < d$

53. Determine whether the statement is always true, sometimes true, or never true.
a. Given that $a > 0$ and $b > 0$, then $ab > 0$.
b. Given that $a < 0$, then $a^2 > 0$.
c. Given that $a > 0$ and $b > 0$, then $a^2 > b$.

54. In your own words, define **a.** a rational number, **b.** an irrational number, and
[W] **c.** a real number.

Projects in Mathematics

Small Business and Gross Income

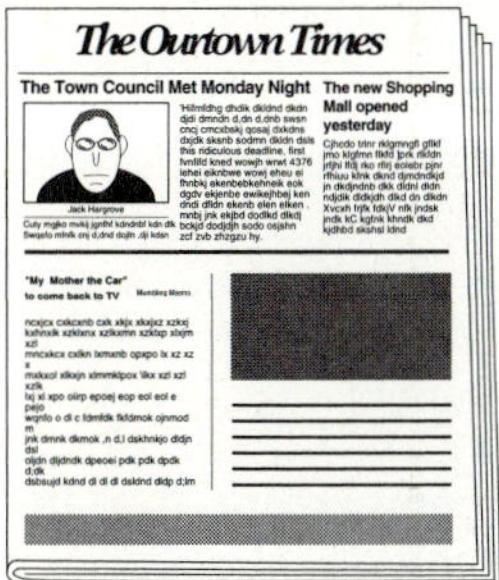

Do some research on your local newspaper. What is the newspaper's circulation? How often is an edition of the newspaper published? What is the price per issue at the newsstand?

Is there a special rate for subscribers to the newspaper? What is the price of a subscription? What is the length of time for which a subscription is paid? How many of the newspaper's readers are subscribers? Use this figure and the newspaper's total circulation to determine the number of copies sold at newsstands.

Use the figures you have gathered in answering the questions above to determine the total annual income, or **gross income,** derived from sales of the newspaper.

Sequences

Suppose you are offered a 30-day job that pays \$.01 the first day, \$.02 the second day, \$.04 the third day, and so on. Each day you work, your earnings are twice your earnings for the previous day. Would you accept this job over a 30-day job that pays \$50,000 per day?

For the job in which earnings double each day, make a guess as to your earnings on the 30th day of work and your total earnings over the 30-day period. Then calculate these figures. You may be surprised at the results!

The list of numbers that indicates your earnings each day is an ordered list of numbers, called a **sequence.**

0.01, 0.02, 0.04, 0.08, 0.16, 0.32, 0.64, 1.28, 2.56, 5.12, . . .

This list is ordered because the position of a number in the list indicates the day on which that amount was earned. For example, the 8th term of the sequence is 1.28, and \$1.28 is earned on the 8th day.

Each number of a sequence is called a **term** of the sequence. A formula can be used to find a specific term of the sequence given above.

$$\text{term } t = (\text{first term})(2^{t-1})$$

For example: $\text{term } 5 = (0.01)(2^{5-1}) = 0.01(2^4) = 0.01(16) = 0.16$

The amount earned on day 5 is \$.16.

Use the formula to find the amount earned on day 15 and on day 20.

Chapter Summary

Key Words

A number written in *decimal notation* has three parts: a whole number part, a decimal point, and a decimal part. The *decimal part* of a number represents a number less than one. A number written in decimal notation is often simply called a *decimal.*

Percent means "parts of 100."

A *rational number* is a number that can be written in the form $\frac{a}{b}$, where a and b are integers and $b \neq 0$. Every rational number can be written either as a terminating or a repeating decimal.

An *irrational number* is a number whose decimal representation never terminates or repeats.

The *real numbers* are all the rational numbers together with all the irrational numbers.

An *inequality* contains the symbol $>$, $<$, $\geq$, or $\leq$, and expresses the relative order of two mathematical expressions.

Essential Rules

To write a decimal in words, write the decimal part as if it were a whole number. Then name the place value of the last digit.

To write a decimal in standard form when it is written in words, write the whole number part, replace the word "and" with a decimal point, and write the decimal part so that the last digit is in the given place-value position.

To compare two decimals, write the decimal part of each number so that each has the same number of decimal places. Then compare the two numbers.

To round a decimal, use the same rules used with whole numbers, except drop the digits to the right of the given place value instead of replacing them with zeros.

To add or subtract decimals, write the decimals so that the decimal points are on a vertical line. Add or subtract as you would with whole numbers. Then write the decimal point in the answer directly below the decimal points in the given numbers.

To estimate the answer to a calculation, round each number to the highest place value of the number; the first digit of each number will be nonzero and all other digits will be zero. If a number is a decimal less than one, round the decimal so that there is one nonzero digit. Perform the calculation using the rounded numbers.

To multiply decimals, multiply the numbers as you would whole numbers. Then write the decimal point in the product so that the number of decimal places in the product is the sum of the decimal places in the factors.

To multiply a decimal by a power of 10, move the decimal point to the right the same number of places as there are zeros in the power of 10. If the power of 10 is written in exponential notation, the exponent indicates how many places to move the decimal point.

To divide decimals, move the decimal point in the divisor to the right so that it is a whole number. Move the decimal point in the dividend the same number of places to the right. Place the decimal point in the quotient directly above the decimal point in the dividend. Then divide as you would with whole numbers.

To divide a decimal by a power of 10, move the decimal point to the left the same number of places as there are zeros in the power of 10. If the power of 10 is written in exponential notation, the exponent indicates how many places to move the decimal point.

To write a fraction as a decimal, divide the numerator of the fraction by the denominator.

To convert a decimal to a fraction, remove the decimal point and place the decimal part over a denominator equal to the place value of the last digit in the decimal.

To find the order relation between a decimal and a fraction, first rewrite the fraction as a decimal. Then compare the two decimals.

To write a percent as a fraction, drop the percent sign and multiply by $\frac{1}{100}$.

To write a percent as a decimal, drop the percent sign and multiply by 0.01.

To write a fraction or a decimal as a percent, multiply by 100%.

Chapter Review Exercises

1. Subtract: $-3.981 - 4.32$

2. Find the quotient of 14.2 and 10^3.

3. Evaluate $a + b + c$ when $a = 80.59$, $b = -3.647$, and $c = 12.3$.

4. Write five and thirty-four thousandths in standard form.

5. Write 32% as a fraction.

6. Evaluate $\frac{x}{y}$ when $x = 0.396$ and $y = 3.6$.

7. Evaluate $a - b$ when $a = 80.32$ and $b = 29.577$.

8. Multiply: $-100(-34.25)$

9. Graph $x \geq -3$.

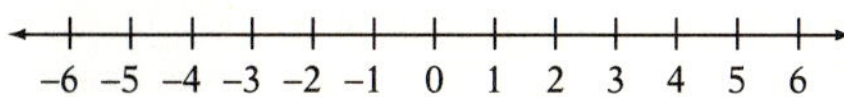

10. Graph all the real numbers between -6 and -2.

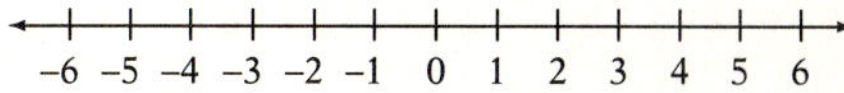

11. Place the correct symbol, < or >, between the two numbers.

$\frac{3}{7}$ 0.429

12. For the inequality $x \geq -1$, what numbers listed below make the inequality true?

a. -6 **b.** -1 **c.** -0.5 **d.** 0.1

13. Estimate the difference between 506.81 and 64.1.

14. Divide and round to the nearest tenth: $-6.8 \div 47.92$

15. Write $\frac{7}{40}$ as a percent.

16. Place the correct symbol, < or >, between the two numbers.

8.039 8.31

17. Find the sum of -247.8 and -193.4.

18. Convert 0.28 to a fraction.

19. Evaluate $60st$ when $s = 5$ and $t = -3.7$.

20. Write 125% as a decimal.

21. Write 126.0439 in words.

22. The boiling point of mercury is 356.58°C. The melting point of mercury is -38.87°C. Find the difference between the boiling point and the melting point of mercury.

23. A student must have a GPA (grade point average) of at least 3.5 to qualify for a certain scholarship. Write an inequality for the GPA a student must have in order to qualify for the scholarship. Does a student who has a GPA of 3.48 qualify for the scholarship?

24. One cup of raisins contains 0.16 mg of thiamin. One cup of peas contains 0.43 mg of thiamin. One cup of orange juice contains 0.23 mg of thiamin. Which contains the greatest amount of thiamin: one cup of raisins, one cup of peas, or one cup of orange juice?

25. Use the formula $P = C + M$, where P is the price of a product to a customer, C is the cost paid by a store for the product, and M is the markup, to find the price of a treadmill that costs a business \$369.99 and has a markup of \$129.50.

Focus on Problem Solving

Some problems in mathematics are solved by using **trial and error.** This method applies repeated tests or experiments until a satisfactory solution is reached.

Many of the Critical Thinking exercises in this text require a trial-and-error method solution. For example, a Critical Thinking exercise on page 224 of this chapter reads:

> Explain how you could cut through a cube so that the face of the resulting solid is **a.** a square, **b.** an equilateral triangle, **c.** a trapezoid, **d.** a hexagon.

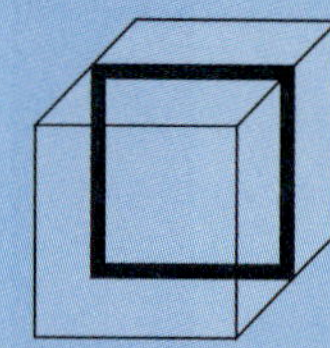

There is no formula to apply to this problem; there is no computation to perform. This problem requires picturing a cube and the results after cutting through it at different places on its surface and at different angles. For part **a**, cutting perpendicular to the top and bottom of the cube (that is, parallel to a face) will result in a square. The other shapes may prove more difficult.

When solving problems of this type, keep an open mind. Sometimes, when using trial and error, we are hampered by narrow vision; we cannot expand our thinking to include other possibilities. Then, when we see someone else's solution, it appears so obvious to us! For example, for the Critical Thinking question above, it is necessary to conceive of cutting through the cube at places other than the top surface; we need to conceive of beginning the cut at one of the corner points of the cube.

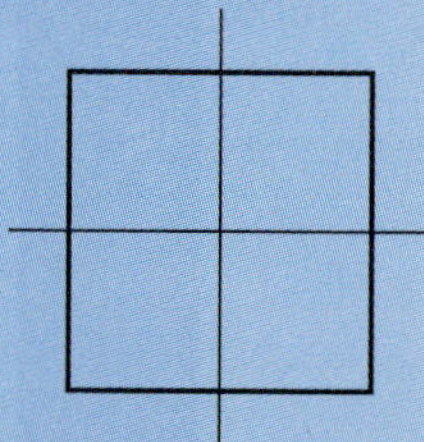

The topic of the Projects in Mathematics feature at the end of this chapter is symmetry. The lines of symmetry of a figure are found by trial and error. For example, in finding lines of symmetry for a square, begin by drawing a square. The horizontal line of symmetry and the vertical line of symmetry may be immediately obvious to you. But there are two others. Do you see that a line drawn through opposite corners of the square is also a line of symmetry?

Many of the questions in this text that require an answer of "always true, sometimes true, or never true" are best solved by trial and error. For example, consider the statement presented in Section 2 of this chapter.

> If two rectangles have the same area, then they have the same perimeter.

Try some numbers. Each of two rectangles, one measuring 6 units by 2 units and another measuring 4 units by 3 units, has an area of 12 square units, but the perimeter of the first is 16 units and the perimeter of the second is 14 units. So the answer "always true" has been eliminated. We still need to determine whether or not there is a case for which it is true. After experimenting with a lot of numbers, you may come to realize that we are trying to determine if it is possible for two different pairs of factors of a number to have the same sum. Is it?

Don't be afraid to make many experiments, and remember that errors (tests that don't work) are part of the process of trial and error.

SECTION 5.1 Variable Expressions in Simplest Form

OBJECTIVE A Simplify a variable expression using the Properties of Addition

POINT OF INTEREST

Historical manuscripts indicate that mathematics is at least 4000 years old. Yet it was only 400 years ago that mathematicians started using variables to stand for numbers. Earlier mathematics had been written in words. The idea that a letter can stand for a number was a turning point in mathematics.

An expression that contains one or more variables is called a **variable expression.** A variable expression is shown at the right. The expression can be rewritten by writing subtraction as addition of the opposite. Note that the expression has 4 addends. The **terms** of a variable expression are the addends of the expression. The expression at the right has 4 terms.

$$4y^3 - 3xy + x - 9$$

$$4y^3 + (-3xy) + x + (-9)$$

The terms $4y^3$, $-3xy$, and x are **variable terms.**

The term -9 is a **constant term,** or simply a constant.

Each variable term consists of a **numerical coefficient** and a **variable part.** The table at the right gives the numerical coefficient and the variable part of each variable term.

Term	Numerical Coefficient	Variable Part
$4y^3$	4	y^3
$-3xy$	-3	xy
x	1	x

In the term x, the numerical coefficient is 1 ($x = 1x$). The numerical coefficient for $-x$ is -1 ($-x = -1x$). The numerical coefficient of $-xy$ is -1 ($-xy = -1xy$). Usually the 1 is not written.

For the variable expression at the right, state: $9x^2 - x - 7yz^2 + 8$

a. the number of terms,
b. the coefficient of the second term,
c. the variable part of the third term, and
d. the constant term.

a. There are 4 terms: $9x^2$, $-x$, $-7yz^2$, and 8.
b. The coefficient of the second term is -1.
c. The variable part of the third term is yz^2.
d. The constant term is 8.

Like terms of a variable expression have the same variable part. Constant terms are also like terms.

For the expression $13ab + 4 - 2ab - 10$, $13ab$ and $-2ab$ are like variable terms, and 4 and -10 are like constant terms.

For the expression at the right, note that $5y^2$ and $-3y$ are not like terms because $y^2 = y \cdot y$, and $y \cdot y \neq y$. However, $6xy$ and $9yx$ are like variable terms because $xy = yx$ by the Commutative Property of Multiplication.

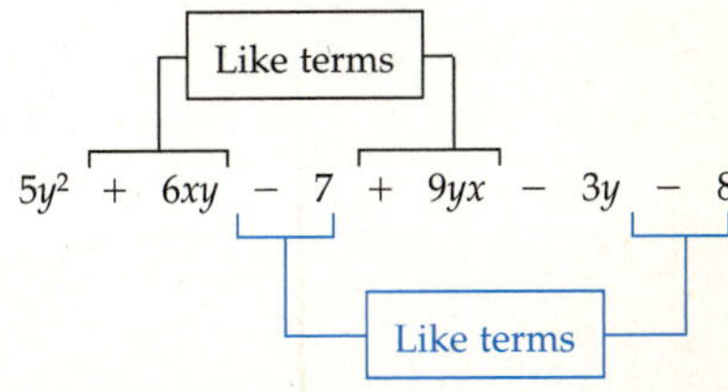

Variable expressions containing like terms are simplified by using the Distributive Property.

The Distributive Property

If a, b, and c are real numbers, then $a(b + c) = ab + ac$.

The Distributive Property can also be written as $ba + ca = (b + c)a$. This form is used to simplify variable expressions.

Simplify: $6c + 7c$

6c and 7c are like terms.
Use the Distributive Property. $6c + 7c = (6 + 7)c$

Simplify. $= 13c$

This example shows that to simplify a variable expression with like terms, add the coefficients of the like terms. The variable part remains unchanged. Adding or subtracting the like terms of a variable expression is called **combining like terms.**

Simplify: $2x - 7x$

Use the Distributive Property to add the numerical coefficients of like variable terms.

$2x - 7x = (2 - 7)x$
$= -5x$

CONSIDER THIS

Combining like terms can be related to many everyday experiences. For instance, 5 bricks plus 7 bricks is 12 bricks. But 5 bricks plus 7 nails is 5 bricks plus 7 nails.

Simplify: $8 + 3t$

The terms 8 and 3t are not like terms and cannot be combined. $8 + 3t$ is in simplest form.

The Properties of Addition can be used to simplify more complex variable expressions.

The Associative Property of Addition

If a, b, and c are real numbers, then $(a + b) + c = a + (b + c)$.

When three or more terms are added, the terms can be grouped (with parentheses, for example) in any order; the sum will be the same. For example,

$(4x + 6x) + 3x = 10x + 3x = 13x$
$4x + (6x + 3x) = 4x + 9x = 13x$

The Commutative Property of Addition

If a and b are real numbers, then $a + b = b + a$.

When two like terms are added, the terms can be added in either order; the sum will be the same. For example,

$$5y + (-7y) = -2y$$
$$(-7y) + 5y = -2y$$

The Addition Property of Zero

If a is a real number, then $a + 0 = 0 + a = a$.

The sum of a term and zero is the term. For example,

$$6x + 0 = 6x \quad \text{and} \quad 0 + 6x = 6x$$

The Inverse Property of Addition

If a is a real number, then $a + (-a) = (-a) + a = 0$.

The sum of a term and its opposite is zero.
The opposite of a number is called its **additive inverse.**
$-a$ is the opposite or additive inverse of a.
a is the opposite or additive inverse of $-a$.

$$9x + (-9x) = 0 \quad \text{and} \quad (-9x) + (9x) = 0$$

Simplify: $6a + 7 - 9a + 3$

Use the Commutative Property of Addition to rearrange terms so that like terms are together.	$6a + 7 - 9a + 3$ $= 6a + (-9a) + 7 + 3$
Use the Associative Property of Addition to group like terms.	$= [6a + (-9a)] + (7 + 3)$
Use the Distributive Property to add like variable terms. Add the constant terms.	$= [6 + (-9)]a + (7 + 3)$ $= -3a + 10$

In the above example, brackets [ ] were used as a grouping symbol to group $6a$ and $-9a$ and to group 6 and -9. Brackets were used because parentheses had already been used in the expression. The expression $[6 + (-9)]a$ is considered easier to read than $(6 + (-9))a$.

Multiply: $(5y)(3y)$

Use the Commutative and Associative Properties of Multiplication.

$$\begin{aligned}(5y)(3y) &= 5 \cdot y \cdot 3 \cdot y \\ &= 5 \cdot 3 \cdot y \cdot y \\ &= (5 \cdot 3)(y \cdot y)\end{aligned}$$

Multiply 5 times 3. Write $y \cdot y$ in exponential form. $= 15y^2$

By the Multiplication Property of One, the product of 1 and x is x.

$$1 \cdot x = x \qquad 1x = x$$

Just as the product of 1 and x is written x, the product of -1 and x is written $-x$.

$$-1 \cdot x = -x \qquad -1x = -x$$

Multiply: $(-2)(-x)$

Write $-x$ as $-1x$. $(-2)(-x) = (-2)(-1x)$

Use the Associative Property of Multiplication. $= [(-2)(-1)]x$

Multiply -2 times -1. $= 2x$

Example 3 Multiply: $-5(7b)$

Solution $-5(7b) = (-5 \cdot 7)b = -35b$

You Try It 3 Multiply: $-6(-3p)$

Your Solution

Example 4 Multiply: $(-4r)(-9t)$

Solution $(-4r)(-9t) = [(-4)(-9)](r \cdot t) = 36rt$

You Try It 4 Multiply: $(-2m)(-8n)$

Your Solution

Example 5 Multiply: $\left(\frac{3}{4}x\right)\left(\frac{8}{9}x\right)$

Solution $\left(\frac{3}{4}x\right)\left(\frac{8}{9}x\right) = \left(\frac{3}{4} \cdot \frac{8}{9}\right)(x \cdot x) = \frac{2}{3}x^2$

You Try It 5 Multiply: $\left(\frac{2}{5}a\right)\left(-\frac{3}{4}a\right)$

Your Solution

Solutions on p. A10

OBJECTIVE C The Distributive Property

Consider the numerical expression $6 \cdot (7 + 9)$. This expression can be evaluated by applying the Order of Operations Agreement.

	$6 \cdot (7 + 9)$
Simplify the expression inside the parentheses.	$= 6 \cdot 16$
Multiply.	$= 96$

There is an alternate method of evaluating this expression.

	$6 \cdot (7 + 9)$
Multiply each number inside the parentheses by 6	$= 6 \cdot 7 + 6 \cdot 9$
and add the products.	$= 42 + 54$
	$= 96$

Each method produced the same result. The second method uses the Distributive Property, which was introduced earlier in this section and is another of the Properties of Real Numbers.

> ***The Distributive Property***
>
> If a, b, and c are real numbers, then $a(b + c) = ab + ac$.

The Distributive Property is used to remove parentheses from a variable expression.

Simplify $3(5a + 4)$ by using the Distributive Property.

Use the Distributive Property.	$3(5a + 4) = 3(5a) + 3(4)$
Simplify.	$= 15a + 12$

Simplify $-4(2a + 3)$ by using the Distributive Property.

Use the Distributive Property.	$-4(2a + 3) = -4(2a) + (-4)(3)$
Simplify.	$= -8a + (-12)$
Rewrite addition of the opposite as subtraction.	$= -8a - 12$

The Distributive Property can also be stated in terms of subtraction.

$$a(b - c) = ab - ac$$

Simplify $5(2x - 4y)$ by using the Distributive Property.

Use the Distributive Property.	$5(2x - 4y) = 5(2x) - 5(4y)$
Simplify.	$= 10x - 20y$

Simplify $-3(2x - 8)$ by using the Distributive Property.

Use the Distributive Property.	$-3(2x - 8) = -3(2x) - (-3)(8)$
Simplify.	$= -6x - (-24)$
Rewrite the subtraction as addition of the opposite.	$= -6x + 24$

The Distributive Property can be extended to more than two addends inside the parentheses. For example

$$\begin{aligned} &4(2a + 3b - 5c) \\ &\quad = 4(2a) + 4(3b) - 4(5c) \\ &\quad = 8a + 12b - 20c \end{aligned}$$

Simplify: $-2(6x - 4y + 7)$

	$-2(6x - 4y + 7)$
Use the Distributive Property.	$= -2(6x) - (-2)(4y) + (-2)(7)$
Simplify.	$= -12x - (-8y) + (-14)$
	$= -12x + 8y - 14$

The Distributive Property is used to remove the parentheses from an expression that has a negative sign in front of the parentheses. Just as $-x = -1 \cdot x$, we can also write $-(x + y) = -1(x + y)$. Therefore

$$-(x + y) = -1(x + y) = -1x - 1y = -x - y$$

When a negative sign precedes parentheses, remove the parentheses and change the sign of *each* term inside the parentheses.

Rewrite the expression $-(4a - 3b + 7)$ without parentheses.

Remove the parentheses and change the sign of each term inside the parentheses.	$-(4a - 3b + 7)$ $= -4a + 3b - 7$

Example 6 Simplify by using the Distributive Property: $6(5c - 12)$

Solution $6(5c - 12) = 6(5c) - 6(12)$
$= 30c - 72$

You Try It 6 Simplify by using the Distributive Property: $-7(2k - 5)$

Your Solution

Example 7 Simplify by using the Distributive Property: $-4(-2a - b)$

Solution $-4(-2a - b) = -4(-2a) - (-4)(b)$
$= 8a + 4b$

You Try It 7 Simplify by using the Distributive Property: $-4(x - 2y)$

Your Solution

Solutions on pp. A10–A11

Example 8 Simplify by using the Distributive Property: $-2(3m - 8n + 5)$

Solution $-2(3m - 8n + 5)$
$= -2(3m) - (-2)(8n) + (-2)(5)$
$= -6m + 16n - 10$

You Try It 8 Simplify by using the Distributive Property: $3(-2v + 3w - 7)$

Your Solution

Example 9 Simplify by using the Distributive Property: $3a(2a + 6)$

Solution $3a(2a + 6) = 3a(2a) + 3a(6)$
$= (3 \cdot 2)(a \cdot a) + (3 \cdot 6)a$
$= 6a^2 + 18a$

You Try It 9 Simplify by using the Distributive Property: $-4z(2z - 7)$

Your Solution

Example 10 Rewrite $-(5x + 3y - 2z)$ without parentheses.

Solution $-(5x + 3y - 2z)$
$= -5x - 3y + 2z$

You Try It 10 Rewrite $-(c - 9d + 1)$ without parentheses.

Your Solution

Solutions on p. A11

OBJECTIVE D General variable expressions

General variable expressions are simplified by repeated use of the Properties of the Real Numbers.

Simplify: $7(2a - 4b) - 3(4a - 2b)$

Use the Distributive Property to remove parentheses.	$7(2a - 4b) - 3(4a - 2b)$ $= 14a - 28b - 12a + 6b$
Use the Commutative Property of Addition to rearrange terms.	$= 14a - 12a - 28b + 6b$
Combine like terms.	$= 2a - 22b$

To simplify variable expressions that contain grouping symbols within other grouping symbols, simplify the inner grouping symbols first.

Simplify: $2x - 4[3 - 2(6x + 5)]$

	$2x - 4[3 - 2(6x + 5)]$
Use the Distributive Property to remove the parentheses.	$= 2x - 4[3 - 12x - 10]$
Combine like terms inside the brackets.	$= 2x - 4[-12x - 7]$
Use the Distributive Property to remove the brackets.	$= 2x + 48x + 28$
Combine like terms.	$= 50x + 28$

Simplify: $2a^2 + 3[4(2a^2 - 5) - 4(3a - 1)]$

	$2a^2 + 3[4(2a^2 - 5) - 4(3a - 1)]$
Use the Distributive Property to remove both sets of parentheses.	$= 2a^2 + 3[8a^2 - 20 - 12a + 4]$
Combine like terms inside the brackets.	$= 2a^2 + 3[8a^2 - 12a - 16]$
Use the Distributive Property to remove the brackets.	$= 2a^2 + 24a^2 - 36a - 48$
Combine like terms.	$= 26a^2 - 36a - 48$

POINT OF INTEREST

Note, in Example 11 below, that we cannot first subtract $4 - 3$ because the Order of Operations Agreement requires multiplication before addition or subtraction.

Example 11 Simplify:
$4 - 3(2a - b) + 4(3a + 2b)$

Solution $4 - 3(2a - b) + 4(3a + 2b)$
$= 4 - 6a + 3b + 12a + 8b$
$= 6a + 11b + 4$

You Try It 11 Simplify:
$6 - 4(2x - y) + 3(x - 4y)$

Your Solution

Example 12 Simplify:
$7y - 4(2y - 3z) - (6y - 4z)$

Solution $7y - 4(2y - 3z) - (6y - 4z)$
$= 7y - 8y + 12z - 6y + 4z$
$= -7y + 16z$

You Try It 12 Simplify:
$8c - 4(3c - 8) - 5(c + 4)$

Your Solution

Example 13 Simplify:
$9v - 4[2(1 - 3v) - 5(2v + 4)]$

Solution $9v - 4[2(1 - 3v) - 5(2v + 4)]$
$= 9v - 4[2 - 6v - 10v - 20]$
$= 9v - 4[-16v - 18]$
$= 9v + 64v + 72$
$= 73v + 72$

You Try It 13 Simplify:
$6p + 5[3(2 - 3p) - 2(5 - 4p)]$

Your Solution

Solutions on p. A11

5.1 EXERCISES

Objective A

List the terms of the variable expression. Then underline the constant term.

1. $3x^2 + 4x - 9$
2. $-7y^2 - 2y + 6$
3. $b + 5$
4. $8n^2 - 1$

List the variable terms of the expression. Then underline the variable part of each term.

5. $9a^2 - 12a + 4b^2$
6. $6x^2y + 7xy^2 + 11$
7. $3x^2 + 16$
8. $-2n^2 + 5n - 8$

State the coefficients of the variable terms.

9. $x^2 - 6x - 7$
10. $-x + 15$
11. $12a^2 + 4ab - 1$
12. $x^2y - x + y$

Simplify by combining like terms.

13. $7a + 9a$
14. $8c + 15c$
15. $12x + 15x$
16. $9b + 24b$
17. $9z - 6z$
18. $12h - 4h$
19. $9x - x$
20. $12y - y$
21. $8z - 15z$
22. $2p - 13p$
23. $w - 7w$
24. $y - 9y$
25. $12v - 12v$
26. $11c - 11c$
27. $9s - 8t$
28. $6m - 5n$
29. $4x - 3y + 2x$
30. $3m - 6n + 4m$
31. $4r + 8p - 2r + 5p$
32. $-12t - 6s + 9t + 4s$
33. $9w - 5v - 12w + 7v$
34. $3c - 8 + 7c - 9$

Simplify by combining like terms.

35. $-4p + 9 - 5p + 2$

36. $-6y - 17 + 4y + 9$

37. $8p + 7 - 6p - 7$

38. $9m - 12 + 2m + 12$

39. $7h + 15 - 7h - 9$

40. $7v^2 - 9v + v^2 - 8v$

41. $9y^2 - 8 + 4y^2 + 9$

42. $r^2 + 4r - 8r - 5r^2$

43. $3w^2 - 7 - 9 + 9w^2$

44. $4c - 7c^2 + 8c - 8c^2$

45. $9w^2 - 15w + w - 9w^2$

46. $12v^2 + 15v - 14v - 12v^2$

47. $7a^2b + 5ab^2 - 2a^2b + 3ab^2$

48. $3xy^2 + 2x^2y - 7xy^2 - 4x^2y$

49. $8a - 9b + 2 - 8a + 9b + 3$

50. $10v + 12w - 9 - v - 12w + 9$

51. $6x^2 - 7x + 1 + 5x^2 + 5x - 1$

52. $4y^2 + 7y + 1 + y^2 - 10y + 9$

Objective B

Multiply.

53. $6(2x)$

54. $3(4y)$

55. $-5(3x)$

56. $-3(6z)$

57. $(3t) \cdot 7$

58. $(9r) \cdot 5$

59. $(-3p) \cdot 7$

60. $(-4w) \cdot 6$

Multiply.

61. $(-2)(-6q)$

62. $(-3)(-5m)$

63. $\frac{1}{2}(4x)$

64. $\frac{2}{3}(6n)$

65. $-\frac{5}{3}(9w)$

66. $-\frac{2}{5}(10v)$

67. $-\frac{1}{2}(-2x)$

68. $-\frac{1}{3}(-3x)$

69. $(2x)(3x)$

70. $(4k)(6k)$

71. $(-3x)(9x)$

72. $(4b)(-12b)$

73. $\left(\frac{1}{2}x\right)(2x)$

74. $\left(\frac{1}{3}h\right)(3h)$

75. $\left(-\frac{2}{3}\right)\left(x \cdot -\frac{3}{2}\right)$

76. $\left(-\frac{4}{3}\right)\left(z \cdot -\frac{3}{4}\right)$

77. $6\left(\frac{1}{6}c\right)$

78. $9\left(\frac{1}{9}v\right)$

79. $-5\left(-\frac{1}{5}a\right)$

80. $-9\left(-\frac{1}{9}s\right)$

81. $\frac{4}{5}w \cdot 15$

82. $\frac{7}{5}y \cdot 30$

83. $2v \cdot 8w$

84. $3m \cdot 7n$

Objective C

Simplify by using the Distributive Property.

85. $2(5z + 2)$

86. $3(4n + 5)$

87. $6(2y + 5z)$

88. $4(7a + 2b)$

89. $3(7x - 9)$

90. $9(3w - 7)$

91. $-(2x - 7)$

92. $-(3x + 4)$

93. $-(-4x - 9)$

Simplify by using the Distributive Property.

94. $-(-5y - 12)$

95. $-5(y + 3)$

96. $-4(x + 5)$

97. $-6(2x - 3)$

98. $-3(7y - 4)$

99. $-5(4n - 8)$

100. $-4(3c - 2)$

101. $-8(-6z + 3)$

102. $-2(-3k + 9)$

103. $-6(-4p - 7)$

104. $-5(-8c - 5)$

105. $5(2a + 3b + 1)$

106. $5(3x + 9y + 8)$

107. $4(3x - y - 1)$

108. $3(2x - 3y + 7)$

109. $9(4m - n + 2)$

110. $-4(3x + 2y - 5)$

111. $-6(-2v + 3w + 7)$

112. $5(4a - 5b + c)$

113. $-4(-2m - n + 3)$

114. $-6(3p - 2r - 9)$

Objective D

Simplify.

115. $5x + 2(x + 1)$

116. $6y + 2(2y + 3)$

117. $9n - 3(2n - 1)$

118. $12x - 2(4x - 6)$

119. $7a - (3a - 4)$

120. $9m - 4(2m - 3)$

Simplify.

121. $7 + 2(2a - 3)$

122. $5 + 3(2y - 8)$

123. $6 + 4(2x + 9)$

124. $4 + 3(7d + 7)$

125. $8 - 4(3x - 5)$

126. $13 - 7(4y + 3)$

127. $2 - 9(2m + 6)$

128. $4 - 7(6w - 9)$

129. $3(6c + 5) + 2(c + 4)$

130. $7(2k - 5) + 3(4k - 3)$

131. $2(a - 2b) + 3(2a + 3b)$

132. $4(3x - 6y) + 5(2x - 3y)$

133. $6(7z - 5) - 3(9z - 6)$

134. $8(2t + 4) - 4(3t - 1)$

135. $-2(6y + 2) + 3(4y - 5)$

136. $-3(2a - 5) - 2(4a + 3)$

137. $-5(x - 2y) - 4(2x + 3y)$

138. $-6(-x - 3y) - 2(-3x + 9y)$

139. $2 - 3(2v - 1) + 2(2v + 4)$

140. $5 - 2(3x + 5) - 3(4x - 1)$

141. $2c - 3(c + 4) - 2(2c - 3)$

142. $5m - 2(3m + 2) - 4(m - 1)$

143. $8a + 3(2a - 1) + 6(4 - 2a)$

144. $9z - 2(2z - 7) + 4(3 - 5z)$

145. $3n - 2[5 - 2(2n - 4)]$

146. $6w + 4[3 - 5(6w - 2)]$

147. $9x - 3[8 - 2(5 - 3x)]$

Simplify.

148. $11y - 7[2(2y - 5) + 3(7 - 5y)]$

149. $-3v - 6[2(3 - 2v) - 5(3v - 7)]$

150. $7y^2 + 2[3y(2y - 4) + 3(2y^2)]$

151. $9z^2 - 3[4(2z + 3) - 3z(2z - 6)]$

Critical Thinking

152. The square and the rectangle at the right can be used to illustrate algebraic expressions. Note at the right the expression for $2x + 1$. The expression below is $3(x + 2)$.

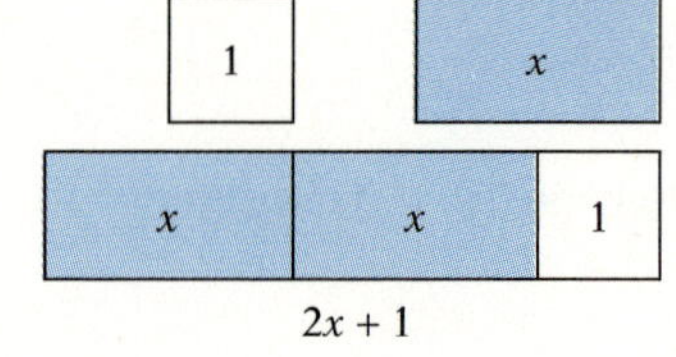

x	1	1	x	1	1	x	1	1

Rearrange these rectangles so that the x's are together and the 1's are together. Write a mathematical expression for the rearranged figure. Using similar squares and rectangles, draw figures that represent the expressions $2 + 3x$, $5x$, $2(2x + 3)$, $4x + 3$, and $4x + 6$. Does the area of the figure for $2(2x + 3)$ equal the area of the figure for $4x + 6$? How does this relate to the Distributive Property? Does the area of the figure for $2 + 3x$ equal the area of the figure for $5x$? How does this relate to combining like terms?

153. Determine whether the statement is true or false. If the statement is false, give an example to show it is false.

a. Division is a commutative operation.
b. Division is an associative operation.
c. Subtraction is an associative operation.
d. Subtraction is a commutative operation.

154. Does every real number have an additive inverse? If not, which real numbers do not have an additive inverse?

155. Does every real number have a multiplicative inverse? If not, which real numbers do not have a multiplicative inverse?

156. [W] Explain why the simplification of the expression $2 + 3(2x + 4)$ shown at the right is incorrect. What is the correct simplification?

Why is this incorrect?

$$2 + 3(2x + 4) = 5(2x + 4)$$
$$= 10x + 20$$

157. [W] Simplifying variable expressions requires combining like terms. Give some examples of how this applies to everyday experience.

SECTION 5.2 Perimeter and Area

OBJECTIVE 

Perimeter of a plane geometric figure

POINT OF INTEREST

Geometry is one of the oldest branches of mathematics. About 350 B.C., the Greek mathematician Euclid wrote the *Elements,* which contained all the known concepts of geometry. Euclid's contribution was to unify various concepts into a single deductive system that was based on a set of axioms.

The word *geometry* comes from the Greek words for *earth* and *measure.* The original purpose of geometry was to measure land. Today geometry is used in many fields, such as physics, medicine, and geology. Geometry is used in applied fields such as mechanical drawing and astronomy. Geometric forms are used in art and design.

Three basic concepts of geometry are point, line, and plane. A **point** is symbolized by drawing a dot. A **line** is determined by two distinct points and extends indefinitely in both directions, as the arrows on the line shown at the right indicate. This line contains points A and B and is represented by $\overleftrightarrow{AB}$. A line can also be represented by a single letter, such as ℓ.

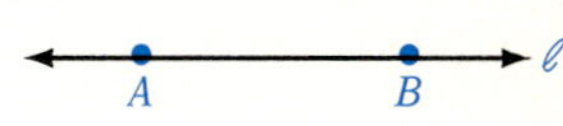

A **ray** starts at a point and extends indefinitely in *one* direction. The point at which a ray starts is called the **endpoint** of the ray. The ray shown at the right is denoted by $\overrightarrow{AB}$. Point A is the endpoint of the ray.

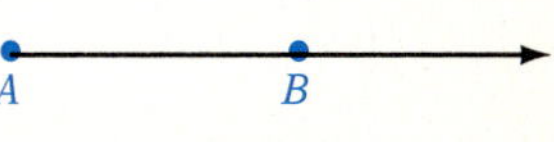

A **line segment** is part of a line and has two endpoints. The line segment shown at the right is denoted by $\overline{AB}$.

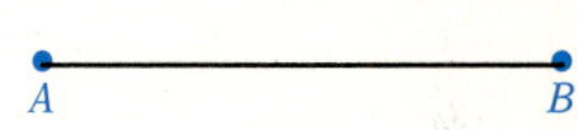

In this section we discuss figures that lie in a plane. A **plane** is a flat surface and can be pictured as a table top or blackboard that extends in all directions. Figures that lie in a plane are called **plane figures.**

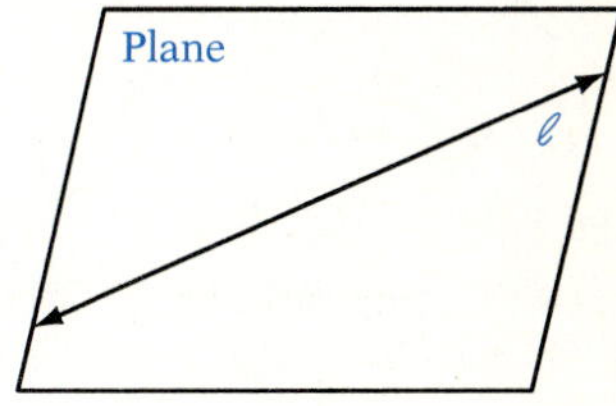

Lines in a plane can be intersecting or parallel. **Intersecting lines** cross at a point in the plane. **Parallel lines** never meet. The distance between them is always the same.

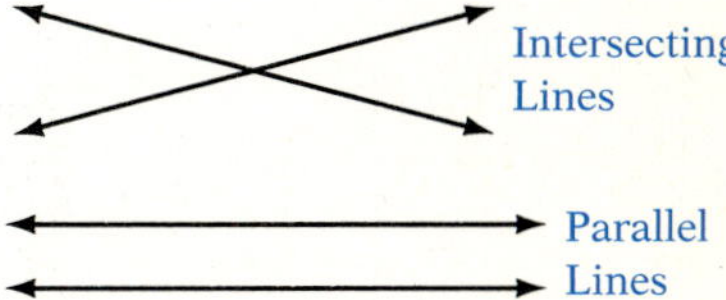

The symbol $\parallel$ means "is parallel to." In the figure at the right, $j \parallel k$ and $\overline{AB} \parallel \overline{CD}$. Note that j contains $\overline{AB}$ and k contains $\overline{CD}$. Parallel lines contain parallel line segments.

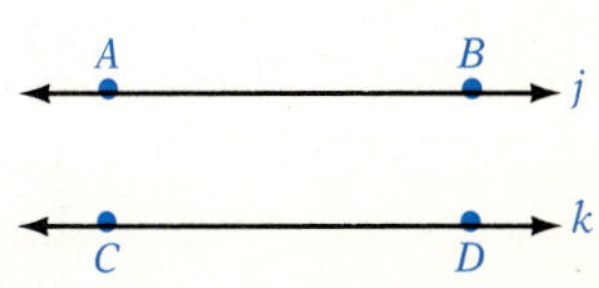

An **angle** is formed by two rays with the same endpoint. The **vertex** of the angle is the point at which the two rays meet. The rays are called the **sides** of the angle.

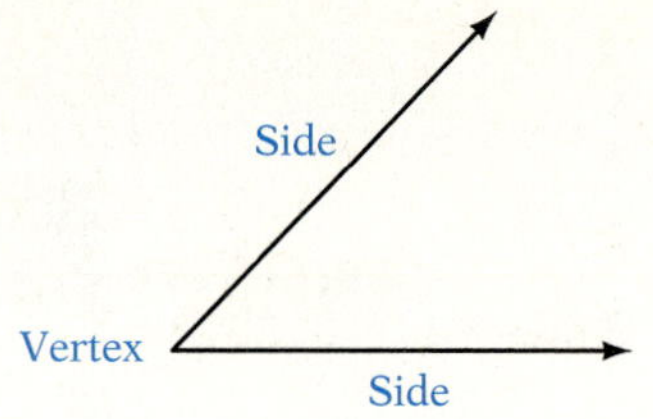

If A and C are points on rays r_1 and r_2, and B is the vertex, then the angle is called $\angle B$ or $\angle ABC$, where $\angle$ is the symbol for angle. Note that the angle is named by the vertex, or the vertex is the second point listed when the angle is named by giving three points. $\angle ABC$ could also be called $\angle CBA$

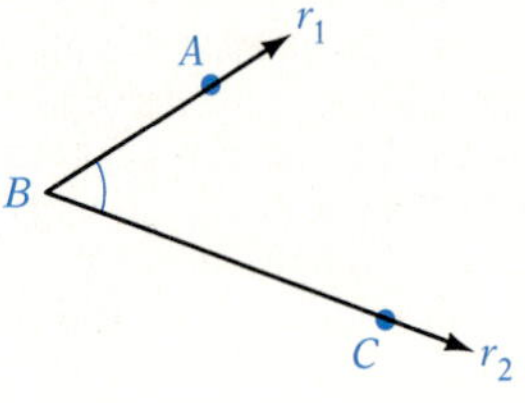

An angle can also be named by a variable written between the rays close to the vertex. In the figure at the right, $\angle x = \angle QRS$ and $\angle y = \angle SRT$. Note that in this figure, more than two rays meet at R. In this case, the vertex cannot be used to name the angle.

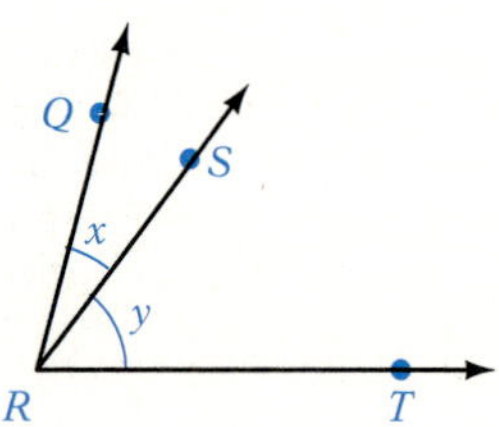

POINT OF INTEREST

The first woman mathematician for which documented evidence exists is Hypatia (370–415). She lived in Alexandria, Egypt, and lectured at the Museum, the forerunner of our modern university. She made important contributions in mathematics, astronomy, and philosophy.

Angles are measured in **degrees.** The symbol for degrees is a small raised circle, °. Probably because early Babylonians believed that Earth revolves around the sun in approximately 360 days, the angle formed by a circle has a measure of 360° (360 degrees).

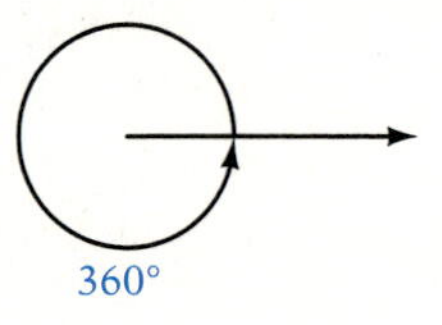

CONSIDER THIS

The corner of a page of this book is a good model for a 90° angle.

A 90° angle is called a **right angle.** The symbol ∟ represents a right angle.

Perpendicular lines are intersecting lines that form right angles.

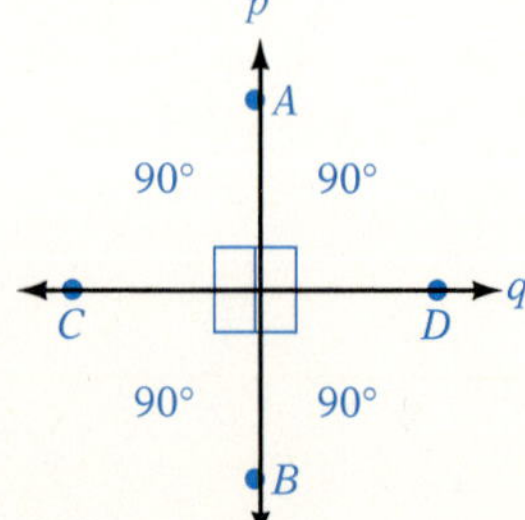

The symbol $\perp$ means "is perpendicular to." In the figure at the right, $p \perp q$ and $\overline{AB} \perp \overline{CD}$. Note that line p contains $\overline{AB}$ and line q contains $\overline{CD}$. Perpendicular lines contain perpendicular line segments.

Complementary angles are two angles whose measures have the sum 90°

$$\angle A + \angle B = 70° + 20° = 90°$$

$\angle A$ and $\angle B$ are complementary angles.

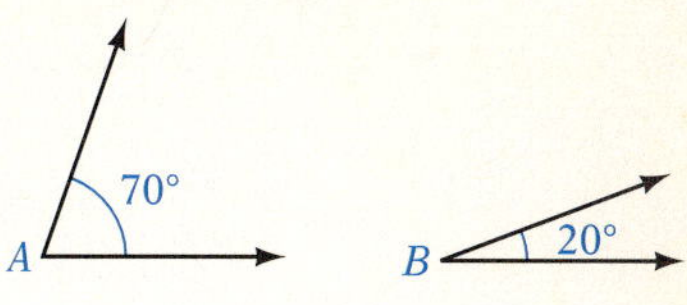

A 180° angle is called a **straight angle.**

$\angle AOB$ is a straight angle.

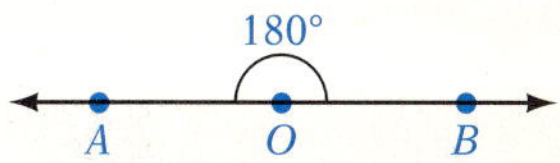

Supplementary angles are two angles whose measures have the sum 180°

$$\angle A + \angle B = 130° + 50° = 180°$$

$\angle A$ and $\angle B$ are supplementary angles.

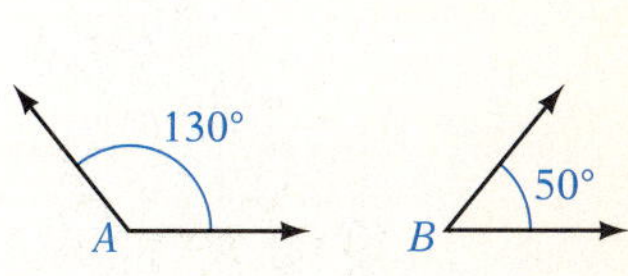

An **acute angle** is an angle whose measure is between 0° and 90°. $\angle B$ above is an acute angle. An **obtuse angle** is an angle whose measure is between 90° and 180°. $\angle A$ above is an obtuse angle.

A **polygon** is a closed figure formed by three or more line segments that lie in a plane. The line segments that form the polygon are called its **sides.** The figures below are examples of polygons.

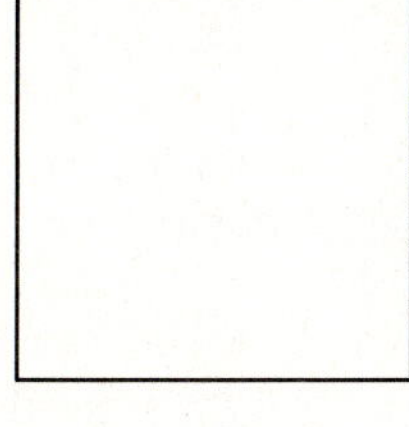

A

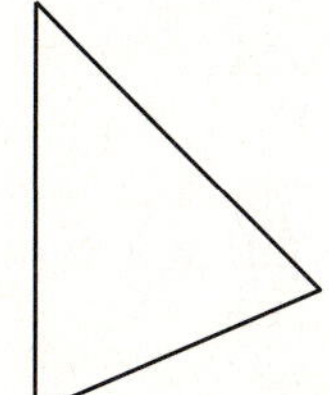

B

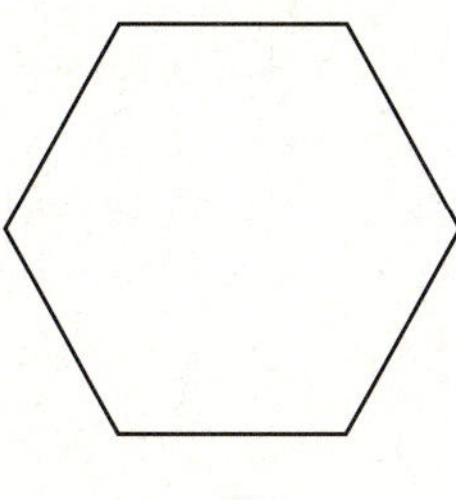

C

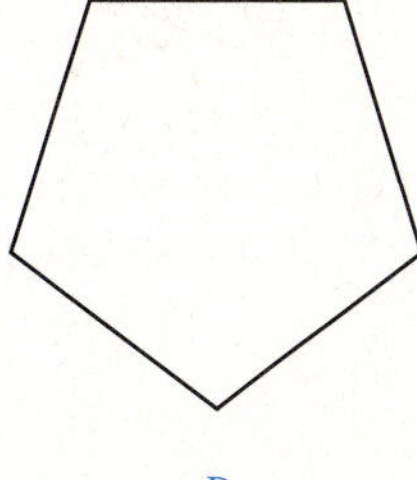

D

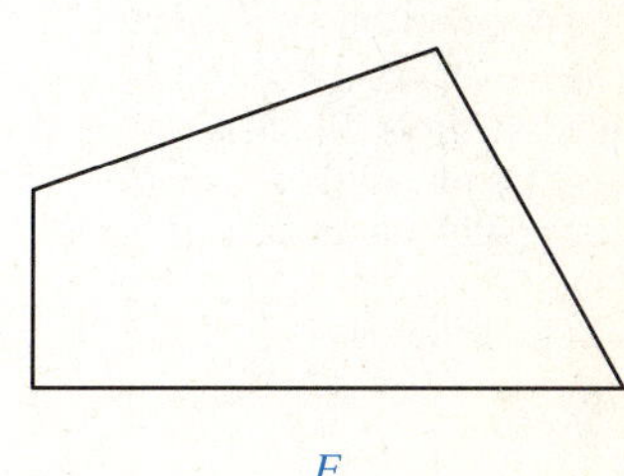

E

POINT OF INTEREST

Although a polygon is defined in terms of its *sides* (see the definition above), the word actually comes from the Latin word *polygonum* which means having many *angles.* This is certainly the case for a polygon.

In a **regular polygon,** each side has the same length and each angle has the same measure. The polygons in Figures *A*, *C*, and *D* above are regular polygons.

The name of a polygon tells the number of its sides. The table below lists the names of polygons that have 3 to 10 sides.

Number of Sides	Polygon
3	Triangle
4	Quadrilateral
5	Pentagon
6	Hexagon
7	Heptagon
8	Octagon
9	Nonagon
10	Decagon

Triangles and quadrilaterals are two of the most common polygons. Triangles are distinguished by the number of equal sides and by the measures of their angles.

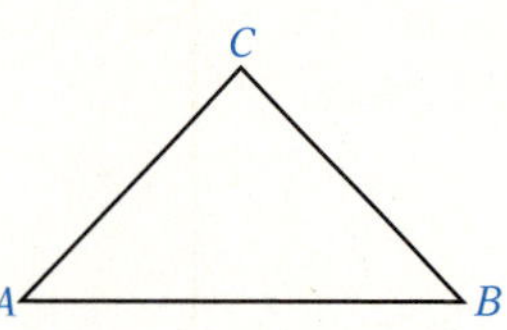

An **isosceles triangle** has two sides of equal length. The angles opposite the equal sides are of equal measure.
$AC = BC$
$\angle A = \angle B$

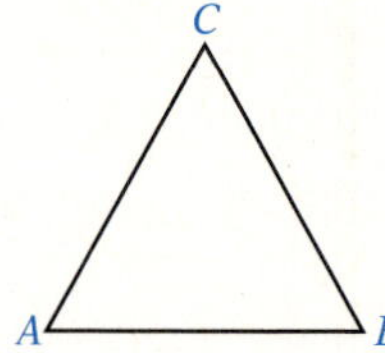

The three sides of an **equilateral triangle** are of equal length. The three angles are of equal measure.
$AB = BC = AC$
$\angle A = \angle B = \angle C$

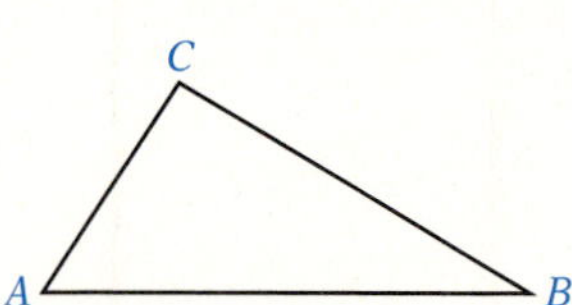

A **scalene triangle** has no two sides of equal length. No two angles are of equal measure.

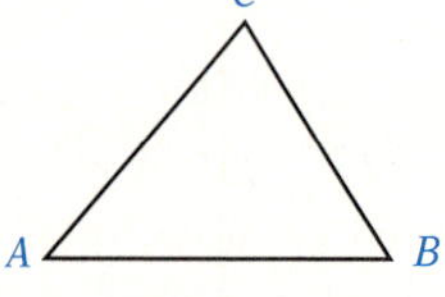

An **acute triangle** has three acute angles.

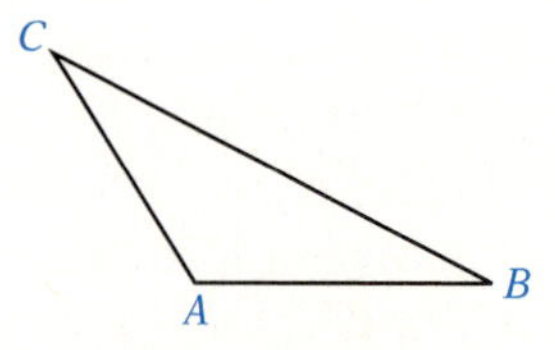

An **obtuse triangle** has one obtuse angle.

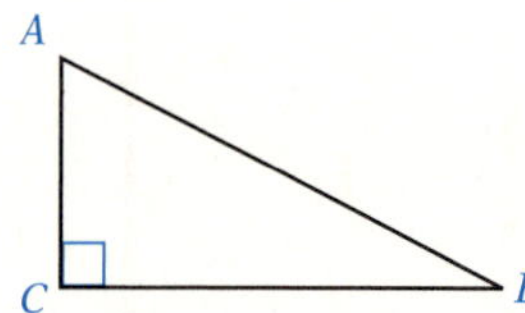

A **right triangle** has a right angle.

POINT OF INTEREST

The diagram below shows the relationships between all quadrilaterals. The description of each quadrilateral is within a drawing of that quadrilateral.

Quadrilaterals are also distinguished by their sides and angles, as shown below. Note that a rectangle, a square, and a rhombus are different forms of a parallelogram.

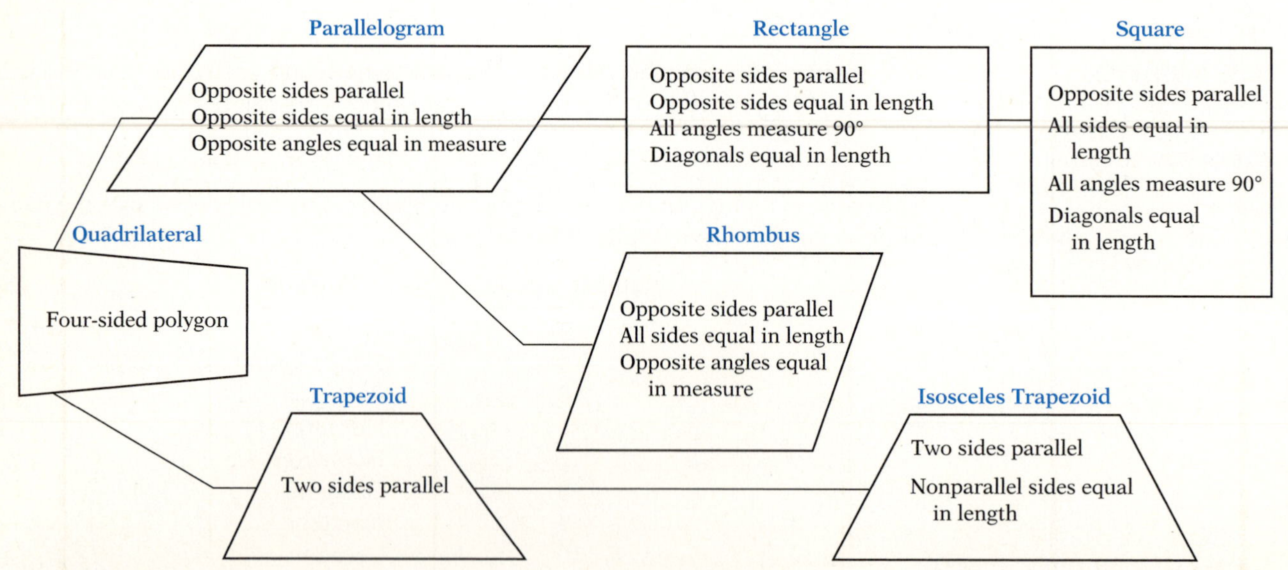

The **perimeter** of a plane geometric figure is a measure of the distance around the figure. Perimeter is used in buying fencing for a lawn or calculating how much baseboard is needed for a room.

The perimeter of a triangle is the sum of the lengths of its three sides.

Perimeter of a Triangle

Let a, b, and c be the lengths of the sides of a triangle. The perimeter, P, of the triangle is given by $P = a + b + c$.

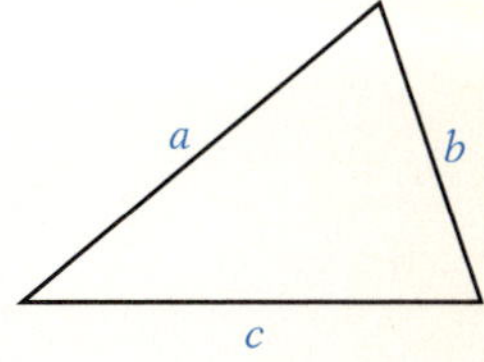

Find the perimeter of the triangle shown at the right.

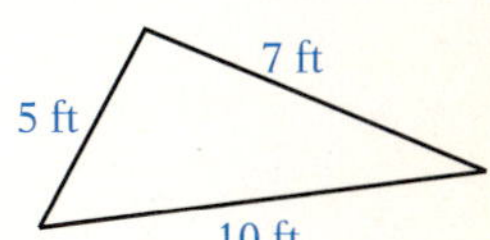

$$P = 5 + 7 + 10 = 22$$

The perimeter is 22 ft.

The perimeter of a quadrilateral is the sum of the lengths of its four sides.

POINT OF INTEREST

Leonardo DaVinci painted the Mona Lisa on a rectangular canvas whose height was approximately 1.6 times its width. Rectangles with these proportions, called golden rectangles, were used extensively in Renaissance art.

A rectangle is a quadrilateral with opposite sides of equal length. Usually the length, L, of a rectangle refers to the length of one of the longer sides of the rectangle, and the width, W, refers to the length of one of the shorter sides. The perimeter can then be represented $P = L + W + L + W$.

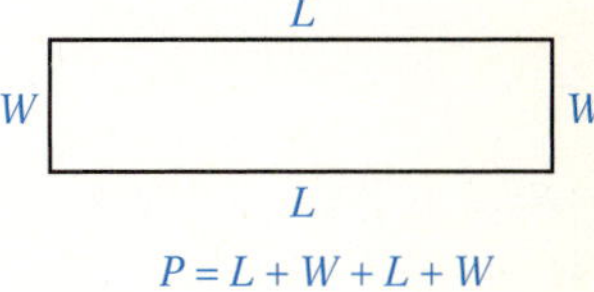

The formula for the perimeter of a rectangle is derived by combining like terms.

$$P = 2L + 2W$$

Perimeter of a Rectangle

Let L represent the length and W the width of a rectangle. The perimeter, P, of the rectangle is given by $P = 2L + 2W$.

Find the perimeter of the rectangle shown at the right.

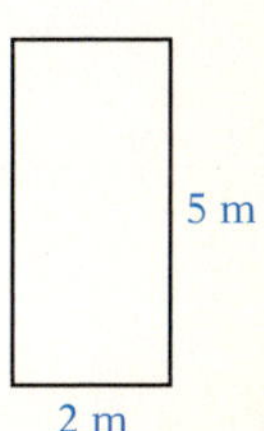

The length is 5 m. Substitute 5 for L. $\quad P = 2L + 2W$

The width is 2 m. Substitute 2 for W. $\quad P = 2(5) + 2(2)$

Solve for P. $\quad P = 10 + 4$

$P = 14$

The perimeter is 14 m.

A square is a rectangle in which each side has the same length. Letting s represent the length of each side of a square, the perimeter of a square is $P = s + s + s + s$.

$P = s + s + s + s$

The formula for the perimeter of a square is derived by combining like terms.

$P = 4s$

> ***Perimeter of a Square***
>
> Let s represent the length of a side of a square. The perimeter, P, of the square is given by $P = 4s$.

Find the perimeter of the square shown at the right.

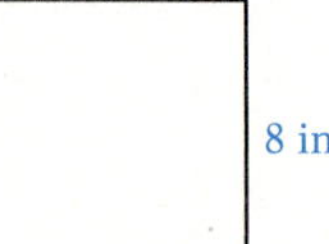

$P = 4s = 4(8) = 32$

The perimeter is 32 in.

A **circle** is a plane figure in which all points are the same distance from point O, called the **center** of the circle.

A **diameter** of a circle is a line segment across the circle through point O. AB is a diameter of the circle at the right. The variable d is used to designate the length of a diameter of a circle.

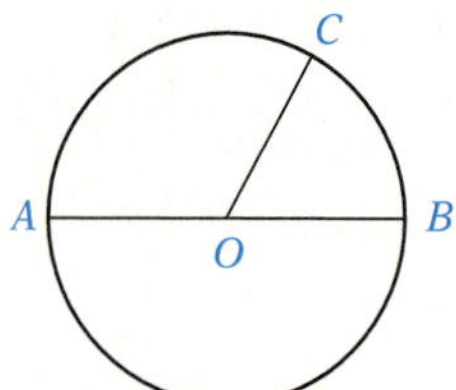

A **radius** of a circle is a line segment from the center of the circle to a point on the circle. OC is a radius of the circle at the right. The variable r is used to designate the length of a radius of a circle.

The length of the diameter is twice the length of the radius.

$d = 2r$ or $r = \frac{1}{2}d$

The distance around a circle is called the **circumference.** The circumference, C, of a circle is equal to the product of π (pi) and the diameter.

$C = \pi d$

Because $d = 2r$, the formula for the circumference can be written in terms of r.

$C = 2\pi r$

> ***The Circumference of a Circle***
>
> The circumference, C, of a circle with diameter d and radius r is given by $C = \pi d$ or $C = 2\pi r$.

POINT OF INTEREST

Archimedes (c. 287–212 B.C.) was the mathematician who gave us the approximate value of π as $\frac{22}{7} = 3\frac{1}{7}$. He actually showed that π was between $3\frac{10}{71}$ and $3\frac{1}{7}$. The approximation $3\frac{10}{71}$ is actually closer to the exact value of π, but it is more difficult to use.

The formula for circumference uses the number π, which is an irrational number. The value of π can be approximated by a fraction or by a decimal.

$$\pi \approx \frac{22}{7} \text{ or } \pi \approx 3.14$$

The π key on a scientific calculator gives a closer approximation of π than 3.14. Use a scientific calculator to find approximate values in calculations involving π.

Find the circumference of a circle with a diameter of 6 in.

The diameter of the circle is given. Use the circumference formula that involves the diameter. $d = 6$.

$C = \pi d$
$C = \pi(6)$

The exact circumference of the circle is 6π in.

$C = 6\pi$

An approximate measure is found by using the π key on a calculator.

$C \approx 18.85$

The approximate circumference is 18.85 in.

Example 1

The dimensions of a triangular sail are 18 ft, 11 ft, and 15 ft. What is the perimeter of the sail?

Strategy

To find the perimeter, use the formula for the perimeter of a triangle. Substitute 18 for a, 11 for b, and 15 for c. Solve for P.

Solution

$P = a + b + c$
$P = 18 + 11 + 15$
$P = 44$

The perimeter of the sail is 44 ft.

You Try It 1

What is the perimeter of a standard sheet of typing paper that measures $8\frac{1}{2}$ in. by 11 in.?

Your Strategy

Your Solution

Example 2

Find the circumference of a circle with a radius of 15 cm. Round to the nearest hundredth.

Strategy

To find the circumference, use the formula that gives the circumference in terms of the radius. An approximation is asked for; use the π key on a calculator. $r = 15$.

Solution

$C = 2\pi r = 2\pi(15) = 30\pi \approx 94.25$

The circumference is about 94.25 cm.

You Try It 2

Find the circumference of a circle with a diameter of 9 in. Give the exact measure.

Your Strategy

Your Solution

Solutions on p. A11

OBJECTIVE B Area of a plane geometric figure

Area is the amount of surface in a region. Area can be used to describe the size of a rug, a parking lot, a farm, or a national park. Area is measured in square units.

A square that measures 1 in. on each side has an area of 1 square inch, written 1 in^2.

A square that measures 1 cm on each side has an area of 1 square centimeter, written 1 cm^2.

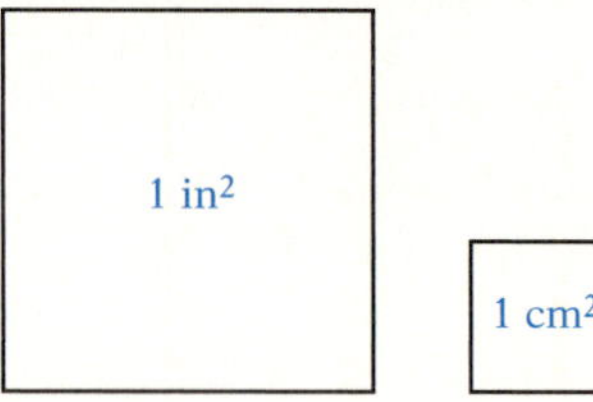

Larger areas can be measured in square feet (ft^2), square meters (m^2), square miles (mi^2), acres (43,560 ft^2), or any other square unit.

The area of a geometic figure is the number of squares that are necessary to cover the figure. In the figures below, two rectangles have been drawn and covered with squares. In the figure on the left, 12 squares, each of area 1 cm^2, were used to cover the rectangle. The area of the rectangle is 12 cm^2. In the figure on the right, 6 squares, each of area 1 in^2, were used to cover the rectangle. The area of the rectangle is 6 in^2.

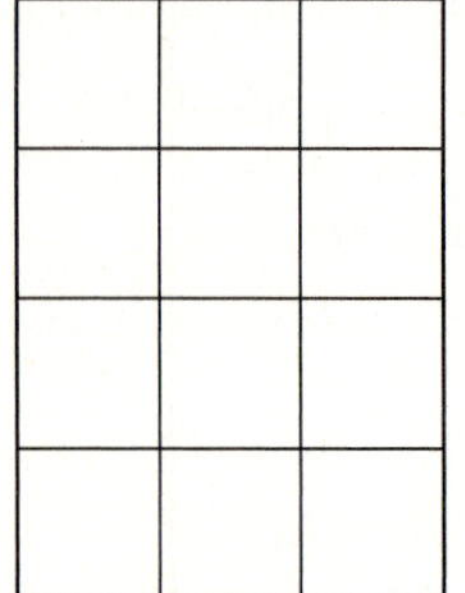

The area of the rectangle is 12 cm^2.

The area of the rectangle is 6 in^2.

Note from the above figures that the area of a rectangle can be found by multiplying the length of the rectangle by its width.

> ***Area of a Rectangle***
>
> Let L represent the length and W the width of a rectangle. The area, A, of the rectangle is given by $A = LW$.

Find the area of the rectangle shown at the right.

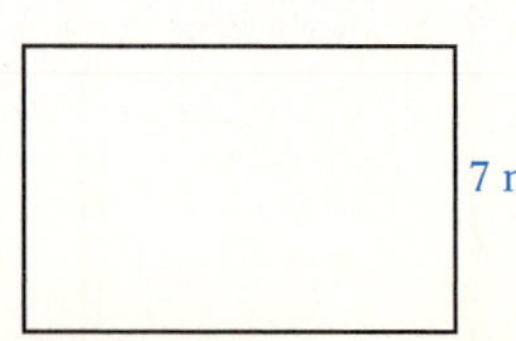

$A = LW = 11(7) = 77$

The area is 77 m^2.

A square is a rectangle in which all sides are the same length. Therefore, both the length and the width of a square can be represented by s, and $A = LW = s \cdot s = s^2$.

Area of a Square

Let s represent the length of a side of a square. The area, A, of the square is given by $A = s^2$.

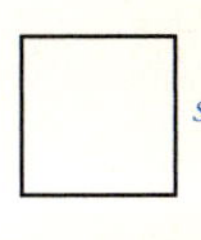

$A = s \cdot s = s^2$

Find the area of the square shown at the right.

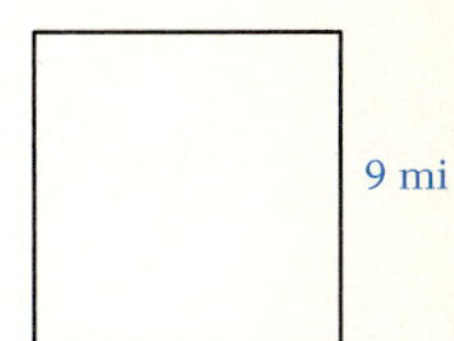

$A = s^2 = 9^2 = 81$

The area is 81 mi^2.

Figure $ABCD$ is a parallelogram. BC is the **base,** b, of the parallelogram. AE, perpendicular to the base, is the **height,** h, of the parallelogram.

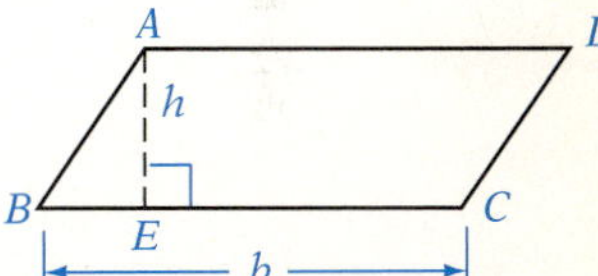

Any side of a parallelogram can be designated as the base. The corresponding height is found by drawing a line segment perpendicular to the base from the opposite side.

A rectangle can be formed from a parallelogram by cutting a right triangle from one end of the parallelogram and attaching it to the other end. The area of the resulting rectangle will equal the area of the original parallelogram.

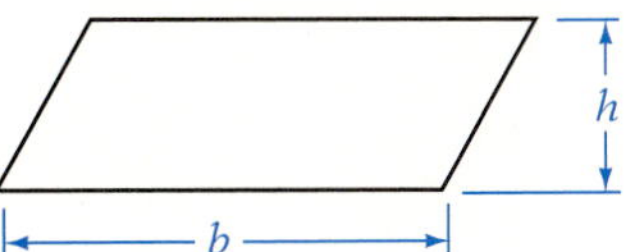

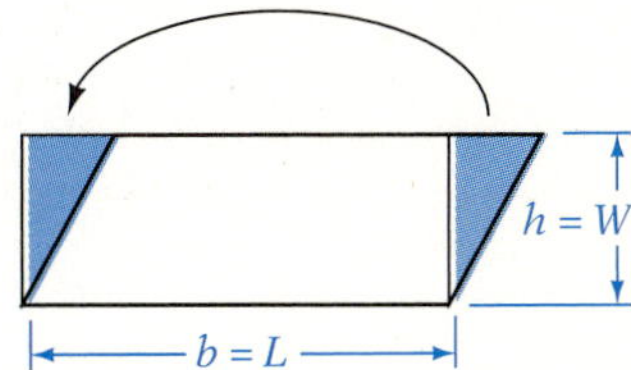

Area of a Parallelogram

Let b represent the length of the base and h the height of a parallelogram. The area, A, of the parallelogram is given by $A = bh$.

Find the area of the parallelogram shown at the right.

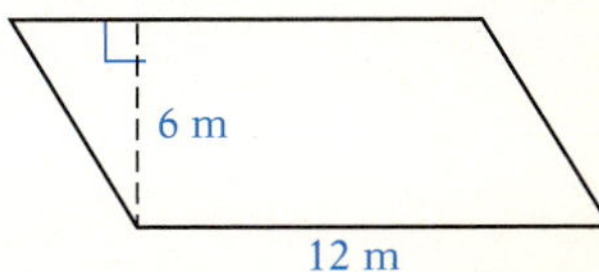

$A = bh = 12 \cdot 6 = 72$

The area is 72 m^2.

Figure ABC is a triangle. AB is the **base,** b, of the triangle. CD, perpendicular to the base, is the **height,** h, of the triangle.

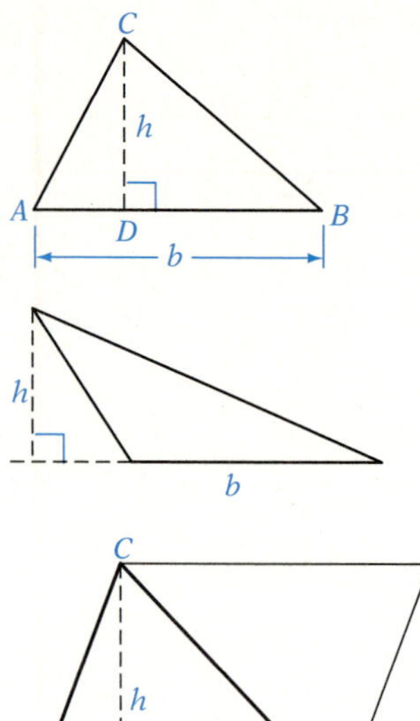

Any side of a triangle can be designated as the base. The corresponding height is found by drawing a line segment perpendicular to the base from the vertex opposite the base.

Consider the triangle with base b and height h shown at the right. By extending a line from C parallel to the base AB and equal in length to the base, a parallelogram is formed. The area of the parallelogram is bh and is twice the area of the triangle. Therefore, the area of the triangle is one-half the area of the parallelogram, or $\frac{1}{2}bh$.

Area of a Triangle

Let b represent the length of the base and h the height of a triangle. The area, A, of the triangle is given by $A = \frac{1}{2}bh$.

Find the area of a triangle with a base of 18 cm and a height of 6 cm.

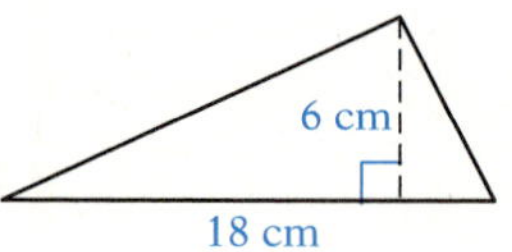

$$A = \frac{1}{2}bh = \frac{1}{2} \cdot 18 \cdot 6 = 54$$

The area is 54 cm^2.

Figure $ABCD$ is a trapezoid. AB is one **base** of the trapezoid and CD is the other base. The **height** is the length of AE, perpendicular to the two bases.

In the trapezoid at the right, the line segment BD divides the trapezoid into two triangles, ABD and BCD. In triangle ABD, b_1 is the base and h is the height. In triangle BCD, b_2 is the base and h is the height. The area of the trapezoid is the sum of the areas of the two triangles.

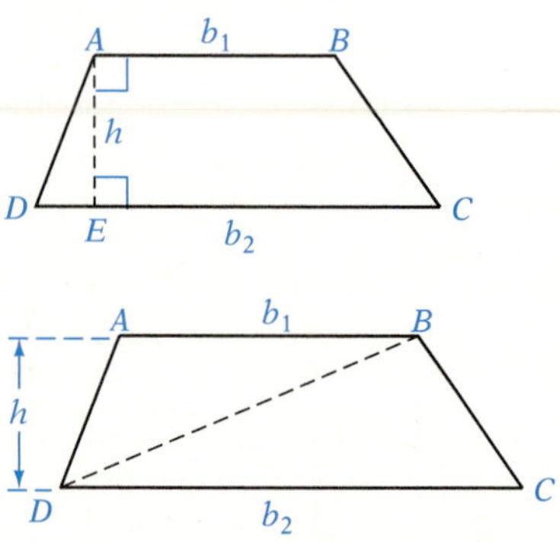

$$\text{Area of trapezoid } ABCD = \text{Area of triangle } ABD + \text{Area of triangle } BCD$$
$$= \frac{1}{2}b_1h + \frac{1}{2}b_2h = \frac{1}{2}h(b_1 + b_2)$$

Area of a Trapezoid

Let b_1 and b_2 represent the lengths of the bases and h the height of a trapezoid. The area, A, of the trapezoid is given by

$A = \frac{1}{2}h(b_1 + b_2)$.

Find the area of a trapezoid that has bases measuring 15 in. and 5 in. and a height of 8 in.

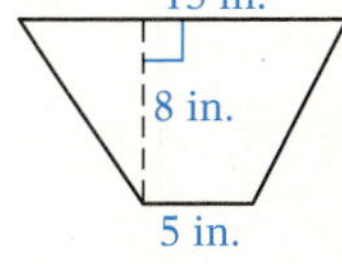

$$A = \frac{1}{2}h(b_1 + b_2)$$
$$= \frac{1}{2} \cdot 8(15 + 5) = 4(20) = 80$$

The area is 80 in^2.

The area of a circle is equal to the product of π and the square of the radius.

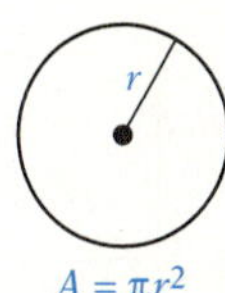

Area of a Circle

The area, A, of a circle with radius r is given by $A = \pi r^2$.

Find the area of a circle that has a radius of 6 cm.

Use the formula for the area of a circle. $r = 6$

$$A = \pi r^2$$
$$A = \pi(6)^2$$
$$A = \pi(36)$$

The exact area of the circle is 36π cm^2.

$$A = 36\pi$$

An approximate measure is found by using the π key on a calculator.

$$A \approx 113.10$$

The approximate area of the circle is 113.10 cm^2.

For your reference, all of the formulas for the perimeter and the area of the figures presented in this section are in the Chapter Summary on page 234.

Example 3

The Parks and Recreation Department of a city plans to plant grass seed in a playground that has the shape of a trapezoid, as shown below. Each bag of grass seed seeds 1500 ft^2. How many bags of grass seed should the department purchase?

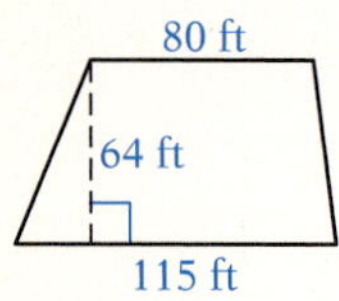

Strategy

To find the number of bags to be purchased:

- Use the formula for the area of a trapezoid to find the area of the playground.
- Divide the area of the playground by the area one bag will seed (1500).

Solution

$$A = \frac{1}{2}h(b_1 + b_2)$$

$$A = \frac{1}{2} \cdot 64(80 + 115)$$

$A = 6240$ The area of the playground is 6240 ft^2.

$6240 \div 1500 = 4.16$

Although only a portion of a fifth bag is needed, 5 bags of grass seed must be purchased.

You Try It 3

An interior designer decides to wallpaper two walls of a room. Each roll of wallpaper will cover 30 ft^2. Each wall measures 8 ft by 12 ft. How many rolls of wallpaper should be purchased?

Your Strategy

Your Solution

Example 4

Find the area of a circle with a diameter of 5 ft. Give the exact measure.

Strategy

To find the area:

- Find the radius of the circle.
- Use the formula for the area of a circle. Leave the answer in terms of π.

Solution

$$r = \frac{1}{2}d = \frac{1}{2}(5) = 2.5$$

$$A = \pi r^2 = \pi(2.5)^2 = \pi(6.25) = 6.25\pi$$

The area of the circle is 6.25π ft^2.

You Try It 4

Find the area of a circle with a radius of 11 cm. Round to the nearest hundredth.

Your Strategy

Your Solution

Solutions on p. A000

5.2 EXERCISES

Objective A

Find the perimeter of the figure.

1. 12 in. 20 in. 24 in.

2.

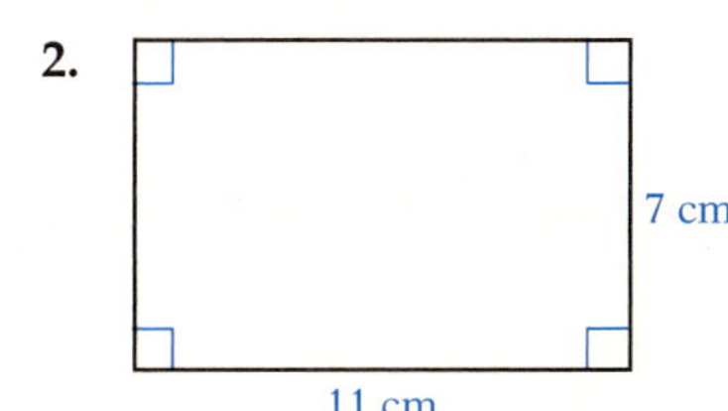

3.

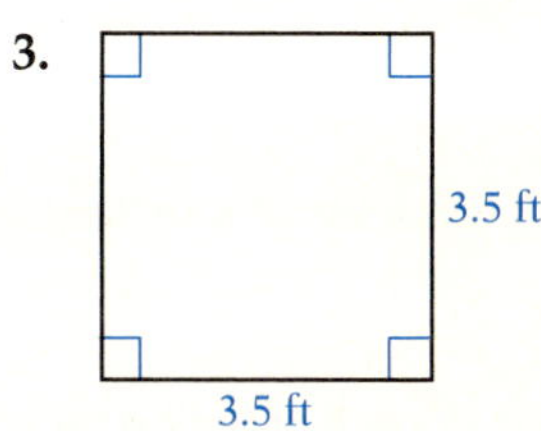

4. 9 m 12 m 8 m 10 m

5.

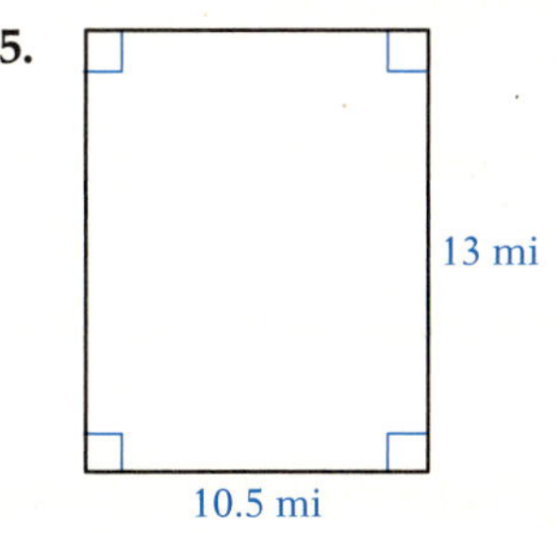

6.

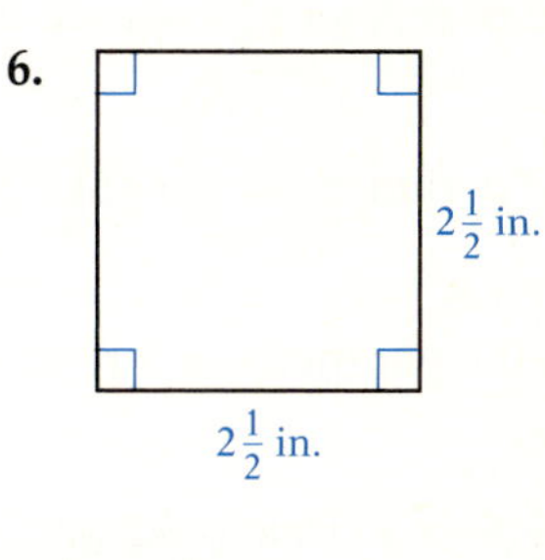

Find the circumference of the figure. Give both the exact value and an approximation to the nearest hundredth.

7. 4 cm

8.

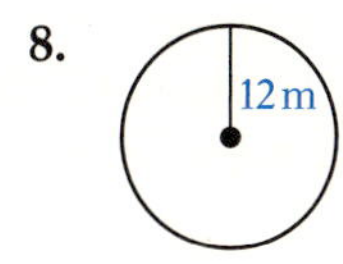

9.

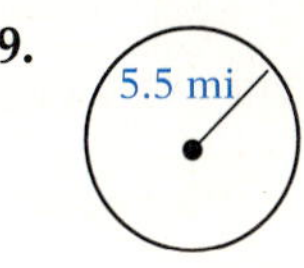

10. 18 in.

11. 17 ft

12.

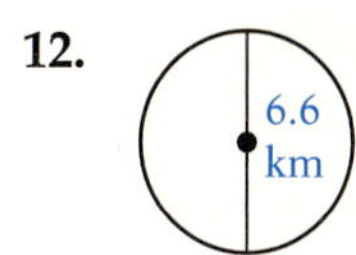

Solve.

13. The lengths of the three sides of a triangle are 3.8 cm, 5.2 cm, and 8.4 cm. Find the perimeter of the triangle.

14. The lengths of the three sides of a triangle are 7.5 m, 6.1 m, and 4.9 m. Find the perimeter of the triangle.

15. The length of each of two sides of an isosceles triangle is $2\frac{1}{2}$ cm. The third side measures 3 cm. Find the perimeter of the triangle.

Find the area of the figure. Give both the exact value and an approximation to the nearest hundredth.

45.
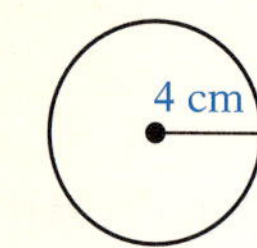

46.
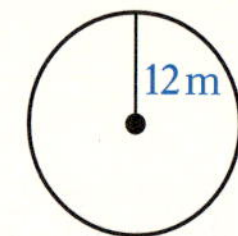

47.

48.
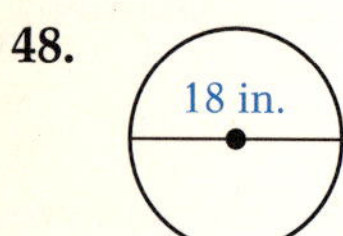

49.
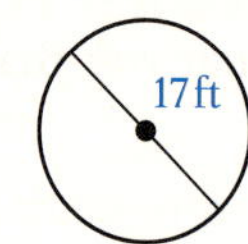

50.

Solve.

51. The length of a side of a square is 12.5 cm. Find the area of the square.

52. Each side of a square measures $3\frac{1}{2}$ in. Find the area of the square.

53. The length of a rectangle is 38 in. and the width is 15 in. Find the area of the rectangle.

54. Find the area of a rectangle that has a length of 6.5 m and a width of 3.8 m.

55. The length of the base of a parallelogram is 16 in., and the height is 12 in. Find the area of the parallelogram.

56. The height of a parallelogram is 3.4 m, and the length of the base is 5.2 m. Find the area of the parallelogram.

57. The length of the base of a triangle is 6 ft. The height is 4.5 ft. Find the area of the triangle.

58. The height of a triangle is 4.2 cm. The length of the base is 5 cm. Find the area of the triangle.

59. The length of one base of a trapezoid is 35 cm, and the length of the other base is 20 cm. If the height is 12 cm, what is the area of the trapezoid?

60. The height of a trapezoid is 5 in. The bases measure 16 in. and 18 in. Find the area of the trapezoid.

Solve.

61. The radius of a circle is 5 in. Find the area of the circle. Give the exact value.

62. Find the area of a circle with a radius of 14 m. Round to the nearest hundredth.

63. Find the area of a circle that has a diameter of 3.4 ft. Round to the nearest hundredth.

64. The diameter of a circle is 6.5 m. Find the area of the circle. Give the exact value.

65. The diameter of the Hale telescope at Mount Palomar, California, is 200 in. Find its area. Give the exact value.

66. An irrigation system waters a circular field that has a 50-foot radius. Find the area watered by the irrigation system. Give the exact value.

67. Find the area of a rectangular flower garden that measures 14 ft by 9 ft.

68. What is the area of a square patio that measures 8.5 m on each side?

69. Artificial turf will be used to cover a playing field. If the field is rectangular with a length of 100 yd and a width of 75 yd, how much artificial turf must be purchased to cover the field?

70. A fabric wall hanging is to fill a space that measures 5 m by 3.5 m. Allowing for 0.1 m of the fabric to be folded back along each edge, how much fabric must be purchased for the wall hanging?

71. You plan to stain the wooden deck attached to your house. The deck measures 10 ft by 8 ft. If a quart of stain will cover 50 ft^2, how many quarts of stain should you buy?

72. Your kitchen floor measures 12 ft by 9 ft. How many tiles, each a square with side $1\frac{1}{2}$ ft, should you purchase to cover the floor?

73. You are wallpapering two walls of a child's room, one measuring 9 ft by 8 ft and the other measuring 11 ft by 8 ft. The wallpaper costs $18.50 per roll, and each roll of the wallpaper will cover 40 ft^2. What is the cost to wallpaper the two walls?

74. An urban renewal project involves reseeding a park that is in the shape of a square, 60 ft on each side. Each bag of grass seed costs $5.75 and will seed 1200 ft^2. How much money should be budgeted for buying grass seed for the park?

Solve.

75. A circle has a radius of 8 in. Find the increase in area when the radius is increased by 2 in. Round to the nearest hundredth.

76. A circle has a radius of 6 cm. Find the increase in area when the radius is doubled. Round to the nearest hundredth.

77. You want to install wall-to-wall carpeting in your living room, which measures 15 ft by 24 ft. If the cost of the carpet you would like to purchase is \$15.95 per square yard, what is the cost of the carpeting for your living room? (*Hint*: 9 ft^2 = 1 yd^2)

78. You want to paint the walls of your bedroom. Two walls measure 15 ft by 9 ft, and the other two walls measure 12 ft by 9 ft. The paint you wish to purchase costs \$12.98 per gallon. Each gallon will cover 400 ft^2 of wall. Find the total amount you will spend on paint.

79. A walkway 2 m wide surrounds a rectangular plot of grass. The plot is 30 m long and 20 m wide. What is the area of the walkway?

80. Pleated draperies for a window must be twice as wide as the width of the window. Draperies are being made for four windows, each 2 ft wide and 4 ft high. Since the drapes will fall slightly below the window sill and extra fabric will be needed for hemming the drapes, 1 ft must be added to the height of the window. How much material must be purchased to make the drapes?

Critical Thinking

81. If both the length and the width of a rectangle are doubled, how many times larger is the area of the resulting rectangle?

82. Derive a formula for the area of a circle in terms of the diameter of the circle.

83. Determine whether the statement is always true, sometimes true, or never true.
- **a.** If two triangles have the same perimeter, then they have the same area.
- **b.** If two rectangles have the same area, then they have the same perimeter.
- **c.** If two squares have the same area, then the sides of the squares have the same length.
- **d.** An equilateral triangle is also an isosceles triangle.
- **e.** All the radii (plural of *radius*) of a circle are equal.
- **f.** All the diameters of a circle are equal.

84. [W] Suppose a circle is cut into 16 equal pieces, which are then arranged as shown at the right. The figure formed resembles a parallelogram. What variable expression could describe the base of the parallelogram? What variable could describe its height? Explain how the formula for the area of a circle is derived from this construction.

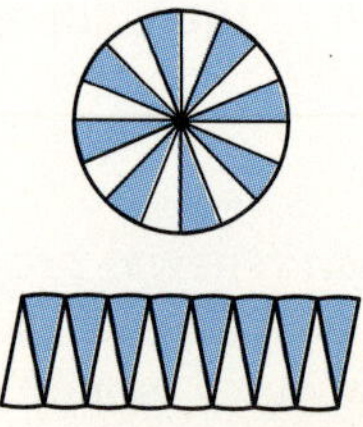

SECTION 5.3 Volume and Surface Area

OBJECTIVE A Volume of a solid

Geometric solids are figures in space. Five common geometric solids are the rectangular solid, the sphere, the cylinder, the cone, and the pyramid.

In a **rectangular solid,** all six sides, called **faces,** are rectangles. The variable L represents the length of a rectangular solid, W its width, and H its height.

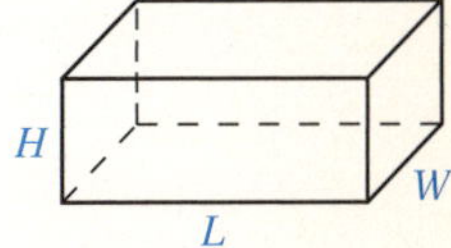

In a **sphere,** all points are the same distance from point O, called the **center** of the sphere. The **diameter,** d, of a sphere is the length of a line segment across the sphere going through point O. The **radius,** r, is the length of a line segment from the center to a point on the sphere. AB is a diameter and OC is a radius of the sphere shown at the right.

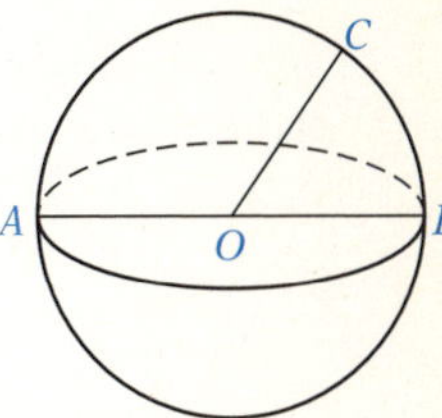

$$d = 2r \text{ or } r = \frac{1}{2}d$$

In a **right circular cylinder,** the bases are circles and are perpendicular to the sides of the cylinder. The variable r represents the radius of a base of a cylinder, and h represents the height. In this text, only right circular cylinders are discussed.

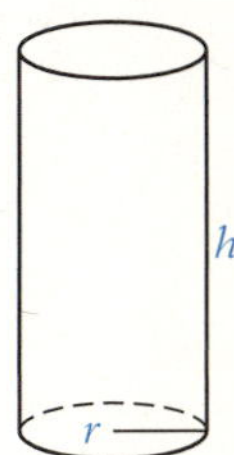

A **right circular cone** is obtained when one base of a right circular cylinder is shrunk to a point, called the **vertex,** V. The variable r represents the radius of the base of the cone, and h represents the height of the cone. The variable l is used to represent the **slant height,** which is the distance from a point on the circumference of the base to the vertex. In this text, only right circular cones are discussed.

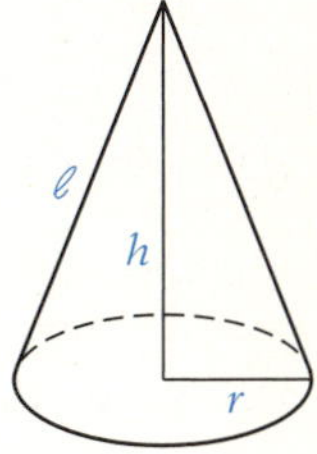

The base of a **regular pyramid** is a regular polygon, and the sides are isosceles triangles. The height, h, is the distance from the vertex, V, to the base and is perpendicular to the base. The variable l represents the **slant height,** which is the height of one of the isosceles triangles on the face of the pyramid. The **regular square pyramid** at the right has a square base. Only square pyramids are discussed in this text.

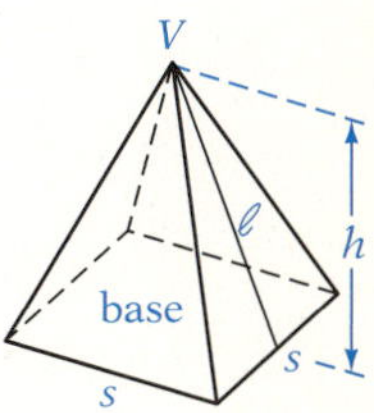

A **cube** is a special type of rectangular solid. Each of the six faces of a cube is a square. The variable s represents the length of the side of the squares.

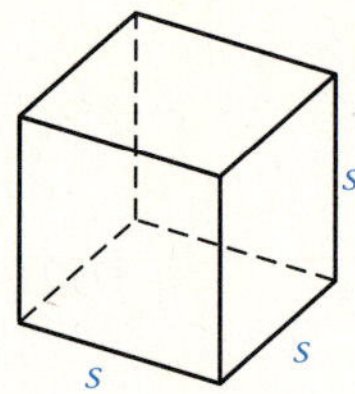

Volume is a measure of the amount of space inside a figure in space. Volume can be used to describe the amount of heating gas used for cooking, the amount of concrete delivered for the foundation of a house, or the amount of water in storage for a city's water supply.

Volume is measured in cubic units. A cube that is 1 ft on each side has a volume of 1 cubic foot, which is written 1 ft^3. A cube that measures 1 cm on each side has a volume of 1 cubic centimeter, written 1 cm^3.

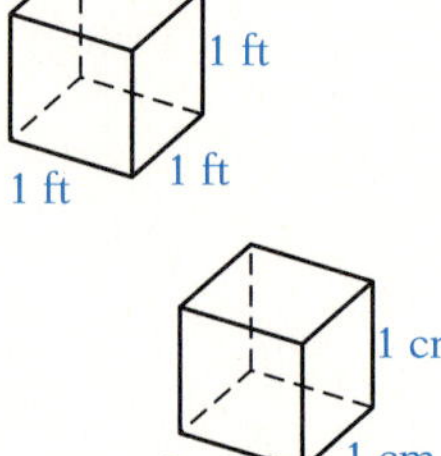

The volume of a solid is the number of cubes that are necessary to exactly fill the solid. The volume of the rectangular solid at the right is 24 cm^3 because it will hold exactly 24 cubes, each 1 cm on a side. Note that the volume can be found by multiplying the length times the width times the height.

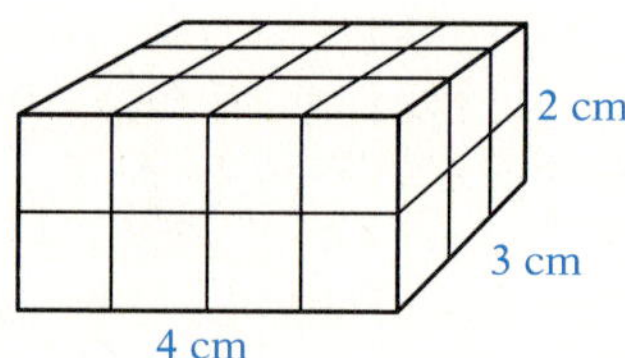

$4 \cdot 3 \cdot 2 = 24$

The formulas for the volumes of these geometric solids are given below.

Volumes of Geometric Solids

The volume, V, of a **rectangular solid** with length L, width W, and height H is given by $V = LWH$.

The volume, V, of a **cube** with side s is given by $V = s^3$.

The volume, V, of a **sphere** with radius r is given by $V = \frac{4}{3}\pi r^3$.

The volume, V, of a **right circular cylinder** is given by $V = \pi r^2 h$, where r is the radius of the base and h is the height.

The volume, V, of a **right circular cone** is given by $V = \frac{1}{3}\pi r^2 h$, where r is the radius of the circular base and h is the height.

The volume, V, of a **regular square pyramid** is given by $V = \frac{1}{3}s^2 h$, where s is the length of a side of the base and h is the height.

Find the volume of a sphere with a diameter of 6 in.

First find the radius of the sphere. $r = \frac{1}{2}d = \frac{1}{2}(6) = 3$

Use the formula for the volume of a sphere.

$$V = \frac{4}{3}\pi r^3$$

$$V = \frac{4}{3}\pi(3)^3$$

$$V = \frac{4}{3}\pi(27)$$

The exact volume of the sphere is 36π in^3. $V = 36\pi$

An approximate measure can be found by using the π key on a calculator. $V \approx 113.10$

The approximate volume is 113.10 in^3.

Example 1

The length of a rectangular solid is 5 m, the width is 3.2 m, and the height is 4 m. Find the volume of the solid.

Strategy

To find the volume, use the formula for the volume of a rectangular solid. $L = 5$, $W = 3.2$, $H = 4$.

Solution

$V = LWH = 5(3.2)(4) = 64$

The volume of the rectangular solid is 64 m^3.

You Try It 1

Find the volume of a cube that measures 2.5 m on a side.

Your Strategy

Your Solution

Example 2

The radius of the base of a cone is 8 cm. The height is 12 cm. Find the volume of the cone. Round to the nearest hundredth.

Strategy

To find the volume, use the formula for the volume of a cone. An approximation is asked for; use the π key on a calculator. $r = 8$, $h = 12$.

Solution

$$V = \frac{1}{3}\pi r^2 h$$

$$V = \frac{1}{3}\pi(8)^2(12) = \frac{1}{3}\pi(64)(12) = 256\pi \approx 804.25$$

The volume is approximately 804.25 cm^3.

You Try It 2

The diameter of the base of a cylinder is 8 ft. The height of the cylinder is 22 ft. Find the exact volume of the cylinder.

Your Strategy

Your Solution

Solutions on p. A12

OBJECTIVE B Surface area of a solid

The **surface area** of a solid is the total area on its surface.

When a rectangular solid is cut open and flattened out, each face is a rectangle. The surface area, S, of the rectangular solid is the sum of the areas of the six rectangles:

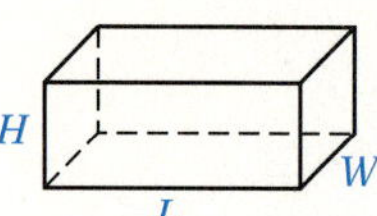

$$S = LW + LH + WH + LW + WH + LH$$

which simplifies to

$$S = 2LW + 2LH + 2WH$$

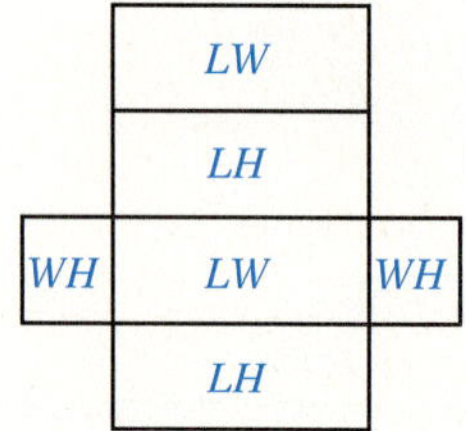

The surface area of a cube is the sum of the areas of the six faces of the cube. The area of each face is s^2. Therefore, the surface area, S, of a cube is given by the formula $S = 6s^2$.

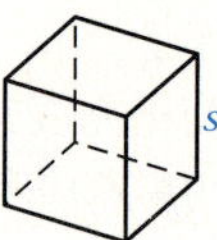

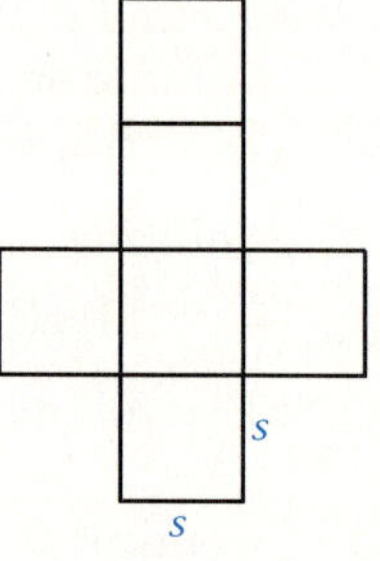

When a cylinder is cut open and flattened out, the top and bottom of the cylinder are circles. The side of the cylinder flattens out to a rectangle. The length of the rectangle is the circumference of the base, which is $2\pi r$; the width is h, the height of the cylinder. Therefore, the area of the rectangle is $2\pi rh$. The surface area, S, of the cylinder is

$$S = \pi r^2 + 2\pi rh + \pi r^2$$

which simplifies to

$$S = 2\pi r^2 + 2\pi rh$$

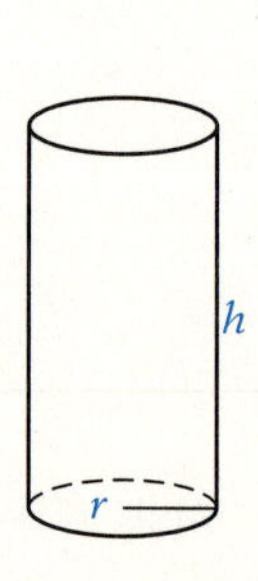

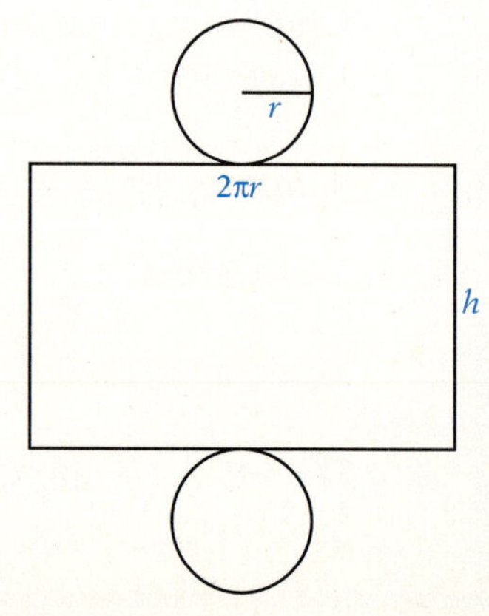

The surface area of a pyramid is the area of the base plus the area of the four isosceles triangles. The length of a side of the square base is s; therefore, the area of the base is s^2. The slant height, l, is the height of each triangle, and s is the base of each triangle. The surface area, S, of a pyramid is

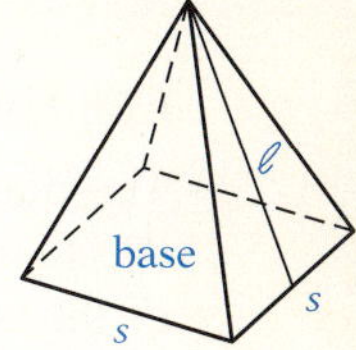

$$S = s^2 + 4\left(\frac{1}{2}sl\right)$$

which simplifies to

$$S = s^2 + 2sl$$

Formulas for the surface area of geometric solids are given below.

Surface Areas of Geometric Solids

The surface area, S, of a **rectangular solid** with length L, width W, and height H is given by $S = 2LW + 2LH + 2WH$.

The surface area, S, of a **cube** with side s is given by $S = 6s^2$.

The surface area, S, of a **sphere** with radius r is given by $S = 4\pi r^2$.

The surface area, S, of a **right circular cylinder** is given by $S = 2\pi r^2 + 2\pi rh$, where r is the radius of the base and h is the height.

The surface area, S, of a **right circular cone** is given by $S = \pi r^2 + \pi rl$, where r is the radius of the circular base and l is the slant height.

The surface area, S, of a **regular square pyramid** is given by $S = s^2 + 2sl$, where s is the length of a side of the base and l is the slant height.

Find the surface area of a sphere with a diameter of 18 cm.

First find the radius of the sphere.

$$r = \frac{1}{2}d = \frac{1}{2}(18) = 9$$

Use the formula for the surface area of a sphere.

$$S = 4\pi r^2$$
$$S = 4\pi(9)^2$$
$$S = 4\pi(81)$$
$$S = 324\pi$$

The exact surface area of the sphere is 324π cm^2.

An approximate measure can be found by using the π key on a calculator.

$$S \approx 1017.88$$

The approximate surface area is 1017.88 cm^2.

Example 3

The diameter of the base of a cone is 5 m and the slant height is 4 m. Find the surface area of the cone. Give the exact measure.

Strategy

To find the surface area of the cone:

▶ Find the radius of the base of the cone.

▶ Use the formula for the surface area of a cone. Leave the answer in terms of π.

Solution

$r = \frac{1}{2}d = \frac{1}{2}(5) = 2.5$

$S = \pi r^2 + \pi r l$
$S = \pi(2.5)^2 + \pi(2.5)(4)$
$S = \pi(6.25) + \pi(2.5)(4)$
$S = 6.25\pi + 10\pi$
$S = 16.25\pi$

The surface area of the cone is 16.25π m^2.

You Try It 3

The diameter of the base of a cylinder is 6 ft, and the height is 8 ft. Find the surface area of the cylinder. Round to the nearest hundredth.

Your Strategy

Your Solution

Example 4

Find the area of a label used to cover a soup can that has a radius of 4 cm and a height of 12 cm. Round to the nearest hundredth.

Strategy

To find the area of the label, use the fact that the surface area of the sides of a cylinder is given by $2\pi rh$. An approximation is asked for; use the π key on a calculator. $r = 4$, $h = 12$.

Solution

Area of the label $= 2\pi rh$
Area of the label $= 2\pi(4)(12) = 96\pi \approx 301.59$

The area is approximately 301.59 cm^2.

You Try It 4

Which has a larger surface area, a cube with a side measuring 10 cm or a sphere with a diameter measuring 8 cm?

Your Strategy

Your Solution

Solutions on p. A12

5.3 EXERCISES

Objective A

Find the volume of the figure. For calculations involving π, give both the exact value and an approximation to the nearest hundredth.

1. 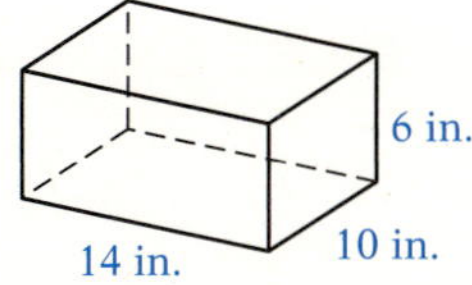

2. 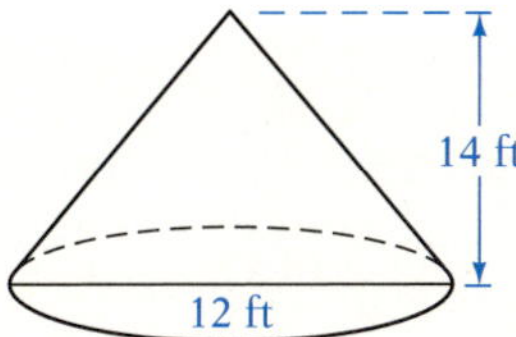

3. 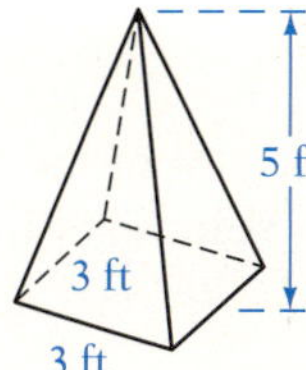

4. 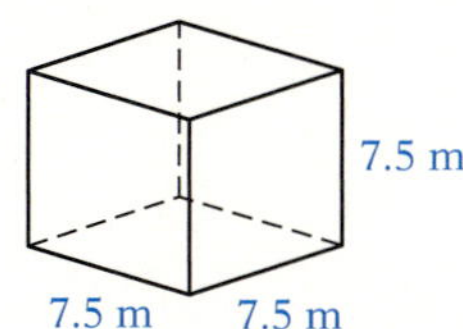

5. 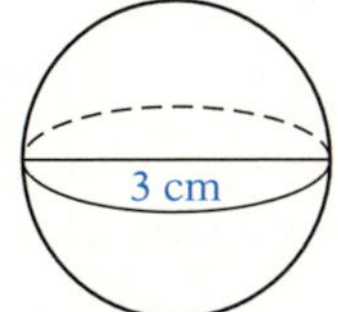

6. 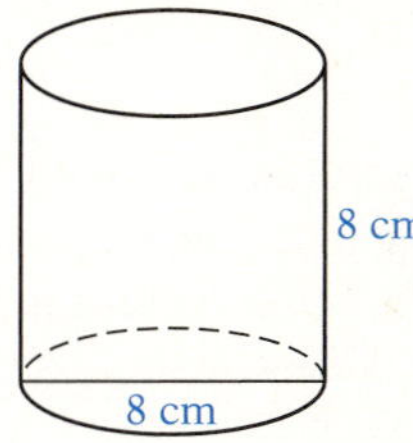

Solve.

7. A rectangular solid has a length of 6.8 m, a width of 2.5 m, and a height of 2 m. Find the volume of the solid.

8. Find the volume of a rectangular solid that has a length of 4.5 ft, a width of 3 ft, and a height of 1.5 ft.

9. Find the volume of a cube whose side measures 2.5 in.

10. The length of a side of a cube is 7 cm. Find the volume of the cube.

11. The diameter of a sphere is 6 ft. Find the volume of the sphere. Give the exact measure.

12. Find the volume of a sphere that has a radius of 1.2 m. Round to the nearest tenth.

13. The diameter of the base of a cylinder is 24 cm. The height of the cylinder is 18 cm. Find the volume of the cylinder. Round to the nearest hundredth.

14. The height of a cylinder is 7.2 m. The radius of the base is 4 m. Find the volume of the cylinder. Give the exact measure.

Solve.

15. The radius of the base of a cone is 5 in. The height of the cone is 9 in. Find the volume of the cone. Give the exact measure.

16. The height of a cone is 15 cm. The diameter of the cone is 10 cm. Find the volume of the cone. Round to the nearest hundredth.

17. The length of a side of the base of a square pyramid is 6 in., and the height is 10 in. Find the volume of the pyramid.

18. The height of a square pyramid is 8 m, and the length of a side of the base is 9 m. What is the volume of the pyramid?

19. An oil storage tank, which is in the shape of a cylinder, is 4 m high and has a diameter of 6 m. The oil tank is two-thirds full. Find the number of cubic meters of oil in the tank. Round to the nearest hundredth.

20. A silo, which is in the shape of a cylinder, is 16 ft in diameter and has a height of 30 ft. The silo is three-fourths full. Find the volume of the portion of the silo that is empty. Round to the nearest hundredth.

Objective B

Find the surface area of the figure.

21.

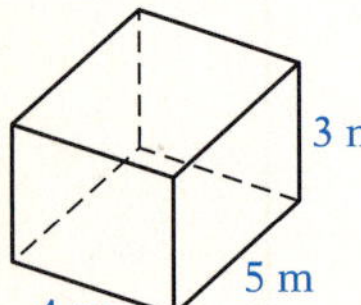

22.

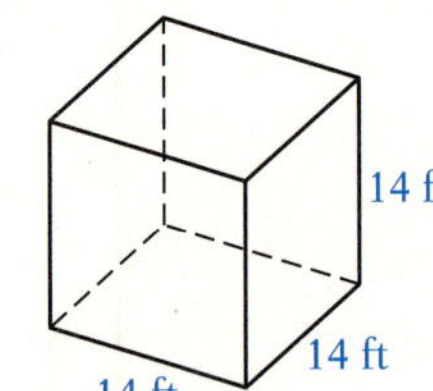

23.

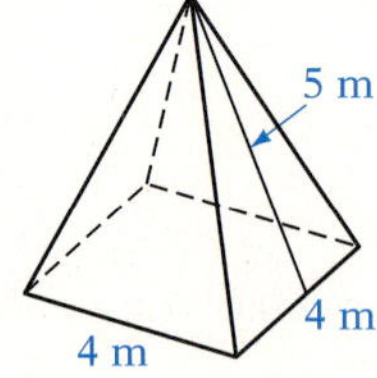

Find the surface area of the figure. Give both the exact value and an approximation to the nearest hundredth.

24.

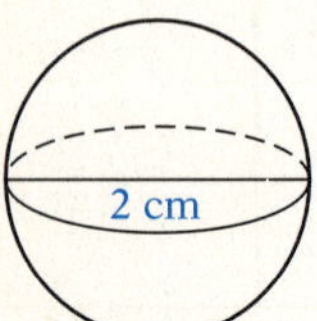

25.

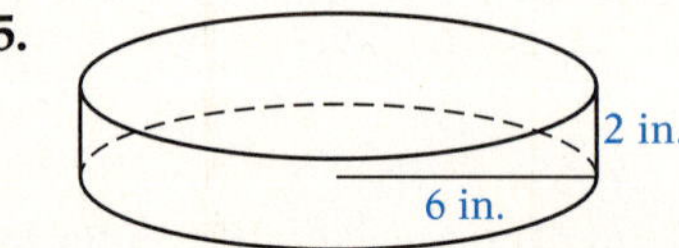

26.

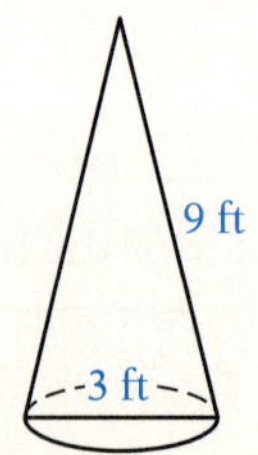

Solve.

27. The height of a rectangular solid is 5 ft. The length is 8 ft, and the width is 4 ft. Find the surface area of the solid.

28. The width of a rectangular solid is 32 cm. The length is 60 cm, and the height is 14 cm. What is the surface area of the solid?

29. The side of a cube measures 3.4 m. Find the surface area of the cube.

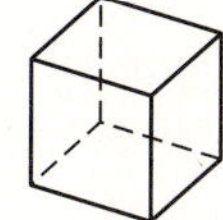

30. Find the surface area of a cube whose side measures 1.5 in.

31. Find the surface area of a sphere with a diameter of 15 cm. Give the exact value.

32. The radius of a sphere is 2 in. Find the surface area of the sphere. Round to the nearest hundredth.

33. The radius of the base of a cylinder is 4 in. The height of the cylinder is 12 in. Find the surface area of the cylinder. Round to the nearest hundredth.

34. The diameter of the base of a cylinder is 1.8 m. The height of the cylinder is 0.7 m. Find the surface area of the cylinder. Give the exact value.

35. The slant height of a cone is 2.5 ft. The radius of the base is 1.5 ft. Find the surface area of the cone. Give the exact value.

36. The diameter of the base of a cone is 21 in. The slant height is 16 in. What is the surface area of the cone? Round to the nearest hundredth.

37. The length of a side of the base of a pyramid is 9 in., and the slant height is 12 in. Find the surface area of the pyramid.

38. The slant height of a pyramid is 18 m, and the length of a side of the base is 16 m. What is the surface area of the pyramid?

39. A can of paint covers 300 ft^2. How many cans of paint should be purchased in order to paint a cylinder that has a height of 30 ft and a radius of 12 ft?

40. A hot air balloon is in the shape of a sphere. Approximately how much fabric was used to construct the balloon if its diameter is 32 ft? Round to the nearest whole number.

Solve.

41. How much glass is needed to make a fish tank that is 12 in. long, 8 in. wide, and 9 in. high? The fish tank is open at the top.

42. Find the area of a label used to cover a can of juice that has a diameter of 16.5 cm and a height of 17 cm. Round to the nearest hundredth.

43. The length of a side of the base of a square pyramid is 5 cm, and the slant height is 8 cm. How much larger is the surface area of this pyramid than the surface area of a cone with a diameter of 5 cm and a slant height of 8 cm? Round to the nearest hundredth.

Critical Thinking

44. Half of a sphere is called a **hemisphere.** Derive formulas for the volume and surface area of a hemisphere.

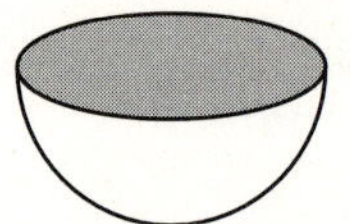

45. Determine whether the statement is always true, sometimes true, or never true.
- **a.** The slant height of a regular pyramid is longer than the height.
- **b.** The slant height of a cone is shorter than the height.
- **c.** The four triangular faces of a regular pyramid are equilateral triangles.

46.
- **a.** What is the effect on the surface area of a rectangular solid if the width and height are doubled?
- **b.** What is the effect on the volume of a rectangular solid if both the length and the width are doubled?
- **c.** What is the effect on the volume of a cube if the length of each side of the cube is doubled?
- **d.** What is the effect on the surface area of a cylinder if the radius and height are doubled?

47. A sphere fits inside a cylinder as shown at the right. The height of the cylinder equals the diameter of the ball. Show that the surface area of the sphere equals the surface area of the sides of the cylinder.

48. [W] Explain how you could cut through a cube so that the face of the resulting solid is
- **a.** a square
- **b.** an equilateral triangle
- **c.** a trapezoid
- **d.** a hexagon

49. [W] Prepare a report on the use of geometric form in architecture. Include examples of both plane geometric figures and geometric solids.

50. [W] Write a paper on the artist M. C. Escher. Explain how he used mathematics and geometry in his works.

SECTION 5.4 Verbal Expressions and Variable Expressions

OBJECTIVE A  Translate verbal expressions into variable expressions

One of the major skills required to use mathematics is translating a verbal expression into a mathematical expression. Following are some verbal phrases that indicate mathematical operations.

POINT OF INTEREST

The way in which mathematical expressions are written has changed over time. Here is how some of the expressions shown at the right may have appeared in the early 16th century.

R p. 9 for $x + 9$. The symbol R was used for a variable to the first power. The symbol p. was used for plus.

R m. 3 for $x - 3$. The symbol R is still used for the variable. The symbol m. is used for minus.

The square of a variable was designated by Q and the cube was designated by C. The expression $x^3 + x^2$ was written C. p. Q.

Addition	more than	8 more than *w*	$w + 8$
	the sum of	the sum of *z* and 9	$z + 9$
	the total of	the total of *r* and *s*	$r + s$
	increased by	*x* increased by 7	$x + 7$
Subtraction	less than	12 less than *b*	$b - 12$
	the difference between	the difference between *x* and 1	$x - 1$
	decreased by	17 decreased by *a*	$17 - a$
Multiplication	times	negative 2 times *c*	$-2c$
	the product of	the product of *x* and *y*	xy
	of	three-fourths of *m*	$\frac{3}{4}m$
	twice	twice *d*	$2d$
Division	divided by	*v* divided by 15	$\frac{v}{15}$
	the quotient of	the quotient of *y* and 3	$\frac{y}{3}$
Power	the square of or the second power of	the square of *x*	x^2
	the cube of or the third power of	the cube of *r*	r^3
	the fifth power of	the fifth power of *a*	a^5

Translating a phrase that contains the word *sum, difference, product,* or *quotient* can sometimes cause a problem. In the examples at the right, note where the operation symbol is placed.

the *sum* of x and y	$x + y$
the *difference* between x and y	$x - y$
the *product* of x and y	$x \cdot y$
the *quotient* of x and y	$\frac{x}{y}$

Translate "3 times the sum of c and 5" into a variable expression.

Identify words that indicate the mathematical operations. — 3 times the sum of c and 5

Use the identified words to write the variable expression. Note that the phrase times the sum of requires parentheses. — $3(c + 5)$

The sum of two numbers is 37. If x represents the smaller number, translate "twice the larger number" into a variable expression.

Write an expression for the larger number by subtracting the smaller number, x, from the sum. — larger number: $37 - x$

Identify the words that indicate the mathematical operations on the larger number. — twice the larger number

Use the identified words to write a variable expression. — $2(37 - x)$

Example 1

Translate "the quotient of r and the sum of r and four" into a variable expression.

Solution

the quotient of r and the sum of r and four

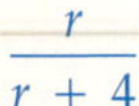

$\frac{r}{r + 4}$

You Try It 1

Translate "twice x divided by the difference between x and 7" into a variable expression.

Your Solution

Example 2

Translate "the sum of the square of y and six" into a variable expression.

Solution

the sum of the square of y and six

$y^2 + 6$

You Try It 2

Translate "the product of negative three and the square of d" into a variable expression.

Your Solution

Solutions on p. A12

OBJECTIVE B Translate and simplify verbal expressions

After translating a verbal expression into a variable expression, it may be possible to simplify the variable expression.

Translate "a number plus five less than the product of eight and the number" into a variable expression. Then simplify.

The letter x is chosen for the unknown number. Any letter could be used.	the unknown number: x
Identify words that indicate the mathematical operations.	x plus 5 less than the product of 8 and x
Use the identified words to write the variable expression.	$x + (8x - 5)$
Simplify the expression.	$x + 8x - 5$ $9x - 5$

Translate "five less than twice the difference between a number and seven" into a variable expression. Then simplify.

	the unknown number: x
Identify words that indicate the mathematical operations.	5 less than twice the difference between x and 7
Use the identified words to write the variable expression.	$2(x - 7) - 5$
Simplify the expression.	$2x - 14 - 5$ $2x - 19$

Example 3

The sum of two numbers is 28. Using x to represent the smaller number, translate "the sum of the smaller number and three times the larger number" into a variable expression. Then simplify.

Solution

The smaller number is x.
The larger number is $28 - x$.
the sum of the smaller number and three times the larger number

$x + 3(28 - x)$ ▶ This is the variable expression.
$x + 84 - 3x$ ▶ Simplify.
$-2x + 84$

You Try It 3

The sum of two numbers is 16. Using x to represent the smaller number, translate "the difference between the larger number and twice the smaller number" into a variable expression. Then simplify.

Your Solution

Solution on p. A12

Example 4

Translate "eight more than the product of four and the total of a number and twelve" into a variable expression. Then simplify.

Solution

Let the unknown number be x.

8 more than the product of 4 and the total of x and 12

$4(x + 12) + 8$ ▶ This is the variable expression.

$4x + 48 + 8$ ▶ Now simplify.

$4x + 56$

You Try It 4

Translate "the difference between fourteen and the sum of a number and seven" into a variable expression. Then simplify.

Your Solution

Solution on p. A12

OBJECTIVE  Applications

Many applications of mathematics require you to identify the unknown quantity, assign a variable to that quantity, and then express the other unknowns in terms of that quantity.

Ten gallons of paint were poured into two containers of different sizes. Express the amount of paint poured into the smaller container in terms of the amount poured into the larger container.

Assign a variable to the amount of paint poured into the larger container.

gallons of paint poured into the larger container: g

Express the amount of paint in the smaller container in terms of g. (g gallons of paint were poured into the larger container.)

The number of gallons of paint in the smaller container is $10 - g$.

Example 5

A cyclist is riding at twice the speed of a runner. Express the speed of the cyclist in terms of the speed of the runner.

Solution

the speed of the runner: r

the speed of the cyclist is twice r: $2r$

You Try It 5

A mixture of candy contains 3 lb more of milk chocolate than of caramel. Express the amount of milk chocolate in the mixture in terms of the amount of caramel in the mixture.

Your Solution

Solution on p. A13

5.4 EXERCISES

Objective A

Translate into a variable expression.

1. three more than t

2. the total of twice q and five

3. five less than the product of six and m

4. seven subtracted from the product of eight and d

5. the difference between three times b and seven

6. the difference between six times c and twelve

7. the product of n and seven

8. the quotient of nine times k and seven

9. twice the sum of three and w

10. six times the difference between y and eight

11. four times the difference between twice r and five

12. seven times the total of p and ten

13. the quotient of v and the difference between v and four

14. x divided by the sum of x and one

15. four times the square of t

16. six times the cube of q

17. the sum of the square of m and the cube of m

18. the difference between the square of d and d

19. The sum of two numbers is 31. Using s to represent the smaller number, translate "five more than the larger number" into a variable expression.

20. The sum of two numbers is 74. Using L to represent the larger number, translate "the quotient of the larger number and the smaller number" into a variable expression.

Objective B

Translate into a variable expression. Then simplify.

21. a number decreased by the total of the number and twelve

22. a number decreased by the difference between six and the number

23. the difference between two-thirds of a number and three-eighths of the number

24. two more than the total of a number and five

Translate into a variable expression. Then simplify.

25. twice the sum of seven times a number and six

26. five times the product of seven and a number

27. the sum of eleven times a number and the product of three and the number

28. a number plus the product of the number and ten

29. nine times the sum of a number and seven

30. a number added to the product of four and the number

31. seven more than the sum of a number and five

32. a number minus the sum of the number and six

33. the product of seven and the difference between a number and four

34. six times the difference between a number and three

35. the difference between ten times a number and the product of three and the number

36. fifteen more than the difference between a number and seven

37. the sum of a number and twice the difference between the number and four

38. the difference between a number and the total of three times the number and five

39. seven times the difference between a number and fourteen

40. the product of three and the sum of a number and twelve

41. the product of eight and the sum of a number and ten

42. the difference between the square of a number and the total of twelve and the square of the number

43. a number increased by the difference between seven times the number and eight

44. the product of ten and the total of a number and one

45. five increased by twice the sum of a number and fifteen

46. eleven less than the difference between eight and a number

47. fourteen decreased by the sum of a number and thirteen

48. eleven minus the sum of a number and six

49. the product of eight times a number and two

50. the sum of four more than the square of a number and the product of three and the square of the number

Translate into a variable expression. Then simplify.

51. the sum of the square of a number and the quotient of the square of the number and two

52. a number plus six added to the difference between three and twice the number

53. eleven more than a number added to the difference between the number and seventeen

54. a number plus nine added to the difference between four times the number and three

55. the sum of a number and ten added to the difference between the number and eleven

56. seven increased by a number added to three times the difference between the number and two

57. The sum of two numbers is 9. Using y to represent the smaller number, translate "five times the larger number" into a variable expression. Then simplify.

58. The sum of two numbers is 14. Using p to represent the smaller number, translate "eight less than the larger number" into a variable expression. Then simplify.

59. The sum of two numbers is 17. Using m to represent the larger number, translate "nine less than three times the smaller number" into a variable expression. Then simplify.

60. The sum of two numbers is 19. Using k to represent the larger number, translate "the difference between twice the smaller number and ten" into a variable expression. Then simplify.

Objective C

Write a variable expression.

61. The distance from Earth to the sun is approximately 390 times the distance from Earth to the moon. Express the distance from Earth to the sun in terms of the distance from Earth to the moon.

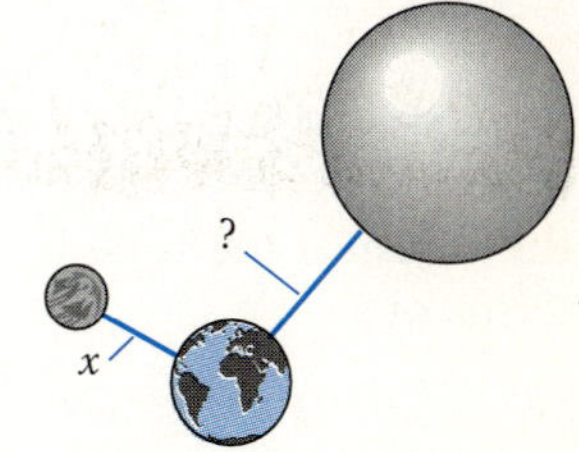

62. The length of an infrared ray is twice the length of an ultraviolet ray. Express the length of the infrared ray in terms of the length of the ultraviolet ray.

63. Mt. Everest is 4430 m higher than Mt. Whitney. Express the height of Mt. Whitney in terms of the height of Mt. Everest.

64. The height of a door is five feet more than the width of the door. Express the height of the door in terms of the width of the door.

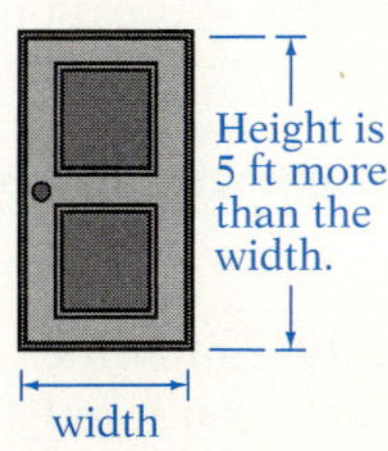

65. A "double-clocked" computer operates at twice its normal speed. Express the "double-clocked" speed in terms of the normal speed.

Write a variable expression.

66. The sale price of a suit is three-fourths of the original price. Express the sale price in terms of the original price.

67. One cyclist pedals six miles per hour faster than another cyclist. Express the speed of the faster cyclist in terms of the speed of the slower cyclist.

68. A mixture contains three times as many peanuts as cashews. Express the amount of peanuts in the mixture in terms of the amount of cashews in the mixture.

69. A string three feet long is cut into two pieces, one shorter than the other. Express the length of the shorter piece in terms of the length of the longer piece.

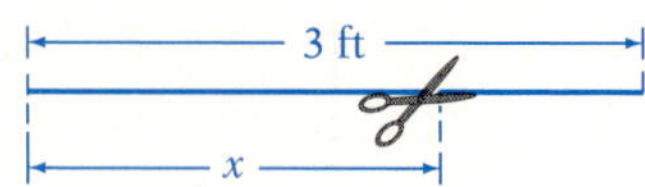

70. The dividend paid on a company's stock is one-twentieth of the price of the stock. Express the dividend paid on the stock in terms of the price of the stock.

71. A twelve-foot board is cut into two pieces of different lengths. Express the length of the longer piece in terms of the length of the shorter piece.

Critical Thinking

72. A wire x inches long is bent into a square. Express the length of a side of the square in terms of x.

x in.

73. The chemical formula for water is H_2O. This formula means that there are two hydrogen atoms and one oxygen atom in each molecule of water. If x represents the number of atoms of oxygen in a glass of pure water, express the number of hydrogen atoms in the glass of water.

74. A block-and-tackle system is designed so that pulling five feet on one end of a rope will move a weight on the other end a distance of three feet. If x represents the distance the rope is pulled, express the distance the weight will move in terms of x.

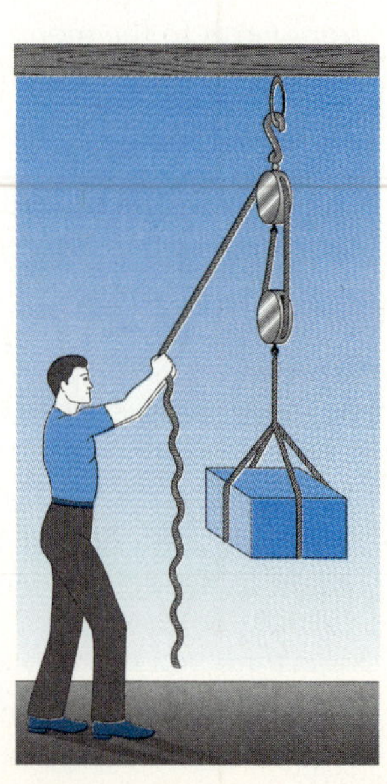

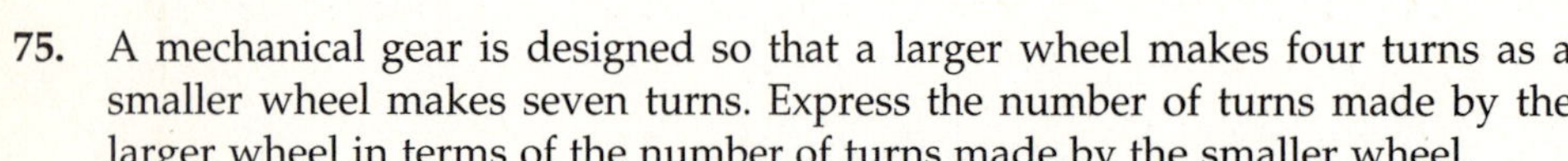

75. A mechanical gear is designed so that a larger wheel makes four turns as a smaller wheel makes seven turns. Express the number of turns made by the larger wheel in terms of the number of turns made by the smaller wheel.

76. Translate the expressions $3x + 4$ and $3(x + 4)$ into verbal phrases.
[W]

77. In your own words, explain how variables are used.
[W]

78. Explain the similarities and the differences between the expressions "the difference between x and 5" and "5 less than x."
[W]

Projects in Mathematics

Lines of Symmetry

Look at the letter A printed at the left. If the letter were folded along line ℓ, the two sides of the letter would match exactly. This letter has **symmetry** with respect to line ℓ. Line ℓ is called the **axis of symmetry.**

Now consider the letter H printed below at the left. Both line ℓ_1 and line ℓ_2 are axes of symmetry for this letter; the letter could be folded along either line and the two sides would match exactly. Does the letter A have more than one axis of symmetry? Find axes of symmetry for other capital letters of the alphabet. Which lowercase letters have one axis of symmetry? Do any of the lowercase letters have more than one axis of symmetry?

Find the number of axes of symmetry for each of the plane geometric figures presented in this chapter.

There are other types of symmetry. Look up the meaning of point symmetry and rotational symmetry. Which plane geometric figures provide examples of these types of symmetry?

Find examples of symmetry in nature, art, and architecture.

Chapter Summary

Key Words

An expression that contains one or more variables is called a *variable expression.* A *term* of a variable expression is one of the addends of the expression. A *variable term* consists of a *numerical coefficient* and a *variable part.* A *constant term* has no variable part. *Like terms* of a variable expression have the same variable part. Constant terms are also like terms. Adding or subtracting the like terms of a variable expression is called *combining like terms.*

The opposite of a term is called its *additive inverse.* The additive inverse of a is $-a$. The reciprocal of a term is called its *multiplicative inverse.* The multiplicative inverse of a nonzero term a is $\frac{1}{a}$.

A *line* is determined by two distinct points and extends indefinitely in both directions. A *line segment* is part of a line that has two endpoints. *Parallel lines* never meet; the distance between them is always the same. *Perpendicular lines* are intersecting lines that form right angles.

An *angle* is formed by two rays with the same endpoint. The *vertex* of an angle is the point at which the two rays meet. An angle is measured in *degrees.* A 90° angle is a *right angle.* An *acute angle* is an angle whose measure is between 0° and 90°. An *obtuse angle* is an angle whose measure is between 90° and 180°.

A *polygon* is a closed figure formed by three or more line segments that lie in a plane. In a *regular polygon,* each side has the same length and each angle has the same measure.

A *triangle* is a plane figure formed by three line segments. An *isosceles triangle* has two sides of equal length. The three sides of an *equilateral triangle* are of equal length. An *acute triangle* has three acute angles. An *obtuse triangle* has one obtuse angle. A *right triangle* has a right angle.

A *quadrilateral* is a four-sided polygon. A parallelogram, a rectangle, a square, a rhombus, and a trapezoid are all quadrilaterals.

A *circle* is a plane figure in which all points are the same distance from the center of the circle. A *diameter* of a circle is a line segment across the circle through the center. A *radius* of a circle is a line segment from the center of the circle to a point on the circle.

The *perimeter* of a plane geometric figure is a measure of the distance around the figure; the distance around a circle is called the *circumference*. *Area* is the amount of surface in a region. *Volume* is a measure of the amount of space inside a figure in space. The *surface area* of a solid is the total area on the surface of the solid.

Essential Rules

Commutative Property of Addition	$a + b = b + a$
Commutative Property of Multiplication	$a \cdot b = b \cdot a$
Associative Property of Addition	$(a + b) + c = a + (b + c)$
Associative Property of Multiplication	$(a \cdot b) \cdot c = a \cdot (b \cdot c)$
Addition Property of Zero	$a + 0 = 0 + a = a$
Multiplication Property of One	$a \cdot 1 = 1 \cdot a = a$
Inverse Property of Addition	$a + (-a) = (-a) + a = 0$
Inverse Property of Multiplication	For $a \neq 0$, $a \cdot \frac{1}{a} = \frac{1}{a} \cdot a = 1$.
Distributive Property	$a(b + c) = ab + ac$

Perimeter	Triangle	$P = a + b + c$
	Rectangle	$P = 2L + 2W$
	Square	$P = 4s$
	Circle	$C = \pi d$ or $C = 2\pi r$
Area	Triangle	$A = \frac{1}{2}bh$
	Rectangle	$A = LW$
	Square	$A = s^2$
	Circle	$A = \pi r^2$
	Parallelogram	$A = bh$
	Trapezoid	$A = \frac{1}{2}h(b_1 + b_2)$
Volume	Rectangular solid	$V = LWH$
	Cube	$V = s^3$
	Sphere	$V = \frac{4}{3}\pi r^3$
	Right circular cylinder	$V = \pi r^2 h$
	Right circular cone	$V = \frac{1}{3}\pi r^2 h$
	Regular square pyramid	$V = \frac{1}{3}s^2 h$
Surface Area	Rectangular solid	$S = 2LW + 2LH + 2WH$
	Cube	$S = 6s^2$
	Sphere	$S = 4\pi r^2$
	Right circular cylinder	$S = 2\pi r^2 + 2\pi rh$
	Right circular cone	$S = \pi r^2 + \pi rl$
	Regular square pyramid	$S = s^2 + 2sl$

Chapter Review Exercises

1. Simplify: $-3x - 7 + 5x - 9$

2. Simplify by using the Distributive Property: $-2(9z + 1)$

3. Multiply: $\frac{2}{3}\left(\frac{3}{2}x\right)$

4. Simplify: $-5(2s - 5t) + 6(3t + s)$

5. Multiply: $-\frac{3}{4}(-8w)$

6. Simplify: $2m - 6n + 7 - 4m + 6n + 9$

7. Simplify: $8 - 2[3(2a - 1) + 2(4 - 2a)]$

8. Simplify: $7a^2 + 9 - 12a^2 + 3a$

9. Simplify: $8(2c - 3d) - 4(c - 5d)$

10. Simplify by using the Distributive Property: $7(2m - 6)$

11. Simplify: $2x^2 + 3[2x(3x - 4) - 4(2x - 3)]$

12. Simplify: $4z^2 + 3z - 9z + 2z^2$

13. Simplify by using the Distributive Property: $-4(3c - 8)$

14. Simplify: $-12x + 7y + 15x - 11y$

15. Simplify: $-7(3a - 4b) - 5(3b - 4a)$

16. Simplify: $9v - 10 + 5v + 8$

17. Translate "nine less than the quotient of four times a number and seven" into a variable expression.

18. Find the volume of a rectangular solid with a length of 6.5 ft, a width of 2 ft, and a height of 3 ft.

19. Find the volume of a sphere that has a diameter of 12 mm. Give the exact value.

20. Translate "the sum of three times a number and twice the difference between the number and seven" into a variable expression. Then simplify.

21. A can of paint will cover 200 ft^2. How many cans of paint should be purchased in order to paint a cylinder that has a height of 15 ft and a radius of 6 ft?

22. The length of a rectangular park is 56 yd. The width is 48 yd. How many yards of fencing are needed to surround the park?

23. Thirty pounds of a blend of coffee beans uses only mocha java and espresso beans. Express the number of pounds of espresso beans in the blend in terms of the number of pounds of mocha java beans in the blend.

24. What is the area of a square patio that measures 9.5 m on each side?

25. A walkway 2 m wide surrounds a rectangular plot of grass. The plot is 40 m long and 25 m wide. What is the area of the walkway?

Cumulative Review Exercises

1. Divide: $4\frac{2}{3} \div 5\frac{3}{5}$

2. Find all the factors of 78.

3. Simplify: $8x^2 - 3x + 4 - 5x^2 + 6x - 1$

4. Evaluate $x^2y - 2z$ when $x = \frac{1}{2}$, $y = \frac{4}{5}$, and $z = -\frac{3}{10}$.

5. Find the product of $2\frac{4}{5}$ and $\frac{6}{7}$.

6. Graph $x > -3$.

−6 −5 −4 −3 −2 −1 0 1 2 3 4 5 6

7. Simplify: $5(2x + 4) - (3x + 2)$

8. Evaluate $2x + 3y^2z$ when $x = 5$, $y = -1$, and $z = -4$.

9. Convert 0.5625 to a fraction.

10. Simplify: $6 \cdot (-2)^3 \div 12 - (-8)$

11. Divide: $-18 \div 0$

12. Evaluate $-6cd$ when $c = -\frac{2}{9}$ and $d = \frac{3}{4}$.

13. Evaluate $(a - b)^2 + 5c$ when $a = -4$, $b = 6$, and $c = -2$.

14. Divide and round to the nearest tenth: $82.93 \div 6.5$

15. Simplify: $9 - 3[4 - 3(2x + 3y) + 2(x + y)]$

16. Write 470,351 in expanded notation.

17. Translate "the quotient of ten and the difference between a number and nine" into a variable expression.

18. The temperature in degrees Celsius is equal to five-ninths the difference between the temperature in degrees Fahrenheit and 32. Use the formula $C = \frac{5}{9}(F - 32)$ to find the temperature in degrees Celsius when the temperature is 5°F.

19. The distance from Neptune to the sun is approximately 30 times the distance from Earth to the sun. Express the distance from Neptune to the sun in terms of the distance from Earth to the sun.

20. Translate "two less than twice the sum of a number and four" into a variable expression. Then simplify.

21. The cost, C, of the shares of stock in a stock purchase is equal to the cost per share, S, times the number of shares purchased, N. Use the equation $C = SN$ to find the cost of purchasing 200 shares of stock selling for $\$15\frac{3}{8}$ per share.

22. Two hundred fifty people are expected to attend a reception. Assuming that each person drinks 12 oz of coffee, how many gallons of coffee should be prepared? Round to the nearest whole number.

23. Find the exact area of a circle whose diameter is 9 cm.

CHAPTER

6

Solving Equations and Inequalities

Focus on Problem Solving

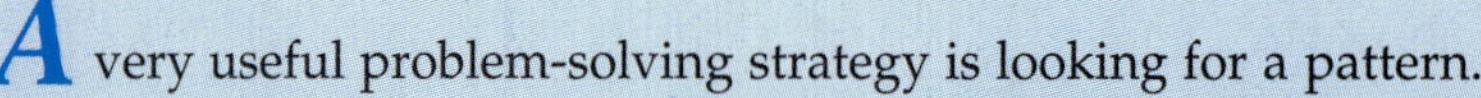

A very useful problem-solving strategy is looking for a pattern.

Problem A legend says that a peasant invented the game of chess and gave it to a very rich king as a present. The king so enjoyed the game that he gave the peasant the choice of anything in the kingdom. The peasant's request was simple. "Place one grain of wheat on the first square, 2 grains on the second square, 4 grains on the third square, 8 on the fourth square, and continue doubling the number of grains until the last square of the chessboard is reached." How many grains of wheat must the king give the peasant?

Solution A chessboard consists of 64 squares. To find the total number of grains of wheat on the 64 squares, we begin by looking at the amount of wheat on the first few squares.

Square 1	Square 2	Square 3	Square 4	Square 5	Square 6	Square 7	Square 8
1	2	4	8	16	32	64	128
1	3	7	15	31	63	127	255

The bottom row of numbers represents the sum of the number of grains of wheat up to and including that square. For instance, the number of grains of wheat on the first 7 squares is $1 + 2 + 4 + 8 + 16 + 32 + 64 = 127$.

One pattern to observe is that the number of grains of wheat on a square can be expressed as a power of 2.

The number of grains on square $n = 2^{n-1}$

For example, the number of grains on square $7 = 2^{7-1} = 2^6 = 64$.

A second pattern of interest is that the number *below* a square (the total number of grains up to and including that square) is one less than the number of grains of wheat *on* the next square. For example, the number *below* square 7 is one less than the number *on* square 8 ($128 - 1 = 127$). From this observation, the number of grains of wheat on the first eight squares is the number on square 8 (128) plus one less than the number on square 8 (127); the total number of grains of wheat on the first eight squares is $128 + 127 = 255$.

From this observation,

$$\begin{aligned}\text{Number of grains of wheat on the chessboard} &= \text{number of grains on square 64} + \text{one less than the number of grains on square 64}\\ &= 2^{64-1} + (2^{64-1} - 1)\\ &= 2^{63} + 2^{63} - 1 \approx 18{,}000{,}000{,}000{,}000{,}000{,}000\end{aligned}$$

To give you an idea of the magnitude of this number, this is more wheat than has been produced in the world since chess was invented.

The same king decided to have a banquet in the long banquet room of the palace to celebrate the invention of chess. The king had 50 square tables, and each table could seat only one person on each side. The king pushed the tables together to form one long banquet table. How many people can sit at this table? *Hint:* Try constructing a pattern by using 2 tables, 3 tables, and 4 tables. [Answer: 102 people]

SECTION 6.1 Introduction to Equations

OBJECTIVE A Solutions of an equation

POINT OF INTEREST

One of the most famous equations ever stated is $E = mc^2$. This equation, stated by Albert Einstein, states a relationship between mass m and energy E. As a side note, the chemical element einsteinium was named in honor of Einstein.

An **equation** expresses the equality of two mathematical expressions. Here is an example of an equation.

$$x + 4 = 9$$

Although some equations are more complicated than this one, this equation does have the main characteristics shared by all equations. There is a *left side* ($x + 4$), an *equal sign* (=), and a *right side* (9).

The display at the right shows some other examples of equations.

$$3x - 7 = 4x + 9$$
$$y = 3x - 6$$
$$2z^2 - 5z + 10 = 0$$

The first equation in the display above is a *first-degree equation in one variable.* The equation has one variable, x, and each occurrence of the variable is the first power. First-degree equations in one variable are the topic of this chapter. The second equation is a *first-degree equation in two variables.* These are discussed in Chapter 7. The third equation is a *second-degree equation in one variable*; the highest exponent on the variable z is 2. These equations are discussed in Chapter 12.

Which of the equations shown at the right are first-degree equations in one variable?

1. $5x + 4 = 9 - 3(2x + 1)$
2. $\sqrt{x} + 9 = 10$
3. $p = -14$
4. $2x - 5 = x^2 - 9$

Equation 1 is a first-degree equation in one variable.
Equation 2 is not a first-degree equation in one variable. First-degree equations do not contain square roots of variable expressions.
Equation 3 is a first-degree equation in one variable.
Equation 4 is not a first-degree equation in one variable. First-degree equations in one variable do not have exponents greater than 1 on the variable.

Just as a statement may be true or false, an equation may be true or false.

The equation at the right is true if the variable is replaced by 5.

$x + 7 = 12$
$5 + 7 = 12$ True equation

The equation is false if the variable is replaced by 2.

$2 + 7 = 12$ False equation

A **solution** of an equation is a number that, when substituted for the variable, results in a true equation.

15 is a solution of $x - 5 = 10$ because $15 - 5 = 10$ is a true equation;
20 is not a solution of $x - 5 = 10$ because $20 - 5 = 10$ is a false equation.

Is 2 a solution of the equation $4 - 3x = 5x - 12$?

Replace the variable by 2.

Evaluate the numerical expressions using the Order of Operations Agreement.

Compare the results. If the results are equal, the given number is a solution. If the results are not equal, the given number is not a solution.

$$\begin{array}{r|l} 4 - 3x = & 5x - 12 \\ \hline 4 - 3(2) & 5(2) - 12 \\ 4 - 6 & 10 - 12 \\ -2 = & -2 \end{array}$$

Yes, 2 is a solution of $4 - 3x = 5x - 12$.

Example 1 Is $\frac{1}{2}$ a solution of $4x - 3 = 6x + 1$?

Solution

$$\begin{array}{r|l} 4x - 3 = & 6x + 1 \\ \hline 4\left(\frac{1}{2}\right) - 3 & 6\left(\frac{1}{2}\right) + 1 \\ 2 - 3 & 3 + 1 \\ -1 \neq & 4 \end{array}$$

No, $\frac{1}{2}$ is not a solution of $4x - 3 = 6x + 1$.

You Try It Is $-\frac{2}{3}$ a solution of $4 - 3x = 6x + 10$?

Your Solution

Solution on p. A13

OBJECTIVE B Equations of the form $x + a = b$

POINT OF INTEREST
Finding solutions of equations has been a principal aim of mathematics for thousands of years. However, the equal sign did not appear in any text until 1557.

To solve an equation means to find its solutions. The simplest equation to solve is an equation of the form **variable = constant**. The constant is the solution.

Consider the equation $x = 7$. The solution is 7 because $7 = 7$ is a true equation.

Note that replacing x in $x + 8 = 12$ by 4 results in a true equation. The solution of the equation $x + 8 = 12$ is 4.

$$\begin{aligned} x + 8 &= 12 \\ 4 + 8 &= 12 \\ 12 &= 12 \end{aligned}$$

If 5 is added to each side of $x + 8 = 12$, the solution is still 4.

$$\begin{aligned} x + 8 &= 12 \\ x + 8 + 5 &= 12 + 5 \\ x + 13 &= 17 \end{aligned}$$

Check: $\begin{array}{r|l} x + 13 = & 17 \\ \hline 4 + 13 & 17 \\ 17 = & 17 \end{array}$

If -3 is added to each side of $x + 8 = 12$, the solution is still 4.

$$\begin{aligned} x + 8 &= 12 \\ x + 8 + (-3) &= 12 + (-3) \\ x + 5 &= 9 \end{aligned}$$

Check: $\begin{array}{r|l} x + 5 = & 9 \\ \hline 4 + 5 & 9 \\ 9 = & 9 \end{array}$

This suggests that adding the same number to each side of an equation does not change the solution of the equation. This is called the Addition Property of Equations.

Addition Property of Equations

The same number or variable expression can be added to each side of an equation without changing the solution of the equation.

CONSIDER THIS

Think of an equation as a balance scale. If the same weight is not added to each side of the equation, the pans no longer balance.

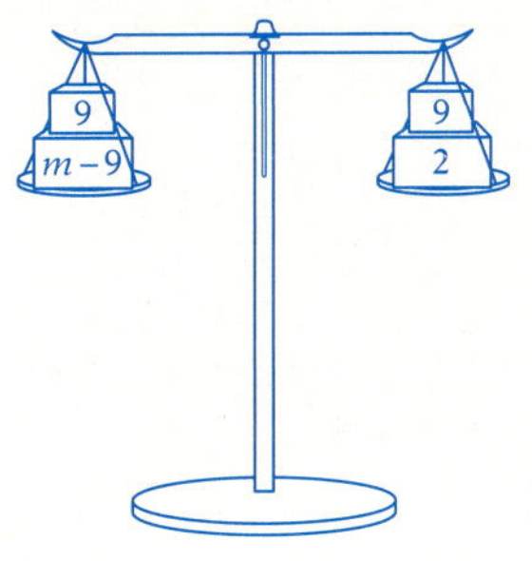

This property is used in solving equations. Note the effect of adding, to each side of the equation $x + 8 = 12$, the opposite of the constant term 8. After simplifying, the equation is in the form *variable* = *constant*. The solution is the constant, 4.

$$x + 8 = 12$$
$$x + 8 + (-8) = 12 + (-8)$$
$$x + 0 = 4$$
$$x = 4$$

Check the solution.

Check: $x + 8 = 12$

$$\begin{array}{c|c} 4 + 8 & 12 \end{array}$$
$$12 = 12$$

The solution checks.

The solution is 4.

The goal in solving an equation is to rewrite it in the form **variable = constant**. **The Addition Property of Equations is used to remove a term from one side of an equation by adding the opposite of that term to each side of the equation.** The resulting equation has the same solution as the original equation.

Solve: $m - 9 = 2$

Remove the constant term -9 from the left side of the equation by adding 9, the opposite of -9, to each side of the equation. Then simplify.

$$m - 9 = 2$$
$$m - 9 + 9 = 2 + 9$$
$$m + 0 = 11$$
$$m = 11$$

You should check the solution.

The solution is 11.

In each of the equations above, the variable appeared on the left side of the equation, and the equation was rewritten in the form *variable* = *constant*. For some equations, it may be more practical to work toward the goal of *constant* = *variable*, as shown in the example below.

Solve: $12 = n - 8$

The variable is on the right side of the equation. The goal is to rewrite the equation in the form *constant* = *variable*.

Remove the constant term from the right side of the equation by adding 8 to each side of the equation. Then simplify.

$$12 = n - 8$$
$$12 + 8 = n - 8 + 8$$
$$20 = n + 0$$
$$20 = n$$

You should check the solution.

The solution is 20.

Because subtraction is defined in terms of addition, the Addition Property of Equations allows the same term to be subtracted from each side of an equation without changing its solution.

Solve: $z + 9 = 6$

The goal is to rewrite the equation in the form *variable* = *constant.*

Add the opposite of 9 to each side of the equation. This is equivalent to subtracting 9 from each side of the equation. Then simplify.

$$\begin{aligned} z + 9 &= 6 \\ z + 9 - 9 &= 6 - 9 \\ z + 0 &= -3 \\ z &= -3 \end{aligned}$$

Check the solution.

Check: $z + 9 = 6$

$$\begin{array}{c|c} -3 + 9 & 6 \end{array}$$

$6 = 6$

The solution checks.

The solution is -3.

Solve: $5 + x - 9 = -10$

Simplify the left side of the equation by combining the constant terms.

$$\begin{aligned} 5 + x - 9 &= -10 \\ x - 4 &= -10 \end{aligned}$$

Add 4 to each side of the equation. Simplify.

$$\begin{aligned} x - 4 + 4 &= -10 + 4 \\ x + 0 &= -6 \\ x &= -6 \end{aligned}$$

-6 checks as a solution.

The solution is -6.

Example 2 Solve: $6 + x = 4$

Solution

$$\begin{aligned} 6 + x &= 4 \\ 6 - 6 + x &= 4 - 6 \\ x &= -2 \end{aligned}$$

The solution is -2.

You Try It 2 Solve: $7 + y = 12$

Your Solution

Example 3 Solve: $7x - 4 - 6x = 3$

Solution

$$\begin{aligned} 7x - 4 - 6x &= 3 \\ x - 4 &= 3 \quad \text{Combine like terms.} \\ x - 4 + 4 &= 3 + 4 \\ x &= 7 \end{aligned}$$

The solution is 7.

You Try It 3 Solve: $-5r + 3 + 6r = 1$

Your Solution

Solutions on p. A13

OBJECTIVE C Equations of the form $ax = b$

POINT OF INTEREST

Equations of the form $ax = b$ have appeared in algebra texts for a long time. The problem below is an adaptation from Fibonacci's text *Liber Abaci,* which dates from 1202.

A merchant purchased 7 eggs for 1 denarius and sold them at a rate of 5 eggs for 1 denarius. The merchant's profit was 18 denarii. How much did the merchant invest?

The resulting equation is

$$\frac{7}{5}x - x = 18$$

The solution is 45 denarii.

Note that replacing x by 3 in $4x = 12$ results in a true equation. The solution of the equation is 3.

$$\begin{aligned} 4x &= 12 \\ 4(3) &= 12 \\ 12 &= 12 \end{aligned}$$

If each side of the equation $4x = 12$ is multiplied by 2, the solution is still 3.

$$\begin{aligned} 4x &= 12 \\ 2(4x) &= 2(12) \\ 8x &= 24 \end{aligned}$$

Check:

$$\begin{array}{c|c} \multicolumn{2}{c}{8x = 24} \\ \hline 8(3) & 24 \\ \multicolumn{2}{c}{24 = 24} \end{array}$$

If each side of the equation $4x = 12$ is multiplied by -3, the solution is still 3.

$$\begin{aligned} 4x &= 12 \\ -3(4x) &= -3(12) \\ -12x &= -36 \end{aligned}$$

Check:

$$\begin{array}{c|c} \multicolumn{2}{c}{-12x = -36} \\ \hline -12(3) & -36 \\ \multicolumn{2}{c}{-36 = -36} \end{array}$$

This suggests that multiplying each side of an equation by the same nonzero number does not change its solution. This is called the Multiplication Property of Equations.

> ***Multiplication Property of Equations***
>
> Each side of an equation can be multiplied by the same nonzero number without changing the solution of the equation.

This property is used in solving equations. Note the effect of multiplying each side of the equation $4x = 12$ by $\frac{1}{4}$, the reciprocal of the coefficient 4. After simplifying, the equation is in the form *variable* = *constant.*

$$\begin{aligned} 4x &= 12 \\ \frac{1}{4} \cdot 4x &= \frac{1}{4} \cdot 12 \\ 1 \cdot x &= 3 \\ x &= 3 \end{aligned}$$

The solution is 3.

The Multiplication Property of Equations is used to remove a coefficient from a variable term of an equation by multiplying each side of the equation by the reciprocal of the coefficient. The resulting equation will have the same solution as the original equation.

Solve: $\frac{3}{4}x = -9$

The goal is to rewrite the equation in the form *variable* = *constant.*

Multiply each side of the equation by $\frac{4}{3}$, the reciprocal of $\frac{3}{4}$. After simplifying, the equation is in the form *variable* = *constant.*

$$\begin{aligned} \frac{3}{4}x &= -9 \\ \frac{4}{3} \cdot \frac{3}{4}x &= \frac{4}{3} \cdot (-9) \\ 1 \cdot x &= -12 \\ x &= -12 \end{aligned}$$

You should check this solution.

The solution is -12.

Because division is defined in terms of multiplication, the Multiplication Property of Equations allows each side of an equation to be divided by the same nonzero number without changing its solution.

Solve: $-2x = 8$

Multiply each side of the equation by the reciprocal of -2. This is equivalent to dividing each side of the equation by -2

$$-2x = 8$$
$$\frac{-2x}{-2} = \frac{8}{-2}$$
$$1 \cdot x = -4$$
$$x = -4$$

Check the solution.

Check: $-2x = 8$

$-2(-4)$	8

$8 = 8$

The solution checks.

The solution is -4.

When using the Multiplication Property of Equations, multiply each side of the equation by the reciprocal of the coefficient when the coefficient is a fraction. Divide each side of the equation by the coefficient when the coefficient is an integer or a decimal.

Example 4 Solve: $\frac{2x}{3} = 12$

Solution

$$\frac{2x}{3} = 12 \qquad \frac{2x}{3} = \frac{2}{3}x$$
$$\frac{3}{2}\left(\frac{2}{3}x\right) = \frac{3}{2}(12)$$
$$x = 18$$

The solution is 18.

You Try It 4 Solve: $10 = \frac{-2x}{5}$

Your Solution

Example 5 Solve and check:
$3y - 7y = 8$

Solution

$$3y - 7y = 8$$
$$-4y = 8 \qquad \text{Combine like terms.}$$
$$\frac{-4y}{-4} = \frac{8}{-4}$$
$$y = -2$$

Check:

$3y - 7y = 8$

$3(-2) - 7(-2)$	8
$-6 - (-14)$	8
$-6 + 14$	8

$8 = 8$

-2 checks as the solution.
The solution is -2.

You Try It 5 Solve and check:
$\frac{1}{3}x - \frac{5}{6}x = 4$

Your Solution

Solutions on p. A13

6.1 EXERCISES

Objective A

1. Is 5 a solution of $x + 2 = 7$?

2. Is 12 a solution of $y - 9 = 3$?

3. Is 2 a solution of $3n = 9$?

4. Is 7 a solution of $2a = 18$?

5. Is 2 a solution of $3b + 5 = 11$?

6. Is 3 a solution of $9 - 2a = 3$?

7. Is 1 a solution of $6 - 2x = 4x$?

8. Is 2 a solution of $8 - y = 3y$?

9. Is 0 a solution of $3n + 2 = 2n + 3$?

10. Is 6 a solution of $2m + 2 = 3m - 4$?

11. Is -1 a solution of $2 + 3x = 5x - 4x$?

12. Is -4 a solution of $4y - 4 = 2 + 5y$?

13. Is -2 a solution of $7 - 4b = 5 - 2b$?

14. Is -4 a solution of $5 - 2t = 1 - 3t$?

15. Is -2 a solution of $4 - 2a = a + 10$?

16. Is -3 a solution of $5 - b = 2 - 2b$?

17. Is 0 a solution of $x + 2(3x - 4) = 7x - 8$?

18. Is -2 a solution of $5y + 3 = y + 2(y + 4)$?

19. Is $-\frac{1}{2}$ a solution of $4t + 1 = -1$?

20. Is $\frac{1}{4}$ a solution of $2x - 1 = 3$?

21. Is $\frac{3}{5}$ a solution of $15y + 1 = 10y - 2$?

22. Is $-\frac{1}{4}$ a solution of $6n + 2 = 2n + 3$?

23. Is $-\frac{2}{5}$ a solution of $2m - 1 = 5m - 3$?

24. Is $\frac{1}{3}$ a solution of $6a + 2 = 9a - 1$?

25. Is $\frac{3}{4}$ a solution of $9b - 6 = 5b - 3$?

26. Is $-\frac{2}{3}$ a solution of $7x - 6 = 4x - 8$?

27. Is $-\frac{1}{5}$ a solution of $2(y - 1) = 7y + 3$?

Objective B

Solve.

28. $x + 3 = 9$
29. $y + 6 = 8$
30. $a - 2 = 3$
31. $b - 3 = 6$
32. $4 + x = 13$
33. $9 + y = 14$
34. $m - 12 = 5$
35. $n - 9 = 3$
36. $x - 3 = -2$
37. $y - 6 = -1$
38. $a + 5 = -2$
39. $b + 3 = -3$
40. $3 + m = -6$
41. $5 + n = -2$
42. $8 + t = -2$
43. $3 + s = -1$
44. $8 = x + 3$
45. $7 = y + 5$
46. $3 = w - 6$
47. $4 = y - 3$
48. $3 = 3 + a$
49. $4 = 4 + b$
50. $-7 = -7 + m$
51. $-9 = -9 + n$
52. $-3 = v + 5$
53. $-1 = w + 2$
54. $-4 = s - 2$
55. $-2 = t - 1$
56. $-5 = 1 + x$
57. $-3 = 4 + y$
58. $3 = -9 + m$
59. $4 = -5 + n$
60. $4 + x - 7 = 3$
61. $12 + y - 4 = 8$
62. $8t + 6 - 7t = -6$
63. $-5z + 5 + 6z = 12$
64. $y + \frac{4}{7} = \frac{6}{7}$
65. $z + \frac{3}{5} = \frac{4}{5}$
66. $x - \frac{3}{8} = \frac{1}{8}$
67. $a - \frac{1}{6} = \frac{5}{6}$

Solve.

68. $c + \frac{2}{3} = \frac{3}{4}$ **69.** $n + \frac{1}{3} = \frac{2}{5}$ **70.** $w - \frac{1}{4} = \frac{3}{8}$ **71.** $t - \frac{1}{3} = \frac{1}{2}$

Objective C

Solve.

72. $3x = 9$ **73.** $8a = 16$ **74.** $4c = -12$ **75.** $5z = -25$

76. $2y = 0$ **77.** $9t = 0$ **78.** $-2r = 16$ **79.** $-6p = 72$

80. $-4m = -28$ **81.** $-12x = -36$ **82.** $-3y = 0$ **83.** $-7a = 0$

84. $12 = 2c$ **85.** $28 = 7x$ **86.** $-56 = 14z$ **87.** $-72 = 18v$

88. $18 = -6n$ **89.** $35 = -5p$ **90.** $-68 = -17t$ **91.** $-60 = -15y$

92. $12x = 30$ **93.** $9v = 15$ **94.** $-6a = 21$ **95.** $-8c = 20$

96. $15m = -40$ **97.** $-18p = 42$ **98.** $-16y = -40$ **99.** $-20z = -30$

100. $28 = -12y$ **101.** $36 = -16z$ **102.** $-52 = -18a$ **103.** $-40 = -30w$

104. $\frac{2}{3}x = 4$ **105.** $\frac{3}{4}y = 9$ **106.** $\frac{1}{3}a = -12$ **107.** $\frac{3y}{5} = -15$

Solve.

108. $-\frac{4c}{7} = 16$

109. $-\frac{5n}{8} = 20$

110. $-\frac{z}{4} = -3$

111. $-\frac{3x}{8} = -15$

112. $8 = \frac{4}{5}y$

113. $9 = \frac{3}{8}x$

114. $10 = -\frac{5}{6}c$

115. $22 = -\frac{11}{12}p$

116. $\frac{2x}{3} = \frac{3}{4}$

117. $\frac{5y}{6} = \frac{7}{12}$

118. $\frac{-3v}{4} = -\frac{7}{8}$

119. $\frac{-4z}{5} = -\frac{8}{9}$

120. $7y - 9y = 10$

121. $8w - 5w = 9$

122. $m - 4m = 21$

123. $2a - 6a = 10$

Critical Thinking

124. Solve the equation $x + a = b$ for x. Is the solution you have written valid for all real numbers a and b?

125. Solve the equation $ax = b$ for x. Is the solution you have written valid for all real numbers a and b?

126. Is it possible for an equation to have no solution? If so, give an example of one.

127. Solve: **a.** $\frac{2}{\frac{1}{x}} = 8$ **b.** $\frac{3}{\frac{2}{x}} = 6$

128. [W] Write out the steps for solving the equation $\frac{2}{3}x = 6$. Identify each Property of Real Numbers or Property of Equations as you use it.

129. [W] Using your own words, state the Addition Property of Equations and the Multiplication Property of Equations.

SECTION 6.2 Applications: The Basic Percent Equation

OBJECTIVE A The basic percent equation

A real estate broker receives a payment that is 6% of a \$175,000 sale. To find the amount the broker receives requires answering the question, "6% of \$175,000 is what?" This sentence can be written using mathematical symbols and then solved for the unknown number. The word **of** is written as · (times), **is** is written as = (equals), and **what** is written as a variable (the unknown number).

6%	of	\$175,000	is	what?
↓	↓	↓	↓	↓
percent	·	base	=	amount
6%		175,000		A

$$0.06 \cdot \$175{,}000 = A$$
$$\$10{,}500 = A$$

The broker receives a payment of \$10,500.

The solution was found by solving the basic percent equation for amount.

The Basic Percent Equation

Percent · base = amount
$PB = A$

Find 2.5% of 800.

Use the basic percent equation.
Percent = 2.5% = 0.025, base = 800,
amount = A

Percent · base = amount
$0.025 \cdot 800 = A$
$20 = A$

2.5% of 800 is 20.

A recent promotional game at a grocery store listed the probability of winning a prize as "1 chance in 2." A percent can be used to describe the chance of winning. This requires answering the question, "What percent of 2 is 1?"

The chance of winning can be found by solving the basic percent equation for percent.

What percent	of	2	is	1?
↓	↓	↓	↓	↓
percent	·	base	=	amount

$$P \cdot 2 = 1$$
$$P = \frac{1}{2}$$

Write the fraction as a percent. $P = \frac{1}{2}(100\%) = 50\%$

There is a 50% chance of winning a prize.

32 is what percent of 20?

Use the basic percent equation.
Percent = P, base = 20,
amount = 32

$$\text{Percent} \cdot \text{base} = \text{amount}$$
$$P \cdot 20 = 32$$
$$P = \frac{32}{20}$$
$$P = 1.6$$

Write 1.6 as a percent.

$$P = 160\%$$

32 is 160% of 20.

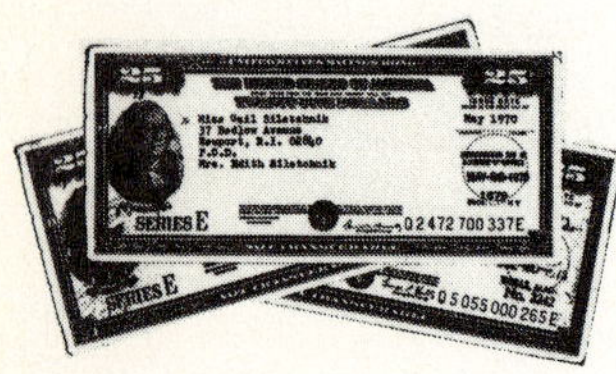

Each year an investor receives a payment that equals 8% of the value of an investment. This year that payment was \$640. To find the value of the investment this year, we must answer the question, "8% of what value is \$640?"

The value of the investment can be found by solving the basic percent equation for the base.

8%	of	what	is	\$640?
↓	↓	↓	↓	↓
Percent	·	base	=	amount
8%		B		640

$$0.08 \cdot B = 640$$
$$B = \frac{640}{0.08}$$
$$B = 8000$$

This year the investment is worth \$8000.

62% of what is 800? Round to the nearest tenth.

Use the basic percent equation.
Percent = 62% = 0.62, base = B,
amount = 800

$$\text{Percent} \cdot \text{base} = \text{amount}$$
$$0.62 \cdot B = 800$$
$$B = \frac{800}{0.62}$$
$$B \approx 1290.3$$

62% of 1290.3 is approximately 800.

Note from the previous problems, if any two parts of the basic percent equation are given, the third part can be found. **In percent problems, the base usually follows the word "of."** Some percent problems may use the word "find." In this case, we can substitute "what is" for find.

Example 1 Find 9.4% of 240.

Strategy To find the amount, solve the basic percent equation.
Percent = 9.4% = 0.094,
base = 240, amount = A

Solution

$$\text{Percent} \cdot \text{base} = \text{amount}$$
$$0.094 \cdot 240 = A$$
$$22.56 = A$$

22.56 is 9.4% of 240.

You Try It 1 Find $33\frac{1}{3}\%$ of 45.

Your Strategy

Your Solution

Solution on p. A13

Example 2 What percent of 30 is 12?

Strategy To find the percent, solve the basic percent equation.
Percent $= P$, base $= 30$,
amount $= 12$

Solution

$$\text{Percent} \cdot \text{base} = \text{amount}$$
$$P \cdot 30 = 12$$
$$P = \frac{12}{30} = 0.4$$
$$P = 40\%$$

12 is 40% of 30.

You Try It 2 25 is what percent of 40?

Your Strategy

Your Solution

Example 3 60 is 2.5% of what?

Strategy To find the base, solve the basic percent equation.
Percent $= 2.5\% = 0.025$,
base $= B$, amount $= 60$

Solution

$$\text{Percent} \cdot \text{base} = \text{amount}$$
$$0.025 \cdot B = 60$$
$$B = \frac{60}{0.025}$$
$$B = 2400$$

60 is 2.5% of 2400.

You Try It 3 $16\frac{2}{3}\%$ of what is 15?

Your Strategy

Your Solution

Solutions on pp. A13–A14

OBJECTIVE B Applications of the basic percent equation

CONSIDER THIS

In application problems involving percent, the basic percent equation frequently results in an equation of the form $ax = b$.

A computer programmer receives a weekly wage of \$650, and \$110.50 is deducted for income tax. Find the percent of the computer programmer's salary deducted for income tax.

Use the basic percent equation.
Percent $= P$, base $= 650$,
amount $= 110.50$

$$\text{Percent} \cdot \text{base} = \text{amount}$$
$$P \cdot 650 = 110.50$$
$$P = \frac{110.50}{650}$$
$$P = 0.17$$

17% of the computer programmer's salary is deducted for income tax.

A department store has 250 employees and must hire an additional 18% for the holiday season. What is the total number of employees needed for the holiday season?

Use the basic percent equation to find the number of additional employees needed. Percent = 18% = 0.18, base = 250, amount = A

$$\text{Percent} \cdot \text{base} = \text{amount}$$
$$0.18 \cdot 250 = A$$
$$45 = A$$

Add the number of employees hired to the present number of employees.

$$250 + 45 = 295$$

The department store will employ 295 employees for the holiday season.

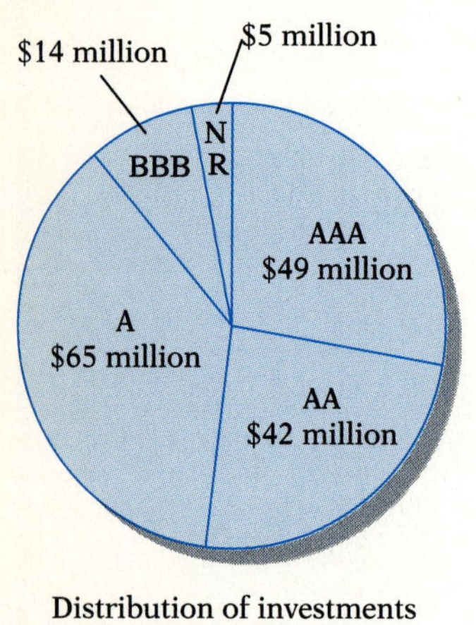

Distribution of investments in municipal bonds

The circle graph at the left represents the amount of money a mutual fund company has invested in different municipal bonds. The complete circle represents the total amount invested in all bonds, $175 million. Each sector of the circle represents the amount of money invested in a different quality of bonds. To find the percent of the total money invested in AAA-rated bonds, solve the basic percent equation for percent.

Percent = P, base = 175 million, amount = 49 million

$$\text{Percent} \cdot \text{base} = \text{amount}$$
$$P \cdot 175 = 49$$
$$P = \frac{49}{175}$$
$$P = 0.28$$
$$P = 28\%$$

28% of the total amount invested is invested in AAA-rated bonds.

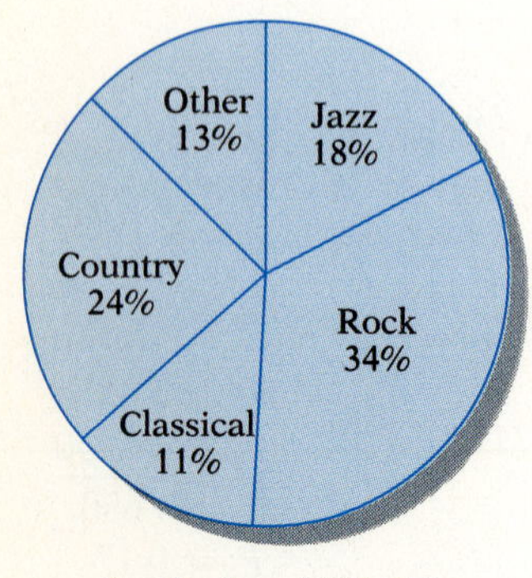

Total sales of $55,000 in compact disks for 1 month

In some circle graphs, each sector represents a percent of the total. The circle graph at the left represents the total sales in dollars of compact disks for a record store for one month. The complete circle represents 100% of the $55,000 in compact disks sold. Each sector expresses the percent of sales for a particular type of music. To find the dollar sales for jazz compact disks for the one-month period, solve the basic percent equation for amount.

Percent = 18% = 0.18, base = 55,000, amount = A

$$\text{Percent} \cdot \text{base} = \text{amount}$$
$$0.18 \cdot 55{,}000 = A$$
$$9900 = A$$

The dollar sales for jazz compact disks were $9900.

Example 4

The circle graph below shows the result of a survey of 300 people who were asked to name their favorite sport. Find the percent of people surveyed who listed tennis as their favorite spectator sport.

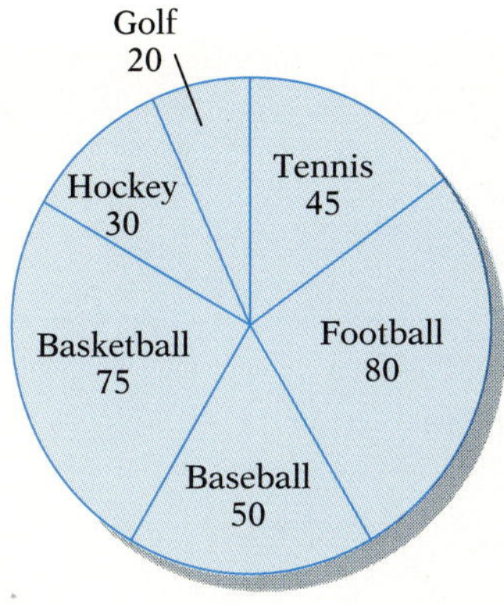

Strategy

To find the percent, solve the basic percent equation for percent. Percent = P, base = 300, amount = 45

Solution

$$\text{Percent} \cdot \text{base} = \text{amount}$$
$$P \cdot 300 = 45$$
$$P = \frac{45}{300}$$
$$P = 0.15 = 15\%$$

15% of the people surveyed selected tennis as their favorite spectator sport.

You Try It 4

The circle graph below represents the percent of a family's annual income of $26,000 that is budgeted for various expenses. Find the annual amount budgeted for food.

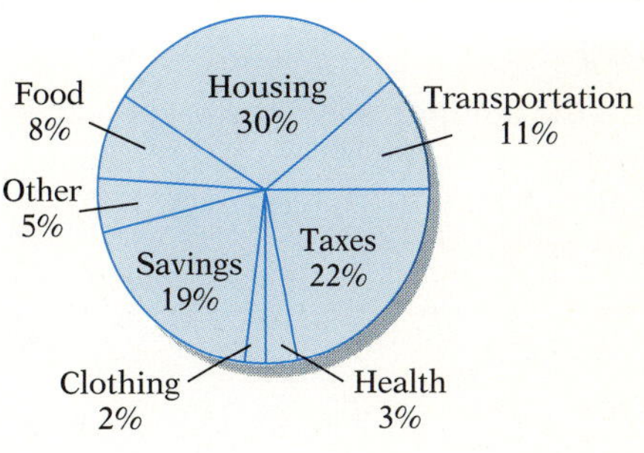

Your Strategy

Your Solution

Example 5

Twelve percent of a company's $60,000 budget is used for advertising. Find the amount of the company's budget spent for advertising.

Strategy

To find the amount, use the basic percent equation.
Percent = 12% = 0.12, base = 60,000, amount = A

Solution

$$\text{Percent} \cdot \text{base} = \text{amount}$$
$$0.12 \cdot 60{,}000 = A$$
$$7200 = A$$

The company spent $7200 for advertising.

You Try It 5

An instructor receives a monthly salary of $2165, and $324.75 is deducted for income tax. Find the percent of the instructor's salary deducted for income tax.

Your Strategy

Your Solution

Solutions on p. A14

Example 6

A taxpayer pays 35% of his net income for state and federal taxes. The taxpayer's net income is $37,500. Find the amount of state and federal taxes paid by the taxpayer.

Strategy

To find the amount, solve the basic percent equation.
Percent = 35% = 0.35, base = 37,500, amount = A

Solution

$$\begin{aligned} \text{Percent} \cdot \text{base} &= \text{amount} \\ 0.35 \cdot 37{,}500 &= A \\ 13{,}125 &= A \end{aligned}$$

The amount of taxes paid is $13,125.

You Try It 6

Seventy percent of the people polled in a public opinion survey approved of the way the mayor was performing the duties of government. If 210 persons approved, how many were polled?

Your Strategy

Your Solution

Example 7

A department store has a blue blazer on sale for $114, which is 60% of the original price. What is the difference between the original price and the sale price?

Strategy

To find the difference between the original price and the sale price:

- ▶ Find the original price. Solve the basic percent equation.
  Percent = 60% = 0.60, amount = 114, base = B
- ▶ Subtract the sale price from the original price.

Solution

$$\begin{aligned} \text{Percent} \cdot \text{base} &= \text{amount} \\ 0.60 \cdot B &= 114 \\ B &= \frac{114}{0.60} \\ B &= 190 \end{aligned}$$

$190 - 114 = 76$

The difference in price is $76.

You Try It 7

An electrician's wage this year is $20.01 per hour, which is 115% of last year's wage. What was the increase in the hourly wage over last year?

Your Strategy

Your Solution

Solutions on p. A14

OBJECTIVE C Percent increase and percent decrease

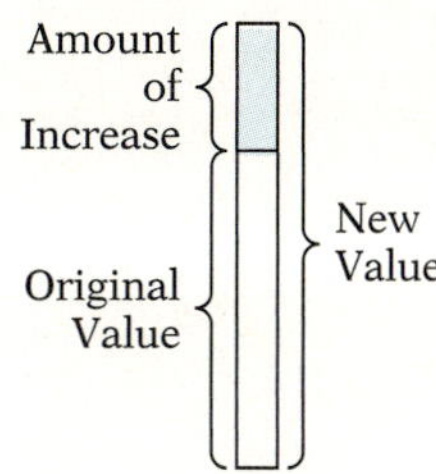

Percent increase is used to show how much a quantity has increased over its original value. The statements "sales volume increased by 11% over last year's sales volume" and "employees received an 8% pay increase" are illustrations of percent increase.

This year a company's production increased 12% over last year's production of 20,000 units. Find this year's production.

Use the basic percent equation to find the increase in production.
Percent = 12% = 0.12, base = 20,000, amount = A

$$\text{Percent} \cdot \text{base} = \text{amount}$$
$$0.12 \cdot 20{,}000 = A$$
$$2400 = A$$

Add the increase (A) in production to the original value to find this year's production.

$$20{,}000 + 2400 = 22{,}400$$

The company's production this year was 22,400 units.

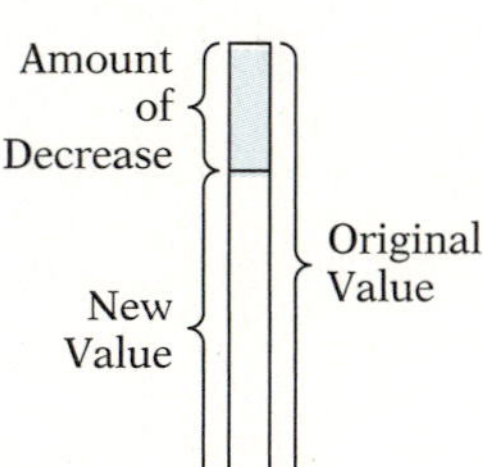

Percent decrease is used to show how much a quantity has decreased from its original value. The statements "the president's approval rating has decreased 9% over last month" and "there has been a 15% decrease in the number of industrial accidents" are illustrations of the use of percent decrease.

The price of a laptop computer has decreased from \$3500 to \$2600 in one year. Find the percent decrease in the price of the computer. Round to the nearest tenth of a percent.

Find the decrease in the price of the computer.

$$\text{Decrease in price} = 3500 - 2600 = 900$$

Use the basic percent equation to find the percent decrease in price.
Percent = P, base = 3500, amount = 900

$$\text{Percent} \cdot \text{base} = \text{amount}$$
$$P \cdot 3500 = 900$$
$$P = \frac{900}{3500}$$

Round the decimal to the nearest thousandth so the percent will be to the nearest tenth.

$$P \approx 0.257$$
$$P \approx 25.7\%$$

The percent decrease in the price of the computer was approximately 25.7%.

Example 8

A sales clerk was earning \$5.60 per hour before an 8% increase in pay. What is the new hourly wage? Round to the nearest cent.

Strategy

To find the new hourly wage:

- ▶ Use the basic percent equation to find the increase in pay.
  Percent = 8% = 0.08, base = 5.60, amount = A
- ▶ Add the amount of increase to the original wage.

Solution

$$\text{Percent} \cdot \text{base} = \text{amount}$$
$$0.08 \cdot 5.60 = A$$
$$0.45 \approx A$$

\$5.60 + \$.45 = \$6.05.

The new hourly wage is \$6.05.

You Try It 8

An automobile manufacturer increased the average mileage on a car from 16.4 mi/gal to 17.2 mi/gal. Find the percent increase in mileage. Round to the nearest tenth of a percent.

Your Strategy

Your Solution

Example 9

Violent crime in a small city decreased from 27 per 1000 people to 24 per 1000 people. Find the percent decrease in violent crime. Round to the nearest tenth of a percent.

Strategy

To find the percent decrease in crime:

- ▶ Find the decrease in the number of crimes.
- ▶ Use the basic percent equation to find the percent decrease in crime.
  Percent = P, base = 27, amount = decrease in the number of crimes

Solution

27 − 24 = 3

$$\text{Percent} \cdot \text{base} = \text{amount}$$
$$P \cdot 27 = 3$$
$$P = \frac{3}{27}$$
$$P \approx 0.111$$

Violent crime decreased by approximately 11.1% during the year.

You Try It 9

The market value of a luxury car decreased 24% during the year. Find the value of a luxury car that cost \$47,000 last year.

Your Strategy

Your Solution

Solutions on p. A14

6.2 EXERCISES

Objective A

Solve.

1. What is 8% of 100?
2. What is 16% of 50?
3. 0.05% of 150 is what?
4. 0.075% of 625 is what?
5. 15 is what percent of 90?
6. 24 is what percent of 60?
7. What percent of 16 is 6?
8. What percent of 24 is 18?
9. 10 is 10% of what?
10. 37 is 37% of what?
11. 2.5% of what is 30?
12. 10.4% of what is 52?
13. Find 10.7% of 485.
14. Find 12.8% of 625.
15. What is 80% of 16.25?
16. What is 26% of 19.5?
17. 54 is what percent of 2000?
18. 8 is what percent of 2500?
19. 16.4 is what percent of 4.1?
20. 5.3 is what percent of 50?
21. 18 is 240% of what?
22. 24 is 320% of what?
23. 25.6 is 12.8% of what?
24. 45.014 is 63.4% of what?
25. 1 is what percent of 40?
26. 0.3 is what percent of 20?
27. What percent of 48 is 18?
28. What percent of 11 is 88?
29. 0.7% of what is 0.56?
30. 0.25% of what is 1?
31. 30% of what is 2.7?
32. 78% of what is 3.9?

Objective B

Solve.

33. A computer programmer's salary increased \$19.52 per week. By what percent did the programmer's salary increase if the salary was \$244 before the raise?

34. A charity organization spent \$2940 for administrative expenses. This amount is 12% of the money it collected. What is the total amount the charity organization collected?

35. A mechanic estimates that the brakes of an RV still have 6000 mi of wear. This amount is 12% of the estimated safe-life use of the brakes. What is the estimated safe-life of the brakes?

36. A company spends \$4500 of its \$90,000 budget for advertising. What percent of the budget is spent for advertising?

37. A sales clerk earns \$2240 per month, and 18% of this amount is deducted for income tax. Find the amount deducted for income tax.

38. An antique shop owner expects to receive $16\frac{2}{3}\%$ of the shop's sales as profit. What is the expected profit in a month when the total sales are \$24,000?

39. Last month a thrift store brought in an income of \$2812.50. The rent for the store is \$900 per month. What percent of last month's income was spent for rent?

40. A used mobile home was purchased for \$18,000. This amount was 64% of the new mobile home cost. What is the cost of a new mobile home?

41. A calculator can be purchased for \$28.50. This amount is 40% of the cost of the calculator 8 years ago. What was the cost of the calculator 8 years ago?

42. A city has a population of 42,000. This amount is 75% of what the population was 5 years ago. What was the city's population 5 years ago?

The circle graph represents the number of playground injuries suffered by children on various pieces of playground equipment.

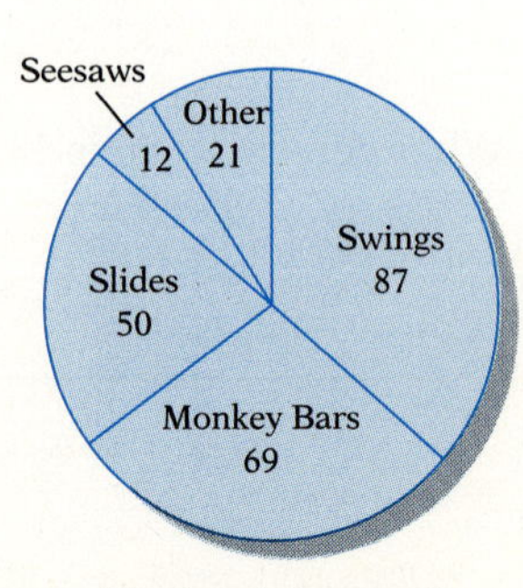

Number of playground injuries in thousands

43. What percent of the total injuries happen on slides? Round to the nearest tenth of a percent.

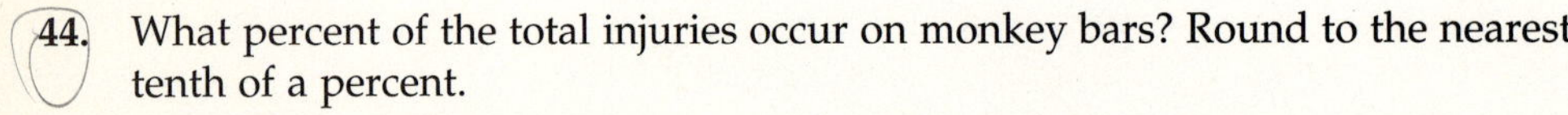

44. What percent of the total injuries occur on monkey bars? Round to the nearest tenth of a percent.

Solve.

45. The snowfall at a ski resort was 198 in. during the year. This amount was 120% of the previous year's snowfall. What was the previous year's snowfall?

46. A fire department answered 24 false alarms out of a total of 200 alarms. What percent of the alarms were false alarms?

47. A store advertised a scientific calculator for $55.80. This amount was 112% of the cost at a competing store. What was the price at the competing store?

48. A car is sold for $8900. The buyer of the car makes a down payment of $1780. What percent of the selling price is the down payment?

49. A farmer is given an income tax credit of 15% of the cost of some farm machinery. What tax credit would the farmer receive on farm equipment that cost $85,000?

An accounting major recorded the number of units required in each discipline to graduate with a degree in accounting. The results are shown in the circle graph.

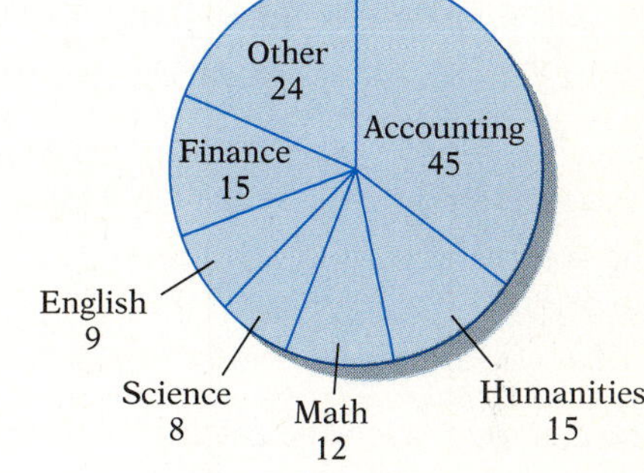

Number of units required to graduate with an accounting degree

50. What percent of the units required to graduate are taken in accounting? Round to the nearest tenth of a percent.

51. What percent of the units required to graduate are taken in mathematics? Round to the nearest tenth of a percent.

52. A customer bought a car for $8500 and paid a sales tax equal to 5.5% of the cost.
a. Find the amount of the sales tax.
b. Find the total cost of the car including the sales tax.

53. During a quality control test, a manufacturer of computer boards found that 56 boards were defective. This amount was 0.7% of the computer boards tested.
a. How many computer boards were tested?
b. How many computer boards tested were not defective?

54. A baseball player with a salary of $1,650,000 is offered a new contract that provides for an 8% salary cut. What is the amount of money offered in the new contract?

55. To receive a license to sell insurance, an insurance account executive must answer 70% of the 250 questions on a test correctly. An account executive answered 177 questions correctly. Did the account executive pass the test?

Objective C

Solve.

56. Enrollment in computer science classes increased from 568 students to 860 students. Find the percent increase in the number of students enrolled in computer science classes. Round to the nearest tenth of a percent.

57. A new model pickup increased in price to \$11,200 from \$10,500, the price for last year's model. Find the percent increase in price. Round to the nearest tenth of a percent.

58. A mutual fund selling for \$15.20 per share increased by 3.2% over a three-month period. Find the new price of the mutual fund. Round to the nearest cent.

59. A contractor built an addition on a house, increasing the 13,500-square foot home by 2500 ft^2. Find the percent increase in size of the remodeled home. Round to the nearest tenth of a percent.

60. During the recession, the number of housing starts in a community decreased from 132 to 86. What percent decrease does this represent? Round to the nearest tenth of a percent.

61. The weight of a chocolate bar was decreased by 20%. Find the new weight if the original weight was 5.5 oz.

62. The cost of a scientific calculator decreased from \$87 to \$62.50. What percent decrease does this represent? Round to the nearest tenth of a percent.

63. The batting average of a baseball player decreased from .298 to .264. What percent decrease does this represent? Round to the nearest tenth of a percent.

64. The 60-member sales staff of a publishing house increased by 15%. Find the number of members on the increased staff.

65. A manufacturer of fax machines increased its monthly output of 2500 machines by 12%. What is the amount of increase?

The double-bar graph represents the annual tuition costs at five western state universities for students who are residents of the state and students who are not.

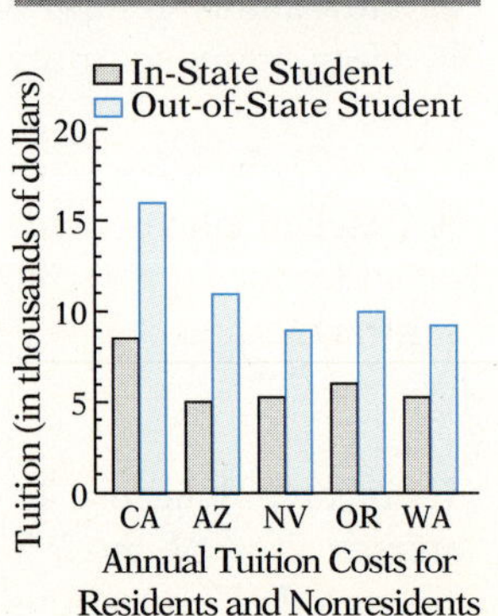

Annual Tuition Costs for Residents and Nonresidents

66. What is the percent increase in tuition for a nonresident student over that of a resident student for Nevada (NV)?

67. What is the percent increase in tuition for a nonresident student over that of a resident student for Arizona (AZ)?

Solve.

68. The average price of gasoline increased from \$1.19 to \$1.32 during a six-month period. Find the percent increase in the price of gasoline. Round to the nearest tenth of a percent.

69. The government requires a 20% increase in the average miles per gallon obtained by the cars manufactured over the next five-year period. One manufacturer had been averaging 16.4 mi/gal. Find the average miles per gallon the manufacturer must obtain to comply with the government order.

70. The compensation of a CEO at a major corporation decreased from \$2,400,000 to \$1,650,000. What percent decrease in compensation does this represent?

71. The income of a mutual fund decreased 30% from last year's income of \$1080. Find this year's income.

72. By careful menu planning, a family was able to reduce its monthly food budget of \$320 by \$50. What percent decrease does this represent?

73. It is estimated that the value of a new car is reduced by 25% after one year of ownership. Using this estimate, how much value does a \$13,200 new car lose after one year?

74. The number of seats in a new airliner is a 30% increase over the number in the older model it is replacing. The older model contained 120 seats. Find the number of seats in the new model.

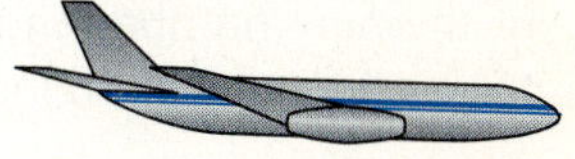

75. An auto manufacturer states that the fuel consumption of its economy car was 35 mi/gal and is now 42 mi/gal. What percent increase in fuel economy does this represent?

76. A new motorcycle that sold for \$1580 last year is now selling for \$1720. Find the percent increase in price for the new model. Round to the nearest tenth of a percent.

77. The dividend of a utility stock increased from \$1.40 to \$1.52 per share. What percent increase does this represent? Round to the nearest tenth of a percent.

The double-bar graph represents the price of a company's stock before restructuring and the price of its stock one year later.

78. Which company had the largest percent increase in stock price?

79. What company had the smallest percent decrease in stock price?

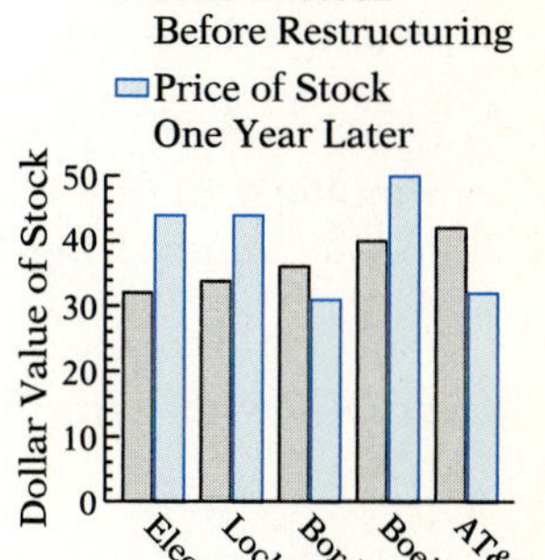

Values of a Company's Stock Before and After Reconstructing

Solve.

80. A new production method reduced the time needed to clean a piece of metal from 8 min to 5 min. What is the amount of decrease? What percent decrease does this represent?

81. A sales manager's average monthly expense for gasoline was $92. After joining a car pool, the manager was able to reduce this expense by 22%. What was the amount of decrease? What is the average monthly gasoline bill now?

82. The price of gold dropped from approximately $650 per ounce to $375 per ounce in six months. What percent decrease does this represent? Round to the nearest tenth of a percent.

83. As a result of an increased number of service lines at a grocery store, the average amount of time a customer waits in line has decreased from 3.8 min to 2.5 min. What percent decrease does this represent? Round to the nearest tenth of a percent.

84. One share of a company's stock sold for $\$44\frac{3}{8}$ at the beginning of the year. At the end of the year the stock sold for $\$65\frac{1}{2}$. Find the percent increase in the price of the stock during the year. Round to the nearest tenth of a percent.

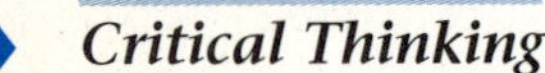

Critical Thinking

85. Find 10% of a number and subtract it from the original number. Now take 10% of the new number and subtract it from the new number. Is this the same as taking 20% of the original number and subtracting it from that number? Explain.

86. Increase a number by 10%. Now decrease the new number by 10%. Is the result the original number? Explain.

87. A $10,000, 8% bond has a quoted price of $10,500. Look up in a business math book the meaning of current yield. Then find the current yield on the $10,000 bond.

88. A wide-screen TV priced at $3000 was on sale for 30% off. When the TV didn't sell, an additional 10% of the sale price was taken off the sale price. Calculate the sale price after the two discounts. Compare the two successive discounts of 30% and 10% with a 40% discount off the original price. Find the equivalent discount of the successive discounts.

89. A welder earning $12 per hour is given a 10% raise. To find the new wage, we can multiply $12 by 0.10 and add the product to $12. Can the new wage be found by multiplying $12 by 1.10? Try both methods and compare your answers.

90. [W] Explain how the term "percentage" is used in different situations.

91. [W] Explain the meaning of "per millage" and explain its relation to percent.

92. [W] Visit a savings and loan institution or credit union to research and write a report on the meaning of points as related to a loan.

SECTION 6.3 General Equations

OBJECTIVE A Equations of the form $ax + b = c$

POINT OF INTEREST

Evariste Galois, despite being killed in a duel at the age of 21, made significant contributions to solving equations. In fact, a branch of mathematics is called Galois theory, which shows what kind of equations can and cannot be solved.

To solve an equation such as $3w - 5 = 16$, both the Addition and Multiplication Properties of Equations are used.

$$3w - 5 = 16$$

Add the opposite of the constant term -5 to each side of the equation.

$$3w - 5 + 5 = 16 + 5$$
$$3w = 21$$

Divide each side of the equation by the coefficient of w. The equation is in the form *variable* = *constant*.

$$\frac{3w}{3} = \frac{21}{3}$$
$$w = 7$$

Check the solution.

Check: $3w - 5 = 16$

$3(7) - 5$	16
$21 - 5$	16

$$16 = 16$$

7 checks as a solution. The solution is 7.

Solve: $7 = 4 - \frac{2}{3}x$

The variable is on the right side of the equation. Work toward the goal of *constant* = *variable*.

$$7 = 4 - \frac{2}{3}x$$

Subtract 4 from each side of the equation.

$$7 - 4 = 4 - 4 - \frac{2}{3}x$$
$$3 = -\frac{2}{3}x$$

Multiply each side of the equation by $-\frac{3}{2}$.

$$-\frac{3}{2} \cdot 3 = \left(-\frac{3}{2}\right)\left(-\frac{2}{3}x\right)$$

The equation is in the form *constant* = *variable*.

$$-\frac{9}{2} = x$$

You should check this solution. The solution is $-\frac{9}{2}$.

Example 1 Solve: $6 - 5x = 16$

Solution

$$6 - 5x = 16$$
$$6 - 6 - 5x = 16 - 6$$
$$-5x = 10$$
$$\frac{-5x}{-5} = \frac{10}{-5}$$
$$x = -2$$

The solution is -2.

You Try It 1 Solve: $-5 - 4t = 7$

Your Solution

Solution on p. A15

Solve.

37. $\frac{4}{9}t - 1 = 5$

38. $\frac{8}{5}p + 7 = 5$

39. $\frac{9w}{4} + 13 = 7$

40. $\frac{6z}{5} - 5 = 10$

41. $\frac{3}{5}y + \frac{1}{4} = \frac{3}{4}$

42. $\frac{5}{6}x - \frac{2}{3} = \frac{5}{3}$

43. $2.86y + 5.61 = 9.042$

44. $3.1x + 2.34 = 3.58$

45. $5.6t - 5.1 = 1.06$

46. $7.2 + 5.2z = 8.76$

47. $6.2 - 3.3t = -12.94$

48. $2.4 - 4.8v = 13.92$

49. $6c - 2 - 3c = 10$

50. $12t + 6 + 3t = 16$

51. $4y + 5 - 12y = -3$

52. $7m - 15 - 10m = 6$

53. $17 = 12p - 5 - 6p$

54. $29 = 4x + 5 - 9x$

Objective B

Solve.

55. $4x + 3 = 2x + 9$

56. $6z + 5 = 3z + 20$

57. $7y - 6 = 3y + 6$

58. $8w - 5 = 5w + 10$

59. $12m + 11 = 5m + 4$

60. $8a + 9 = 2a - 9$

61. $7c - 5 = 2c - 25$

62. $7r - 1 = 5r - 13$

63. $2n - 3 = 5n - 18$

64. $4t - 7 = 10t - 25$

65. $3z + 5 = 19 - 4z$

66. $2m + 3 = 23 - 8m$

Solve.

67. $5v - 3 = 4 - 2v$

68. $3r - 8 = 2 - 2r$

69. $7 - 4a = 2a$

70. $5 - 3x = 5x$

71. $12 - 5y = 3y - 12$

72. $8 - 3m = 8m - 14$

73. $7r = 8 + 2r$

74. $-2w = 4 - 5w$

75. $5a + 3 = 3a + 10$

76. $7y + 3 = 5y + 12$

77. $9w - 2 = 5w + 4$

78. $7n - 3 = 3n + 6$

79. $3x - 5 = 9x + 4$

80. $4p + 5 = 10p - 9$

81. $7c - 5 = 11c + 9$

82. $4n + 5 = 10n + 13$

83. $x - 7 = 5x - 21$

84. $3z + 2 = 9z - 12$

85. $3y - 4 = 9y - 24$

86. $2c + 5 = 10c - 7$

87. $5n - 1 + 2n = 4n + 8$

88. $3y + 1 + y = 2y + 11$

89. $3z - 2 - 7z = 4z + 6$

90. $2a + 3 - 9a = 3a + 33$

91. $4t - 8 + 12t = 3 - 4t - 11$

92. $6x - 5 + 9x = 7 - 4x - 12$

Objective C

Solve.

93. $3(4y + 5) = 25$

94. $5(3z - 2) = 8$

95. $-2(4x + 1) = 22$

96. $-3(2x - 5) = 30$

97. $5(2k + 1) - 7 = 28$

98. $7(3t - 4) + 8 = -6$

99. $3(3v - 4) + 2v = 10$

100. $4(3x + 1) - 5x = 25$

101. $3y + 2(y + 1) = 12$

102. $7x + 3(x + 2) = 33$

103. $7v - 3(v - 4) = 20$

104. $15m - 4(2m - 5) = 34$

105. $6 + 3(3x - 3) = 24$

106. $9 + 2(4p - 3) = 24$

107. $9 - 3(4a - 2) = 9$

108. $17 - 8(x - 3) = 1$

109. $3(2z - 5) = 4z + 1$

110. $4(3z - 1) = 5z + 17$

111. $2 - 3(5x + 2) = 2(3 - 5x)$

112. $5 - 2(3y + 1) = 3(2 - 3y)$

113. $4r + 11 = 5 - 2(3r + 3)$

114. $3v + 6 = 9 - 4(2v - 2)$

115. $7n - 2 = 5 - (9 - n)$

116. $8x - 5 = 7 - 2(5 - x)$

Objective D

Translate into an equation and solve.

117. The sum of a number and twelve is twenty. Find the number.

118. The difference between nine and a number is seven. Find the number.

119. Three-fifths of a number is negative thirty. Find the number.

120. The quotient of a number and six is twelve. Find the number.

121. Four more than three times a number is thirteen. Find the number.

122. The sum of twice a number and five is fifteen. Find the number.

123. The difference between nine times a number and six is twelve. Find the number.

124. Six less than four times a number is twenty-two. Find the number.

125. The sum of a number and twice the number is nine. Find the number.

126. Eleven more than the product of negative four and a number is three. Find the number.

127. Seventeen less than the product of five and a number is two. Find the number.

128. Eight less than the product of eleven and a number is negative nineteen. Find the number.

129. Seven more than the product of six and a number is eight less than the product of three and the number. Find the number.

130. Fifteen less than the product of four and a number is eleven less than the product of six and the number. Find the number.

Translate into an equation and solve.

131. Thirty equals nine less than the product of seven and a number. Find the number.

132. Twenty-three equals the difference between eight and the product of five and a number. Find the number.

133. The sum of two numbers is twenty-one. Twice the smaller number is three more than the larger number. Find the two numbers.

134. The sum of two numbers is thirty. Three times the smaller number is twice the larger number. Find the two numbers.

135. The sum of two numbers is twenty-three. The larger number is five more than twice the smaller number. Find the two numbers.

136. The sum of two numbers is twenty-five. The larger number is seven less than four times the smaller number. Find the two numbers.

Critical Thinking

137. Solve: $x \div 28 = 1481$ remainder 25

138. If $3 + 2(4a - 3) = 5$ and $4 - 3(2 - 3b) = 11$, which is larger, a or b?

139. Does the sentence "Solve $2x - 3(4x + 1)$" make sense? Why or why not?

140. [W] Explain in your own words the steps you would take to solve the equation $\frac{2}{3}x - 4 = 10$. State the Property of Real Numbers or the Property of Equations you would use at each step.

141. [W] The equation $x = x + 1$ has no solution, whereas the solution of the equation $2x + 3 = 3$ is zero. Is there a difference between no solution and a solution of zero? Explain your answer.

142. [W] Explain the difference between the word *equation* and the word *expression.*

SECTION 6.4 Applications: Geometry

OBJECTIVE A Lines and angles

Recall that a **line segment** is part of a line and has two endpoints. The line segment shown at the right is denoted by $\overline{AB}$.

The distance between the endpoints of $\overline{AC}$ is denoted by AC. If B is a point on $\overline{AC}$, then AC (the distance from A to C) is the sum of AB (the distance from A to B) and BC (the distance from B to C).

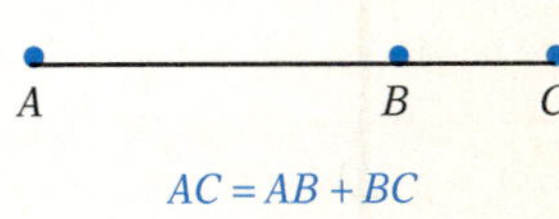

Given $AB = 22$ cm and $AC = 31$ cm, find BC.

Write an equation for the distances between points on the line segment. $AC = AB + BC$

Substitute the given distances for AB and AC into the equation. $31 = 22 + BC$

Solve for BC. $9 = BC$

$BC = 9$ cm

Recall that an **angle** is formed by two rays with the same endpoint. The **vertex** of the angle is the point at which the two rays meet. The rays are called the **sides** of the angle. An angle is measured in **degrees**.

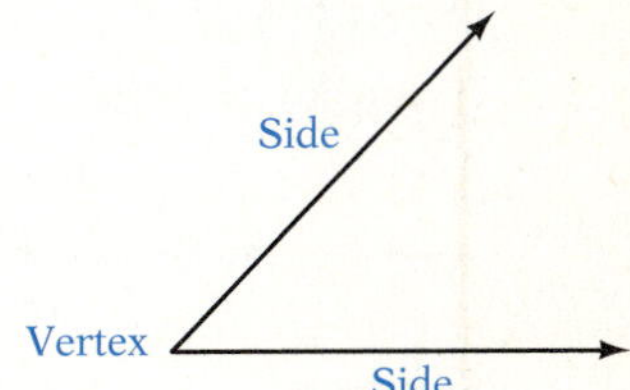

A **protractor** is used to measure an angle. Place the center of the protractor at the vertex of the angle with the edge of the protractor along a side of the angle. The angle shown in the figure below measures 58°.

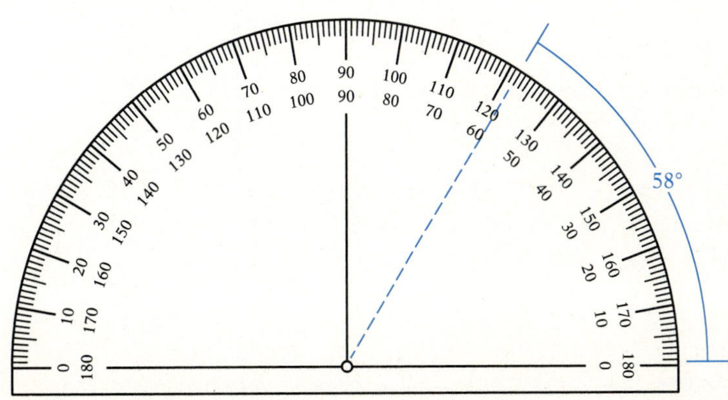

Two angles that share a common side are **adjacent angles**. In the figure at the right, $\angle DAC$ and $\angle CAB$ are adjacent angles

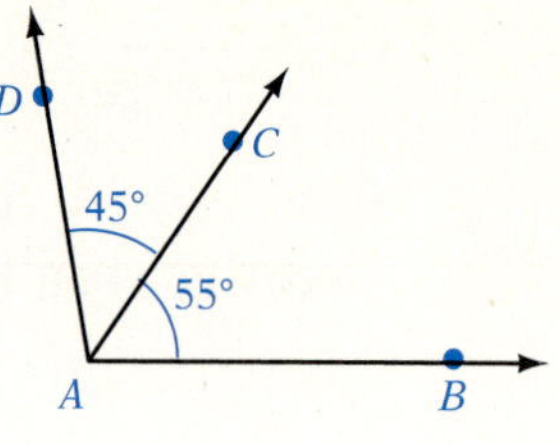

$\angle DAC = 45°$ and $\angle CAB = 55°$.

$$\angle DAB = \angle DAC + \angle CAB$$
$$= 45° + 55° = 100°$$

CONSIDER THIS
Answers to application problems must include units, such as degrees, feet, dollars, or hours.

In the figure at the right, $\angle EDG = 80°$. $\angle FDG$ is three times the measure of $\angle EDF$. Find the measure of $\angle EDF$.

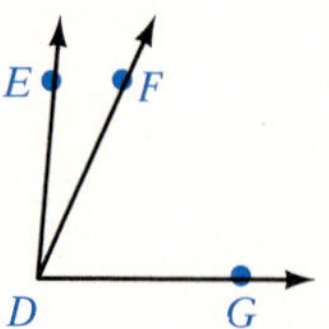

Let x = the measure of $\angle EDF$. Then $3x$ = the measure of $\angle FDG$. Write an equation and solve for x, the measure of $\angle EDF$.

$$\angle EDF + \angle FDG = \angle EDG$$
$$x + 3x = 80$$
$$4x = 80$$
$$x = 20$$

$\angle EDF = 20°$

Example 1

Given $XY = 9$ m and YZ is twice XY, find XZ.

Solution

$XZ = XY + YZ$
$XZ = XY + 2(XY)$ ▶ $YZ = 2(XY)$
$XZ = 9 + 2(9)$
$XZ = 9 + 18$
$XZ = 27$

$XZ = 27$ m

You Try It 1

Given $BC = 16$ ft and $AB = \frac{1}{4}(BC)$, find AC.

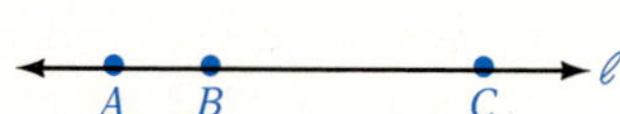

Your Solution

Solution on p. A16

Example 2

Given $MN = 15$ mm, $NO = 18$ mm, and $MP = 48$ mm, find OP.

Solution

$$MN + NO + OP = MP$$
$$15 + 18 + OP = 48$$
$$33 + OP = 48$$
$$OP = 15$$

$OP = 15$ mm

You Try It 2

Given $QR = 24$ cm, $ST = 17$ cm, and $QT = 62$ cm, find RS.

Your Solution

Example 3

Find the complement of a 38° angle.

Strategy

Complementary angles are two angles whose sum is 90°. To find the complement, let x represent the complement of a 38° angle. Write an equation and solve for x.

Solution

$$x + 38° = 90°$$
$$x = 52°$$

The complement of a 38° angle is a 52° angle.

You Try It 3

Find the supplement of a 129° angle.

Your Strategy

Your Solution

Example 4

Find the measure of $\angle x$.

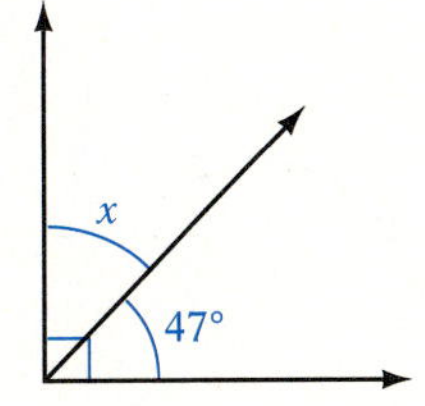

Strategy

To find the measure of $\angle x$, write an equation using the fact that the sum of the measure of $\angle x$ and 47° is 90°. Solve for $\angle x$.

Solution

$$\angle x + 47° = 90°$$
$$\angle x = 43°$$

The measure of $\angle x$ is 43°.

You Try It 4

Find the measure of $\angle a$.

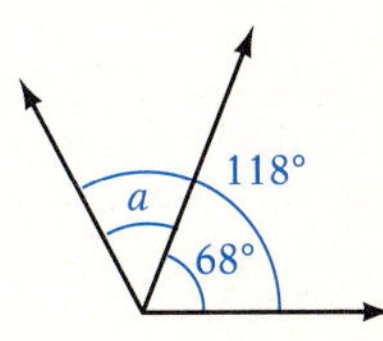

Your Strategy

Your Solution

Solutions on p. A16

OBJECTIVE B Angles formed by intersecting lines

Four angles are formed by the intersection of two lines. If the two lines are perpendicular, each of the four angles is a right angle. If the two lines are not perpendicular, then two of the angles formed are acute angles and two of the angles are obtuse angles. The two acute angles are always opposite each other, and the two obtuse angles are always opposite each other.

In the figure at the right, $\angle w$ and $\angle y$ are acute angles. $\angle x$ and $\angle z$ are obtuse angles.

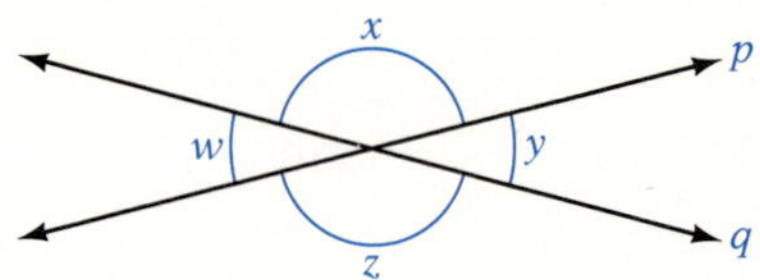

Two angles that are on opposite sides of the intersection of two lines are called **vertical angles**. Vertical angles have the same measure. $\angle w$ and $\angle y$ are vertical angles. $\angle x$ and $\angle z$ are vertical angles.

Vertical angles have the same measure.

$$\angle w = \angle y$$
$$\angle x = \angle z$$

Two angles that share a common side are called **adjacent angles**. For the figure shown above, $\angle x$ and $\angle y$ are adjacent angles, as are $\angle y$ and $\angle z$, $\angle z$ and $\angle w$, and $\angle w$ and $\angle x$. Adjacent angles of intersecting lines are supplementary angles.

Adjacent angles of intersecting lines are supplementary angles.

$$\angle x + \angle y = 180°$$
$$\angle y + \angle z = 180°$$
$$\angle z + \angle w = 180°$$
$$\angle w + \angle x = 180°$$

Given that $\angle c = 65°$, find the measures of angles a, b, and d.

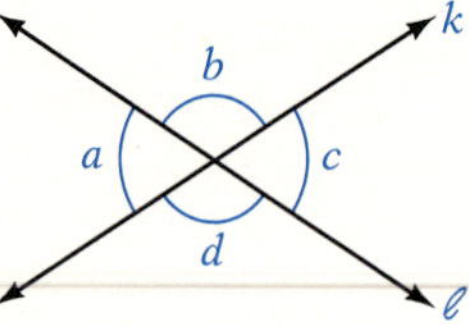

$\angle a = \angle c$ because $\angle a$ and $\angle c$ are vertical angles.

$\angle a = 65°$

$\angle b$ is supplementary to $\angle c$ because $\angle b$ and $\angle c$ are adjacent angles of intersecting lines.

$$\angle b + \angle c = 180°$$
$$\angle b + 65° = 180°$$
$$\angle b = 115°$$

$\angle d = \angle b$ because $\angle d$ and $\angle b$ are vertical angles.

$\angle d = 115°$

A line that intersects two other lines at different points is called a **transversal**.

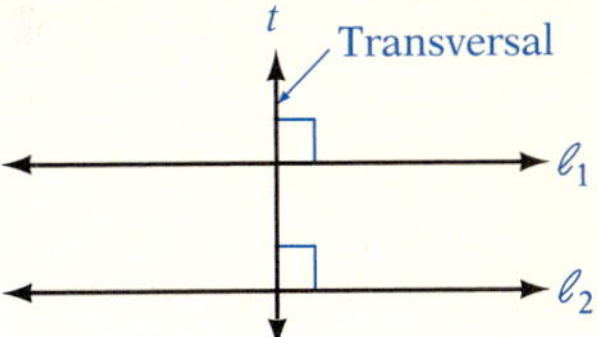

If the lines cut by a transversal *t* are parallel lines and the transversal is perpendicular to the parallel lines, all eight angles formed are right angles.

If the lines cut by a transversal *t* are parallel lines and the transversal is not perpendicular to the parallel lines, all four acute angles have the same measure and all four obtuse angles have the same measure. For the figure at the right,

$$\angle b = \angle d = \angle x = \angle z$$
$$\angle a = \angle c = \angle w = \angle y$$

Alternate interior angles are two angles that are on opposite sides of the transversal and between the lines. In the figure above, $\angle c$ and $\angle w$ are alternate interior angles; $\angle d$ and $\angle x$ are alternate interior angles. Alternate interior angles have the same measure.

Alternate interior angles have the same measure.

$$\angle c = \angle w$$
$$\angle d = \angle x$$

Alternate exterior angles are two angles that are on opposite sides of the transversal and outside the parallel lines. In the figure above, $\angle a$ and $\angle y$ are alternate exterior angles; $\angle b$ and $\angle z$ are alternate exterior angles. Alternate exterior angles have the same measure.

Alternate exterior angles have the same measure.

$$\angle a = \angle y$$
$$\angle b = \angle z$$

Corresponding angles are two angles that are on the same side of the transversal and are both acute angles or are both obtuse angles. For the figure above, the following pairs of angles are corresponding angles: $\angle a$ and $\angle w$, $\angle d$ and $\angle z$, $\angle b$ and $\angle x$, $\angle c$ and $\angle y$. Corresponding angles have the same measure.

Corresponding angles have the same measure.

$$\angle a = \angle w$$
$$\angle d = \angle z$$
$$\angle b = \angle x$$
$$\angle c = \angle y$$

Example 7

Given that $\angle y = 55°$, find the measures of angles a, b, and d.

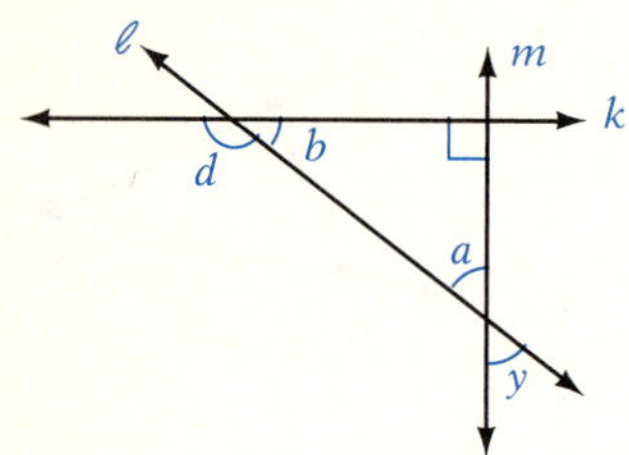

Strategy

- To find the measure of angle a, use the fact that $\angle a$ and $\angle y$ are vertical angles.
- To find the measure of angle b, use the fact that the sum of the measures of the interior angles of a triangle is 180°.
- To find the measure of angle d, use the fact that the sum of an interior and an exterior angle is 180°.

Solution

$\angle a = \angle y = 55°$

$$\angle a + \angle b + 90° = 180°$$
$$55° + \angle b + 90° = 180°$$
$$\angle b + 145° = 180°$$
$$\angle b = 35°$$

$$\angle d + \angle b = 180°$$
$$\angle d + 35° = 180°$$
$$\angle d = 145°$$

You Try It 7

Given that $\angle a = 45°$ and $\angle x = 100°$, find the measures of angles b, c, and y.

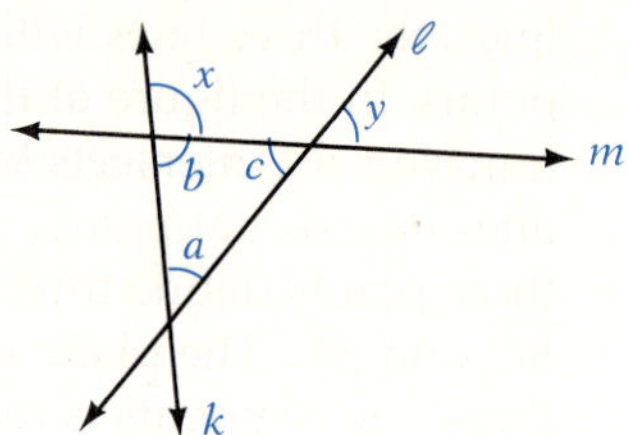

Your Strategy

Your Solution

Example 8

Two angles of a triangle measure 53° and 78°. Find the measure of the third angle.

Strategy

To find the measure of the third angle, use the fact that the sum of the measures of the interior angles of a triangle is 180°. Write an equation using x to represent the measure of the third angle. Solve the equation for x.

Solution

$$x + 53° + 78° = 180°$$
$$x + 131° = 180°$$
$$x = 49°$$

The measure of the third angle is 49°.

You Try It 8

One angle in a triangle is a right angle, and one angle measures 34°. Find the measure of the third angle.

Your Strategy

Your Solution

Solutions on p. A17

6.4 EXERCISES

Objective A

Use a protractor to measure the angle. State whether the angle is acute, obtuse, or right.

1.

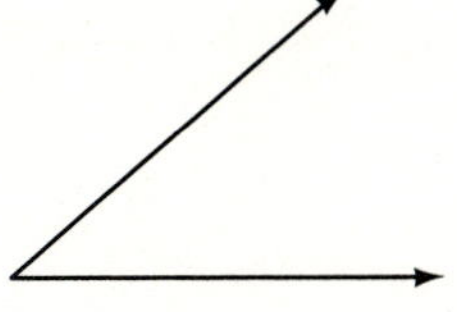

2.

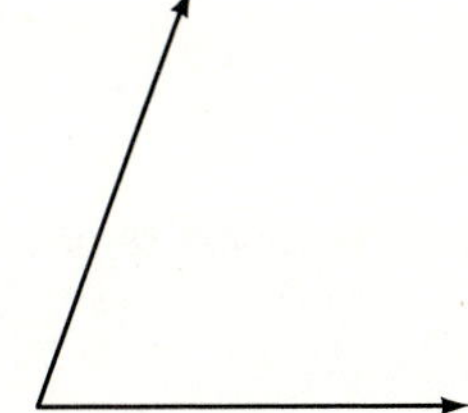

3.

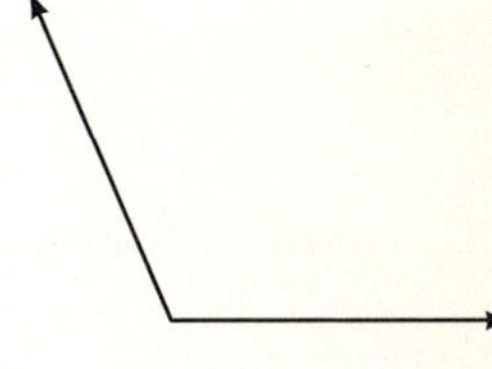

4.

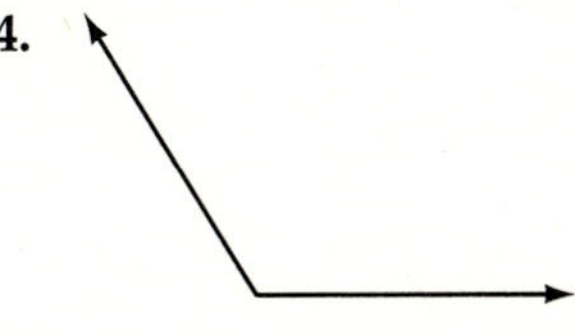

5.

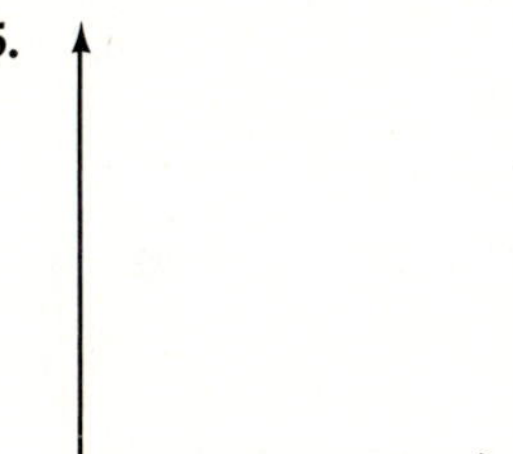

6.

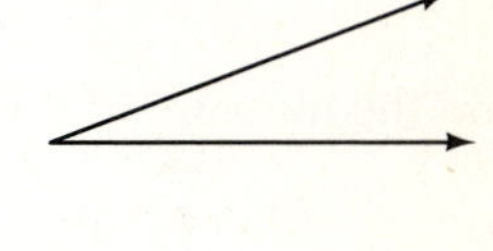

Solve.

7. Find the complement of a 62° angle.

8. Find the complement of a 31° angle.

9. Find the supplement of a 162° angle.

10. Find the supplement of a 72° angle.

11. Given $AB = 12$ cm, $CD = 9$ cm, and $AD = 35$ cm, find the length of BC.

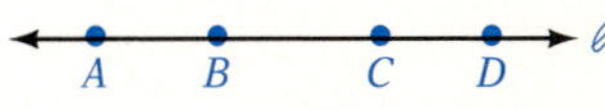

12. Given $AB = 21$ mm, $BC = 14$ mm, and $AD = 54$ mm, find the length of CD.

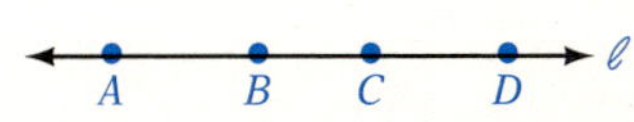

13. Given $QR = 7$ ft and RS is three times the length of QR, find the length of QS.

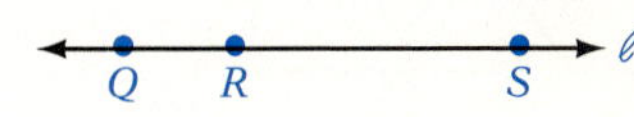

14. Given $QR = 15$ in. and RS is twice the length of QR, find the length of QS.

15. Given $EF = 20$ m and FG equals one-half the length of EF, find the length of EG.

Solve.

16. Given $EF = 18$ cm and FG is one-third the length of EF, find the length of EG.

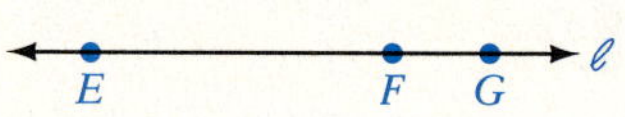

17. Given $\angle LOM = 53°$ and $\angle LON = 139°$, find the measure of $\angle MON$.

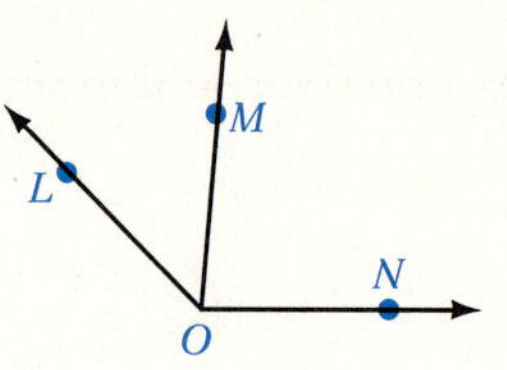

18. Given $\angle MON = 38°$ and $\angle LON = 85°$, find the measure of $\angle LOM$.

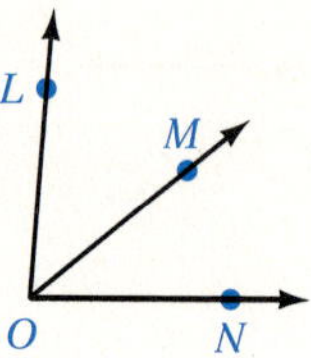

Find the measure of $\angle x$.

19.

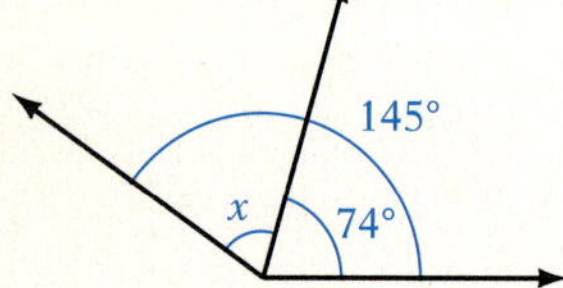

20.

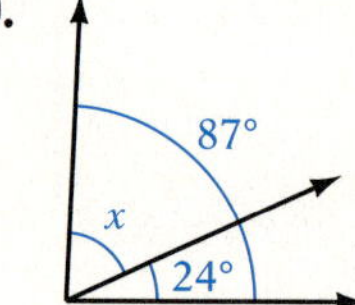

Given that $\angle LON$ is a right angle, find the measure of $\angle x$.

21.

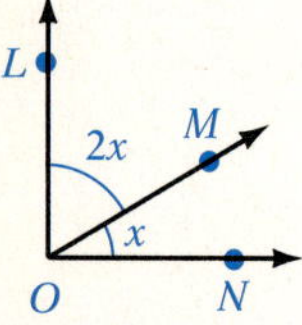

22.

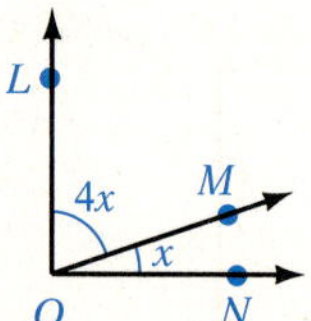

23.

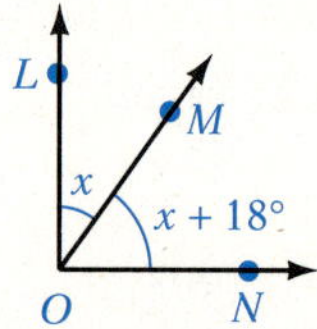

24.

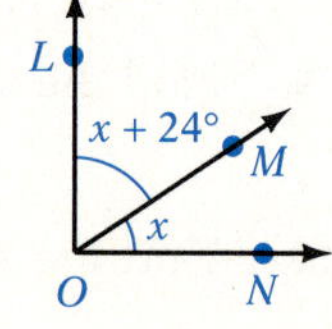

Find the measure of $\angle a$.

25.

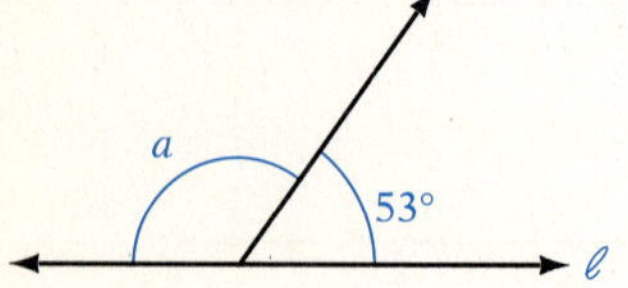

26.

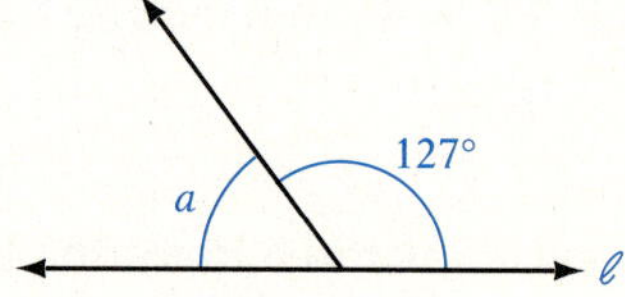

27.

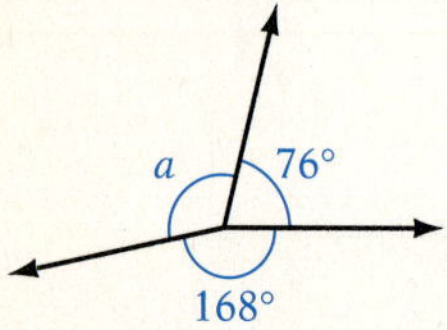

28.

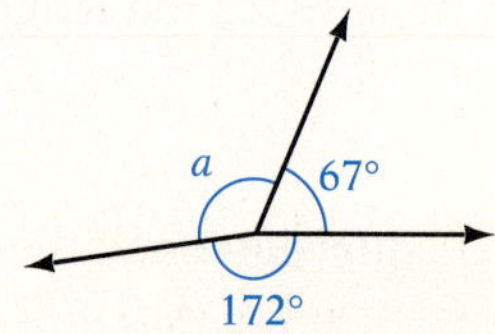

Find x.

29.

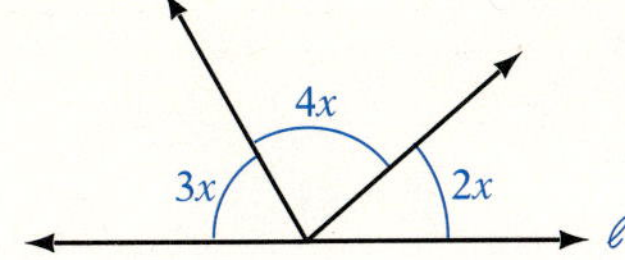

30.

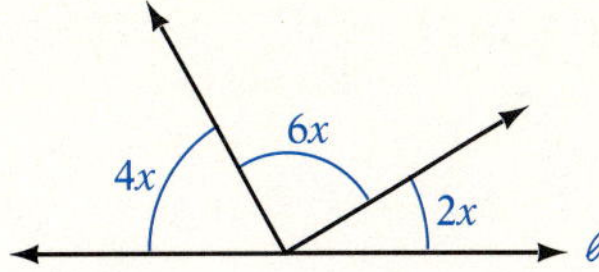

31.

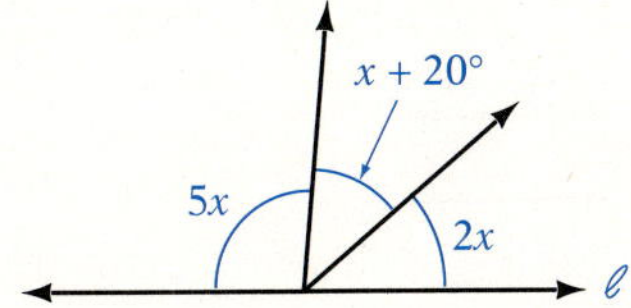

32.

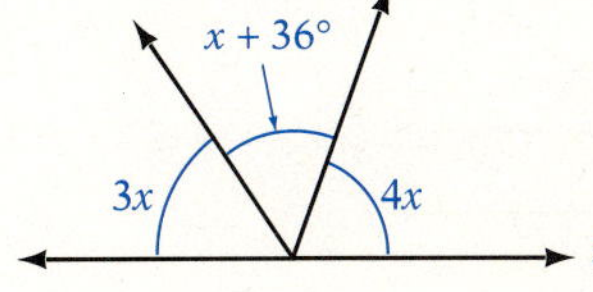

33.

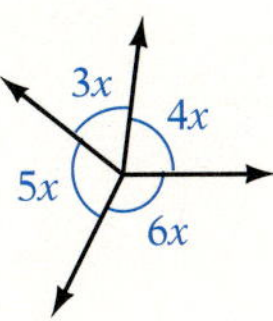

34.

2x x 2x 3x

Solve.

35. Given $\angle a = 51°$, find the measure of $\angle b$.

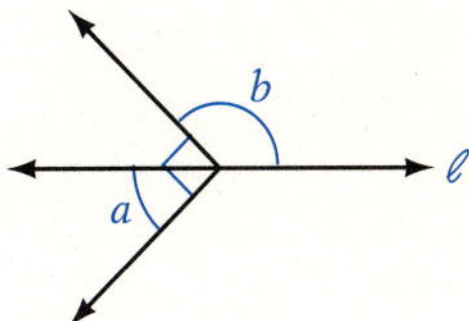

36. Given $\angle a = 38°$, find the measure of $\angle b$.

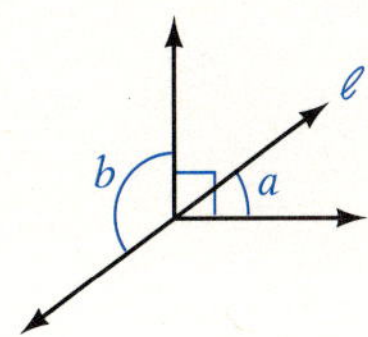

Objective B

Find the measure of $\angle x$.

37.

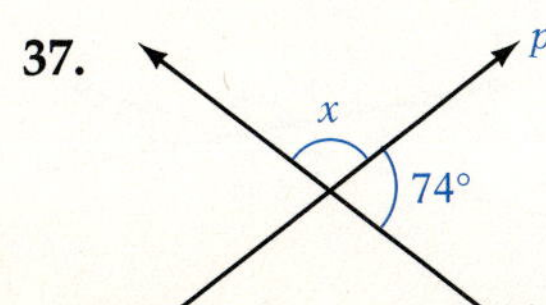

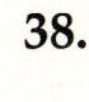

38.

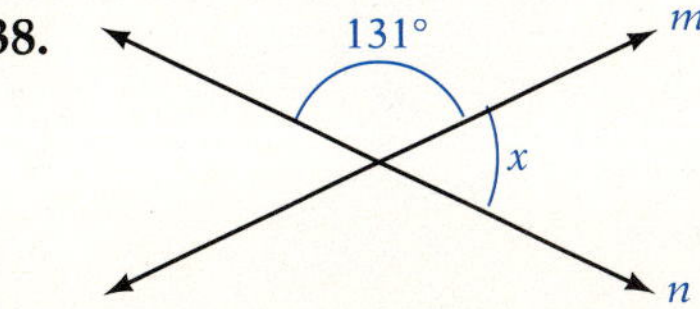

Find x.

39. 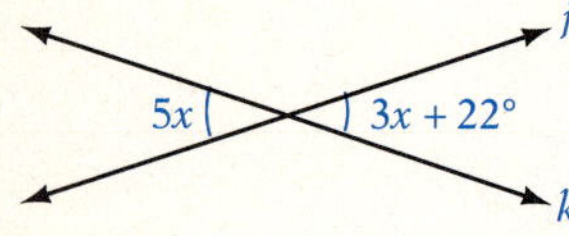

40. 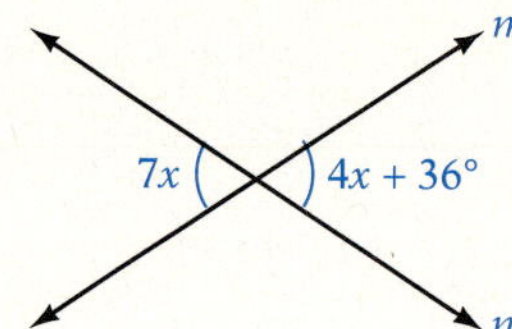

Given that $\ell_1 \parallel \ell_2$, find the measures of angles a and b.

41. 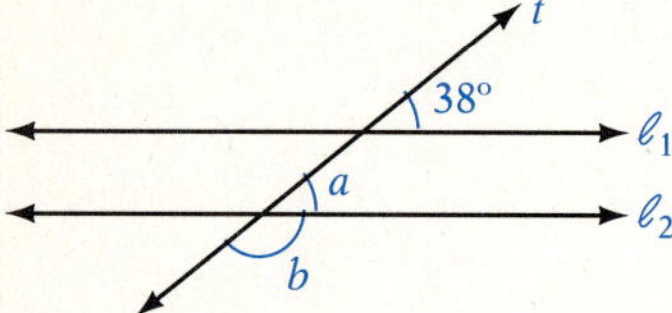

42. 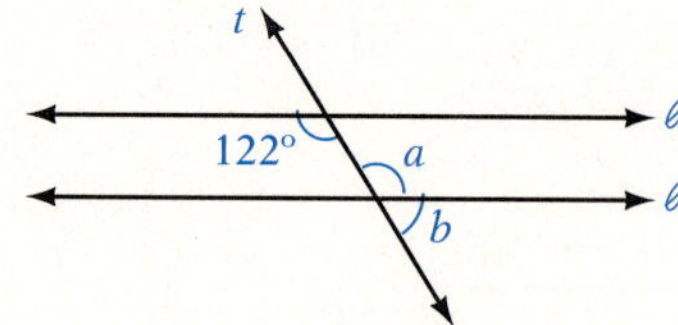

43. 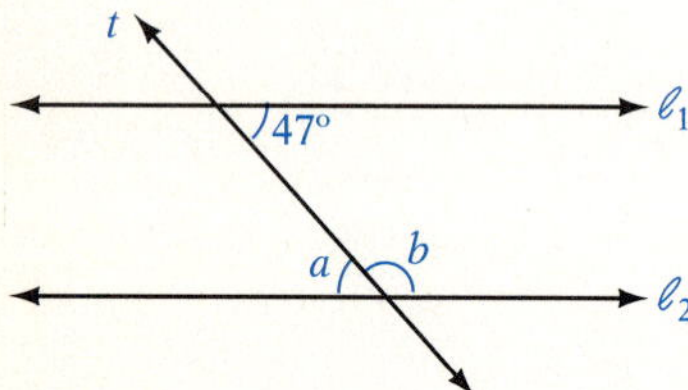

44. 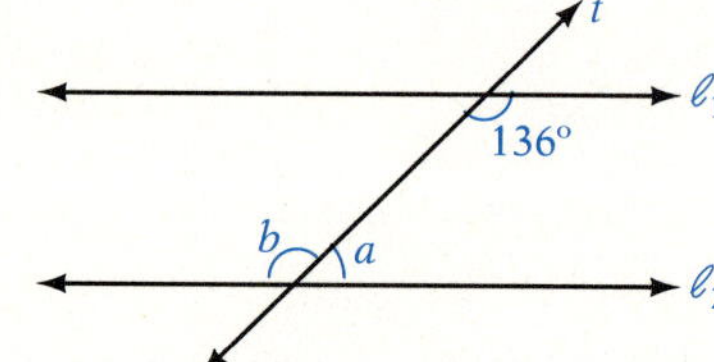

Given that $\ell_1 \parallel \ell_2$, find x.

45. 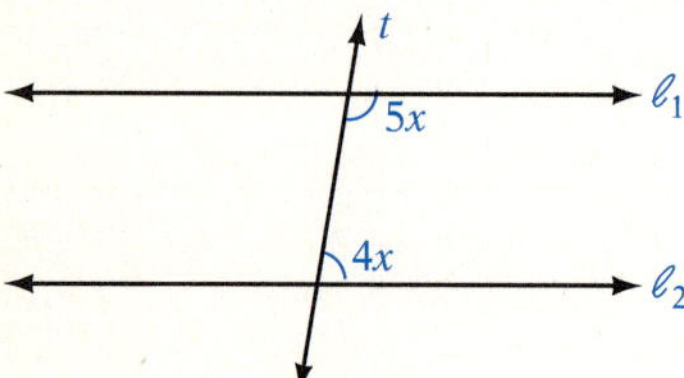

46. 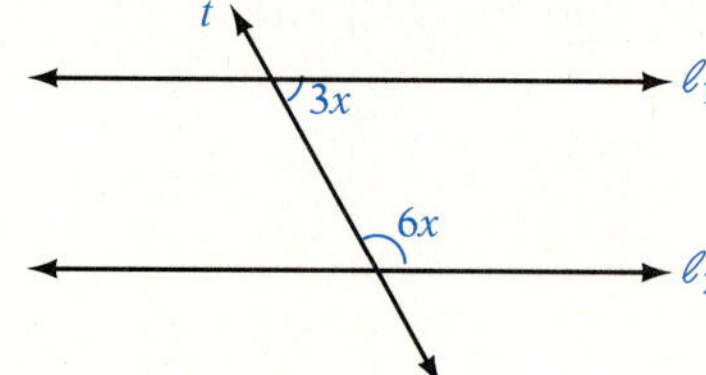

47. 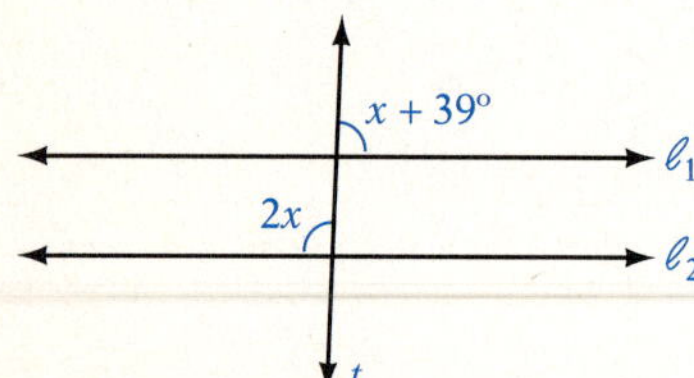

48. 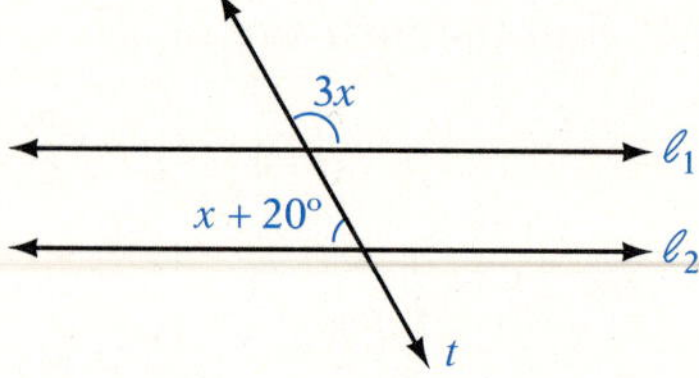

Objective C

Solve.

49. Given that $\angle a = 95°$ and $\angle b = 70°$, find the measures of angles x and y.

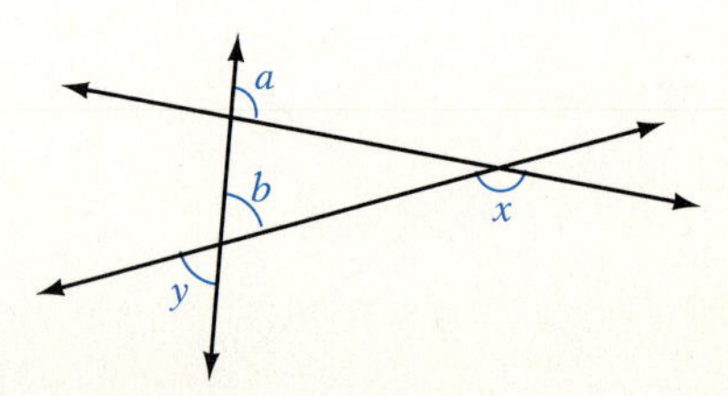

Solve.

50. Given that $\angle a = 35°$ and $\angle b = 55°$, find the measures of angles x and y.

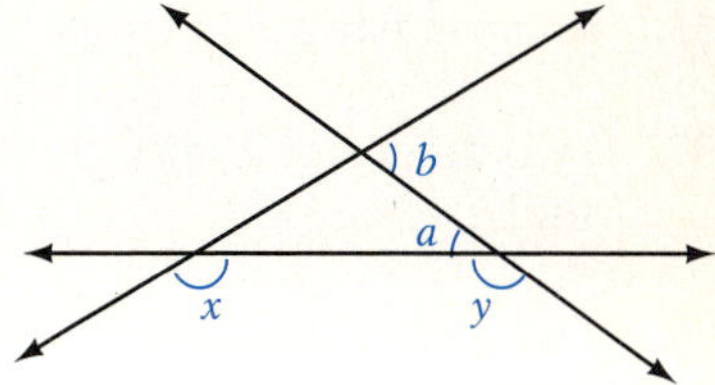

51. Given that $\angle y = 45°$, find the measures of angles a and b.

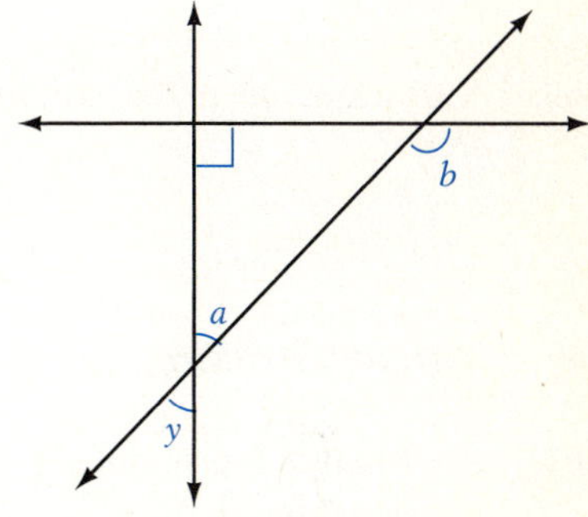

52. Given that $\angle y = 130°$, find the measures of angles a and b.

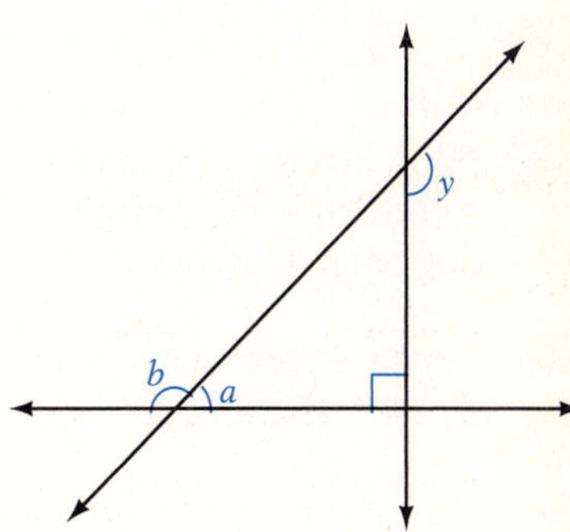

53. Given that $\overline{AO} \perp \overline{OB}$, express in terms of x the measure of $\angle BOC$.

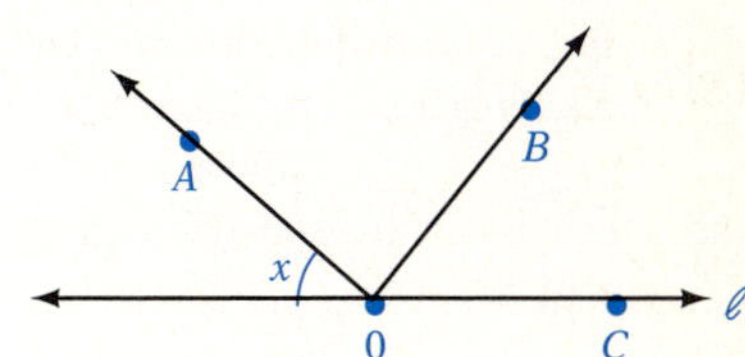

54. Given that $\overline{AO} \perp \overline{OB}$, express in terms of x the measure of $\angle AOC$.

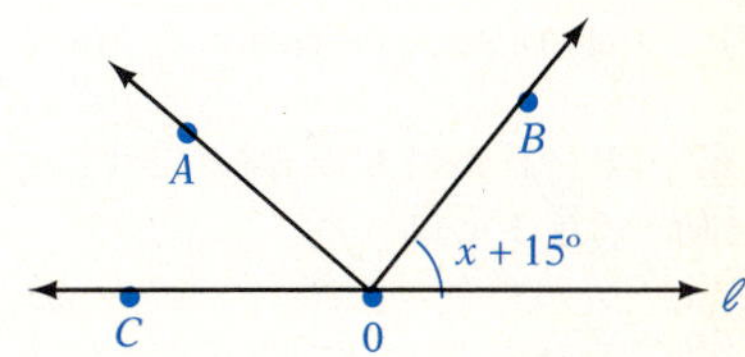

55. One angle in a triangle is a right angle, and one angle is equal to 30°. What is the measure of the third angle?

Solve.

56. A triangle has a 45° angle and a right angle. Find the measure of the third angle.

57. Two angles of a triangle measure 42° and 103°. Find the measure of the third angle.

58. Two angles of a triangle measure 62° and 45°. Find the measure of the third angle.

59. A triangle has a 13° angle and a 65° angle. What is the measure of the third angle?

60. A triangle has a 105° angle and a 32° angle. What is the measure of the third angle?

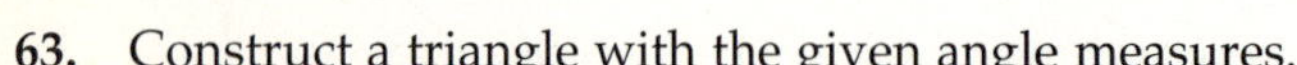

Critical Thinking

61. **a.** What is the smallest possible whole number of degrees in an angle of a triangle?
b. What is the largest possible whole number of degrees in an angle of a triangle?

62. Cut out a triangle and then tear off two of the angles, as shown at the right. Position the pieces you tore off so that angle *a* is adjacent to angle *b* and angle *c* is adjacent to angle *b*. Describe what you observe. What does this demonstrate?

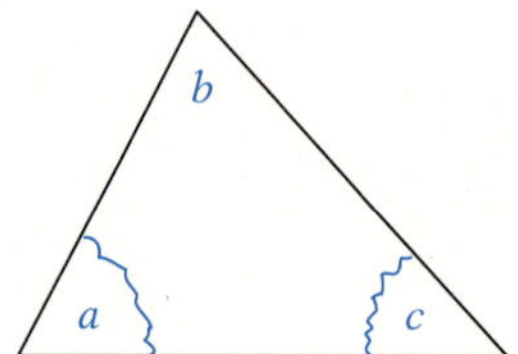

63. Construct a triangle with the given angle measures.
a. 45°, 45°, and 90° **b.** 30°, 60°, and 90° **c.** 40°, 40°, and 100°

64. Determine whether the statement is always true, sometimes true, or never true.
a. Two lines that are parallel to a third line are parallel to each other.
b. A triangle contains two acute angles.
c. Vertical angles are complementary angles.

65. For the figure at the right, find the sum of the measures of angles x, y, and z.

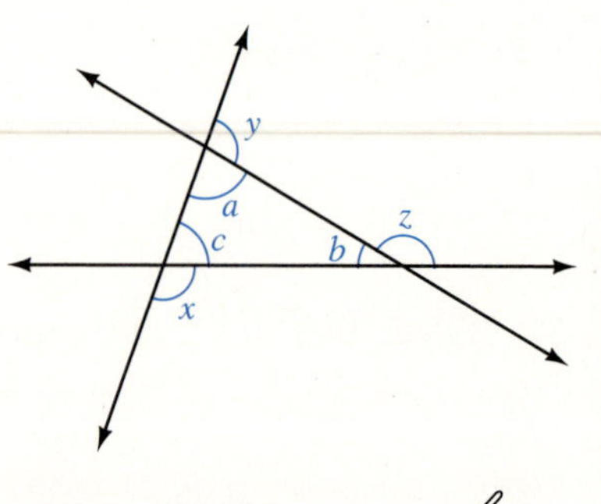

66. [W] For the figure at the right, explain why $\angle a + \angle b = \angle x$. Write a rule that describes the relationship between an exterior angle of a triangle and the opposite interior angles. Use the rule to write an equation involving angles a, c, and z.

67. [W] If $\overline{AB}$ and $\overline{CD}$ intersect at point O, and $\angle AOC = \angle BOC$, explain why $\overline{AB} \perp \overline{CD}$.

68. [W] What are the meanings of the words *acute* and *obtuse* in describing a person?

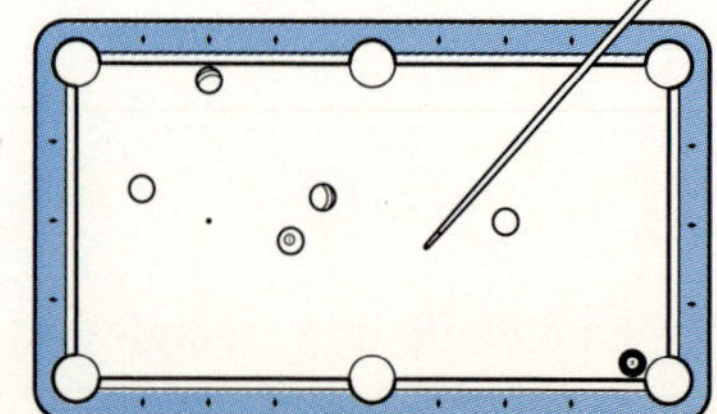

69. [W] Do some research on the principle of reflection. Explain how this principle applies to the operation of a periscope and to the game of billiards.

SECTION 6.5 Mixture, Investment, and Motion Problems

OBJECTIVE A Value mixture problems

A value mixture problem involves combining two ingredients that have different prices into a single blend. For example, a coffee merchant may blend two types of coffee into a single blend.

The solution of a value mixture problem is based on the equation $V = AC$, where V is the value of an ingredient, A is the amount of the ingredient, and C is the cost per unit of the ingredient.

Find the value of 12 lb of walnuts costing $1.60 per pound.

The amount is 12 lb. The cost per pound is $1.60.
The value of the 12 lb of walnuts is $19.20.

$$V = AC$$
$$V = 12(1.60) = 19.20$$

A coffee merchant wants to make 9 lb of a blend of coffee costing $6 per pound. The blend is made using a $7 grade and a $4 grade of coffee. How many pounds of each of these grades should be used?

CONSIDER THIS
Use the information given in the problem to fill in the amount and unit cost columns of the table. Fill in the value column by multiplying the two expressions you wrote in each row. Use the expressions in the last column to write the equation.

Strategy for Solving a Value Mixture Problem

For each ingredient in the mixture, write a numerical or variable expression for the amount of the ingredient used, the unit cost of the ingredient, and the value of the amount used. For the blend, write a numerical or variable expression for the amount, the unit cost of the blend, and the value of the amount. The results can be recorded in a table.

The sum of the amounts is 9 lb.

Amount of $7 coffee: x
Amount of $4 coffee: $9 - x$

	Amount, A	·	Unit cost, C	=	Value, V
$7 grade	x	·	$7	=	$7x$
$4 grade	$9 - x$	·	$4	=	$4(9 - x)$
$6 blend	9	·	$6	=	$6(9)$

Determine how the values of each ingredient are related. Use the fact that the sum of the values of each ingredient is equal to the value of the blend.

The sum of the values of the $7 grade and the $4 grade is equal to the value of the $6 blend.

$$7x + 4(9 - x) = 6(9)$$
$$7x + 36 - 4x = 54$$
$$3x + 36 = 54$$
$$3x = 18$$
$$x = 6$$
$$9 - x = 9 - 6 = 3$$

The merchant must use 6 lb of the $7 coffee and 3 lb of the $4 coffee.

Example 1

How many ounces of a silver alloy that costs \$4 an ounce must be mixed with 10 oz of an alloy that costs \$6 an ounce to make a mixture that costs \$4.32 an ounce?

Strategy

▶ Ounces of \$4 alloy: x

	Amount	Cost	Value
\$4 alloy	x	\$4	$4x$
\$6 alloy	10	\$6	6(10)
\$4.32 mixture	$10 + x$	\$4.32	$4.32(10 + x)$

▶ The sum of the values before mixing equals the value after mixing.

Solution

$$\begin{aligned} 4x + 6(10) &= 4.32(10 + x) \\ 4x + 60 &= 43.2 + 4.32x \\ -0.32x + 60 &= 43.2 \\ -0.32x &= -16.8 \\ x &= 52.5 \end{aligned}$$

52.5 oz of the \$4 silver alloy must be used.

You Try It 1

A gardener has 20 lb of a lawn fertilizer that costs \$.80 per pound. How many pounds of a fertilizer that costs \$.55 per pound should be mixed with the 20 lb of lawn fertilizer to produce a mixture that costs \$.75 per pound?

Your Strategy

Your Solution

Solution on p. A17

OBJECTIVE B Percent mixture problems

The amount of a substance in a solution can be given as a percent of the total solution. For example, a 5% salt water solution means that 5% of the total solution is salt. The remaining 95% is water.

The solution of a percent mixture problem is based on the equation $Q = Ar$, where Q is the quantity of a substance in the solution, r is the percent of concentration, and A is the amount of solution.

A 500-milliliter bottle contains a 4% solution of hydrogen peroxide. Find the amount of hydrogen peroxide in the solution.

Given: $A = 500$
$r = 4\% = 0.04$
Unknown: Q

$Q = Ar$
$Q = 500(0.04)$
$Q = 20$

The bottle contains 20 ml of hydrogen peroxide.

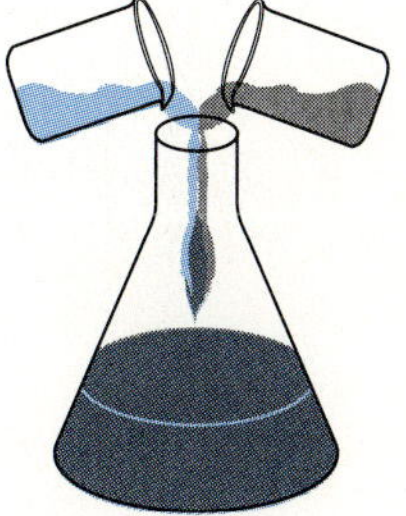

How many gallons of a 20% salt solution must be mixed with 6 gal of a 30% salt solution to make a 22% salt solution?

Strategy for Solving a Percent Mixture Problem

For each solution, write a numerical or variable expression for the amount of solution, percent of concentration, and the quantity of the substance in the solution. The results can be recorded in a table.

CONSIDER THIS
Use the information given in the problem to fill in the amount and percent columns of the table. Fill in the quantity column by multiplying the two expressions you wrote in each row. Use the expressions in the last column to write the equation.

The unknown quantity of 20% solution: x

	Amount of solution, A	$\cdot$	Percent of concentration, r	=	Quantity of substance, Q
20% solution	x	$\cdot$	0.20	=	$0.20x$
30% solution	6	$\cdot$	0.30	=	$0.30(6)$
22% solution	$x + 6$	$\cdot$	0.22	=	$0.22(x + 6)$

Determine how the quantities of the substance in each solution are related. Use the fact that the sum of the quantities of the substances being mixed is equal to the quantity of the substance after mixing.

The sum of the quantities of the substance in the 20% solution and the 30% solution is equal to the quantity of the substance in the 22% solution.

$$\begin{aligned} 0.20x + 0.30(6) &= 0.22(x + 6) \\ 0.20x + 1.80 &= 0.22x + 1.32 \\ -0.02x + 1.80 &= 1.32 \\ -0.02x &= -0.48 \\ x &= 24 \end{aligned}$$

24 gal of the 20% solution are required.

Example 2

A chemist wishes to make 2 L of an 8% acid solution by mixing a 10% acid solution and a 5% acid solution. How many liters of each solution should the chemist use?

Strategy

- Liters of 10% solution: x
  Liters of 5% solution: $2 - x$

	Amount	Percent	Quantity
10%	x	0.10	$0.10x$
5%	$2 - x$	0.05	$0.05(2 - x)$
8%	2	0.08	0.08(2)

- The sum of the quantities before mixing is equal to the quantity after mixing.

Solution

$$\begin{aligned} 0.10x + 0.05(2 - x) &= 0.08(2) \\ 0.10x + 0.10 - 0.05x &= 0.16 \\ 0.05x + 0.10 &= 0.16 \\ 0.05x &= 0.06 \\ x &= 1.2 \end{aligned}$$

$2 - x = 2 - 1.2 = 0.8$

The chemist needs 1.2 L of the 10% solution and 0.8 L of the 5% solution.

You Try It 2

A pharmacist dilutes 5 L of a 12% solution by adding water. How many liters of water are added to make an 8% solution?

Your Strategy

Your Solution

Solution on p. A17

OBJECTIVE C Investment problems

POINT OF INTEREST

You may be familiar with the simple interest formula $I = Prt$. If so, you know that t represents time. In this section, time is always 1 (one year), so the formula $I = Prt$ simplifies to:

$I = Pr(1)$
$I = Pr$

The annual simple interest that an investment earns is given by the equation $I = Pr$, where I is the simple interest, P is the principal, or the amount invested, and r is the simple interest rate.

The annual interest rate on a \$2500 investment is 11%. Find the annual simple interest earned on the investment.

Given: $P = \$2500$
$r = 11\% = 0.11$
Unknown: I

$I = Pr$
$I = 2500(0.11)$
$I = 275$

The annual simple interest is \$275.

An investor has a total of \$10,000 to deposit into two simple interest accounts. On one account, the annual simple interest rate is 7%. On the second account, the annual simple interest rate is 11%. How much should be invested in each account so that the total annual interest earned is \$1000?

Strategy for Solving a Problem Involving Money Deposited in Two Simple Interest Accounts

For each amount invested, write a numerical or variable expression for the principal, the interest rate, and the interest earned. The results can be recorded in a table.

The sum of the amounts at each interest rate is \$10,000.

Amount invested at 7%: x
Amount invested at 11%: $10,000 - x$

CONSIDER THIS

Use the information given in the problem to fill in the principal and interest rate columns of the table. Fill in the interest earned column by multiplying the two expressions you wrote in each row.

	Principal, P	·	Interest rate, r	=	Interest earned, I
Amount at 7%	x	·	0.07	=	$0.07x$
Amount at 11%	$10,000 - x$	·	0.11	=	$0.11(10,000 - x)$

Determine how the amounts of interest earned on each amount are related. For example, the total interest earned by both accounts may be known or it may be known that the interest earned on one account is equal to the interest earned by the other account.

The total annual interest earned is \$1000

$$0.07x + 0.11(10,000 - x) = 1000$$
$$0.07x + 1100 - 0.11x = 1000$$
$$-0.04x + 1100 = 1000$$
$$-0.04x = -100$$
$$x = 2500$$
$$10,000 - x = 10,000 - 2500 = 7500$$

The amount invested at 7% is \$2500.
The amount invested at 11% is \$7500.

Example 3

An investment counselor invested 75% of a client's money in a 9% annual simple interest money market fund. The remainder was invested in 7% annual simple interest government securities. Find the amount invested in each if the total annual interest earned is $3825.

Strategy

▶ Amount invested: x
Amount invested at 7%: $0.25x$
Amount invested at 9%: $0.75x$

	Principal	Rate	Interest
Amount at 7%	$0.25x$	0.07	$0.0175x$
Amount at 9%	$0.75x$	0.09	$0.0675x$

▶ The sum of the interest earned by the two investments equals the total annual interest earned ($3825).

Solution

$$0.0175x + 0.0675x = 3825$$
$$0.085x = 3825$$
$$x = 45{,}000$$

$$0.25x = 0.25(45{,}000) = 11{,}250$$

$$0.75x = 0.75(45{,}000) = 33{,}750$$

The amount invested at 7% is $11,250.
The amount invested at 9% is $33,750.

You Try It 3

An investment of $5000 is made at an annual simple interest rate of 8%. How much additional money must be invested at 11% so that the total interest earned will be 9% of the total investment?

Your Strategy

Your Solution

Solution on p. A18

OBJECTIVE D Uniform motion problems

A train that travels at a constant speed in a straight line at 50 mph is in *uniform motion.* **Uniform motion** means that the speed of an object does not change.

The solution of a uniform motion problem is based on the equation $d = rt$, where d is the distance traveled, r is the rate of travel, and t is the time spent traveling.

A train traveled at a speed of 55 mph for 3 h. The distance traveled by the train can be found by the equation $d = rt$.

The rate is 55 mph. The time is 3 h.

$$d = rt$$
$$d = 55(3)$$
$$d = 165$$

The train traveled 165 mi.

A car leaves a town traveling at 40 mph. Two hours later, a second car leaves the same town, on the same road, traveling at 60 mph. In how many hours will the second car be passing the first car?

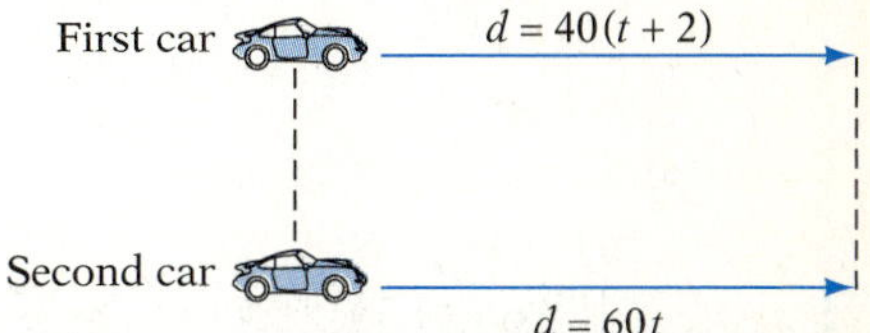

Strategy for Solving a Uniform Motion Problem

For each object, write a numerical or variable expression for the distance, rate, and time. The results can be recorded in a table.

The first car traveled 2 h longer than the second car.

Time for the second car: t
Time for the first car: $t + 2$

CONSIDER THIS
Use the information given in the problem to fill in the rate and time columns of the table. Fill in the distance column by multiplying the two expressions you wrote in each row.

	Rate, r	·	Time, t	=	Distance, d
First car	40	·	$t + 2$	=	$40(t + 2)$
Second car	60	·	t	=	$60t$

Determine how the distances traveled by each object are related. For example, the total distance traveled by both objects may be known or it may be known that the two objects traveled the same distance.

The two cars traveled the same distance.

$$40(t + 2) = 60t$$
$$40t + 80 = 60t$$
$$80 = 20t$$
$$4 = t$$

The second car will pass the first car in 4 h.

Example 4

Two cars, the second car traveling 10 mph faster than the first car, start at the same time from the same point and travel in opposite directions. In 3 h they are 300 mi apart. Find the rate of the first car.

Strategy

▶ Rate of 1st car: r
Rate of 2nd car: $r + 10$

	Rate	Time	Distance
1st car	r	3	$3r$
2nd car	$r + 10$	3	$3(r + 10)$

▶ The total distance traveled by the two cars is 300 mi.

Solution

$$3r + 3(r + 10) = 300$$
$$3r + 3r + 30 = 300$$
$$6r + 30 = 300$$
$$6r = 270$$
$$r = 45$$

The first car is traveling 45 mph.

You Try It 4

Two trains, one traveling at twice the speed of the other, start at the same time from stations 288 mi apart and travel toward each other. In 3 h the trains pass each other. Find the rate of each train.

Your strategy

Your Solution

Example 5

How far can a bicycling club ride out into the country at a speed of 12 mph and return over the same road at 8 mph if they travel a total of 10 h?

Strategy

▶ Time spent riding out: t
Time spent riding back: $10 - t$

	Rate	Time	Distance
Out	12	t	$12t$
Back	8	$10 - t$	$8(10 - t)$

▶ The distance out equals the distance back.

Solution

$$12t = 8(10 - t)$$
$$12t = 80 - 8t$$
$$20t = 80$$
$$t = 4 \text{ (The time is 4 h.)}$$

The distance out $= 12t = 12(4) = 48$ mi.
The club can ride 48 mi into the country.

You Try It 5

On a survey mission, a pilot flew out to a parcel of land and back in 5 h. The rate out was 150 mph. The rate returning was 100 mph. How far away was the parcel of land?

Your Strategy

Your Solution

Solutions on p. A18

6.5 EXERCISES

Objective A

Solve.

1. A high-protein diet supplement that costs $6.75 per pound is mixed with a vitamin supplement that costs $3.25 per pound. How many pounds of each should be used to make 5 lb of a mixture that costs $4.65 per pound?

2. A 20-ounce alloy of platinum that costs $220 per ounce is mixed with an alloy that costs $400 per ounce. How many ounces of the $400 alloy should be mixed with the 20-ounce alloy to make an alloy that costs $300 per ounce?

3. Find the cost per pound of a coffee mixture made from 8 lb of coffee that costs $9.20 per pound and 12 lb of coffee that costs $5.50 per pound.

4. How many pounds of tea that costs $4.20 per pound must be mixed with 12 lb of tea that costs $2.25 per pound to make a mixture that costs $3.40 per pound?

5. A goldsmith combined an alloy that costs $4.30 per ounce with an alloy that costs $1.80 per ounce. How many ounces of each were used to make a mixture of 200 oz that costs $2.50 per ounce?

6. How many liters of a solvent that costs $80 per liter must be mixed with 6 L of a solvent that costs $25 per liter to make a solvent that costs $36 per liter?

7. Find the cost per pound of a trail mix made from 40 lb of raisins that cost $4.40 per pound and 100 lb of granola that costs $2.30 per pound.

8. Find the cost per ounce of a mixture of 200 oz of cologne that costs $5.50 per ounce and 500 oz of cologne that costs $2.00 per ounce.

9. How many kilograms of hard candy that costs $7.50 per kilogram must be mixed with 24 kg of jelly beans that cost $3.25 per kilogram to make a mixture that costs $4.50 per kilogram?

10. A grocery store offers a cheese and fruit sampler that combines cheddar cheese that costs $8 per kilogram with kiwi fruit that costs $3 per kilogram. How many kilograms of each were used to make a 5-kilogram mixture that costs $4.50 per kilogram?

Solve.

11. A ground meat mixture is formed by combining meat that costs \$2.20 per pound with meat that costs \$4.20 per pound. How many pounds of each were used to make a 50-pound mixture that costs \$3.00 per pound?

12. A lumber company combined oak wood chips that cost \$3.10 per pound with pine wood chips that cost \$2.50 per pound. How many pounds of each were used to make an 80-pound mixture that costs \$2.65 per pound?

13. A caterer makes an ice cream punch by combining fruit juice that costs \$2.25 per gallon with ice cream that costs \$3.25 per gallon. How many gallons of each should be used to make 100 gal of punch that costs \$2.50 per gallon?

14. The manager of a specialty food store combined almonds that cost \$4.50 per pound with walnuts that cost \$2.50 per pound. How many pounds of each were used to make a 100-pound mixture that costs \$3.24 per pound?

15. Find the cost per gallon of a carbonated fruit drink made from 12 gal of fruit juice that costs \$4.00 per gallon and 30 gal of carbonated water that costs \$2.25 per gallon.

16. Find the cost per pound of a sugar-coated breakfast cereal made from 40 lb of sugar that costs \$1.00 per pound and 120 lb of corn flakes that cost \$.60 per pound.

Objective B

Solve.

17. A chemist wants to make 50 ml of a 16% acid solution. How many milliliters each of a 13% acid solution and an 18% acid solution should be mixed to produce the desired solution?

18. A blend of coffee was made by combining some coffee that was 40% java beans with 80 lb of coffee that was 30% java beans to make a mixture that is 32% java. How many pounds of the 40% java coffee were used?

19. Thirty ounces of pure silver are added to 50 oz of a silver alloy that is 20% silver. What is the percent concentration of the silver in the resulting mixture?

Solve.

20. Two hundred liters of a punch that contains 35% fruit juice is mixed with 300 L of another punch. The resulting fruit punch is 20% fruit juice. Find the percent of fruit juice in the 300 L of punch.

21. The manager of a garden shop mixes grass seed that is 60% rye grass with 70 lb of grass seed that is 80% rye grass to make a mixture that is 74% rye grass. How much of the 60% mixture is used?

22. Ten grams of sugar are added to a 40-gram serving of a breakfast cereal that is 30% sugar. What is the percent concentration of sugar in the resulting mixture?

23. A dermatologist mixes 50 g of a cream that is 0.5% hydrocortisone with 150 g of another hydrocortisone cream. The resulting mixture is 0.68% hydrocortisone. Find the percent of hydrocortisone in the 150-gram cream.

24. A carpet manufacturer blends two fibers, one 20% wool and the second 50% wool. How many pounds of each fiber should be woven together to produce 500 lb of a fabric that is 35% wool?

25. A hair dye is made by blending some 7% hydrogen peroxide solution with some 4% hydrogen peroxide solution. How many milliliters of each should be mixed to make a 300-milliliter solution that is 5% hydrogen peroxide?

26. How many grams of pure salt must be added to 40 g of a 20% salt solution to make a solution that is 36% salt?

27. How many ounces of pure water must be added to 50 oz of a 15% saline solution to make a saline solution that is 10% salt?

28. A paint is blended by using a paint that contains 21% green pigment and a paint that contains 15% green pigment. How many gallons of each must be mixed to produce 60 gal of paint that is 19% green pigment?

29. A goldsmith mixes 8 oz of a 30% gold alloy with 12 oz of a 25% gold alloy. What is the percent concentration of the resulting alloy?

30. A physicist mixes 40 L of oxygen with 50 L of air that contains 64% oxygen. What is the percent concentration of the resulting air?

31. A 50-ounce box of cereal is 40% bran flakes. How many ounces of pure bran flakes must be added to this box to produce a mixture that is 50% bran flakes?

Solve.

32. A pastry chef has 150 ml of a chocolate topping that is 50% chocolate. How many milliliters of pure chocolate must be added to this topping to make a topping that is 75% chocolate?

Objective C

Solve.

33. An investment of $3000 is made at an annual simple interest rate of 5%. How much additional money must be invested at an annual simple interest rate of 9% so that the total annual interest earned is 7.5% of the total investment?

34. A total of $6000 is invested in two simple interest accounts. The annual simple interest rate on one account is 9%; on the second account, the annual simple interest rate is 6%. How much should be invested in each account so that both accounts earn the same amount of annual interest?

35. An engineer invested a portion of $15,000 in a 7% annual simple interest account and the remainder in a 6.5% annual simple interest government bond. The amount of interest earned for one year was $1020. How much was invested in each account?

36. An investment club invested part of $20,000 in preferred stock that pays 8% annual simple interest and the remainder in a municipal bond that pays 7% annual simple interest. The amount of interest earned each year is $1520. How much was invested in each account?

37. A grocery checker deposited an amount of money into a high-yield mutual fund that returns a 9% annual simple interest rate. A second deposit, $2500 more than the first, was placed in a certificate of deposit that returns a 5% annual simple interest rate. The total interest earned on both investments for one year was $475. How much money was deposited in the mutual fund?

38. A deposit was made into a 7% annual simple interest account. Another deposit, $1500 less than the first deposit, was placed in a 9% annual simple interest certificate of deposit. The total interest earned on both accounts for one year was $505. How much money was deposited in the certificate of deposit?

39. A corporation gave a university $300,000 to support product safety research. The university deposited some of the money in a 10% simple interest account and the remainder in an 8.5% simple interest account. How much should be deposited in each account so that the annual interest earned is $28,500?

Solve.

40. A financial consultant advises a client to invest part of $30,000 in municipal bonds that earn 6.5% annual simple interest and the remainder of the money in 8.5% corporate bonds. How much should be invested in each account so that the total annual interest earned each year is $2190?

41. To provide for retirement income, an auto mechanic purchases a $5000 bond that earns 7.5% annual simple interest. How much money must be invested in additional bonds that have an interest rate of 8% so that the total annual interest earned from the two investments is $615?

42. The portfolio manager for an investment group invested $40,000 in a certificate of deposit that earns 7.25% annual simple interest. How much money must be invested in additional certificates that have an interest rate of 8.5% so that the total annual interest earned from the two investments is $5025?

43. A charity deposited a total of $54,000 into two simple interest accounts. The annual simple interest rate on one account is 8%. The annual simple interest rate on the second account is 12%. How much was invested in each account if the total interest earned is 9% of the total investment?

44. A college sports foundation deposited a total of $24,000 into two simple interest accounts. The annual simple interest rate on one account is 7%. The annual simple interest rate on the second account is 11%. How much is invested in each account if the total annual interest earned is 10% of the total investment?

45. An investment banker invested 55% of the bank's available cash in an account that earns 8.25% annual simple interest. The remainder of the cash was placed in an account that earns 10% annual simple interest. The interest earned in one year was $58,743.75. Find the total amount invested.

46. A financial planner recommended that 40% of a client's cash account be placed in preferred stock that earns 9% annual simple interest. The remainder of the client's cash was placed in treasury bonds that earn 7% annual interest. The total annual interest earned from the two investments was $2496. What was the total amount invested?

47. The manager of a mutual fund placed 30% of the fund's available cash in a 6% simple interest account, 25% in 8% corporate bonds, and the remainder in a money market fund that earns 7.5% annual simple interest. The total annual interest from the investments was $35,875. What was the total amount invested?

Solve.

48. The manager of a trust decided to invest 30% of a client's cash in government bonds that earn 6.5% annual simple interest. Another 30% was placed in utility stocks that earn 7% annual simple interest. The remainder of the cash was placed in an account earning 8% annual simple interest. The total annual interest earned from the investments was $5437.50. What was the total amount invested?

Objective D

Solve.

49. A 555-mile, 5-hour plane trip was flown at two speeds. For the first part of the trip, the average speed was 105 mph. For the remainder of the trip, the average speed was 115 mph. For how long did the plane fly at each speed?

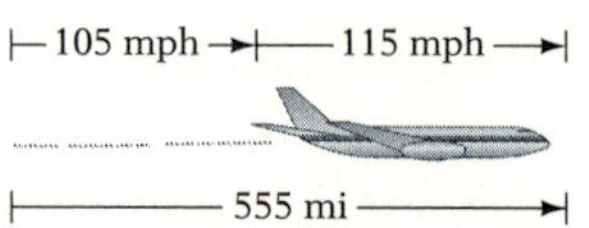

50. An executive drove from home at an average speed of 30 mph to an airport where a helicopter was waiting. The executive boarded the helicopter and flew to the corporate offices at an average speed of 60 mph. The entire distance was 150 mi. The entire trip took 3 h. Find the distance from the airport to the corporate offices.

51. After a sailboat had been on the water for 3 h, a change in wind direction reduced the average speed of the boat by 5 mph. The entire distance sailed was 57 mi. The total time spent sailing was 6 h. How far did the sailboat travel in the first 3 h?

52. A car and a bus set out at 2 P.M. from the same point headed in the same direction. The average speed of the car is 30 mph slower than twice the speed of the bus. In 2 h the car is 20 mi ahead of the bus. Find the rate of the car.

53. A passenger train leaves a train depot 2 h after a freight train leaves the same depot. The freight train is traveling 20 mph slower than the passenger train. Find the rate of each train if the passenger train overtakes the freight train in 3 h.

54. Two cyclists start at the same time from opposite ends of a course that is 45 mi long. One cyclist is riding at 14 mph and the second cyclist is riding at 16 mph. How long after they begin will they meet?

55. A cyclist and a jogger set out at 11 A.M. from the same point headed in the same direction. The average speed of the cyclist is twice the average speed of the jogger. In 1 h the cyclist is 8 mi ahead of the jogger. Find the rate of the cyclist.

Solve.

56. Two cyclists start from the same point and ride in opposite directions. One cyclist rides twice as fast as the other. In 3 h they are 72 mi apart. Find the rate of each cyclist.

57. Two small planes start from the same point and fly in opposite directions. The first plane is flying 25 mph slower than the second plane. In 2 h the planes are 430 mi apart. Find the rate of each plane.

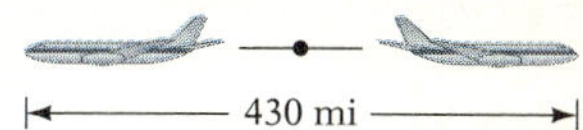

58. A motorboat leaves a harbor and travels at an average speed of 8 mph toward a small island. Two hours later a cabin cruiser leaves the same harbor and travels at an average speed of 16 mph toward the same island. In how many hours after the cabin cruiser leaves will it be alongside the motorboat?

59. Two joggers start at the same time from opposite ends of a 10-mile course. One jogger is running at 4 mph and the other is running at 6 mph. How long after they begin will they meet?

60. On a 195-mile trip, a car traveled at an average speed of 45 mph and then reduced its speed to an average of 30 mph for the remainder of the trip. The trip took a total of 5 h. How long did the car travel at each speed?

61. A long-distance runner started on a course running at an average speed of 6 mph. One hour later, a second runner began the same course at an average speed of 8 mph. How long after the second runner started will the second runner overtake the first runner?

62. A family drove to a resort at an average speed of 30 mph and later returned over the same road at an average speed of 50 mph. Find the distance to the resort if the total driving time was 8 h.

63. Three campers left their campsite by canoe and paddled downstream at an average rate of 8 mph. They then turned around and paddled back upstream at an average rate of 4 mph to return to their campsite. How long did it take the campers to canoe downstream if the total trip took 1 h?

64. A car traveling at 48 mph overtakes a cyclist who, riding at 12 mph, had a 3-hour head start. How far from the starting point does the car overtake the cyclist?

Critical Thinking

65. Find the cost per ounce of a mixture of 30 oz of an alloy that costs \$4.50 per ounce, 40 oz of an alloy that costs \$3.50 per ounce, and 30 oz of an alloy that costs \$3.00 per ounce.

Solve.

66. A grocer combined walnuts that cost \$1.60 per pound and cashews that cost \$2.50 per pound with 20 lb of peanuts that cost \$1.00 per pound. Find the amount of walnuts and the amount of cashews used to make a 50-pound mixture that costs \$1.72 per pound.

67. How many ounces of water must be evaporated from 50 oz of a 12% salt solution to produce a 15% salt solution?

68. A chemist mixed pure acid with water to make 10 L of a 30% acid solution. How much pure acid and how much water did the chemist use?

69. A radiator contains 15 gal of a 20% antifreeze solution. How many gallons must be drained from the radiator and replaced by pure antifreeze so that the radiator will contain 15 gal of a 40% antifreeze solution?

70. A sales representative invests in a stock paying 9% dividends. A research consultant invests \$5000 more than the sales representative in bonds paying 8% annual simple interest. The research consultant's income from the investment is equal to the sales representative's. Find the amount of the research consultant's investment.

71. A financial manager invested 20% of a client's money in bonds paying 9% annual simple interest, 35% in an 8% simple interest account, and the remainder in 9.5% corporate bonds. Find the amount invested in each if the total annual interest earned is \$5325.

CORPORATION BONDS
Volume, \$46,830,000

Bonds	Cur Yld	Vol	Close	Net Chg.
AMR 9s16	8.9	20	$101\frac{1}{2}$	$-\ \frac{1}{4}$
ANR $10\frac{5}{8}$95	10.4	10	102	$+\ \frac{1}{2}$
Advst 9\$08	cv	15	89	...
AetnLf $8\frac{1}{8}$07	8.2	75	$98\frac{3}{4}$	$-\ \frac{1}{8}$
AlaP $7\frac{7}{8}$\$02	7.9	8	$99\frac{1}{2}$	$+\ \frac{1}{2}$
AlaP $7\frac{3}{4}$\$02	7.9	3	$97\frac{3}{4}$	$+\ \frac{3}{8}$
AlaP $8\frac{7}{8}$\$03	8.7	8	102	...
AlaP $8\frac{1}{4}$\$03	8.0	22	$102\frac{3}{4}$	...
AlaP $9\frac{3}{4}$\$04	9.3	1	105	$+\ \frac{1}{8}$
AlaP $8\frac{7}{8}$06	8.5	31	104	$+\ 1\frac{5}{8}$
AlaP $8\frac{3}{4}$07	8.6	31	102	$+\ \frac{1}{4}$
AlaP $9\frac{1}{2}$08	9.1	4	$104\frac{3}{4}$	$+\ \frac{5}{8}$
AlskAr $6\frac{7}{8}$14	cv	34	$79\frac{1}{2}$	$-\ \frac{1}{2}$
AlskAr zr06	...	22	$33\frac{1}{4}$	...
AlldC zr98	...	20	$59\frac{3}{8}$	$-\ \frac{3}{8}$
AlldC zr92	...	5	$97\frac{17}{32}$	$-\ \frac{1}{32}$
AlldC zr96	...	4	$76\frac{3}{4}$	...
AlldC zr2000	...	4	$50\frac{1}{2}$	$+\ \frac{5}{8}$
AlldC zr9	...	10	$91\frac{7}{8}$	$-\ \frac{5}{8}$
AlldC zr95	...	50	$78\frac{1}{4}$	$-\ \frac{1}{8}$
AlldC zr01	...	10	$45\frac{3}{8}$	$-\ \frac{1}{8}$
AlldC zr 03	...	15	$38\frac{1}{4}$	$-\ \frac{1}{2}$
Allwst $7\frac{1}{4}$14	cv	13	87	$+\ \frac{1}{2}$
AmStor 01	cv	12	$98\frac{3}{4}$	$+\ \frac{1}{4}$
ATT $5\frac{5}{8}$95	5.7	170	$97\frac{7}{8}$	$+\ \frac{1}{8}$
ATT $5\frac{1}{2}$97	5.9	50	$93\frac{3}{4}$	$-\ 1\frac{1}{4}$

72. A plant manager invested \$3000 more in stocks than in bonds. The stocks paid 8% annual simple interest, and the bonds paid 9.5% annual simple interest. Both investments yielded the same income. Find the total annual interest received on both investments.

73. At 10 A.M., two campers left their campsite by canoe and paddled downstream at an average speed of 12 mph. They then turned around and paddled back upstream at an average rate of 4 mph. The total trip took 1 h. At what time did the campers turn around downstream?

74. At 7 A.M., two joggers start from opposite ends of an 8-mi course. One jogger is running at a rate of 4 mph, and the other is running at a rate of 6 mph. At what time will the joggers meet?

75. A bicyclist rides for 2 h at a speed of 10 mph and then returns at a speed of 20 mph. Find the cyclist's average speed for the trip.

76. A car travels a 1-mile track at an average speed of 30 mph. At what average speed must the car travel the next mile so that the average speed for the 2 mi is 60 mph?

SECTION 6.6 Inequalities

OBJECTIVE A The Addition and Multiplication Properties of Inequalities

The solution set of an inequality is a set of numbers, each element of which, when substituted for the variable, makes the inequality true.

The inequality at the right is true if the variable is replaced by 7, 9.3, or $\frac{15}{2}$.

$$x + 5 > 8$$

$$\left.\begin{aligned} 7 + 5 &> 8 \\ 9.3 + 5 &> 8 \\ \frac{15}{2} + 5 &> 8 \end{aligned}\right\} \text{True inequalities}$$

The inequality $x + 5 > 8$ is false if the variable is replaced by 0, 1.5, or $-\frac{1}{2}$.

$$\left.\begin{aligned} 0 + 5 &> 8 \\ 1.5 + 5 &> 8 \\ -\frac{1}{2} + 5 &> 8 \end{aligned}\right\} \text{False inequalities}$$

There are many values of the variable x that will make the inequality $x + 5 > 8$ true. The solution set of $x + 5 > 8$ is any number greater than 3.

At the right is the graph of the solution set of $x + 5 > 8$.

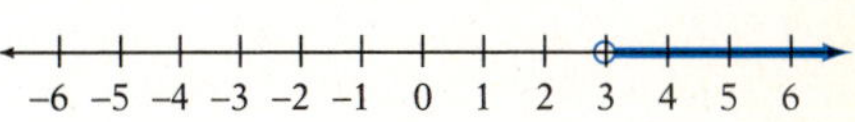

In solving an inequality, the goal is to rewrite the given inequality in the form *variable* $>$ *constant* or *variable* $<$ *constant*. The Addition Property of Inequalities is used to rewrite an inequality in this form.

> ***Addition Property of Inequalities***
>
> The same term can be added to each side of an inequality without changing the solution set of the inequality.
>
> If $a > b$, then $a + c > b + c$.
> If $a < b$, then $a + c < b + c$.

The Addition Property of Inequalities also holds true for an inequality containing the symbol $\geq$ or $\leq$.

The Addition Property of Inequalities is used when, in order to rewrite an inequality in the form *variable* $>$ *constant* or *variable* $<$ *constant*, we must remove a term from one side of the inequality. Add the same term to each side of the inequality.

Solve: $x - 4 < -3$

$$x - 4 < -3$$

Add 4 to each side of the inequality. $\quad x - 4 + 4 < -3 + 4$

Simplify. $\quad x < 1$

At the right is the graph of the solution set of $x - 4 < -3$.

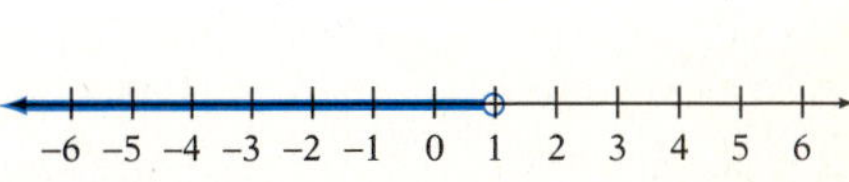

Because subtraction is defined in terms of addition, the Addition Property of Inequalities allows the same term to be subtracted from each side of an inequality.

Solve: $5x - 6 \le 4x - 4$

	$5x - 6 \le 4x - 4$
Subtract $4x$ from each side of the inequality.	$5x - 4x - 6 \le 4x - 4x - 4$
Simplify.	$x - 6 \le -4$
Add 6 to each side of the inequality.	$x - 6 + 6 \le -4 + 6$
Simplify.	$x \le 2$

The Multiplication Property of Inequalities is used when, in order to rewrite an inequality in the form *variable* > *constant* or *variable* < *constant*, we must remove a coefficient from one side of the inequality.

CONSIDER THIS
$c > 0$ means c is a positive number.

Multiplication Property of Inequalities

Each side of an inequality can be multiplied by the same positive number without changing the solution set of the inequality.

If $a > b$ and $c > 0$, then $ac > bc$.
If $a < b$ and $c > 0$, then $ac < bc$.

CONSIDER THIS
The inequality symbol is not reversed.

$$5 > 4$$
$$5(2) > 4(2)$$ • Multiply by *positive* 2.
$$10 > 8$$ • A true inequality

CONSIDER THIS
$c < 0$ means c is a negative number.

If each side of an inequality is multiplied by the same negative number and the inequality symbol is reversed, then the solution set of the inequality is not changed.

If $a > b$ and $c < 0$, then $ac < bc$.
If $a < b$ and $c < 0$, then $ac > bc$.

CONSIDER THIS
The inequality symbol is reversed.

$$6 < 9$$
$$6(-3) > 9(-3)$$ • Multiply by *negative* 3 and *reverse* the inequality.
$$-18 > -27$$ • A true inequality

The Multiplication Property of Inequalities also holds true for an inequality containing the symbol $\ge$ or $\le$.

Solve and graph the solution set: $-\frac{3}{2}x \le 6$

$$-\frac{3}{2}x \le 6$$

Multiply each side of the inequality by $-\frac{2}{3}$.

Because $-\frac{2}{3}$ is a negative number, the inequality symbol must be reversed.

$$-\frac{2}{3}\left(-\frac{3}{2}x\right) \ge -\frac{2}{3}(6)$$
$$x \ge -4$$

The graph of the solution set is shown at the right.

CONSIDER THIS

Whenever an inequality is multiplied or divided by a negative number, the inequality symbol must be reversed. Compare the next two examples.

$2x < -4$, $\frac{2x}{2} < \frac{-4}{2}$, $x < -2$ — Divide each side by *positive* 2. Inequality *is not* reversed.

$-2x < 4$, $\frac{-2x}{-2} > \frac{4}{-2}$, $x > -2$ — Divide each side by *negative* 2. Inequality *is* reversed.

Because division is defined in terms of multiplication, the Multiplication Property of Inequalities allows each side of an inequality to be divided by the same nonzero number. When each side of an inequality is divided by a negative number, the inequality symbol must be reversed.

Solve: $-4 < 6x$

$$-4 < 6x$$

Divide each side of the equality by 6

$$\frac{-4}{6} < \frac{6x}{6}$$

Simplify. Note that $\frac{-4}{6} = -\frac{2}{3}$.

$$-\frac{2}{3} < x$$

The solution can be written $-\frac{2}{3} < x$ or $x > -\frac{2}{3}$

Example 1

Solve and graph the solution set of $3 < x + 5$.

Solution

$$3 < x + 5$$
$$3 - 5 < x + 5 - 5$$
$$-2 < x$$

You Try It 1

Solve and graph the solution set of $x + 2 < -2$.

Your Solution

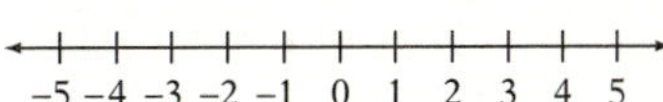

Example 2

Solve: $7x - 14 \le 6x - 16$

Solution

$$7x - 14 \le 6x - 16$$
$$7x - 6x - 14 \le 6x - 6x - 16$$
$$x - 14 \le -16$$
$$x - 14 + 14 \le -16 + 14$$
$$x \le -2$$

You Try It 2

Solve: $5x + 3 > 4x + 5$

Your Solution

Solutions on p. A18

Example 3 Solve and graph the solution set of $-7x > 14$.

Solution $-7x > 14$

$$\frac{-7x}{-7} < \frac{14}{-7}$$

$$x < -2$$

−5 −4 −3 −2 −1 0 1 2 3 4 5

You Try It 3 Solve and graph the solution set of $-3x > -9$.

Your Solution

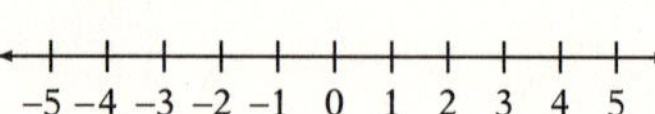

Example 4 Solve: $-\frac{5}{8}x \le \frac{5}{12}$

Solution

$$-\frac{5}{8}x \le \frac{5}{12}$$

$$-\frac{8}{5}\left(-\frac{5}{8}x\right) \ge -\frac{8}{5}\left(\frac{5}{12}\right)$$

$$x \ge -\frac{2}{3}$$

You Try It 4 Solve: $-\frac{3}{4}x \ge 18$

Your Solution

Solutions on p. A19

OBJECTIVE B General inequalities

CONSIDER THIS
Solving these inequalities is similar to solving the equations solved in Section 3 *except* if you multiply or divide the inequality by a negative number, you must reverse the inequality symbol.

Solving an inequality frequently requires application of both the Addition and the Multiplication Properties of Inequalities.

Solve: $4y - 3 \ge 6y + 5$

$$4y - 3 \ge 6y + 5$$

Subtract $6y$ from each side of the inequality.

$$4y - 6y - 3 \ge 6y - 6y + 5$$

Simplify.

$$-2y - 3 \ge 5$$

Add 3 to each side of the inequality.

$$-2y - 3 + 3 \ge 5 + 3$$

Simplify.

$$-2y \ge 8$$

Divide each side of the inequality by -2. Because -2 is a negative number, the inequality symbol must be reversed.

$$\frac{-2y}{-2} \le \frac{8}{-2}$$

$$y \le -4$$

When an inequality contains parentheses, solving the inequality requires the use of the Distributive Property.

Solve: $-2(x - 7) > 3 - 4(2x - 3)$

$$-2(x - 7) > 3 - 4(2x - 3)$$

Use the Distributive Property to remove parentheses.

$$-2x + 14 > 3 - 8x + 12$$

Simplify.

$$-2x + 14 > -8x + 15$$

Add $8x$ to each side of the inequality.

$$-2x + 8x + 14 > -8x + 8x + 15$$

Simplify.

$$6x + 14 > 15$$

Subtract 14 from each side of the inequality.

$$6x + 14 - 14 > 15 - 14$$

Simplify.

$$6x > 1$$

Divide each side of the inequality by 6

$$\frac{6x}{6} > \frac{1}{6}$$

$$x > \frac{1}{6}$$

Example 5 Solve: $7x - 3 \le 3x + 17$

Solution

$$7x - 3 \le 3x + 17$$
$$7x - 3x - 3 \le 3x - 3x + 17$$
$$4x - 3 \le 17$$
$$4x - 3 + 3 \le 17 + 3$$
$$4x \le 20$$
$$\frac{4x}{4} \le \frac{20}{4}$$
$$x \le 5$$

You Try It 5 Solve: $5 - 4x > 9 - 8x$

Your Solution

Example 6

Solve: $3(3 - 2x) \ge -5x - 2(3 - x)$

Solution

$$3(3 - 2x) \ge -5x - 2(3 - x)$$
$$9 - 6x \ge -5x - 6 + 2x$$
$$9 - 6x \ge -3x - 6$$
$$9 - 6x + 3x \ge -3x + 3x - 6$$
$$9 - 3x \ge -6$$
$$9 - 9 - 3x \ge -6 - 9$$
$$-3x \ge -15$$
$$\frac{-3x}{-3} \le \frac{-15}{-3}$$
$$x \le 5$$

You Try It 6

Solve: $8 - 4(3x + 5) \le 6(x - 8)$

Your Solution

Solutions on p. A19

OBJECTIVE C Applications

Example 7

A rectangle is 10 ft wide and $(2x + 4)$ ft long. Express as an integer the maximum length of the rectangle when the area is less than 200 ft^2. (The area of a rectangle is equal to its length times its width.)

Strategy

To find the maximum length:

- ▶ Replace the variables in the area formula by the given values and solve for x.
- ▶ Replace the variable in the expression $2x + 4$ with the value found for x.

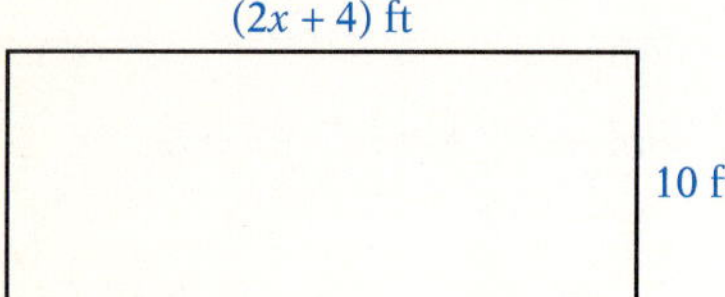

Solution

Length times width	is less than	200 ft^2

$$(2x + 4)10 < 200$$
$$20x + 40 < 200$$
$$20x + 40 - 40 < 200 - 40$$
$$20x < 160$$
$$\frac{20x}{20} < \frac{160}{20}$$
$$x < 8$$

The length is $(2x + 4)$ ft. Because $x < 8$, $2x + 4 < 2(8) + 4 = 20$. Therefore, the length is less than 20 ft.

The maximum length is 19 ft.

You Try It 7

Company A rents cars for \$8 a day plus 10¢ for every mile driven. Company B rents cars for \$10 a day and 8¢ per mile driven. You want to rent a car for one week. What is the maximum number of miles you can drive a Company A car if it is to cost you less than a Company B car?

Your Strategy

Your Solution

Solution on p. A19

6.6 EXERCISES

Objective A

Solve and graph the solution.

1. $x + 1 < 3$

2. $y + 2 < 2$

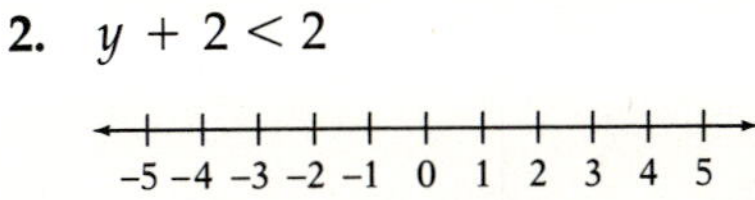

3. $x - 5 > -2$

4. $x - 3 > -2$

5. $7 \leq n + 4$

6. $3 \leq 5 + x$

7. $3x < 12$

8. $8x \leq -24$

9. $15 \leq 5y$

10. $-48 < 24x$

11. $16x \leq 16$

12. $3x > 0$

Solve.

13. $y - 3 \geq -12$

14. $x + 8 \geq -14$

15. $3x - 5 < 2x + 7$

16. $5x + 4 < 4x - 10$

17. $8x - 7 \geq 7x - 2$

18. $3n - 9 \geq 2n - 8$

19. $2x + 4 < x - 7$

20. $9x + 7 < 8x - 7$

21. $4x - 8 \leq 2 + 3x$

22. $5b - 9 < 3 + 4b$

23. $x + \frac{5}{8} \geq -\frac{2}{3}$

24. $y + \frac{5}{12} \geq -\frac{3}{4}$

Solve.

25. $6x - \frac{1}{3} \le 5x - \frac{1}{2}$

26. $3x + \frac{5}{8} > 2x + \frac{5}{6}$

27. $4b - \frac{7}{12} \ge 3b - \frac{9}{16}$

28. $3.8x < 2.8x - 3.8$

29. $1.2x < 0.2x - 7.3$

30. $x + 5.8 \le 4.6$

31. $n - 3.82 \le 3.95$

32. $x - 3.5 < 2.1$

33. $x - 0.23 \le 0.47$

34. $-5y \ge 0$

35. $-3z < 0$

36. $7x > 2$

37. $6x \le -1$

38. $2x \le -5$

39. $\frac{5}{6}n < 15$

40. $\frac{3}{4}x < 12$

41. $\frac{2}{3}y \ge 4$

42. $10 \le \frac{5}{8}x$

43. $4 \ge \frac{2}{3}x$

44. $-\frac{3}{7}x \le 6$

45. $-\frac{2}{11}b \ge -6$

46. $-\frac{3}{8}x \ge \frac{9}{14}$

47. $-\frac{3}{5}x < -\frac{6}{7}$

48. $-\frac{4}{5}x < -\frac{8}{15}$

49. $-\frac{3}{4}y \ge -\frac{5}{8}$

50. $-\frac{8}{9}x \ge -\frac{16}{27}$

51. $1.5x \le 6.30$

52. $2.3x \le 5.29$

53. $-3.5d > 7.35$

Objective B

Solve.

54. $4x - 8 < 2x$

55. $7x - 4 < 3x$

56. $2x - 8 > 4x$

57. $3y + 2 > 7y$

58. $8 - 3x \le 5x$

59. $10 - 3x \le 7x$

60. $3x + 2 > 5x - 8$

61. $2n - 9 \ge 5n + 4$

62. $5x - 2 < 3x - 2$

63. $8x - 9 > 3x - 9$

64. $0.1(180 + x) > x$

65. $x > 0.2(50 + x)$

Solve.

66. $2(2y - 5) \le 3(5 - 2y)$

67. $2(5x - 8) \le 7(x - 3)$

68. $5(2 - x) > 3(2x - 5)$

69. $4(3d - 1) > 3(2 - 5d)$

70. $5(x - 2) > 9x - 3(2x - 4)$

71. $3x - 2(3x - 5) > 4(2x - 1)$

72. $4 - 3(3 - n) \le 3(2 - 5n)$

73. $15 - 5(3 - 2x) \le 4(x - 3)$

74. $2x - 3(x - 4) \ge 4 - 2(x - 7)$

75. $4 + 2(3 - 2y) \le 4(3y - 5) - 6y$

Objective C

Solve.

76. The circumference of an official major league baseball is between 9.00 in. and 9.25 in. Find the possible diameters of a major league baseball. Round to the nearest hundredth of an inch.

77. To be eligible for a basketball tournament, a basketball team must win at least 60% of its remaining games. If the team has 17 games remaining, how many games must the team win to qualify for the tournament?

78. Computer software engineers are fond of saying that software takes at least twice as long to develop as they think it will. Applying that saying, how many hours will it take to develop a software product that an engineer thinks can be finished in 50 h?

79. To pass a course with a B grade, a student must have an average of 80 points on five tests. The student's grades on the first four tests were 75, 83, 86, and 78. What scores can the student receive on the fifth test to earn a B grade?

80. A car sales representative receives a commission that is the greater of $250 or 8% of the selling price of a car. What dollar amounts for the sale price of a car will make the commission offer more attractive?

Solve.

81. The sales agent for a jewelry company is offered a flat monthly salary of \$3200 or a salary of \$1000 plus an 11% commission on the selling price of each item sold by the agent. If the agent chooses the \$3200, what dollar amount does the agent expect to sell in one month?

82. A computer bulletin board service offers service for a flat fee of \$10 per month or \$4 per month plus \$.10 for each minute the service is used. How many minutes must a person use this service to exceed \$10?

83. A site licensing fee for a computer program is \$1500. Paying this fee allows the company to use the program at any computer terminal within the company. Alternatively, the company can choose to pay \$200 for each individual computer it has. How many individual computers must a company have for the site license to be more economical?

84. For a product to be labeled orange juice, a state agency requires that at least 80% of the drink be real orange juice. How many ounces of artificial flavor can be added to 32 oz of real orange juice if this drink can be labeled orange juice?

85. Grade A hamburger cannot contain more than 20% fat. How much fat can a butcher mix with 300 lb of lean meat to meet the 20% requirement?

86. A shuttle service taking skiers to a ski area charges \$8 per person each way. Four skiers are debating whether to take the shuttle bus or rent a car for \$45 plus \$.25 per mile. Assuming that the skiers will share the cost of the car and that they want the least expensive method of transportation, find how far away the ski area is if they should choose the shuttle service.

87. Company A rents a car for \$25 per day and \$.08 per mile. Company B rents a car for \$15 per day and \$.14 per mile. Find the maximum number of miles you can drive company B's car before the cost exceeds the cost of company A's car.

Critical Thinking

88. Determine whether the statement is always true, sometimes true, or never true, for real numbers a, b, and c.
   a. If $a > b$, then $-a > -b$.
   b. If $a < b$, then $ac < bc$.
   c. If $a > b$, then $a + c > b + c$.
   d. If $a \neq 0$, $b \neq 0$, and $a > b$, then $\frac{1}{a} > \frac{1}{b}$.

89. List the positive integers which are solutions of $7 - 2b \leq 15 - 5b$.

90. Determine the solution set of $2 - 3(x + 4) < 5 - 3x$.

91. Determine the solution set of $3x + 2(x - 1) > 5(x + 1)$.

92. [W] In your own words, state the Multiplication Property of Inequalities.

Projects in Mathematics

Buying a Car

Besides the initial expense of buying a car, there are continuing expenses involved in owning a car. These ongoing expenses include car insurance, gas and oil, general maintenance, and monthly car payments.

A student takes an after-school job to earn money to buy and maintain a car. How many hours per week must the student work to support a car? Assume that the student earns \$4.50 per hour. Here are some assumptions about the monthly cost in several categories.

Monthly Payment: Assume that the car cost \$2500 with a down payment of \$300. The remainder is financed for 3 years at an annual simple interest rate of 11%.

Amount financed $= 2500 - 300 = 2200$
The amount of interest $= \$2200 \cdot 0.11 \cdot 3 = \$726$
The total amount to be repaid $= \$2200 + \$726 = \$2926$
Monthly payment $= \dfrac{2926}{36} \approx \$81.28$

Insurance: Assume \$1020 per year $= \dfrac{\$1020}{12 \text{ months}} = \$85$ per month

Auto Repair Bill	
Parts	\$ 179
Labor	\$ 78
Sales tax	\$ 15

01373

Gas: Assume that the student travels 600 mi per month. At 25 mi/gal of gas, the student uses 24 gal of gas per month. The gas costs \$1.20 per gallon.

$24 \cdot \$1.20 = \$28.80$ per month for gas

Miscellaneous: Assume \$.10 per mile for upkeep.

$600 \cdot \$.10 = \$60$ per month for upkeep

Total monthly expenses $= \$81.28 + \$85 + \$28.80 + \$60 = \$255.08$

To find the number of hours per month that the student must work to finance the car, divide the total monthly expenses by the hourly rate.

Number of hours per month $= \dfrac{\$255.08}{\$4.50} \approx 56.7$ h

Number of hours per week $= \dfrac{56.7 \text{ h}}{1 \text{ month}} \cdot \dfrac{1 \text{ month}}{4 \text{ weeks}} \approx 14.2$ h/week

The student has to work approximately 14 h per week to pay the monthly car expenses.

If you own a car, make out your own expense record. If you do not own a car, decide on the kind of car that you would want to purchase, and calculate the total monthly expenses that you would have. An insurance company can give you rates on different kinds of insurance. An automobile club can give you approximations of miscellaneous expenses.

Chapter Summary

Key Words

An *equation* expresses the equality of two mathematical expressions. The *solution* of an equation is a number that, when substituted for the variable, results in a true equation. To *solve an equation* means to find a solution of the equation.

Two angles that are on opposite sides of the intersection of two lines are *vertical angles*; vertical angles have the same measure. Two angles that share a common side are *adjacent angles*; adjacent angles of intersecting lines are *supplementary angles.*

A line that intersects two other lines at two different points is a *transversal.* If the lines cut by a transversal are parallel lines, equal angles are formed: *alternate interior angles, alternate exterior angles,* and *corresponding angles.*

The *solution set* of an inequality is a set of numbers, each element of which, when substituted for the variable, results in a true inequality.

Essential Rules

Addition Property of Equations
The same quantity can be added to each side of an equation without changing its solution.

If $a = b$, then $a + c = b + c$.

Multiplication Property of Equations
Each side of an equation can be multiplied by the same nonzero number without changing its solution.

If $a = b$ and $c \neq 0$, then $ac = bc$.

Addition Property of Inequalities
The same term can be added to each side of an inequality without changing its solution.

If $a > b$, then $a + c > b + c$.
If $a < b$, then $a + c < b + c$.

Multiplication Property of Inequalities
Each side of an inequality can be multiplied by the same positive number without changing the solution set of the inequality.

If $a > b$ and $c > 0$, then $ac > bc$.
If $a < b$ and $c > 0$, then $ac < bc$.

If each side of an inequality is multiplied by the same *negative* number and the inequality symbol is reversed, then the solution set of the inequality is not changed.

If $a > b$ and $c < 0$, then $ac < bc$.
If $a < b$ and $c < 0$, then $ac > bc$.

Basic Percent Equation

Percent · base = amount
$PB = A$

Interior Angles of a Triangle

The sum of the interior angles = 180°.

Value Mixture Equation

Value = amount · unit cost
$V = AC$

Percent Mixture Equation

Quantity = amount · percent concentration
$Q = Ar$

Annual Simple Interest Equation

Simple Interest = principal · simple interest rate
$I = Pr$

Uniform Motion Equation

Distance = rate · time
$d = rt$

Chapter Review Exercises

1. Solve: $3m = -12$

2. Solve: $7 = 8a - 5$

3. Solve and graph the solution set of $x - 3 > -1$.

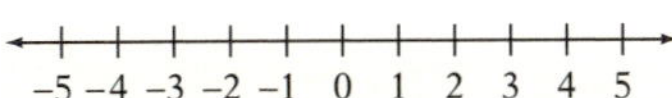

4. Given that $BC = 11$ cm and AB is three times the length of BC, find the length of AC.

5. Solve: $3(2n - 3) = 2n + 3$

6. 4.5 is what percent of 80?

7. Solve: $3x + 4 \geq -8$

8. Solve: $z + 5 = 2$

9. Given that $\angle a = 74°$ and $\angle b = 52°$, find the measures of angles x and y.

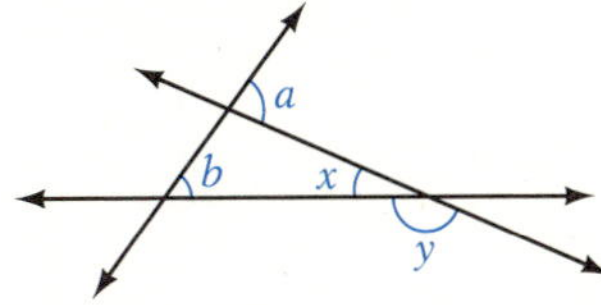

10. Given that $\ell_1 \parallel \ell_2$, find the measures of angles a and b.

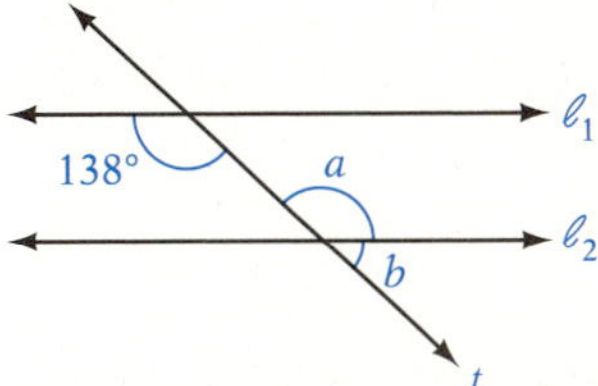

11. Solve: $3a + 8 = 12 - 5a$

12. Is -2 a solution of $x(x - 3) = 3(2 - x) - 2$?

13. 32% of what number is 180?

14. Solve: $-15x \leq 45$

15. Solve: $12 - 4(x - 1) \leq 5(x - 4)$

16. Solve: $4(2 - 2t) = 8 + 2(5 - 3t)$

17. Translate "the difference between seven and the product of five and a number is thirty-seven" into an equation and solve.

18. An airline overbooks flights by selling 12% more tickets than there are seats available. How many tickets would this airline sell for an airplane that has 175 seats?

19. A computer system that sold for \$2400 one year ago can now be purchased for \$1800. What percent decrease does this represent?

20. A college sports foundation deposits a total of \$24,000 into two simple interest accounts. The annual simple interest rate on one account is 4%; on the second account, the annual simple interest rate is 9%. How much is invested in each account if the total interest earned is 7% of the total investment?

21. A dairy mixed 5 gal of cream that is 30% butterfat with 8 gal of milk that is 4% butterfat. What is the percent of butterfat in the resulting mixture?

22. Cranberry juice that costs \$1.79 per quart is combined with apple juice that costs \$1.19 per quart. How many quarts of each are used to make 10 qt of cranapple juice costing \$1.61 per quart?

23. A jet plane traveling at 600 mph overtakes a propeller-driven plane that had a two-hour head start. The propeller-driven plane is traveling at 200 mph. How far from the starting point does the jet overtake the propeller-driven plane?

24. Florist A charges a \$3 delivery fee plus \$21 per bouquet delivered. Florist B charges a \$15 delivery fee plus \$18 per bouquet delivered. A scout troop wants to supply each resident of a nursing home with a bouquet for Grandparent's Day. Find the number of residents of the nursing home if Florist B is more economical than Florist A.

Florist A

Florist B

Cumulative Review Exercises

1. Simplify: $2^3 \cdot 3 - 4(3 - 4 \cdot 5)$

2. Find the GCF of 18 and 45.

3. Evaluate $-|-23|$.

4. Write eleven and nine hundredths in standard form.

5. Simplify: $-7y^2 + 4 + 5y^2 - 8$

6. Find the product of $\frac{2}{9}$ and $\frac{3}{4}$.

7. Find the measure of $\angle x$.

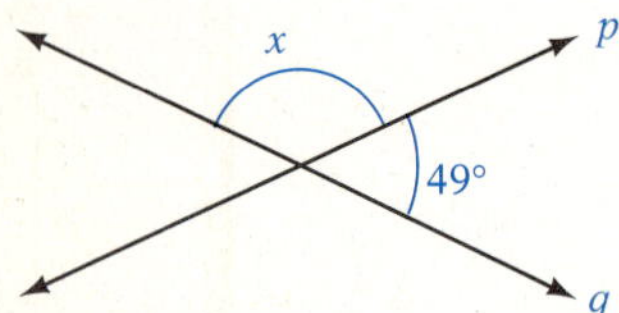

8. Place the correct symbol, < or >, between the two numbers.

$$-\frac{1}{2} \quad -\frac{2}{3}$$

9. Identify the property that justifies the statement.

$$(x + 3)5 = 5(x + 3)$$

10. For the inequality $x < -4$, what numbers listed below make the inequality true?

a. -8 **b.** -6.1 **c.** -4 **d.** 0

11. Solve: $4x + 2 = 6x - 8$

12. Evaluate $4a^2b - a^3$ when $a = -2$ and $b = 3$.

13. Simplify: $3(b - 2) - 6(4b + 5)$

14. What percent of 15 is 3?

15. Evaluate $\frac{x}{y - z}$ when $x = \frac{1}{2}$, $y = \frac{4}{5}$, and $z = \frac{3}{10}$.

16. Solve: $3(2x + 5) = 18$

17. The pressure, P, in pounds per square inch, at a certain depth in the ocean can be approximated by the equation $P = 15 + \frac{1}{2}D$, where D is the depth in feet. Use this equation to find the pressure when the depth is 20 ft.

18. Your rent increased from \$500 per month to \$545 per month. Find the percent increase in the monthly rent.

19. Of the 375 students in a high school graduating class, 300 students went on to college. What percent of the students in the class went on to college?

20. A child's circular wading pool is 21 ft in diameter and 1.5 ft deep. What volume of water does the pool hold? Round to the nearest hundredth.

21. Find the perimeter of a rectangle that has a length of 4.25 m and a width of 3.5 m.

22. Six less than a number is greater than twenty-five. Find the smallest integer that satisfies the inequality.

23. A pharmacist has 15 L of an 80% alcohol solution. How many liters of pure water should be added to the alcohol solution to make a diluted alcohol solution that is 75% alcohol?

CHAPTER

7

Graphs and Linear Equations

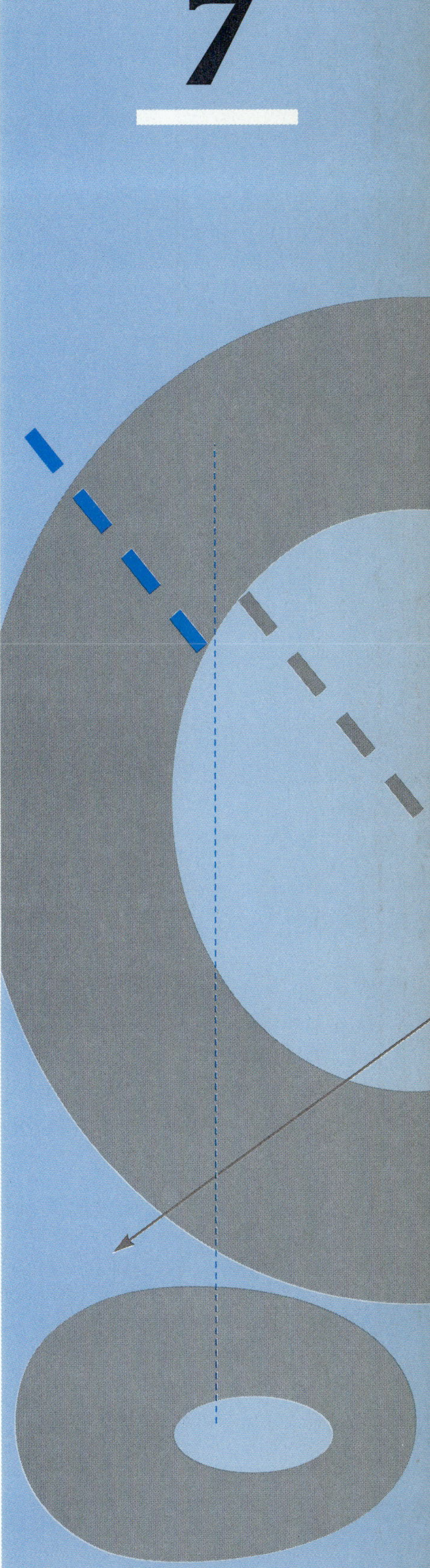

7.1 Graphs of Straight Lines

A The rectangular coordinate system
B Solutions of linear equations in two variables
C Graph equations of the form $y = mx + b$
D Graph equations of the form $Ax + By = C$

7.2 Slopes of Straight Lines

A The slope of a straight line
B Graph a line using the slope and y-intercept

7.3 Equations of Straight Lines

A Find the equation of a line given a point and the slope
B Find the equation of a line given two points
C Application problems

7.4 Solving Systems of Linear Equations

A Solve a system of linear equations by graphing
B Solve a system of linear equations by the substitution method
C Solve a system of linear equations by the addition method

7.5 Application Problems in Two Variables

A Solve rate–of–wind or current problems
B Application problems using two variables

Focus on Problem Solving

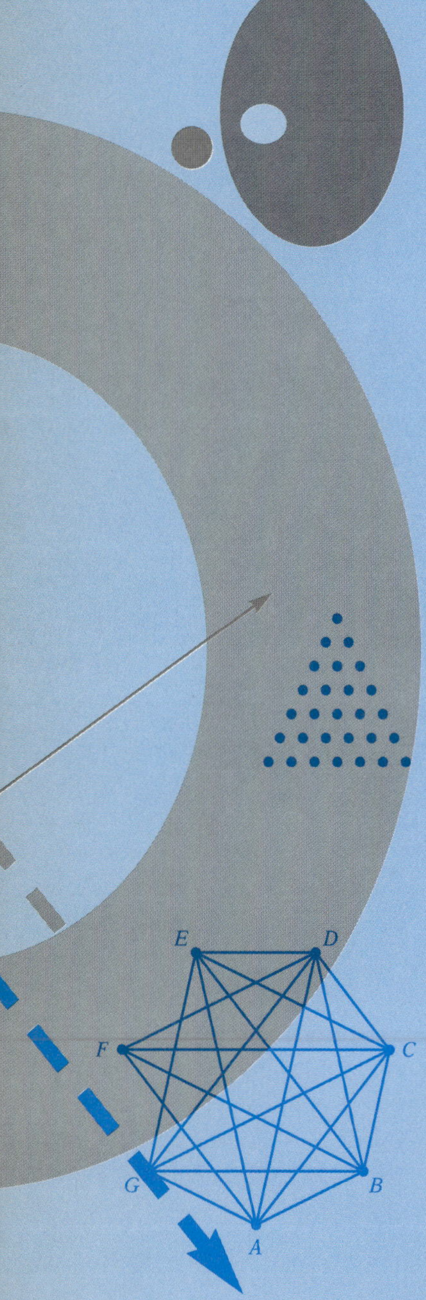

Problem solving in the previous chapters concentrates on solving a certain problem. After the problem is solved, there is an important question to be asked: "Does the solution to this problem apply to other types of problems?"

To illustrate this extension to problem solving, we will consider *triangular numbers,* which were studied by ancient Greek mathematicians. The numbers 1, 3, 6, 10, 15, 21 are the first six triangular numbers. What is the next triangular number?

To answer this question, note in the diagram below that a triangle can be formed using the number of dots that correspond to a triangular number.

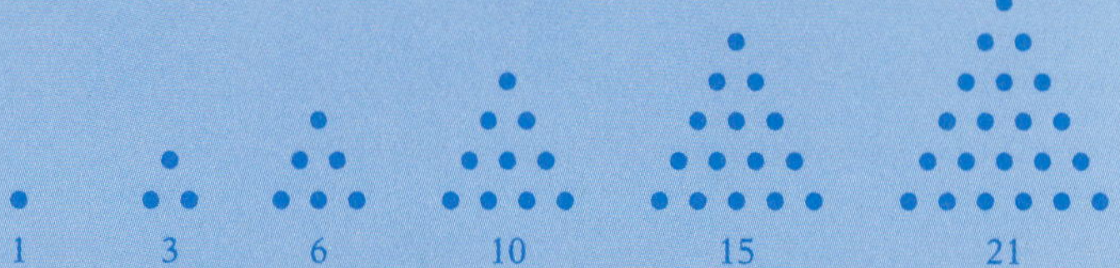

Observe that the number of dots in a row is one more than the number in the row above. The total number of dots can be found by addition.

$$1 = 1 \qquad 1 + 2 = 3 \qquad 1 + 2 + 3 = 6 \qquad 1 + 2 + 3 + 4 = 10$$
$$1 + 2 + 3 + 4 + 5 = 15 \qquad 1 + 2 + 3 + 4 + 5 + 6 = 21$$

The pattern suggests that the next triangular number (the 7th one) is the sum of the first 7 natural numbers.

$$1 + 2 + 3 + 4 + 5 + 6 + 7 = 28$$

The 7th triangular number is 28. The diagram at the left shows the 7th triangular number.

Using the pattern for triangular numbers, the 10th triangular number is

$$1 + 2 + 3 + 4 + 5 + 6 + 7 + 8 + 9 + 10 = 55$$

Now consider a situation that may seem to be totally unrelated to triangular numbers. Suppose you are in charge of scheduling softball games for a league. There are seven teams in the league, and each team must play every other team once. How many games must be scheduled?

We label the teams A, B, C, D, E, F, and G. (See the figure at the left.) A line between two teams indicates that the two teams play each other. Beginning with A, there are 6 lines for the 6 teams that A must play.

Now consider B. There are 6 teams that B must play, but the line between A and B has already been drawn, so there are only 5 remaining games to schedule for B. Now move on to C. The lines between C and A and C and B have already been drawn, so there are 4 additional lines to be drawn to represent the teams C will play. Moving on to D, the lines between D and A, D and B, and D and C have already been drawn, so there are 3 more lines to be drawn to represent the teams D will play.

Note that as we move from team to team, one fewer line needs to be drawn. When we reach F, there is only one line to be drawn, the one between F and G. The total number of lines drawn is $6 + 5 + 4 + 3 + 2 + 1 = 21$, the sixth triangular number. For a league with 7 teams, the number of games that must be scheduled so that each team plays every other team once is the 6th triangular number. If there were 10 teams in the league, the number of games that must be scheduled would be the 9th triangular number, which is 45.

SECTION 7.1 Graphs of Straight Lines

OBJECTIVE A The rectangular coordinate system

POINT OF INTEREST

Although Descartes is given credit for introducing analytic geometry, others were working on the same concept, notably Pierre Fermat. Nowhere in Descartes's work is there a coordinate system as we draw it with two axes. Descartes did not use the word *coordinate* in his work. This word was introduced by Gottfried Leibnitz, who also first used the words *abscissa* and *ordinate.*

Before the 15th century, geometry and algebra were considered separate branches of mathematics. That all changed when René Descartes, a French mathematician who lived from 1596 to 1650, founded **analytic geometry**. In this geometry, a *coordinate system* is used to study relationships between variables.

A **rectangular coordinate system** is formed by two number lines, one horizontal and one vertical, that intersect at the zero point of each line. The point of intersection is called the **origin**. The two lines are called **coordinate axes**, or simply **axes**.

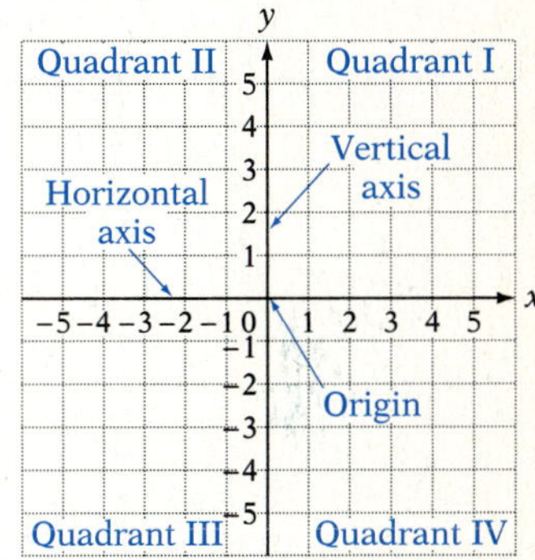

The axes determine a **plane**, which can be thought of as a large, flat sheet of paper. The two axes divide the plane into four regions called **quadrants**, which are numbered counterclockwise from I to IV.

Each point in the plane can be identified by a pair of numbers called an **ordered pair**. The first number of the pair measures a horizontal distance and is called the **abscissa**, or ***x*-coordinate**. The second number of the pair measures a vertical distance and is called the **ordinate**, or ***y*-coordinate**. The ordered pair (a, b) associated with a point is also called the **coordinates** of the point.

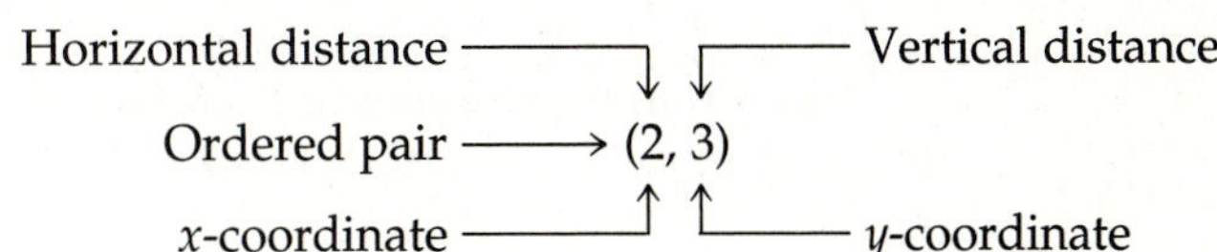

To **graph** or **plot** a point in the plane, place a dot at the location given by the ordered pair. The **graph of an ordered pair** is the dot drawn at the coordinates of the point in the plane. The points whose coordinates are $(3, 4)$ and $(-2.5, -3)$ are graphed in the figures below.

CONSIDER THIS

Locating points in the plane may be difficult at first. You should practice plotting points such as $(-2, 3)$, $(4, -3)$, $(-2, -4)$, $(0, 3)$, and $(3, 0)$. Another good exercise is for you to quickly determine in what quadrant a given point lies. For instance, in what quadrant is the point whose coordinates are $(3, -1)$?

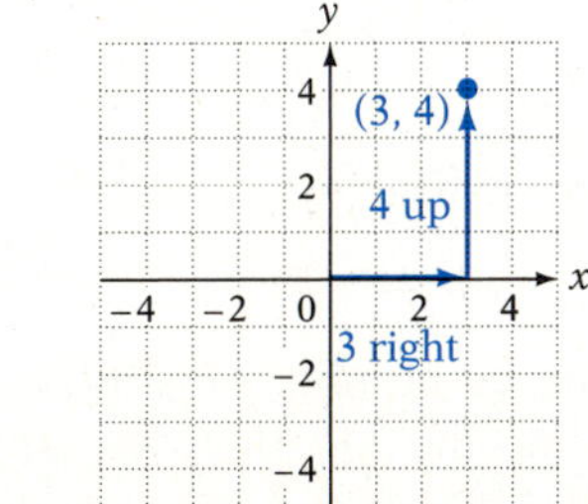

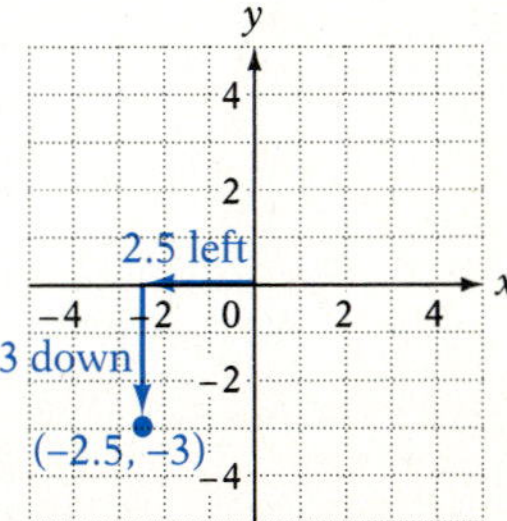

Example 1 Graph the ordered pairs $(-2, -3)$, $(3, -2)$, $(1, 3)$, and $(4, 1)$.

Solution

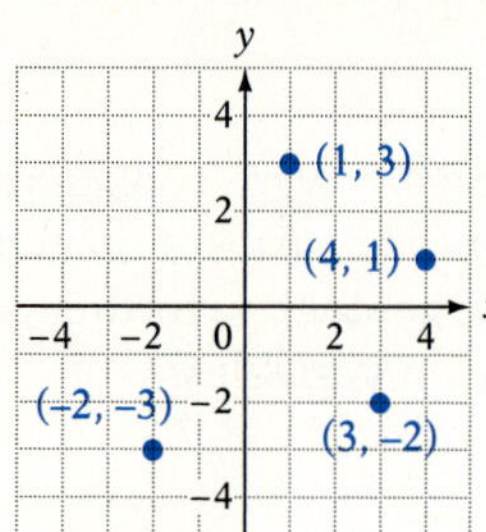

You Try It 1 Graph the ordered pairs $(-1, 3)$, $(1, 4)$, $(-4, 0)$, and $(-2, -1)$.

Your Solution

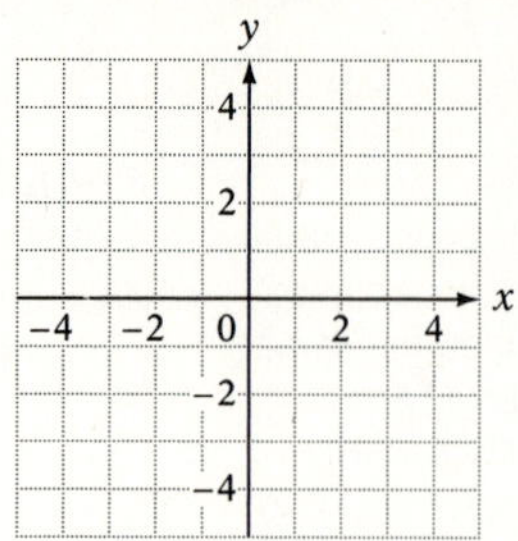

Example 2 Find the coordinates of each point.

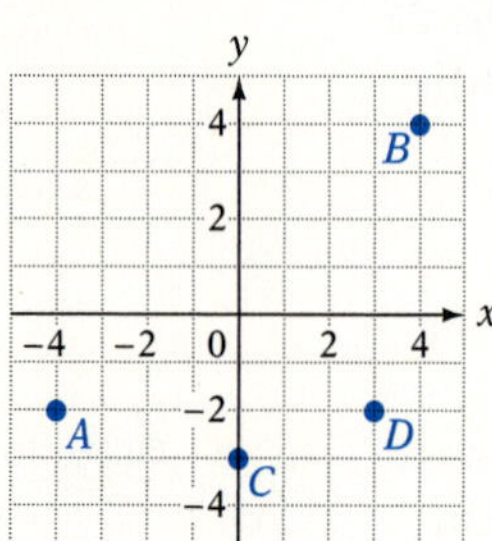

Solution $A(-4, -2)$ $C(0, -3)$
$B(4, 4)$ $D(3, -2)$

You Try It 2 Find the coordinates of each point.

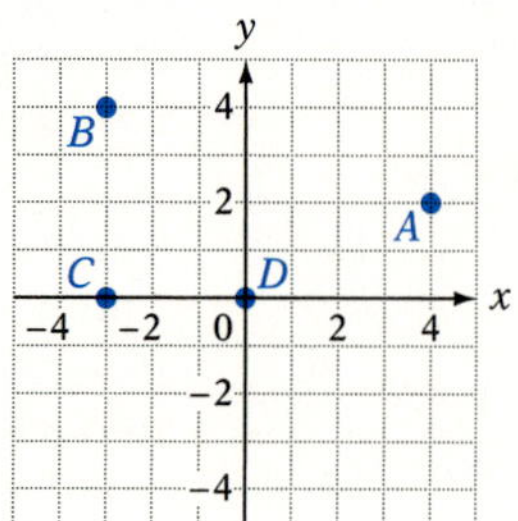

Your Solution

Solutions on pp. A19–A20

OBJECTIVE B Solutions of linear equations in two variables

A **solution of an equation in two variables** is an ordered pair (x, y) whose coordinates make the equation a true statement.

Is $(-3, 7)$ a solution of $y = -2x + 1$?

Replace x by -3; replace y by 7. Simplify. Compare the results. If the results are equal, the given ordered pair is a solution. If the results are not equal, the given ordered pair is not a solution.

y	$= -2x + 1$
7	$-2(-3) + 1$
7	$6 + 1$

$7 = 7$

$(-3, 7)$ is a solution of the equation $y = -2x + 1$.

Besides $(-3, 7)$, there are many other ordered-pair solutions of $y = -2x + 1$. For example, $(0, 1)$, $\left(-\frac{3}{2}, 4\right)$, and $(4, -7)$ are also solutions. In general, an equation in two variables has an infinite number of solutions. By choosing any value of x and substituting that value into the equation, we can calculate the corresponding value of y.

Find the ordered pair solution of $y = \frac{2}{3}x - 3$ that corresponds to $x = 6$.

	$y = \frac{2}{3}x - 3$
Replace x by 6.	$y = \frac{2}{3}(6) - 3$
Solve for y.	$y = 4 - 3$
	$y = 1$

The ordered-pair solution is (6, 1).

CONSIDER THIS

It may help to think of an ordered pair as the address (location) of a point in the plane.

The solutions of an equation in two variables can be graphed in an xy-coordinate system.

Graph the ordered-pair solutions of $y = -2x + 1$ when $x = -2, -1, 0, 1$, and 2.

Use the values of x to determine the ordered-pair solutions. It is convenient to record these in a table.

x	$y = -2x + 1$	y	(x, y)
-2	$-2(-2) + 1$	5	$(-2, 5)$
-1	$-2(-1) + 1$	3	$(-1, 3)$
0	$-2(0) + 1$	1	$(0, 1)$
1	$-2(1) + 1$	-1	$(1, -1)$
2	$-2(2) + 1$	-3	$(2, -3)$

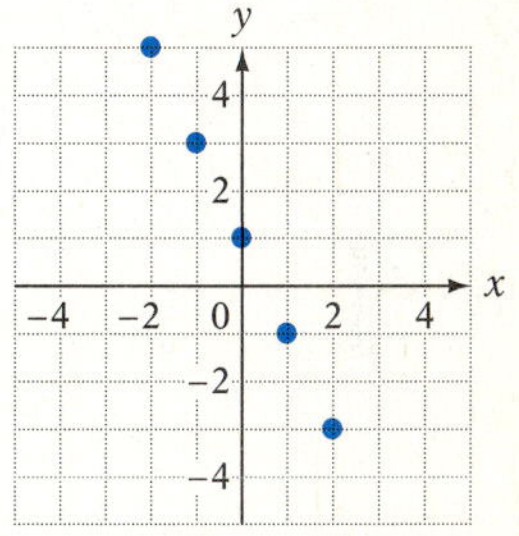

CONSIDER THIS

Problems such as the one at the right prepare you to graph straight lines, the subject of the next objective. Keep in mind the relationship between a solution of an equation and the pictorial representation of that solution, that is, its graph.

Example 3

Is $(-3, 2)$ a solution of $y = 2x + 2$?

Solution

$y = 2x + 2$

2	$2(-3) + 2$
	$-6 + 2$

$2 \neq -4$

No, $(-3, 2)$ is not a solution of $y = 2x + 2$.

You Try It 3

Is $(2, -4)$ a solution of $y = -\frac{1}{2}x - 3$?

Your Solution

Example 4

Graph the ordered-pair solutions of $y = 2x - 1$ when $x = -2, -1, 0, 1$, and 3.

Solution

x	y
-2	-5
-1	-3
0	-1
1	1
3	5

(3, 5)
(1, 1)
(0, –1)
(–1, –3)
(–2, –5)

You Try It 4

Graph the ordered-pair solutions of $y = 2x + 3$ when $x = -3, -1, 0$, and 1.

Your Solution

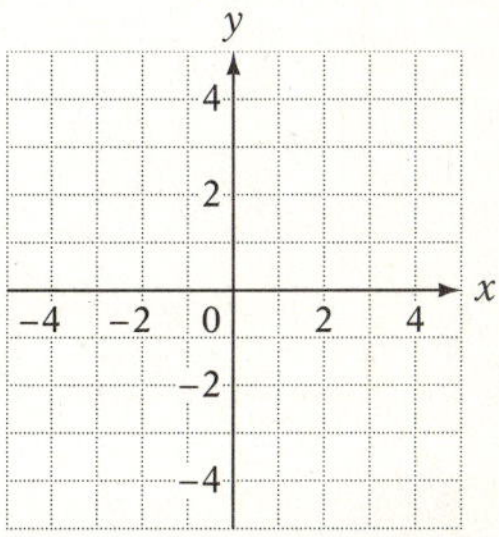

Solutions on p. A20

OBJECTIVE C Graph equations of the form $y = mx + b$

The **graph of an equation in two variables** is a graph of the ordered-pair solutions of the equation.

Consider $y = 2x + 1$. Choosing $x = -2, -1, 0, 1$, and 2 and determining the corresponding values of y produces some of the ordered pairs of the equation. These are recorded in the table at the right. See the graph of the ordered pairs in Fig. 1.

x	$y = 2x + 1$	y	(x, y)
-2	$2(-2) + 1$	-3	$(-2, -3)$
-1	$2(-1) + 1$	-1	$(-1, -1)$
0	$2(0) + 1$	1	$(0, 1)$
1	$2(1) + 1$	3	$(1, 3)$
2	$2(2) + 1$	5	$(2, 5)$

Choosing values of x that are not integers produces more ordered pairs to graph, such as $\left(-\frac{5}{2}, -4\right)$ and $\left(\frac{3}{2}, 4\right)$, as shown in Fig. 2. Choosing still other values of x would result in more and more ordered pairs being graphed. The result would be so many dots that the graph would appear as the straight line shown in Fig. 3, which is the graph of $y = 2x + 1$.

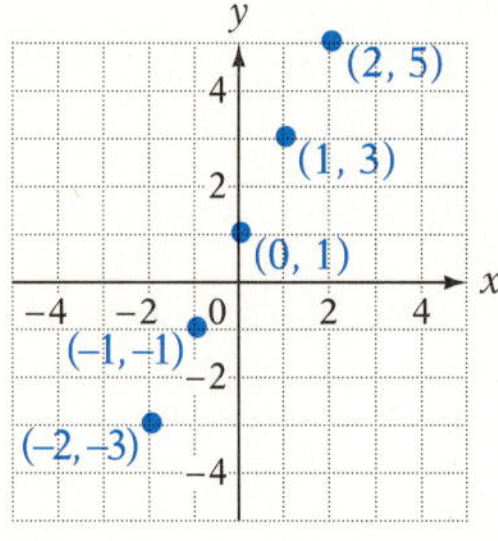

Fig. 1

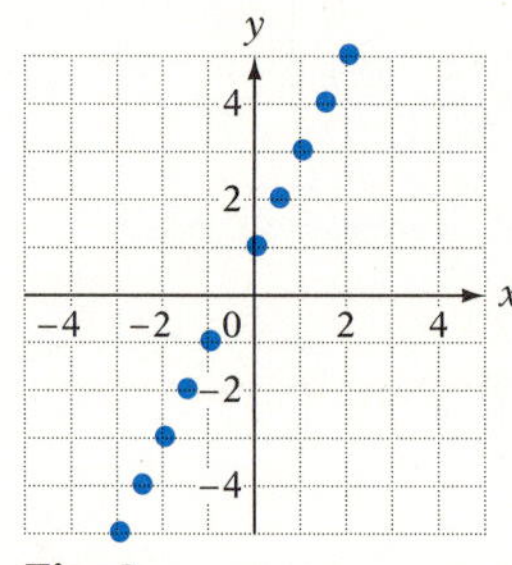

Fig. 2

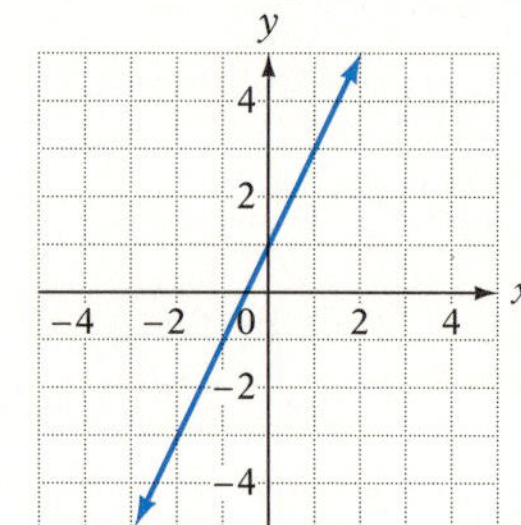

Fig. 3

Equations in two variables have characteristic graphs. The equation $y = 2x + 1$ is an example of a *linear equation* because its graph is a straight line.

Linear Equation in Two Variables

Any equation of the form $y = mx + b$, where m and b are constants, is a **linear equation in two variables**. The graph of a linear equation in two variables is a straight line.

Examples of linear equations are shown at the right. Note from $y = 3 - 2x$ that m is the coefficient of x and b is the constant.

$y = 2x + 1$ $(m = 2, b = 1)$

$y = 3 - 2x$ $(m = -2, b = 3)$

$y = -\frac{3}{4}x$ $\left(m = -\frac{3}{4}, b = 0\right)$

The equation $y = x^2 + 4x + 3$ is not a linear equation in two variables because there is a term with a variable squared. The equation $y = \frac{3}{x - 4}$ is not a linear equation because a variable occurs in the denominator of the fraction.

To graph a linear equation, choose some values of x and then find the corresponding values of y. Because a straight line is determined by two points, it is sufficient to find only two ordered-pair solutions. However, it is recommended that at least three ordered-pair solutions be used to ensure accuracy.

Graph $y = -\frac{3}{2}x + 2$.

This is a linear equation with $m = -\frac{3}{2}$ and $b = 2$. Find at least three solutions. Because m is a fraction, choose values of x that will simplify the calculations. We have chosen -2, 0, and 4 for x. (Any values of x could have been selected.)

x	$y = -\frac{3}{2}x + 2$	y	(x, y)
-2	$-\frac{3}{2}(-2) + 2$	5	$(-2, 5)$
0	$-\frac{3}{2}(0) + 2$	2	$(0, 2)$
4	$-\frac{3}{2}(4) + 2$	-4	$(4, -4)$

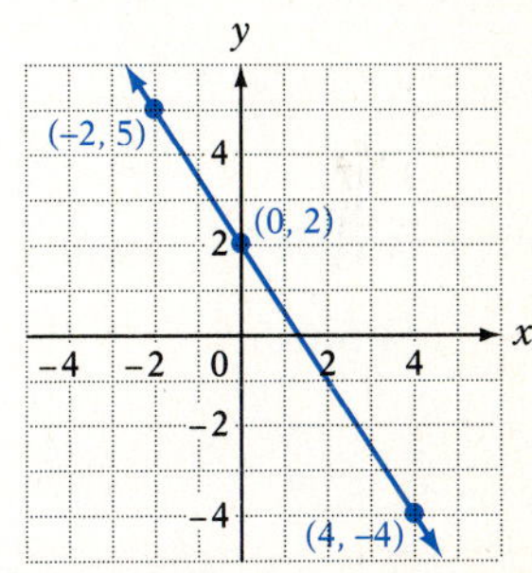

Remember that a graph is a drawing of the ordered-pair solutions of the equation. Therefore, every point on the graph is a solution of the equation and every solution of the equation is a point on the graph.

CONSIDER THIS

It may be helpful to estimate some of the coordinates of the points on the graph. Then use these points to verify that the coordinates are solutions of the equation.

The graph at the right is the graph of $y = x + 2$. Note that $(-4, -2)$ and $(1, 3)$ are points on the graph and that these points are solutions of $y = x + 2$. The point whose coordinates are $(4, 1)$ is not a point on the graph and is not a solution of the equation.

Example 5 Graph $y = 3x - 2$.

Solution

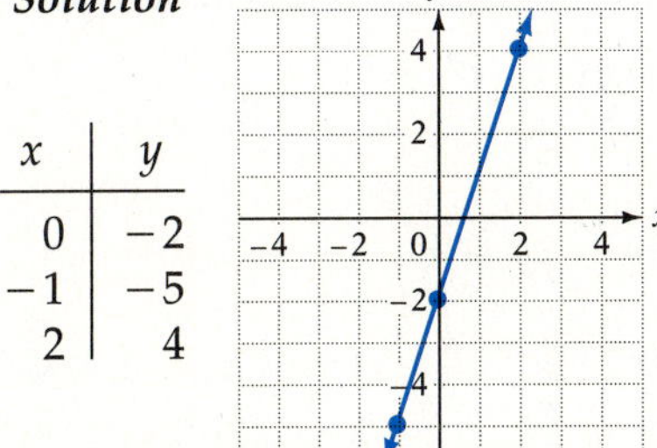

x	y
0	-2
-1	-5
2	4

You Try It 5 Graph $y = 3x + 1$.

Your Solution

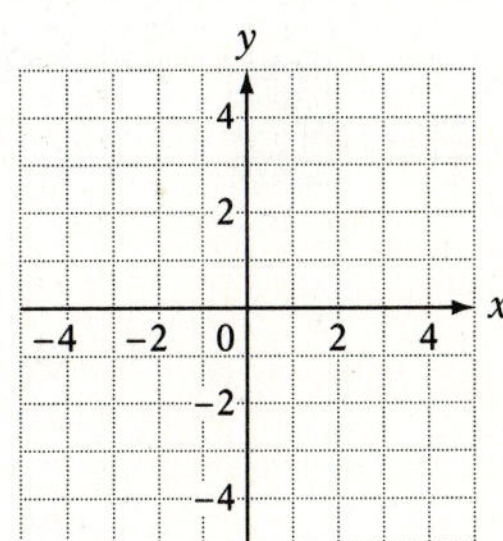

Solution on p. A20

Example 6 Graph $y = 2x$.

Solution

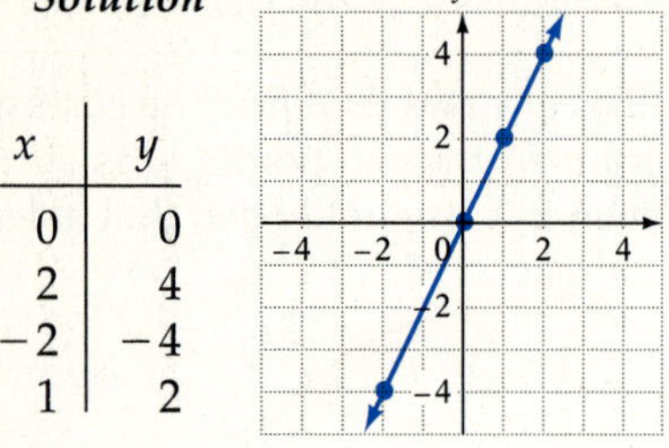

x	y
0	0
2	4
−2	−4
1	2

You Try It 6 Graph $y = -2x$.

Your Solution

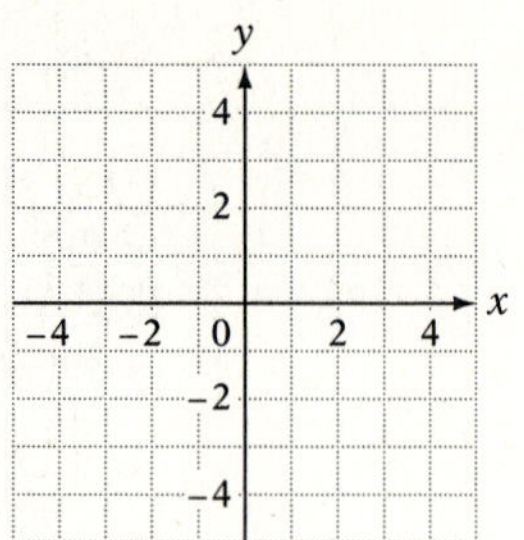

Example 7 Graph $y = \frac{1}{2}x - 1$.

Solution

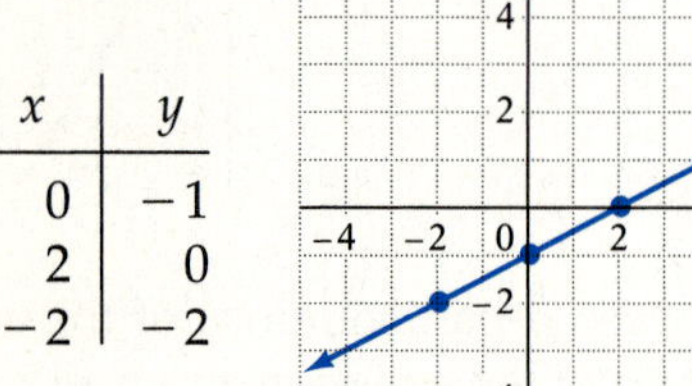

x	y
0	−1
2	0
−2	−2

You Try It 7 Graph $y = \frac{1}{3}x - 3$.

Your Solution

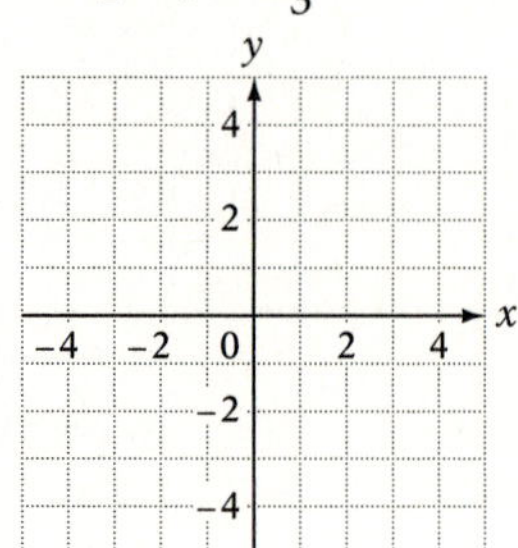

Solutions on p. A20

OBJECTIVE D Graph equations of the form $Ax + By = C$

The equation $Ax + By = C$, where A, B, and C are constants, is also a linear equation. Examples of these equations are shown at the right.

$2x + 3y = 6 \quad (A = 2, B = 3, C = 6)$
$x - 2y = -4 \quad (A = 1, B = -2, C = -4)$
$2x + y = 0 \quad (A = 2, B = 1, C = 0)$
$4x - 5y = 2 \quad (A = 4, B = -5, C = 2)$

To graph an equation of the form $Ax + By = C$, first solve the equation for y. Then follow the same procedure used for graphing $y = mx + b$.

Graph $3x + 4y = 12$.

The equation is in the form $Ax + By = C$. Use the Addition Property of Equations to subtract the term $3x$ from each side of the equation. Simplify. Note that on the right side of the equation, the term containing x is first.

$$3x + 4y = 12$$
$$3x - 3x + 4y = -3x + 12$$
$$4y = -3x + 12$$

Use the Multiplication Property of Equations to divide each side of the equation by the coefficient of y. Note that we must divide each term on the right side by 4.

$$\frac{4y}{4} = \frac{-3x + 12}{4}$$
$$y = \frac{-3x}{4} + \frac{12}{4}$$

Simplify. The equation is in the form $y = mx + b$.

$$y = -\frac{3}{4}x + 3$$

Find at least three solutions. Display the ordered pairs in a table.

x	y
0	3
4	0
−4	6

Graph the ordered pairs on a rectangular coordinate system and draw a straight line through the points.

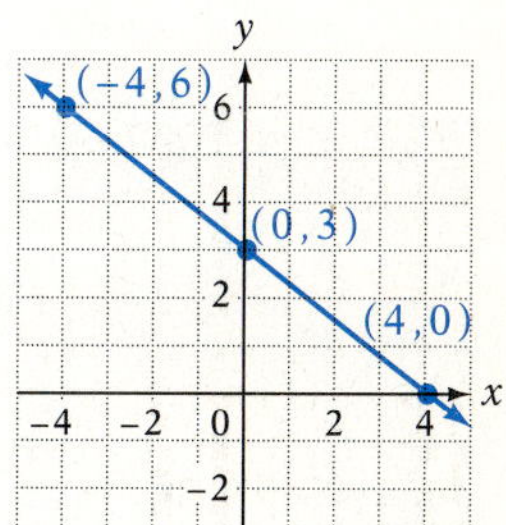

The graph of an equation with one of the variables missing is either a horizontal or a vertical line.

The equation $y = 2$ could be written $0 \cdot x + y = 2$. Because $0 \cdot x = 0$ for any value of x, the value of y is always 2 no matter what value of x is chosen. For instance, replace x by -4, -1, 0, or 3. In each case, $y = 2$.

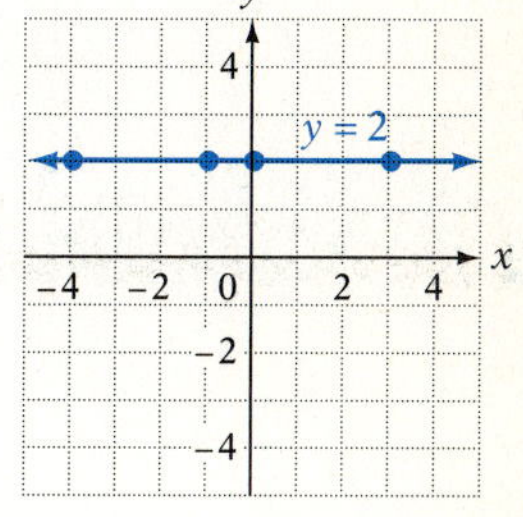

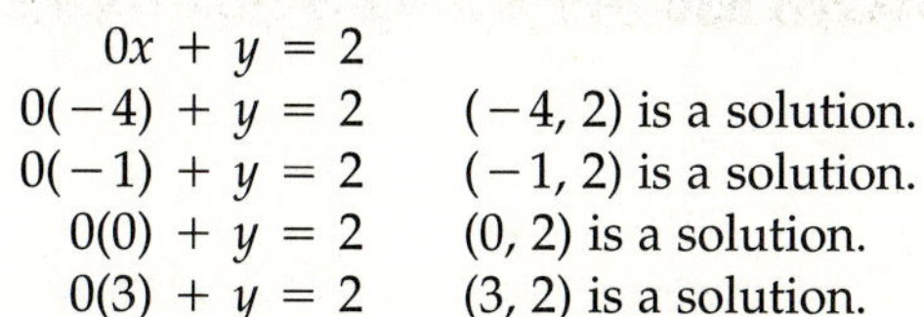

$$0x + y = 2$$

$0(-4) + y = 2$ $\quad(-4, 2)$ is a solution.

$0(-1) + y = 2$ $\quad(-1, 2)$ is a solution.

$0(0) + y = 2$ $\quad(0, 2)$ is a solution.

$0(3) + y = 2$ $\quad(3, 2)$ is a solution.

The solutions are plotted in the graph above, and a line is drawn through the plotted points. Note that the line is horizontal.

Graph of a Horizontal Line

The graph of $y = b$ is a horizontal line passing through $(0, b)$.

The equation $x = -2$ could be written $x + 0 \cdot y = -2$. Because $0 \cdot y = 0$ for any value of y, the value of x is always -2 no matter what value of y is chosen. For instance, replace y by -2, 0, 2, or 3. In each case, $x = -2$.

$x + 0y = -2$
$x + 0(-2) = -2$ $(-2, -2)$ is a solution.
$x + 0(0) = -2$ $(-2, 0)$ is a solution.
$x + 0(2) = -2$ $(-2, 2)$ is a solution.
$x + 0(3) = -2$ $(-2, 3)$ is a solution.

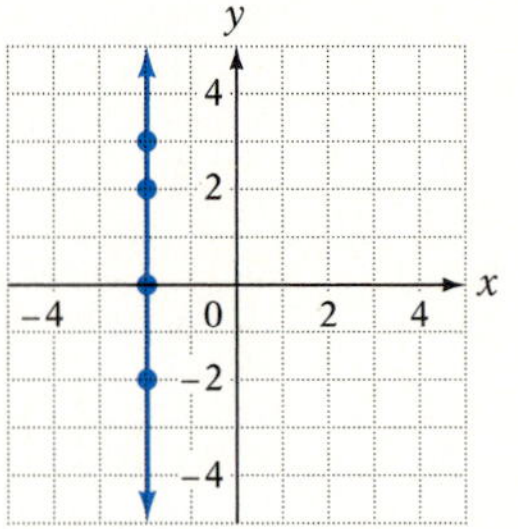

The solutions are plotted in the graph at the right, and a line is drawn through the plotted points. Note that the line is vertical.

Graph of a Vertical Line

The graph of $x = a$ is a vertical line passing through $(a, 0)$.

Graph $x = -3$ and $y = 2$ in the same coordinate grid.

The graph of $x = -3$ is a vertical line passing through $(-3, 0)$.

The graph of $y = 2$ is a horizontal line passing through $(0, 2)$.

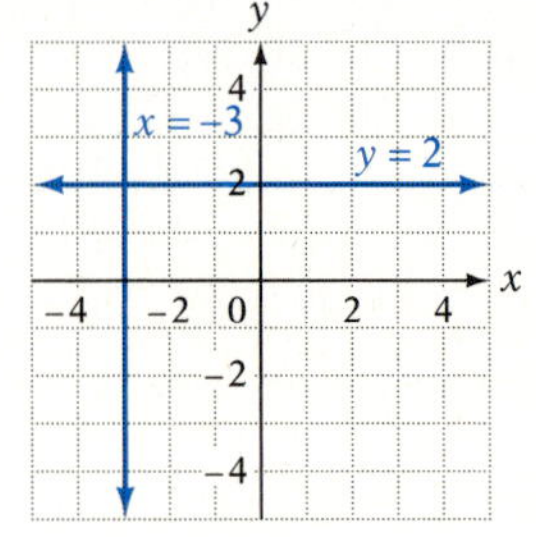

The graph of the equation $2x + 3y = 6$ is shown at the right. The graph crosses the x-axis at the point (3, 0). This point is called the ***x*-intercept**. The graph also crosses the y-axis at the point (0, 2). This point is called the ***y*-intercept.**

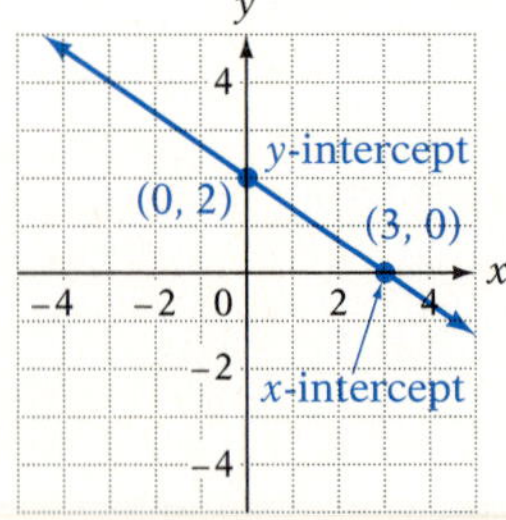

Find the x-intercept and the y-intercept of the graph of the equation $2x - 3y = 12$.

To find the x-intercept, let $y = 0$. (Any point on the x-axis has y-coordinate 0.)

$$2x - 3y = 12$$
$$2x - 3(0) = 12$$
$$2x = 12$$
$$x = 6$$

The x-intercept is (6, 0).

To find the y-intercept, let $x = 0$. (Any point on the y-axis has x-coordinate 0.)

$$2x - 3y = 12$$
$$2(0) - 3y = 12$$
$$-3y = 12$$
$$y = -4$$

The y-intercept is $(0, -4)$.

Find the x- and y-intercept for $x + 4y = -4$. Graph the line.

x-intercept:

$$x + 4y = -4$$
$$x + 4(0) = -4$$
$$x + 0 = -4$$
$$x = -4$$

The x-intercept is $(-4, 0)$.

y-intercept:

$$x + 4y = -4$$
$$0 + 4y = -4$$
$$4y = -4$$
$$y = -1$$

The y-intercept is $(0, -1)$.

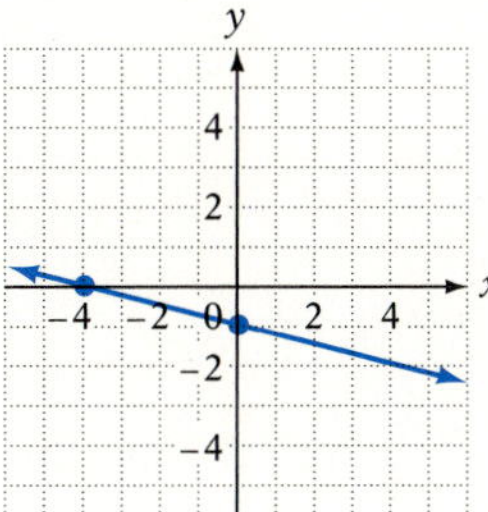

Example 8 Graph $2x - 5y = 10$.

Solution

$$2x - 5y = 10$$
$$-5y = -2x + 10$$
$$y = \frac{2}{5}x - 2$$

x	y
0	−2
5	0
−5	−4

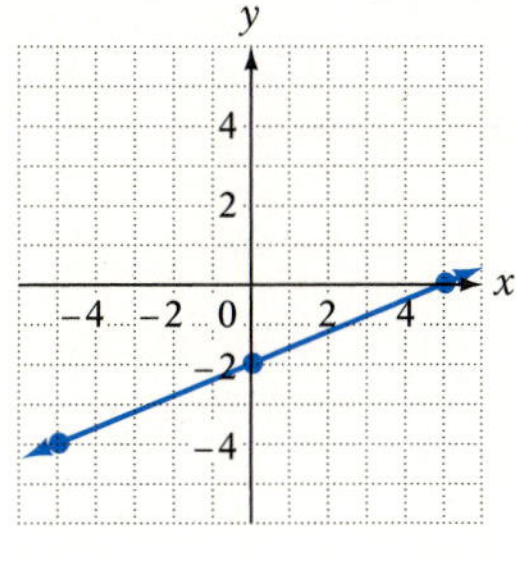

You Try It 8 Graph $5x - 2y = 10$.

Your Solution

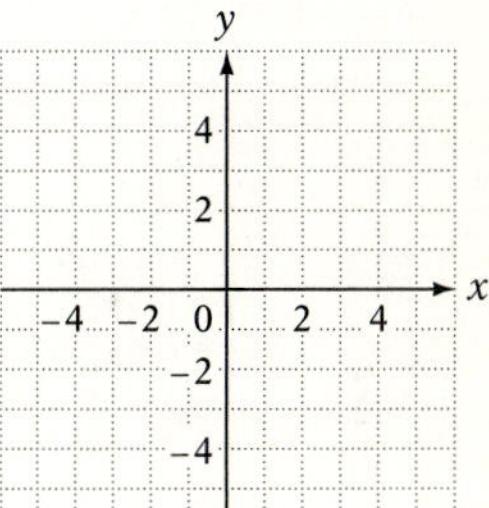

Example 9 Graph $x + 2y = 6$.

Solution

$$x + 2y = 6$$
$$2y = -x + 6$$
$$y = -\frac{1}{2}x + 3$$

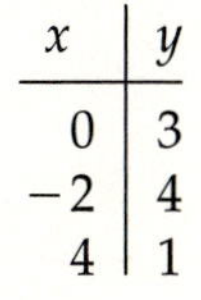

x	y
0	3
−2	4
4	1

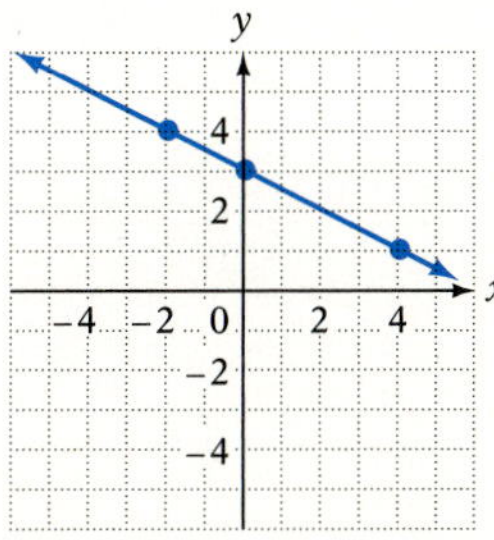

You Try It 9 Graph $x - 3y = 9$.

Your Solution

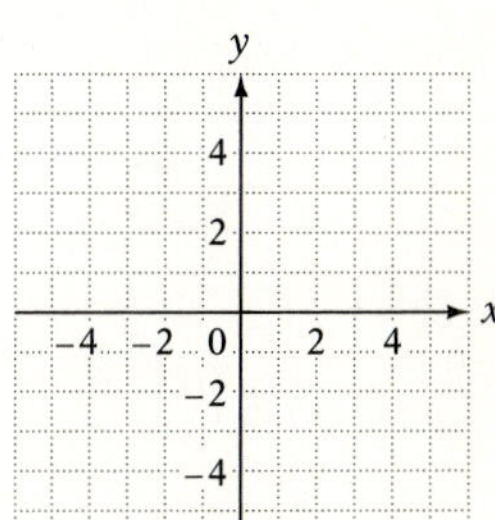

Solutions on p. A20

Example 10 Graph $y = -2$.

Solution The graph of an equation of the form $y = b$ is a horizontal line passing through the point $(0, b)$.

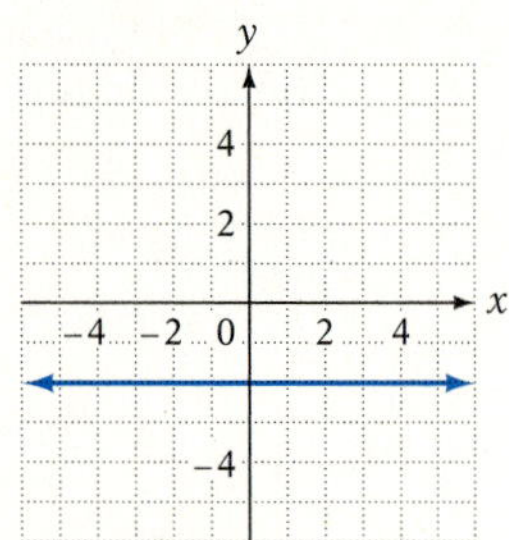

You Try It 10 Graph $y = 3$.

Your Solution

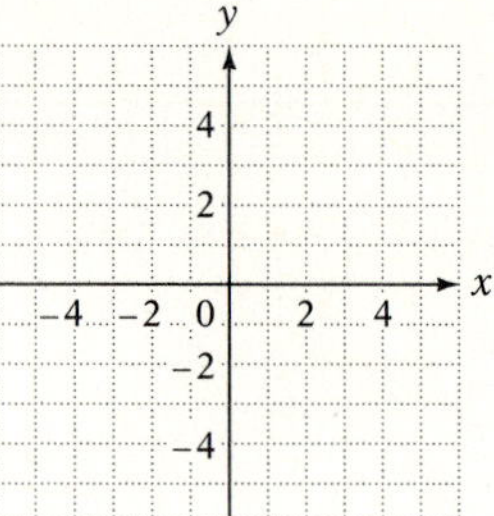

Example 11 Graph $x = 3$.

Solution The graph of an equation of the form $x = a$ is a vertical line passing through the point $(a, 0)$.

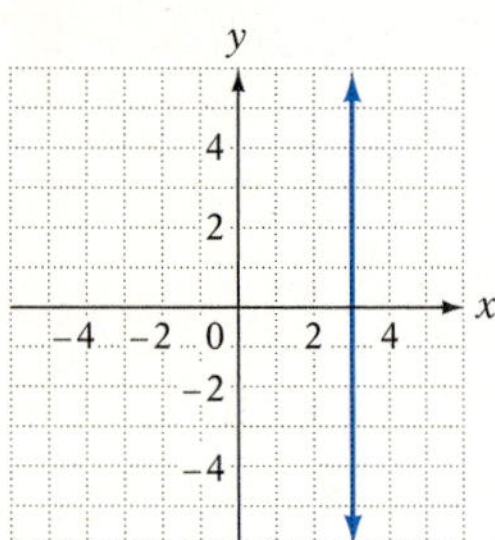

You Try It 11 Graph $x = -4$.

Your Solution

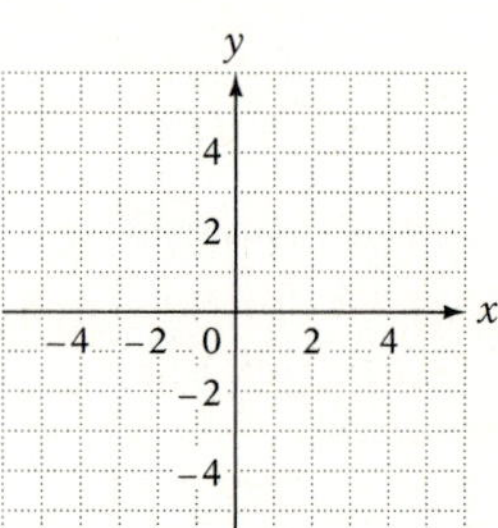

Example 12 Find the x- and y-intercepts for $x - 2y = 4$. Graph the line.

Solution

x-intercept:

$$x - 2y = 4$$
$$x - 2(0) = 4$$
$$x = 4$$

$(4, 0)$

y-intercept:

$$x - 2y = 4$$
$$0 - 2y = 4$$
$$-2y = 4$$
$$y = -2$$

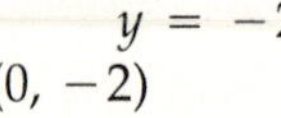

$(0, -2)$

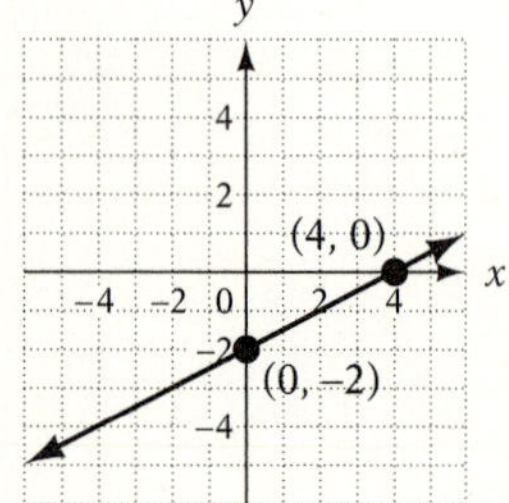

You Try It 12 Find the x- and y-intercepts for $4x - y = 4$. Graph the line.

Your Solution

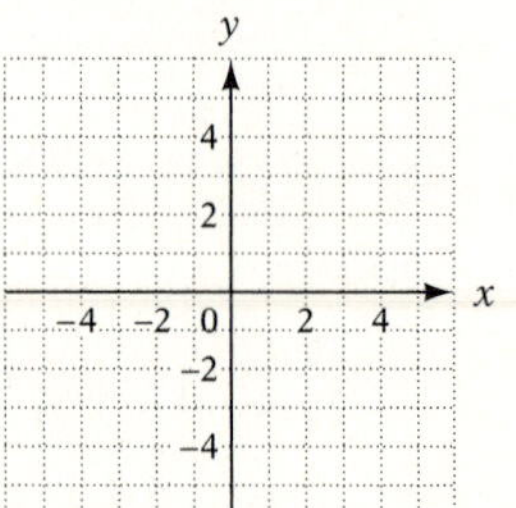

Solutions on p. A20

7.1 EXERCISES

Objective A

1. Graph the ordered pairs $(-2, 1)$, $(3, -5)$, $(-2, 4)$, and $(0, 3)$.

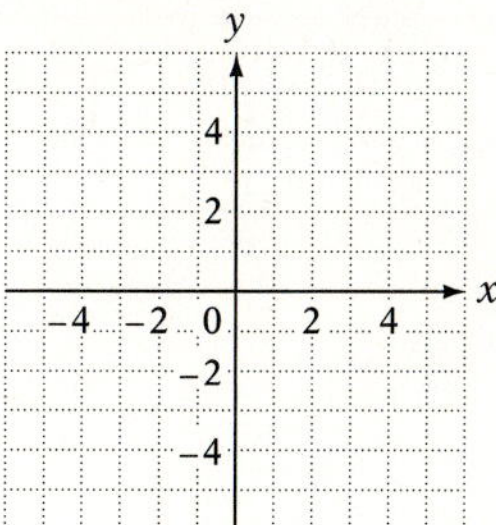

2. Graph the ordered pairs $(5, -1)$, $(-3, -3)$, $(-1, 0)$, and $(1, -1)$.

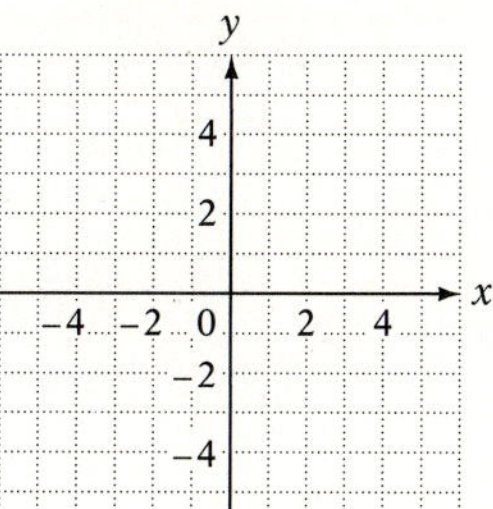

3. Graph the ordered pairs $(0, 0)$, $(0, -5)$, $(-3, 0)$, and $(0, 2)$.

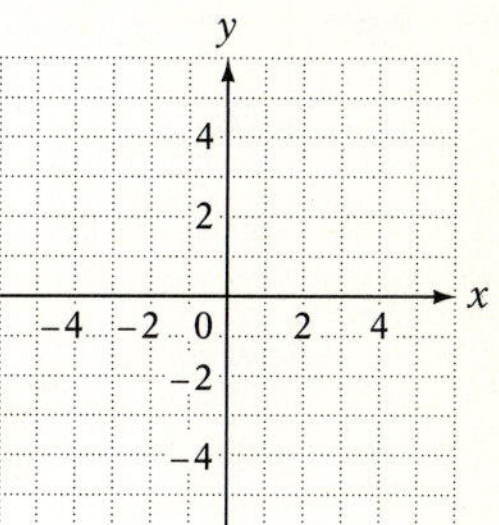

4. Graph the ordered pairs $(-4, 5)$, $(-3, 1)$, $(3, -4)$, and $(5, 0)$.

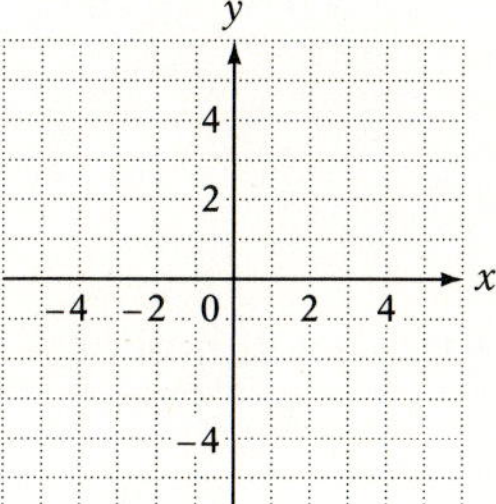

5. Graph the ordered pairs $(-1, 4)$, $(-2, -3)$, $(0, 2)$, and $(4, 0)$.

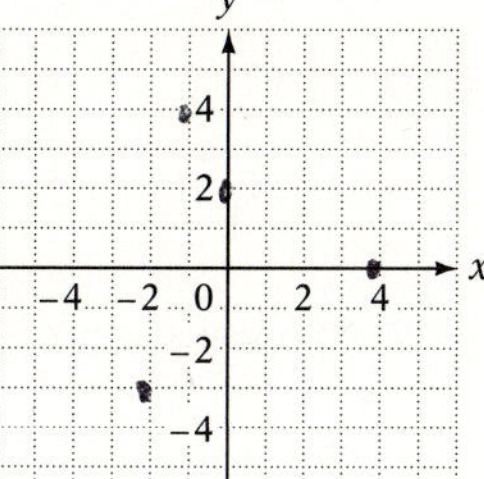

6. Graph the ordered pairs

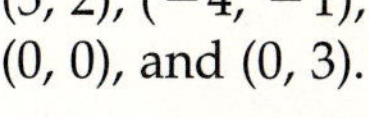

$(5, 2)$, $(-4, -1)$, $(0, 0)$, and $(0, 3)$.

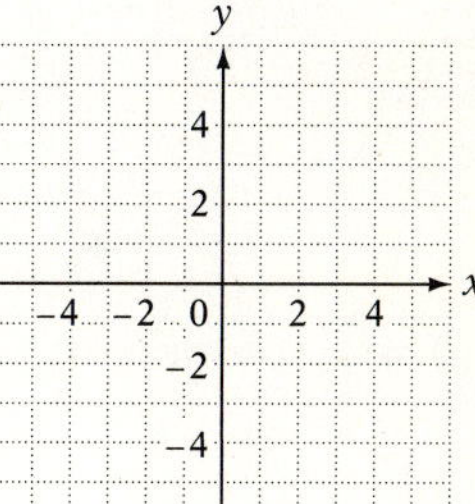

7. Find the coordinates of each point.

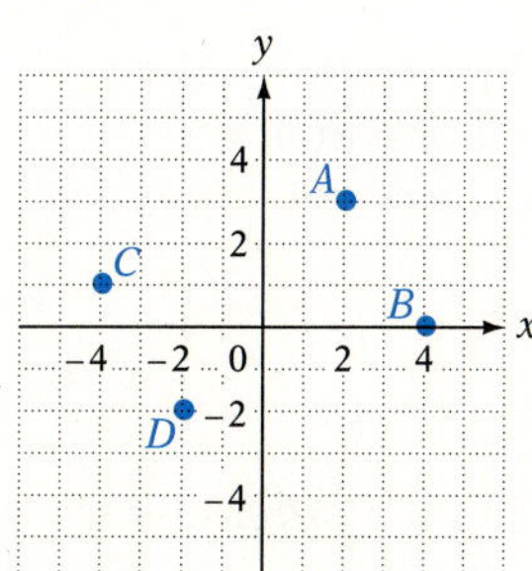

8. Find the coordinates of each point.

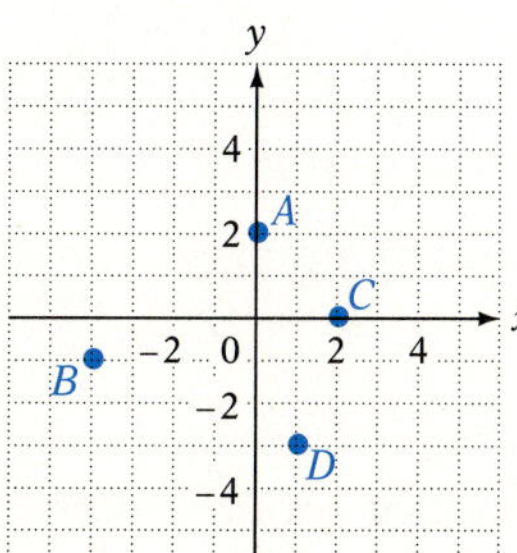

9. Find the coordinates of each point.

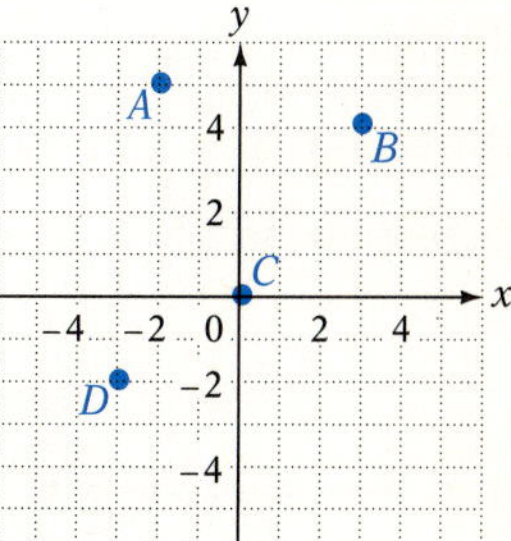

10. Find the coordinates of each point.

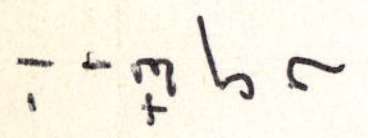

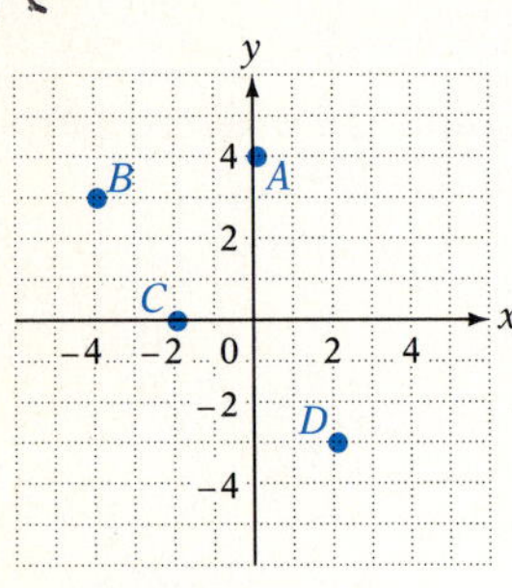

11. **a.** Name the abscissas of points A and C.
b. Name the ordinates of points B and D.

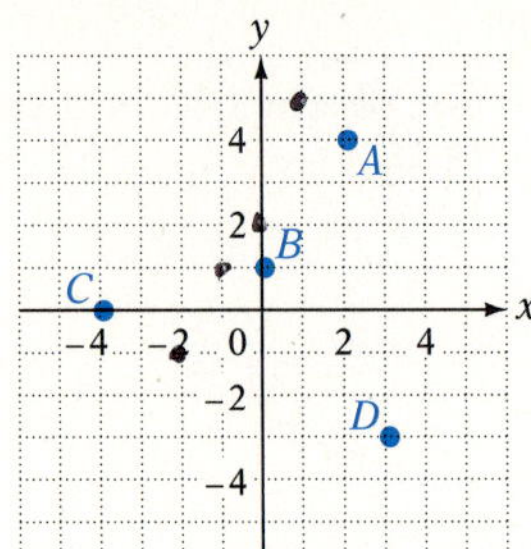

12. **a.** Name the abscissas of points A and C.
b. Name the ordinates of points B and D.

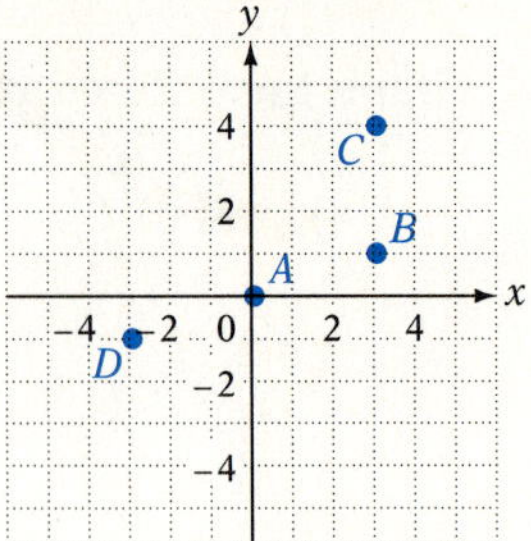

Objective B

13. Is (3, 4) a solution of $y = -x + 7$?

14. Is (2, −3) a solution of $y = x + 5$?

15. Is (−1, 2) a solution of $y = \frac{1}{2}x - 1$?

16. Is (1, −3) a solution of $y = -2x - 1$?

17. Is (4, 1) a solution of $y = \frac{1}{4}x + 1$?

18. Is (−5, 3) a solution of $y = -\frac{2}{5}x + 1$?

19. Is (0, 4) a solution of $y = \frac{3}{4}x + 4$?

20. Is (−2, 0) a solution of $y = -\frac{1}{2}x - 1$?

21. Is (0, 0) a solution of $y = 3x + 2$?

22. Is (0, 0) a solution of $y = -\frac{3}{4}x$?

23. Find the ordered-pair solution of $y = 3x - 2$ corresponding to $x = 3$.

24. Find the ordered-pair solution of $y = 4x + 1$ corresponding to $x = -1$.

25. Find the ordered-pair solution of $y = \frac{2}{3}x - 1$ corresponding to $x = 6$.

26. Find the ordered-pair solution of $y = \frac{3}{4}x - 2$ corresponding to $x = 4$.

27. Find the ordered-pair solution of $y = -3x + 1$ corresponding to $x = 0$.

28. Find the ordered-pair solution of $y = \frac{2}{5}x - 5$ corresponding to $x = 0$.

29. Find the ordered-pair solution of $y = \frac{2}{5}x + 2$ corresponding to $x = -5$.

30. Find the ordered-pair solution of $y = -\frac{1}{6}x - 2$ corresponding to $x = 12$.

Graph the ordered-pair solutions of each equation for the given values of x.

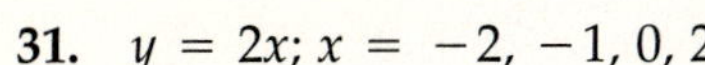

31. $y = 2x;\ x = -2, -1, 0, 2$

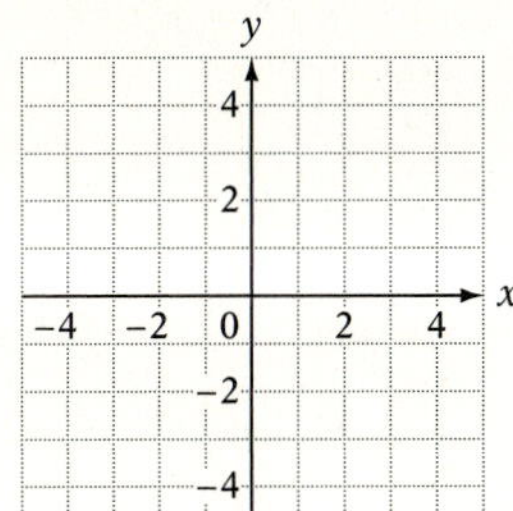

32. $y = -2x;\ x = -2, -1, 0, 2$

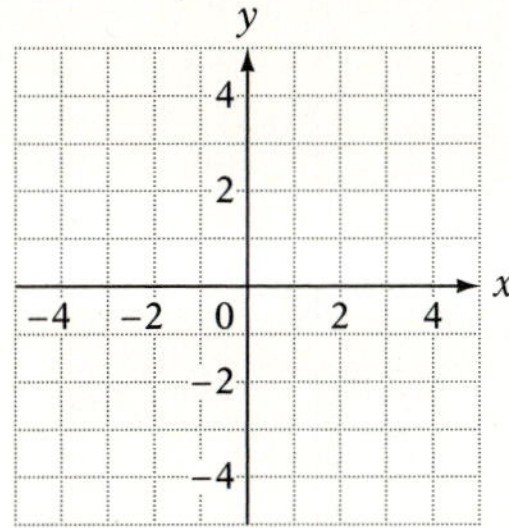

33. $y = x + 2;\ x = -4, -2, 0, 3$

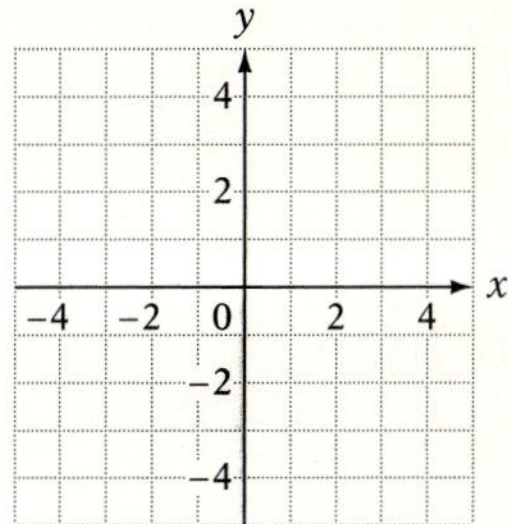

34. $y = \frac{1}{2}x - 1;\ x = -2, 0, 2, 4$

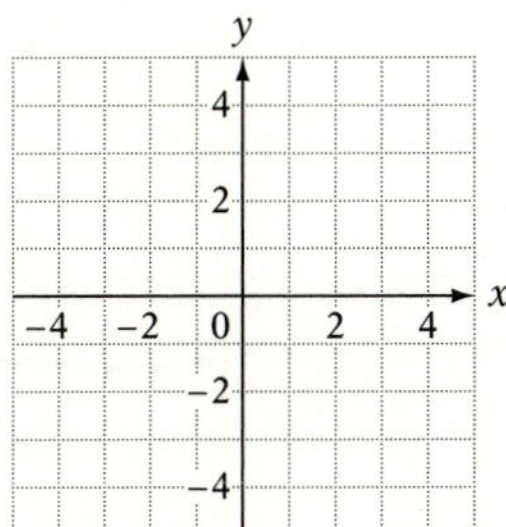

35. $y = \frac{2}{3}x + 1;\ x = -3, 0, 3$

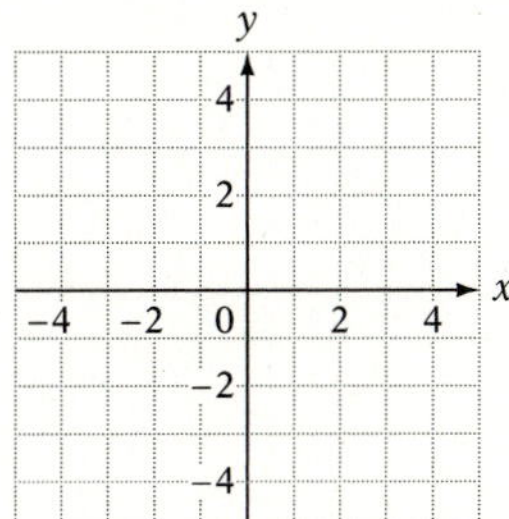

36. $y = -\frac{1}{3}x - 2;\ x = -3, 0, 3$

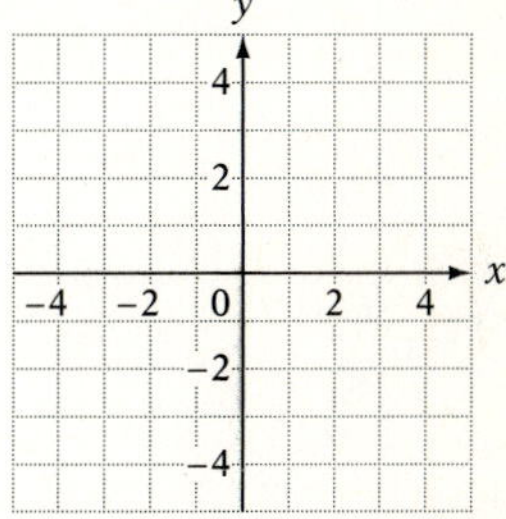

37. $2x + 3y = 6;\ x = -3, 0, 3$

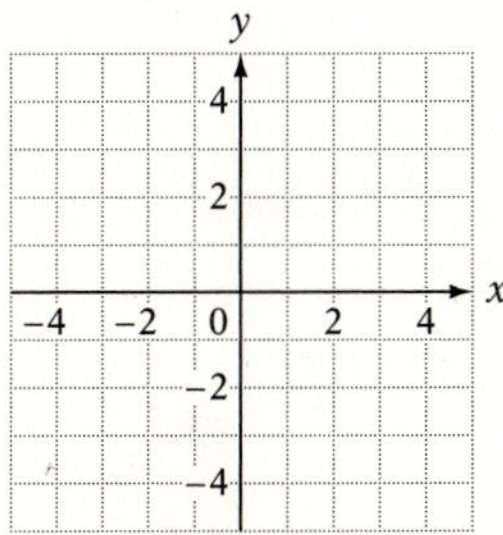

38. $x - 2y = 4;\ x = -2, 0, 2$

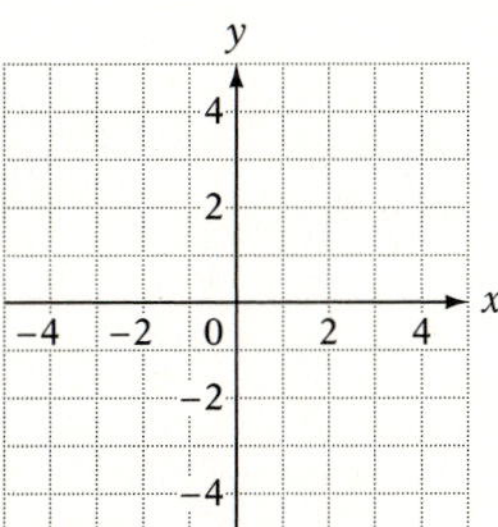

39. $2x + y = 3;\ x = -1, 0, 1, 2$

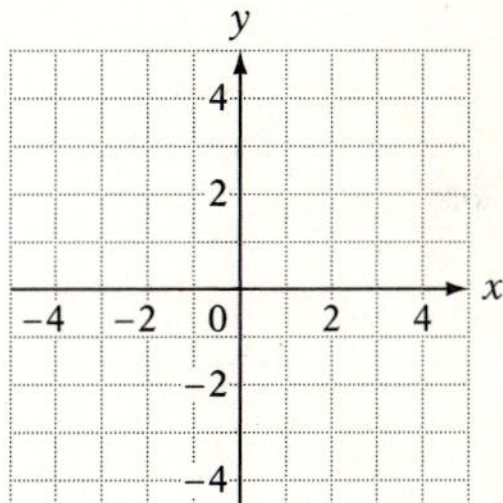

Objective C

Graph.

40. $y = 2x - 3$

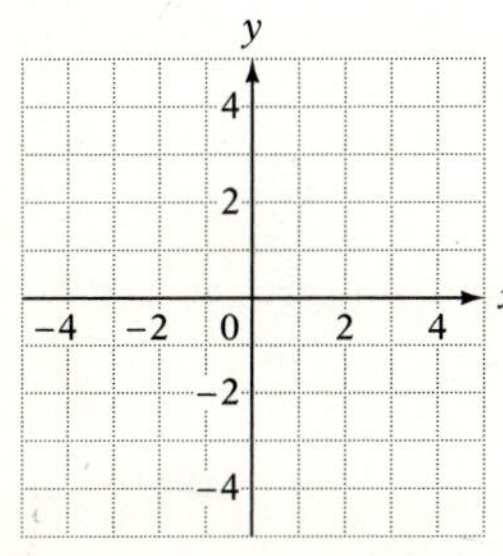

41. $y = -2x + 2$

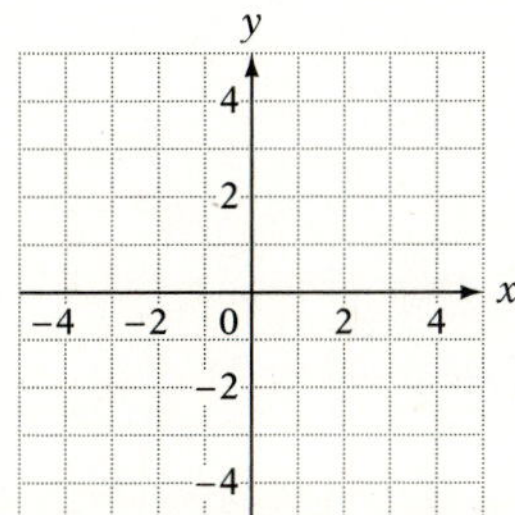

42. $y = \frac{1}{3}x$

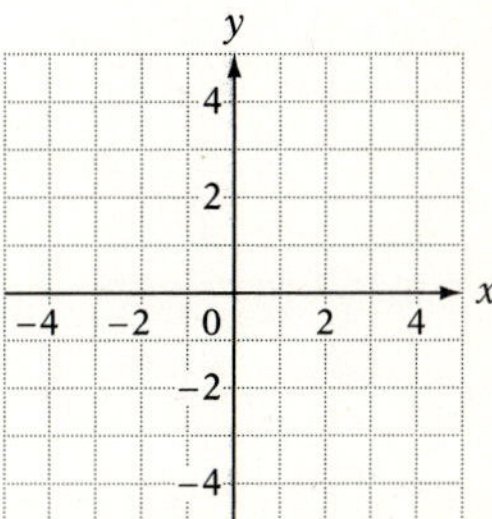

Graph.

43. $y = -3x$

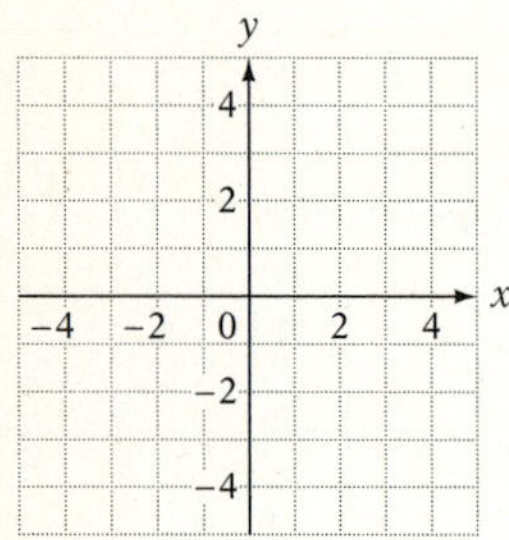

44. $y = \frac{2}{3}x - 1$

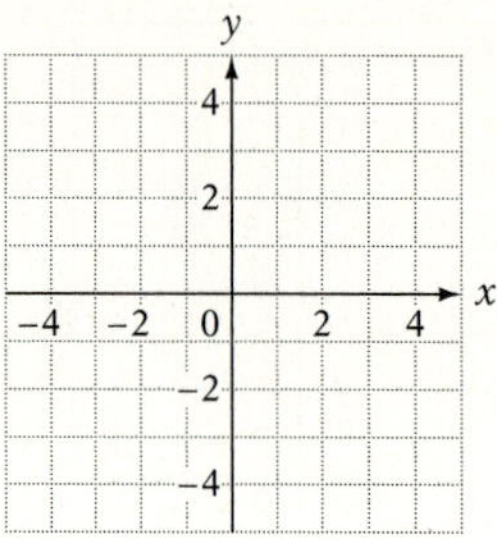

45. $y = \frac{3}{4}x + 2$

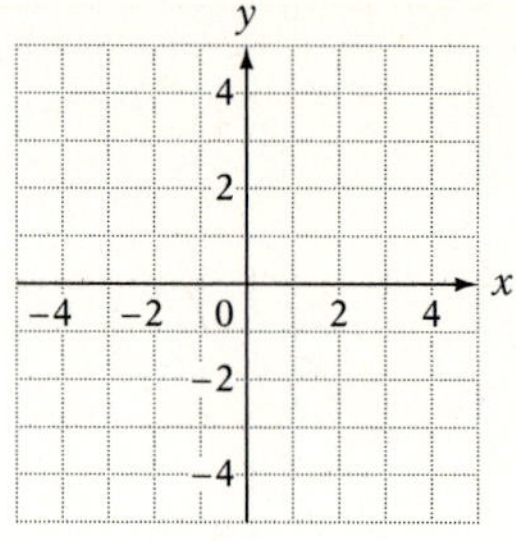

46. $y = -x + 2$

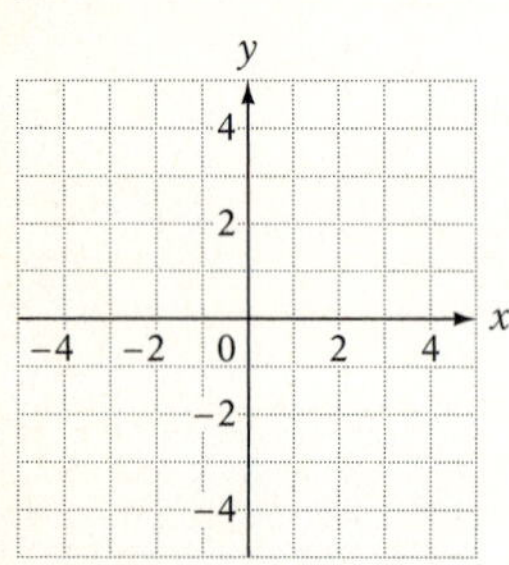

47. $y = -x - 1$

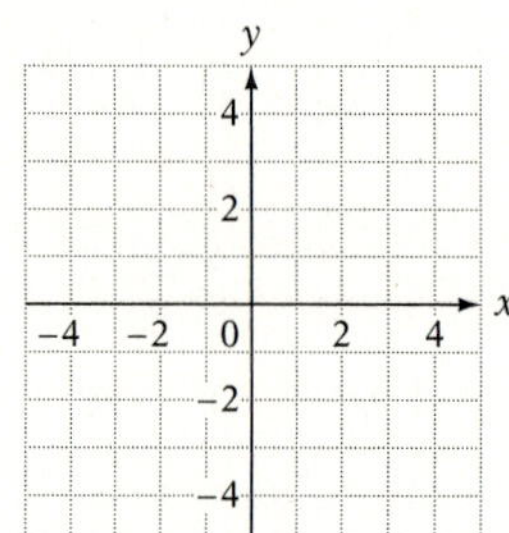

48. $y = -\frac{2}{3}x + 1$

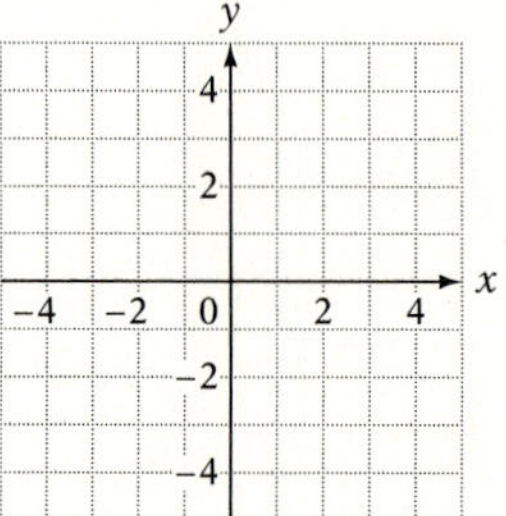

Objective D

Graph.

49. $3x + y = 3$

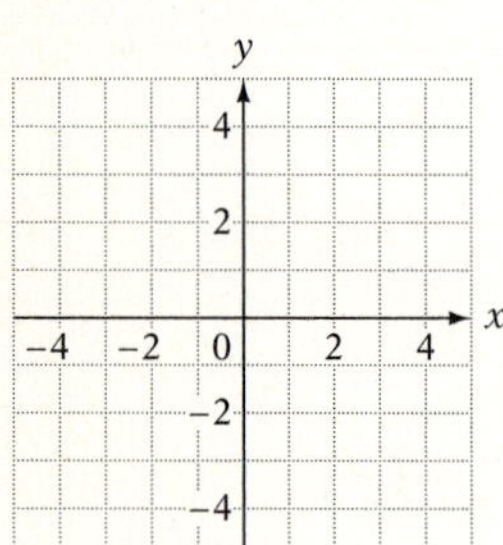

50. $2x + y = 4$

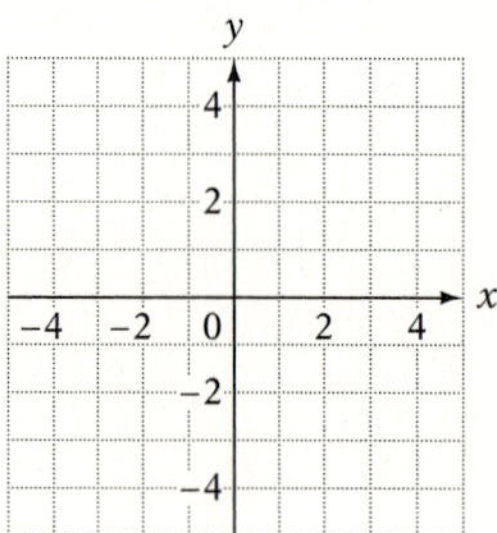

51. $2x + 3y = 6$

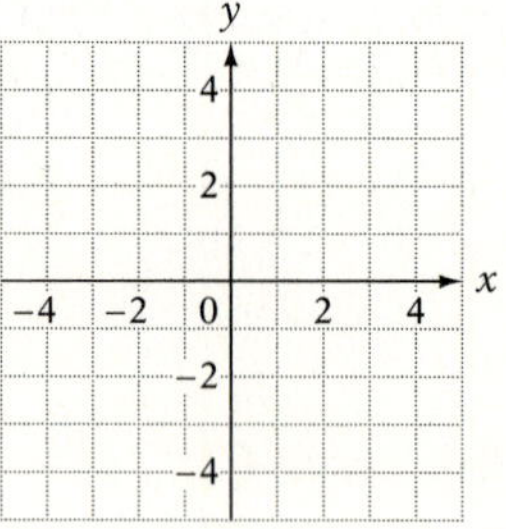

52. $3x + 2y = 4$

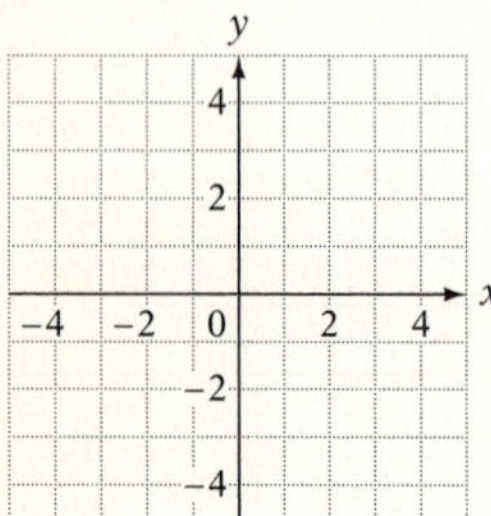

53. $x = 3$

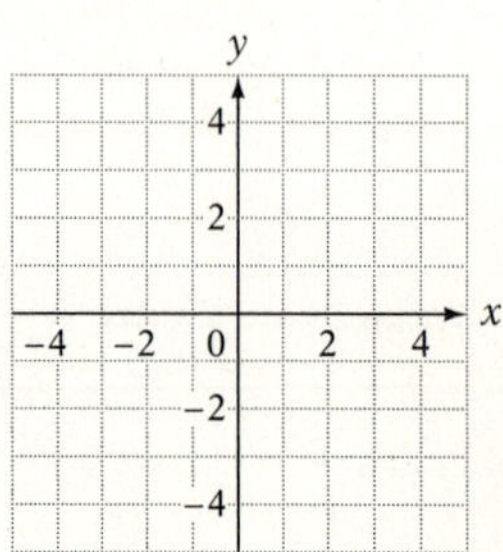

54. $y = -4$

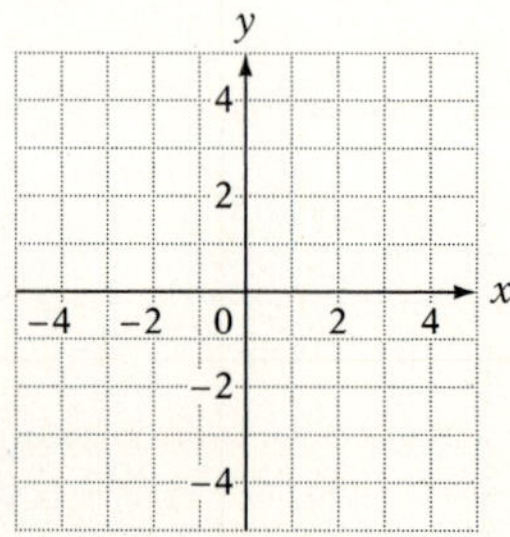

Find the x- and y-intercepts and graph.

55. $3x - 2y = 8$

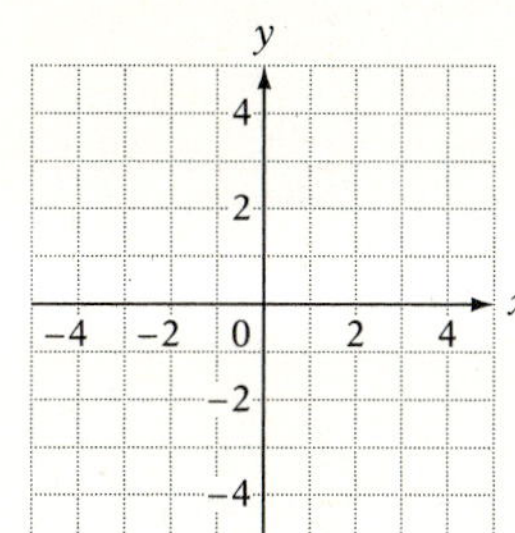

56. $2x + 5y = 10$

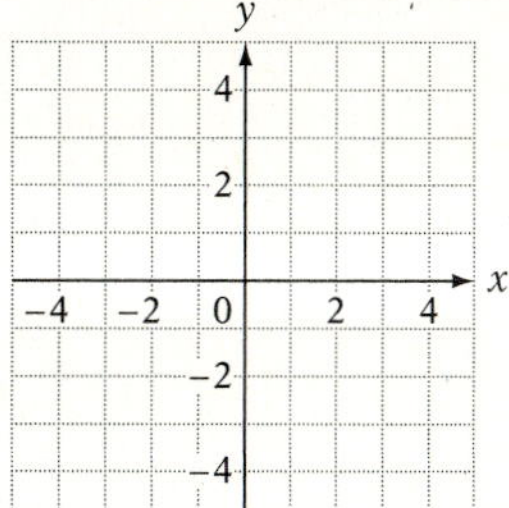

57. $3x + 4y = 12$

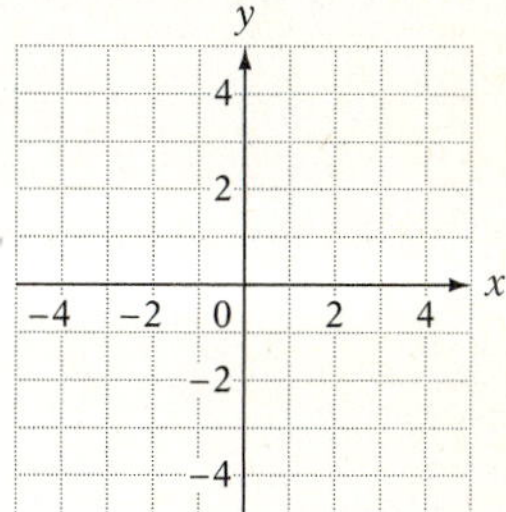

58. $x - 2y = 4$

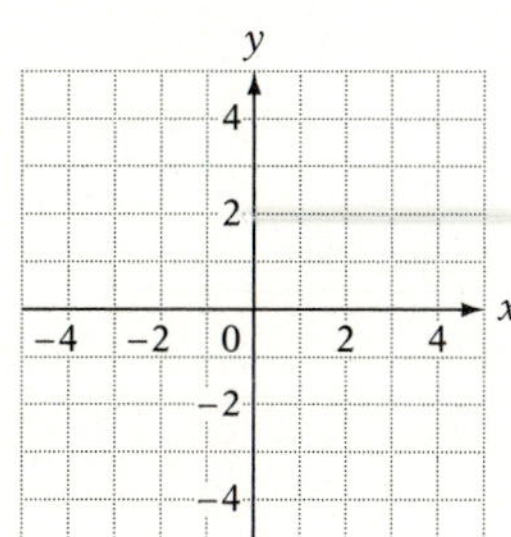

59. $x - 3y = 6$

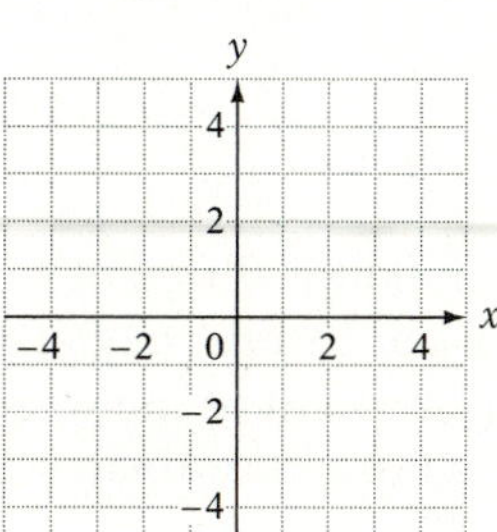

60. $2x + 3y = 3$

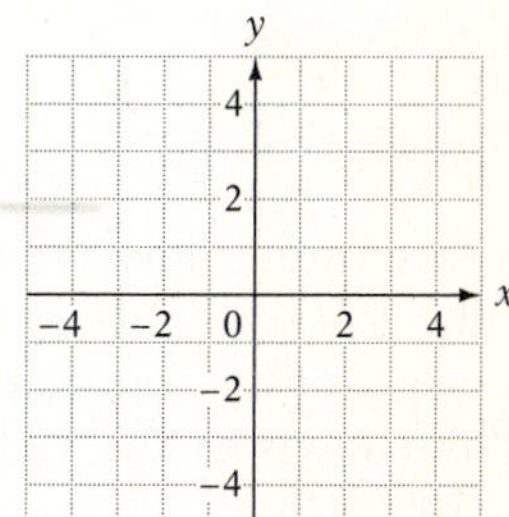

Critical Thinking

In which quadrant is the given point located?

61. $(2, -4)$

62. $(-3, 2)$

63. $(-1, -6)$

What is the distance from the given point to the horizontal axis?

64. $(-5, 1)$

65. $(3, -4)$

66. $(-6, 0)$

What is the distance from the given point to the vertical axis?

67. $(-2, 4)$

68. $(1, -3)$

69. $(5, 0)$

70. Name the coordinates of a point plotted at the origin of the rectangular coordinate system.

71. Describe the signs of the coordinates of a point plotted in (a) Quadrant I, (b) Quadrant II, (c) Quadrant III, and (d) Quadrant IV.

72. **a.** Show that the equation $y + 3 = 2(x + 4)$ is a linear equation by writing it in the form $y = mx + b$.
b. Find the ordered-pair solution corresponding to $x = -4$.

73. **a.** Show that the equation $y + 4 = -\frac{1}{2}(x + 2)$ is a linear equation by writing it in the form $y = mx + b$.

b. Find the ordered-pair solution corresponding to $x = -2$.

74. For the linear equation $y = 2x - 3$, what is the increase in y when x is increased by 1?

75. For the linear equation $y = -x - 4$, what is the decrease in y when x is increased by 1?

76. Write an equation of a line that has (0, 0) as both the x-intercept and the y-intercept.

77. A computer screen has a coordinate system that is different from the xy-coordinate system we have discussed. In one mode, the origin of the coordinate system is the top left point of the screen, as shown at the right. Plot the points whose coordinates are (200, 400), (0, 100), and (100, 0).

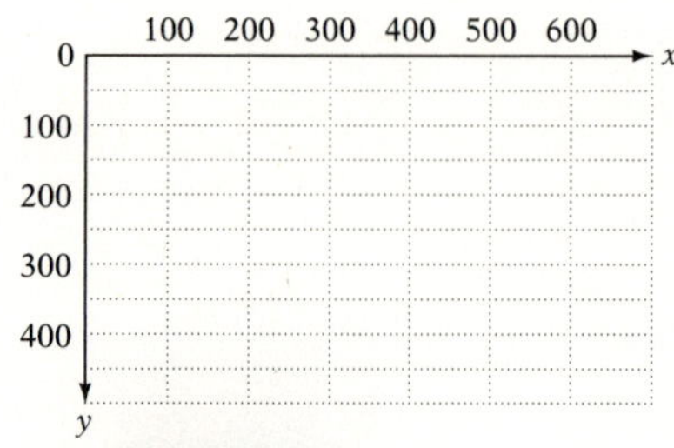

78. [W] There is a coordinate system for locations on Earth which uses *longitude* and *latitude*. Write a report on how location is determined on the surface of Earth. Include in your report the longitude and latitude coordinates of your school.

79. [W] Identify two quantities that may be related and collect at least 10 pairs of values. Here are some examples: height and weight, time studying for a test and the test grade, age of a car and its cost. Draw a scatter diagram for the data. Is there any trend? That is, as the values on the horizontal axis increase, do the values on the vertical axis increase or decrease?

80. [W] Write a paragraph explaining how to plot points in a rectangular coordinate system.

81. [W] Explain (a) why the y-coordinate of any point on the x-axis is zero and (b) why the x-coordinate of any point on the y-axis is zero.

SECTION 7.2 Slopes of Straight Lines

OBJECTIVE  The slope of a straight line

The graphs of $y = \frac{2}{3}x + 1$ and $y = 2x + 1$ are shown in Fig. 1. Each graph crosses the y-axis at the point (0, 1), but the graphs have different slants. The **slope** of a line is a measure of the slant of a line. The symbol for slope is m.

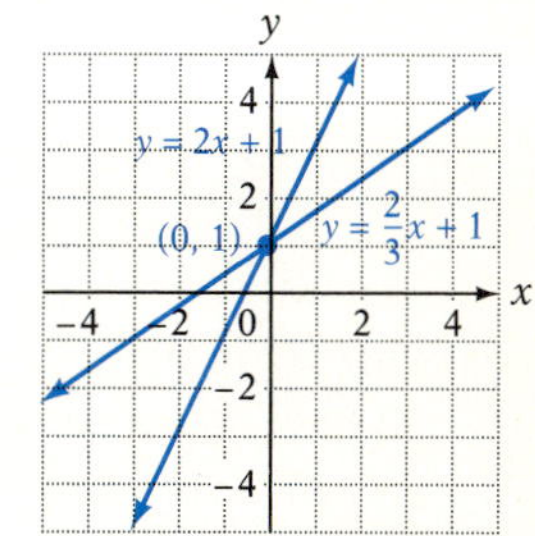

Fig. 1

The slope of a line is the ratio of the change in the y values to the change in the x values for any pair of points on the line. The line containing the points $(-2, -3)$ and (6, 1) is graphed in Fig. 2. The change in the y values is the difference between the two ordinates.

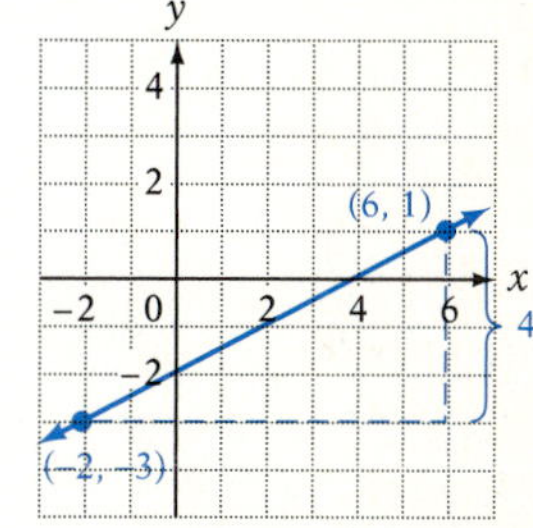

Fig. 2

Change in $y = 1 - (-3) = 4$

The change in the x values is the difference between the two abscissas (Fig. 3).

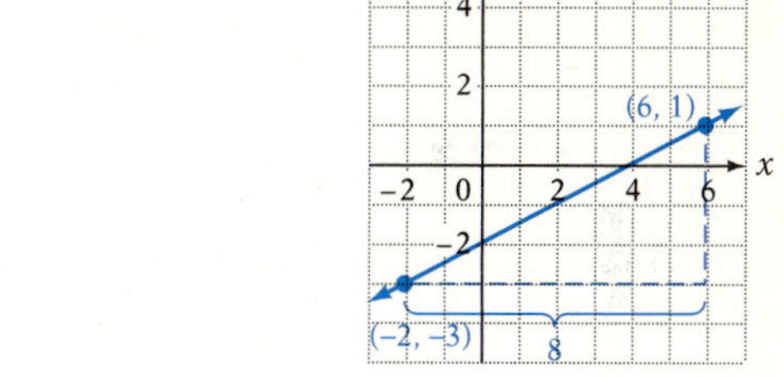

Fig. 3

Change in $x = 6 - (-2) = 8$

$$\text{Slope} = m = \frac{\text{change in } y}{\text{change in } x} = \frac{4}{8} = \frac{1}{2}$$

Slope Formula

If $P_1(x_1, y_1)$ and $P_2(x_2, y_2)$ are two points on a line and $x_1 \neq x_2$, then $m = \frac{y_2 - y_1}{x_2 - x_1}$ (Fig. 4). If $x_1 = x_2$, the slope is undefined.

Fig. 4

Find the slope of the line containing the points $(-1, 1)$ and (2, 3).

Let P_1 be $(-1, 1)$ and P_2 be (2, 3). Then,

$x_1 = -1, y_1 = 1, x_2 = 2, y_2 = 3.$

$$m = \frac{y_2 - y_1}{x_2 - x_1} = \frac{3 - 1}{2 - (-1)} = \frac{2}{3}$$

The slope is $\frac{2}{3}$.

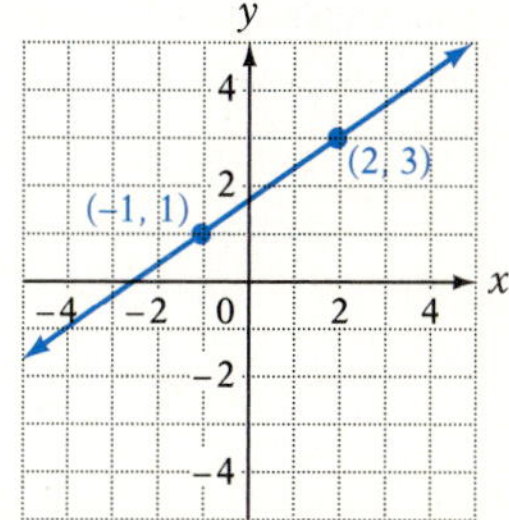

A line that slants upward to the right has a **positive slope**.

CONSIDER THIS

Positive slope means that the value of y increases as the value of x increases.

Find the slope of the line containing the points $(-3, 4)$ and $(2, -2)$.

Let $P_1 = (-3, 4)$ and $P_2 = (2, -2)$.

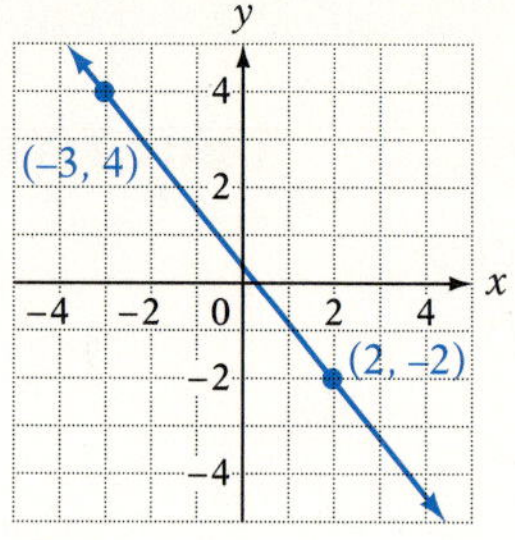

$$m = \frac{y_2 - y_1}{x_2 - x_1} = \frac{-2 - 4}{2 - (-3)} = \frac{-6}{5} = -\frac{6}{5}$$

A line that slants downward to the right has a **negative slope**.

Find the slope of the line containing the points $(-1, 3)$ and $(4, 3)$.

Let $P_1 = (-1, 3)$ and $P_2 = (4, 3)$.

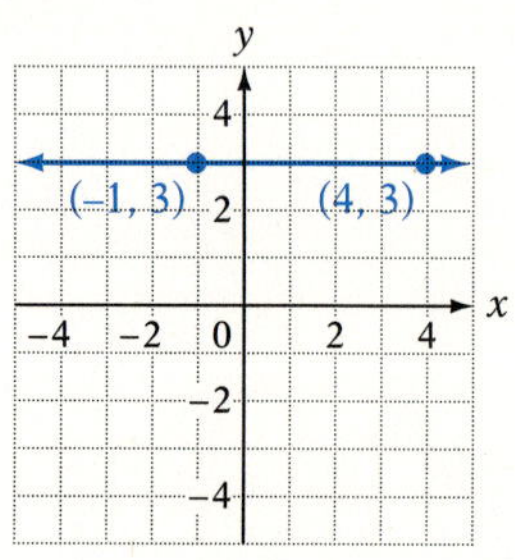

$$m = \frac{y_2 - y_1}{x_2 - x_1} = \frac{3 - 3}{4 - (-1)} = \frac{0}{5} = 0$$

A horizontal line has **zero slope**.

Find the slope of the line containing the points $(2, -2)$ and $(2, 4)$.

Let $P_1 = (2, -2)$ and $P_2 = (2, 4)$.

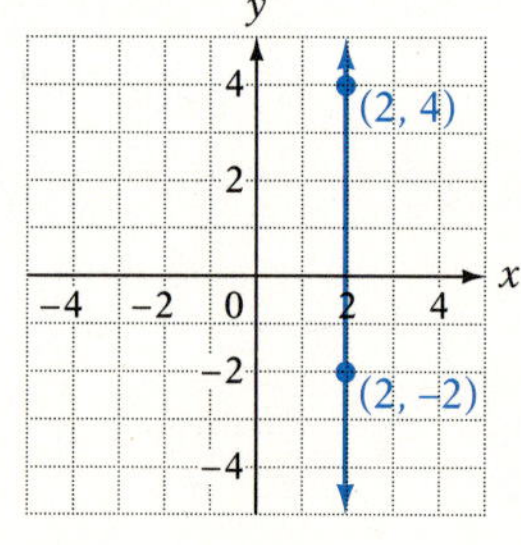

$$m = \frac{y_2 - y_1}{x_2 - x_1} = \frac{4 - (-2)}{2 - 2} = \frac{6}{0}$$

Division by zero is not defined.

The slope of a vertical line is undefined.

CONSIDER THIS

Just as the phrase "the sum of" means to add, any expression that contains the word "per" translates mathematically to slope. Suggest some everyday situations that involve slope. Some examples might include miles per gallon or feet per second.

There are many applications of the concept of slope. Here are two.

In 1988, when Florence Griffith-Joyner set the world record for the 100-meter dash, her average speed was approximately 9.5 m/s. The graph at the right shows the distance she ran during her record-setting run. Reading the graph, note that after 4 s she had run 38 m and that after 6 s she had run 57 m. The slope of the line between these two points is

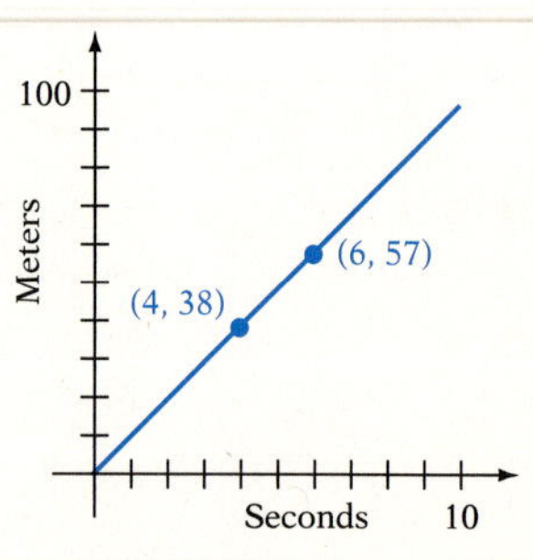

$$m = \frac{57 - 38}{6 - 4} = \frac{19}{2} = 9.5$$

Note that the slope of the line is the same as the rate she was running, 9.5 m/s. The average speed of an object is related to slope.

Here is an example of slope taken from economics. According to the Department of Commerce, from 1987 to 1994, U.S. exports of goods to other countries had been increasing at a rate of approximately $2.5 billion per year. The graph at the right shows the value of exports for each year. From the graph, we learn that in 1988 exports were $27 billion and in 1992 exports were $37 billion. The slope of the line between these two points is

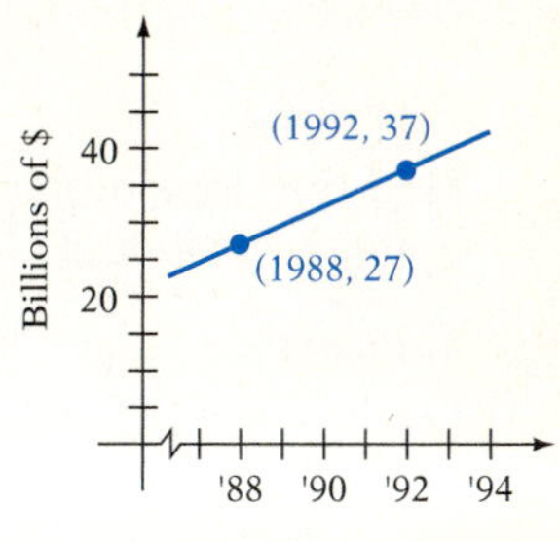

$$m = \frac{37 - 27}{1992 - 1988} = \frac{10}{4} = 2.5$$

Observe that the slope of the line is the same as the rate at which exports are increasing, $2.5 billion per year.

In general, any quantity that is expressed by using the word *per* is represented mathematically as slope. In the first example, the slope was 9.5 meters *per* second; in the second example, the slope was $2.5 billion *per* year.

CONSIDER THIS

In You Try It 5, you are required to write a complete sentence that explains the meaning of the slope. This requires that you understand the final result in the context of the application.

Example 1 Find the slope of the line containing the points $(-2, -1)$ and $(3, 4)$.

Solution Let $P_1 = (-2, -1)$ and $P_2 = (3, 4)$.

$$m = \frac{y_2 - y_1}{x_2 - x_1} = \frac{4 - (-1)}{3 - (-2)} = \frac{5}{5} = 1$$

The slope is 1.

You Try It 1 Find the slope of the line containing the points $(-1, 2)$ and $(1, 3)$.

Your Solution

Example 2 Find the slope of the line containing the points $(-3, 1)$ and $(2, -2)$.

Solution Let $P_1 = (-3, 1)$ and $P_2 = (2, -2)$.

$$m = \frac{y_2 - y_1}{x_2 - x_1} = \frac{-2 - 1}{2 - (-3)} = \frac{-3}{5}$$

The slope is $-\frac{3}{5}$.

You Try It 2 Find the slope of the line containing the points $(1, 2)$ and $(4, -5)$.

Your Solution

Solutions on p. A20

Example 3

Find the slope of the line containing the points $(-1, 4)$ and $(-1, 0)$.

Solution

Let $P_1 = (-1, 4)$ and $P_2 = (-1, 0)$.

$$m = \frac{y_2 - y_1}{x_2 - x_1} = \frac{0 - 4}{-1 - (-1)} = \frac{-4}{0}$$

The slope of the line is undefined.

You Try It 3

Find the slope of the line containing the points $(2, 3)$ and $(2, 7)$.

Your Solution

Example 4

Find the slope of the line containing the points $(-1, 2)$ and $(4, 2)$.

Solution

Let $P_1 = (-1, 2)$ and $P_2 = (4, 2)$.

$$m = \frac{y_2 - y_1}{x_2 - x_1} = \frac{2 - 2}{4 - (-1)} = \frac{0}{5} = 0$$

The slope of the line is zero.

You Try It 4

Find the slope of the line containing the points $(1, -3)$ and $(-5, -3)$.

Your Solution

Example 5

The graph below shows the decline in the price of a used Macintosh IIci computer from January to June (shown as the numbers 1 through 6). Find the slope of the line. Write a sentence that states the meaning of the slope.

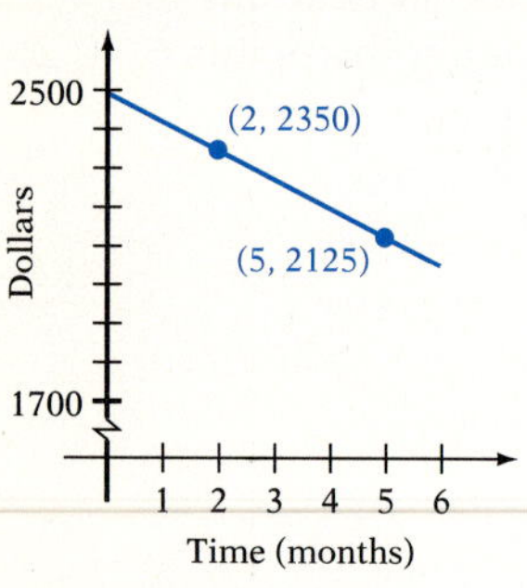

Solution

$$m = \frac{2125 - 2350}{5 - 2}$$

$$= \frac{-225}{3} = -75$$

A slope of -75 means that the price of a Macintosh IIci is *decreasing* at a rate of \$75 per month.

You Try It 5

The graph below shows the approximate decline in the value of a used car over a five-year period. Find the slope of the line. Write a sentence that states the meaning of the slope.

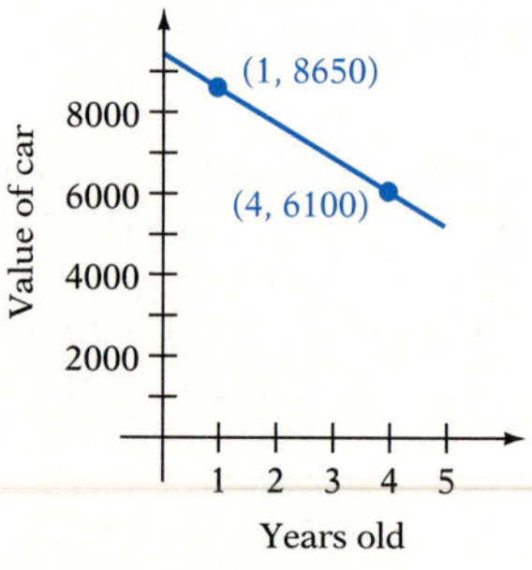

Your Solution

Solutions on p. A21

OBJECTIVE B Graph a line using the slope and y-intercept

The graph of the equation $y = \frac{2}{3}x + 1$ is shown at the right. The points $(-3, -1)$ and $(3, 3)$ are on the graph. The slope of the line is

$$m = \frac{3 - (-1)}{3 - (-3)} = \frac{4}{6} = \frac{2}{3}$$

Note that the slope of the line has the same value as the coefficient of x.

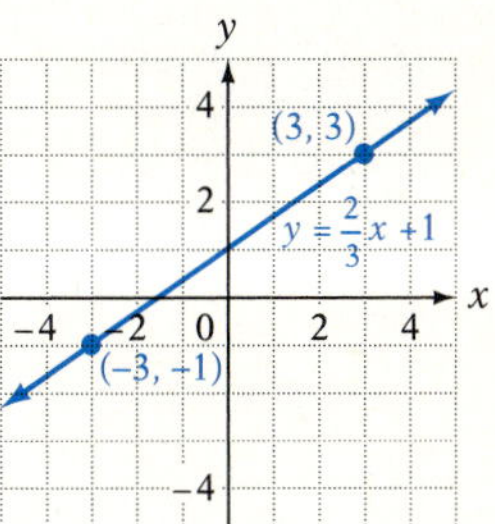

For any equation of the form $y = mx + b$, the slope of the line is m, the coefficient of x. The ***y*-intercept** is $(0, b)$. Thus, an equation of the form $y = mx + b$ is called the **slope-intercept form of a straight line.**

Find the slope and the y-intercept of the line $y = -\frac{3}{4}x + 1$.

$$y = \boxed{m}\, x + \boxed{b}$$

$$y = \boxed{-\frac{3}{4}}\, x + \boxed{1}$$

Slope $= m = -\frac{3}{4}$ — y-intercept $= (0, b) = (0, 1)$

The slope is $-\frac{3}{4}$. The y = intercept is $(0, 1)$.

When the equation of a straight line is in the form $y = mx + b$, the graph can be drawn using the slope and y-intercept. First locate the y-intercept. Use the slope to find a second point on the line. Then draw a line through the two points.

Graph $y = 2x - 3$.

$y = 2x + (-3)$

y-intercept $= (0, b) = (0, -3)$

$m = 2 = \frac{2}{1} = \frac{\text{change in } y}{\text{change in } x}$

Beginning at the y-intercept, move right 1 unit (change in x) and then up 2 units (change in y).

$(1, -1)$ is a second point on the graph.

Draw a line through the two points $(0, -3)$ and $(1, -1)$.

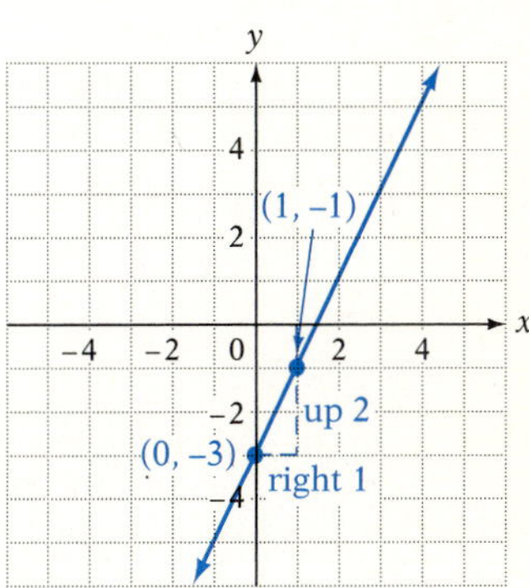

Example 6 Graph $y = -\frac{2}{3}x + 1$ by using the slope and y-intercept.

Solution y-intercept $= (0, b) = (0, 1)$

$m = -\frac{2}{3} = \frac{-2}{3}$ (Move right 3 units, then down 2 units.)

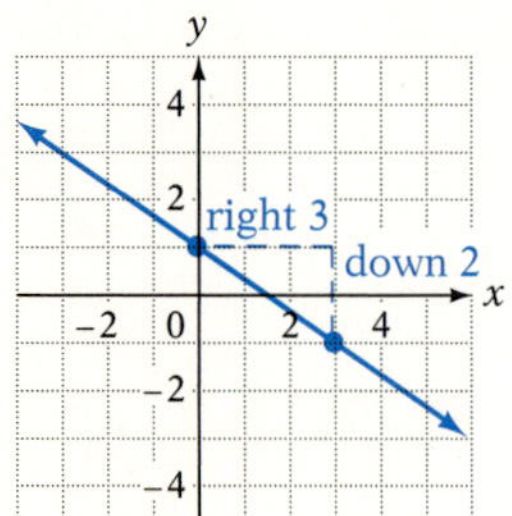

You Try It 6 Graph $y = -\frac{1}{4}x - 1$ by using the slope and y-intercept.

Your Solution

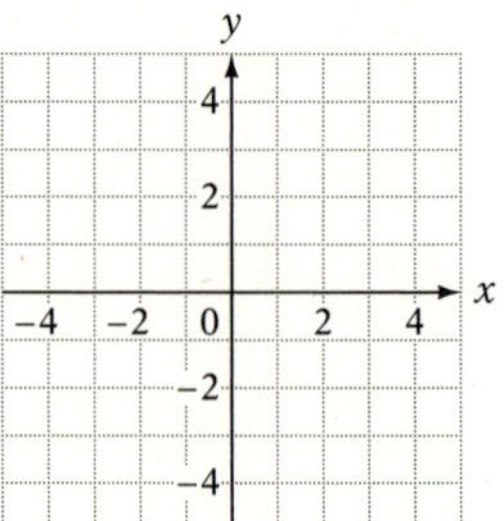

Example 7 Graph $y = -\frac{3}{4}x$ by using the slope and y-intercept.

Solution y-intercept $= (0, b) = (0, 0)$

$m = -\frac{3}{4} = \frac{-3}{4}$

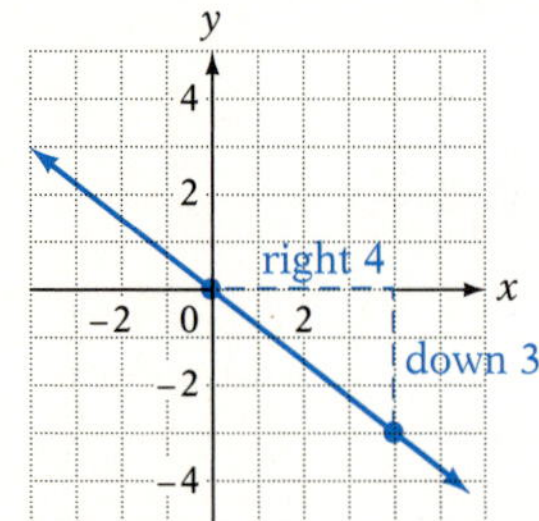

You Try It 7 Graph $y = -\frac{3}{5}x$ by using the slope and y-intercept.

Your Solution

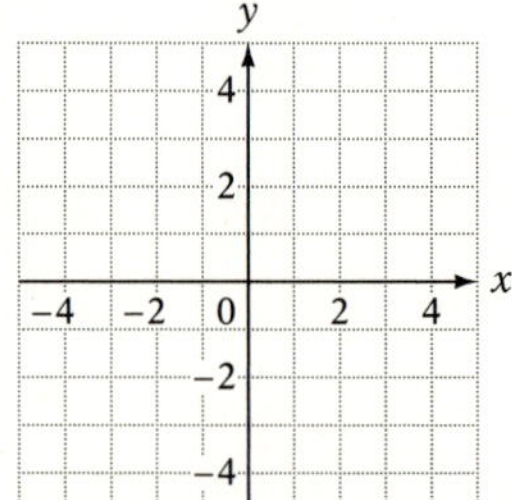

Example 8 Graph $2x - 3y = 6$ by using the slope and y-intercept.

Solution Solve the equation for y.

$$2x - 3y = 6$$
$$-3y = -2x + 6$$
$$y = \frac{2}{3}x - 2$$

y-intercept $= (0, -2)$ $\quad m = \frac{2}{3}$

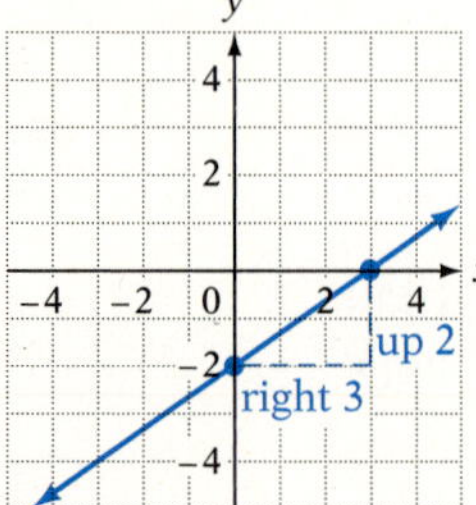

You Try It 8 Graph $x - 2y = 4$ by using the slope and y-intercept.

Your Solution

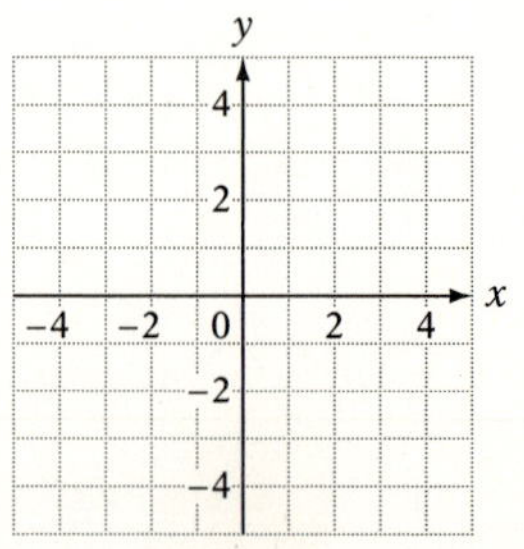

Solutions on p. A21

7.2 EXERCISES

Objective A

Find the slope of the line containing the points.

1. $P_1(4, 2), P_2(3, 4)$
2. $P_1(2, 1), P_2(3, 4)$
3. $P_1(-1, 3), P_2(2, 4)$
4. $P_1(-2, 1), P_2(2, 2)$
5. $P_1(2, 4), P_2(4, -1)$
6. $P_1(1, 3), P_2(5, -3)$
7. $P_1(-2, 3), P_2(2, 1)$
8. $P_1(5, -2), P_2(1, 0)$
9. $P_1(8, -3), P_2(4, 1)$
10. $P_1(0, 3), P_2(2, -1)$
11. $P_1(3, -4), P_2(3, 5)$
12. $P_1(-1, 2), P_2(-1, 3)$
13. $P_1(4, -2), P_2(3, -2)$
14. $P_1(5, 1), P_2(-2, 1)$
15. $P_1(0, -1), P_2(3, -2)$
16. $P_1(3, 0), P_2(2, -1)$
17. $P_1(-2, 3), P_2(1, 3)$
18. $P_1(4, -1), P_2(-3, -1)$
19. $P_1(-2, 4), P_2(-1, -1)$
20. $P_1(6, -4), P_2(4, -2)$
21. $P_1(-2, -3), P_2(-2, 1)$
22. $P_1(5, 1), P_2(5, -2)$
23. $P_1(-1, 5), P_2(5, 1)$
24. $P_1(-1, 5), P_2(7, 1)$

25. The graph below shows the cost, in dollars, to make a transatlantic telephone call. Find the slope of the line. Write a sentence that states the meaning of the slope.

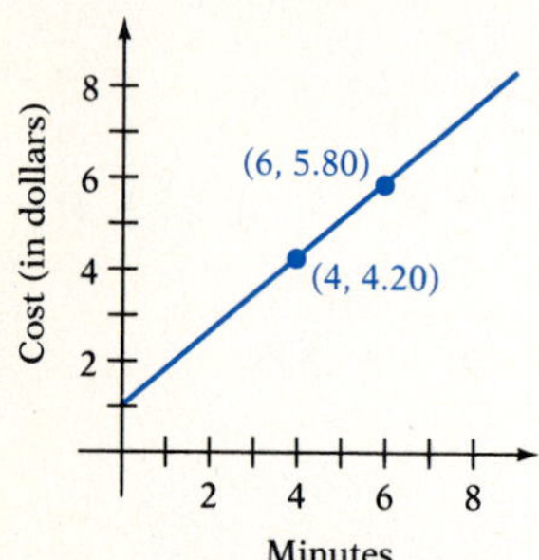

26. The graph below shows the pressure in pounds per square inch on a diver. Find the slope of the line. Write a sentence that states the meaning of the slope.

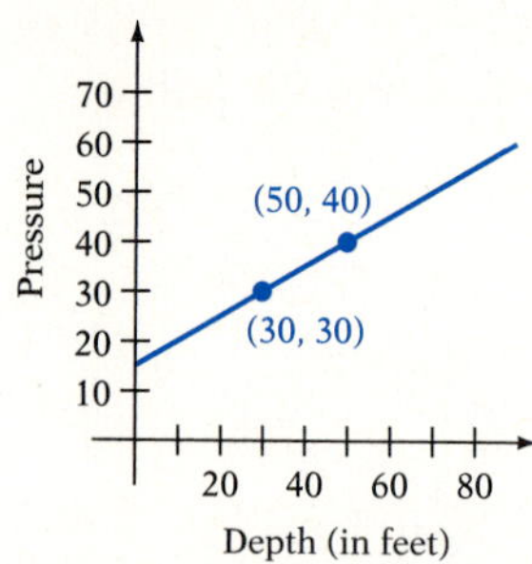

27. The graph below shows the approximate decline in the price of Merck stock during the month of March. Find the slope of the line. Write a sentence that states the meaning of the slope.

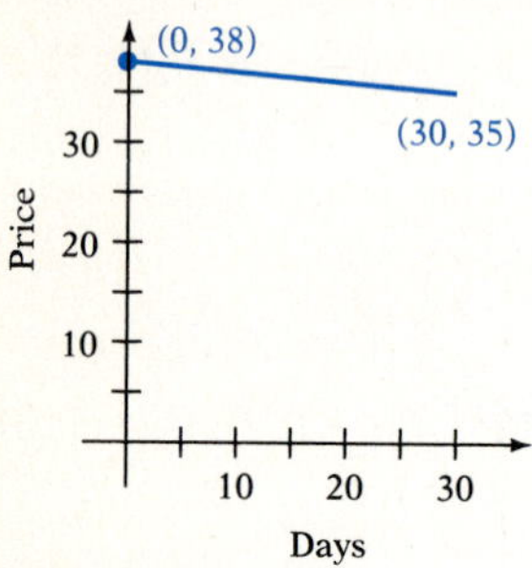

28. The graph below shows the decrease in the median price of a home in a certain city over a 6-month period. Find the slope of the line. Write a sentence that states the meaning of the slope.

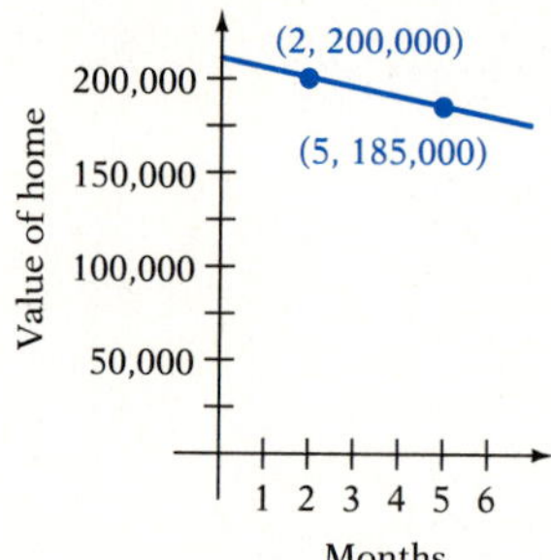

Objective B

Graph by using the slope and y-intercept.

29. $y = 3x + 1$

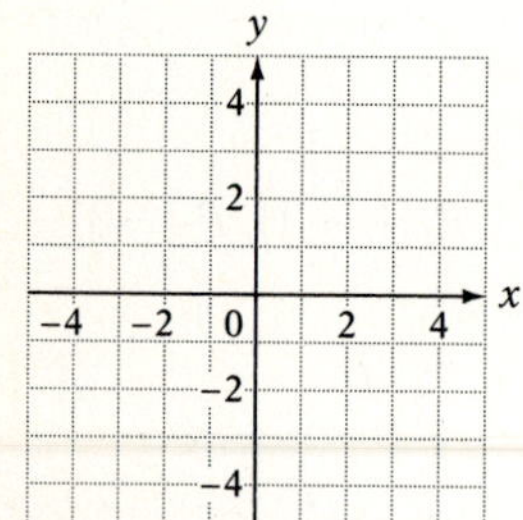

30. $y = -2x - 1$

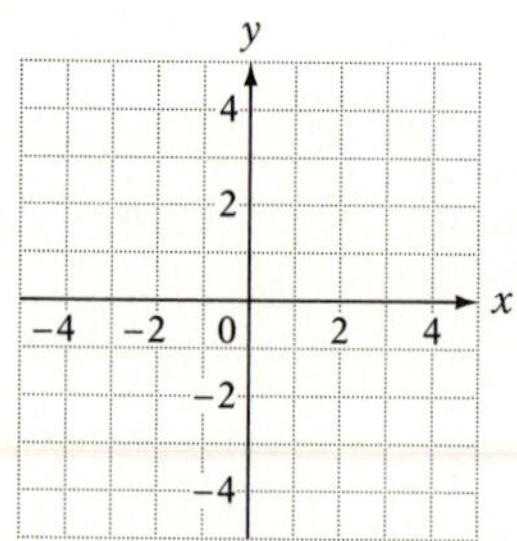

31. $y = \frac{2}{5}x - 2$

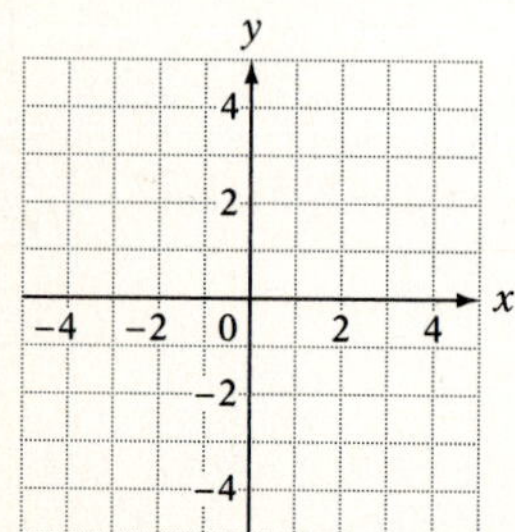

32. $y = \frac{3}{4}x + 1$

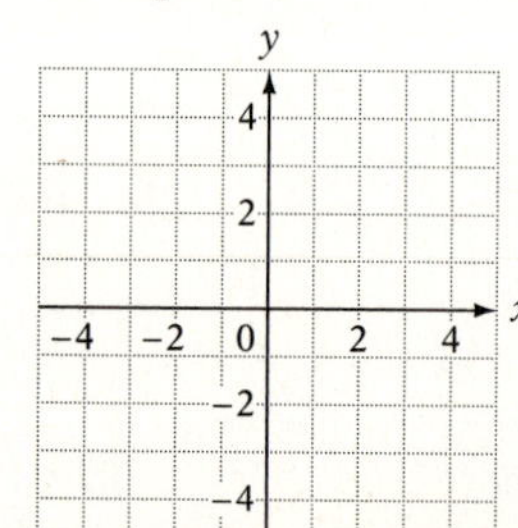

Graph by using the slope and y-intercept.

33. $y = \frac{2}{3}x$

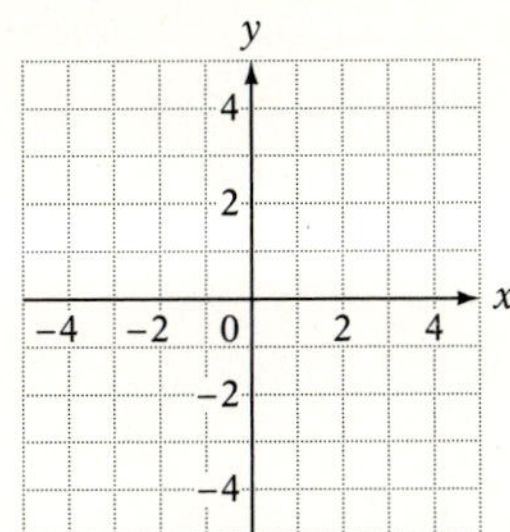

34. $y = \frac{1}{2}x$

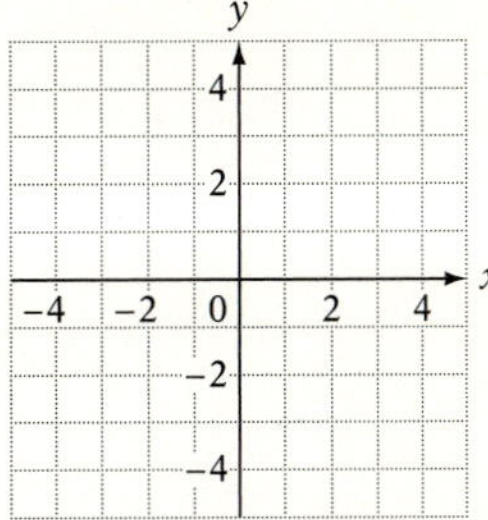

35. $y = -x + 1$

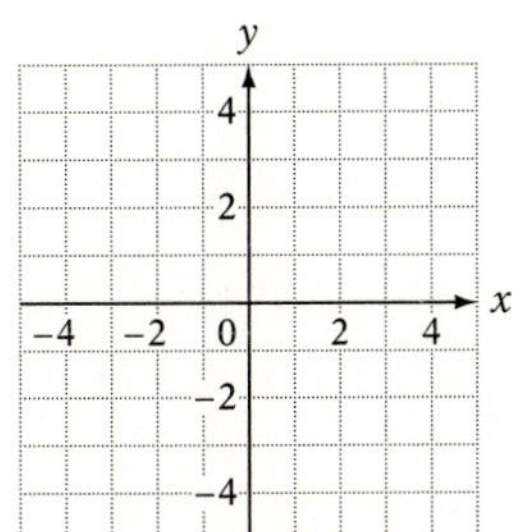

36. $y = -x - 3$

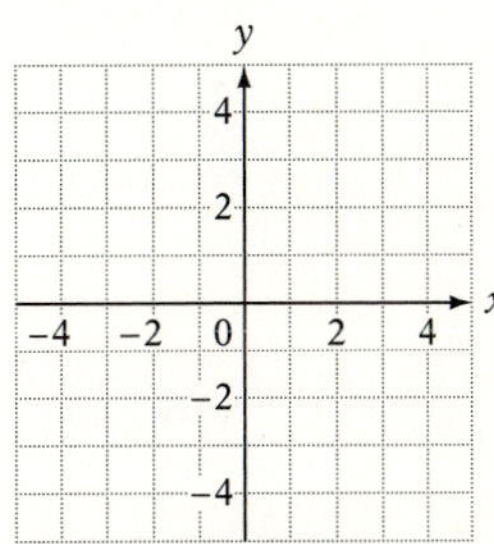

37. $2x + y = 3$

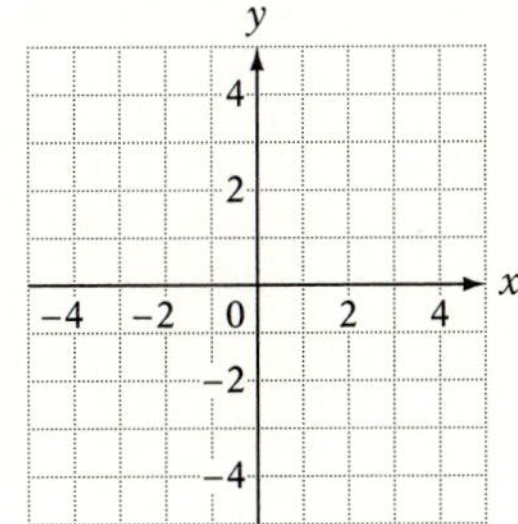

38. $3x - y = 1$

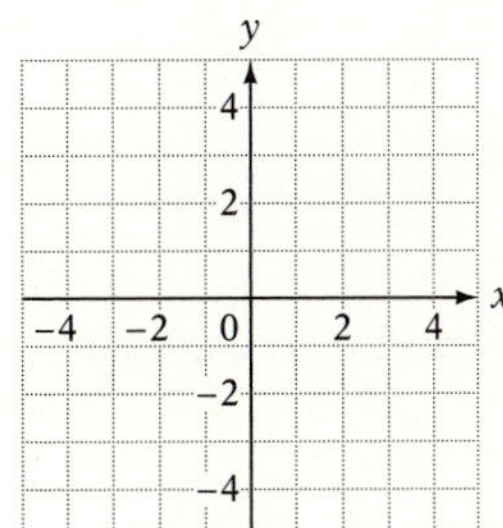

39. $x - 2y = 4$

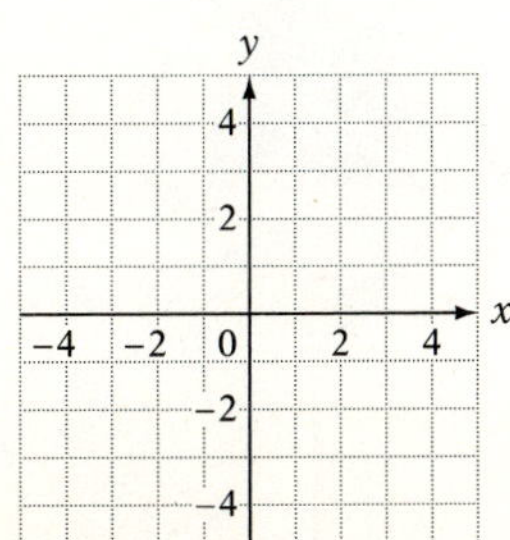

40. $x + 3y = 6$

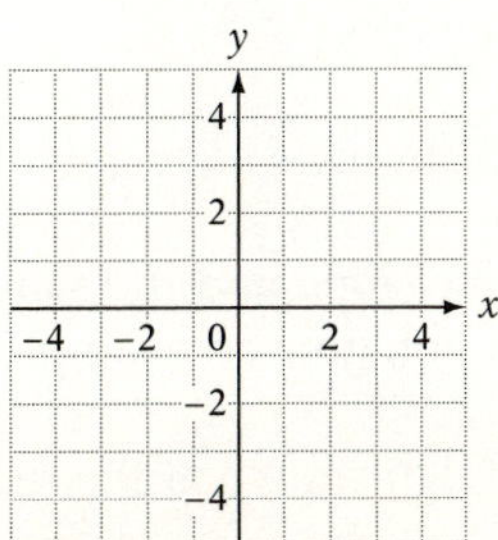

Graph by using the slope and y-intercept.

41. $3x - 4y = 12$

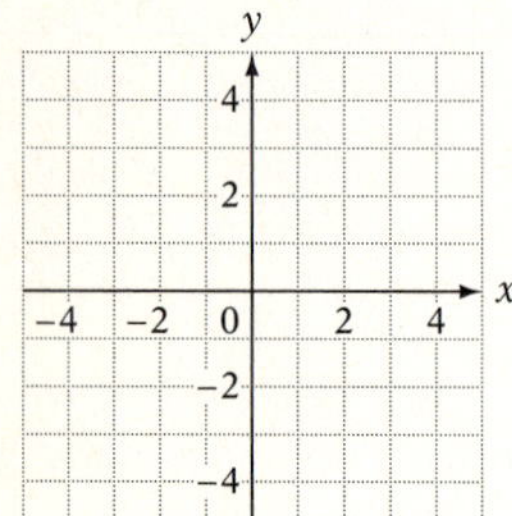

42. $5x - 2y = 10$

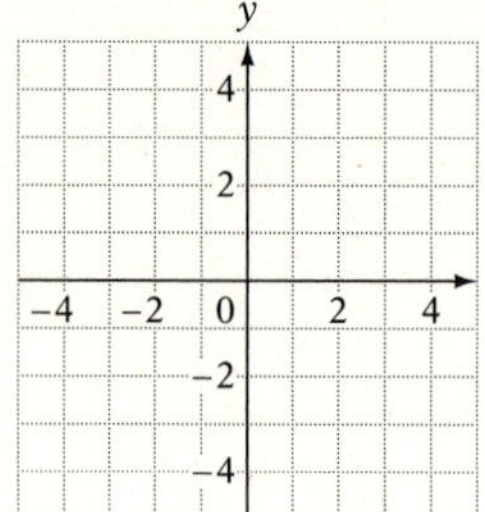

43. $y = -4x + 2$

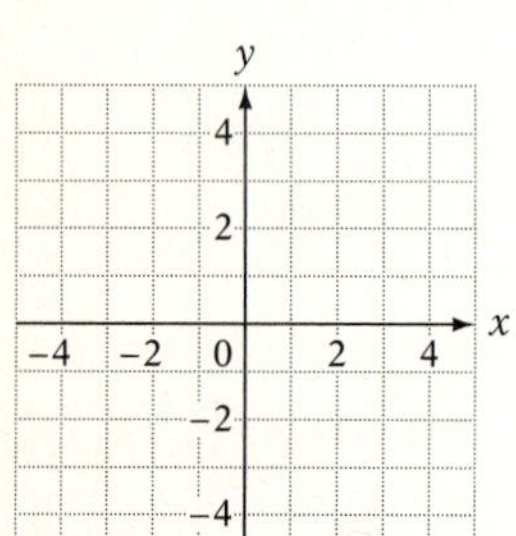

44. $4x - 5y = 20$

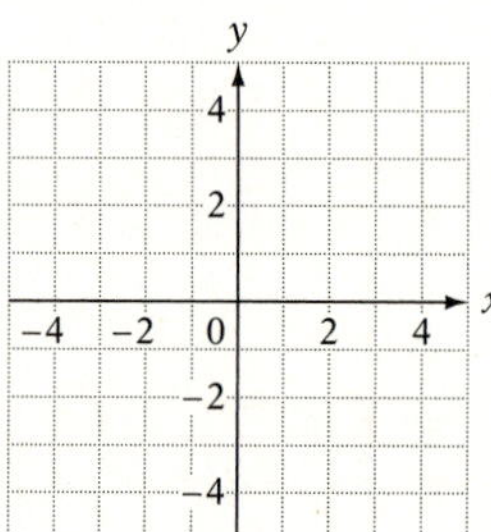

Critical Thinking

45. What effect does increasing the coefficient of x have on the graph of $y = mx + b$?

46. What effect does decreasing the coefficient of x have on the graph of $y = mx + b$?

47. What effect does increasing the constant term have on the graph of $y = mx + b$?

48. What effect does decreasing the constant term have on the graph of $y = mx + b$?

49. **a.** Show that the equation $\frac{x}{4} + \frac{y}{3} = 1$ is a linear equation by writing it in the form $y = mx + b$.

b. Find the x- and y-intercepts.

50. Do the graphs of all straight lines have a y-intercept? If not, give an example.

51. If two lines have the same slope and the same y-intercept, must the graphs of the lines be the same? If not, give an example.

52. [W] What does the highway sign shown at the right have to do with slope?

SECTION 7.3 Equations of Straight Lines

OBJECTIVE A Find the equation of a line given a point and the slope

When the slope of a line and the y-intercept are known, the equation of the line can be written using the slope-intercept form, $y = mx + b$.

Find the equation of the line with slope $-\frac{1}{2}$ and y-intercept (0, 3).

Use the slope-intercept form. $\quad y = mx + b$

$m = -\frac{1}{2}$; $(0, b) = (0, 3)$, so $b = 3$. $\quad y = -\frac{1}{2}x + 3$

The equation of the line is $y = -\frac{1}{2}x + 3$.

When the slope of a line and the coordinates of a point other than the y-intercept are known, the equation of the line can be found by using the formula for slope.

A method for finding the equation of a line, given the slope and the coordinates of a point on the line, involves use of the point–slope formula. The point–slope formula is derived from the formula for slope.

Let (x_1, y_1) be the coordinates of the given point on the line, and let (x, y) be the coordinates of any other point on the line.

Formula for slope $\quad \frac{y - y_1}{x - x_1} = m$

Multiply both sides of the equation by $(x - x_1)$. $\quad \frac{y - y_1}{x - x_1}(x - x_1) = m(x - x_1)$

Simplify. $\quad y - y_1 = m(x - x_1)$

Point–Slope Formula

The equation of the line that has slope m and contains the point whose coordinates are (x_1, y_1) can be found by the point–slope formula:

$$y - y_1 = m(x - x_1)$$

CONSIDER THIS

A model of the point–slope formula with empty parentheses may help you substitute coordinates correctly.

$y - y_1 = m(x - x_1)$

$y - (\ ) = (\ )[x - (\ )]$

Find the equation of the line that passes through point (2, 3) and has slope -2.

Use the point–slope formula. $y - y_1 = m(x - x_1)$

$m = -2;\ (x_1, y_1) = (2, 3)$ $y - 3 = -2(x - 2)$

Solve for y. $y - 3 = -2x + 4$

$y = -2x + 7$

The equation of the line is $y = -2x + 7$.

Example 1

Find the equation of the line whose slope is $-\frac{2}{3}$ and whose y-intercept is $(0, -1)$.

Solution

Because the slope and y-intercept are known, use the slope–intercept form, $y = mx + b$.

$y = -\frac{2}{3}x - 1$ • $m = -\frac{2}{3};\ b = -1$

You Try It 1

Find the equation of the line whose slope is $\frac{5}{3}$ and whose y-intercept is $(0, 2)$.

Your Solution

Example 2

Use the point–slope formula to find the equation of the line that passes through the point $(-2, -1)$ and has slope $\frac{3}{2}$.

Solution

$$y - y_1 = m(x - x_1)$$

$$y - (-1) = \frac{3}{2}[x - (-2)]$$ • $m = \frac{3}{2};\ (x_1, y_1) = (-2, -1)$

$$y + 1 = \frac{3}{2}(x + 2)$$

$$y + 1 = \frac{3}{2}x + 3$$

$$y = \frac{3}{2}x + 2$$

The equation of the line is $y = \frac{3}{2}x + 2$.

You Try It 2

Use the point–slope formula to find the equation of the line that passes through the point $(4, -2)$ and has slope $\frac{3}{4}$.

Your Solution

Solutions on p. A21

OBJECTIVE B Find the equation of a line given two points

The point–slope formula is used to find the equation of a line when a point on the line and the slope of the line are known. But this formula can also be used to find the equation of a line given two points on the line. In this case:

1. Use the slope formula to determine the slope of the line between the points.
2. Use the point–slope formula, the slope you just calculated, and one of the given points to find the equation of the line.

Find the equation of the line that passes through the points $(-3, -1)$ and $(3, 3)$.

Use the slope formula to determine the slope of the line between the points.

$(x_1, y_1) = (-3, -1); (x_2, y_2) = (3, 3)$ $\quad m = \dfrac{y_2 - y_1}{x_2 - x_1} = \dfrac{3 - (-1)}{3 - (-3)} = \dfrac{4}{6} = \dfrac{2}{3}$

Use the point–slope formula, the slope you just calculated, and one of the known points to find the equation of the line.

Use the point–slope formula.

$m = \frac{2}{3}; (x_1, y_1) = (-3, -1)$

$$\begin{aligned} y - y_1 &= m(x - x_1) \\ y - (-1) &= \frac{2}{3}[x - (-3)] \\ y + 1 &= \frac{2}{3}(x + 3) \\ y + 1 &= \frac{2}{3}x + 2 \\ y &= \frac{2}{3}x + 1 \end{aligned}$$

You can verify that the equation $y = \frac{2}{3}x + 1$ passes through the points $(-3, -1)$ and $(3, 3)$ by substituting the coordinates of these points into the equation.

$y = \frac{2}{3}x + 1$	
-1	$\frac{2}{3}(-3) + 1$
-1	$-2 + 1$
$-1 = -1$	

$y = \frac{2}{3}x + 1$	
3	$\frac{2}{3}(3) + 1$
3	$2 + 1$
$3 = 3$	

The equation of the line that passes through the two points is $y = \frac{2}{3}x + 1$.

Example 3

Find the equation of the line that passes through the points $(-4, 0)$ and $(2, -3)$.

Solution

Find the slope of the line between the two points.

$$m = \frac{y_2 - y_1}{x_2 - x_1} = \frac{-3 - 0}{2 - (-4)} = \frac{-3}{6} = -\frac{1}{2}$$

Use the point–slope formula.

$$\begin{aligned} y - y_1 &= m(x - x_1) \\ y - 0 &= -\frac{1}{2}[x - (-4)] \quad \bullet\ m = -\frac{1}{2}; (x_1, y_1) = (-4, 0) \\ y &= -\frac{1}{2}(x + 4) \\ y &= -\frac{1}{2}x - 2 \end{aligned}$$

The equation of the line is $y = -\frac{1}{2}x - 2$.

You Try It 3

Find the equation of the line that passes through the points $(-6, -1)$ and $(3, 1)$.

Your Solution

Solution on p. A22

OBJECTIVE C Application problems

A **linear model** is a first-degree equation that describes a relationship between quantities. In many cases, a linear model is used to approximate collected data. The data are graphed as points in a coordinate system, and then a line is drawn that approximates the data. The graph of the points is called a **scatter diagram;** the line is called a **line of best fit.**

Consider an experiment to determine the weight required to stretch a spring a certain distance. Data from such an experiment are shown in the table below.

Distance, in.	2.5	4	2	3.5	1	4.5
Weight, lb	63.0	104	47	85	27	115

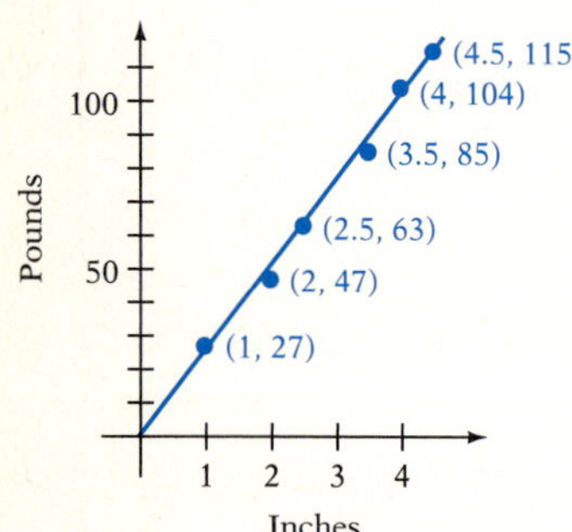

The accompanying graph shows the scatter diagram (the collection of plotted points) and the line of best fit (the line that approximately goes through the plotted points). The equation of the line of best fit is $y = 25.7x - 1.3$, where x is the number of inches the spring is stretched and y is the weight in pounds.

The table below shows the values that the model would predict to the nearest tenth. Good linear models should predict values that are close to the actual values. A more thorough analysis of lines of best fit is undertaken in statistics courses.

Distance (x), in.	2.5	4	2	3.5	1	4.5
Weight predicted using $y = 25.7x - 1.3$	63.0	101.5	50.1	88.7	24.4	114.4

Example 4

The data in the table below show the size of a house in square feet and the cost to build the house. The line $y = 70.3x + 41{,}100$, where x is the number of square feet and y is the cost of the house, is the best fit.

Square ft	1250	1400	1348	2675	2900
Cost	128,000	140,000	136,100	233,450	241,500

Graph the data and the line of best fit in the coordinate system below. Write a sentence that describes the meaning of the slope of the line of best fit.

Solution

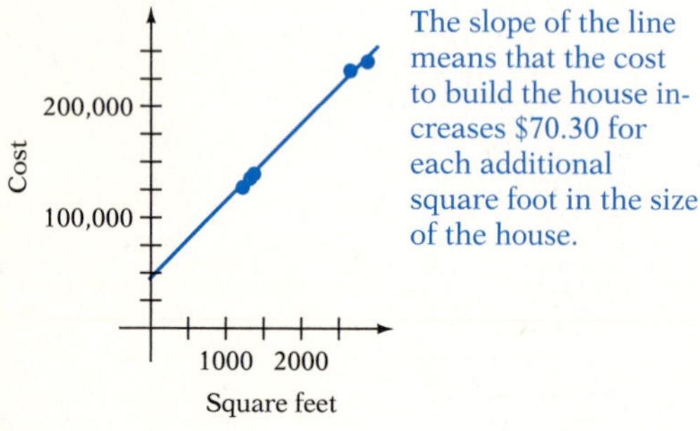

You Try It 4

The data in the table below show a reading test grade and the final exam grade in a history class. The line of best fit is $y = 8.3x - 7.8$, where x is the reading test score and y is the history test score.

Reading	8.5	9.4	10.0	11.4	12.0
History	64	68	76	87	92

Graph the data and the line of best fit in the coordinate system below. Write a sentence that describes the meaning of the slope of the line of best fit.

Your Solution

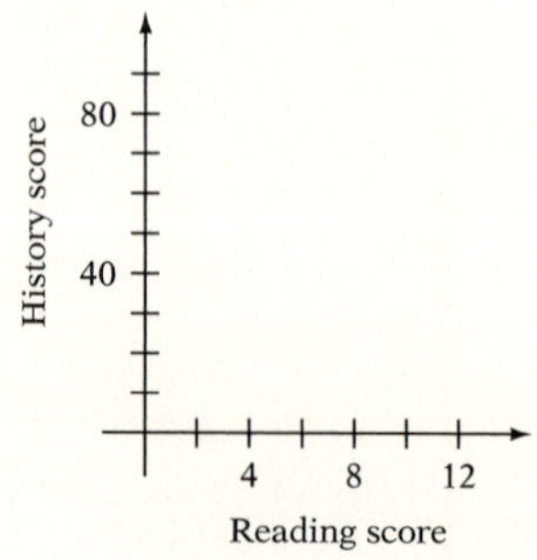

Solution on p. A22

7.3 EXERCISES

Objective A

Solve. Use the point–slope formula.

1. Find the equation of the line that contains the point $(0, 2)$ and has slope 2.

2. Find the equation of the line that contains the point $(0, -1)$ and has slope -2.

3. Find the equation of the line that contains the point $(-1, 2)$ and has slope -3.

4. Find the equation of the line that contains the point $(2, -3)$ and has slope 3.

5. Find the equation of the line that contains the point $(3, 1)$ and has slope $\frac{1}{3}$.

6. Find the equation of the line that contains the point $(-2, 3)$ and has slope $\frac{1}{2}$.

7. Find the equation of the line that contains the point $(4, -2)$ and has slope $\frac{3}{4}$.

8. Find the equation of the line that contains the point $(2, 3)$ and has slope $-\frac{1}{2}$.

9. Find the equation of the line that contains the point $(5, -3)$ and has slope $-\frac{3}{5}$.

10. Find the equation of the line that contains the point $(5, -1)$ and has slope $\frac{1}{5}$.

11. Find the equation of the line that contains the point $(2, 3)$ and has slope $\frac{1}{4}$.

12. Find the equation of the line that contains the point $(-1, 2)$ and has slope $-\frac{1}{2}$.

Objective B

Solve.

13. Find the equation of the line that passes through the points $(1, -1)$ and $(-2, -7)$.

14. Find the equation of the line that passes through the points $(2, 3)$ and $(3, 2)$.

15. Find the equation of the line that passes through the points $(-2, 1)$ and $(1, -5)$.

16. Find the equation of the line that passes through the points $(-1, -3)$ and $(2, -12)$.

17. Find the equation of the line that passes through the points $(0, 0)$ and $(-3, -2)$.

18. Find the equation of the line that passes through the points $(0, 0)$ and $(-5, 1)$.

19. Find the equation of the line that passes through the points $(2, 3)$ and $(-4, 0)$.

20. Find the equation of the line that passes through the points $(3, -1)$ and $(0, -3)$.

21. Find the equation of the line that passes through the points $(-4, 1)$ and $(4, -5)$.

22. Find the equation of the line that passes through the points $(-5, 0)$ and $(10, -3)$.

23. Find the equation of the line that passes through the points $(-2, 1)$ and $(2, 4)$.

24. Find the equation of the line that passes through the points $(3, -2)$ and $(-3, -3)$.

Objective C

Solve.

25. The data in the table below show the tread depth of a tire and the number of miles that have been driven on that tire. The line of best fit is $y = -0.2x + 10$, where x is the number of miles driven in thousands and y is the depth of the tread in millimeters.

Miles driven, x	25	35	40	20	45
Tread depth, y	4.8	3.5	2.1	5.5	1.0

Graph the data and the line of best fit in the coordinate system at the right. Write a sentence that describes the meaning of the slope of the line of best fit.

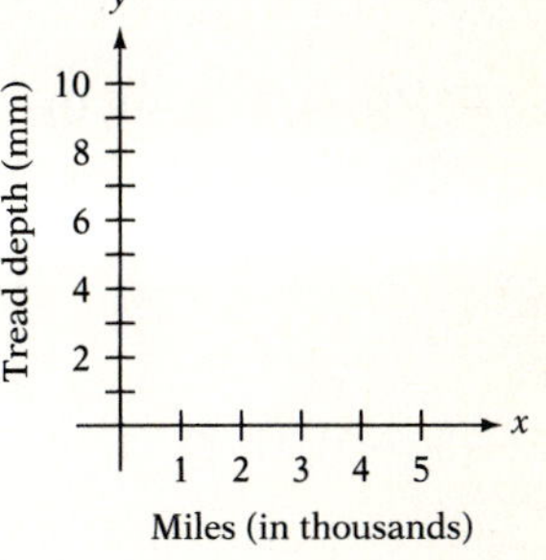

26. The data in the table below are estimates of the number of employed persons per retired person during the decade from 2020 to 2030. This information is important to the Social Security Administration as it plans for paying future retirement benefits. The line of best fit is $y = -0.13x + 3.35$, where x is the year (with 2020 as 0) and y is the number of employed persons per retired person.

Year, x	2 (2022)	4 (2024)	6 (2026)	8 (2028)	10 (2030)
Employed per retired, y	3.1	2.9	2.5	2.4	2.1

Graph the data and the line of best fit in the coordinate system at the right. Write a sentence that describes the meaning of the slope of the line of best fit.

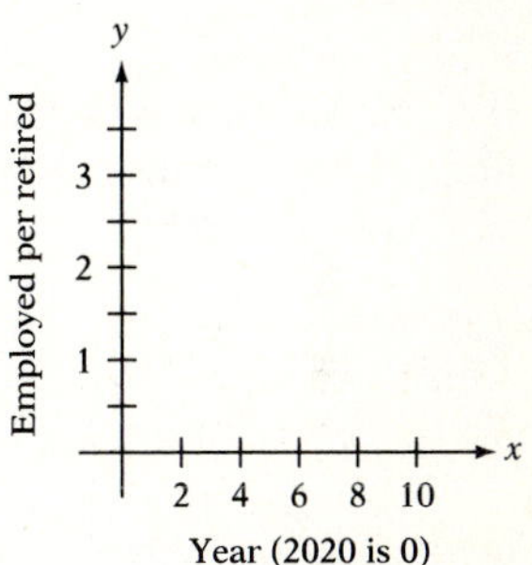

27. The data in the table below show the recorded and estimated sales of microprocessors by Maple Instruments for 1991 through 1995. The line of best fit is $y = 1.6x + 2.4$, where x is the year (with 1991 as 1) and y is the revenue in billions.

Year, x	1 (1991)	2 (1992)	3 (1993)	4 (1994)	5 (1995)
Revenue, y	3.9	5.5	7.7	8.7	10.3

Graph the data and the line of best fit in the coordinate system at the right. Write a sentence that describes the meaning of the slope of the line of best fit.

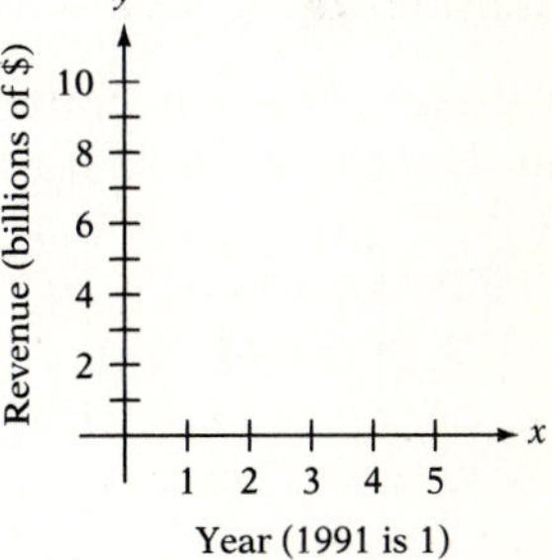

28. The data in the table below show the recorded and projected numbers of ATMs (automated teller machines) in service from 1985 through 1995. The line of best fit is $y = 3.6x + 61.6$, where x is the year (with 1985 as 0) and y is the number of ATMs in thousands.

Year, x	0 (1985)	3 (1988)	5 (1990)	7 (1992)	10 (1995)
ATMs, y	61	73	81	86	98

Graph the data and the line of best fit in the coordinate system at the right. Write a sentence that describes the meaning of the slope of the line of best fit.

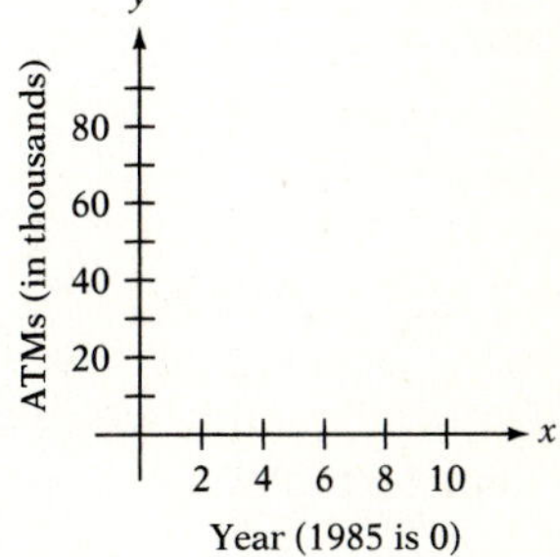

Critical Thinking

In Exercises 29–32, the first two given points are on a line. Determine whether the third point is on the line.

29. $(-3, 2), (4, 1); (-1, 0)$

30. $(2, -2), (3, 4); (-1, 5)$

31. $(-3, -5), (1, 3); (4, 9)$

32. $(-3, 7), (0, -2); (1, -5)$

33. If $(-2, 4)$ are the coordinates of a point on the line whose equation is $y = mx + 1$, what is the slope of the line?

34. If $(3, 1)$ are the coordinates of a point on the line whose equation is $y = mx - 3$, what is the slope of the line?

35. If $(0, -3)$, $(6, -7)$, and $(3, n)$ are coordinates of points on the same line, determine n.

36. If $(-4, 11)$, $(2, -4)$, and $(6, n)$ are coordinates of points on the same line, determine n.

The formula $y - y_1 = \frac{y_2 - y_1}{x_2 - x_1}(x - x_1)$, where $x_1 \neq x_2$, is called the **two-point formula** for a straight line. This formula can be used to find the equation of a line given two points. Use this formula for Exercises 37–38.

37. Find the equation of the line passing through $(-2, 3)$ and $(4, -1)$.

38. Find the equation of the line passing through $(3, -1)$ and $(4, -3)$.

39. [W] Explain why the condition $x_1 \neq x_2$ is placed on the two-point formula given above.

40. [W] Explain how the two-point formula given above can be derived from the point–slope formula.

SECTION 7.4 Solving Systems of Linear Equations

OBJECTIVE A Solve a system of linear equations by graphing

Equations considered together are called a **system of equations.** A system of equations is shown at the right.

$$2x + y = 3$$
$$x + y = 1$$

A **solution of a system of equations** is an ordered pair that is a solution of each equation of the system.

For example, $(2, -1)$ is a solution of the system of equations given above because it is a solution of each equation in the system.

$2x + y = 3$	
$2(2) + (-1)$	3
$4 + (-1)$	3
$3 = 3$	

$x + y = 1$	
$2 + (-1)$	1
$1 = 1$	

However, $(3, -3)$ is not a solution of this system because it is not a solution of each equation in the system.

$2x + y = 3$	
$2(3) + (-3)$	3
$6 + (-3)$	3
$3 = 3$	

$x + y = 1$	
$3 + (-3)$	1
$0 \neq 1$	

The solution of a system of linear equations in two variables can be found by graphing the two equations on the same coordinate system. Three situations are possible.

1. The graphs of the lines may intersect at one point. The point of intersection of the lines is the ordered pair that is a solution of each equation of the system. It is the solution of the system of equations. This system of equations is **independent.**

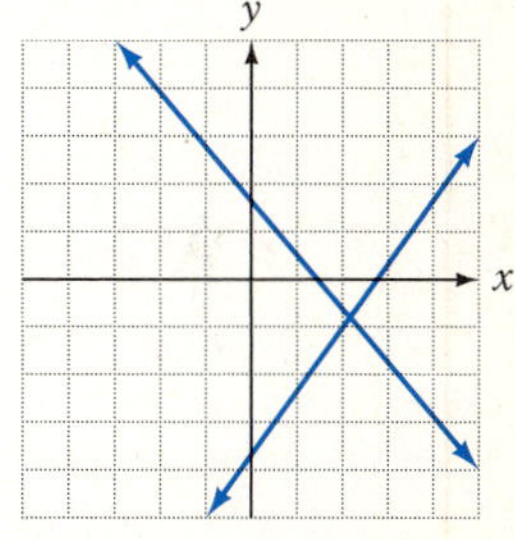

2. The graphs of the lines may be parallel and not intersect at all. This system of equations is **inconsistent** and has no solution.

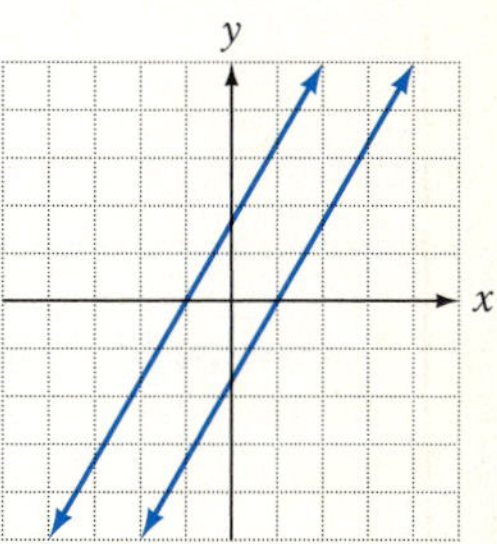

3. The graphs of the lines can be the same line. The lines intersect at infinitely many points; therefore, there are infinitely many solutions. This system of equations is **dependent.** The solutions are the ordered pairs that are solutions of either one of the two equations in the system.

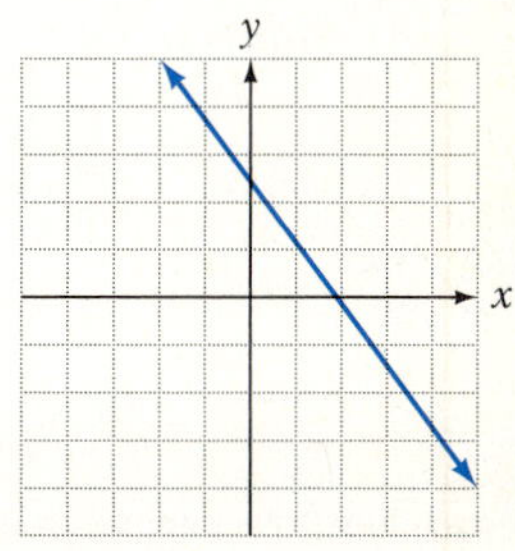

Solve by graphing: $2x + 3y = 6$
$2x + y = -2$

Graph each line.

The equations intersect at one point. This system of equations is independent.

Find the point of intersection.

The solution is $(-3, 4)$.

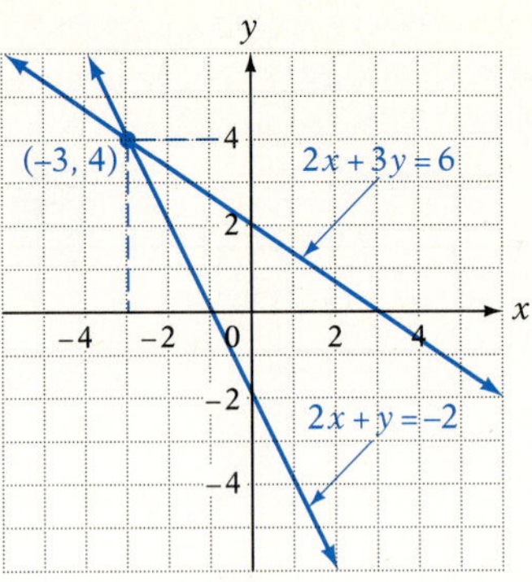

Solve by graphing: $2x - y = 1$
$6x - 3y = 12$

Graph each line.

The lines are parallel and therefore do not intersect.

This system of equations is inconsistent and has no solution.

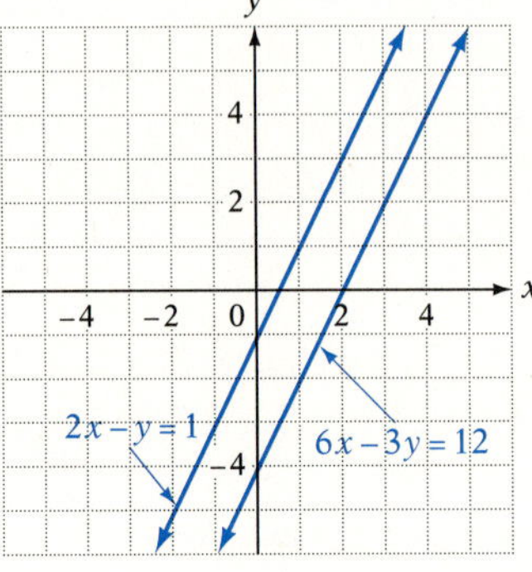

Solve by graphing: $2x + 3y = 6$
$6x + 9y = 18$

Graph each line.

The two equations represent the same line. This system of equations is dependent and, therefore, has an infinite number of solutions.

The solutions are the ordered pairs that are solutions of the equation $2x + 3y = 6$.

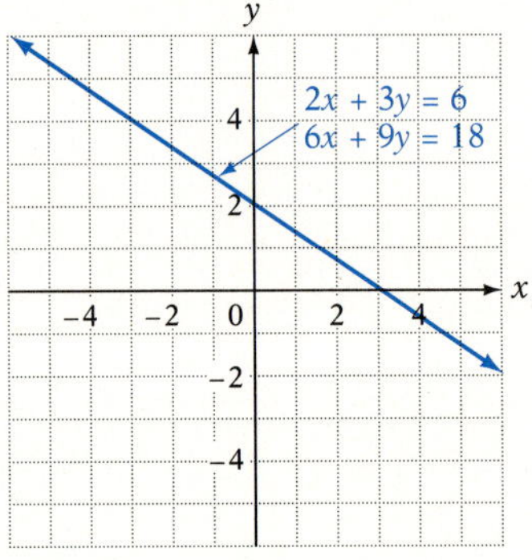

By choosing values for x, substituting these values into the equation $2x + 3y = 6$, and finding the corresponding values for y, some specific ordered pair solutions can be found. For example, $(3, 0)$, $(0, 2)$, and $(6, -2)$ are solutions of this system of equations.

Example 1 Is $(1, -3)$ a solution of the system
$3x + 2y = -3$
$x - 3y = 6$?

Solution

$$\begin{array}{r|l} 3x + 2y & = -3 \\ \hline 3\cdot 1 + 2(-3) & -3 \\ 3 + (-6) & -3 \\ -3 & = -3 \end{array} \qquad \begin{array}{r|l} x - 3y & = 6 \\ \hline 1 - 3(-3) & 6 \\ 1 - (-9) & 6 \\ 10 & \neq 6 \end{array}$$

No, $(1, -3)$ is not a solution of the system of equations.

You Try It 1 Is $(-1, -2)$ a solution of the system
$2x - 5y = 8$
$-x + 3y = -5$?

Your Solution

Solution on p. A22

Example 2 Solve by graphing:

$$x - 2y = 2$$
$$x + y = 5$$

Solution

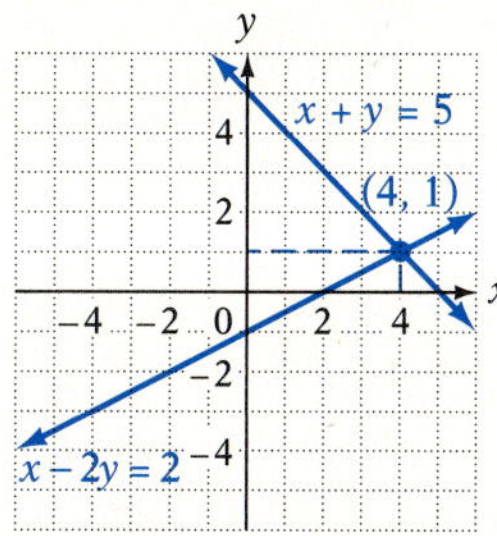

The solution is (4, 1).

You Try It 2 Solve by graphing:

$$x + 3y = 3$$
$$-x + y = 5$$

Your Solution

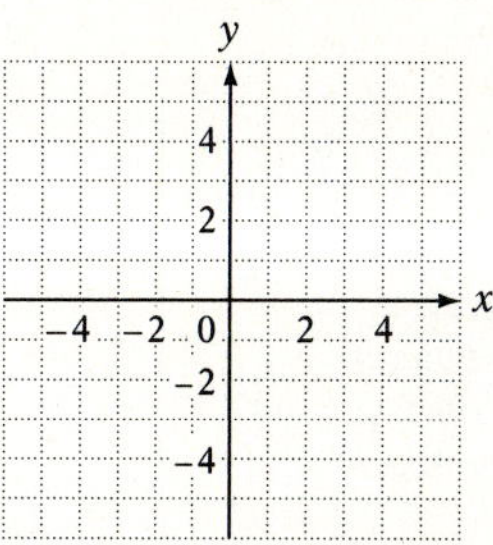

Example 3 Solve by graphing:

$$4x - 2y = 6$$
$$y = 2x - 3$$

Solution

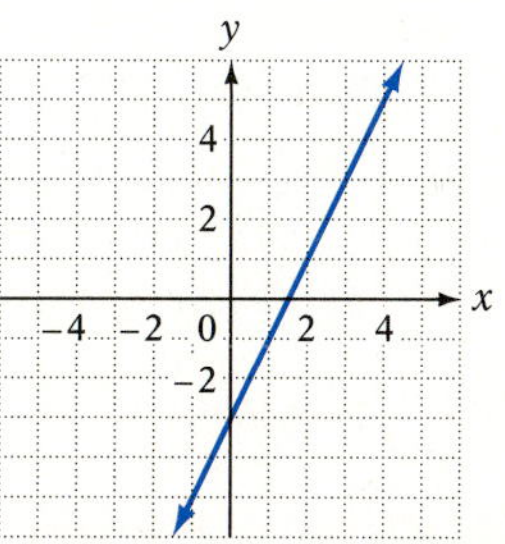

The two equations represent the same line. Any ordered pair that is a solution of one equation is also a solution of the other equation.

You Try It 3 Solve by graphing:

$$y = 3x - 1$$
$$6x - 2y = -6$$

Your Solution

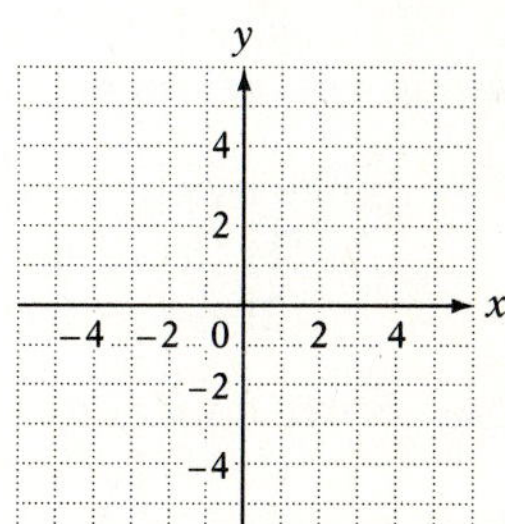

Solutions on p. A22

OBJECTIVE B Solve a system of linear equations by the substitution method

Solving a system of equations graphically may give only an approximate solution. For example, the point $\left(\frac{1}{4}, \frac{1}{2}\right)$ would be difficult to read from the graph. An algebraic method, the **substitution method,** can be used to find an exact solution of a system.

In the system of equations at the right, equation (2) states that $y = 3x - 9$. Substitute $3x - 9$ for y in equation (1).

$$(1) \quad 2x + 5y = -11$$
$$(2) \quad y = 3x - 9$$

Solve for x.

$$2x + 5(3x - 9) = -11$$
$$2x + 15x - 45 = -11$$
$$17x = 34$$
$$x = 2$$

Substitute the value of x into equation (2) and solve for y.

$$(2) \quad y = 3x - 9$$
$$y = 3 \cdot 2 - 9$$
$$y = 6 - 9$$
$$y = -3$$

The solution is (2, −3).

CONSIDER THIS

In previous work, you have replaced a variable with a constant when evaluating a variable expression. Here you are replacing a variable with a variable expression.

Solve: $5x + y = 4$ (1)
$2x - 3y = 5$ (2)

$$5x + y = 4$$
$$2x - 3y = 5$$

Solve equation (1) for y. Equation (1) is chosen because it is the easier equation to solve for one variable in terms of the other.

$$5x + y = 4$$
$$y = -5x + 4$$

Substitute $-5x + 4$ for y in equation (2). Solve for x.

$$2x - 3(-5x + 4) = 5$$
$$2x + 15x - 12 = 5$$
$$17x - 12 = 5$$
$$17x = 17$$
$$x = 1$$

Substitute the value of x in equation (1) and solve for y.

$$5x + y = 4$$
$$5(1) + y = 4$$
$$5 + y = 4$$
$$y = -1$$

The solution is $(1, -1)$.

Example 4 Solve by substitution:
(1) $3x + 4y = -2$
(2) $-x + 2y = 4$

Solution Solve equation (2) for x.

$$-x + 2y = 4$$
$$-x = -2y + 4$$
$$x = 2y - 4$$

Substitute in equation (1).

$$3(2y - 4) + 4y = -2$$
$$6y - 12 + 4y = -2$$
$$10y - 12 = -2$$
$$10y = 10$$
$$y = 1$$

Substitute in equation (2).

$$-x + 2(1) = 4$$
$$-x + 2 = 4$$
$$-x = 2$$
$$x = -2$$

The solution is $(-2, 1)$.

You Try It 4 Solve by substitution:
(1) $7x - y = 4$
(2) $3x + 2y = 9$

Your Solution

Example 5 Solve by substitution:

$$4x + 2y = 5$$
$$y = -2x + 1$$

Solution

$$4x + 2y = 5$$
$$4x + 2(-2x + 1) = 5$$
$$4x - 4x + 2 = 5$$
$$2 = 5$$

This is not a true equation. The lines are parallel and therefore do not intersect. This system does not have a solution.

You Try It 5 Solve by substitution:

$$3x - y = 4$$
$$y = 3x + 2$$

Your Solution

Solutions on p. A22

Example 6 Solve by substitution:

$$6x - 2y = 4$$
$$y = 3x - 2$$

Solution

$$6x - 2y = 4$$
$$6x - 2(3x - 2) = 4$$
$$6x - 6x + 4 = 4$$
$$4 = 4$$

This is a true equation. The two equations represent the same line. Any ordered pair that is a solution of one equation is also a solution of the other equation.

You Try It 6 Solve by substitution:

$$y = -2x + 1$$
$$6x + 3y = 3$$

Your Solution

Solution on p. A23

OBJECTIVE C Solve a system of linear equations by the addition method

Another algebraic method for solving a system of equations is called the **addition method.** It is based on the Addition Property of Equations.

CONSIDER THIS

The addition method may look like magic to you. It may help to show that this method relies on nothing more than the Addition Property of Equations and the substitution property.

$3x + 2y = 4$

$4x - 2y = 10$

Add $4x - 2y$ to each side of the first equation.

$3x + 2y + (4x - 2y)$
$= (4x - 2y) + 4$

Substitute 10 for $4x - 2y$ on the right side of the equation.

$3x + 2y + (4x - 2y) = 10 + 4$

Now simplify.

$7x = 14$

Note, for the system of equations at the right, the effect of adding equation (2) to equation (1). Since $2y$ and $-2y$ are opposites, adding the equations results in an equation with only one variable.

$$(1) \quad 3x + 2y = 4$$
$$(2) \quad 4x - 2y = 10$$
$$7x + 0y = 14$$
$$7x = 14$$

The solution of the resulting equation is the first coordinate of the ordered pair solution of the system.

$$7x = 14$$
$$x = 2$$

The second coordinate is found by substituting the value of x into equation (1) or (2) and then solving for y. Equation (1) is used here.

$$(1) \quad 3x + 2y = 4$$
$$3 \cdot 2 + 2y = 4$$
$$6 + 2y = 4$$
$$2y = -2$$
$$y = -1$$

The solution is $(2, -1)$.

Sometimes adding the two equations does not eliminate one of the variables. In this case, use the Multiplication Property of Equations to rewrite one or both of the equations so that when the equations are added, one of the variables is eliminated.

To do this, first choose which variable to eliminate. The coefficients of that variable must be opposites. Multiply each equation by a constant that will produce coefficients that are opposites.

Solve: $3x + 2y = 7$
$5x - 4y = 19$

(1) $3x + 2y = 7$
(2) $5x - 4y = 19$

Eliminate y. Multiply equation (1) by 2.

$2(3x + 2y) = 2 \cdot 7$
$5x - 4y = 19$

Now the coefficients of the y terms are opposites.

$6x + 4y = 14$
$5x - 4y = 19$

Add the equations.
Solve for x.

$11x + 0y = 33$
$11x = 33$
$x = 3$

Substitute the value of x into one of the equations and solve for y. Equation (2) is used here.

(2) $5x - 4y = 19$
$5 \cdot 3 - 4y = 19$
$15 - 4y = 19$
$-4y = 4$
$y = -1$

The solution is $(3, -1)$.

CONSIDER THIS

Demonstrate that the same value of y is determined by substituting the value of x into equation (1).

Solve: $5x + 6y = 3$
$2x - 5y = 16$

(1) $5x + 6y = 3$
(2) $2x - 5y = 16$

Eliminate x. Multiply equation (1) by 2 and equation (2) by -5. Note how the constants are selected.

$2(5x + 6y) = 2 \cdot 3$
$-5(2x - 5y) = -5 \cdot 16$

The negative is used so that the coefficients will be opposites.

Now the coefficients of the x terms are opposites.

$10x + 12y = 6$
$-10x + 25y = -80$

Add the equations.
Solve for y.

$0x + 37y = -74$
$37y = -74$
$y = -2$

Substitute the value of y into one of the equations and solve for x. Equation (1) is used here.

(1) $5x + 6y = 3$
$5x + 6(-2) = 3$
$5x - 12 = 3$
$5x = 15$
$x = 3$

The solution is $(3, -2)$.

Solve: $2x + y = 2$
$4x + 2y = 5$

(1) $2x + y = 2$
(2) $4x + 2y = 5$

Eliminate y. Multiply equation (1) by -2.

$$-2(2x + y) = -2 \cdot 2$$
$$4x + 2y = 5$$

$$-4x - 2y = -4$$
$$4x + 2y = 5$$

Add the equations.

$$0x + 0y = 1$$
$$0 \neq 1$$

This is not a true equation. The lines are parallel and therefore do not intersect. This system does not have a solution.

Example 7 Solve by the addition method:
$2x + 4y = 7$
$5x - 3y = -2$

Solution Eliminate x.

$$5(2x + 4y) = 5 \cdot 7$$
$$-2(5x - 3y) = -2(-2)$$

$$10x + 20y = 35$$
$$-10x + 6y = 4$$

Add the equations.

$$26y = 39$$
$$y = \frac{39}{26} = \frac{3}{2}$$

Replace y in equation (1).

$$2x + 4\left(\frac{3}{2}\right) = 7$$
$$2x + 6 = 7$$
$$2x = 1$$
$$x = \frac{1}{2}$$

The solution is $\left(\frac{1}{2}, \frac{3}{2}\right)$.

You Try It 7 Solve by the addition method:
$x - 2y = 1$
$2x + 4y = 0$

Your Solution

Solution on p. A23

Example 8 Solve by the addition method:

$$6x + 9y = 15$$
$$4x + 6y = 10$$

Solution Eliminate x.

$$4(6x + 9y) = 4 \cdot 15$$
$$-6(4x + 6y) = -6 \cdot 10$$

$$24x + 36y = 60$$
$$-24x - 36y = -60$$

Add the equations.

$$0x + 0y = 0$$
$$0 = 0$$

This is a true equation. The two equations represent the same line. Any ordered pair that is a solution of one equation is also a solution of the other equation.

You Try It 8 Solve by the addition method:

$$2x - 3y = 4$$
$$-4x + 6y = -8$$

Your Solution

Example 9 Solve by the addition method:

$$2x = y + 8$$
$$3x + 2y = 5$$

Solution Write equation (1) in the form $Ax + By = C$.

$$2x = y + 8$$
$$2x - y = 8$$

Eliminate y.

$$2(2x - y) = 2 \cdot 8$$
$$3x + 2y = 5$$

$$4x - 2y = 16$$
$$3x + 2y = 5$$

Add the equations.

$$7x = 21$$
$$x = 3$$

Replace x in equation (1).

$$2x = y + 8$$
$$2 \cdot 3 = y + 8$$
$$6 = y + 8$$
$$-2 = y$$

The solution is $(3, -2)$.

You Try It 9 Solve by the addition method:

$$4x + 5y = 11$$
$$3y = x + 10$$

Your Solution

Solutions on p. A23

7.4 EXERCISES

Objective A

1. Is (2, 3) a solution of the system
$$3x + 4y = 18$$
$$2x - y = 1?$$

2. Is (2, −1) a solution of the system
$$x - 2y = 4$$
$$2x + y = 3?$$

3. Is (1, −2) a solution of the system
$$3x - y = 5$$
$$2x + 5y = -8?$$

4. Is (−1, −1) a solution of the system
$$x - 4y = 3$$
$$3x + y = 2?$$

5. Is (4, 3) a solution of the system
$$5x - 2y = 14$$
$$x + y = 8?$$

6. Is (2, 5) a solution of the system
$$3x + 2y = 16$$
$$2x - 3y = 4?$$

7. Is (−1, 3) a solution of the system
$$4x - y = -5$$
$$2x + 5y = 13?$$

8. Is (4, −1) a solution of the system
$$x - 4y = 9$$
$$2x - 3y = 11?$$

9. Is (0, 0) a solution of the system
$$4x + 3y = 0$$
$$2x - y = 1?$$

10. Is (2, 0) a solution of the system
$$3x - y = 6$$
$$x + 3y = 2?$$

Solve by graphing.

11. $x - y = 3$
$x + y = 5$

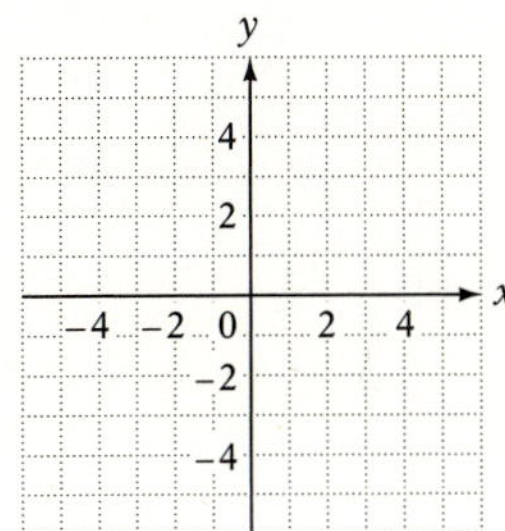

12. $2x - y = 4$
$x + y = 5$

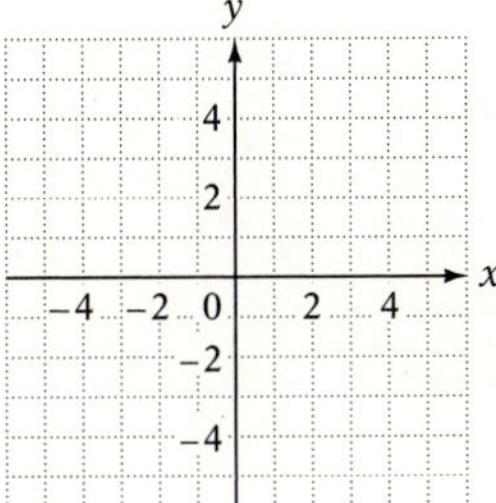

13. $x + 2y = 6$
$x - y = 3$

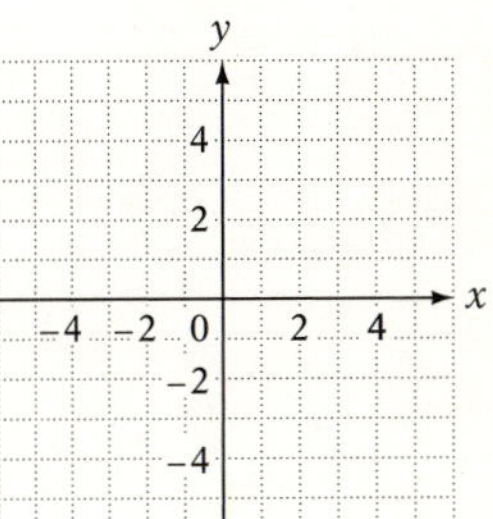

14. $3x - y = 3$
$2x + y = 2$

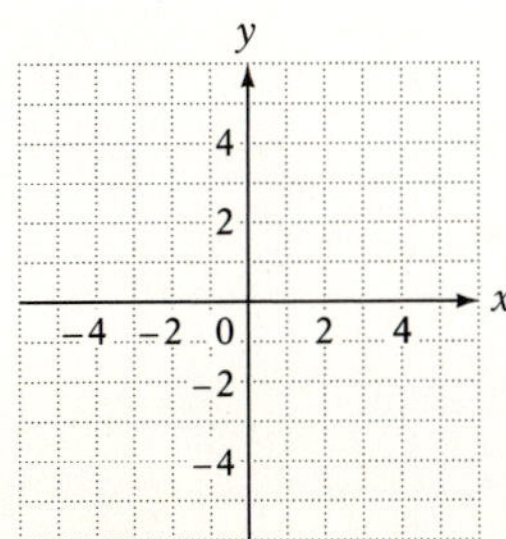

15. $3x - 2y = 6$
$y = 3$

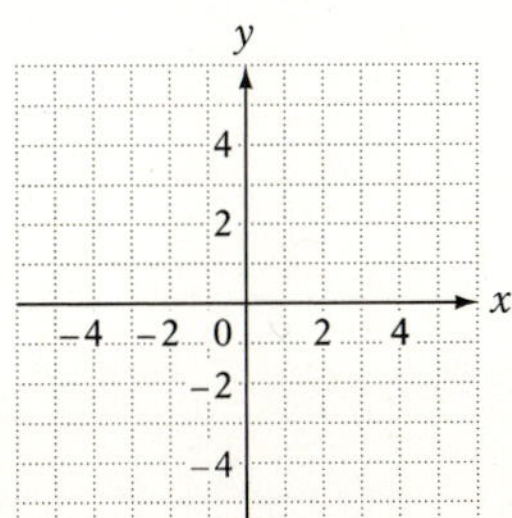

16. $x = 2$
$3x + 2y = 4$

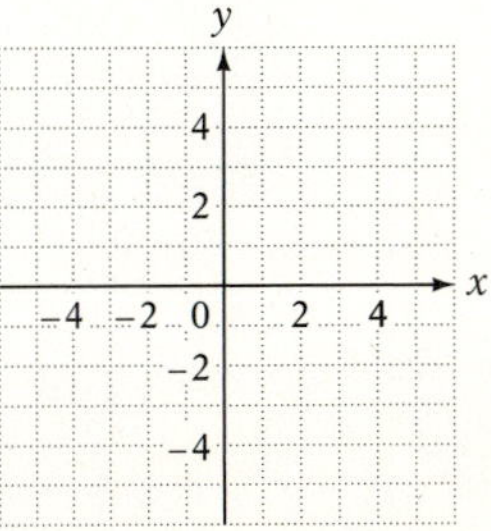

Solve by graphing.

17. $x = 3$
$y = -2$

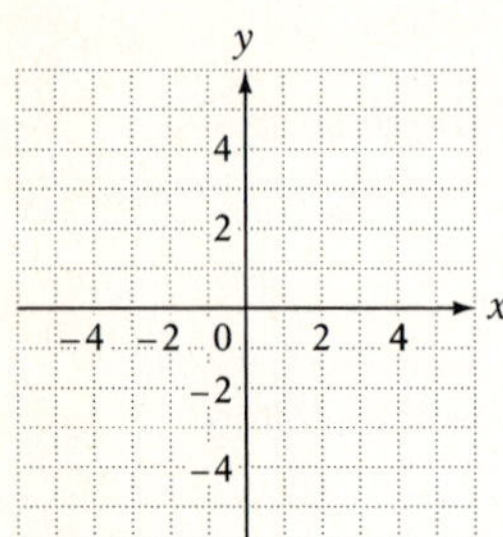

18. $x + 1 = 0$
$y - 3 = 0$

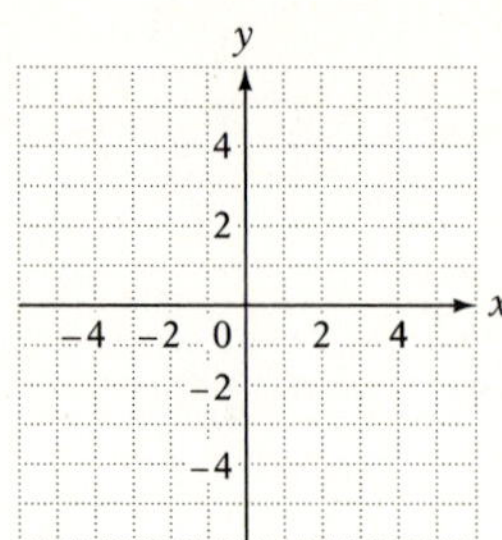

19. $y = 2x - 6$
$x + y = 0$

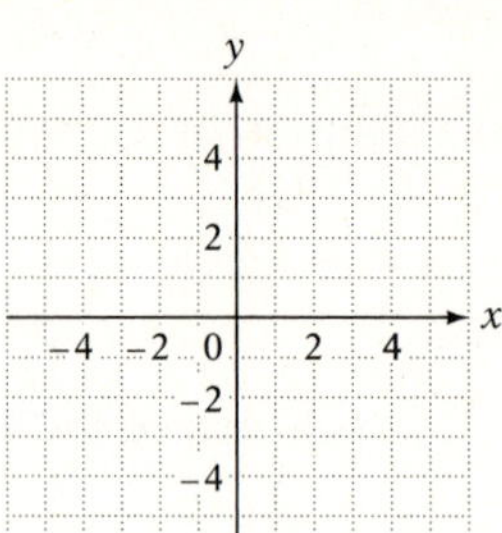

20. $5x - 2y = 11$
$y = 2x - 5$

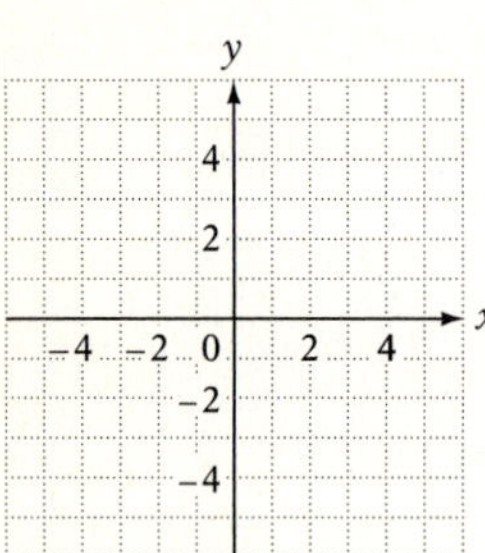

21. $2x + y = -2$
$6x + 3y = 6$

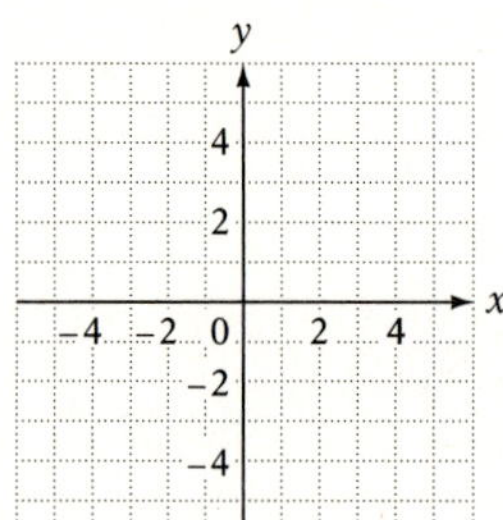

22. $x + y = 5$
$3x + 3y = 6$

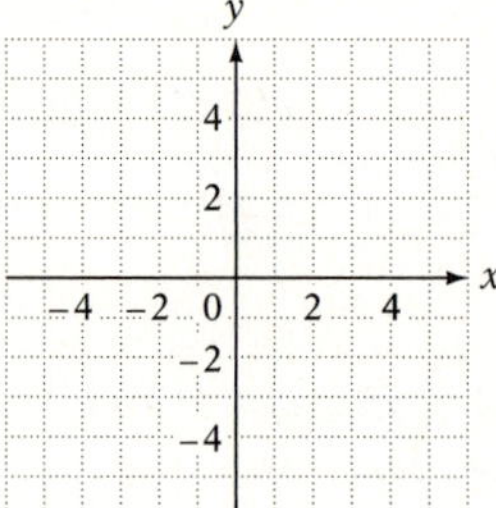

23. $4x - 2y = 4$
$y = 2x - 2$

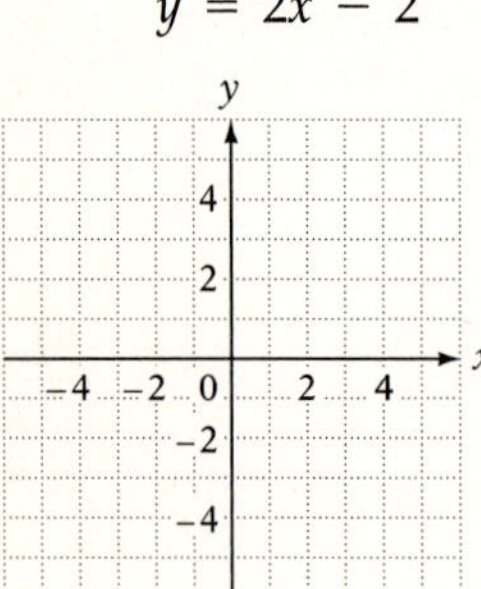

24. $2x + 6y = 6$
$y = -\frac{1}{3}x + 1$

25. $4x + 6y = 12$
$6x + 9y = 18$

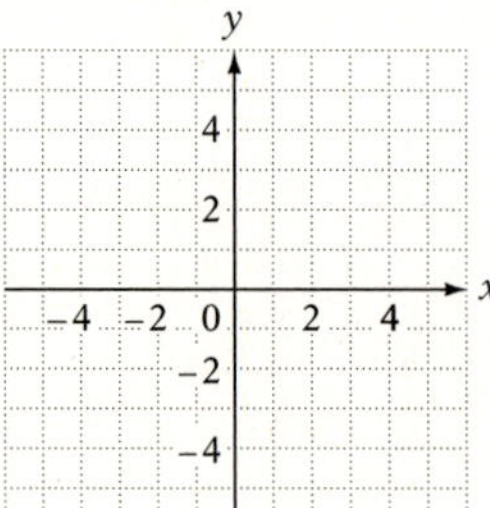

26. $x - 3y = 3$
$2x - 6y = 12$

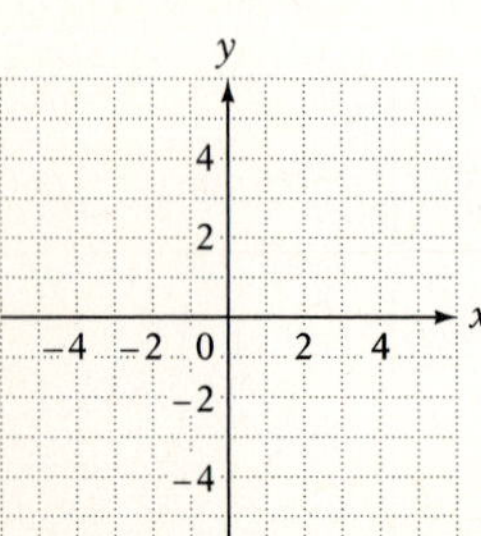

27. $3x + 4y = 0$
$2x - 5y = 0$

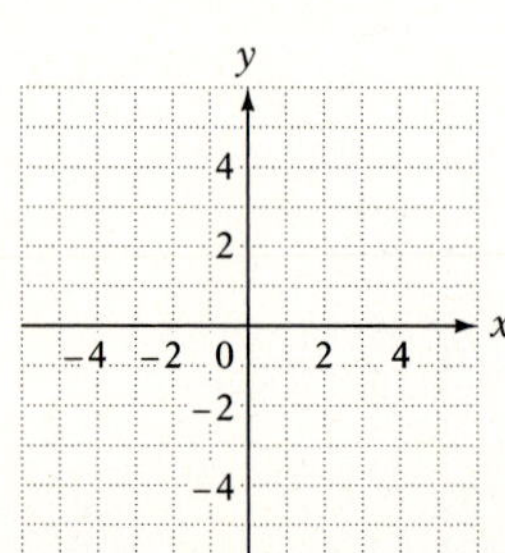

28. $2x - 3y = 0$
$y = -\frac{1}{3}x$

Objective B

Solve by substitution.

29. $2x + 3y = 7$
$x = 2$

30. $y = 3$
$3x - 2y = 6$

31. $y = x - 3$
$x + y = 5$

32. $y = x + 2$
$x + y = 6$

33. $x = y - 2$
$x + 3y = 2$

34. $x = y + 1$
$x + 2y = 7$

35. $2x + 3y = 9$
$y = x - 2$

36. $3x + 2y = 11$
$y = x + 3$

37. $3x - y = 2$
$y = 2x - 1$

38. $2x - y = -5$
$y = x + 4$

39. $x = 2y - 3$
$2x - 3y = -5$

40. $x = 3y - 1$
$3x + 4y = 10$

41. $y = 4 - 3x$
$3x + y = 5$

42. $y = 2 - 3x$
$6x + 2y = 7$

43. $x = 3y + 3$
$2x - 6y = 12$

44. $x = 2 - y$
$3x + 3y = 6$

45. $3x + 5y = -6$
$x = 5y + 3$

46. $y = 2x + 3$
$4x - 3y = 1$

47. $4x - 3y = -1$
$y = 2x - 3$

48. $3x - 7y = 28$
$x = 3 - 4y$

49. $7x + y = 14$
$2x - 5y = -33$

50. $4x + 3y = 0$
$2x - y = 0$

51. $5x + 2y = 0$
$x - 3y = 0$

52. $6x - 3y = 6$
$2x - y = 2$

Solve by substitution.

53. $y = 2x + 11$
$y = 5x - 19$

54. $y = 2x - 8$
$y = 3x - 13$

55. $y = -4x + 2$
$y = -3x - 1$

56. $x = 3y + 7$
$x = 2y - 1$

57. $x = 4y - 2$
$x = 6y + 8$

58. $x = 3 - 2y$
$x = 5y - 10$

59. $y = 2x - 7$
$y = 4x + 5$

60. $3x - y = 11$
$2x + 5y = -4$

61. $-x + 6y = 8$
$2x + 5y = 1$

Objective C

Solve by the addition method.

62. $x + y = 4$
$x - y = 6$

63. $2x + y = 3$
$x - y = 3$

64. $x + y = 4$
$2x + y = 5$

65. $x - 3y = 2$
$x + 2y = -3$

66. $2x - y = 1$
$x + 3y = 4$

67. $x - 2y = 4$
$3x + 4y = 2$

68. $4x - 5y = 22$
$x + 2y = -1$

69. $3x - y = 11$
$2x + 5y = 13$

70. $2x - y = 1$
$4x - 2y = 2$

71. $x + 3y = 2$
$3x + 9y = 6$

72. $4x + 3y = 15$
$2x - 5y = 1$

73. $3x - 7y = 13$
$6x + 5y = 7$

74. $2x - 3y = 1$
$4x - 6y = 2$

75. $2x + 4y = 6$
$3x + 6y = 9$

76. $5x - 2y = -1$
$x + 3y = -5$

Solve by the addition method.

77. $3x - 2y = 0$
$6x + 5y = 0$

78. $5x + 2y = 0$
$3x + 5y = 0$

79. $2x - 3y = 16$
$3x + 4y = 7$

80. $3x + 4y = 10$
$4x + 3y = 11$

81. $5x + 3y = 7$
$2x + 5y = 1$

82. $-2x + 7y = 9$
$3x + 2y = -1$

83. $7x - 2y = 13$
$5x + 3y = 27$

84. $12x + 5y = 23$
$2x - 7y = 39$

85. $8x - 3y = 11$
$6x - 5y = 11$

86. $4x - 8y = 36$
$3x - 6y = 27$

87. $5x + 15y = 20$
$2x + 6y = 8$

88. $y = 2x - 3$
$3x + 4y = -1$

89. $3x = 2y + 7$
$5x - 2y = 13$

90. $2y = 4 - 9x$
$9x - y = 25$

91. $2x + 9y = 16$
$5x = 1 - 3y$

92. $3x - 4 = y + 18$
$4x + 5y = -21$

93. $2x + 3y = 7 - 2x$
$7x + 2y = 9$

94. $5x - 3y = 3y + 4$
$4x + 3y = 11$

95. $3x + y = 1$
$5x + y = 2$

96. $2x - y = 1$
$2x - 5y = -1$

97. $4x + 3y = 3$
$x + 3y = 1$

98. $2x - 5y = 4$
$x + 5y = 1$

99. $3x - 4y = 1$
$4x + 3y = 1$

100. $2x - 7y = -17$
$3x + 5y = 17$

Critical Thinking

For what value of k does the system of equations have no solution?

101. $2x - 3y = 7$
$kx - 3y = 4$

102. $8x - 4y = 1$
$2x - ky = 3$

103. $x = 4y + 4$
$kx - 8y = 4$

For what value of k is the system of equations dependent?

104. $2x + 3y = 7$
$4x + 6y = k$

105. $y = \frac{2}{3}x - 3$
$y = kx - 3$

106. $x = ky - 1$
$y = 2x + 2$

For what values of k is the system of equations independent?

107. $x + y = 7$
$kx + y = 3$

108. $x + 2y = 4$
$kx + 3y = 2$

109. $2x + ky = 1$
$x + 2y = 2$

Write the system of equations given its graph.

110.

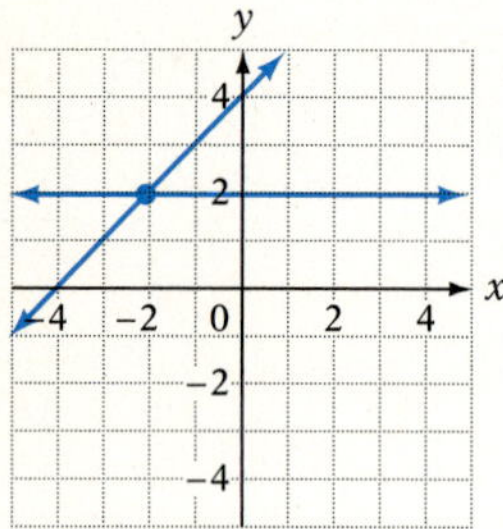

111.

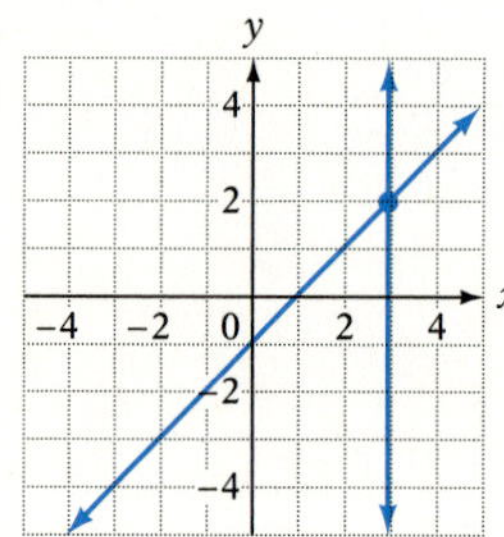

112. [W] Explain how you can determine from the graph of a system of two equations in two variables whether it is an independent system.

113. [W] Explain how you can determine from the graph of a system of two equations in two variables whether it is an inconsistent system.

114. [W] When you solve a system of equations by the substitution method, how do you determine whether the system is dependent?

115. [W] When you solve a system of equations by the substitution method, how do you determine whether the system is inconsistent?

116. [W] The graph at the right shows the Cumulative Total Return Performance for the Lipper Science and Technology Average versus the S&P 500 Stock Index from December 1990 to December 1991. Explain what the solution of this system of equations means.

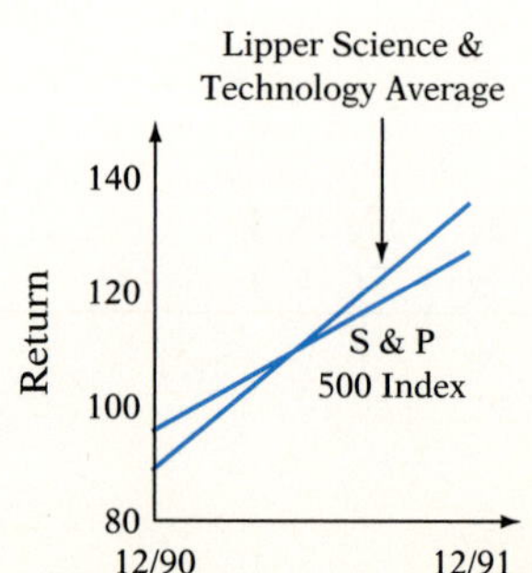

SECTION 7.5 Application Problems in Two Variables

OBJECTIVE A Solve rate-of-wind or current problems

Motion problems in which an object moves with or against wind or current normally require two variables to solve.

Flying with the wind, a small plane can fly 600 mi in 3 h. Against the wind, the plane can fly the same distance in 4 h. Find the rate of the plane in calm air and the rate of the wind.

Strategy for Solving Rate-of-Wind or Current Problems

Choose one variable to represent the rate of the object in calm conditions and a second variable to represent the rate of the wind or current. Using these variables, express the rate of the object moving with and against the wind or current. Use the equation $d = rt$ to write expressions for the distance traveled by the object. The results can be recorded in a table.

Rate of plane in calm air: p
Rate of wind: w

	Rate	·	Time	=	Distance
With the wind	$p + w$	·	3	=	$3(p + w)$
Against the wind	$p - w$	·	4	=	$4(p - w)$

Determine how the expressions for distance are related.

The distance traveled with the wind is 600 mi. $\quad 3(p + w) = 600$
The distance traveled against the wind is 600 mi. $\quad 4(p - w) = 600$

Solve the system of equations.

$$3(p + w) = 600 \quad \rightarrow \quad \frac{1}{3} \cdot 3(p + w) = \frac{1}{3} \cdot 600 \quad \rightarrow \quad p + w = 200$$

$$4(p - w) = 600 \quad \quad \frac{1}{4} \cdot 4(p - w) = \frac{1}{4} \cdot 600 \quad \quad p - w = 150$$

$$2p = 350$$
$$p = 175$$

$$p + w = 200$$
$$175 + w = 200$$
$$w = 25$$

The rate of the plane in calm air is 175 mph.
The rate of the wind is 25 mph.

Example 1

A 450-mile trip from one city to another takes 3 h when a plane is flying with the wind. The return trip, against the wind, takes 5 h. Find the rate of the plane in still air and the rate of the wind.

Strategy

- Rate of the plane in still air: p
  Rate of the wind: w

	Rate	*Time*	*Distance*
With wind	$p + w$	3	$3(p + w)$
Against wind	$p - w$	5	$5(p - w)$

- The distance traveled with the wind is 450 mi. The distance traveled against the wind is 450 mi.

Solution

$$3(p + w) = 450 \qquad \frac{1}{3} \cdot 3(p + w) = \frac{1}{3} \cdot 450$$

$$5(p - w) = 450 \qquad \frac{1}{5} \cdot 5(p - w) = \frac{1}{5} \cdot 450$$

$$p + w = 150$$
$$p - w = 90$$

$$2p = 240$$
$$p = 120$$

$$p + w = 150$$
$$120 + w = 150$$
$$w = 30$$

The rate of the plane in still air is 120 mph.
The rate of the wind is 30 mph.

You Try It 1

A canoeist paddling with the current can travel 15 mi in 3 h. Against the current, it takes 5 h to travel the same distance. Find the rate of the current and the rate of the canoeist in calm water.

Your Strategy

Your Solution

Solution on p. A24

OBJECTIVE B Application problems using two variables

The application problems in this section are varieties of problems solved earlier in the text. Each of the strategies for the problems in this section will result in a system of equations.

POINT OF INTEREST

The Babylonians had a method for solving a system of equations. Here is an adaptation of a problem from an ancient (about 1500 B.C.) Babylonian text. "There are two silver blocks. The sum of $\frac{1}{7}$ of the first block and $\frac{1}{11}$ of the second block is one sheqel (a weight). The first block diminished by $\frac{1}{7}$ of its weight equals the second diminished by $\frac{1}{11}$ of its weight. What are the weights of the two blocks?"

first block: $\frac{35}{8}$ sheqel

second block: $\frac{33}{8}$ sheqel

A jeweler purchased 5 oz of a gold alloy and 20 oz of a silver alloy for a total cost of \$540. The next day, at the same prices per ounce, the jeweler purchased 4 oz of the gold alloy and 25 oz of the silver alloy for a total cost of \$450. Find the cost per ounce of the gold and silver alloys.

Strategy for Solving an Application Problem in Two Variables

Choose one variable to represent one of the unknown quantities and a second variable to represent the other unknown quantity. Write numerical or variable expressions for all the remaining quantities. These results can be recorded in two tables, one for each of the conditions.

Cost per ounce of gold: g
Cost per ounce of silver: s

First Day

	Amount	·	*Unit Cost*	=	*Value*
Gold	5	·	g	=	$5g$
Silver	20	·	s	=	$20s$

Second Day

	Amount	·	*Unit Cost*	=	*Value*
Gold	4	·	g	=	$4g$
Silver	25	·	s	=	$25s$

Determine a system of equations. Each table will give one equation of the system.

The total value of the purchase on the first day was \$540. $5g + 20s = 540$
The total value of the purchase on the second day was \$450. $4g + 25s = 450$

Solve the system of equations.

$$\begin{aligned} 5g + 20s &= 540 \\ 4g + 25s &= 450 \end{aligned} \qquad \begin{aligned} 4(5g + 20s) &= 4 \cdot 540 \\ -5(4g + 25s) &= -5 \cdot 450 \end{aligned} \qquad \begin{aligned} 20g + 80s &= 2160 \\ -20g - 125s &= -2250 \\ \hline -45s &= -90 \\ s &= 2 \end{aligned}$$

$$\begin{aligned} 5g + 20s &= 540 \\ 5g + 20(2) &= 540 \qquad s = 2 \\ 5g + 40 &= 540 \\ 5g &= 500 \\ g &= 100 \end{aligned}$$

The cost per ounce of the gold alloy was \$100.
The cost per ounce of the silver alloy was \$2.

Example 2

A store owner purchased 20 incandescent light bulbs and 30 fluorescent bulbs for a total cost of $40. A second purchase, at the same prices, included 30 incandescent bulbs and 10 fluorescent bulbs for a total cost of $25. Find the cost of an incandescent bulb and of a fluorescent bulb.

Strategy

Cost of an incandescent bulb: b
Cost of a fluorescent bulb: f

First purchase

	Amount	*Unit Cost*	*Value*
Incandescent	20	b	$20b$
Fluorescent	30	f	$30f$

Second purchase

	Amount	*Unit Cost*	*Value*
Incandescent	30	b	$30b$
Fluorescent	10	f	$10f$

The total of the first purchase was $40.
The total of the second purchase was $25.

Solution

$$20b + 30f = 40 \qquad 3(20b + 30f) = 3 \cdot 40$$
$$30b + 10f = 25 \qquad -2(30b + 10f) = -2 \cdot 25$$

$$\begin{aligned} 60b + 90f &= 120 \\ -60b - 20f &= -50 \\ \hline 70f &= 70 \\ f &= 1 \end{aligned}$$

$$\begin{aligned} 20b + 30f &= 40 \\ 20b + 30(1) &= 40 \\ 20b &= 10 \\ b &= \frac{1}{2} \end{aligned}$$

The cost of an incandescent bulb was $.50.
The cost of a fluorescent bulb was $1.00.

You Try It 2

A citrus grower purchased 25 orange trees and 20 grapefruit trees for $290. The next week, at the same prices, the grower bought 20 orange trees and 30 grapefruit trees for $330. Find the cost of an orange tree and the cost of a grapefruit tree.

Your Strategy

Your Solution

Solution on p. A24

7.5 EXERCISES

Objective A

Solve.

1. A plane flying with the jet stream flew from Los Angeles to Chicago, a distance of 2250 mi, in 5 h. Flying against the jet stream, the plane could fly only 1750 mi in the same amount of time. Find the rate of the plane in calm air and the rate of the wind.

2. A rowing team rowing with the current traveled 40 km in 2 h. Rowing against the current, the team could travel only 16 km in 2 h. Find the rate of rowing in calm water and the rate of the current.

3. A motorboat traveling with the current went 35 mi in 3.5 h. Traveling against the current, the boat went 12 mi in 3 h. Find the rate of the boat in calm water and the rate of the current.

4. A small plane flew 270 mi in 3 h into a headwind. Flying with the wind, the plane traveled 260 mi in 2 h. Find the rate of the plane in calm air and the rate of the wind.

5. A plane flying with a tailwind flew 300 mi in 2 h. Against the wind, it took 3 h to travel the same distance. Find the rate of the plane in calm air and the rate of the wind.

6. A rowing team rowing with the current traveled 17 mi in 2 h. Against the current, the team rowed 7 mi in the same amount of time. Find the rate of the rowing team in calm water and the rate of the current.

7. A seaplane pilot flying with the wind flew from an ocean port to a lake, a distance of 240 mi, in 2 h. Flying against the wind, the trip from the lake to the ocean port took 2 h and 40 min. Find the rate of the plane in calm air and the rate of the wind.

8. Rowing with the current, a canoeist paddled 14 mi in 2 h. Against the current, the canoeist could paddle only 10 mi in the same amount of time. Find the rate of the canoeist in calm water and the rate of the current.

Objective B

Solve.

9. A computer software store received two shipments of software. The value of the first shipment, which contained 12 identical word processing programs and 10 identical spreadsheet programs, was $6190. The second shipment, at the same prices, contained 5 word processing programs and 8 spreadsheet programs. The value of the second shipment was $3825. Find the cost of a word processing program and of a spreadsheet program.

10. A baker purchased 12 lb of wheat flour and 15 lb of rye flour for a total cost of $18.30. A second purchase, at the same prices, included 15 lb of wheat flour and 10 lb of rye flour. The cost of the second purchase was $16.75. Find the cost per pound of the wheat flour and of the rye flour.

Solve.

11. An investor owned 300 shares of an oil company and 200 shares of a movie company. The quarterly dividend from the two stocks was $165. After the investor sold 100 shares of the oil company and bought an additional 100 shares of the movie company, the quarterly dividend became $185. Find the dividend per share for each stock.

12. The charge for 25 min of prime time and 35 min of nonprime time to a customer for using a computerized financial news network was $10.75. A second customer used 30 min of prime time and 45 min of nonprime time for a cost of $13.35. Find the cost per minute for using the financial news network during prime and during nonprime time.

13. A college football team scored 30 points in one game with only touchdowns and field goals. If the number of touchdowns had been field goals and the number of field goals had been touchdowns, the score would have been 33. Find the number of touchdowns and field goals that were actually scored. Use 6 points for a touchdown and 3 points for a field goal.

14. A professional basketball team scored 87 points in two-point baskets and three-point baskets. If the number of two-point baskets had been three-point baskets and the number of three-point baskets had been two-point baskets, the score would have been 93. Find the number of two-point and three-point baskets that were actually scored.

15. Two angles are supplementary. The larger angle is 15° more than twice the measure of the smaller angle. Find the measures of the two angles.

Critical Thinking

Solve.

16. The value of the nickels and dimes in a coin bank is $.25. If the number of nickels and the number of dimes were doubled, the value of the coins would be $.50. How many nickels and how many dimes are in the bank?

17. An investor has $5000 to invest in two accounts. The first account earns 8% annual simple interest and the second account earns 10% annual simple interest. How much money should be invested in each account so that the annual interest earned is $600?

18. Solve the following problem, which dates from a Chinese manuscript called the Jinzhang that is approximately 2100 years old. "The price of 1 acre of good land is 300 pieces of gold; the price of 7 acres of bad land is 500 pieces of gold. One has purchased altogether 100 acres. The price was 10,000 pieces of gold. How much good land and how much bad land was bought?" Adapted from *A History of Mathematics, An Introduction*, Victor J. Katz (New York: HarperCollins, 1993, p. 15).

19. A coin bank contains only nickels or dimes, but there are no more than 27 coins. The value of the coins is $2.10. How many different combinations of nickels and dimes could be in the coin bank?

Projects in Mathematics

A graphical representation of data can sometimes be misleading. Consider the graphs shown below. An investment firm's financial advisor claims that an investment with the firm will grow as shown in the graph on the left, whereas an investment with a competitor will grow as shown in the graph on the right. Apparently, you would accumulate more money by choosing the investment on the left.

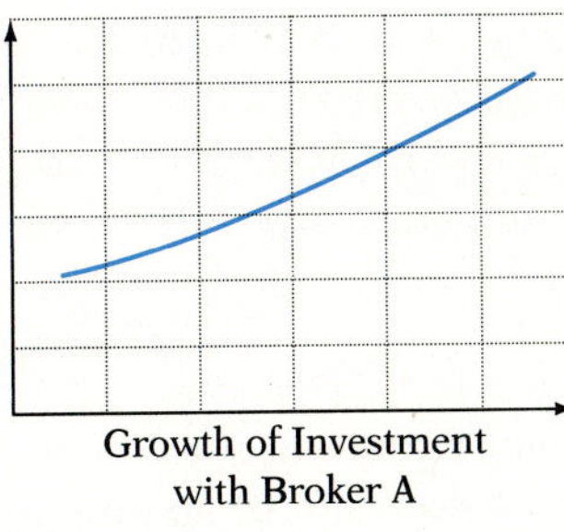

Growth of Investment with Broker A

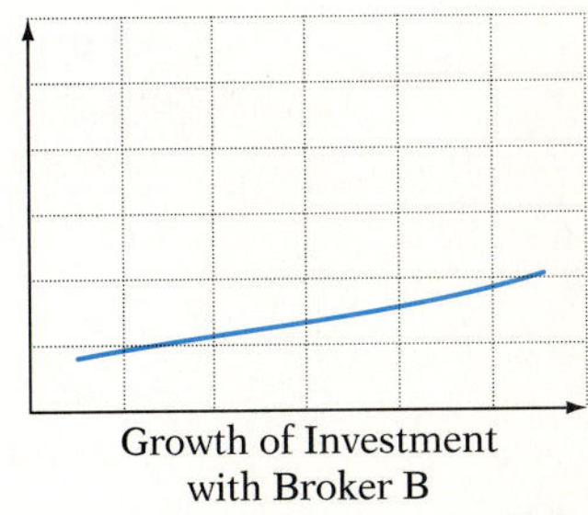

Growth of Investment with Broker B

However, these graphs have a serious flaw. There are no labels on the horizontal and vertical axes. Therefore, it is impossible to tell which investment increased more or over what time interval. When labels are not placed on the axes of a graph, the data that the graph represents are meaningless. It is one way advertisers use a visual image to distort the true meaning of data.

The graphs below are the same as those drawn above except that scales have been drawn along each axis. Now it is possible to tell how each investment has performed. Note that each one has exactly the same performance.

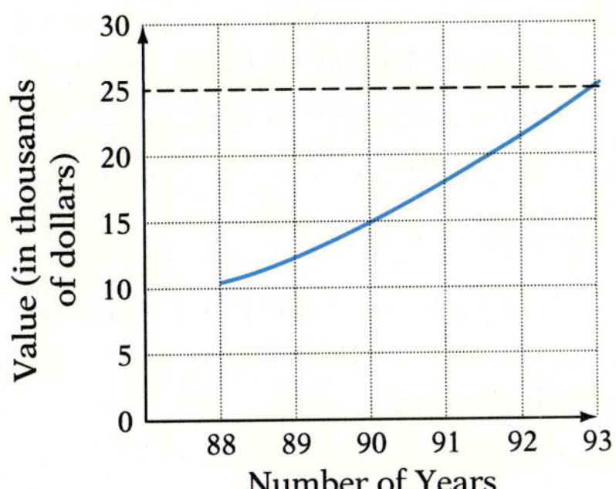

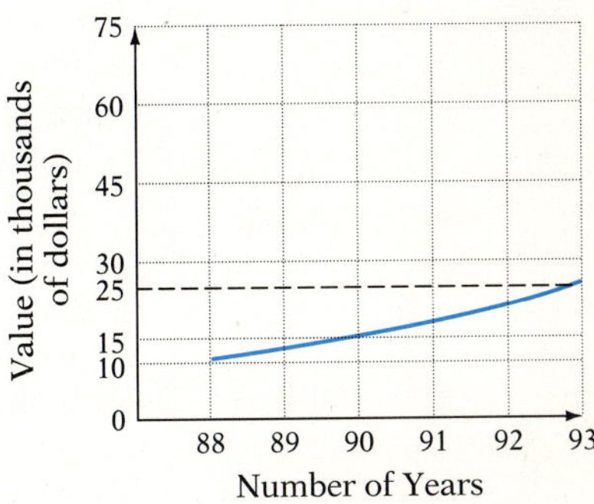

Drawing a circle graph as an oval is another way of distorting data. This is especially true if a three-dimensional representation is presented. In the circle graph at the left, region *A* appears larger than region *B*. However, that isn't true. Measure the angle of each sector to see for yourself.

As you read newspapers and magazines, find examples of graphs that may distort actual data. Discuss how these graphs should be drawn to be more accurate.

Chapter Summary

Key Words

A **rectangular coordinate system** is formed by two number lines, one horizontal and one vertical, that intersect at the zero point of each line. The point of intersection is called the **origin**. The two number lines are called the **coordinate axes,** or simply **axes**. A rectangular coordinate system divides the **plane** into four regions called **quadrants**.

An **ordered pair** (a, b) is used to locate a point in a plane. The first number in an ordered pair measures a horizontal distance and is called the **abscissa**. The second number measures a vertical distance and is called the **ordinate**. The **coordinates** of a point are the numbers in the ordered pair associated with the point.

An equation of the form $y = mx + b$, where m is the coefficient of x and b is a constant, is a **linear equation in two variables**. The equation $Ax + By = C$ is also a linear equation. A **solution of an equation in two variables** is an ordered pair of numbers that makes the equation a true statement.

The **graph of an equation in two variables** is a drawing of the ordered-pair solutions of the equation. For a linear equation in two variables, the graph is a straight line.

The **graph of** $y = b$ is a horizontal line passing through point $(0, b)$. The **graph of** $x = a$ is a vertical line passing through point $(a, 0)$.

The point at which a graph crosses the x-axis is called the **x-intercept**. The point at which a graph crosses the y-axis is called the **y-intercept**.

The **slope** of a line is the measure of the slant of a line. The symbol for slope is m. A line that slants upward to the right has a **positive slope**. A line that slants downward to the right has a **negative slope**. A horizontal line has **zero slope**. The slope of a vertical line is undefined.

A set of equations considered together is called a **system of equations**. A **solution of a system of equations** in two variables is an ordered pair that is a solution of each equation of the system.

Essential Rules

Slope of a straight line	slope $= m = \dfrac{y_2 - y_1}{x_2 - x_1}$
Slope–intercept form of a straight line	$y = mx + b$
Point–slope form of a straight line	$y - y_1 = m(x - x_1)$
Methods to solve linear systems	the graphing method, the substitution method, and the addition method

Chapter Review Exercises

1. Graph the ordered pairs $(-3, 1)$ and $(0, 2)$.

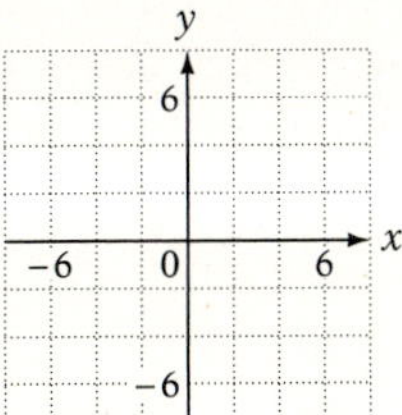

2. Graph $3x - 2y = 6$.

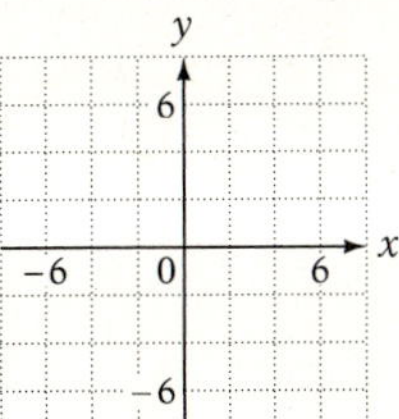

3. Find the x- and y-intercepts for $6x - 4y = 12$.

4. Find the slope of the line containing the points $(2, -3)$ and $(4, 1)$.

5. Find the equation of the line that contains the point $(0, -1)$ and has slope 3.

6. Find the equation of the line that contains the point $(2, 3)$ and has slope $\frac{1}{2}$.

7. Graph the line that has slope $-\frac{2}{3}$ and y-intercept $(0, 4)$.

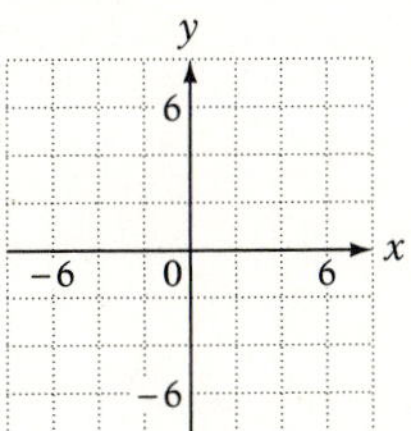

8. Solve by graphing: $3x + 2y = 2$
$x - 2y = 6$

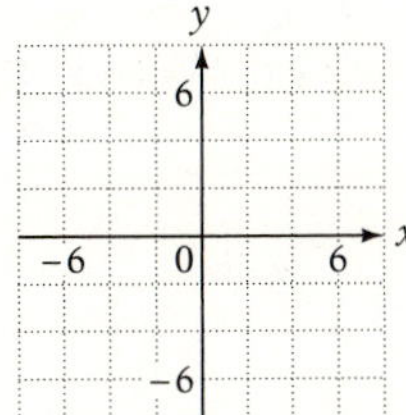

9. Is $(-2, 3)$ a solution of the system
$2x + 5y = 11$
$x + 3y = 7$?

10. Solve by substitution: $3x + 5y = 1$
$2x - y = 5$

11. Solve by the addition method:
$2x - 5y = 6$
$4x + 3y = -1$

12. Solve by the addition method:
$5x + 6y = -7$
$3x + 4y = -5$

13. The graph at the right shows the increase in the cost of tuition for a college for the years 1989 through 1994 (with 1989 as 0). Find the slope of the line. Write a sentence that states the meaning of the slope.

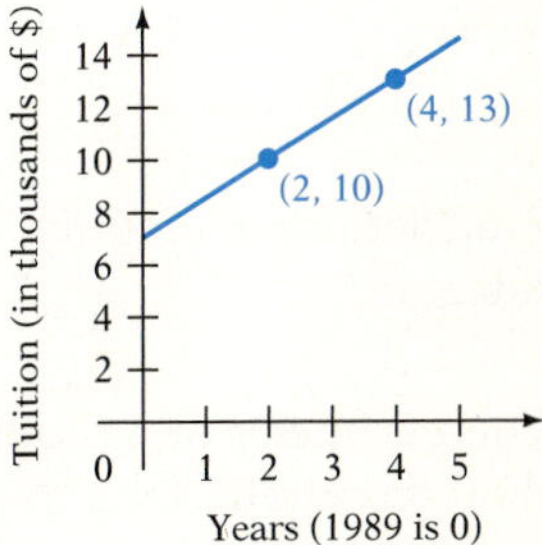

14. Flying with the wind, a small plane flew 280 mi in 2 h. Flying against the wind, the plane flew 160 mi in 2 h. Find the rate of the plane in calm air and the rate of the wind.

15. The total value of the nickels and dimes in a coin bank is \$3. If the nickels were dimes and the dimes were nickels, the total value of the coins would be \$3.75. Find the number of nickels and the number of dimes in the bank.

Cumulative Review Exercises

1. Simplify: $4(2 - 3x) - 5(x - 4)$

2. Write $6\frac{2}{3}\%$ as a fraction.

3. Solve: $2x - \frac{2}{3} = \frac{7}{3}$

4. Simplify: $12 - 18 \div 3 \cdot (-2)^2$

5. Evaluate $\frac{a - b}{a^2 - c}$ when $a = -2$, $b = 3$, and $c = -4$.

6. Solve: $-\frac{4}{5}x > 16$

7. Given that $\ell_1 \parallel \ell_2$, find the measures of $\angle a$ and $\angle b$.

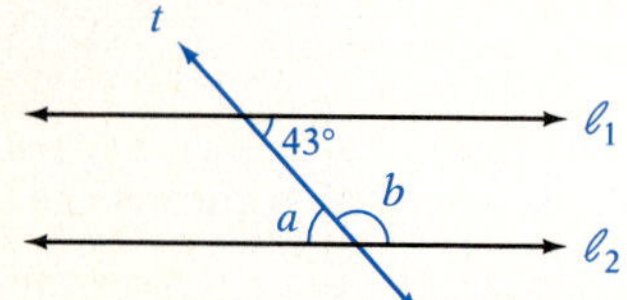

8. Given $\angle LON$ is a right angle, find the measure of $\angle x$.

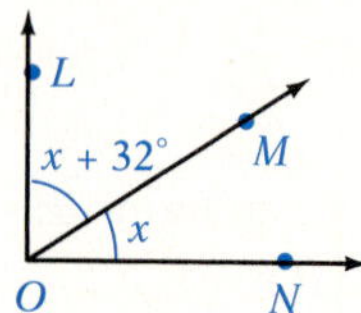

9. Multiply: $-2y^2(-3y^2 - 4y + 8)$

10. Simplify: $\left(-\frac{1}{2}\right)^3 \div \left(\frac{3}{8} - \frac{5}{6}\right) + 2$

11. Evaluate $-|-6.8|$.

12. Place the correct symbol, < or >, between the two numbers. $\frac{5}{9}$ 0.5

13. Subtract: $-2 - (-3) - 5 - (-11)$

14. Solve: $3x - 2 = 12 - 5x$

15. Solve: $-\frac{5}{7}x = -\frac{10}{21}$

16. Multiply: $-\frac{3}{4}(-20x^2)$

17. Graph: $y = \frac{1}{2}x - 1$

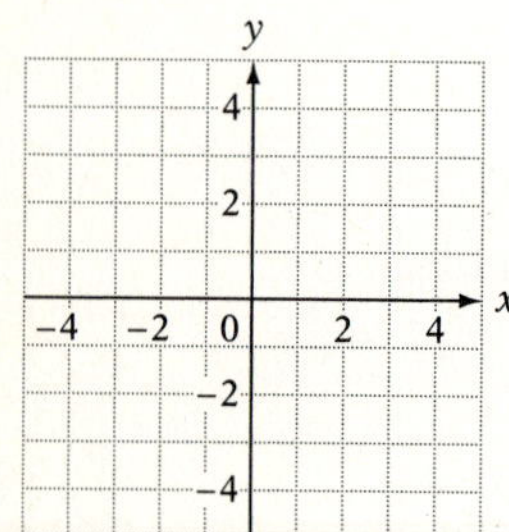

18. Solve by graphing: $3x + 2y = 6$
$3x - 2y = 6$

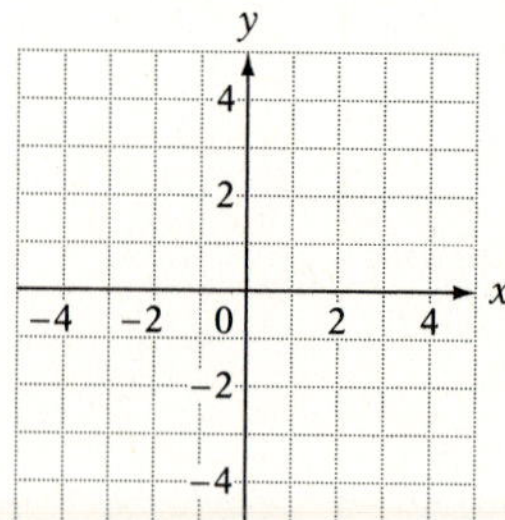

19. Translate and simplify "the product of four and the sum of two consecutive integers."

20. The perimeter of a rectangle is 64 cm. The length of the rectangle is 8 m more than the width. Find the length of the rectangle.

21. A family drove to a resort at an average speed of 42 mph and later returned over the same road at an average speed of 56 mph. Find the distance to the resort if the total driving time was 7 h.

CHAPTER

8

Polynomials

Focus on Problem Solving

In the Critical Thinking exercises in this text, you are sometimes asked to determine whether a statement is true or false. For instance, the statement "every real number has a reciprocal" is false because 0 is a real number and 0 does not have a reciprocal.

Finding an example, such as 0, which has no reciprocal, to show that the statement "every real number has a reciprocal" is not always true is called *finding a counterexample.* A counterexample is an example that shows that a statement is not always true.

Consider the statement "the product of two numbers is greater than either factor."

A counterexample to this statement is the numbers $\frac{2}{3}$ and $\frac{3}{4}$. The product of these numbers is $\frac{1}{2}$, and $\frac{1}{2}$ is *smaller* than $\frac{2}{3}$ or $\frac{3}{4}$. There are many other counterexamples to the given statement.

Here are some counterexamples to the statement that the square root of a number is smaller than the number.

$$\sqrt{\frac{1}{4}} = \frac{1}{2} \text{ but } \frac{1}{2} > \frac{1}{4} \qquad \sqrt{1} = 1 \text{ but } 1 = 1$$

For each of the next five statements, find at least one counterexample to show that the conjecture is false.

1. The product of two integers is always a positive number.
2. The sum of two prime numbers is never a prime number.
3. For all real numbers, $|x + y| = |x| + |y|$.
4. If x and y are nonzero real numbers and $x > y$, then $x^2 > y^2$.
5. The quotient of any two nonzero real numbers is less than either one of the numbers.

When a problem is posed, it may not be known whether the problem statement is true or false. For instance, Christian Goldbach (1690–1764) stated that every even integer greater than 2 can be written as the sum of two prime numbers. No one has been able to find a counterexample to this statement, but no one has been able to prove that it is always true.

For the statements below, answer true if the statement is always true or give a counterexample if there is an instance when the statement is false.

6. The reciprocal of a positive number is always smaller than the number.
7. If $x < 0$, then $|x| = -x$.
8. For any two real numbers x and y, $x + y > x - y$.
9. For any positive integer n, $n^2 + n + 17$ is a prime number.
10. The list of numbers, 1, 11, 111, 1111, 11111, . . . contains infinitely many composite numbers. *Hint:* A number is divisible by 3 if the sum of the digits of the number is divisible by 3.

SECTION 8.1 Addition and Subtraction of Polynomials

OBJECTIVE A Addition of polynomials

A **monomial** is a number, a variable, or a product of numbers and variables. For instance,

7	b	$\frac{2}{3}a$	$12xy^2$
A number	A variable	A product of a number and a variable	A product of a number and variables

The expression $3\sqrt{x}$ is not a monomial because $\sqrt{x}$ cannot be written as a product of variables. The expression $\frac{2x}{y^2}$ is not a monomial because it is a *quotient* of variables.

A **polynomial** is a variable expression in which the terms are monomials.

A polynomial of *one* term is a **monomial**. — $-7x^2$ is a monomial.
A polynomial of *two* terms is a **binomial**. — $4x + 2$ is a binomial.
A polynomial of *three* terms is a **trinomial**. — $7x^2 + 5x - 7$ is a trinomial.

CONSIDER THIS

An analogy may help to understand these terms. *Polynomial* is similar to the word *car*. Chevrolet and Ford are types of cars. Monomials and binomials are types of polynomials.

The terms of a polynomial in one variable are usually arranged so that the exponents of the variable decrease from left to right. This is called **descending order**.

$$5x^3 - 4x^2 + 6x - 1$$
$$7z^4 + 4z^3 + z - 6$$
$$2y^4 + y^3 - 2y^2 + 4y - 1$$

The **degree** of a polynomial in one variable is its largest exponent (on the variable). The degree of $4x^3 - 5x^2 + 7x - 8$ is 3; the degree of $2y^4 + y^2 - 1$ is 4.

To add polynomials, add the coefficients of like terms. Either a horizontal format or a vertical format can be used.

Add $(3x^3 - 7x + 2) + (7x^2 + 2x - 7)$. Use a horizontal format.

Use the Commutative and Associative Properties of Addition to rearrange and group like terms.

$$(3x^3 - 7x + 2) + (7x^2 + 2x - 7)$$
$$= 3x^3 + 7x^2 + (-7x + 2x) + (2 - 7)$$

Then combine terms.

$$= 3x^3 + 7x^2 - 5x - 5$$

Add $(-4x^2 + 6x - 9) + (12 - 8x + 2x^3)$. Use a vertical format.

Arrange the terms of each polynomial in descending order with like terms in the same column. Combine the terms in each column.

$$\begin{array}{r} -4x^2 + 6x - 9 \\ \underline{2x^3 \qquad\quad - 8x + 12} \\ 2x^3 - 4x^2 - 2x + 3 \end{array}$$

Example 1

Use a horizontal format to add $(8x^2 - 4x - 9) + (2x^2 + 9x - 9)$.

Solution

$$(8x^2 - 4x - 9) + (2x^2 + 9x - 9)$$
$$= (8x^2 + 2x^2) + (-4x + 9x) + (-9 - 9)$$
$$= 10x^2 + 5x - 18$$

You Try It 1

Use a horizontal format to add $(-4x^3 + 2x^2 - 8) + (4x^3 + 6x^2 - 7x + 5)$.

Your Solution

Solution on p. A24

Example 2

Use a vertical format to add
$(-5x^3 + 4x^2 - 7x + 9) + (2x^3 + 5x - 11)$.

Solution

$$\begin{array}{r} -5x^3 + 4x^2 - 7x + 9 \\ 2x^3 + 5x - 11 \\ \hline -3x^3 + 4x^2 - 2x - 2 \end{array}$$

You Try It 2

Use a vertical format to add
$(6x^3 + 2x + 8) + (-9x^3 + 2x^2 - 12x - 8)$.

Your Solution

Solution on p. A24

OBJECTIVE B Subtraction of polynomials

The **opposite** of the polynomial $(3x^2 - 7x + 8)$ is $-(3x^2 - 7x + 8)$.

To simplify the opposite of a polynomial, change the sign of each term inside the parentheses.

$$-(3x^2 - 7x + 8) = -3x^2 + 7x - 8$$

To subtract two polynomials, add the opposite of the second polynomial to the first.

Subtract $(4y^2 - 6y + 7) - (2y^3 - 5y - 4)$. Use a horizontal format.

Add the opposite of the second polynomial to the first. Combine like terms.

$$\begin{aligned} (4y^2 - 6y + 7) - (2y^3 - 5y - 4) &= (4y^2 - 6y + 7) + (-2y^3 + 5y + 4) \\ &= -2y^3 + 4y^2 + (-6y + 5y) + (7 + 4) \\ &= -2y^3 + 4y^2 - y + 11 \end{aligned}$$

Subtract $(9 + 4y + 3y^3) - (2y^2 + 4y - 21)$. Use a vertical format.

Arrange $(9 + 4y + 3y^3)$ and the opposite of $(2y^2 + 4y - 21)$ in descending order with like terms in the same column. Then add.

$$\begin{array}{r} 3y^3 + 4y + 9 \\ - 2y^2 - 4y + 21 \\ \hline 3y^3 - 2y^2 + 30 \end{array}$$

Example 3

Use a horizontal format to subtract
$(7c^2 - 9c - 12) - (9c^2 + 5c - 8)$.

Solution

$$\begin{aligned} &(7c^2 - 9c - 12) - (9c^2 + 5c - 8) \\ &\quad = (7c^2 - 9c - 12) + (-9c^2 - 5c + 8) \\ &\quad = -2c^2 - 14c - 4 \end{aligned}$$

You Try It 3

Use a horizontal format to subtract
$(-4w^3 + 8w - 8) - (3w^3 - 4w^2 - 2w - 1)$.

Your Solution

Example 4

Use a vertical format to subtract
$(3k^2 - 4k + 1) - (k^3 + 3k^2 - 6k - 8)$.

Solution

$$\begin{array}{r} 3k^2 - 4k + 1 \\ -k^3 - 3k^2 + 6k + 8 \\ \hline -k^3 + 2k + 9 \end{array}$$

Add the opposite of $(k^3 + 3k^2 - 6k - 8)$ to the first polynomial.

You Try It 4

Use a vertical format to subtract
$(13y^3 - 6y - 7) - (4y^2 - 6y - 9)$.

Your Solution

Solutions on p. A25

8.1 EXERCISES

Objective A

Write the polynomial in descending order.

1. $8x^2 - 2x + 3x^3 - 6$
2. $7y - 8 + 2y^2 + 4y^3$
3. $2a - 3a^2 + 5a^3 + 1$
4. $b - 3b^2 + b^4 - 2b^3$
5. $5 + r^5 - 6r^2 + r^3$
6. $8x^2 - 2x^6 + x^4 + 7$
7. $1 - y^4$
8. $4 - b^2$
9. $3 - 9a^2 + a$

Identify the degree of the polynomial.

10. $9x^5 + 3x^3 - 4x$
11. $3y^4 - 2y^2 + 10$
12. $6b^8 - 5b^6 + 7b^4$
13. $a^6 - 2a^4 + a^2 - 8$
14. $5r + 1$
15. $9x - 2$

Add. Use a vertical format.

16. $(x^2 + 7x) + (-3x^2 - 4x)$
17. $(3y^2 - 2y) + (5y^2 + 6y)$
18. $(y^2 + 4y) + (-4y - 8)$
19. $(3x^2 + 9x) + (6x - 24)$
20. $(2x^2 + 6x + 12) + (3x^2 + x + 8)$
21. $(x^2 + x + 5) + (3x^2 - 10x + 4)$
22. $(x^3 - 7x + 4) + (2x^2 + x - 10)$
23. $(3y^3 + y^2 + 1) + (-4y^3 - 6y - 3)$
24. $(2a^3 - 7a + 1) + (-3a^2 - 4a + 1)$
25. $(5r^3 - 6r^2 + 3r) + (r^2 - 2r - 3)$

Add. Use a horizontal format.

26. $(4x^2 + 2x) + (x^2 + 6x)$
27. $(-3y^2 + y) + (4y^2 + 6y)$
28. $(4x^2 - 5xy) + (3x^2 + 6xy - 4y^2)$
29. $(2x^2 - 4y^2) + (6x^2 - 2xy + 4y^2)$
30. $(2a^2 - 7a + 10) + (a^2 + 4a + 7)$
31. $(-6x^2 + 7x + 3) + (3x^2 + x + 3)$
32. $(5x^3 + 7x - 7) + (10x^2 - 8x + 3)$
33. $(3y^3 + 4y + 9) + (2y^2 + 4y - 21)$
34. $(2r^2 - 5r + 7) + (3r^3 - 6r)$
35. $(3y^3 + 4y + 14) + (-4y^2 + 21)$
36. $(3x^2 + 7x + 10) + (-2x^3 + 3x + 1)$
37. $(7x^3 + 4x - 1) + (2x^2 - 6x + 2)$

Objective B

Subtract. Use a vertical format.

38. $(x^2 - 6x) - (x^2 - 10x)$

39. $(y^2 + 4y) - (y^2 + 10y)$

40. $(2y^2 - 4y) - (-y^2 + 2)$

41. $(-3a^2 - 2a) - (4a^2 - 4)$

42. $(x^2 - 2x + 1) - (x^2 + 5x + 8)$

43. $(3x^2 + 2x - 2) - (5x^2 - 5x + 6)$

44. $(4x^3 + 5x + 2) - (-3x^2 + 2x + 1)$

45. $(5y^2 - y + 2) - (-2y^3 + 3y - 3)$

46. $(2y^3 + 6y - 2) - (y^3 + y^2 + 4)$

47. $(-2x^2 - x + 4) - (-x^3 + 3x - 2)$

Subtract. Use a horizontal format.

48. $(y^2 - 10xy) - (2y^2 + 3xy)$

49. $(x^2 - 3xy) - (-2x^2 + xy)$

50. $(3x^2 + x - 3) - (x^2 + 4x - 2)$

51. $(5y^2 - 2y + 1) - (-3y^2 - y - 2)$

52. $(-2x^3 + x - 1) - (-x^2 + x - 3)$

53. $(2x^2 + 5x - 3) - (3x^3 + 2x - 5)$

54. $(4a^3 - 2a + 1) - (a^3 - 2a + 3)$

55. $(b^2 - 8b + 7) - (4b^3 - 7b - 8)$

56. $(4y^3 - y - 1) - (2y^2 - 3y + 3)$

57. $(3x^2 - 2x - 3) - (2x^3 - 2x^2 + 4)$

Critical Thinking

State whether the polynomial is a monomial, a binomial, or a trinomial.

58. $8x^4 - 6x^2$

59. $4a^2b^2 + 9ab + 10$

60. $7x^3y^4$

State whether or not the expression is a monomial.

61. $3\sqrt{x}$

62. $\frac{4}{x}$

63. x^2y^2

State whether or not the expression is a polynomial.

64. $\frac{1}{5}x^3 + \frac{1}{2}x$

65. $\frac{1}{5x^2} + \frac{1}{2x}$

66. $x + \sqrt{5}$

67. [W] In your own words, explain the terms monomial, binomial, trinomial, and polynomial. Give an example of each.

SECTION 8.2 Multiplication of Polynomials

OBJECTIVE A Multiplication of monomials

Recall that in the exponential expression x^5, x is the base and 5 is the exponent. The exponent indicates the number of times the base is used as a factor.

The product of exponential expressions with the *same* base can be simplified by writing each expression in factored form and writing the result with an exponent.

$$x^3 \cdot x^2 = \overbrace{(x \cdot x \cdot x)}^{\text{3 factors}} \cdot \overbrace{(x \cdot x)}^{\text{2 factors}}$$
$$\text{5 factors}$$
$$= x \cdot x \cdot x \cdot x \cdot x$$
$$= x^5$$

Note that adding the exponents results in the same product.

$$x^3 \cdot x^2 = x^{3+2} = x^5$$

Rule for Multiplying Exponential Expressions

If m and n are integers, then $x^m \cdot x^n = x^{m+n}$.

Multiply: $a^2 \cdot a^6 \cdot a$

The bases are the same. Add the exponents.

$$a^2 \cdot a^6 \cdot a = a^{2+6+1} = a^9$$

Multiply: $(2xy)(3x^2y)$

Use the Commutative and Associative Properties of Multiplication to rearrange and group factors.

$$(2xy)(3x^2y) = (2 \cdot 3)(x \cdot x^2)(y \cdot y)$$

Multiply variables with like bases by adding the exponents.

$$= 6x^{1+2}y^{1+1}$$
$$= 6x^3y^2$$

CONSIDER THIS
In the Rule for Multiplying Exponential Expressions, the bases must be the same. The expression x^3y^2 cannot be simplified.

Example 1
Multiply: $(-4y)(5y^3)$

Solution
$(-4y)(5y^3) = (-4 \cdot 5)(y \cdot y^3) = -20y^4$

You Try It 1
Multiply: $(3x^2)(6x^3)$

Your Solution

Example 2
Multiply: $(2x^2y)(-5xy^4)$

Solution
$(2x^2y)(-5xy^4) = [2(-5)](x^2 \cdot x)(y \cdot y^4) = -10x^3y^5$

You Try It 2
Multiply: $(-3xy^2)(-4x^2y^3)$

Your Solution

Solutions on p. A25

OBJECTIVE B Powers of monomials

A power of a monomial can be simplified by rewriting the expression in factored form and then using the Rule for Multiplying Exponential Expressions.

$$(x^2)^3 = x^2 \cdot x^2 \cdot x^2 = x^{2+2+2} = x^6$$

$$(x^4y^3)^2 = (x^4y^3)(x^4y^3) = x^4 \cdot y^3 \cdot x^4 \cdot y^3 = (x^4 \cdot x^4)(y^3 \cdot y^3) = x^{4+4}y^{3+3} = x^8y^6$$

POINT OF INTEREST

One of the first symbolic representations of powers was given by Diophantus (c. A.D. 250) in his book *Arithmetica.* He used Δ^Y for x^2 and κ^Y for x^3. The symbol Δ^Y was the first two letters of the Greek word *dunamis* meaning power; κ^Y was from the Greek word *kubos* meaning cube. He also combined these symbols to denote higher powers. For instance, $\Delta\kappa^Y$ was the symbol for x^5.

Note that multiplying each exponent inside the parentheses by the exponent outside the parentheses gives the same result.

$$(x^2)^3 = x^{2\cdot 3} = x^6 \qquad (x^4y^3)^2 = x^{4\cdot 2}y^{3\cdot 2} = x^8y^6$$

Rule for Simplifying Powers of Exponential Expressions

If m and n are integers, then $(x^m)^n = x^{m\cdot n}$.

Rule for Simplifying Powers of Products

If m, n, and p are integers, then $(x^m \cdot y^n)^p = x^{m\cdot p}y^{n\cdot p}$.

Simplify: $(x^5)^2$

Multiply the exponents. $\qquad (x^5)^2 = x^{5\cdot 2} = x^{10}$

Simplify: $(3a^2b)^3$

Multiply each exponent inside the parentheses by the exponent outside the parentheses.

$$(3a^2b)^3 = 3^{1\cdot 3}a^{2\cdot 3}b^{1\cdot 3} = 3^3a^6b^3 = 27a^6b^3$$

Example 3

Simplify: $(2xy^3)^4$

Solution

$(2xy^3)^4 = 2^4x^4y^{12} = 16x^4y^{12}$

You Try It 3

Simplify: $(3x)(2x^2y)^3$

Your Solution

Example 4

Simplify: $(-2x)(-3xy^2)^3$

Solution

$(-2x)(-3xy^2)^3 = (-2x)(-3)^3x^3y^6$
$= (-2x)(-27)x^3y^6 = [-2(-27)](x \cdot x^3)y^6$
$= 54x^4y^6$

You Try It 4

Simplify: $(3x^2)^2(-2xy^2)^3$

Your Solution

Solutions on p. A25

OBJECTIVE C Multiplication of polynomials

To multiply a polynomial by a monomial, use the Distributive Property and the Rule for Multiplying Exponential Expressions.

Multiply: $-2x(x^2 - 4x - 3)$

$-2x(x^2 - 4x - 3)$

Use the Distributive Property. $= -2x(x^2) - (-2x)(4x) - (-2x)(3)$

Use the Rule for Multiplying Exponential Expressions. $= -2x^3 + 8x^2 + 6x$

Multiplication of two polynomials requires the repeated application of the Distributive Property.

$$\begin{aligned} &(y - 2)(y^2 + 3y + 1) \\ &= (y - 2)(y^2) + (y - 2)(3y) + (y - 2)(1) \\ &= y^3 - 2y^2 + 3y^2 - 6y + y - 2 \\ &= y^3 + y^2 - 5y - 2 \end{aligned}$$

A more convenient method of multiplying two polynomials is to use a vertical format, similar to that used for multiplying whole numbers.

CONSIDER THIS

Before doing the example at the right, you might want to review the procedure for multiplying whole numbers, such as 473 × 28, and relate this to the multiplication of polynomials.

Multiply each term in the trinomial by -2.
Multiply each term in the trinomial by y.
Write like terms in the same column.
Add the terms in each column.

$$\begin{array}{rrrr} & y^2 & + 3y & + 1 \\ \times & & y & - 2 \\ \hline & -2y^2 & - 6y & - 2 \\ y^3 + & 3y^2 & + y & \\ \hline y^3 + & y^2 & - 5y & - 2 \end{array}$$

Multiply: $(a^2 - 3)(a + 5)$

Multiply each term of $a^2 - 3$ by 5.
Multiply each term of $a^2 - 3$ by a.
Arrange the terms in descending order.
Add the terms in each column.

$$\begin{array}{rrrr} & a^2 & & - 3 \\ & \times\ a & & + 5 \\ \hline & 5a^2 & & - 15 \\ a^3 & & - 3a & \\ \hline a^3 & + 5a^2 & - 3a & - 15 \end{array}$$

Example 5

Multiply: $(5x + 4)(-2x)$

Solution

$(5x + 4)(-2x) = -10x^2 - 8x$

You Try It 5

Multiply: $(-2y + 3)(-4y)$

Your Solution

Example 6

Multiply: $x^3(2x^2 - 3x + 2)$

Solution

$x^3(2x^2 - 3x + 2) = 2x^5 - 3x^4 + 2x^3$

You Try It 6

Multiply: $-a^2(3a^2 + 2a - 7)$

Your Solution

Solutions on p. A25

Example 7

Multiply: $(2b^3 - b + 1)(2b + 3)$

Solution

$$\begin{array}{r} 2b^3 - b + 1 \\ \times \quad 2b + 3 \\ \hline 6b^3 \quad - 3b + 3 \\ 4b^4 + \quad - 2b^2 + 2b \quad \\ \hline 4b^4 + 6b^3 - 2b^2 - b + 3 \end{array}$$

You Try It 7

Multiply: $(2y^3 + 2y^2 - 3)(3y - 1)$

Your Solution

Example 8

Multiply: $(x^2 - 1)(x + 3)$

Solution

$$\begin{array}{r} x^2 - 1 \\ \times \quad x + 3 \\ \hline 3x^2 \quad - 3 \\ x^3 \quad - x \quad \\ \hline x^3 + 3x^2 - x - 3 \end{array}$$

You Try It 8

Multiply: $(a^3 - 2)(a + 7)$

Your Solution

Solutions on p. A25

OBJECTIVE D Multiplication of two binomials

The product of two binomials can be found using a method called **FOIL**, which is based on the Distributive Property. The letters of FOIL stand for **F**irst, **O**uter, **I**nner, and **L**ast.

Multiply: $(2x + 3)(x + 5)$

Multiply the First terms.	$(2x + 3)(x + 5)$	$2x \cdot x = 2x^2$
Multiply the Outer terms.	$(2x + 3)(x + 5)$	$2x \cdot 5 = 10x$
Multiply the Inner terms.	$(2x + 3)(x + 5)$	$3 \cdot x = 3x$
Multiply the Last terms.	$(2x + 3)(x + 5)$	$3 \cdot 5 = 15$
Add the products.	$(2x + 3)(x + 5)$	F O I L $= 2x^2 + 10x + 3x + 15$
Combine like terms.		$= 2x^2 + 13x + 15$

CONSIDER THIS

FOIL is not really a different way of multiplying. It is based on the Distributive Property.

$(2x + 3)(x + 5)$
$= 2x(x + 5) + 3(x + 5)$
F O I L
$= 2x^2 + 10x + 3x + 15$
$= 2x^2 + 13x + 15$

Multiply: $(4x - 3)(3x - 2)$

$$\begin{aligned}(4x - 3)(3x - 2) &= 4x(3x) + 4x(-2) + (-3)(3x) + (-3)(-2)\\ &= 12x^2 - 8x - 9x + 6\\ &= 12x^2 - 17x + 6\end{aligned}$$

Multiply: $(3x - 2y)(x + 4y)$

$$\begin{aligned}(3x - 2y)(x + 4y) &= 3x(x) + 3x(4y) + (-2y)(x) + (-2y)(4y)\\ &= 3x^2 + 12xy - 2xy - 8y^2\\ &= 3x^2 + 10xy - 8y^2\end{aligned}$$

Example 9

Multiply: $(y + 4)(y - 7)$

Solution

$$\begin{aligned}(y + 4)(y - 7) &= y^2 - 7y + 4y - 28\\ &= y^2 - 3y - 28\end{aligned}$$

You Try It 9

Multiply: $(b - 5)(b + 8)$

Your Solution

Example 10

Multiply: $(2a - 1)(3a - 2)$

Solution

$$\begin{aligned}(2a - 1)(3a - 2) &= 6a^2 - 4a - 3a + 2\\ &= 6a^2 - 7a + 2\end{aligned}$$

You Try It 10

Multiply: $(4y - 5)(2y - 3)$

Your Solution

Example 11

Multiply: $(3x - 2)(4x + 3)$

Solution

$$\begin{aligned}(3x - 2)(4x + 3) &= 12x^2 + 9x - 8x - 6\\ &= 12x^2 + x - 6\end{aligned}$$

You Try It 11

Multiply: $(3b + 2)(3b - 5)$

Your Solution

Example 12

Multiply: $(2x - 3y)(3x + 4y)$

Solution

$$\begin{aligned}&(2x - 3y)(3x + 4y)\\ &\quad = 6x^2 + 8xy - 9xy - 12y^2\\ &\quad = 6x^2 - xy - 12y^2\end{aligned}$$

You Try It 12

Multiply: $(4x - y)(2x + 3y)$

Your Solution

Solutions on p. A25

OBJECTIVE E Binomials that have special products

Using FOIL, a pattern for the product of the sum and difference of two terms and for the square of a binomial can be found.

The Sum and Difference of Two Terms

$$(a + b)(a - b) = a^2 - ab + ab - b^2$$
$$= a^2 - b^2$$

Square of first term (a^2)
Square of second term (b^2)

The Square of a Binomial

$$(a + b)^2 = (a + b)(a + b) = a^2 + ab + ab + b^2$$
$$= a^2 + 2ab + b^2$$

Square of first term (a^2)
Twice the product of the two terms ($2ab$)
Square of last term (b^2)

Multiply: $(2x + 3)(2x - 3)$

$(2x + 3)(2x - 3)$ is the sum and difference of two terms.

$$(2x + 3)(2x - 3) = (2x)^2 - 3^2$$
$$= 4x^2 - 9$$

Simplify: $(3x - 2)^2$

$(3x - 2)^2$ is the square of a binomial.

$$(3x - 2)^2 = (3x)^2 + 2(3x)(-2) + (-2)^2$$
$$= 9x^2 - 12x + 4$$

Example 13
Multiply: $(4z - 2w)(4z + 2w)$

Solution

$(4z - 2w)(4z + 2w) = 16z^2 - 4w^2$

You Try It 13
Multiply: $(2a + 5c)(2a - 5c)$

Your Solution

Example 14
Simplify: $(2r - 3s)^2$

Solution

$(2r - 3s)^2 = 4r^2 - 12rs + 9s^2$

You Try It 14
Simplify: $(3x + 2y)^2$

Your Solution

Solutions on p. A25

8.2 EXERCISES

Objective A

Multiply.

1. $x(2x)$
2. $(-3y)(y)$
3. $(3x)(4x)$
4. $(7y^3)(7y^2)$
5. $(-2a^3)(-3a^4)$
6. $(5a^6)(-2a^5)$
7. $(x^2y)(xy^4)$
8. $(x^2y^4)(xy^7)$
9. $(-2x^4)(5x^5y)$
10. $(-3a^3)(2a^2b^4)$
11. $(x^2y^4)(x^5y^4)$
12. $(a^2b^4)(ab^3)$
13. $(2xy)(-3x^2y^4)$
14. $(-3a^2b)(-2ab^3)$
15. $(x^2yz)(x^2y^4)$
16. $(-ab^2c)(a^2b^5)$
17. $(a^2b^3)(ab^2c^4)$
18. $(x^2y^3z)(x^3y^4)$
19. $(-a^2b^2)(a^3b^6)$
20. $(xy^4)(-xy^3)$
21. $(-6a^3)(a^2b)$
22. $(2a^2b^3)(-4ab^2)$
23. $(-5y^4z)(-8y^6z^5)$
24. $(3x^2y)(-4xy^2)$
25. $(x^2y)(yz)(xyz)$
26. $(xy^2z)(x^2y)(z^2y^2)$
27. $(-2x^2y^3)(3xy)(-5x^3y^4)$
28. $(4a^2b)(-3a^3b^4)(a^5b^2)$
29. $(3ab^2)(-2abc)(4ac^2)$
30. $(3a^2b)(-6bc)(2ac^2)$

Objective B

Simplify.

31. $(2^2)^3$
32. $(3^2)^2$
33. $(-2)^2$
34. $(-3)^3$
35. $(-2^2)^3$
36. $(-2^3)^3$
37. $(x^3)^3$
38. $(y^4)^2$
39. $(x^7)^2$
40. $(y^5)^3$
41. $(-x^2)^2$
42. $(-x^2)^3$
43. $(2x)^2$
44. $(3y)^3$
45. $(-2x^2)^3$
46. $(-3y^3)^2$
47. $(x^2y^3)^2$
48. $(x^3y^4)^5$
49. $(3x^2y)^2$
50. $(-2ab^3)^4$
51. $(x^2y)(x^2y)^3$
52. $(a^3b)(ab)^3$
53. $(ab^2)^2(ab)^2$
54. $(x^2y)^2(x^3y)^3$
55. $(-2x)(-2x^3y)^3$
56. $(-3y)(-4x^2y^3)^3$
57. $(ab^2)(-2a^2b)^3$
58. $(a^2b^2)(-3ab^4)^2$
59. $(-2a^3)(3a^2b)^3$
60. $(-3b^2)(2ab^2)^3$
61. $(-3ab)^2(-2ab)^3$
62. $(-3ab^2)^3(-3ab)^3$

Objective C

Multiply.

63. $x(x - 2)$

64. $y(3 - y)$

65. $-x(x + 7)$

66. $-y(7 - y)$

67. $3a^2(a - 2)$

68. $4b^2(b + 8)$

69. $-5x^2(x^2 - x)$

70. $-6y^2(y + 2y^2)$

71. $2x(6x^2 - 3x)$

72. $3y(4y - y^2)$

73. $(2x - 4)3x$

74. $(3y - 2)y$

75. $(3x + 4)x$

76. $(2x + 1)2x$

77. $-xy(x^2 - y^2)$

78. $-x^2y(2xy - y^2)$

79. $x(2x^3 - 3x + 2)$

80. $y(-3y^2 - 2y + 6)$

81. $-a(-2a^2 - 3a - 2)$

82. $-b(5b^2 + 7b - 35)$

83. $x^2(3x^4 - 3x^2 - 2)$

84. $y^3(-4y^3 - 6y + 7)$

85. $2y^2(-3y^2 - 6y + 7)$

86. $4x^2(3x^2 - 2x + 6)$

87. $(a^2 + 3a - 4)(-2a)$

88. $(b^3 - 2b + 2)(-5b)$

89. $-3y^2(-2y^2 + y - 2)$

90. $-5x^2(3x^2 - 3x - 7)$

91. $xy(x^2 - 3xy + y^2)$

92. $ab(2a^2 - 4ab - 6b^2)$

93. $(x^2 + 3x + 2)(x + 1)$

94. $(x^2 - 2x + 7)(x - 2)$

95. $(a^2 - 3a + 4)(a - 3)$

96. $(x^2 - 3x + 5)(2x - 3)$

97. $(-2b^2 - 3b + 4)(b - 5)$

98. $(-a^2 + 3a - 2)(2a - 1)$

99. $(-2x^2 + 7x - 2)(3x - 5)$

100. $(-a^2 - 2a + 3)(2a - 1)$

101. $(x^2 + 5)(x - 3)$

102. $(y^2 - 2y)(2y + 5)$

103. $(x^3 - 3x + 2)(x - 4)$

104. $(y^3 + 4y^2 - 8)(2y - 1)$

105. $(5y^2 + 8y - 2)(3y - 8)$

106. $(3y^2 + 3y - 5)(4y - 3)$

107. $(5a^3 - 5a + 2)(a - 4)$

108. $(3b^3 - 5b^2 + 7)(6b - 1)$

Objective D

Multiply.

109. $(x + 1)(x + 3)$

110. $(y + 2)(y + 5)$

111. $(a - 3)(a + 4)$

112. $(b - 6)(b + 3)$

113. $(y + 3)(y - 8)$

114. $(x + 10)(x - 5)$

115. $(y - 7)(y - 3)$

116. $(a - 8)(a - 9)$

117. $(2x + 1)(x + 7)$

118. $(y + 2)(5y + 1)$

119. $(3x - 1)(x + 4)$

120. $(7x - 2)(x + 4)$

121. $(4x - 3)(x - 7)$

122. $(2x - 3)(4x - 7)$

123. $(3y - 8)(y + 2)$

124. $(5y - 9)(y + 5)$

125. $(2a - b)(3a + 2b)$

126. $(5a - 3b)(2a + 4b)$

127. $(2x + y)(x - 2y)$

128. $(3x - 7y)(3x + 5y)$

129. $(2x + 3y)(5x + 7y)$

130. $(5x + 3y)(7x + 2y)$

131. $(3a - 2b)(2a - 7b)$

132. $(5a - b)(7a - b)$

133. $(a - 9b)(2a + 7b)$

134. $(2a + 5b)(7a - 2b)$

135. $(10a - 3b)(10a - 7b)$

136. $(12a - 5b)(3a - 4b)$

137. $(5x + 12y)(3x + 4y)$

138. $(11x + 2y)(3x + 7y)$

139. $(2x - 15y)(7x + 4y)$

Objective E

Simplify.

140. $(y - 5)(y + 5)$

141. $(y + 6)(y - 6)$

142. $(2x + 3)(2x - 3)$

143. $(4x - 7)(4x + 7)$

144. $(x + 1)^2$

145. $(y - 3)^2$

146. $(3a - 5)^2$

147. $(6x - 5)^2$

148. $(3x - 7)(3x + 7)$

149. $(9x - 2)(9x + 2)$

150. $(2a + b)^2$

151. $(x + 3y)^2$

152. $(x - 2y)^2$

153. $(2x - 3y)^2$

154. $(4 - 3y)(4 + 3y)$

155. $(2 + 5x)(2 - 5x)$

156. $(5x + 2y)^2$

157. $(2a - 9b)^2$

158. $(3 - 5y)^2$

159. $(2 + 7x)^2$

Critical Thinking

Simplify.

160. $(6x)(2x^2) + (4x^2)(5x)$

161. $(2a^7)(7a^2) - (6a^3)(5a^6)$

162. $(3a^2b^2)(2ab) - (9ab^2)(a^2b)$

163. $(3x^2y^2)^2 - (2xy)^4$

164. $a^n \cdot a^n$

165. $(a^n)^2$

166. $(a^2)^n$

167. $a^2 \cdot a^n$

168. $(a + b)^2 - (a - b)^2$

169. $(x + 3y)^2 + (x + 3y)(x - 3y)$

170. $(3a^2 - 4a + 2)^2$

171. $(x + 4)^3$

172. $3x^2(2x^3 + 4x - 1) - 6x^3(x^2 - 2)$

173. $(3b + 2)(b - 6) + (4 + 2b)(3 - b)$

174. $x^n(x^n + 1)$

175. $(x^n + 1)(x^n - 1)$

176. $(x^n + 1)(x^n + 1)$

177. $(x^n - 1)^2$

For Exercises 178–181, answer true or false. If the answer is false, correct the right side of the equation.

178. $(-a)^5 = -a^5$

179. $(-b)^8 = b^8$

180. $(x^2)^5 = x^{2+5} = x^7$

181. $x^3 + x^3 = 2x^{3+3} = 2x^6$

182. Evaluate $(2^3)^2$ and $2^{(3^2)}$. Are the results the same? If not, which expression has the larger value?

183. What is the Order of Operations for the expression x^{m^n}?

184. The length of a rectangle is $4ab$. The width is $2ab$. Find the perimeter of the rectangle in terms of ab.

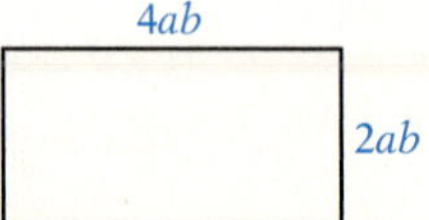

185. Add $x^2 + 2x - 3$ to the product of $2x - 5$ and $3x + 1$.

186. Subtract $4x^2 - x - 5$ from the product of $x^2 + x + 3$ and $x - 4$.

187. If a polynomial of degree 3 is multiplied by a polynomial of degree 2, what is the degree of the resulting polynomial?

188. [W] Is it possible to multiply a polynomial of degree 2 by a polynomial of degree 2 and have the product be a polynomial of degree 3? If so, give an example. If not, explain why not.

SECTION 8.3 Division of Polynomials

OBJECTIVE  Division of monomials

The quotient of two exponential expressions with the *same* base can be simplified by writing each expression in factored form, dividing by the common factors, and then writing the result with an exponent.

$$\frac{x^5}{x^2} = \frac{\overset{1}{\cancel{x}} \cdot \overset{1}{\cancel{x}} \cdot x \cdot x \cdot x}{\underset{1}{\cancel{x}} \cdot \underset{1}{\cancel{x}}} = x^3$$

$$\frac{x^4}{x^6} = \frac{\overset{1}{\cancel{x}} \cdot \overset{1}{\cancel{x}} \cdot \overset{1}{\cancel{x}} \cdot \overset{1}{\cancel{x}}}{\underset{1}{\cancel{x}} \cdot \underset{1}{\cancel{x}} \cdot \underset{1}{\cancel{x}} \cdot \underset{1}{\cancel{x}} \cdot x \cdot x} = \frac{1}{x^2}$$

Note that subtracting the smaller exponent from the larger exponent results in the same quotient.

$$\frac{x^5}{x^2} = \frac{x^{5-2}}{1} = x^3$$

$$\frac{x^4}{x^6} = \frac{1}{x^{6-4}} = \frac{1}{x^2}$$

Rule for Dividing Exponential Expressions

If m and n are integers and $x \neq 0$, then $\dfrac{x^m}{x^n} = x^{m-n}$ if $m > n$

and $\dfrac{x^m}{x^n} = \dfrac{1}{x^{n-m}}$ if $m < n$.

Simplify: $\dfrac{x^7}{x^2}$

$7 > 2$
The bases are the same.
Subtract the exponent in the denominator from the exponent in the numerator.

$$\frac{x^7}{x^2} = x^{7-2} = x^5$$

Simplify: $\dfrac{a^3}{a^9}$

$3 < 9$
The bases are the same.
Subtract the exponent in the numerator from the exponent in the denominator.

$$\frac{a^3}{a^9} = \frac{1}{a^{9-3}} = \frac{1}{a^6}$$

Simplify: $\dfrac{x^4y^2}{-xy^3}$

Negative signs are placed in front of a fraction.

$$\frac{x^4y^2}{-xy^3} = -\frac{x^4y^2}{xy^3}$$

Divide variables with like bases by subtracting the exponents.

$$= -\frac{x^{4-1}}{y^{3-2}} = -\frac{x^3}{y}$$

Simplify: $\frac{10x^3y^5}{4x^6y^2}$

Factor the coefficients. Divide by the common factors.

$$\frac{10x^3y^5}{4x^6y^2} = \frac{\overset{1}{\cancel{2}} \cdot 5x^3y^5}{\underset{1}{\cancel{2}} \cdot 2x^6y^2}$$

Divide variables with like bases by subtracting the exponents.

$$= \frac{5y^{5-2}}{2x^{6-3}}$$

$$= \frac{5y^3}{2x^3}$$

Example 1

Simplify: $\frac{-16x^4}{4x^9}$

Solution

$$\frac{-16x^4}{4x^9} = -\frac{\overset{1}{\cancel{2}} \cdot \overset{1}{\cancel{2}} \cdot 2 \cdot 2x^4}{\underset{1}{\cancel{2}} \cdot \underset{1}{\cancel{2}}x^9} = -\frac{4}{x^5}$$

You Try It 1

Simplify: $\frac{42y^{12}}{-14y^{17}}$

Your Solution

Example 2

Simplify: $\frac{-28x^4y^3}{6xy}$

Solution

$$\frac{-28x^4y^3}{6xy} = -\frac{\overset{1}{\cancel{2}} \cdot 2 \cdot 7x^4y^3}{\underset{1}{\cancel{2}} \cdot 3xy} = -\frac{14x^3y^2}{3}$$

You Try It 2

Simplify: $\frac{12r^4s^2}{-8r^3s}$

Your Solution

Example 3

Simplify: $\frac{(-3ab)^2}{9a^3b}$

Solution

$(-3)^2 = 9$

$$\frac{(-3ab)^2}{9a^3b} = \frac{(-3)^2a^2b^2}{9a^3b} = \frac{\overset{1}{\cancel{3}} \cdot \overset{1}{\cancel{3}}a^2b^2}{\underset{1}{\cancel{3}} \cdot \underset{1}{\cancel{3}}a^3b} = \frac{b}{a}$$

You Try It 3

Simplify: $\frac{(2x^2y)^3}{-4xy^5}$

Your Solution

Solutions on p. A25

OBJECTIVE B Division of a polynomial by a monomial

Note that $\frac{8+4}{2}$ can be simplified by first adding the terms in the numerator and then dividing the result. It can also be simplified by first dividing each term in the numerator by the denominator and then adding the results.

$$\frac{8+4}{2} = \frac{12}{2} = 6$$

$$\frac{8+4}{2} = \frac{8}{2} + \frac{4}{2} = 4 + 2 = 6$$

To divide a polynomial by a monomial, divide each term in the numerator by the denominator, and write the sum of the quotients.

$$\frac{a+b}{c} = \frac{a}{c} + \frac{b}{c}$$

Divide: $\frac{6x^2 + 4x}{2x}$

Divide each term of the polynomial by the monomial.
Simplify each expression.

$$\frac{6x^2 + 4x}{2x} = \frac{6x^2}{2x} + \frac{4x}{2x}$$
$$= 3x + 2$$

Check: $2x(3x + 2) = 6x^2 + 4x$

Example 4

Divide: $\frac{6x^3 - 3x^2 + 9x}{3x}$

Solution

$$\frac{6x^3 - 3x^2 + 9x}{3x} = \frac{6x^3}{3x} - \frac{3x^2}{3x} + \frac{9x}{3x}$$
$$= 2x^2 - x + 3$$

You Try It 4

Divide: $\frac{4x^3y + 8x^2y^2 - 4xy^3}{2xy}$

Your Solution

Example 5

Divide: $\frac{12x^2y - 6xy + 4x^2}{2xy}$

Solution

$$\frac{12x^2y - 6xy + 4x^2}{2xy} = \frac{12x^2y}{2xy} - \frac{6xy}{2xy} + \frac{4x^2}{2xy}$$
$$= 6x - 3 + \frac{2x}{y}$$

You Try It 5

Divide: $\frac{24x^2y^2 - 18xy + 6y}{6xy}$

Your Solution

Solutions on pp. A25–A26

OBJECTIVE C Division of polynomials

To divide polynomials, use a method similar to that used for division of whole numbers. The same equation used to check division of whole numbers is used to check polynomial division.

Dividend = (quotient × divisor) + remainder

CONSIDER THIS

It may be helpful to review division of whole numbers and compare the procedure used to divide polynomials.

Divide: $(x^2 - 5x + 8) \div (x - 3)$

Step 1

$$\begin{array}{r} x \\ x-3\overline{)x^2 - 5x + 8} \\ \underline{x^2 - 3x}\quad\downarrow \\ -2x + 8 \end{array}$$

Think: $x\overline{)x^2} = \frac{x^2}{x} = x$

Multiply: $x(x - 3) = x^2 - 3x$

Subtract: $(x^2 - 5x) - (x^2 - 3x) = -2x$

Step 2

$$\begin{array}{r} x - 2 \\ x-3\overline{)x^2 - 5x + 8} \\ \underline{x^2 - 3x} \\ -2x + 8 \\ \underline{-2x + 6} \\ 2 \end{array}$$

Think: $x\overline{)-2x} = \frac{-2x}{x} = -2$

Multiply: $-2(x - 3) = -2x + 6$

Subtract: $(-2x + 8) - (-2x + 6) = 2$

The remainder is 2.

Check: $(x - 2)(x - 3) + 2 = x^2 - 3x - 2x + 6 + 2 = x^2 - 5x + 8$

$(x^2 - 5x + 8) \div (x - 3) = x - 2 + \frac{2}{x - 3}$

CONSIDER THIS

The quotient $15 \div 4$ can be written as $3\frac{3}{4}$, which is $3 + \frac{3}{4}$. This is the same form in which a remainder for a quotient of polynomials is written.

Divide: $(6x + 2x^3 + 26) \div (x + 2)$

Arrange the terms in descending order. There is no x^2 term in $2x^3 + 6x + 26$. Insert $0x^2$ for the missing term so that like terms will be in columns.

$$\begin{array}{r} 2x^2 - 4x + 14 \\ x+2\overline{)2x^3 + 0x^2 + 6x + 26} \\ \underline{2x^3 + 4x^2} \qquad\qquad \\ -4x^2 + 6x \qquad \\ \underline{-4x^2 - 8x} \qquad \\ 14x + 26 \\ \underline{14x + 28} \\ -2 \end{array}$$

$(2x^3 + 6x + 26) \div (x + 2) = 2x^2 - 4x + 14 - \frac{2}{x + 2}$

Example 6

Divide: $(x^2 - 1) \div (x + 1)$

Solution

Insert $0x$ for the missing term.

$$\begin{array}{r} x - 1 \\ x+1\overline{)x^2 + 0x - 1} \\ \underline{x^2 + x} \qquad \\ -x - 1 \\ \underline{-x - 1} \\ 0 \end{array}$$

$(x^2 - 1) \div (x + 1) = x - 1$

You Try It 6

Divide: $(2x^3 + x^2 - 8x - 3) \div (2x - 3)$

Your Solution

Solution on p. A26

8.3 EXERCISES

Objective A

Simplify.

1. $\frac{3x^2}{x}$
2. $\frac{4y^2}{2y}$
3. $\frac{2x^2}{-2x}$
4. $\frac{-8y^2}{4y}$
5. $\frac{12x^4}{3x}$
6. $\frac{5x^2}{15x}$
7. $\frac{-16x}{4x^3}$
8. $\frac{27y}{-12y^3}$
9. $\frac{a^4b^5}{a^3b^9}$
10. $\frac{a^5b^7}{a^8b^2}$
11. $\frac{x^3y^4}{x^3y}$
12. $\frac{x^4y^5}{x^2y^5}$
13. $\frac{(3x)^2}{15x^2}$
14. $\frac{(4y)^2}{2y}$
15. $\frac{(6b)^3}{(-3b^2)^2}$
16. $\frac{(-3a^2)^3}{(9a)^2}$
17. $\frac{-36a^4b^7}{60a^5b^9}$
18. $\frac{-50a^2b^7}{45ab^2}$
19. $\frac{12a^2b^3}{-27a^2b^2}$
20. $\frac{-16xy^4}{96x^4y^4}$
21. $\frac{-8x^2y^4}{44y^2z^5}$
22. $\frac{22a^2b^4}{-132b^3c^2}$
23. $\frac{-(8a^2b^4)^3}{64a^3b^8}$
24. $\frac{-(14ab^4)^2}{28a^4b^2}$
25. $\frac{-20a^3b^4}{-45ab^7}$
26. $\frac{-14x^6y^4}{-70x^3y}$
27. $\frac{x^4y^7z}{x^2y^5z^4}$
28. $\frac{x^2y^4z^2}{x^3yz^5}$
29. $\frac{(-2ab^2)^3}{-8ab^7}$
30. $\frac{(-3x^3y)^2}{-12xy^5}$

Objective B

Divide.

31. $\frac{2x + 2}{2}$
32. $\frac{5y + 5}{5}$
33. $\frac{10a - 25}{5}$
34. $\frac{16b - 40}{8}$
35. $\frac{3a^2 + 2a}{a}$
36. $\frac{6y^2 + 4y}{y}$
37. $\frac{4b^3 - 3b}{b}$
38. $\frac{12x^2 - 7x}{x}$
39. $\frac{3x^2 - 6x}{3x}$
40. $\frac{10y^2 - 6y}{2y}$
41. $\frac{5x^2 - 10x}{-5x}$
42. $\frac{3y^2 - 27y}{-3y}$
43. $\frac{x^3 + 3x^2 - 5x}{x}$
44. $\frac{a^3 - 5a^2 + 7a}{a}$
45. $\frac{x^6 - 3x^4 - x^2}{x^2}$
46. $\frac{a^8 - 5a^5 - 3a^3}{a^2}$
47. $\frac{5x^2y^2 + 10xy}{5xy}$
48. $\frac{8x^2y^2 - 24xy}{8xy}$
49. $\frac{9y^6 - 15y^3}{-3y^3}$
50. $\frac{4x^4 - 6x^2}{-2x^2}$
51. $\frac{3x^2 - 2x + 1}{x}$
52. $\frac{8y^2 + 2y - 3}{y}$
53. $\frac{-3x^2 + 7x - 6}{x}$
54. $\frac{2y^2 - 6y + 9}{y}$
55. $\frac{16a^2b - 20ab + 24ab^2}{4ab}$
56. $\frac{22a^2b + 11ab - 33ab^2}{11ab}$
57. $\frac{9x^2y + 6xy - 3xy^2}{xy}$

Objective C

Divide.

58. $(x^2 + 2x + 1) \div (x + 1)$

59. $(x^2 + 10x + 25) \div (x + 5)$

60. $(a^2 - 6a + 9) \div (a - 3)$

61. $(b^2 - 14b + 49) \div (b - 7)$

62. $(x^2 - x - 6) \div (x - 3)$

63. $(y^2 + 2y - 35) \div (y + 7)$

64. $(2x^2 + 5x + 2) \div (x + 2)$

65. $(2y^2 - 13y + 21) \div (y - 3)$

66. $(4x^2 - 16) \div (2x + 4)$

67. $(2y^2 + 7) \div (y - 3)$

68. $(x^2 + 1) \div (x - 1)$

69. $(x^2 + 4) \div (x + 2)$

70. $(6x^2 - 7x) \div (3x - 2)$

71. $(6y^2 + 2y) \div (2y + 4)$

72. $(5x^2 + 7x) \div (x - 1)$

73. $(6x^2 - 5) \div (x + 2)$

74. $(a^2 + 5a + 10) \div (a + 2)$

75. $(b^2 - 8b - 9) \div (b - 3)$

76. $(2y^2 - 9y + 8) \div (2y + 3)$

77. $(x^4 - x^2 - 6) \div (x^2 + 2)$

78. $(x^4 + 3x^2 - 10) \div (x^2 - 2)$

Critical Thinking

Simplify.

79. $\left(\dfrac{9x^2y^4}{3xy^2}\right) - \left(\dfrac{12x^5y^6}{6x^4y^4}\right)$

80. $\left(\dfrac{6x^2 + 9x}{3x}\right) + \left(\dfrac{8xy^2 + 4y^2}{4y^2}\right)$

81. $\left(\dfrac{6x^4yz^3}{2x^2y^3}\right)\left(\dfrac{2x^2z^3}{4y^2z}\right) \div \left(\dfrac{6x^2y^3}{x^4y^2z}\right)$

82. $\left(\dfrac{5x^2yz^3}{3x^4yz^3}\right) \div \left(\dfrac{10x^2y^5z^4}{2y^3z}\right) \div \left(\dfrac{5y^4z^2}{x^2y^6z}\right)$

83. The product of a monomial and $4b$ is $12a^2b$. Find the monomial.

84. The product of a monomial and $6x$ is $24xy^2$. Find the monomial.

85. The quotient of a polynomial and $2x + 1$ is $2x - 4 + \dfrac{7}{2x + 1}$. Find the polynomial.

86. The quotient of a polynomial and $x - 3$ is $x^2 - x + 8 + \dfrac{22}{x - 3}$. Find the polynomial.

87. If $m = n$ and $a \neq 0$, then $\dfrac{a^m}{a^n} =$ ______.

88. If $m = n + 1$ and $a \neq 0$, then $\dfrac{a^m}{a^n} =$ ______.

89. In your own words, explain how to divide exponential expressions.
[W]

SECTION 8.4 Negative and Zero Exponents

OBJECTIVE A Expressions containing negative and zero exponents

Note that when an exponential expression is divided by itself, the exponent is zero.

$$\frac{x^3}{x^3} = x^{3-3} = x^0$$

When the same expression is simplified by factoring and dividing by common factors, the result is 1.

$$\frac{x^3}{x^3} = \frac{\overset{1}{\cancel{x}} \cdot \overset{1}{\cancel{x}} \cdot \overset{1}{\cancel{x}}}{\underset{1}{\cancel{x}} \cdot \underset{1}{\cancel{x}} \cdot \underset{1}{\cancel{x}}} = 1$$

To ensure that the two answers are equal, a number or variable to the zero power must be equal to 1.

$$x^0 = 1, x \neq 0$$

Negative integers, as well as positive, can be used as exponents. The rules that have been developed for exponential expressions can be extended to include negative integers.

$x^m \cdot x^n = x^{m+n}$ $\qquad$ $a^{-2} \cdot a^3 = a^{-2+3} = a$

$(x^m)^n = x^{m \cdot n}$ $\qquad$ $(a^{-2})^3 = a^{-2 \cdot 3} = a^{-6}$

$(x^m \cdot y^n)^p = x^{m \cdot p}y^{n \cdot p}$ $\qquad$ $(a^{-2}b^4)^{-3} = a^{-2(-3)}b^{4(-3)} = a^6b^{-12}$

The Rule for Dividing Exponential Expressions can be stated as a single rule.

$\frac{x^m}{x^n} = x^{m-n}$ $\qquad$ $\frac{a^2}{a^5} = a^{2-5} = a^{-3}$

POINT OF INTEREST

In the 15th century, the expression $12^{2\overline{m}}$ was used to mean $12x^{-2}$. The use of $\overline{m}$ reflected an Italian influence. In Italy, m was used for minus and p was used for plus. It was understood that $2\overline{m}$ referred to a power of an unnamed variable. Isaac Newton, in the 17th century, advocated the current use of negative exponents.

The meaning of a negative exponent can be developed using the rules presented in this chapter.

The exponential expressions at the right are multiplied by adding the exponents. The product is a^0, or 1.

$$a^4 \cdot a^{-4} = a^{4+(-4)} = a^0 = 1$$

Recall that a number times its reciprocal is equal to 1.

$$a^4 \cdot \frac{1}{a^4} = 1$$

Since $a^4 \cdot a^{-4} = 1$, and $a^4 \cdot \frac{1}{a^4} = 1$, then a^{-4} must equal $\frac{1}{a^4}$.

> ***Rule for Negative Exponents***
>
> If n is a positive integer and $x \neq 0$, then $x^{-n} = \frac{1}{x^n}$ and $\frac{1}{x^{-n}} = x^n$.

An exponential expression is in simplest form when written with positive exponents.

A number with a negative exponent can be written with a positive exponent and then evaluated.

CONSIDER THIS

Note from the example that 2^{-3} is a positive number. A negative exponent does not change the sign of a number.

Write 2^{-3} with a positive exponent. Then evaluate.

Write the expression with a positive exponent. $2^{-3} = \frac{1}{2^3}$

Evaluate. $= \frac{1}{8}$

Simplify: $\frac{x^{-4}y^6}{xy^{-2}}$

Divide variables with like bases by subtracting the exponents. $\frac{x^{-4}y^6}{xy^{-2}} = x^{-4-1}y^{6-(-2)}$

$= x^{-5}y^8$

Write the expression with positive exponents. $= \frac{y^8}{x^5}$

Example 1

Write $\frac{3^{-3}}{3^2}$ with a positive exponent.

Then evaluate.

Solution

$\frac{3^{-3}}{3^2} = 3^{-5} = \frac{1}{3^5} = \frac{1}{243}$

You Try It 1

Write $\frac{2^{-2}}{2^3}$ with a positive exponent.

Then evaluate.

Your Solution

Example 2

Simplify: $(-2x)(3x^{-2})^{-3}$

Solution

$(-2x)(3x^{-2})^{-3} = (-2x)(3^{-3}x^6)$

$= \frac{-2x \cdot x^6}{3^3} = -\frac{2x^7}{27}$

You Try It 2

Simplify: $(-2x^2)(x^{-3}y^{-4})^{-2}$

Your Solution

Example 3

Simplify: $\frac{(2ab^{-2})^{-2}}{ab^2}$

Solution

$\frac{(2ab^{-2})^{-2}}{ab^2} = \frac{2^{-2}a^{-2}b^4}{ab^2} = 2^{-2}a^{-3}b^2 = \frac{b^2}{2^2a^3} = \frac{b^2}{4a^3}$

You Try It 3

Simplify: $\frac{(3x^{-2}y)^3}{9xy^0}$

Your Solution

Solutions on p. A26

OBJECTIVE B Scientific notation

Very large and very small numbers are encountered in science and engineering. For example, the mass of the electron is 0.0000000000000000000000000000009 g. Numbers such as this one are difficult to read and write, so a more convenient system for writing such numbers has been developed. It is called **scientific notation**.

To express a number in scientific notation, write the number as the product of a number between 1 and 10 and a power of 10. The form for scientific notation is $\boldsymbol{a \cdot 10^n}$, where a is a number between 1 and 10.

For numbers greater than 10, move the decimal point to the right of the first digit. The exponent n is positive and equal to the number of places the decimal point has been moved.

$965{,}000 = 9.65 \cdot 10^5$

$3{,}600{,}000 = 3.6 \cdot 10^6$

$92{,}000{,}000{,}000 = 9.2 \cdot 10^{10}$

POINT OF INTEREST

An electron microscope uses wavelengths that are approximately $4 \cdot 10^{-12}$ meters to make images of viruses.

The human eye can detect wavelengths between $4.3 \cdot 10^{-7}$ meters and $6.9 \cdot 10^{-7}$ meters. Although these are very short, they are approximately 10^5 times longer than the waves used in an electron microscope.

For numbers less than 1, move the decimal point to the right of the first nonzero digit. The exponent n is negative. The absolute value of the exponent is equal to the number of places the decimal point has been moved.

$0.0002 = 2 \cdot 10^{-4}$

$0.0000000974 = 9.74 \cdot 10^{-8}$

$0.000000000086 = 8.6 \cdot 10^{-11}$

Converting a number written in scientific notation to decimal notation requires moving the decimal point.

When the exponent is positive, move the decimal point to the right the same number of places as the exponent.

$1.32 \cdot 10^4 = 13{,}200$

$1.4 \cdot 10^8 = 140{,}000{,}000$

When the exponent is negative, move the decimal point to the left the same number of places as the absolute value of the exponent.

$1.32 \cdot 10^{-2} = 0.0132$

$1.4 \cdot 10^{-4} = 0.00014$

Numerical calculations involving numbers that have more digits than the hand-held calculator is able to handle can be performed using scientific notation.

Simplify: $\dfrac{220{,}000 \cdot 0.000000092}{0.0000011}$

Write the numbers in scientific notation.

$$\frac{220{,}000 \cdot 0.000000092}{0.0000011} = \frac{2.2 \cdot 10^5 \cdot 9.2 \cdot 10^{-8}}{1.1 \cdot 10^{-6}}$$

Simplify.

$$= \frac{(2.2)(9.2) \cdot 10^{5+(-8)-(-6)}}{1.1}$$

$$= 18.4 \cdot 10^3 = 18{,}400$$

Example 4 Write 0.000041 in scientific notation.

Solution $0.000041 = 4.1 \cdot 10^{-5}$

You Try It 4 Write 942,000,000 in scientific notation.

Your Solution

Solution on p. A26

Example 5 Write $3.3 \cdot 10^7$ in decimal notation.

Solution $3.3 \cdot 10^7 = 33{,}000{,}000$

You Try It 5 Write $2.7 \cdot 10^{-5}$ in decimal notation.

Your Solution

Example 6 Simplify:

$$\frac{2{,}400{,}000{,}000 \cdot 0.0000063}{0.00009 \cdot 480}$$

Solution

$$\frac{2{,}400{,}000{,}000 \cdot 0.0000063}{0.00009 \cdot 480}$$

$$= \frac{2.4 \cdot 10^9 \cdot 6.3 \cdot 10^{-6}}{9 \cdot 10^{-5} \cdot 4.8 \cdot 10^2}$$

$$= \frac{(2.4)(6.3) \cdot 10^{9+(-6)-(-5)-2}}{(9)(4.8)}$$

$$= 0.35 \cdot 10^6 = 350{,}000$$

You Try It 6 Simplify:

$$\frac{5{,}600{,}000 \cdot 0.000000081}{900 \cdot 0.000000028}$$

Your Solution

Solutions on p. A26

OBJECTIVE C Applications

Example 7 How many miles does light travel in one day? The speed of light is 186,000 mi/s. Write the answer in scientific notation.

Strategy To find the distance traveled:

- ▶ Write the speed of light in scientific notation.
- ▶ Write the number of seconds in one day in scientific notation.
- ▶ Use the equation $d = rt$, where r is the speed of light and t is the number of seconds in one day.

Solution

$186{,}000 = 1.86 \cdot 10^5$

$24 \cdot 60 \cdot 60 = 86{,}400 = 8.64 \cdot 10^4$

$d = rt$

$d = (1.86 \cdot 10^5)(8.64 \cdot 10^4)$

$d = 1.86 \cdot 8.64 \cdot 10^9$

$d = 16.0704 \cdot 10^9$

$d = 1.60704 \cdot 10^{10}$

Light travels $1.60704 \cdot 10^{10}$ mi in one day.

You Try It 7 How long does it take light to travel to Earth from the sun? The sun is $9.3 \cdot 10^7$ mi from Earth and light travels $1.86 \cdot 10^5$ mi/s. Write the answer in scientific notation.

Your Strategy

Your Solution

Solution on p. A26

8.4 EXERCISES

Objective A

Write with a positive exponent, and then evaluate.

1. 5^{-2}
2. 3^{-3}
3. $\frac{3^{-2}}{3}$
4. $\frac{5^{-3}}{5}$
5. $\frac{2^{-3}}{2^3}$
6. $\frac{3^{-2}}{3^2}$

Simplify.

7. x^{-2}
8. y^{-10}
9. a^{-6}
10. b^{-4}
11. x^2y^{-3}
12. $a^{-2}b$
13. $x^{-1}y^{-2}$
14. $x^{-3}y^{-4}$
15. $x^{-3}x^4$
16. $x \cdot x^{-2}$
17. $a^{-3}a^{-4}$
18. $a^{-2}a^{-5}$
19. $\frac{x^{-2}}{x^2}$
20. $\frac{x^{-1}}{x}$
21. $\frac{a^{-3}}{a^5}$
22. $\frac{a^{-10}}{a^{10}}$
23. $\frac{x^{-2}y}{x}$
24. $\frac{x^4y^{-3}}{x^2}$
25. $\frac{a^{-2}b}{b^0}$
26. $\frac{a^2b^{-4}}{b^0}$
27. $(a^2)^{-2}$
28. $(b^3)^{-3}$
29. $(a^{-3})^2$
30. $(y^{-4})^3$
31. $(x^{-2})^{-3}$
32. $(y^{-3})^{-4}$
33. $(a^{-6})^{-3}$
34. $(a^{-1})^{-3}$
35. $(ab^{-1})^0$
36. $(a^2b)^0$
37. $(x^{-2}y^2)^2$
38. $(x^{-3}y^{-1})^2$
39. $\frac{x^2y^{-4}}{xy}$
40. $\frac{ab^{-5}}{a^7b}$
41. $\frac{x^{-2}y}{x^2y^{-3}}$
42. $\frac{a^{-3}b}{a^{-1}b^2}$
43. $(x^2y^{-1})^{-2}$
44. $(a^{-1}b^{-1})^{-4}$
45. $(-2xy^{-2})^3$
46. $(-3x^{-1}y^2)^2$
47. $(-5a^2)(a^{-5})^2$
48. $(2a^{-3})(a^7b^{-1})^3$
49. $(3ab^{-2})(2a^{-1}b)^{-3}$
50. $\frac{a^{-1}b^{-1}}{ab}$
51. $\frac{a^{-3}b^{-4}}{a^2b^2}$
52. $\frac{3x^{-2}y^2}{6xy^2}$
53. $\frac{2x^{-2}y}{8xy^{-3}}$
54. $\frac{2x^{-1}y^{-4}}{4xy^{-2}}$
55. $\frac{(x^{-1}y)^2}{xy^2}$
56. $\frac{(x^{-2}y)^2}{x^{-3}y^3}$
57. $\frac{(x^{-3}y^{-2})^2}{x^6y^8}$
58. $\frac{(a^{-2}y^3)^{-3}}{a^{-2}y}$

Objective B

Write in scientific notation.

59. 0.00000467
60. 0.00000000017
61. 4,300,000
62. 9,800,000,000

Write in decimal notation.

63. $1.23 \cdot 10^{-7}$
64. $6.2 \cdot 10^{-12}$
65. $6.34 \cdot 10^5$
66. $4.35 \cdot 10^9$

Simplify. Write the answer in decimal notation.

67. (0.0000065)(3,200,000,000,000)

68. (480,000)(0.0000000096)

69. $(3 \cdot 10^{-12})(5 \cdot 10^{16})$

70. $(8.9 \cdot 10^{-5})(3.2 \cdot 10^{-6})$

71. $\dfrac{9 \cdot 10^{-3}}{6 \cdot 10^{5}}$

72. $\dfrac{2.7 \cdot 10^{4}}{3 \cdot 10^{-6}}$

73. $\dfrac{4800}{0.00000024}$

74. $\dfrac{0.00056}{0.000000000004}$

Objective C

Solve. Write the answer in scientific notation.

75. How many kilometers does light travel in one day? The speed of light is 300,000 km/s.

76. How many meters does light travel in 8 h? The speed of light is 300,000,000 m/s.

77. The mass of the Earth is $5.9 \cdot 10^{27}$ g. The mass of the sun is $2 \cdot 10^{33}$ g. How many times heavier is the sun than the earth?

78. The distance to the sun is $9.3 \cdot 10^{7}$ mi. A satellite leaves the Earth traveling at a constant speed of $1 \cdot 10^{5}$ mph. How long would it take for the satellite to reach the sun if such a trip were possible?

79. One light year, an astronomical unit of distance, is the distance that light travels in one year. Light travels $1.86 \cdot 10^{5}$ mi/s. Find the measure of one light year in miles. Use a 360-day year.

80. The light from the star Alpha Centauri takes 4.3 years to reach the Earth. Light travels at $1.86 \cdot 10^{5}$ mi/s. How far is Alpha Centauri from the Earth? Use a 360-day year.

Critical Thinking

Evaluate.

81. $8^{-2} + 2^{-5}$

82. $9^{-2} + 3^{-3}$

Write in decimal notation.

83. 2^{-4}

84. 25^{-2}

85. Evaluate 2^{x} and 2^{-x} when $x = -2, -1, 0, 1,$ and 2.

86. Evaluate 3^{x} and 3^{-x} when $x = -2, -1, 0, 1,$ and 2.

87. [W] If x is a nonzero real number, is x^{-2} always positive, always negative, or positive or negative depending on whether x is positive or negative? Explain your answer.

88. [W] If x is a nonzero real number, is x^{-3} always positive, always negative, or positive or negative depending on whether x is positive or negative? Explain your answer.

Projects in Mathematics

Intensity of Illumination

The rate at which light falls upon a one-square-unit area of surface is called the **intensity of illumination**. Intensity of illumination is measured in **lumens** (lm). A lumen is defined in the following illustration.

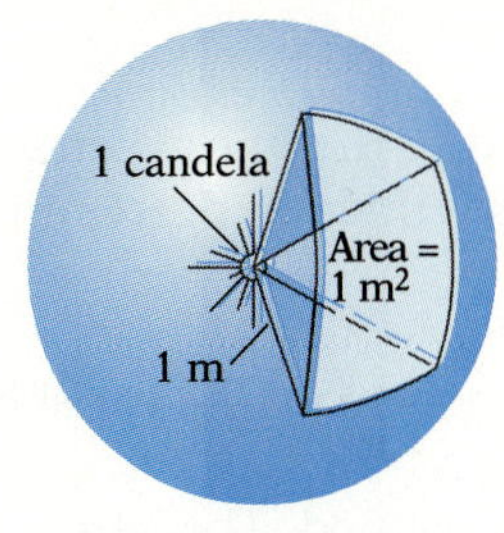

Picture a source of light whose intensity equals one candela (1 cd) placed at the center of a hollow sphere that has a radius of 1 m. The rate at which light falls upon 1 m^2 of the inner surface of the sphere is equal to one lumen. If a light source whose intensity equals 4 cd is placed at the center of the sphere, each square meter of the inner surface receives four times as much illumination, or 4 lm.

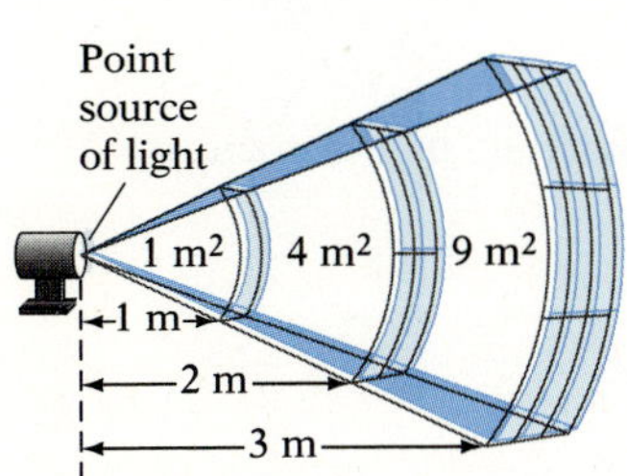

Light rays diverge as they leave a light source. The light that falls upon an area of 1 m^2 at a distance of 1 m from the source of light spreads out over an area of 4 m^2 when it is 2 m from the source. The same light spreads out over an area of 9 m^2 when it is 3 m from the light source and over an area of 16 m^2 when it is 4 m from the light source. Therefore, as a surface moves farther away from the source of light, the intensity of illumination on the surface decreases from its value at 1 m to $\left(\frac{1}{2}\right)^2$, or $\frac{1}{4}$, that value at 2 m; to $\left(\frac{1}{3}\right)^2$, or $\frac{1}{9}$, that value at 3 m; and to $\left(\frac{1}{4}\right)^2$, or $\frac{1}{16}$, that value at 4 m. The formula for the intensity of illumination is

$$I = \frac{s}{r^2}$$

where I is the intensity of illumination, s is the strength of the light source in candelas, and r is the distance in meters between the light source and the illuminated surface.

A 30-cd lamp is 0.5 m above a desk. Find the illumination on the desk.

$$I = \frac{s}{r^2} = \frac{30}{(0.5)^2} = 120$$

The illumination on the desk is 120 lm.

Solve.

1. A 100-cd light hangs 5 m above a floor. What is the intensity of illumination on the floor beneath it?
2. A 25-cd source of light is 2 m above a desk. Find the intensity of illumination on the desk.
3. How strong a light source is needed to cast 80 lm of light on a surface 5 m from the source?
4. Two lights cast the same intensity of illumination on a wall. One light is 6 m from the wall and its strength is 36 cd. The second light is 8 m from the wall. Find the strength in candelas of the second light.

Chapter Summary

Key Words

A **monomial** is a number, a variable, or a product of numbers and variables.

A **polynomial** is a variable expression in which the terms are monomials.

A **monomial** is a polynomial of one term. A **binomial** is a polynomial of two terms. A **trinomial** is a polynomial of three terms.

When the terms of a polynomial in one variable are arranged so that the exponents of the variable decrease from left to right, the polynomial is written in **descending order**.

The **degree of a polynomial** in one variable is the largest exponent on the variable.

The product of two binomials can be found using the **FOIL** method. The letters of FOIL stand for **F**irst, **O**uter, **I**nner, and **L**ast.

A number written in **scientific notation** is a number written in the form $a \cdot 10^n$, where a is a number between 1 and 10.

Essential Rules

Rule for Multiplying Exponential Expressions
If m and n are integers, then $x^m \cdot x^n = x^{m+n}$.

Rule for Simplifying Powers of Exponential Expressions
If m and n are integers, then $(x^m)^n = x^{m \cdot n}$.

Rule for Simplifying Powers of Products
If m, n, and p are integers, then $(x^m \cdot y^n)^p = x^{m \cdot p}y^{n \cdot p}$.

The Sum and Difference of Two
$(a + b)(a - b) = a^2 - b^2$

The Square of a Binomial
$(a + b)^2 = a^2 + 2ab + b^2$
$(a - b)^2 = a^2 - 2ab + b^2$

Rule for Dividing Exponential Expressions
If m and n are integers and $x \neq 0$, then

$\dfrac{x^m}{x^n} = x^{m-n}$ if $m > n$ and

$\dfrac{x^m}{x^n} = \dfrac{1}{x^{n-m}}$ if $m < n$.

Rule for Negative Exponents
If n is a positive integer and $x \neq 0$, then
$x^{-n} = \dfrac{1}{x^n}$ and $\dfrac{1}{x^{-n}} = x^n$.

Chapter Review Exercises

1. Write the polynomial in descending order.
$5x - 2x^2 + 3 - 8x^3$

2. Identify the degree of the polynomial.
$6x^4 - x^2 + 7$

3. Add:
$(3x^3 - 2x^2 - 4) + (8x^2 - 8x + 7)$

4. Subtract:
$(3a^2 - 2a - 7) - (5a^3 + 2a - 10)$

5. Multiply: $(-2xy^2)(3x^2y^4)$

6. Simplify: $(x^2y^3)^4$

7. Multiply: $2x(2x^2 - 3x)$

8. Multiply: $-3y^2(-2y^2 + 3y - 6)$

9. Multiply: $(x - 3)(x^2 - 4x + 5)$

10. Multiply: $(-2x^3 + x^2 - 7)(2x - 3)$

11. Multiply: $(a - 2b)(a + 5b)$

12. Multiply: $(2x - 7y)(5x - 4y)$

13. Multiply: $(4y - 3)(4y + 3)$

14. Simplify: $(2x - 5)^2$

15. Simplify: $\dfrac{12x^2}{-3x^8}$

16. Simplify: $\dfrac{(3xy^3)^3}{3x^4y^3}$

17. Divide: $\dfrac{16x^5 - 8x^3 + 20x}{4x}$

18. Divide: $(x^2 + 6x - 7) \div (x - 1)$

19. Divide: $(4x^2 - 7) \div (2x - 3)$

20. Simplify: $(a^2b^{-3})^2$

21. Simplify: $(-2ab^{-3})(3a^{-2}b^4)$

22. Write 0.00000005 in scientific notation.

23. Write $3.9 \cdot 10^{-2}$ in decimal notation.

24. A space vehicle travels $2.4 \cdot 10^5$ mi from the Earth to the moon at an average velocity of $2 \cdot 10^4$ mph. How long does it take the space vehicle to reach the moon? Write the answer in scientific notation.

Cumulative Review Exercises

1. Solve: $12 = -\frac{1}{4}x$

2. 35.2 is what percent of 160?

3. Multiply: $(12x)\left(-\frac{3}{4}\right)$

4. Simplify: $-2[4x - 2(3 - 2x) - 8x]$

5. Find the ordered-pair solution of $y = 2x - 1$ corresponding to $x = -2$.

6. Is (2, 0) a solution of the system $5x - 3y = 10$, $4x + 7y = 8$?

7. Find the slope of the line containing the points (2, 3) and (−2, 3).

8. Solve: $3(x - 7) \geq 5x - 12$

9. Solve: $2x - 4[3x - 2(1 - 3x)] = 2(3 - 4x)$

10. Multiply: $(2a - 7)(5a^2 - 2a + 3)$

11. Simplify: $\frac{(-2a^2b^3)^2}{8a^4b^8}$

12. Divide: $(a^2 - 4a - 21) \div (a + 3)$

13. Write 0.068 as a percent.

14. Simplify: $\frac{x^{-4}y^{-3}}{xy^2}$

15. Evaluate $-2a^2 \div 2b - c$ when $a = -4$, $b = 2$, and $c = -1$.

16. Simplify: $\frac{8.4 \cdot 10^9}{2.1 \cdot 10^6}$

17. Graph the line that has slope $-\frac{1}{2}$ and y-intercept 2.

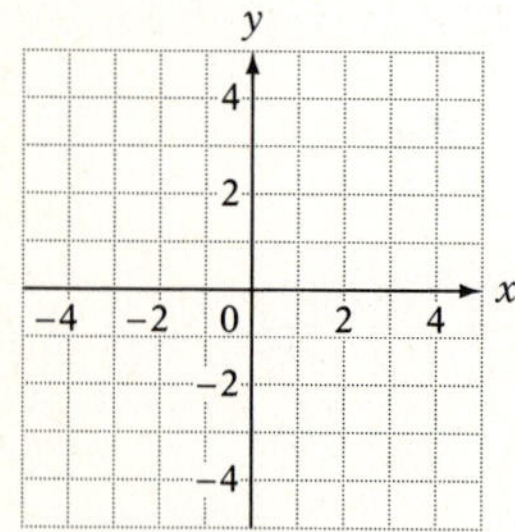

18. Graph: $2x - 3y = 6$

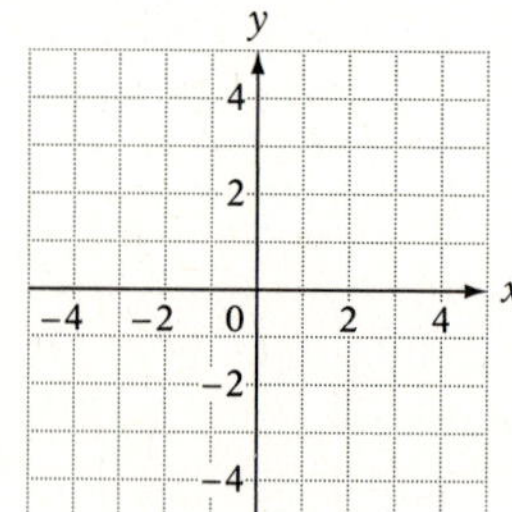

19. Find the equation of the line that contains the point (−2, −3) and that has slope $\frac{1}{2}$.

20. A right triangle has a 53° angle. Find the measures of the other two angles.

21. Find the surface area of a sphere with a diameter of 14 in. Give the exact value.

22. The slant height of a pyramid is 15 cm and the base is 10 cm. Find the surface area of the pyramid.

23. A company increased its dividend of \$1.50 by 8%. Find the new dividend paid by the company.

CHAPTER

9

Factoring

Focus on Problem Solving

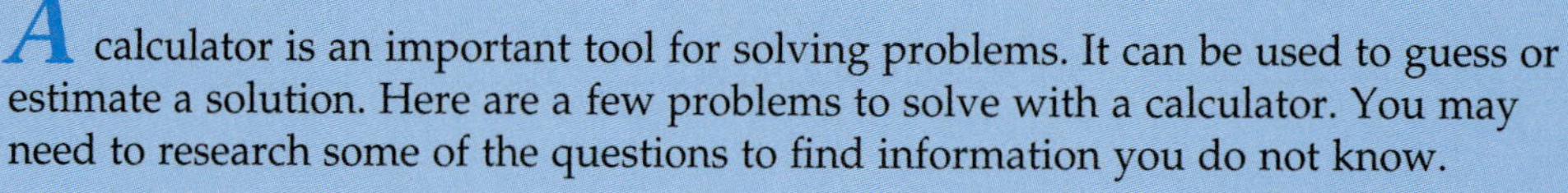

A calculator is an important tool for solving problems. It can be used to guess or estimate a solution. Here are a few problems to solve with a calculator. You may need to research some of the questions to find information you do not know.

1. Choose any single-digit positive number. Multiply the number by 1507. Now multiply the result by 7519. What is the answer? Choose another positive single-digit number and again multiply by 1507 and 7519. What is the answer? What pattern do you see? Why does this work?
2. Are there enough people in the United States so that if they held hands in a line they would stretch around the world at the equator? To answer this question, begin by asking yourself what information you need. What assumptions must you make?
3. The gross national product in 1990 was about \$5,400,000,000,000,000. Is this more or less than the amount of money that would be placed on the last square of a standard checkerboard if 1¢ is placed on the first square, 2¢ is placed on the second square, 4¢ is placed on the third square, 8¢ is placed on the fourth square, and so on until the 64th square is reached?
4. Which of the reciprocals of the first 16 natural numbers have a terminating decimal representation and which have a repeating decimal representation?
5. What is the largest natural number n for which $4^n > 1 \cdot 2 \cdot 3 \cdots n$?
6. If \$1000 bills were stacked one on top of another, is the height of one billion dollars less than or more than the height of the Washington Monument?
7. What is the value of $1 + \cfrac{1}{1 + \cfrac{1}{1 + \cfrac{1}{1 + \cfrac{1}{1 + 1}}}}$?
8. Calculate 15^2, 35^2, 65^2, and 85^2. Study the results. Make a conjecture about a relationship between a number ending in 5 and its square. Use your conjecture to find 75^2 and 95^2. Does your conjecture work for 125^2?
9. Find the sum of the first 1000 natural numbers. (*Hint:* You could just start adding $1 + 2 + 3 + 4 + \ldots$, but even if you performed one operation each second, it would take over 15 minutes to find the sum. Instead, try pairing the numbers and then adding the numbers in each pair. Pair 1 and 1000, 2 and 999, 3 and 998, and so on. What is the sum of each pair? How many pairs are there? Use this information to answer the original question.)
10. To qualify for a home loan, a bank requires that the monthly mortgage payment be less than 25% of a borrower's monthly take-home income. A laboratory technician has deductions for taxes, insurance, and retirement that amount to 25% of the technician's monthly gross income. What minimum gross monthly income must this technician earn to receive a bank loan that has a \$1200 per month mortgage payment?

SECTION 9.1 Monomial Factors

OBJECTIVE A The greatest common factor (GCF) of two or more monomials

The **greatest common factor (GCF)** of two or more integers is the greatest integer that is a factor of all the integers.

$24 = 2 \cdot 2 \cdot 2 \cdot 3$
$60 = 2 \cdot 2 \cdot 3 \cdot 5$
$\text{GCF} = 2 \cdot 2 \cdot 3 = 12$

The GCF of two or more monomials is the product of the GCF of the coefficients and the common variable factors.

$6x^3y = 2 \cdot 3 \cdot x \cdot x \cdot x \cdot y$
$8x^2y^2 = 2 \cdot 2 \cdot 2 \cdot x \cdot x \cdot y \cdot y$
$\text{GCF} = 2 \cdot x \cdot x \cdot y = 2x^2y$

Note that the exponent of each variable in the GCF is the same as the *smallest* exponent of that variable in either of the monomials.

The GCF of $6x^3y$ and $8x^2y^2$ is $2x^2y$.

Find the GCF of $12a^4b$ and $18a^2b^2c$.

$12a^4b = 2 \cdot 2 \cdot 3 \cdot a^4 \cdot b$
$18a^2b^2c = 2 \cdot 3 \cdot 3 \cdot a^2 \cdot b^2 \cdot c$
$\text{GCF} = 2 \cdot 3 \cdot a^2 \cdot b = 6a^2b$

The common variable factors are a^2 and b. c is not a common variable factor.

Example 1

Find the GCF of $4a^3b$, $12ab^2$, and $6a^2b^2$.

Solution

$4a^3b = 2 \cdot 2 \cdot a^3 \cdot b$
$12ab^2 = 2 \cdot 2 \cdot 3 \cdot a \cdot b^2$
$6a^2b^2 = 2 \cdot 3 \cdot a^2 \cdot b^2$

The GCF is $2ab$.

You Try It 1

Find the GCF of $8a^4bc$, $12ab^3$, and $20abc^2$.

Your Solution

Solution on p. A27

OBJECTIVE B Factor a monomial from a polynomial

The Distributive Property is used to multiply factors of a polynomial. To **factor** a polynomial means to write the polynomial as a product of other polynomials.

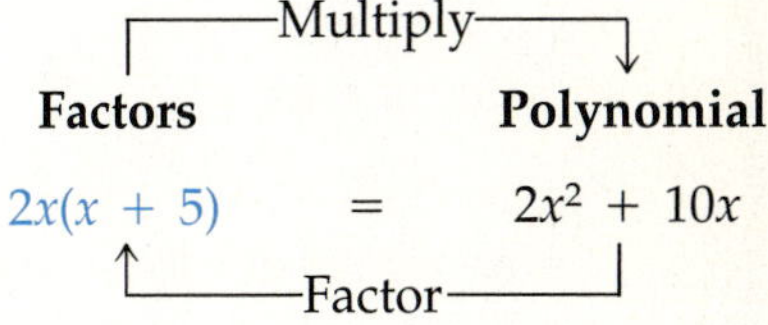

In the example above, $2x$ is the GCF of the terms $2x^2$ and $10x$. It is a **common monomial factor** of the terms. $x + 5$ is a **binomial factor** of $2x^2 + 10x$.

Factor $8a^2b^2 - 6ab$.

Find the GCF of the terms of the polynomial.

$8a^2b^2 = 2 \cdot 2 \cdot 2 \cdot a^2 \cdot b^2$
$6ab = 2 \cdot 3 \cdot a \cdot b$
The GCF is $2ab$.

Divide each term of the polynomial by the GCF.

$\dfrac{8a^2b^2}{2ab} = 4ab \qquad \dfrac{-6ab}{2ab} = -3$

Use the quotients to rewrite the polynomial, expressing each term as a product with the GCF as one of the factors.

$8a^2b^2 - 6ab = 2ab(4ab) + 2ab(-3)$

Use the Distributive Property to write the polynomial as a product of factors.

$= 2ab(4ab - 3)$

Check: $2ab(4ab - 3) = 8a^2b^2 - 6ab$

Factor $5x^3 - 35x^2 + 10x$.

Find the GCF of the terms of the polynomial.

$5x^3 = 5 \cdot x^3$
$35x^2 = 5 \cdot 7 \cdot x^2$
$10x = 2 \cdot 5 \cdot x$
The GCF is $5x$.

Divide each term of the polynomial by the GCF.

$\dfrac{5x^3}{5x} = x^2 \qquad \dfrac{-35x^2}{5x} = -7x \qquad \dfrac{10x}{5x} = 2$

Use the quotients to rewrite the polynomial, expressing each term as a product with the GCF as one of the factors.

$5x^3 - 35x^2 + 10x = 5x(x^2) + 5x(-7x) + 5x(2)$

Use the Distributive Property to write the polynomial as a product of factors.

$= 5x(x^2 - 7x + 2)$

Example 2

Factor $6x^3 - 12x^2 + 30x$.

Solution

$6x^3 = 2 \cdot 3 \cdot x^3$
$12x^2 = 2 \cdot 2 \cdot 3 \cdot x^2$
$30x = 2 \cdot 3 \cdot 5 \cdot x$
The GCF is $6x$.

$6x^3 - 12x^2 + 30x$
$= 6x(x^2) + 6x(-2x) + 6x(5)$
$= 6x(x^2 - 2x + 5)$

You Try It 2

Factor $18a^4 + 27a^3 - 9a^2$.

Your Solution

Solution on p. A27

9.1 EXERCISES

Objective A

Find the greatest common factor.

1. x^7, x^3
2. y^6, y^{12}
3. x^2y^4, xy^6
4. a^5b^3, a^3b^8
5. $x^2y^4z^6, xy^8z^2$
6. ab^2c^3, a^3b^2c
7. $a^3b^2c^3, ab^4c^3$
8. x^3y^2z, x^4yz^5
9. $3x^4, 12x^2$
10. $12x, 30x^2$
11. $16a^3, 18a$
12. $8y^3, 12y^6$
13. $14a^3, 49a^7$
14. $12y^2, 27y^4$
15. $3x^2y^2, 5ab^2$
16. $8x^2y^3, 7ab^4$
17. $9a^2b^4, 24a^4b^2$
18. $15a^4b^2, 9ab^5$
19. $ab^3, 4a^2b, 12a^3b^3$
20. $12x^2y, x^4y, 16x$
21. $18ab^2, a^3b^3, 27a^2b$
22. $2x^2y, 4xy, 8x$
23. $16x^2, 8x^4y^2, 12xy$
24. $3x^2y^2, 6x, 9x^3y^3$
25. $4a^2b^3, 8a^3, 12ab^4$
26. $8a^3bc, 2ab^3c, 6a^2b^2$
27. $6m^2n^2, 9m^3n^3, 18m^4n^4$
28. $21x^2y, 14y^2z, 35xz^2$

Objective B

Factor.

29. $5a + 5$
30. $7b - 7$
31. $6 - 18x$
32. $8 - 4y$
33. $2x^2 - 20$
34. $3y^2 - 24$
35. $16 - 8a^2$
36. $12 + 12y^2$
37. $7x^2 - 3x$
38. $12y^2 - 5y$
39. $3a^2 + 5a^5$
40. $12b^3 + 4b^4$
41. $9x - 5x^2$
42. $14y^2 + 11y$
43. $6b^3 - 5b^2$
44. $2x^4 - 4x$
45. $3y^4 - 9y$
46. $10x^4 - 12x^2$
47. $12a^5 - 32a^2$
48. $8a^8 - 4a^5$

Factor.

49. $m^4n^2 - m^2n^4$

50. $3x^2y^4 - 6xy$

51. $12a^2b^5 - 9ab$

52. $x^2y - xy^3$

53. $a^2b + a^4b^2$

54. $2a^5b + 3xy^3$

55. $5x^2y - 7ab^3$

56. $6a^2b^3 - 12b^2$

57. $8x^2y^3 - 4x^2$

58. $a^3 - 3a^2 + 5a$

59. $b^3 - 5b^2 - 7b$

60. $5x^2 - 15x + 35$

61. $8y^2 - 12y + 32$

62. $3x^3 + 6x^2 + 9x$

63. $5y^3 - 20y^2 + 10y$

64. $2x^4 - 4x^3 + 6x^2$

65. $3y^4 - 9y^3 - 6y^2$

66. $2x^3 + 6x^2 - 14x$

67. $3y^3 - 9y^2 + 24y$

68. $2y^5 - 3y^4 + 7y^3$

69. $6a^5 - 3a^3 - 2a^2$

70. $x^3y - 3x^2y^2 + 7xy^3$

71. $2a^2b - 5a^2b^2 + 7ab^2$

72. $5y^3 + 10y^2 - 25y$

73. $4b^5 + 6b^3 - 12b$

74. $3a^2b^2 - 9ab^2 + 15b^2$

75. $8x^2y^2 - 4x^2y + x^2$

Critical Thinking

A whole number is a perfect number if it equals the sum of all its factors less than itself. For example, 6 is a perfect number because all the factors of 6 that are less than 6 are 1, 2, and 3, and $1 + 2 + 3 = 6$.

76. Find the one perfect number between 20 and 30.

77. Find the one perfect number between 490 and 500.

78. In the equation $P = 2L + 2W$, what is the effect on P when the quantity $L + W$ doubles?

79. Write an expression in factored form for the shaded portion in the diagrams.

a.

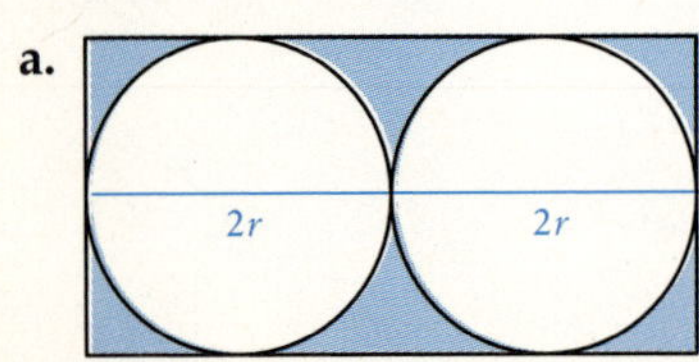

b.

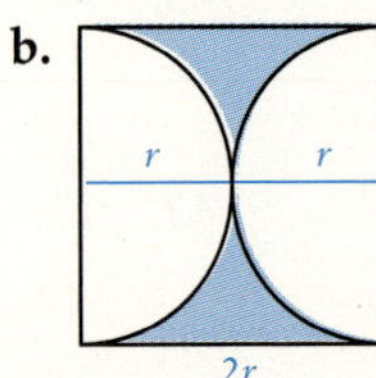

SECTION 9.2 Factoring Trinomials

OBJECTIVE A Factor a trinomial of the form $x^2 + bx + c$

Trinomials of the form $x^2 + bx + c$, where b and c are integers, are shown at the right.

$x^2 + 8x + 12$, $b = 8$, $c = 12$
$x^2 - 7x + 12$, $b = -7$, $c = 12$
$x^2 - 2x - 15$, $b = -2$, $c = -15$

To **factor** a trinomial of this form means to express the trinomial as the product of two binomials.

Trinomials expressed as the product of binomials are shown at the right.

$x^2 + 8x + 12 = (x + 6)(x + 2)$
$x^2 - 7x + 12 = (x - 3)(x - 4)$
$x^2 - 2x - 15 = (x + 3)(x - 5)$

The factors of a trinomial are found by using the FOIL method. Consider the following binomial products, noting the relationship between the constant terms of the binomials and the terms of the trinomials.

Signs in the binomials are the same

$(x + 6)(x + 2) = x^2 + 2x + 6x + (6)(2) = x^2 + 8x + 12$
sum of 6 and 2 (gives 8)
product of 6 and 2 (gives 12)

$(x - 3)(x - 4) = x^2 - 4x - 3x + (-3)(-4) = x^2 - 7x + 12$
sum of -3 and -4
product of -3 and -4

CONSIDER THIS
Keep in mind that the relationships between the binomial factors and the terms of the polynomial are based on the rules for multiplying binomials.

Signs in the binomials are opposite

$(x + 3)(x - 5) = x^2 - 5x + 3x + (3)(-5) = x^2 - 2x - 15$
sum of 3 and -5
product of 3 and -5

$(x - 4)(x + 6) = x^2 + 6x - 4x + (-4)(6) = x^2 + 2x - 24$
sum of -4 and 6
product of -4 and 6

Relationships for Factoring Trinomials

1. When the constant term of the trinomial is positive, the constant terms of the binomials have the same sign. They are both positive when the coefficient of the x term in the trinomial is positive. They are both negative when the coefficient of the x term in the trinomial is negative.

2. When the constant term of the trinomial is negative, the constant terms of the binomials have opposite signs.

3. In the trinomial, the coefficient of x is the sum of the constant terms of the binomials.

4. In the trinomial, the constant term is the product of the constant terms of the binomials.

The following trinomial factoring patterns summarize the relationships stated on the previous page.

Trinomial	Factoring Pattern
$x^2 + bx + c$	$(x + \blacksquare)(x + \blacksquare)$
$x^2 - bx + c$	$(x - \blacksquare)(x - \blacksquare)$
$x^2 + bx - c$	$(x + \blacksquare)(x - \blacksquare)$ or $(x - \blacksquare)(x + \blacksquare)$
$x^2 - bx - c$	$(x + \blacksquare)(x - \blacksquare)$ or $(x - \blacksquare)(x + \blacksquare)$

Factor $x^2 + 7x + 10$.

The constant term is positive. The coefficient of x is positive. The binomial constants will be positive. $(x + \blacksquare)(x + \blacksquare)$

Find two positive factors of 10 whose sum is 7.

Factors	Sum
+1, +10	11
+2, +5	7

Write the factors of the trinomial. $(x + 2)(x + 5)$

Check:

$$(x + 2)(x + 5) = x^2 + 5x + 2x + 10$$
$$= x^2 + 7x + 10$$

CONSIDER THIS

The phrase *nonfactorable over the integers* does not mean that the polynomial does not factor but that the polynomial does not factor if integers are used. For instance, the only way to write 7 as a product involving *integers* is $1 \cdot 7$ or $(-1)(-7)$. However, $\frac{21}{5} \cdot \frac{5}{3} = 7$. The ability to factor depends on the numbers that are allowed as factors.

When only integers are used, some trinomials do not factor. For example, to factor $x^2 + 5x + 3$, it would be necessary to find two positive integers whose product is 3 and whose sum is 5. This is not possible, since the only positive factors of 3 are 1 and 3, and the sum of 1 and 3 is 4. This trinomial is **nonfactorable over the integers.** Binomials of the form $x + a$ or $x - a$ are also nonfactorable over the integers.

Example 1

Factor $x^2 - 8x + 15$.

Solution

$(x - \blacksquare)(x - \blacksquare)$

Factors	Sum
−1, −15	−16
−3, −5	−8

$(x - 3)(x - 5)$

$x^2 - 8x + 15 = (x - 3)(x - 5)$

You Try It 1

Factor $x^2 - 9x + 20$.

Your Solution

Example 2

Factor $x^2 + 6x - 27$.

Solution

$(x + \blacksquare)(x - \blacksquare)$

Factors	Sum
+1, −27	−26
−1, +27	26
+3, −9	−6
−3, +9	6

$(x - 3)(x + 9)$

$x^2 + 6x - 27 = (x - 3)(x + 9)$

You Try It 2

Factor $x^2 + 3x - 18$.

Your Solution

Solutions on p. A27

OBJECTIVE B Factor a trinomial of the form $ax^2 + bx + c$

Trinomials of the form $ax^2 + bx + c$, where a, b, and c are integers and $a \neq 0$, are shown at the right.

$3x^2 - x + 4, a = 3, b = -1, c = 4$

$6x^2 + 8x - 6, a = 6, b = 8, c = -6$

To factor a trinomial of this form, a trial-and-error method is used. Trial factors are written, using the factors of a and c to write the binomials. Then FOIL is used to check for b, the coefficient of the middle term.

To reduce the number of factors to try, remember the following:

1. Use the signs of the constant and the coefficient of x in the trinomial to determine the signs of the terms in the binomial factors.

Trinomial	Factoring Pattern
$ax^2 + bx + c$	$(\blacksquare x + \blacksquare)(\blacksquare x + \blacksquare)$
$ax^2 - bx + c$	$(\blacksquare x - \blacksquare)(\blacksquare x - \blacksquare)$
$ax^2 - bx - c$	$(\blacksquare x + \blacksquare)(\blacksquare x - \blacksquare)$ or $(\blacksquare x - \blacksquare)(\blacksquare x + \blacksquare)$
$ax^2 + bx - c$	$(\blacksquare x + \blacksquare)(\blacksquare x - \blacksquare)$ or $(\blacksquare x - \blacksquare)(\blacksquare x + \blacksquare)$

2. If the terms of the trinomial do not have a common factor, then the two terms in either one of the binomial factors will not have a common factor.

Factor $2x^2 - 7x + 3$.

The terms have no common factor.
The constant term is positive.
The coefficient of x is negative.
The binomial constants will be negative.

$(\blacksquare x - \blacksquare)(\blacksquare x - \blacksquare)$

Write the factors of 2 (the coefficient of x^2). These factors will be the coefficients of the x terms in the binomial factors.

Factors of 2: 1, 2

Write the negative factors of 3 (the constant term). These factors will be the constants in the binomial factors.

Factors of 3: $-1, -3$

Write trial factors. Writing the 1 when it is the coefficient of x may be helpful. Use the Outer and Inner products of FOIL to determine the middle term of the trinomial.

Trial Factors	Middle Term
$(1x - 1)(2x - 3)$	$-3x - 2x = -5x$
$(1x - 3)(2x - 1)$	$-x - 6x = -7x$

Write the factors of the trinomial.

$(x - 3)(2x - 1)$

Check: $(x - 3)(2x - 1)$
$= 2x^2 - x - 6x + 3$
$= 2x^2 - 7x + 3$

Factor $6x^2 - x - 2$.

The terms have no common factor. The constant term is negative. The signs of the binomial constants will be opposites.

$(\blacksquare x + \blacksquare)(\blacksquare x - \blacksquare)$ or $(\blacksquare x - \blacksquare)(\blacksquare x + \blacksquare)$

Write the factors of 6. These factors will be the coefficients of the x terms in the binomial factors.

Factors of 6: 1, 6
2, 3

Write the factors of -2. These factors will be the constants in the binomial factors.

Factors of -2: $-1, +2$
$+1, -2$

Write the trial factors. Use the Outer and Inner terms of FOIL to determine the middle term of the trinomial. It is not necessary to test trial factors that have a common factor. For example, $6x + 2$ need not be tested because it has a common factor of 2. Once a trial solution has the correct middle term, other trial factors need not be tried.

Trial Factors	Middle Term
$(1x - 1)(6x + 2)$	Common factor
$(1x + 2)(6x - 1)$	$-x + 12x = 11x$
$(1x + 1)(6x - 2)$	Common factor
$(1x - 2)(6x + 1)$	$x - 12x = -11x$
$(2x - 1)(3x + 2)$	$4x - 3x = x$
$(2x + 2)(3x - 1)$	Common factor
$(2x + 1)(3x - 2)$	$-4x + 3x = -x$
$(2x - 2)(3x + 1)$	Common factor

CONSIDER THIS
Always check your proposed factorization to ensure accuracy.

Write the factors of the trinomial.

$(2x + 1)(3x - 2)$

Check: $(2x + 1)(3x - 2)$
$= 6x^2 - 4x + 3x - 2$
$= 6x^2 - x - 2$

Example 3

Factor $3x^2 + x - 2$.

Solution

$(\blacksquare x + \blacksquare)(\blacksquare x - \blacksquare)$ or $(\blacksquare x - \blacksquare)(\blacksquare x + \blacksquare)$

Factors of 3: 1, 3 Factors of -2: $+1, -2$
$-1, +2$

Trial Factors	Middle Term
$(1x + 1)(3x - 2)$	$-2x + 3x = x$
$(1x - 2)(3x + 1)$	$x - 6x = -5x$
$(1x - 1)(3x + 2)$	$2x - 3x = -x$
$(1x + 2)(3x - 1)$	$-x + 6x = 5x$

$(x + 1)(3x - 2)$

$3x^2 + x - 2 = (x + 1)(3x - 2)$

You Try It 3

Factor $2x^2 - x - 3$.

Your Solution

Solution on p. A27

9.2 EXERCISES

Objective A

Factor.

1. $x^2 + 3x + 2$
2. $x^2 + 5x + 6$
3. $x^2 - x - 2$
4. $x^2 + x - 6$
5. $a^2 + a - 12$
6. $a^2 - 2a - 35$
7. $a^2 - 3a + 2$
8. $a^2 - 5a + 4$
9. $a^2 + a - 2$
10. $a^2 - 2a - 3$
11. $b^2 - 6b + 9$
12. $b^2 + 8b + 16$
13. $b^2 + 7b - 8$
14. $y^2 - y - 6$
15. $y^2 + 6y - 55$
16. $z^2 - 4z - 45$
17. $y^2 - 5y + 6$
18. $y^2 - 8y + 15$
19. $z^2 - 14z + 45$
20. $z^2 - 14z + 49$
21. $z^2 - 12z - 160$
22. $p^2 + 2p - 35$
23. $p^2 + 12p + 27$
24. $p^2 - 6p + 8$
25. $x^2 + 20x + 100$
26. $x^2 + 18x + 81$
27. $b^2 + 9b + 20$
28. $b^2 + 13b + 40$
29. $x^2 - 11x - 42$
30. $x^2 + 9x - 70$
31. $b^2 - b - 20$
32. $b^2 + 3b - 40$
33. $y^2 - 14y - 51$
34. $y^2 - y - 72$
35. $p^2 - 4p - 21$
36. $p^2 + 16p + 39$
37. $y^2 - 8y + 32$
38. $y^2 - 9y + 81$
39. $x^2 - 20x + 75$
40. $p^2 + 24p + 63$

Objective B

Factor.

41. $2x^2 + 3x + 1$
42. $5x^2 + 6x + 1$
43. $2y^2 + 7y + 3$
44. $3y^2 + 7y + 2$
45. $2a^2 - 3a + 1$
46. $3a^2 - 4a + 1$

Factor.

47. $2t^2 - t - 10$

48. $2t^2 + 5t - 12$

49. $3p^2 - 16p + 5$

50. $6p^2 + 5p + 1$

51. $12y^2 - 7y + 1$

52. $6y^2 - 5y + 1$

53. $6z^2 - 7z + 3$

54. $9z^2 + 3z + 2$

55. $6t^2 - 11t + 4$

56. $10t^2 + 11t + 3$

57. $8x^2 + 33x + 4$

58. $7x^2 + 50x + 7$

59. $5x^2 - 62x - 7$

60. $9x^2 - 13x - 4$

61. $12y^2 + 19y + 5$

62. $18t^2 - 9t - 5$

63. $12t^2 + 28t - 5$

64. $6b^2 + 71b - 12$

65. $8b^2 + 65b + 8$

66. $9x^2 + 12x + 4$

67. $25x^2 - 30x + 9$

68. $6b^2 - 13b + 6$

69. $20b^2 + 37b + 15$

70. $33b^2 + 34b - 35$

71. $15b^2 - 43b + 22$

72. $18y^2 - 39y + 20$

73. $24y^2 + 41y + 12$

Critical Thinking

Determine the positive integer values of k for which the following polynomials are factorable over the integers.

74. $y^2 + 4y + k$

75. $z^2 + 7z + k$

76. $a^2 - 6a + k$

77. $c^2 - 7c + k$

78. $x^2 - 3x + k$

79. $y^2 + 5y + k$

80. [W] In your own words, explain how the signs of the last terms of the two binomial factors of a trinomial are determined.

SECTION 9.3 Special Factoring

OBJECTIVE A Factor the difference of two perfect squares or a perfect square trinomial

The product of a term and itself is called a **perfect square**. The exponents of variables of perfect squares are always even numbers.

Term		Perfect Square
2	$2 \cdot 2 =$	4
x	$x \cdot x =$	x^2
$3y^3$	$3y^3 \cdot 3y^3 =$	$9y^6$

The **square root** of a perfect square is one of the two equal factors of the perfect square. $\sqrt{\ }$, called a radical sign, is the symbol for square root. To find the exponent of the square root of a variable term, divide the exponent by 2.

$\sqrt{4} = 2$

$\sqrt{x^2} = x$

$\sqrt{9y^6} = 3y^3$

Sum and Difference of Two Terms		Difference of Two Perfect Squares
$(a + b)(a - b)$	$=$	$a^2 - b^2$

The factors of the difference of two perfect squares are the sum and difference of the square roots of the perfect squares.

$a^2 + b^2$ is the *sum* of two perfect squares. It is nonfactorable over the integers.

Factor $x^2 - 16$.

Write $x^2 - 16$ as the difference of two perfect squares.

$$x^2 - 16 = x^2 - 4^2$$

The factors are the sum and difference of the square roots of the perfect squares.

$$= (x + 4)(x - 4)$$

Check: $(x + 4)(x - 4) = x^2 - 4x + 4x - 16$
$= x^2 - 16$

Example 1

Factor $16x^2 - y^2$.

Solution

$16x^2 - y^2 = (4x)^2 - y^2 = (4x + y)(4x - y)$

You Try It 1

Factor $25a^2 - b^2$.

Your Solution

Example 2

Factor $z^6 - 25$.

Solution

$z^6 - 25 = (z^3)^2 - 5^2 = (z^3 + 5)(z^3 - 5)$

You Try It 2

Factor $n^8 - 36$.

Your Solution

Solutions on p. A27

OBJECTIVE C Factor completely

A polynomial is factored completely when it is written as a product of factors that are nonfactorable over the integers.

CONSIDER THIS

The first step in *any* factoring problem is to determine whether the terms of the polynomials have a **common factor**. If they do, factor it out first.

Factor $3x^3 - 23x^2 + 14x$.

Find the GCF of the terms of the polynomial. — The GCF is x.

Factor out the GCF. — $3x^3 - 23x^2 + 14x = x(3x^2 - 23x + 14)$

Factor the trinomial. — $x(\blacksquare x - \blacksquare)(\blacksquare x - \blacksquare)$

Write the factors of 3. — Factors of 3: 1, 3

Write the negative factors of 14. — Factors of 14: $-1, -14$; $-2, -7$

CONSIDER THIS

If the terms of the polynomial do not have a common factor, the trial factors will not have a common factor.

Write trial factors. Writing the 1 when it is the coefficient of x may be helpful. Determine the middle term of the trinomial.

Trial Factors	Middle Term
$(1x - 1)(3x - 14)$	$-14x - 3x = -17x$
$(1x - 14)(3x - 1)$	$-x - 42x = -43x$
$(1x - 2)(3x - 7)$	$-7x - 6x = -13x$
$(1x - 7)(3x - 2)$	$-2x - 21x = -23x$

Write the product of the GCF and the factors of the trinomial. — $x(x - 7)(3x - 2)$

Check: $x(x - 7)(3x - 2)$
$= x(3x^2 - 2x - 21x + 14)$
$= x(3x^2 - 23x + 14)$
$= 3x^3 - 23x^2 + 14x$

Factor $x^2 + 9xy + 20y^2$.

The terms have no common factor.
There are two variables. — $(x + \blacksquare y)(x + \blacksquare y)$
Find two positive factors of 20 whose sum is 9.

Factors	Sum
$+1, +20$	21
$+2, +10$	12
$+4, +5$	9

Write the factors of the trinomial. — $(x + 4y)(x + 5y)$

Check: $(x + 4y)(x + 5y)$
$= x^2 + 5xy + 4xy + 20y^2$
$= x^2 + 9xy + 20y^2$

Factoring Strategy

To factor a polynomial completely, ask the following questions about the polynomial.

1. Is there a common factor? If so, factor out the common factor.
2. Is the polynomial the difference of two perfect squares? If so, factor.
3. Is the polynomial a perfect square trinomial? If so, factor.
4. Is the polynomial a trinomial that is the product of two binomials? If so, factor.
5. Is each factor nonfactorable over the integers? If not, factor.

Example 7

Factor $4x^2 - 40xy + 84y^2$.

Solution

The GCF is 4.

$4x^2 - 40xy + 84y^2 = 4(x^2 - 10xy + 21y^2)$

Factor the trinomial.

$4(x - \blacksquare y)(x - \blacksquare y)$

Factors	Sum
−1, −21	−22
−3, −7	−10

$4(x - 3y)(x - 7y)$

$4x^2 - 40xy + 84y^2 = 4(x - 3y)(x - 7y)$

You Try It 7

Factor $3x^2 - 9xy - 12y^2$.

Your Solution

Example 8

Factor $3x^2 - 48$.

Solution

The GCF is 3.

$3x^2 - 48 = 3(x^2 - 16)$

Factor the difference of two perfect squares.

$= 3(x + 4)(x - 4)$

$3x^2 - 48 = 3(x + 4)(x - 4)$

You Try It 8

Factor $12x^3 - 75x$.

Your Solution

Example 9

Factor $x^2(x - 3) + 4(3 - x)$.

Solution

The common binomial factor is $x - 3$.

$x^2(x - 3) + 4(3 - x) =$

$x^2(x - 3) - 4(x - 3) = (x - 3)(x^2 - 4)$

Factor the difference of two perfect squares.

$= (x - 3)(x + 2)(x - 2)$

$x^2(x - 3) + 4(3 - x) = (x - 3)(x + 2)(x - 2)$

You Try It 9

Factor $a^2(b - 7) + (7 - b)$.

Your Solution

Solutions on p. A28

Example 10

Factor $2x^2y + 19xy - 10y$.

Solution

The GCF is y.

$2x^2y + 19xy - 10y = y(2x^2 + 19x - 10)$

Factor the trinomial.

$y(\blacksquare x + \blacksquare)(\blacksquare x - \blacksquare)$ or $y(\blacksquare x - \blacksquare)(\blacksquare x + \blacksquare)$

Factors of 2: 1, 2 Factors of -10: $+1, -10$
$-1, +10$
$+2, -5$
$-2, +5$

Trial Factors	Middle Term
$(1x + 1)(2x - 10)$	Common factor
$(1x - 10)(2x + 1)$	$x - 20x = -19x$
$(1x - 1)(2x + 10)$	Common factor
$(1x + 10)(2x - 1)$	$-x + 20x = 19x$
$(1x + 2)(2x - 5)$	$-5x + 4x = -x$
$(1x - 5)(2x + 2)$	Common factor
$(1x - 2)(2x + 5)$	$5x - 4x = x$
$(1x + 5)(2x - 2)$	Common factor

$y(x + 10)(2x - 1)$

$2x^2y + 19xy - 10y = y(x + 10)(2x - 1)$

You Try It 10

Factor $4a^2b^2 + 26a^2b - 14a^2$.

Your Solution

Example 11

Factor $12x - 32x^2 - 12x^3$.

Solution

The GCF is $4x$.

$12x - 32x^2 - 12x^3 = 4x(3 - 8x - 3x^2)$

Factor the trinomial.

$4x(\blacksquare + \blacksquare x)(\blacksquare - \blacksquare x)$ or $4x(\blacksquare - \blacksquare x)(\blacksquare + \blacksquare x)$

Factors of 3: 1, 3 Factors of -3: $+1, -3$
$-1, +3$

Trial Factors	Middle Term
$(1 + 1x)(3 - 3x)$	Common factor
$(1 - 3x)(3 + 1x)$	$x - 9x = -8x$
$(1 - 1x)(3 + 3x)$	Common factor
$(1 + 3x)(3 - 1x)$	$-x + 9x = 8x$

$4x(1 - 3x)(3 + x)$

$12x - 32x^2 - 12x^3 = 4x(1 - 3x)(3 + x)$

You Try It 11

Factor $12y + 12y^2 - 45y^3$.

Your Solution

Solutions on p. A28

9.3 EXERCISES

Objective A

Factor.

1. $x^2 - 4$
2. $x^2 - 9$
3. $a^2 - 81$
4. $a^2 - 49$
5. $4x^2 - 1$
6. $9x^2 - 16$
7. $x^6 - 9$
8. $y^{12} - 64$
9. $25x^2 - 1$
10. $4x^2 - 1$
11. $1 - 49x^2$
12. $1 - 64x^2$
13. $t^2 + 36$
14. $x^2 + 64$
15. $x^4 - y^2$
16. $b^4 - 16a^2$
17. $9x^2 - 16y^2$
18. $25z^2 - y^2$
19. $x^2y^2 - 4$
20. $a^2b^2 - 25$
21. $y^2 + 2y + 1$
22. $y^2 + 14y + 49$
23. $a^2 - 2a + 1$
24. $x^2 + 8x - 16$
25. $z^2 - 18z - 81$
26. $x^2 - 12x + 36$
27. $x^2 + 2xy + y^2$
28. $x^2 + 6xy + 9y^2$
29. $4a^2 + 4a + 1$
30. $25x^2 + 10x + 1$
31. $64a^2 - 16a + 1$
32. $9a^2 + 6a + 1$
33. $16b^2 + 8b + 1$
34. $4a^2 - 20a + 25$
35. $4b^2 + 28b + 49$
36. $9a^2 - 42a + 49$
37. $25a^2 + 30ab + 9b^2$
38. $4a^2 - 12ab + 9b^2$
39. $49x^2 + 28xy + 4y^2$
40. $4y^2 - 36yz + 81z^2$

Objective B

Factor.

41. $x(a + b) + 2(a + b)$
42. $a(x + y) + 4(x + y)$
43. $x(b + 2) - y(b + 2)$
44. $a(y - 4) - b(y - 4)$
45. $z(x - 3) - (x - 3)$
46. $a(y + 7) - (y + 7)$
47. $x(b - 2c) + y(b - 2c)$
48. $2x(x - 3) - (x - 3)$
49. $a(x - 2) + 5(2 - x)$
50. $a(x - 7) + b(7 - x)$
51. $b(y - 2) - 2a(y - 2)$
52. $x(a - 3) - 2y(a - 3)$
53. $b(y - 3) + 3(3 - y)$
54. $c(a - 2) - b(2 - a)$
55. $a(x - y) - 2(y - x)$

Objective C

Factor.

56. $2x^2 + 6x + 4$

57. $3x^2 + 15x + 18$

58. $3a^2 + 3a - 18$

59. $4x^2 - 4x - 8$

60. $ab^2 + 2ab - 15a$

61. $ab^2 + 7ab - 8a$

62. $xy^2 - 5xy + 6x$

63. $xy^2 + 8xy + 15x$

64. $z^3 - 7z^2 + 12z$

65. $2a^3 + 6a^2 + 4a$

66. $3y^3 - 15y^2 + 18y$

67. $4y^3 + 12y^2 - 72y$

68. $3x^2 + 3x - 36$

69. $2x^3 - 2x^2 - 4x$

70. $5z^2 - 15z - 140$

71. $6z^2 + 12z - 90$

72. $2a^3 + 8a^2 - 64a$

73. $3a^3 - 9a^2 - 54a$

74. $x^2 - 5xy + 6y^2$

75. $x^2 + 4xy - 21y^2$

76. $a^2 - 9ab + 20b^2$

77. $a^2 - 15ab + 50b^2$

78. $x^2 - 3xy - 28y^2$

79. $s^2 + 2st - 48t^2$

80. $5x^2 - 5$

81. $2x^2 - 18$

82. $x^3 + 4x^2 + 4x$

83. $y^3 - 10y^2 + 25y$

84. $x^4 + 2x^3 - 35x^2$

85. $a^4 - 11a^3 + 24a^2$

86. $5b^2 + 75b + 180$

87. $6y^2 - 48y + 72$

88. $3a^2 + 36a + 10$

89. $5a^2 - 30a + 4$

90. $2x^2y + 16xy - 66y$

91. $3a^2b + 21ab - 54b$

92. $x^2y^2 - 7xy^2 - 8y^2$

93. $a^2b^2 + 3a^2b - 88a^2$

94. $a^2b^2 - 10ab^2 + 25b^2$

95. $a^2b^2 + 6ab^2 + 9b^2$

96. $12a^3b - a^2b^2 - ab^3$

97. $2x^3y - 7x^2y^2 + 6xy^3$

Factor.

98. $4x^2 + 6x + 2$

99. $12x^2 + 33x - 9$

100. $15y^2 - 50y + 35$

101. $30y^2 + 10y - 20$

102. $2x^3 - 11x^2 + 5x$

103. $2x^3 - 3x^2 - 5x$

104. $3a^2b - 16ab + 16b$

105. $2a^2b - ab - 21b$

106. $3z^2 + 95z + 10$

107. $8z^2 - 36z + 1$

108. $3x^2 + xy - 2y^2$

109. $6x^2 + 10xy + 4y^2$

110. $3a^2 + 5ab - 2b^2$

111. $2a^2 - 9ab + 9b^2$

112. $4y^2 - 11yz + 6z^2$

113. $2y^2 + 7yz + 5z^2$

114. $12 - x - x^2$

115. $2 + x - x^2$

116. $28 + 3z - z^2$

117. $15 - 2z - z^2$

118. $8 - 7x - x^2$

119. $12 + 11x - x^2$

120. $9x^2 + 33x - 60$

121. $16x^2 - 16x - 12$

122. $80y^2 - 36y + 4$

123. $24y^2 - 24y - 18$

124. $8z^3 + 14z^2 + 3z$

125. $6z^3 - 23z^2 + 20z$

126. $6x^2y - 11xy - 10y$

127. $8x^2y - 27xy + 9y$

128. $24x^2 - 52x + 24$

129. $60x^2 + 95x + 20$

130. $35a^4 + 9a^3 - 2a^2$

131. $15a^4 + 26a^3 + 7a^2$

132. $15b^2 - 115b + 70$

133. $25b^2 + 35b - 30$

Factor.

134. $3x^2 - 26xy + 35y^2$

135. $4x^2 + 16xy + 15y^2$

136. $216y^2 - 3y - 3$

137. $360y^2 + 4y - 4$

138. $21 - 20x - x^2$

139. $18 + 17x - x^2$

140. $15a^2 + 11ab - 14b^2$

141. $15a^2 - 31ab + 10b^2$

142. $33z - 8z^2 - z^3$

143. $12a^3 - 12a^2 + 3a$

144. $18a^3 + 24a^2 + 8a$

145. $x^4 - 25x^2$

146. $a^4 - 16$

147. $a(2x - 2) + b(2x - 2)$

148. $4a(x - 3) - 2b(x - 3)$

149. $x^2(x - 2) - (x - 2)$

150. $y^2(a - b) - (a - b)$

151. $a(x^2 - 4) + b(x^2 - 4)$

Critical Thinking

Find all integers k such that the trinomial is a perfect square trinomial.

152. $4x^2 - kx + 9$

153. $25x^2 - kx + 1$

154. $36x^2 + kxy + y^2$

155. $64x^2 + kxy + y^2$

156. $x^2 + 6x + k$

157. $x^2 - 4x + k$

158. $x^2 - 2x + k$

159. $x^2 + 10x + k$

160. The prime factorization of a number is $2^3 \cdot 3^2$. How many of its whole number factors are perfect squares?

161. The product of two numbers is 48. One is a perfect square. The other is a prime number. Find the sum of the two numbers.

162. What is the smallest whole number by which 300 can be multiplied so that the product will be a perfect square?

163. Select any odd integer greater than 1, square it, and then subtract 1. Is the result evenly divisible by 8? Prove that this procedure always produces an integer divisible by 8. (Suggestion: Any odd integer greater than 1 can be expressed as $2n + 1$, where n is a natural number.)

SECTION 9.4 Solving Equations by Factoring

OBJECTIVE A Solve an equation by factoring

Recall that the Multiplication Property of Zero states that the product of a number and zero is zero. This property is restated below.

If a is a real number, then $a \cdot 0 = 0 \cdot a = 0$.

Now consider $x \cdot y = 0$. For this to be a true equation, then either $x = 0$ or $y = 0$.

> ***Principle of Zero Products***
>
> If the product of two factors is zero, then at least one of the factors must be zero.
>
> If $a \cdot b = 0$, then $a = 0$ or $b = 0$.

The Principle of Zero Products is used to solve some equations.

Solve: $(x - 2)(x - 3) = 0$

Use the Principle of Zero Products to set each factor equal to zero.

$$(x - 2)(x - 3) = 0$$
$$x - 2 = 0 \qquad x - 3 = 0$$

Solve each equation for x.

$$x = 2 \qquad x = 3$$

Check:

$$\begin{array}{r|l} (x - 2)(x - 3) & 0 \\ \hline (2 - 2)(2 - 3) & 0 \\ 0(-1) & 0 \\ 0 & 0 \text{ True} \end{array} \qquad \begin{array}{r|l} (x - 2)(x - 3) & 0 \\ \hline (3 - 2)(3 - 3) & 0 \\ (1)(0) & 0 \\ 0 & 0 \text{ True} \end{array}$$

The solutions are 2 and 3.

An equation of the form $ax^2 + bx + c = 0$, $a \neq 0$, is a **quadratic equation**. A quadratic equation is in **standard form** when the polynomial is in descending order and equal to zero. The quadratic equations at the right are in standard form.

$$3x^2 + 2x + 1 = 0$$
$$4x^2 - 3x + 2 = 0$$

Solve: $2x^2 + x = 6$ $\qquad 2x^2 + x = 6$

Write the equation in standard form. $\qquad 2x^2 + x - 6 = 0$

Factor. $\qquad (2x - 3)(x + 2) = 0$

Use the Principle of Zero Products. $\qquad 2x - 3 = 0 \qquad x + 2 = 0$

Rewrite each equation in the form *variable = constant.* $\qquad 2x = 3 \qquad x = -2$

$x = \frac{3}{2}$

$\frac{3}{2}$ and -2 check as solutions. The solutions are $\frac{3}{2}$ and -2.

Example 1

Solve: $x(x - 3) = 0$

Solution

$x(x - 3) = 0$

$x = 0 \qquad x - 3 = 0$

$x = 3$

The solutions are 0 and 3.

You Try It 1

Solve: $2x(x + 7) = 0$

Your Solution

Example 2

Solve: $2x^2 - 50 = 0$

Solution

$$2x^2 - 50 = 0$$
$$2(x^2 - 25) = 0$$
$$2(x + 5)(x - 5) = 0$$

$x + 5 = 0 \qquad x - 5 = 0$

$x = -5 \qquad x = 5$

The solutions are -5 and 5.

You Try It 2

Solve: $4x^2 - 9 = 0$

Your Solution

Example 3

Solve: $(x - 3)(x - 10) = -10$

Solution

$(x - 3)(x - 10) = -10$

$x^2 - 13x + 30 = -10$ • Multiply $(x - 3)(x - 10)$.

$x^2 - 13x + 40 = 0$ • Add 10 to each side of the equation. The equation is now in standard form.

$(x - 8)(x - 5) = 0$

$x - 8 = 0 \qquad x - 5 = 0$

$x = 8 \qquad x = 5$

The solutions are 8 and 5.

You Try It 3

Solve: $(x + 2)(x - 7) = 52$

Your Solution

Solutions on p. A29

OBJECTIVE B Application problems

Example 4

The sum of the squares of two consecutive positive even integers is equal to 100. Find the two integers.

Strategy

First positive even integer: n
Second positive even integer: $n + 2$
The sum of the square of the first positive even integer and the square of the second positive even integer is 100.

Solution

$$
\begin{aligned}
n^2 + (n + 2)^2 &= 100 \\
n^2 + n^2 + 4n + 4 &= 100 \\
2n^2 + 4n + 4 &= 100 \\
2n^2 + 4n - 96 &= 0 \\
2(n^2 + 2n - 48) &= 0 \\
2(n - 6)(n + 8) &= 0
\end{aligned}
$$

$n - 6 = 0 \qquad n + 8 = 0$
$n = 6 \qquad n = -8$

Because -8 is not a positive even integer, it is not a solution.

$n = 6$
$n + 2 = 6 + 2 = 8$

The two integers are 6 and 8.

You Try It 4

The sum of the squares of two consecutive positive integers is 61. Find the two integers.

Your Strategy

Your Solution

Solution on p. A29

Example 5

A stone is thrown into a well with an initial speed of 4 ft/s. The well is 420 ft deep. How many seconds later will the stone hit the bottom of the well? Use the equation $d = vt + 16t^2$, where d is the distance in feet, v is the initial speed, and t is the time in seconds.

Strategy

To find the time for the stone to drop to the bottom of the well, replace the variables d and v by their given values and solve for t.

Solution

$$d = vt + 16t^2$$
$$420 = 4t + 16t^2$$
$$0 = -420 + 4t + 16t^2$$
$$16t^2 + 4t - 420 = 0$$
$$4(4t^2 + t - 105) = 0$$
$$4(4t + 21)(t - 5) = 0$$

$$4t + 21 = 0 \qquad t - 5 = 0$$
$$4t = -21 \qquad t = 5$$
$$t = -\frac{21}{4}$$

Because time cannot be negative, $-\frac{21}{4}$ is not a solution.

The time is 5 s.

You Try It 5

The length of a rectangle is 4 in. longer than twice the width. The area of the rectangle is 96 in^2. Find the length and width of the rectangle.

Your Strategy

Your Solution

Solution on p. A29

9.4 EXERCISES

Objective A

Solve.

1. $(y+3)(y+2)=0$
2. $(y-3)(y-5)=0$
3. $(z-7)(z-3)=0$
4. $(z+8)(z-9)=0$
5. $x(x-5)=0$
6. $x(x+2)=0$
7. $a(a-9)=0$
8. $a(a+12)=0$
9. $y(2y+3)=0$
10. $t(4t-7)=0$
11. $2a(3a-2)=0$
12. $4b(2b+5)=0$
13. $(b+2)(b-5)=0$
14. $(b-8)(b+3)=0$
15. $x^2-81=0$
16. $x^2-121=0$
17. $4x^2-49=0$
18. $16x^2-1=0$
19. $9x^2-1=0$
20. $16x^2-49=0$
21. $x^2+6x+8=0$
22. $x^2-8x+15=0$
23. $z^2+5z-14=0$
24. $z^2+z-72=0$
25. $x^2-5x+6=0$
26. $x^2-3x-10=0$
27. $y^2+4y-21=0$
28. $2y^2-y-1=0$
29. $2a^2-9a-5=0$
30. $3a^2+14a+8=0$
31. $6z^2+5z+1=0$
32. $6y^2-19y+15=0$
33. $x^2-3x=0$
34. $a^2-5a=0$
35. $x^2-7x=0$
36. $2a^2-8a=0$
37. $a^2+5a=-4$
38. $a^2-5a=24$
39. $y^2-5y=-6$
40. $y^2-7y=8$
41. $2t^2+7t=4$
42. $3t^2+t=10$
43. $3t^2-13t=-4$
44. $5t^2-16t=-12$
45. $x(x-12)=-27$
46. $x(x-11)=12$
47. $y(y-7)=18$
48. $y(y+8)=-15$

Solve.

49. $p(p + 3) = -2$
50. $p(p - 1) = 20$
51. $y(y + 4) = 45$
52. $y(y - 8) = -15$
53. $x(x + 3) = 28$
54. $p(p - 14) = 15$
55. $(x + 8)(x - 3) = -30$
56. $(x + 4)(x - 1) = 14$
57. $(z - 5)(z + 4) = 52$
58. $(z - 8)(z + 4) = -35$
59. $(z - 6)(z + 1) = -10$
60. $(a + 3)(a + 4) = 72$
61. $(a - 4)(a + 7) = -18$
62. $(2x + 5)(x + 1) = -1$
63. $(z + 3)(z - 10) = -42$
64. $(y + 3)(2y + 3) = 5$
65. $(y + 5)(3y - 2) = -14$

Objective B

Solve.

66. The square of a positive number is seven more than six times the positive number. Find the number.

67. The square of a negative number is fifteen more than twice the negative number. Find the number.

68. The sum of the squares of two consecutive positive integers is sixty-one. Find the two integers.

69. The sum of the squares of two consecutive positive even integers is fifty-two. Find the two integers.

70. The product of two consecutive positive integers is two hundred forty. Find the integers.

71. The product of two consecutive positive even integers is one hundred sixty-eight. Find the integers.

72. The length of the base of a triangle is three times the height. The area of the triangle is 54 ft^2. Find the base and height of the triangle.

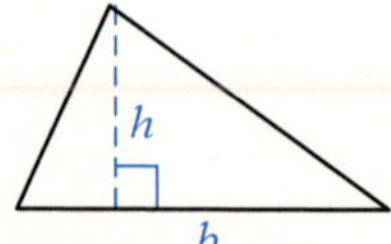

73. The height of a triangle is 4 m more than twice the length of the base. The area of the triangle is 35 m^2. Find the height of the triangle.

74. The length of a rectangle is four times the width. The area is 400 m^2. Find the length and width of the rectangle.

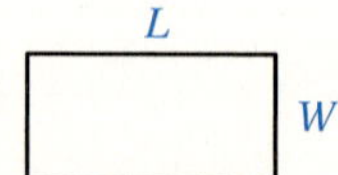

75. The length of a rectangle is two more than twice the width. The area is 144 ft^2. Find the length and width of the rectangle.

76. The length of each side of a square is extended 2 cm. The area of the resulting square is 64 cm^2. Find the length of a side of the original square.

Solve.

77. The length of each side of a square is extended 4 m. The area of the resulting square is 64 m^2. Find the length of a side of the original square.

78. A circle has a radius of 10 in. Find the increase in area when the radius is increased by 2 in. Round to the nearest tenth.

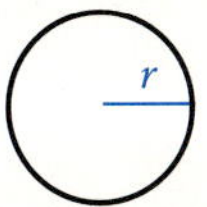

79. The radius of a circle is increased by 3 ft, increasing the area by 100 ft^2. Find the radius of the original circle. Round to the nearest tenth.

80. The page of a book measures 6 in. by 9 in. A uniform border around the page leaves 28 in^2 for type. Find the dimensions of the type area.

81. A small garden measures 8 ft by 10 ft. A uniform border around the garden increases the total area to 168 ft^2. Find the width of the border.

Use the formula $d = vt + 16t^2$, where d is the distance in feet, v is the initial velocity, and t is the time in seconds.

82. An object is released from a plane at an altitude of 1600 ft. The initial velocity is 0 ft/s. How many seconds later will the object hit the ground?

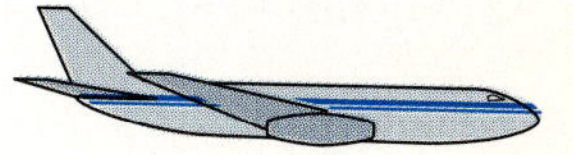

83. An object is released from the top of a building 320 ft high. The initial velocity is 16 ft/s. How many seconds later will the object hit the ground?

Use the formula $S = \frac{n^2 + n}{2}$, where S is the sum of the first n natural numbers.

84. How many consecutive natural numbers beginning with 1 will give a sum of 78?

85. How many consecutive natural numbers beginning with 1 will give a sum of 120?

Use the formula $N = \frac{t^2 - t}{2}$, where N is the number of football games that must be scheduled in a league with t teams if each team is to play every other team once.

86. A league has 28 games scheduled. How many teams are in the league if each team plays every other team once?

87. A league has 45 games scheduled. How many teams are in the league if each team plays every other team once?

Use the formula $h = vt - 16t^2$, where h is the height an object will attain (neglecting air resistance) in t seconds and v is its initial velocity.

88. A foul ball leaves a bat and travels straight up with an initial velocity of 64 ft/s. How many seconds later will the ball be 64 ft above the ground?

89. A golf ball is thrown onto a cement surface and rebounds straight up. The initial velocity of the rebound is 96 ft/s. How many seconds later will the golf ball return to the ground?

Critical Thinking

Solve.

90. $2y(y + 4) = -5(y + 3)$

91. $2y(y + 4) = 3(y + 4)$

92. $(a - 3)^2 = 36$

93. $(b + 5)^2 = 16$

94. $p^3 = 9p^2$

95. $p^3 = 7p^2$

96. $(2z - 3)(z + 5) = (z + 1)(z + 3)$

97. $(x + 3)(2x - 1) = (3 - x)(5 - 3x)$

98. Find $3n^2$ if $n(n + 5) = -4$.

99. Find $2n^3$ if $n(n + 3) = 4$.

100. The length of a rectangle is 7 cm, and the width is 4 cm. If both the length and the width are increased by equal amounts, the area of the rectangle is increased by 42 cm^3. Find the length and width of the larger rectangle.

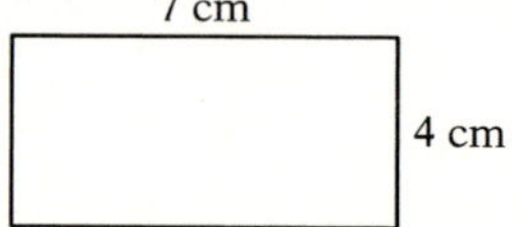

101. A rectangular piece of cardboard is 10 in. longer than it is wide. Squares 2 in. on a side are cut from each corner, and then the sides are folded up to make an open box with a volume of 192 in^3. Find the length and width of the piece of cardboard.

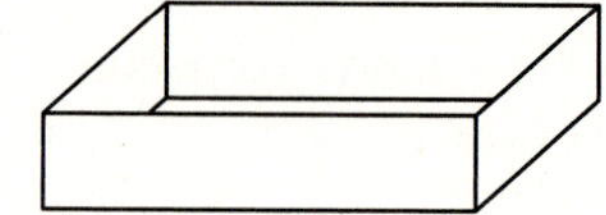

102. [W] Explain the error made in solving the equation at the right. Solve the equation correctly.

$$(x + 2)(x - 3) = 6$$
$$x + 2 = 6 \qquad x - 3 = 6$$
$$x = 4 \qquad x = 9$$

103. [W] Explain the error made in solving the equation at the right. Solve the equation correctly.

$$x^2 = x$$
$$\frac{x^2}{x} = \frac{x}{x}$$
$$x = 1$$

104. [W] In your own words, explain why it is possible to solve a quadratic equation using the Principle of Zero Products.

Projects in Mathematics

Continued Fractions

The following complex fraction is called a **continued fraction**.

$$1 + \cfrac{1}{1 + \cfrac{1}{1 + \cfrac{1}{1 + \cfrac{1}{1 + \cdots}}}}$$

The dots indicate that the pattern repeats forever.

A **convergent** for a continued fraction is an approximation of the repeated pattern. For instance,

$$c_2 = 1 + \cfrac{1}{1 + \cfrac{1}{1 + 1}} \qquad c_3 = 1 + \cfrac{1}{1 + \cfrac{1}{1 + \cfrac{1}{1 + 1}}} \qquad c_4 = 1 + \cfrac{1}{1 + \cfrac{1}{1 + \cfrac{1}{1 + \cfrac{1}{1 + 1}}}}$$

1. Calculate c_5 for the continued fraction above.

This continued fraction is related to the golden rectangle, which has been used in architectural designs as diverse as the Parthenon in Athens, built around 440 B.C., and the United Nations building. In a golden rectangle,

$$\frac{\text{length}}{\text{width}} = \frac{\text{length} + \text{width}}{\text{length}}$$

An example of a golden rectangle is shown at the right.

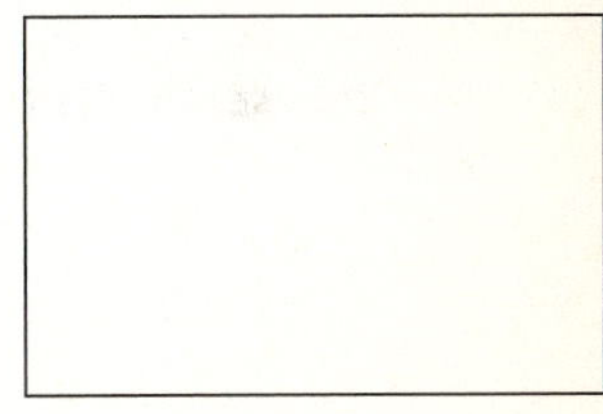

Here is another continued fraction that was discovered by Leonard Euler (1707–1793). Calculating the convergents of this continued fraction yields approximations that are closer and closer to π.

$$\pi = 3 + \cfrac{1^2}{6 + \cfrac{3^2}{6 + \cfrac{5^2}{6 + \cfrac{7^2}{6 + \cdots}}}}$$

2. Calculate $c_5 = 3 + \cfrac{1^2}{6 + \cfrac{3^2}{6 + \cfrac{5^2}{6 + \cfrac{7^2}{6 + \cfrac{9^2}{6 + 11^2}}}}}$.

Chapter Summary

Key Words

The *greatest common factor* (GCF) of two or more integers is the greatest integer that is a factor of all the integers.

To *factor* a polynomial means to write the polynomial as a product of other polynomials.

To *factor* a trinomial of the form $ax^2 + bx + c$ means to express the trinomial as the product of two binomials.

A product of a term and itself is a *perfect square.*

A polynomial that does not factor using only integers is *nonfactorable over the integers.*

An equation of the form $ax^2 + bx + c = 0$, where $a \neq 0$, is a *quadratic equation.*

A quadratic equation is in *standard form* when the polynomial is in descending order and equal to zero. $ax^2 + bx + c = 0$ is in standard form.

Essential Rules

The Principle of Zero Products
If the product of two factors is zero, then at least one of the factors must be zero.

If $a \cdot b = 0$, then $a = 0$ or $b = 0$.

The difference of two squares is the product of the sum and difference of two terms.

Difference of Two Perfect Squares		**Sum and Difference of Two Terms**
$a^2 - b^2$	$=$	$(a + b)(a - b)$

A perfect square trinomial is the square of a binomial.

Perfect Square Trinomial		**Square of a Binomial**
$a^2 + 2ab + b^2$	$=$	$(a + b)^2$
$a^2 - 2ab + b^2$	$=$	$(a - b)^2$

General Factoring Strategy

1. Is there a common factor? If so, factor out the common factor.
2. Is the polynomial the difference of two perfect squares? If so, factor.
3. Is the polynomial a perfect square trinomial? If so, factor.
4. Is the polynomial a trinomial that is the product of two binomials? If so, factor.
5. Is each binomial factor nonfactorable over the integers? If not, factor.

Chapter Review Exercises

1. Find the GCF of $12a^2b^3$ and $16ab^6$.

2. Factor: $6x^3 - 8x^2 + 10x$

3. Factor: $p^2 + 5p + 6$

4. Factor: $a^2 - 19a + 48$

5. Factor: $x^2 + 2x - 15$

6. Factor: $a(x - 2) + b(x - 2)$

7. Factor: $5x^2 - 45x - 15$

8. Factor: $2y^4 - 14y^3 - 16y^2$

9. Factor: $2x^2 + 4x - 5$

10. Factor: $6x^2 + 19x + 8$

11. Factor: $x(p + 1) - (p + 1)$

12. Factor: $8x^2 + 20x - 48$

13. Factor: $6x^2y^2 + 9xy^2 + 12y^2$

14. Factor: $y^3 - 9y$

15. Factor: $4x^2 - 49y^2$

16. Factor: $x(a^2 - b^2) - y(a^2 - b^2)$

17. Factor: $4a^2 - 12ab + 9b^2$

18. Solve: $(2a - 3)(a + 7) = 0$

19. Solve: $x(x - 8) = -15$

20. Solve: $4x^2 - 1 = 0$

21. The length of a rectangle is 3 cm longer than twice the width. The area of the rectangle is 90 cm^2. Find the length and width of the rectangle.

22. The length of the base of a triangle is three times the height. The area of the triangle is 24 in^2. Find the length of the base of the triangle.

23. The product of two consecutive negative integers is one hundred fifty-six. Find the two integers.

Cumulative Review Exercises

1. Simplify: $(-2x^2y)^3(2xy^2)^2$

2. Simplify: $\dfrac{-15x^7}{5x^5}$

3. Solve by substitution: $3x - 5y = -23$
$x + 2y = -4$

4. Multiply: $(3b - 2)(5b - 7)$

5. Multiply: $1\frac{3}{4} \cdot 2\frac{5}{8}$

6. Write $8.1 \cdot 10^{-4}$ in decimal notation.

7. Solve: $2x - 9 = 3x + 7$

8. 120% of what number is 54?

9. Simplify: $-3[x - 2(3 - 2x) - 5] + 2x$

10. Simplify: $(x^{-4}y^3)^2$

11. Factor: $p^2 - 9p - 10$

12. Solve: $-2 + 4[3x - 2(4 - x) - 3] = 4x + 2$

13. Multiply: $(-3x^2y)(-2x^3y^4)$

14. Solve: $2x - 3(2 - 3x) \geq 2x - 5$

15. Factor: $3x^3 + 2x^2 - 8x$

16. Solve by substitution: $6x + y = 7$
$x - 3y = 17$

17. Multiply: $(x + 2)(x^2 - 5x + 4)$

18. Divide: $(x^3 - 8) \div (x - 2)$

19. Graph: $2x + 3y = 6$

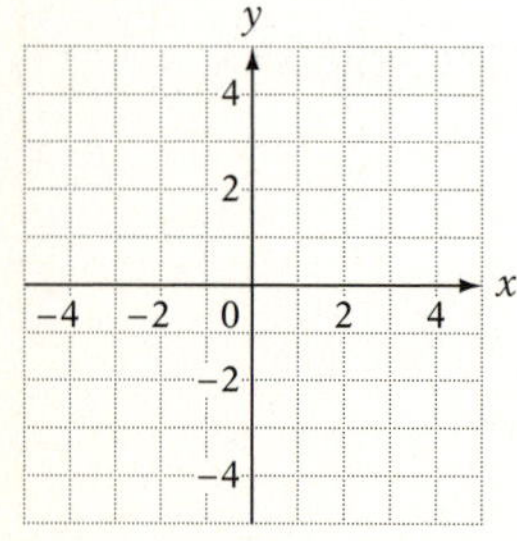

20. Find x.

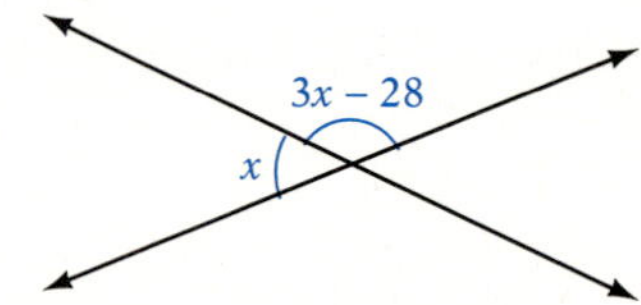

21. A silversmith mixes 60 g of an alloy that is 40% silver with 120 g of another silver alloy. The resulting alloy is 60% silver. Find the percent of silver in the 120-gram alloy.

22. A canoeist can paddle at a rate of 7 mph in calm water. Traveling with the current, the canoe traveled 40 mi in the same time as it traveled 16 mi against the current. Find the rate of the current.

23. The difference between two complementary angles is 16°. Find the measure of the two angles.

24. Find the length of a radius of a circle whose circumference is 12 in. Round to the nearest hundredth.

CHAPTER

10

Algebraic Fractions

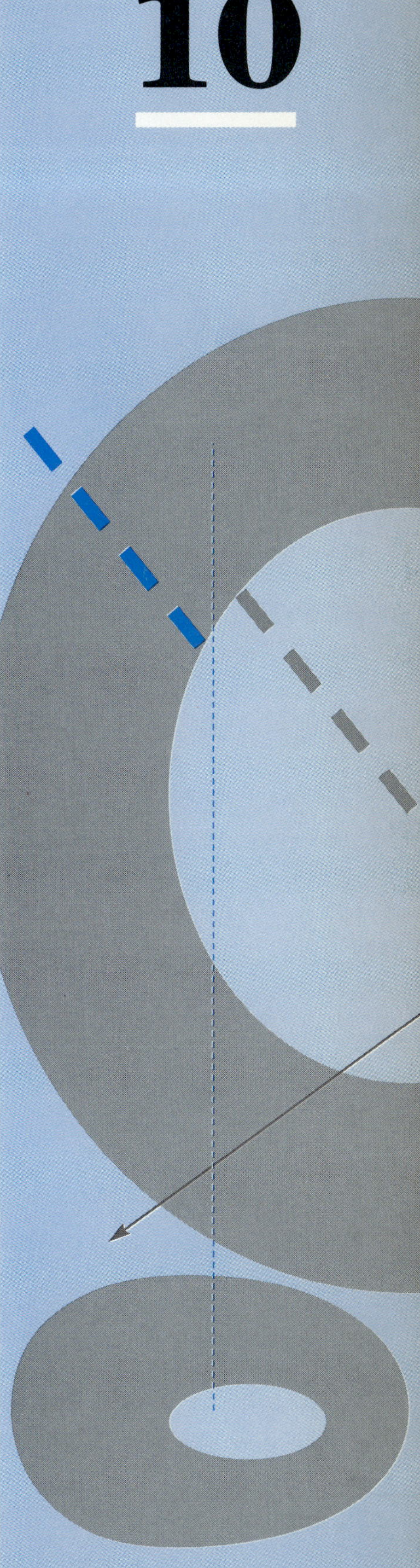

Focus on Problem Solving

Problems in mathematics or real life involve a question or a need and information or circumstances about that need. Solving problems in the sciences usually involves a question, observations, and measurements of some kind.

One of the challenges of problem solving in the sciences is to separate the relevant information about a problem from other information. Following is an example from the physical sciences in which some relevant information was omitted.

Hooke's Law states that the distance that a weight will stretch a spring is directly proportional to the weight on the spring. That is, $d = kF$, where d is the distance the spring is stretched and F is the force. In an experiment to verify this law, some physics students were continually getting the wrong results. Finally, the instructor discovered that the heat produced when the lights were turned on was affecting the experiment. In this case, some relevant information affecting the experiment was being omitted.

A lawyer drove 8 mi to the train station. After a 35-minute ride of 18 mi, the lawyer walked 10 min to the office. Find the total time it took the lawyer to get to work.

From this situation, answer the following before reading on.

- **a.** What is asked for?
- **b.** Is there enough information to answer the question?
- **c.** Is information given that is not necessary?

Here are the answers.

- **a.** We want the total time for the lawyer to get to work.
- **b.** No. We do not know the time it takes him to get to the train station.
- **c.** Yes. The distance to the train station and the distance of the train ride are not necessary to answer the question.

In the following problems

- **a.** What is asked for?
- **b.** Is there enough information to answer the question?
- **c.** Is some information not needed?

1. A customer bought 6 boxes of strawberries and paid with a $20 bill. What was the change?
2. A board is cut into two pieces. One piece is 3 ft longer than the other piece. What is the length of the original board?
3. A family rented a car for their vacation and drove 680 mi. The cost of the rental car was $21 per day with 150 free miles per day and $.15 for each mile over 150. How many miles did the family drive per day?
4. An investor bought 8 acres of land for $80,000. One and one-half acres were set aside for a park, and the remainder was developed into one-half-acre lots. How many lots were available for sale?
5. You wrote checks of $43.67, $122.88, and $432.22 after making a deposit of $768.55. How much do you have left in your checking account?

SECTION 10.1 Multiplication and Division of Algebraic Fractions

OBJECTIVE A Simplify algebraic fractions

A fraction in which the numerator or denominator is a variable expression is called an **algebraic fraction**. Examples of algebraic fractions are shown at the right.

$$\frac{5}{z}, \quad \frac{x^2+1}{2x-1}, \quad \frac{y^2-3}{3xy+1}$$

Care must be exercised with algebraic fractions to ensure that when the variables are replaced with numbers, the denominator does not equal zero.

Consider the algebraic fraction at the right. The value of x cannot be 2 because the denominator would then be zero.

$$\frac{3x+1}{2x-4}$$

$$\frac{3 \cdot 2+1}{2 \cdot 2-4} = \frac{7}{0} \quad \text{Not a real number}$$

A rational expression is in simplest form when the numerator and denominator have no common factors. The Multiplication Property of One is used to write a rational expression in simplest form.

CONSIDER THIS

Simplifying an algebraic fraction is closely related to simplifying an arithmetic fraction—the common factors are removed.

Evaluate $\frac{x^2-4}{x^2-2x-8}$ and $\frac{x-2}{x-4}$ for various values of x and note that, except for -2 and 4, the expressions are equal.

Simplify: $\frac{x^2-4}{x^2-2x-8}$

Factor the numerator and denominator.

$$\frac{x^2-4}{x^2-2x-8} = \frac{(x-2)(x+2)}{(x-4)(x+2)}$$

$$= \frac{x-2}{x-4} \cdot \boxed{\frac{x+2}{x+2}} = \frac{x-2}{x-4} \cdot 1$$

The restrictions, $x \neq -2, 4$, are necessary to prevent division by zero.

$$= \frac{x-2}{x-4}, \quad x \neq -2, 4$$

This simplification is usually shown with slashes through the common factors. The last simplification would be shown as follows:

Factor the numerator and denominator.

$$\frac{x^2-4}{x^2-2x-8} = \frac{(x-2)\overset{1}{\cancel{(x+2)}}}{(x-4)\underset{1}{\cancel{(x+2)}}}$$

Divide by the common factors. The restrictions, $x = -2, 4$, are necessary to prevent division by zero.

$$= \frac{x-2}{x-4}, \; x \neq -2, 4$$

For the remaining examples, we will omit the restrictions on the variables that prevent division by zero and assume the values of the variables are such that the denominator will not equal zero.

Recall that $(b - a) = (-a + b) = -(a - b)$.

Therefore, $\frac{b-a}{a-b} = \frac{-(a-b)}{a-b} = \frac{-1}{1} = -1$.

Simplify: $\frac{6-2x}{x-3}$

Factor the numerator. $\frac{6-2x}{x-3} = \frac{2(3-x)}{x-3}$

Divide by the common factors. $= \frac{2\overset{-1}{\cancel{(3-x)}}}{\underset{1}{\cancel{x-3}}}$

$\frac{3-x}{x-3} = \frac{-(x-3)}{x-3} = \frac{-1}{1}$

Write the fraction in simplest form. $= -2$

Simplify: $\frac{10+3x-x^2}{x^2-4x-5}$

Factor the numerator and the denominator. $\frac{10+3x-x^2}{x^2-4x-5} = \frac{(5-x)(2+x)}{(x-5)(x+1)}$

Divide by the common factors. $= \frac{\overset{-1}{\cancel{(5-x)}}(2+x)}{\underset{1}{\cancel{(x-5)}}(x+1)}$

$\frac{5-x}{x-5} = \frac{-(x-5)}{x-5} = \frac{-1}{1}$

Write the fraction in simplest form. $= -\frac{x+2}{x+1}$

Example 1

Simplify: $\frac{4x^3y^4}{6x^4y}$

Solution

$\frac{4x^3y^4}{6x^4y} = \frac{\overset{1}{\cancel{2}} \cdot 2x^3y^4}{\underset{1}{\cancel{2}} \cdot 3x^4y} = \frac{2y^3}{3x}$ Use rules of exponents.

You Try It 1

Simplify: $\frac{6x^5y}{12x^2y^3}$

Your Solution

Example 2

Simplify: $\frac{9-x^2}{x^2+x-12}$

Solution

$\frac{9-x^2}{x^2+x-12} = \frac{\overset{-1}{\cancel{(3-x)}}(3+x)}{\underset{1}{\cancel{(x-3)}}(x+4)} = -\frac{x+3}{x+4}$

You Try It 2

Simplify: $\frac{x^2+2x-24}{16-x^2}$

Your Solution

Solutions on p. A29

OBJECTIVE B Multiplication of algebraic fractions

The product of two fractions is a fraction whose numerator is the product of the numerators of the two fractions and whose denominator is the product of the denominators of the two fractions.

$$\frac{a}{b} \cdot \frac{c}{d} = \frac{ac}{bd}$$

$$\frac{3x}{y} \cdot \frac{2}{z} = \frac{6x}{yz}$$

$$\frac{x+2}{x} \cdot \frac{3}{x-2} = \frac{3x+6}{x^2-2x}$$

Multiply: $\frac{x^2+3x}{x^2-3x-4} \cdot \frac{x^2-5x+4}{x^2+2x-3}$

$$\frac{x^2+3x}{x^2-3x-4} \cdot \frac{x^2-5x+4}{x^2+2x-3}$$

Factor the numerator and denominator of each fraction.

$$= \frac{x(x+3)}{(x-4)(x+1)} \cdot \frac{(x-4)(x-1)}{(x+3)(x-1)}$$

Multiply. Divide by the common factors.

$$= \frac{x\overset{1}{\cancel{(x+3)}}\overset{1}{\cancel{(x-4)}}\overset{1}{\cancel{(x-1)}}}{\underset{1}{\cancel{(x-4)}}(x+1)\underset{1}{\cancel{(x+3)}}\underset{1}{\cancel{(x-1)}}}$$

Write the product in simplest form.

$$= \frac{x}{x+1}$$

Example 3

Multiply: $\frac{10x^2-15x}{12x-8} \cdot \frac{3x-2}{20x-25}$

Solution

$$\frac{10x^2-15x}{12x-8} \cdot \frac{3x-2}{20x-25}$$

$$= \frac{5x(2x-3)}{4(3x-2)} \cdot \frac{(3x-2)}{5(4x-5)}$$

$$= \frac{\overset{1}{\cancel{5}}x(2x-3)\overset{1}{\cancel{(3x-2)}}}{2 \cdot 2\underset{1}{\cancel{(3x-2)}}\underset{1}{\cancel{5}}(4x-5)} = \frac{x(2x-3)}{4(4x-5)}$$

You Try It 3

Multiply: $\frac{12x^2+3x}{10x-15} \cdot \frac{8x-12}{9x+18}$

Your Solution

Example 4

Multiply: $\frac{x^2+x-6}{x^2+7x+12} \cdot \frac{x^2+3x-4}{4-x^2}$

Solution

$$\frac{x^2+x-6}{x^2+7x+12} \cdot \frac{x^2+3x-4}{4-x^2}$$

$$= \frac{(x+3)(x-2)}{(x+3)(x+4)} \cdot \frac{(x+4)(x-1)}{(2-x)(2+x)}$$

$$= \frac{\overset{1}{\cancel{(x+3)}}\overset{-1}{\cancel{(x-2)}}\overset{1}{\cancel{(x+4)}}(x-1)}{\underset{1}{\cancel{(x+3)}}\underset{1}{\cancel{(x+4)}}\underset{1}{\cancel{(2-x)}}(2+x)} = -\frac{x-1}{x+2}$$

You Try It 4

Multiply: $\frac{x^2+2x-15}{9-x^2} \cdot \frac{x^2-3x-18}{x^2-7x+6}$

Your Solution

Solutions on p. A30

OBJECTIVE C Division of algebraic fractions

The **reciprocal** of a fraction is the fraction with the numerator and denominator interchanged.

Fraction	Reciprocal
$\dfrac{a}{b}$	$\dfrac{b}{a}$
$x^2 = \dfrac{x^2}{1}$	$\dfrac{1}{x^2}$
$\dfrac{x+2}{x}$	$\dfrac{x}{x+2}$

To divide two fractions, multiply by the reciprocal of the divisor.

$$\frac{a}{b} \div \frac{c}{d} = \frac{a}{b} \cdot \frac{d}{c} = \frac{ad}{bc}$$

$$\frac{4}{x} \div \frac{y}{5} = \frac{4}{x} \cdot \frac{5}{y} = \frac{20}{xy}$$

$$\frac{x+4}{x} \div \frac{x-2}{4} = \frac{x+4}{x} \cdot \frac{4}{x-2} = \frac{4(x+4)}{x(x-2)}$$

The basis for the division rule is shown at the right.

$$\frac{a}{b} \div \frac{c}{d} = \frac{\frac{a}{b}}{\frac{c}{d}} \cdot \frac{\frac{d}{c}}{\frac{d}{c}} = \frac{\frac{a}{b} \cdot \frac{d}{c}}{\frac{c}{d} \cdot \frac{d}{c}} = \frac{\frac{a}{b} \cdot \frac{d}{c}}{1} = \frac{a}{b} \cdot \frac{d}{c}$$

$$\frac{a}{b} \div \frac{c}{d} = \frac{a}{b} \cdot \frac{d}{c}$$

Example 5

Divide: $\dfrac{xy^2 - 3x^2y}{z^2} \div \dfrac{6x^2 - 2xy}{z^3}$

Solution

$$\frac{xy^2 - 3x^2y}{z^2} \div \frac{6x^2 - 2xy}{z^3}$$

$$= \frac{xy^2 - 3x^2y}{z^2} \cdot \frac{z^3}{6x^2 - 2xy}$$

$$= \frac{xy(y - 3x) \cdot z^3}{z^2 \cdot 2x(3x - y)} = -\frac{yz}{2}$$

You Try It 5

Divide: $\dfrac{a^2}{4bc^2 - 2b^2c} \div \dfrac{a}{6bc - 3b^2}$

Your Solution

Example 6

Divide: $\dfrac{2x^2 + 5x + 2}{2x^2 + 3x - 2} \div \dfrac{3x^2 + 13x + 4}{2x^2 + 7x - 4}$

Solution

$$\frac{2x^2 + 5x + 2}{2x^2 + 3x - 2} \div \frac{3x^2 + 13x + 4}{2x^2 + 7x - 4}$$

$$= \frac{2x^2 + 5x + 2}{2x^2 + 3x - 2} \cdot \frac{2x^2 + 7x - 4}{3x^2 + 13x + 4}$$

$$= \frac{(2x+1)(x+2)}{(2x-1)(x+2)} \cdot \frac{(2x-1)(x+4)}{(3x+1)(x+4)} = \frac{2x+1}{3x+1}$$

You Try It 6

Divide: $\dfrac{3x^2 + 26x + 16}{3x^2 - 7x - 6} \div \dfrac{2x^2 + 9x - 5}{x^2 + 2x - 15}$

Your Solution

Solutions on p. A30

10.1 EXERCISES

Objective A

Simplify.

1. $\dfrac{9x^3}{12x^4}$

2. $\dfrac{16x^2y}{24xy^3}$

3. $\dfrac{(x+3)^2}{(x+3)^3}$

4. $\dfrac{(2x-1)^5}{(2x-1)^4}$

5. $\dfrac{3n-4}{4-3n}$

6. $\dfrac{5-2x}{2x-5}$

7. $\dfrac{6y(y+2)}{9y^2(y+2)}$

8. $\dfrac{12x^2(3-x)}{18x(3-x)}$

9. $\dfrac{6x(x-5)}{8x^2(5-x)}$

10. $\dfrac{14x^3(7-3x)}{21x(3x-7)}$

11. $\dfrac{10-5y}{y-2}$

12. $\dfrac{18-6a}{a-3}$

13. $\dfrac{a^2+4a}{ab+4b}$

14. $\dfrac{x^2-3x}{2x-6}$

15. $\dfrac{4-6x}{3x^2-2x}$

16. $\dfrac{5xy-3y}{9-15x}$

17. $\dfrac{y^2-3y+2}{y^2-4y+3}$

18. $\dfrac{x^2+5x+6}{x^2+8x+15}$

19. $\dfrac{x^2+3x-10}{x^2+2x-8}$

20. $\dfrac{a^2+7a-8}{a^2+6a-7}$

21. $\dfrac{x^2+x-12}{x^2-6x+9}$

22. $\dfrac{x^2+8x+16}{x^2-2x-24}$

23. $\dfrac{x^2-3x-10}{25-x^2}$

24. $\dfrac{4-y^2}{y^2-3y-10}$

25. $\dfrac{2x^3+2x^2-4x}{x^3+2x^2-3x}$

26. $\dfrac{3x^3-12x}{6x^3-24x^2+24x}$

27. $\dfrac{6x^2-7x+2}{6x^2+5x-6}$

28. $\dfrac{2n^2-9n+4}{2n^2-5n-12}$

29. $\dfrac{x^2+3x-28}{24-2x-x^2}$

30. $\dfrac{x^2+2x-3}{5-4x-x^2}$

31. $\dfrac{8+2y-y^2}{y^2+y-20}$

32. $\dfrac{1-x^2}{x^2+8x-9}$

Objective B

Multiply.

33. $\frac{8x^2}{9y^3} \cdot \frac{3y^2}{4x^3}$

34. $\frac{4a^2b^3}{15x^5y^2} \cdot \frac{25x^3y}{16ab}$

35. $\frac{12x^3y^4}{7a^2b^3} \cdot \frac{14a^3b^4}{9x^2y^2}$

36. $\frac{18a^4b^2}{25x^2y^3} \cdot \frac{50x^5y^6}{27a^6b^2}$

37. $\frac{3x - 6}{5x - 20} \cdot \frac{10x - 40}{27x - 54}$

38. $\frac{8x - 12}{14x + 7} \cdot \frac{42x + 21}{32x - 48}$

39. $\frac{x^2 + 5x + 4}{x^3y^2} \cdot \frac{x^2y^3}{x^2 + 2x + 1}$

40. $\frac{x^2 + x - 2}{xy^2} \cdot \frac{x^3y}{x^2 + 5x + 6}$

41. $\frac{x^4y^2}{x^2 + 3x - 28} \cdot \frac{x^2 - 49}{xy^4}$

42. $\frac{x^5y^3}{x^2 + 13x + 30} \cdot \frac{x^2 + 2x - 3}{x^7y^2}$

43. $\frac{x^2 - 2x - 24}{x^2 - 5x - 6} \cdot \frac{x^2 + 5x + 6}{x^2 + 6x + 8}$

44. $\frac{x^2 - 8x + 7}{x^2 + 3x - 4} \cdot \frac{x^2 + 3x - 10}{x^2 - 9x + 14}$

45. $\frac{x^2 + 2x - 35}{x^2 + 4x - 21} \cdot \frac{x^2 + 3x - 18}{x^2 + 9x + 18}$

46. $\frac{y^2 + y - 20}{y^2 + 2y - 15} \cdot \frac{y^2 + 4y - 21}{y^2 + 3y - 28}$

47. $\frac{2x^2 + 5x + 2}{2x^2 + 7x + 3} \cdot \frac{x^2 - 7x - 30}{x^2 - 6x - 40}$

48. $\frac{x^2 - 4x - 32}{x^2 - 8x - 48} \cdot \frac{3x^2 + 17x + 10}{3x^2 - 22x - 16}$

49. $\frac{2x^2 + x - 3}{2x^2 - x - 6} \cdot \frac{2x^2 - 9x + 10}{2x^2 - 3x + 1}$

50. $\frac{3y^2 + 14y + 8}{2y^2 + 7y - 4} \cdot \frac{2y^2 + 9y - 5}{3y^2 + 16y + 5}$

51. $\frac{6x^2 - 11x + 4}{6x^2 + x - 2} \cdot \frac{12x^2 + 11x + 2}{8x^2 + 14x + 3}$

52. $\frac{6 - x - 2x^2}{4x^2 + 3x - 10} \cdot \frac{3x^2 + 7x - 20}{2x^2 + 5x - 12}$

Objective C

Divide.

53. $\dfrac{4x^2y^3}{15a^2b^3} \div \dfrac{6xy}{5a^3b^5}$

54. $\dfrac{9x^3y^4}{16a^4b^2} \div \dfrac{45x^4y^2}{14a^7b}$

55. $\dfrac{6x - 12}{8x + 32} \div \dfrac{18x - 36}{10x + 40}$

56. $\dfrac{28x + 14}{45x - 30} \div \dfrac{14x + 7}{30x - 20}$

57. $\dfrac{6x^3 + 7x^2}{12x - 3} \div \dfrac{6x^2 + 7x}{36x - 9}$

58. $\dfrac{5a^2y + 3a^2}{2x^3 + 5x^2} \div \dfrac{10ay + 6a}{6x^3 + 15x^2}$

59. $\dfrac{x^2 + 4x + 3}{x^2y} \div \dfrac{x^2 + 2x + 1}{xy^2}$

60. $\dfrac{x^3y^2}{x^2 - 3x - 10} \div \dfrac{xy^4}{x^2 - x - 20}$

61. $\dfrac{x^2 - 49}{x^4y^3} \div \dfrac{x^2 - 14x + 49}{x^4y^3}$

62. $\dfrac{x^2y^5}{x^2 - 11x + 30} \div \dfrac{xy^6}{x^2 - 7x + 10}$

63. $\dfrac{4ax - 8a}{c^2} \div \dfrac{2y - xy}{c^3}$

64. $\dfrac{3x^2y - 9xy}{a^2b} \div \dfrac{3x^2 - x^3}{ab^2}$

65. $\dfrac{x^2 - 5x + 6}{x^2 - 9x + 18} \div \dfrac{x^2 - 6x + 8}{x^2 - 9x + 20}$

66. $\dfrac{x^2 + 3x - 40}{x^2 + 2x - 35} \div \dfrac{x^2 + 2x - 48}{x^2 + 3x - 18}$

67. $\dfrac{x^2 + 2x - 15}{x^2 - 4x - 45} \div \dfrac{x^2 + x - 12}{x^2 - 5x - 36}$

68. $\dfrac{y^2 - y - 56}{y^2 + 8y + 7} \div \dfrac{y^2 - 13y + 40}{y^2 - 4y - 5}$

69. $\dfrac{8 + 2x - x^2}{x^2 + 7x + 10} \div \dfrac{x^2 - 11x + 28}{x^2 - x - 42}$

70. $\dfrac{x^2 - x - 2}{x^2 - 7x + 10} \div \dfrac{x^2 - 3x - 4}{40 - 3x - x^2}$

Divide.

71. $\dfrac{2x^2 - 3x - 20}{2x^2 - 7x - 30} \div \dfrac{2x^2 - 5x - 12}{4x^2 + 12x + 9}$

72. $\dfrac{6n^2 + 13n + 6}{4n^2 - 9} \div \dfrac{6n^2 + n - 2}{4n^2 - 1}$

73. $\dfrac{9x^2 - 16}{6x^2 - 11x + 4} \div \dfrac{6x^2 + 11x + 4}{8x^2 + 10x + 3}$

74. $\dfrac{15 - 14x - 8x^2}{4x^2 + 4x - 15} \div \dfrac{4x^2 + 13x - 12}{3x^2 + 13x + 4}$

75. $\dfrac{8x^2 + 18x - 5}{10x^2 - 9x + 2} \div \dfrac{8x^2 + 22x + 15}{10x^2 + 11x - 6}$

76. $\dfrac{10 + 7x - 12x^2}{8x^2 - 2x - 15} \div \dfrac{6x^2 - 13x + 5}{10x^2 - 13x + 4}$

Critical Thinking

For what values of x is the algebraic fraction undefined? (*Hint:* Set the denominator equal to zero and solve for x.)

77. $\dfrac{x}{(x + 6)(x - 1)}$

78. $\dfrac{x}{(x - 2)(x + 5)}$

79. $\dfrac{8}{x^2 - 1}$

80. $\dfrac{7}{x^2 - 16}$

81. $\dfrac{x - 4}{x^2 - x - 6}$

82. $\dfrac{x + 5}{x^2 - 4x - 5}$

83. $\dfrac{3x}{x^2 + 6x + 9}$

84. $\dfrac{3x - 8}{3x^2 - 10x - 8}$

85. $\dfrac{4x + 7}{6x^2 - 5x - 4}$

Simplify.

86. $\dfrac{y^2}{x} \cdot \dfrac{x}{2} \div \dfrac{y}{x}$

87. $\dfrac{ab}{3} \cdot \dfrac{a}{b^2} \div \dfrac{a}{4}$

88. $\left(\dfrac{2x}{y}\right)^3 \div \left(\dfrac{x}{3y}\right)^2$

89. $\left(\dfrac{c}{3}\right)^2 \div \left(\dfrac{c}{2} \cdot \dfrac{c}{4}\right)$

90. $\left(\dfrac{a - 3}{b}\right)^2\left(\dfrac{b}{3 - a}\right)^3$

91. $\left(\dfrac{x - 4}{y^2}\right)^3 \cdot \left(\dfrac{y}{4 - x}\right)^2$

92. $\dfrac{x^2 + 3x - 40}{x^2 + 2x - 35} \div \dfrac{x^2 + 2x - 48}{x^2 + 3x - 18} \cdot \dfrac{x^2 - 36}{x^2 - 9}$

93. $\dfrac{x^2 + x - 6}{x^2 + 7x + 12} \cdot \dfrac{x^2 + 3x - 4}{x^2 + x - 2} \div \dfrac{x^2 - 16}{x^2 - 4}$

94. [W] Given the expression $\dfrac{9}{x^2 + 1}$, choose some values of x and evaluate the expression for those values. Is it possible to choose a value of x for which the value of the expression is greater than 10? If so, what is that value of x? If not, explain why it is not possible.

95. [W] Given the expression $\dfrac{1}{y - 3}$, choose some values of y and evaluate the expression for those values. Is it possible to choose a value of y for which the value of the expression is greater than 10,000,000? If so, what is that value of y? If not, explain why it is not possible.

SECTION 10.2 Addition and Subtraction of Algebraic Fractions

OBJECTIVE A The least common multiple (LCM) of two or more polynomials

The **least common multiple (LCM)** of two or more numbers is the smallest number that contains the prime factorization of each number.

The LCM of 12 and 18 is 36. 36 contains the prime factors of 12 and the prime factors of 18.

$$12 = 2 \cdot 2 \cdot 3$$
$$18 = 2 \cdot 3 \cdot 3$$
$$\text{LCM} = 36 = \overbrace{2 \cdot 2 \cdot 3}^{\text{Factors of 12}} \cdot 3 \quad\text{and}\quad 2 \cdot \underbrace{2 \cdot 3 \cdot 3}_{\text{Factors of 18}}$$

The least common multiple of two or more polynomials is the simplest polynomial that contains the factors of each polynomial.

To find the LCM of two or more polynomials, first factor each polynomial completely. The LCM is the product of each factor the greatest number of times it occurs in any one factorization.

CONSIDER THIS
The LCM must contain the factors of each polynomial. As shown with the braces at the right, the LCM contains the factors of $4x^2 + 4x$ and the factors of $x^2 + 2x + 1$.

Find the LCM of $4x^2 + 4x$ and $x^2 + 2x + 1$.

The LCM of the polynomials is the product of the LCM of the numerical coefficients and each variable factor the greatest number of times it occurs in any one factorization.

$$4x^2 + 4x = 4x(x + 1) = 2 \cdot 2 \cdot x(x + 1)$$
$$x^2 + 2x + 1 = (x + 1)(x + 1)$$

$$\text{LCM} = \overbrace{2 \cdot 2 \cdot x(x + 1)}^{\text{Factors of } 4x^2 + 4x}(x + 1) = 4x(x + 1)(x + 1)$$

(Factors of $x^2 + 2x + 1$: $(x + 1)(x + 1)$)

Example 1
Find the LCM of $4x^2y$ and $6xy^2$.

Solution
$4x^2y = 2 \cdot 2 \cdot x \cdot x \cdot y \qquad 6xy^2 = 2 \cdot 3 \cdot x \cdot y \cdot y$
$\text{LCM} = 2 \cdot 2 \cdot 3 \cdot x \cdot x \cdot y \cdot y = 12x^2y^2$

You Try It 1
Find the LCM of $8uv^2$ and $12uw$.

Your Solution

Example 2
Find the LCM of $x^2 - x - 6$ and $9 - x^2$.

Solution
$x^2 - x - 6 = (x - 3)(x + 2)$
$9 - x^2 = -(x^2 - 9) = -(x + 3)(x - 3)$
$\text{LCM} = (x - 3)(x + 2)(x + 3)$

You Try It 2
Find the LCM of $m^2 - 6m + 9$ and $m^2 - 2m - 3$.

Your Solution

Solutions on p. A30

OBJECTIVE B Express two fractions in terms of the LCM of their denominators

When adding and subtracting fractions, it is frequently necessary to express two or more fractions in terms of a common denominator. This common denominator is the LCM of the denominators of the fractions.

CONSIDER THIS

The LCM must contain the factors of each polynomial. The LCM of the denominators at the right contains the factors of $4x^2$ and the factors of $6x^2 - 12x$.

Write the fractions $\frac{x+1}{4x^2}$ and $\frac{x-3}{6x^2-12x}$ in terms of the LCM of the denominators.

Find the LCM of the denominators. The LCM is $12x^2(x - 2)$.

For each fraction, multiply the numerator and denominator by the factors whose product with the denominator is the LCM.

$$\frac{x+1}{4x^2} = \frac{x+1}{4x^2} \cdot \frac{3(x-2)}{3(x-2)} = \frac{3x^2-3x-6}{12x^2(x-2)} \leftarrow \text{LCM}$$

$$\frac{x-3}{6x^2-12x} = \frac{x-3}{6x(x-2)} \cdot \frac{2x}{2x} = \frac{2x^2-6x}{12x^2(x-2)} \leftarrow \text{LCM}$$

Example 3

Write the fractions $\frac{x+2}{3x^2}$ and $\frac{x-1}{8xy}$ in terms of the LCM of the denominators.

Solution

The LCM is $24x^2y$.

$$\frac{x+2}{3x^2} = \frac{x+2}{3x^2} \cdot \frac{8y}{8y} = \frac{8xy+16y}{24x^2y}$$

$$\frac{x-1}{8xy} = \frac{x-1}{8xy} \cdot \frac{3x}{3x} = \frac{3x^2-3x}{24x^2y}$$

You Try It 3

Write the fractions $\frac{x-3}{4xy^2}$ and $\frac{2x+1}{9y^2z}$ in terms of the LCM of the denominators.

Your Solution

Example 4

Write the fractions $\frac{2x-1}{2x-x^2}$ and $\frac{x}{x^2+x-6}$ in terms of the LCM of the denominators.

Solution

$$\frac{2x-1}{2x-x^2} = \frac{2x-1}{-(x^2-2x)} = -\frac{2x-1}{x^2-2x}$$

The LCM is $x(x - 2)(x + 3)$.

$$\frac{2x-1}{2x-x^2} = -\frac{2x-1}{x(x-2)} \cdot \frac{x+3}{x+3} = -\frac{2x^2+5x-3}{x(x-2)(x+3)}$$

$$\frac{x}{x^2+x-6} = \frac{x}{(x-2)(x+3)} \cdot \frac{x}{x} = \frac{x^2}{x(x-2)(x+3)}$$

You Try It 4

Write the fractions $\frac{x+4}{x^2-3x-10}$ and $\frac{2x}{25-x^2}$ in terms of the LCM of the denominators.

Your Solution

Solutions on p. A30

OBJECTIVE C Addition and subtraction of algebraic fractions with like denominators

When adding algebraic fractions in which the denominators are the same, add the numerators. The denominator of the sum is the common denominator.

$$\frac{a}{b} + \frac{c}{b} = \frac{a+c}{b}$$

$$\frac{5x}{18} + \frac{7x}{18} = \frac{12x}{18} = \frac{2x}{3}$$

Note that the sum is written in simplest form.

$$\frac{x}{x^2-1} + \frac{1}{x^2-1} = \frac{x+1}{x^2-1} = \frac{\overset{1}{\cancel{(x+1)}}}{(x-1)\underset{1}{\cancel{(x+1)}}} = \frac{1}{x-1}$$

When subtracting algebraic fractions with like denominators, subtract the numerators. The denominator of the difference is the common denominator. Write the answer in simplest form.

$$\frac{a}{b} - \frac{c}{b} = \frac{a-c}{b}$$

CONSIDER THIS

Evaluate the expression $\frac{2x}{x-2} - \frac{4}{x-2}$ for several values of x and show that the calculation always yields 2 (provided $x \neq 2$).

$$\frac{2x}{x-2} - \frac{4}{x-2} = \frac{2x-4}{x-2} = \frac{2\overset{1}{\cancel{(x-2)}}}{\underset{1}{\cancel{x-2}}} = 2$$

$$\frac{3x-1}{x^2-5x+4} - \frac{2x+3}{x^2-5x+4} = \frac{(3x-1)-(2x+3)}{x^2-5x+4} = \frac{x-4}{x^2-5x+4} = \frac{\overset{1}{\cancel{(x-4)}}}{\underset{1}{\cancel{(x-4)}}(x-1)} = \frac{1}{x-1}$$

Example 5

Add: $\frac{7}{x^2} + \frac{9}{x^2}$

Solution

$$\frac{7}{x^2} + \frac{9}{x^2} = \frac{7+9}{x^2} = \frac{16}{x^2}$$

You Try It 5

Add: $\frac{3}{xy} + \frac{12}{xy}$

Your Solution

Example 6

Subtract: $\frac{3x^2}{x^2-1} - \frac{x+4}{x^2-1}$

Solution

$$\frac{3x^2}{x^2-1} - \frac{x+4}{x^2-1} = \frac{3x^2-(x+4)}{x^2-1} = \frac{3x^2-x-4}{x^2-1}$$

$$= \frac{(3x-4)\overset{1}{\cancel{(x+1)}}}{(x-1)\underset{1}{\cancel{(x+1)}}} = \frac{3x-4}{x-1}$$

You Try It 6

Subtract: $\frac{2x^2}{x^2-x-12} - \frac{7x+4}{x^2-x-12}$

Your Solution

Solutions on p. A30

Example 7

Simplify:

$$\frac{2x^2+5}{x^2+2x-3}-\frac{x^2-3x}{x^2+2x-3}+\frac{x-2}{x^2+2x-3}$$

Solution

$$\frac{2x^2+5}{x^2+2x-3}-\frac{x^2-3x}{x^2+2x-3}+\frac{x-2}{x^2+2x-3}$$

$$=\frac{(2x^2+5)-(x^2-3x)+(x-2)}{x^2+2x-3}$$

$$=\frac{2x^2+5-x^2+3x+x-2}{x^2+2x-3}$$

$$=\frac{x^2+4x+3}{x^2+2x-3}=\frac{\overset{1}{\cancel{(x+3)}}(x+1)}{\underset{1}{\cancel{(x+3)}}(x-1)}=\frac{x+1}{x-1}$$

You Try It 7

Simplify:

$$\frac{x^2-1}{x^2-8x+12}-\frac{2x+1}{x^2-8x+12}+\frac{x}{x^2-8x+12}$$

Your Solution

Solution on p. A31

OBJECTIVE D

Addition and subtraction of algebraic fractions with different denominators

Before two fractions with unlike denominators can be added or subtracted, each fraction must be expressed in terms of a common denominator. This common denominator is the LCM of the denominators of the fractions.

Add: $\frac{x-3}{x^2-2x}+\frac{6}{x^2-4}$

Find the LCM of the denominators. The LCM is $x(x-2)(x+2)$.

Write each fraction in terms of the LCM of the denominators. Multiply the factors in the numerator.

$$\frac{x-3}{x^2-2x}=\frac{x-3}{x(x-2)}\cdot\frac{x+2}{x+2}=\frac{x^2-x-6}{x(x-2)(x+2)}$$

$$\frac{6}{x^2-4}=\frac{6}{(x-2)(x+2)}\cdot\frac{x}{x}=\frac{6x}{x(x-2)(x+2)}$$

Add the fractions.

$$\frac{x-3}{x^2-2x}+\frac{6}{x^2-4}$$

$$=\frac{x^2-x-6}{x(x-2)(x+2)}+\frac{6x}{x(x-2)(x+2)}$$

$$=\frac{x^2-x-6+6x}{x(x-2)(x+2)}$$

$$=\frac{x^2+5x-6}{x(x-2)(x+2)}$$

Factor the numerator to determine whether there are common factors in the numerator and denominator.

$$=\frac{(x+6)(x-1)}{x(x-2)(x+2)}$$

Example 8

Simplify: $\frac{y}{x} - \frac{4y}{3x} + \frac{3y}{4x}$

Solution

The LCM of the denominators is $12x$.

$\frac{y}{x} = \frac{y}{x} \cdot \frac{12}{12} = \frac{12y}{12x}$ $\qquad \frac{4y}{3x} = \frac{4y}{3x} \cdot \frac{4}{4} = \frac{16y}{12x}$

$\frac{3y}{4x} = \frac{3y}{4x} \cdot \frac{3}{3} = \frac{9y}{12x}$

$$\frac{y}{x} - \frac{4y}{3x} + \frac{3y}{4x} = \frac{12y}{12x} - \frac{16y}{12x} + \frac{9y}{12x}$$

$$= \frac{12y - 16y + 9y}{12x} = \frac{5y}{12x}$$

You Try It 8

Simplify: $\frac{z}{8y} - \frac{4z}{3y} + \frac{5z}{4y}$

Your Solution

Example 9

Subtract: $\frac{2x}{x-3} - \frac{5}{3-x}$

Solution

The LCM of $x - 3$ and $3 - x$ is $x - 3$.
Remember: $3 - x = -(x - 3)$

$$\frac{2x}{x-3} = \frac{2x}{x-3} \cdot \frac{1}{1} = \frac{2x}{x-3}$$

$$\frac{5}{3-x} = \frac{5}{-(x-3)} \cdot \frac{-1}{-1} = \frac{-5}{x-3}$$

$$\frac{2x}{x-3} - \frac{5}{3-x} = \frac{2x}{x-3} - \frac{-5}{x-3}$$

$$= \frac{2x - (-5)}{x-3} = \frac{2x+5}{x-3}$$

You Try It 9

Subtract: $\frac{5x}{x-2} - \frac{3}{2-x}$

Your Solution

Example 10

Subtract: $\frac{2x}{2x-3} - \frac{1}{x+1}$

Solution

The LCM is $(2x - 3)(x + 1)$.

$$\frac{2x}{2x-3} = \frac{2x}{2x-3} \cdot \frac{x+1}{x+1} = \frac{2x^2 + 2x}{(2x-3)(x+1)}$$

$$\frac{1}{x+1} = \frac{1}{x+1} \cdot \frac{2x-3}{2x-3} = \frac{2x-3}{(2x-3)(x+1)}$$

$$\frac{2x}{2x-3} - \frac{1}{x+1} = \frac{2x^2 + 2x}{(2x-3)(x+1)} - \frac{2x-3}{(2x-3)(x+1)}$$

$$= \frac{(2x^2 + 2x) - (2x - 3)}{(2x-3)(x+1)} = \frac{2x^2 + 3}{(2x-3)(x+1)}$$

You Try It 10

Subtract: $\frac{4x}{3x-1} - \frac{9}{x+4}$

Your Solution

Solutions on p. A31

Example 11

Add: $\dfrac{x+3}{x^2-2x-8}+\dfrac{3}{4-x}$

Solution

The LCM is $(x-4)(x+2)$.

$$\frac{x+3}{x^2-2x-8}=\frac{x+3}{(x-4)(x+2)}$$

$$\frac{3}{4-x}=\frac{3}{-(x-4)}\cdot\frac{-1\cdot(x+2)}{-1\cdot(x+2)}=\frac{-3(x+2)}{(x-4)(x+2)}$$

$$\frac{x+3}{x^2-2x-8}+\frac{3}{4-x}=\frac{x+3}{(x-4)(x+2)}+\frac{-3(x+2)}{(x-4)(x+2)}$$

$$=\frac{(x+3)+(-3)(x+2)}{(x-4)(x+2)}=\frac{x+3-3x-6}{(x-4)(x+2)}$$

$$=\frac{-2x-3}{(x-4)(x+2)}$$

You Try It 11

Add: $\dfrac{2x-1}{x^2-25}+\dfrac{2}{5-x}$

Your Solution

Example 12

Simplify: $\dfrac{3x+2}{2x^2-x-1}-\dfrac{3}{2x+1}+\dfrac{4}{x-1}$

Solution

The LCM is $(2x+1)(x-1)$.

$$\frac{3x+2}{2x^2-x-1}=\frac{3x+2}{(2x+1)(x-1)}$$

$$\frac{3}{2x+1}=\frac{3}{2x+1}\cdot\frac{x-1}{x-1}=\frac{3x-3}{(2x+1)(x-1)}$$

$$\frac{4}{x-1}=\frac{4}{x-1}\cdot\frac{2x+1}{2x+1}=\frac{8x+4}{(2x+1)(x-1)}$$

$$\frac{3x+2}{2x^2-x-1}-\frac{3}{2x+1}+\frac{4}{x-1}$$

$$=\frac{3x+2}{(2x+1)(x-1)}-\frac{3x-3}{(2x+1)(x-1)}+\frac{8x+4}{(2x+1)(x-1)}$$

$$=\frac{(3x+2)-(3x-3)+(8x+4)}{(2x+1)(x-1)}$$

$$=\frac{3x+2-3x+3+8x+4}{(2x+1)(x-1)}=\frac{8x+9}{(2x+1)(x-1)}$$

You Try It 12

Simplify: $\dfrac{2x-3}{3x^2-x-2}+\dfrac{5}{3x+2}-\dfrac{1}{x-1}$

Your Solution

Solutions on p. A31

10.2 EXERCISES

Objective A

Find the LCM of the expressions.

1. $8x^3y$
   $12xy^2$

2. $6ab^2$
   $18ab^3$

3. $10x^4y^2$
   $15x^3y$

4. $12a^2b$
   $18ab^3$

5. $8x^2$
   $4x^2 + 8x$

6. $6y^2$
   $4y + 12$

7. $2x^2y$
   $3x^2 + 12x$

8. $4xy^2$
   $6xy^2 + 12y^2$

9. $9x(x + 2)$
   $12(x + 2)^2$

10. $8x^2(x - 1)^2$
    $10x^3(x - 1)$

11. $3x + 3$
    $2x^2 + 4x + 2$

12. $4x - 12$
    $2x^2 - 12x + 18$

13. $(x - 1)(x + 2)$
    $(x - 1)(x + 3)$

14. $(2x - 1)(x + 4)$
    $(2x + 1)(x + 4)$

15. $(2x + 3)^2$
    $(2x + 3)(x - 5)$

16. $(x - 7)(x + 2)$
    $(x - 7)^2$

17. $x - 1$
    $x - 2$
    $(x - 1)(x - 2)$

18. $(x + 4)(x - 3)$
    $x + 4$
    $x - 3$

19. $x^2 - x - 6$
    $x^2 + x - 12$

20. $x^2 + 3x - 10$
    $x^2 + 5x - 14$

21. $x^2 + 5x + 4$
    $x^2 - 3x - 28$

22. $x^2 - 10x + 21$
    $x^2 - 8x + 15$

23. $x^2 - 2x - 24$
    $x^2 - 36$

24. $x^2 + 7x + 10$
    $x^2 - 25$

25. $x^2 - 7x - 30$
    $x^2 - 5x - 24$

26. $2x^2 - 7x + 3$
    $2x^2 + x - 1$

27. $3x^2 - 11x + 6$
    $3x^2 + 4x - 4$

28. $2x^2 - 9x + 10$
    $2x^2 + x - 15$

29. $6 + x - x^2$
    $x + 2$
    $x - 3$

30. $15 + 2x - x^2$
    $x - 5$
    $x + 3$

31. $x^2 + 3x - 18$
    $3 - x$
    $x + 6$

32. $x^2 - 5x + 6$
    $1 - x$
    $x - 6$

Objective B

Write each fraction in terms of the LCM of the denominators.

33. $\frac{4}{x}, \frac{3}{x^2}$

34. $\frac{5}{ab^2}, \frac{6}{ab}$

35. $\frac{x}{3y^2}, \frac{z}{4y}$

36. $\frac{5y}{6x^2}, \frac{7}{9xy}$

37. $\frac{y}{x(x-3)}, \frac{6}{x^2}$

38. $\frac{a}{y^2}, \frac{6}{y(y+5)}$

39. $\frac{9}{(x-1)^2}, \frac{6}{x(x-1)}$

40. $\frac{a^2}{y(y+7)}, \frac{a}{(y+7)^2}$

41. $\frac{3}{x-3}, \frac{5}{x(3-x)}$

42. $\frac{b}{y(y-4)}, \frac{b^2}{4-y}$

43. $\frac{3}{(x-5)^2}, \frac{2}{5-x}$

44. $\frac{3}{7-y}, \frac{2}{(y-7)^2}$

45. $\frac{3}{x^2+2x}, \frac{4}{x^2}$

46. $\frac{2}{y-3}, \frac{3}{y^3-3y^2}$

47. $\frac{x-2}{x+3}, \frac{x}{x-4}$

48. $\frac{x^2}{2x-1}, \frac{x+1}{x+4}$

49. $\frac{3}{x^2+x-2}, \frac{x}{x+2}$

50. $\frac{3x}{x-5}, \frac{4}{x^2-25}$

51. $\frac{5}{2x^2-9x+10}, \frac{x-1}{2x-5}$

52. $\frac{x-3}{3x^2+4x-4}, \frac{2}{x+2}$

53. $\frac{x}{x^2+x-6}, \frac{2x}{x^2-9}$

54. $\frac{x-1}{x^2+2x-15}, \frac{x}{x^2+6x+5}$

55. $\frac{x}{9-x^2}, \frac{x-1}{x^2-6x+9}$

56. $\frac{2x}{10+3x-x^2}, \frac{x+2}{x^2-8x+15}$

57. $\frac{3x}{x-5}, \frac{x}{x+4}, \frac{3}{20+x-x^2}$

58. $\frac{x+1}{x+5}, \frac{x+2}{x-7}, \frac{3}{35+2x-x^2}$

Objective C

Simplify.

59. $\frac{3}{y^2} + \frac{8}{y^2}$

60. $\frac{6}{ab} - \frac{2}{ab}$

61. $\frac{3}{x + 4} - \frac{10}{x + 4}$

62. $\frac{x}{x + 6} - \frac{2}{x + 6}$

63. $\frac{3x}{2x + 3} + \frac{5x}{2x + 3}$

64. $\frac{6y}{4y + 1} - \frac{11y}{4y + 1}$

65. $\frac{2x + 1}{x - 3} + \frac{3x + 6}{x - 3}$

66. $\frac{4x + 3}{2x - 7} + \frac{3x - 8}{2x - 7}$

67. $\frac{5x - 1}{x + 9} - \frac{3x + 4}{x + 9}$

68. $\frac{6x - 5}{x - 10} - \frac{3x - 4}{x - 10}$

69. $\frac{x - 7}{2x + 7} - \frac{4x - 3}{2x + 7}$

70. $\frac{2n}{3n + 4} - \frac{5n - 3}{3n + 4}$

71. $\frac{x}{x^2 + 2x - 15} - \frac{3}{x^2 + 2x - 15}$

72. $\frac{3x}{x^2 + 3x - 10} - \frac{6}{x^2 + 3x - 10}$

73. $\frac{2x + 3}{x^2 - x - 30} - \frac{x - 2}{x^2 - x - 30}$

74. $\frac{3x - 1}{x^2 + 5x - 6} - \frac{2x - 7}{x^2 + 5x - 6}$

75. $\frac{4y + 7}{2y^2 + 7y - 4} - \frac{y - 5}{2y^2 + 7y - 4}$

76. $\frac{x + 1}{2x^2 - 5x - 12} + \frac{x + 2}{2x^2 - 5x - 12}$

77. $\frac{2x^2 + 3x}{x^2 - 9x + 20} + \frac{2x^2 - 3}{x^2 - 9x + 20} - \frac{4x^2 + 2x + 1}{x^2 - 9x + 20}$

78. $\frac{2x^2 + 3x}{x^2 - 2x - 63} - \frac{x^2 - 3x + 21}{x^2 - 2x - 63} - \frac{x - 7}{x^2 - 2x - 63}$

Objective D

Simplify.

79. $\frac{4}{x} + \frac{5}{y}$

80. $\frac{7}{a} + \frac{5}{b}$

81. $\frac{12}{x} - \frac{5}{2x}$

82. $\frac{5}{3a} - \frac{3}{4a}$

83. $\frac{1}{2x} - \frac{5}{4x} + \frac{7}{6x}$

84. $\frac{7}{4y} + \frac{11}{6y} - \frac{8}{3y}$

85. $\frac{2}{x} - \frac{3}{2y} + \frac{3}{5x} - \frac{1}{4y}$

86. $\frac{5}{2a} + \frac{7}{3b} - \frac{2}{b} - \frac{3}{4a}$

87. $\frac{2x + 1}{3x} + \frac{x - 1}{5x}$

88. $\frac{4x - 3}{6x} + \frac{2x + 3}{4x}$

89. $\frac{x - 3}{6x} + \frac{x + 4}{8x}$

90. $\frac{2x - 3}{2x} + \frac{x + 3}{3x}$

91. $\frac{x + 4}{2x} - \frac{x - 1}{x^2}$

92. $\frac{x - 2}{3x^2} - \frac{x + 4}{x}$

93. $\frac{x - 10}{4x^2} + \frac{x + 1}{2x}$

94. $\frac{x + 5}{3x^2} + \frac{2x + 1}{2x}$

95. $\frac{4x - 3}{3x^2y} + \frac{2x + 1}{4xy^2}$

96. $\frac{5x + 7}{6xy^2} - \frac{4x - 3}{8x^2y}$

97. $\frac{x - 2}{8x^2} - \frac{x + 7}{12xy}$

98. $\frac{3x - 1}{6y^2} - \frac{x + 5}{9xy}$

Simplify.

99. $\frac{4}{x - 2} + \frac{5}{x + 3}$

100. $\frac{2}{x - 3} + \frac{5}{x - 4}$

101. $\frac{2x}{x + 1} + \frac{1}{x - 3}$

102. $\frac{3x}{x - 4} + \frac{2}{x + 6}$

103. $\frac{4x}{2x - 1} - \frac{5}{x - 6}$

104. $\frac{6x}{x + 5} - \frac{3}{2x + 3}$

105. $\frac{x}{x^2 - 9} + \frac{3}{x - 3}$

106. $\frac{y}{y^2 - 16} + \frac{1}{y - 4}$

107. $\frac{2x}{x^2 - x - 6} - \frac{3}{x + 2}$

108. $\frac{5x}{x^2 + 2x - 8} - \frac{2}{x + 4}$

109. $\frac{3x - 1}{x^2 - 10x + 25} - \frac{3}{x - 5}$

110. $\frac{2a + 3}{a^2 - 7a + 12} - \frac{2}{a - 3}$

111. $\frac{x + 4}{x^2 - x - 42} + \frac{3}{7 - x}$

112. $\frac{x + 3}{x^2 - 3x - 10} + \frac{2}{5 - x}$

113. $\frac{1}{x + 1} + \frac{x}{x - 6} - \frac{5x - 2}{x^2 - 5x - 6}$

114. $\frac{x}{x - 4} + \frac{5}{x + 5} - \frac{11x - 8}{x^2 + x - 20}$

115. $\frac{3x + 1}{x - 1} - \frac{x - 1}{x - 3} + \frac{x + 1}{x^2 - 4x + 3}$

116. $\frac{4x + 1}{x - 8} - \frac{3x + 2}{x + 4} - \frac{49x + 4}{x^2 - 4x - 32}$

117. $\frac{2x + 9}{3 - x} + \frac{x + 5}{x + 7} - \frac{2x^2 + 3x - 3}{x^2 + 4x - 21}$

118. $\frac{3x + 5}{x + 5} - \frac{x + 1}{2 - x} - \frac{4x^2 - 3x - 1}{x^2 + 3x - 10}$

Critical Thinking

Simplify.

119. $\frac{a}{a-b}+\frac{b}{b-a}+1$

120. $\frac{y}{x-y}+2-\frac{x}{y-x}$

121. $b-3+\frac{5}{b+4}$

122. $2y-1+\frac{6}{y+5}$

123. $\frac{(n+1)^2}{(n-1)^2}-1$

124. $1-\frac{(y-2)^2}{(y+2)^2}$

125. $\frac{x^2+x-6}{x^2+2x-8}\cdot\frac{x^2+5x+4}{x^2+2x-3}-\frac{2}{x-1}$

126. $\frac{x^2+9x+20}{x^2+4x-5}\div\frac{x^2-49}{x^2+6x-7}-\frac{x}{x-7}$

127. $\frac{x^2-9}{x^2+6x+9}\div\frac{x^2+x-20}{x^2-x-12}+\frac{1}{x+1}$

128. $\frac{x^2-25}{x^2+10x+25}\cdot\frac{x^2-7x+10}{x^2-x-2}+\frac{1}{x+1}$

129. Find the sum of the following:

$$\frac{1}{1\cdot 2}+\frac{1}{2\cdot 3}$$
$$\frac{1}{1\cdot 2}+\frac{1}{2\cdot 3}+\frac{1}{3\cdot 4}$$
$$\frac{1}{1\cdot 2}+\frac{1}{2\cdot 3}+\frac{1}{3\cdot 4}+\frac{1}{4\cdot 5}$$

Note the pattern in these sums, and find the sum of 50 terms, of 100 terms, and of 1000 terms.

Rewrite the expression as the sum of two algebraic fractions in simplest form.

130. $\frac{5b+4a}{ab}$

131. $\frac{6x+7y}{xy}$

132. $\frac{3x^2+4xy}{x^2y^2}$

133. $\frac{2mn^2+8m^2n}{m^3n^3}$

134. [W] When is the LCM of two expressions equal to their product?

135. [W] In your own words, explain the procedure for adding algebraic fractions with different denominators.

SECTION 10.3 Solving Equations Containing Fractions

OBJECTIVE A Equations containing fractions

To solve an equation containing fractions, **clear denominators** by multiplying each side of the equation by the LCM of the denominators. Then solve for the variable.

Solve: $\frac{3x-1}{4} + \frac{2}{3} = \frac{7}{6}$

The LCM is 12.

$$\frac{3x-1}{4} + \frac{2}{3} = \frac{7}{6}$$

Multiply each side of the equation by the LCM of the denominators.

$$12\left(\frac{3x-1}{4} + \frac{2}{3}\right) = 12 \cdot \frac{7}{6}$$

Simplify using the Distributive Property and the properties of fractions.

$$12\left(\frac{3x-1}{4}\right) + 12 \cdot \frac{2}{3} = 12 \cdot \frac{7}{6}$$

$$\frac{\overset{3}{\cancel{12}}}{1}\left(\frac{3x-1}{\underset{1}{\cancel{4}}}\right) + \frac{\overset{4}{\cancel{12}}}{1} \cdot \frac{2}{\underset{1}{\cancel{3}}} = \frac{\overset{2}{\cancel{12}}}{1} \cdot \frac{7}{\underset{1}{\cancel{6}}}$$

Solve for x.

$$9x - 3 + 8 = 14$$
$$9x + 5 = 14$$
$$9x = 9$$
$$x = 1$$

1 checks as a solution. The solution is 1.

Occasionally, a value of the variable that appears to be a solution will make one of the denominators zero. In this case, that value of the variable is not a solution of the equation.

CONSIDER THIS

This example illustrates the importance of checking a solution of an equation containing fractions when each side is multiplied by a variable expression. The Multiplication Property of Equations states that each side can be multiplied by a *nonzero* number. If x were allowed to be 2, each side of the equation is multiplied by zero.

Solve: $\frac{2x}{x-2} = 1 + \frac{4}{x-2}$

The LCM is $x - 2$.

$$\frac{2x}{x-2} = 1 + \frac{4}{x-2}$$

Multiply each side of the equation by the LCM of the denominators.

$$(x-2) \cdot \frac{2x}{x-2} = (x-2)\left(1 + \frac{4}{x-2}\right)$$

Simplify using the Distributive Property and properties of fractions.

$$(x-2) \cdot \frac{2x}{x-2} = (x-2) \cdot 1 + (x-2) \cdot \frac{4}{x-2}$$

$$\frac{\overset{1}{\cancel{x-2}}}{1} \cdot \frac{2x}{\underset{1}{\cancel{x-2}}} = (x-2) + \frac{\overset{1}{\cancel{x-2}}}{1} \cdot \frac{4}{\underset{1}{\cancel{x-2}}}$$

Solve for x.

$$2x = x - 2 + 4$$
$$2x = x + 2$$
$$x = 2$$

When x is replaced by 2, the denominators of $\frac{2x}{x-2}$ and $\frac{4}{x-2}$ are zero. The equation has no solution.

Example 1

Solve: $\frac{4}{x+4} = \frac{2}{x}$

Solution

$$\frac{4}{x+4} = \frac{2}{x} \qquad \text{The LCM is } x(x+4).$$

$$\frac{x(x+4)}{1} \cdot \frac{4}{x+4} = \frac{x(x+4)}{1} \cdot \frac{2}{x}$$

$$\frac{x\overset{1}{\cancel{(x+4)}}}{1} \cdot \frac{4}{\underset{1}{\cancel{x+4}}} = \frac{\overset{1}{\cancel{x}}(x+4)}{1} \cdot \frac{2}{\underset{1}{\cancel{x}}}$$

$$4x = (x+4)2$$
$$4x = 2x + 8$$
$$2x = 8$$
$$x = 4$$

4 checks as a solution.

The solution is 4.

You Try It 1

Solve: $\frac{2}{x+6} = \frac{3}{x}$

Your Solution

Example 2

Solve: $\frac{3x}{x-4} = 5 + \frac{12}{x-4}$

Solution

$$\frac{3x}{x-4} = 5 + \frac{12}{x-4} \qquad \text{The LCM is } x - 4.$$

$$\frac{(x-4)}{1} \cdot \frac{3x}{x-4} = \frac{(x-4)}{1}\left(5 + \frac{12}{x-4}\right)$$

$$\frac{\overset{1}{\cancel{(x-4)}}}{1} \cdot \frac{3x}{\underset{1}{\cancel{x-4}}} = \frac{(x-4)}{1} \cdot 5 + \frac{\overset{1}{\cancel{(x-4)}}}{1} \cdot \frac{12}{\underset{1}{\cancel{x-4}}}$$

$$3x = (x-4)5 + 12$$
$$3x = 5x - 20 + 12$$
$$3x = 5x - 8$$
$$-2x = -8$$
$$x = 4$$

4 does not check as a solution.

The equation has no solution.

You Try It 2

Solve: $\frac{5x}{x+2} = 3 - \frac{10}{x+2}$

Your Solution

Solutions on p. A32

OBJECTIVE B Variation problems

The equation $y = kx$, where k is a constant value, is an example of a **direct variation.** The equation $y = kx$ is read "y varies directly as x" or "y is proportional to x." The constant k is called the **constant of variation** or the **constant of proportionality.**

A clerical worker makes \$10 per hour. The worker's total wage (w) is directly proportional to the number of hours (h) worked. The equation of variation is $w = 10h$. The constant of proportionality is 10.

A direct variation equation can be written in the form $y = kx^n$, where n is a positive number. For example, the equation $y = kx^2$ is read "y varies directly as the square of x."

Given that V varies directly as r and that $V = 20$ when $r = 4$, find the constant of variation and the equation of variation.

Write the basic variation equation.	$V = kr$
Replace V and r by the given values.	$20 = k \cdot 4$
Solve for the constant of variation.	$5 = k$
	The constant of variation is 5.
Write the direct variation equation by substituting the value of k into the direct variation equation.	$V = 5r$

The tension (T) in a spring varies directly as the distance (x) it is stretched. If $T = 8$ lb when $x = 2$ in., find T when $x = 4$ in.

Write the basic direct variation equation.	$T = kx$
Replace T and x by the given values.	$8 = k \cdot 2$
Solve for the constant of variation.	$4 = k$
Write the direct variation equation by substituting the value of k into the basic direct variation equation.	$T = 4x$
To find T when $x = 4$, substitute 4 for x in the equation and solve for T.	$T = 4 \cdot 4 = 16$
	The tension is 16 lb.

The equation $y = \frac{k}{x}$, where k is a constant, is an example of an **inverse variation.**

The equation $y = \frac{k}{x}$ is read "y varies inversely as x" or "y is inversely proportional to x."

In general, an inverse variation equation can be written $y = \frac{k}{x^n}$, where n is a positive number. For example, the equation $y = \frac{k}{x^2}$ is read "y varies inversely as the square of x."

Given that P varies inversely as x and that $P = 5$ when $x = 2$, find P when $x = 10$.

Write the basic inverse variation equation. $P = \frac{k}{x}$

Replace P and x by the given values. $5 = \frac{k}{2}$

Solve for the constant of variation. $10 = k$

Write the inverse variation equation by substituting the value of k into the basic inverse variation equation. $P = \frac{10}{x}$

To find P when $x = 10$, substitute 10 for x in the equation and solve for P. $P = \frac{10}{10}$

$P = 1$

Example 3

The amount (A) of medication prescribed for a person is directly related to the person's weight (W). For a 50-kilogram person, 2 ml of medication are prescribed. How many milliliters of medication are required for a person who weighs 75 kg?

Strategy

To find the required amount of medication:

- ▶ Write the basic direct variation equation, replace the variables by the given values, and solve for k.
- ▶ Write the direct variation equation, replacing k by its value. Substitute 75 for W and solve for A.

Solution

$A = kW$

$2 = k \cdot 50$

$\frac{1}{25} = k$

$A = \frac{1}{25}W = \frac{1}{25} \cdot 75 = 3$

The required amount of medication is 3 ml.

You Try It 3

A company that produces personal computers has determined that the number of computers it can sell (s) is inversely proportional to the price (P) of the computer. Two thousand computers can be sold when the price is \$2500. How many computers can be sold if the price of a computer is \$2000?

Your Strategy

Your Solution

Solution on p. A32

OBJECTIVE C Literal equations

A **literal equation** is an equation that contains more than one variable. Examples of literal equations are shown at the right.

$$2x + 3y = 6$$
$$4w - 2x + z = 0$$

Formulas are used to express a relationship among physical quantities. A **formula** is a literal equation that states a relationship between measurements. Examples of formulas are shown at the right.

$\frac{1}{R_1} + \frac{1}{R_2} = \frac{1}{R}$ (Physics)

$s = a + (n - 1)d$ (Mathematics)

$A = P + Prt$ (Business)

The Addition and Multiplication Properties of Equations can be used to solve a literal equation for one of the variables. The goal is to rewrite the equation so that the letter being solved for is alone on one side of the equation and all the other numbers and variables are on the other side.

Solve $A = P(1 + i)$ for i.

The goal is to rewrite the equation so that i is on one side of the equation and all other variables are on the other side.

Use the Distributive Property to remove parentheses.

$$A = P(1 + i)$$
$$A = P + Pi$$

Subtract P from each side of the equation.

$$A - P = P - P + Pi$$
$$A - P = Pi$$

Divide each side of the equation by P.

$$\frac{A - P}{P} = \frac{Pi}{P}$$

$$\frac{A - P}{P} = i$$

Example 4

Solve $3x - 4y = 12$ for x.

Solution

$$3x - 4y = 12$$
$$3x - 4y + 4y = 4y + 12$$
$$3x = 4y + 12$$
$$\frac{3x}{3} = \frac{4y + 12}{3}$$
$$x = \frac{4}{3}y + 4$$

You Try It 4

Solve $5x - 2y = 10$ for x.

Your Solution

Solution on p. A32

Example 5

Solve $i = \frac{E}{R + r}$ for R.

Solution

$$i = \frac{E}{R + r}$$

$$(R + r)i = (R + r)\frac{E}{R + r}$$

$$Ri + ri = E$$

$$Ri + ri - ri = E - ri$$

$$Ri = E - ri$$

$$\frac{Ri}{i} = \frac{E - ri}{i}$$

$$R = \frac{E - ri}{i}$$

You Try It 5

Solve $s = \frac{A + L}{2}$ for L.

Your Solution

Example 6

Solve $L = a(1 + ct)$ for c.

Solution

$$L = a(1 + ct)$$

$$L = a + act$$

$$L - a = a - a + act$$

$$L - a = act$$

$$\frac{L - a}{at} = \frac{act}{at}$$

$$\frac{L - a}{at} = c$$

You Try It 6

Solve $S = a + (n - 1)d$ for n.

Your Solution

Example 7

Solve $S = C - rC$ for C.

Solution

$$S = C - rC$$

$$S = (1 - r)C \quad \text{• Factor}$$

$$\frac{S}{1 - r} = \frac{(1 - r)C}{1 - r}$$

$$\frac{S}{1 - r} = C$$

You Try It 7

Solve $S = C + rC$ for C.

Your Solution

Solution on pp. A32–A33

10.3 EXERCISES

Objective A

Solve.

1. $\frac{2x}{3} - \frac{5}{2} = -\frac{1}{2}$

2. $\frac{x}{3} - \frac{1}{4} = \frac{1}{12}$

3. $\frac{x}{3} = \frac{x}{6} + 2$

4. $\frac{a}{4} = \frac{a}{2} + 3$

5. $\frac{x}{2} = 3 + \frac{x}{5}$

6. $\frac{n}{3} = 2 + \frac{n}{4}$

7. $\frac{x}{3} - \frac{1}{4} = \frac{x}{4} - \frac{1}{6}$

8. $\frac{2y}{9} - \frac{1}{6} = \frac{y}{9} + \frac{1}{6}$

9. $\frac{2x - 5}{8} + \frac{1}{4} = \frac{x}{8} + \frac{3}{4}$

10. $\frac{3x + 4}{12} - \frac{1}{3} = \frac{5x + 2}{12} - \frac{1}{2}$

11. $\frac{5x}{3} = 4 + 3x$

12. $6 = \frac{3x}{2} + 9$

13. $3 = \frac{4 + a}{a}$

14. $2 = \frac{5 + x}{x}$

15. $\frac{3 + y}{y} = 6$

16. $\frac{2 + x}{x} = 8$

17. $\frac{6}{2a + 1} = 2$

18. $\frac{12}{3x - 2} = 3$

19. $\frac{9}{2x - 5} = -2$

20. $\frac{6}{4 - 3x} = 3$

21. $\frac{y + 3}{y - 2} = 6$

22. $\frac{x + 4}{x - 1} = 5$

23. $7 = \frac{a - 2}{a + 1}$

24. $4 = \frac{x - 5}{x + 1}$

Solve.

25. $2 + \frac{5}{x} = 7$

26. $3 + \frac{8}{n} = 5$

27. $1 - \frac{9}{x} = 4$

28. $3 - \frac{12}{x} = 7$

29. $\frac{2}{y} + 5 = 9$

30. $\frac{6}{x} + 3 = 11$

31. $\frac{3}{x - 2} = \frac{4}{x}$

32. $\frac{5}{x + 3} = \frac{3}{x - 1}$

33. $\frac{5}{x + 2} = \frac{5}{x}$

34. $\frac{8}{x - 3} = \frac{8}{x}$

35. $\frac{2}{3x - 1} = \frac{3}{4x + 1}$

36. $\frac{5}{3x - 4} = \frac{-3}{1 - 2x}$

37. $\frac{-3}{2x + 5} = \frac{2}{x - 1}$

38. $\frac{4}{5y - 1} = \frac{2}{2y - 1}$

39. $\frac{4x}{x - 4} + 5 = \frac{5x}{x - 4}$

40. $\frac{2x}{x + 2} - 5 = \frac{7x}{x + 2}$

41. $2 + \frac{3}{a - 3} = \frac{a}{a - 3}$

42. $\frac{x}{x + 4} = 3 - \frac{4}{x + 4}$

43. $\frac{4}{x^2 + 2x - 8} = \frac{2}{x - 2} + \frac{2}{x + 4}$

44. $\frac{2}{x^2 + 2x - 3} = \frac{2}{x + 3} + \frac{3}{x - 1}$

45. $\frac{10}{x^2 - x - 6} = \frac{2}{x - 3} + \frac{3}{x + 2}$

46. $\frac{15}{x^2 - 3x - 4} = \frac{3}{x - 4} + \frac{1}{x + 1}$

Objective B

Solve.

47. Given that P varies directly as R and $P = 20$ when $R = 5$, find P when $R = 6$.

48. Given that T varies directly as S and $T = 36$ when $S = 9$, find T when $S = 7$.

49. Given that M is directly proportional to P and $M = 15$ when $P = 30$, find M when $P = 20$.

50. Given that A is directly proportional to B and $A = 6$ when $B = 18$, find A when $B = 21$.

51. D varies directly as T and $D = 120$ when $T = 2$. Find T when $D = 300$.

52. P varies directly as x and $P = 68$ when $x = 4$. Find x when $P = 85$.

53. Given that T varies inversely as x and $T = 10$ when $x = 4$, find T when $x = 5$.

54. Given that M varies inversely as Q and $M = 12$ when $Q = 5$, find M when $Q = 6$.

55. C is inversely proportional to H and $C = 8$ when $H = 3$. Find C when $H = 12$.

56. W is inversely proportional to V and $W = 9$ when $V = 1$. Find W when $V = 3$.

57. Given that Y varies inversely as X and $Y = 16$ when $X = 4$, find X when $Y = 12$.

58. Given that L varies inversely as P and $L = 12$ when $P = 5$, find P when $L = 5$.

59. If A varies directly as the square of r and $A = 3.14$ when $r = 1$, find A when $r = 2$.

60. If A varies directly as the square of r and $A = \frac{22}{7}$ when $r = 1$, find A when $r = 7$.

61. If P varies inversely as the square of x and $P = 2$ when $x = 4$, find P when $x = 12$.

62. If M varies inversely as the square of N and $M = 4$ when $N = 2$, find M when $N = 6$.

63. The distance (d) a spring will stretch varies directly as the force (f) applied to the spring. If a force of 6 lb is required to stretch the spring 3 in., what force is required to stretch the spring 4 in.?

64. The pressure (p) on a diver under water varies directly as the depth (d). If the pressure is 4.5 lb per square inch when the depth is 10 ft, what is the pressure per square inch when the depth is 15 ft?

Solve.

65. The number of bushels of wheat (b) produced by a farm is directly proportional to the number of acres (A) planted in wheat. If a 20-acre farm yields 450 bushels of wheat, what is the yield of a farm that has 30 acres of wheat?

66. The profit (P) realized by a company varies directly as the number of products it sells (s). If a company makes a profit of \$4000 on the sale of 250 products, what is the profit when the company sells 5000 products?

67. The stopping distance (s) of a car varies directly as the square of its speed (v). If a car traveling 50 mph requires 170 ft to stop, find the stopping distance for a car traveling 60 mph.

68. The distance (s) a ball has rolled down an inclined plane is directly proportional to the square of the time (t) it has been rolling. If the ball rolls 6 ft in one second, how far will it roll in 3 s?

69. The distance (s) a body falls from rest varies directly as the square of the time (t) of the fall. An object falls 64 ft in 2 s. How far will it fall in 5 s?

70. The resistance (R) to the flow of electric current in a wire is inversely proportional to the square of the diameter (d) of a wire. If a wire of diameter 0.01 cm has a resistance of 0.5 ohm, what is the resistance in a wire that is 0.02 cm in diameter?

71. The speed (v) of a gear varies inversely as the number of teeth (t). If a gear that has 45 teeth makes 24 revolutions per minute, how many revolutions will a gear that has 36 teeth make?

72. For a constant temperature, the pressure (P) of a gas varies inversely as the volume (V). If the pressure is 30 lb/in^2 when the volume is 500 ft^3, find the pressure per square inch when the volume is 200 ft^3.

73. The number of items (n) that can be purchased for a given amount of money is inversely proportional to the cost (C) of the item. If 60 items can be purchased when the cost per item is \$.25, how many items can be purchased when the cost per item is \$.20?

74. The length (L) of a rectangle of fixed area varies inversely as the width (W). If the length of a rectangle is 8 ft when the width is 5 ft, find the length of the rectangle when the width is 4 ft.

75. The intensity (I) of a light source is inversely proportional to the square of the distance (d) from the source. If the intensity is 12 lumens at a distance of 10 ft, what is the intensity when the distance is 5 ft?

76. The repulsive force (f) between the north poles of two magnets is inversely proportional to the square of the distance (d) between them. If the repulsive force is 20 lb when the distance is 4 in., find the repulsive force when the distance is 2 in.

Objective C

Solve for x.

77. $x + 3y = 6$

78. $x + 2y = 8$

79. $x - 4y = 12$

80. $x - 3y = 9$

81. $3x - y + 7 = 0$

82. $2x - y + 5 = 0$

83. $x + 3y = 6$

84. $x + 6y = 10$

85. $4(2 - 3ax) = 11$

86. $7(3 - 2ax) = 14$

87. $2(1 - 3ax) = x + 4$

88. $3(2 + ax) = 4x + 8$

Solve the formula for the given variable.

89. $A = \frac{1}{2}bh; h$
(Geometry)

90. $P = a + b + c; b$
(Geometry)

91. $d = rt; t$
(Physics)

92. $E = IR; R$
(Physics)

93. $PV = nRT; T$
(Chemistry)

94. $A = bh; h$
(Geometry)

95. $P = 2L + 2W; L$
(Geometry)

96. $F = \frac{9}{5}C + 32; C$
(Temperature Conversion)

97. $A = \frac{1}{2}h(b_1 + b_2); b_1$
(Geometry)

98. $C = \frac{5}{9}(F - 32); F$
(Temperature Conversion)

99. $V = \frac{1}{3}Ah; h$
(Geometry)

100. $P = R - C; C$
(Business)

101. $R = \frac{C - S}{t}; S$
(Business)

102. $P = \frac{R - C}{n}; R$
(Business)

103. $A = P + Prt; P$
(Business)

104. $T = fm - gm; m$
(Engineering)

105. $A = Sw + w; w$
(Physics)

106. $a = S - Sr; S$
(Mathematics)

107. $V = \pi r^2 h; h$
(Geometry)

108. $S = C + rC; C$
(Business)

109. $F_1 x = F_2(d - x); x$
(Physics)

Critical Thinking

Solve.

110. $\frac{2}{3}(x + 2) - \frac{x + 1}{6} = \frac{1}{2}$

111. $\frac{3}{4}a = \frac{1}{2}(3 - a) + \frac{a - 2}{4}$

112. $\frac{x}{2x^2 - x - 1} = \frac{3}{x^2 - 1} + \frac{3}{2x + 1}$

113. $\frac{x + 1}{x^2 + x - 2} = \frac{x + 2}{x^2 - 1} + \frac{3}{x + 2}$

When markup is based on selling price, the selling price of a product is given by the formula $S = \frac{C}{1 - r}$, where C is the cost of the product, and r is the markup rate.

114. **a.** Solve the formula $S = \frac{C}{1 - r}$ for r.

b. Use your answer to part **a** to find the markup rate on a tennis racket when the cost is \$112 and the selling price is \$140.

c. Use your answer to part **a** to find the markup rate on a radio when the cost is \$50.40 and the selling price is \$72.

Break-even analysis is a method used to determine the sales volume required for a company to break even, or experience neither a profit nor a loss on the sale of its product. The break-even point represents the number of units that must be made and sold for income from sales to equal the cost of the product. The break-even point can be calculated using the formula $B = \frac{F}{S - V}$, where F is the fixed costs, S is the selling price per unit, and V is the variable costs per unit.

115. **a.** Solve the formula $B = \frac{F}{S - V}$ for S.

b. Use your answer to part **a** to find the required selling price per desk for a company to break even. The fixed costs are \$20,000, the variable costs per desk are \$80, and the company plans to make and sell 200 desks.

c. Use your answer to part **a** to find the required selling price per camera for a company to break even. The fixed costs are \$15,000, the variable costs per camera are \$50, and the company plans to make and sell 600 cameras.

The surface area of a right circular cylinder is given by the following formula: $S = 2\pi rh + 2\pi r^2$, where r is the radius of the base, and h is the height of the cylinder.

116. **a.** Solve the formula $S = 2\pi rh + 2\pi r^2$ for h.

b. Use your answer to part **a** to find the height of a right circular cylinder when the surface area is 12π in^2 and the radius is 1 in.

c. Use your answer to part **a** to find the height of a right circular cylinder when the surface area is 23π in^2 and the radius is 2 in.

SECTION 10.4 Rates and Proportions

OBJECTIVE A Unit rates

Quantities such as 3 feet, 12 cents, and 9 cars are number quantities written with **units**.

3 feet
12 cents
9 cars

These are only some examples of units. Shirts, dollars, trees, miles, and gallons are further examples.

A **ratio** is the quotient of two quantities that have the same unit.

The length of a living room is 16 ft, and the width is 12 ft. The ratio of the length to the width is written

$\frac{16\text{ ft}}{12\text{ ft}} = \frac{16}{12} = \frac{4}{3}$

A ratio is in simplest form when the two numbers do not have a common factor. Note that the units are not written.

A **rate** is the quotient of two quantities that have *different* units. A rate is written as a fraction.

A distance runner ran 26 mi in 4 h. The distance-to-time rate is written

$\frac{26\text{ mi}}{4\text{ h}} = \frac{13\text{ mi}}{2\text{ h}}$

A rate is in **simplest form** when the numbers that form the rate have no common factors. Note that the units are written as part of the rate.

A **unit rate** is a rate in which the number in the denominator is 1.

$\frac{\$3.25}{1\text{ lb}}$ or \$3.25/lb is read "\$3.25 per pound."

To write a rate as a unit rate, divide the number in the numerator by the number in the denominator.

A car traveled 344 mi on 16 gal of gasoline. To find the miles per gallon (unit rate), divide the numerator of the rate by the denominator of the rate.

$\frac{344\text{ mi}}{16\text{ gal}}$ is the rate.

$$16\overline{)344.0} = 21.5$$

21.5 mi/gal is the unit rate.

Example 1 Write "300 ft in 8 s" as a unit rate.

Solution $\frac{300 \text{ ft}}{8 \text{ s}}$

$$8\overline{)300.0} = 37.5$$

37.5 ft/s

You Try It 1 Write "260 mi in 8 h" as a unit rate.

Your Solution

Solution on p. A33

OBJECTIVE B Proportions

A **proportion** is an equation that states the equality of two ratios or rates.

$\frac{30 \text{ mi}}{4 \text{ h}} = \frac{15 \text{ mi}}{2 \text{ h}}$ Note that the units of the numerator are the same and the units of the denominators are the same.

$\frac{3}{5} = \frac{9}{15}$

In a proportion, the "cross products" are equal. For the proportion $\frac{2}{3} = \frac{8}{12}$,

$$\frac{2}{3} \;\; \frac{8}{12} \Rightarrow 3 \times 8 = 24$$
$$\Rightarrow 2 \times 12 = 24$$

Sometimes one of the numbers in a proportion is unknown. In this case, it is necessary to *solve* the proportion.

Solve the proportion $\frac{4}{x} = \frac{2}{3}$.

$$\frac{4}{x} = \frac{2}{3}$$

Find the cross products. $4 \cdot 3 = x \cdot 2$

Solve for x. $12 = 2x$

$6 = x$

The solution is 6.

Example 2 Solve $\frac{x}{12} = \frac{25}{60}$ and check.

Solution

$$x \cdot 60 = 12 \cdot 25$$
$$60x = 300$$
$$x = 5$$

Check: $\frac{5}{12} \;\; \frac{25}{60} \Rightarrow 12 \times 25 = 300$

$\Rightarrow 5 \times 60 = 300$

The solution is 5.

You Try It 2 Solve $\frac{x}{14} = \frac{3}{7}$ and check.

Your Solution

Solution on p. A33

Example 3 Solve: $\frac{4}{9} = \frac{x}{16}$
Round to the nearest tenth.

Solution

$$4 \cdot 16 = 9 \cdot x$$
$$64 = 9x$$
$$7.1 \approx x$$

Note: A rounded answer is an approximation. Therefore, the answer to a check will not be exact.

The solution is 7.1.

You Try It 3 Solve: $\frac{5}{8} = \frac{x}{20}$

Your Solution

Example 4 Solve: $\frac{8}{x+3} = \frac{4}{x}$

Solution

$$8 \cdot x = 4(x + 3)$$
$$8x = 4x + 12$$
$$4x = 12$$
$$x = 3$$

The solution is 3.

You Try It 4 Solve: $\frac{2}{x+3} = \frac{6}{5x+5}$

Your Solution

Solutions on p. A33

OBJECTIVE  Applications

Example 5

A grocery store sells 3 tomatoes for 98 cents. What is the cost per tomato to the nearest tenth of a cent?

Strategy

To find the cost per tomato, divide the cost for 3 tomatoes (98¢) by the number of tomatoes (3).

Solution

$$3\overline{)98.00} = 32.66 \approx 32.7$$

The cost is 32.7¢ per tomato.

You Try It 5

A cyclist rode 47 mi in 3 h. What is the miles-per-hour rate to the nearest tenth?

Your Strategy

Your Solution

Solution on p. A33

Example 6

The dosage of a certain medication is 2 oz for every 50 lb of body weight. How many ounces of this medication are required for a person who weighs 175 lb?

Strategy

To find the number of ounces required, write and solve a proportion using n to represent the number of ounces of medication for a 175-pound person.

Solution

$$\frac{2 \text{ oz}}{50 \text{ lb}} = \frac{n \text{ oz}}{175 \text{ lb}}$$

$$2 \cdot 175 = 50 \cdot n$$

$$350 = 50n$$

$$7 = n$$

For a 175-pound person, 7 oz of medication are required.

You Try It 6

Three tablespoons of a liquid plant fertilizer are to be added to every 4 gal of water. How many tablespoons of fertilizer are required for 10 gal of water?

Your Strategy

Your Solution

Example 7

An investment of \$500 earns \$60 each year. At the same rate, how much additional money must be invested to earn \$90 each year?

Strategy

To find the additional amount of money that must be invested, write and solve a proportion using x to represent the additional money. Then $500 + x$ is the total amount invested.

Solution

$$\frac{\$60}{\$500} = \frac{\$90}{\$500 + x}$$

$$60(500 + x) = 500(90)$$

$$30{,}000 + 60x = 45{,}000$$

$$60x = 15{,}000$$

$$x = 250$$

An additional \$250 must be invested.

You Try It 7

Three ounces of medication are required for a 150-pound adult. At the same rate, how many additional ounces of medication are required for a 200-pound adult?

Your Strategy

Your Solution

Solutions on p. A34

OBJECTIVE D Similar triangles

Similar objects have the same shape but not necessarily the same size. A baseball is similar to a basketball. A model airplane is similar to an actual airplane.

Similar objects have corresponding parts; for example, the propellers on the model airplane correspond to the propellers on the actual airplane. The relationship between the sizes of each of the corresponding parts can be written as a ratio, and each ratio will be the same. If the propellers on the model plane are $\frac{1}{50}$ the size of the propellers on the actual plane, then the model wing is $\frac{1}{50}$ the size of the actual wing, the model fuselage is $\frac{1}{50}$ the size of the actual fuselage, and so on.

The two triangles ABC and DEF are similar. The ratios of corresponding sides are equal.

$$\frac{AB}{DE} = \frac{2}{6} = \frac{1}{3}, \frac{BC}{EF} = \frac{3}{9} = \frac{1}{3}, \text{ and } \frac{AC}{DF} = \frac{4}{12} = \frac{1}{3}$$

The ratio of corresponding sides $= \frac{1}{3}$.

Since the ratios of corresponding sides are equal, three proportions can be formed.

$$\frac{AB}{DE} = \frac{BC}{EF}, \frac{AB}{DE} = \frac{AC}{DF}, \text{ and } \frac{BC}{EF} = \frac{AC}{DF}.$$

The ratio of corresponding heights equals the ratio of corresponding sides.

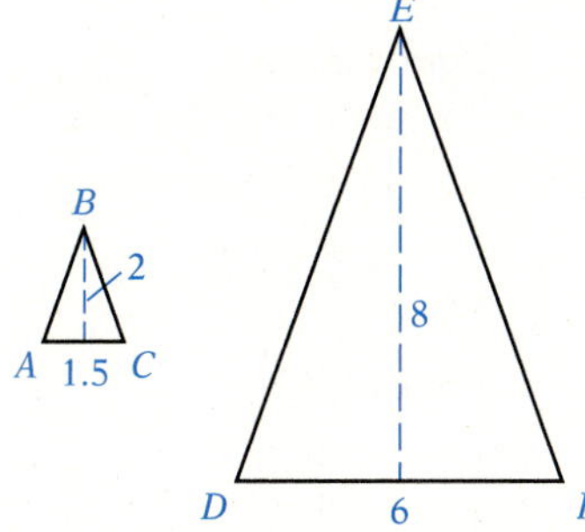

Ratio of corresponding sides $= \frac{1.5}{6} = \frac{1}{4}$.

Ratio of heights $= \frac{2}{8} = \frac{1}{4}$.

Congruent objects have the same shape *and* the same size.

The two triangles at the right are congruent. They have exactly the same size.

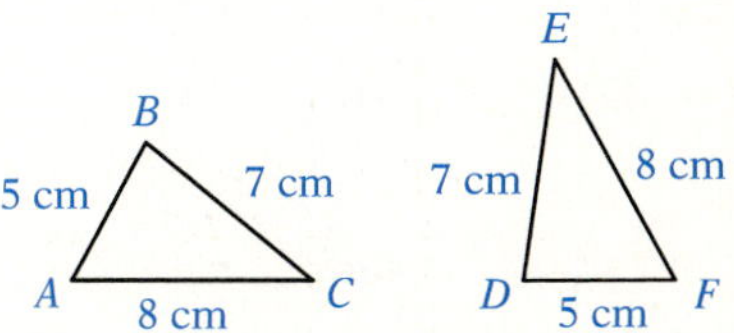

For triangles, congruent means that the corresponding sides *and* angles of the triangle must be equal, unlike similar triangles that just have corresponding angles equal but not necessarily the corresponding sides.

Example 8

Triangles ABC and DEF are similar. Find FG, the height of triangle DEF.

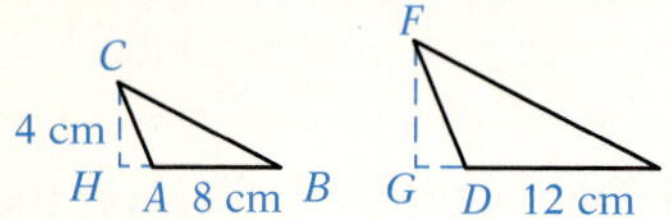

Strategy

To find FG, solve a proportion to find the height.

Solution

$$\frac{AB}{DE} = \frac{CH}{FG}$$

$$\frac{8}{12} = \frac{4}{FG}$$

$$8(FG) = 12(4)$$

$$8(FG) = 48$$

$$FG = 6$$

The height FG of triangle DEF is 6 cm.

You Try It 8

Triangles ABC and DEF are similar. Find FG, the height of triangle DEF.

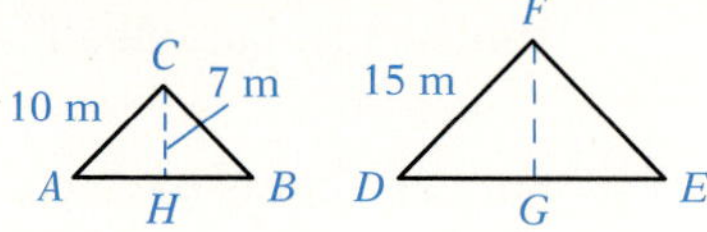

Your Strategy

Your Solution

Example 9

Triangles ABC and DEF are similar. Find the area of triangle DEF.

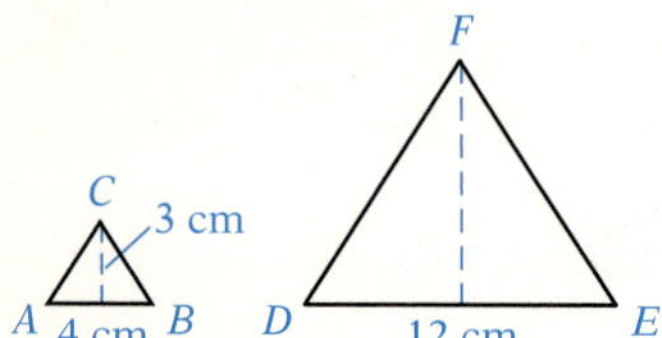

Strategy

To find the area of triangle DEF:

- Solve a proportion to find the height of triangle DEF. Let h represent the height.
- Use the equation for the area of a triangle.

Solution

$$\frac{AB}{DE} = \frac{\text{height of triangle } ABC}{h}$$

$$\frac{4}{12} = \frac{3}{h}$$

$$4 \cdot h = 12 \cdot 3$$

$$4h = 36$$

$$h = 9$$

$$A = \frac{1}{2}bh = \frac{1}{2}(12)(9) = 54$$

The area of triangle DEF is 54 cm^2.

You Try It 9

Triangles ABC and DEF are similar. Find the perimeter of triangle ABC.

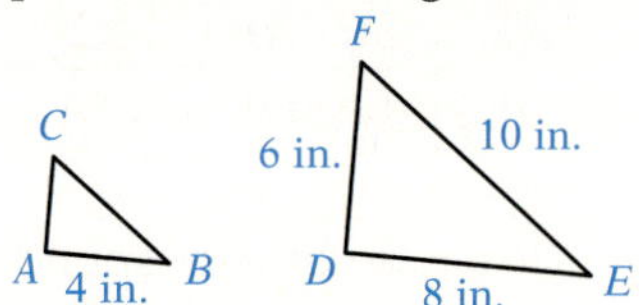

Your Strategy

Your Solution

Solutions on p. A34

10.4 EXERCISES

Objective A

Write as a unit rate.

1. 10 ft in 4 s
2. 816 mi in 6 days
3. \$1300 earned in 4 weeks
4. 1100 trees on 10 acres
5. 3750 words on 15 pages
6. \$32.97 earned in 7 h
7. 628.8 mi in 12 h
8. 388.8 mi in 8 h
9. 344.4 mi on 12.3 gal
10. 409.40 mi on 11.5 gal
11. \$349.80 for 212 lb
12. \$11.05 for 3.4 lb
13. A chef earns \$100 in 8 h. What is the chef's wage per hour?
14. A mechanic earns \$18,000 in 12 months. What is the mechanic's wage per month?
15. Five quarts of oil cost \$4.80. What is the cost per quart?
16. Twenty feet of lumber cost \$138.40. What is the cost per foot?
17. An investor purchased 350 shares of stock for \$12,600. Find the cost per share.
18. You own 300 shares of Lumex stock and receive a dividend of \$729.00. Find the dividend per share.
19. A manufacturer's cost to produce 1000 ballpoint pens was \$250. The company sold the pens to a retail store for \$450. What was the manufacturer's profit per pen?
20. A florist purchased 150 roses for \$405. During one week, the florist sold all the roses and received \$825. What was the florist's profit per rose?

Objective B

Solve. Round to the nearest tenth.

21. $\frac{x}{12} = \frac{3}{4}$
22. $\frac{6}{x} = \frac{2}{3}$
23. $\frac{4}{9} = \frac{x}{27}$
24. $\frac{16}{9} = \frac{64}{x}$
25. $\frac{n}{45} = \frac{17}{135}$

Solve. Round to the nearest tenth.

26. $\frac{n}{5} = \frac{7}{8}$
27. $\frac{4}{n} = \frac{9}{5}$
28. $\frac{8}{5} = \frac{n}{6}$
29. $\frac{n}{11} = \frac{32}{4}$
30. $\frac{3}{4} = \frac{8}{n}$

31. $\frac{5}{12} = \frac{n}{8}$
32. $\frac{36}{20} = \frac{12}{n}$
33. $\frac{15}{n} = \frac{65}{100}$
34. $\frac{n}{15} = \frac{21}{12}$
35. $\frac{40}{n} = \frac{15}{8}$

36. $\frac{32}{n} = \frac{1}{3}$
37. $\frac{x+3}{12} = \frac{5}{6}$
38. $\frac{3}{5} = \frac{x-4}{10}$
39. $\frac{18}{x+4} = \frac{9}{5}$
40. $\frac{2}{11} = \frac{20}{x-3}$

41. $\frac{2}{x} = \frac{4}{x+1}$
42. $\frac{16}{x-2} = \frac{8}{x}$
43. $\frac{x+3}{4} = \frac{x}{8}$
44. $\frac{x-6}{3} = \frac{x}{5}$
45. $\frac{2}{x-1} = \frac{6}{2x+1}$

46. $\frac{9}{x+2} = \frac{3}{x-2}$
47. $\frac{x+3}{5} = \frac{3}{4}$
48. $\frac{2x}{7} = \frac{x-2}{14}$
49. $\frac{4}{x} = \frac{7}{x+2}$
50. $\frac{5}{x+3} = \frac{2}{x-1}$

Objective C

Solve.

51. A salt water solution is made by dissolving 2 lb of salt in 5 gal of water. At this rate, how many pounds of salt are required for 25 gal of water?

52. A building contractor estimates that three overhead lights are needed for every 250 ft^2 of floor space. Using this estimate, how many light fixtures are necessary for a 10,000-square-foot office building?

53. A pre-election survey showed that 2 out of every 5 voters would vote in an election. At this rate, how many people would be expected to vote in a city of 25,000?

54. A quality control inspector found 2 defective electric blenders in a shipment of 100 blenders. At this rate, how many blenders would be defective in a shipment of 5000?

55. An exit poll showed that 4 out of every 7 voters cast a ballot in favor of an amendment to a city charter. At this rate, how many people voted in favor of the amendment if 35,000 people voted?

Solve.

56. A simple syrup is made by dissolving 2 c of sugar in $\frac{2}{3}$ c of boiling water. At this rate, how many cups of sugar are required for 2 c of boiling water?

57. A landscape designer estimates that 6 pieces of lumber are necessary for each 25 ft^2 of patio wood decking. Using this estimate, how many square feet of decking can be made from 36 pieces of lumber?

58. A carpet manufacturer uses 2 lb of wool for every 3 lb of nylon in a certain grade of carpet. At this rate, how many pounds of nylon are required for 250 lb of wool?

59. The license fee for a car that cost \$5500 was \$66. At the same rate, what is the license fee for a car that costs \$7500?

60. The real estate tax for a home that costs \$75,000 is \$750. At this rate, what is the cost of a home for which the real estate tax is \$562.50?

61. A soft drink is made by mixing 4 parts syrup with every 3 parts carbonated water. How many milliliters of syrup are in 280 ml of soft drink?

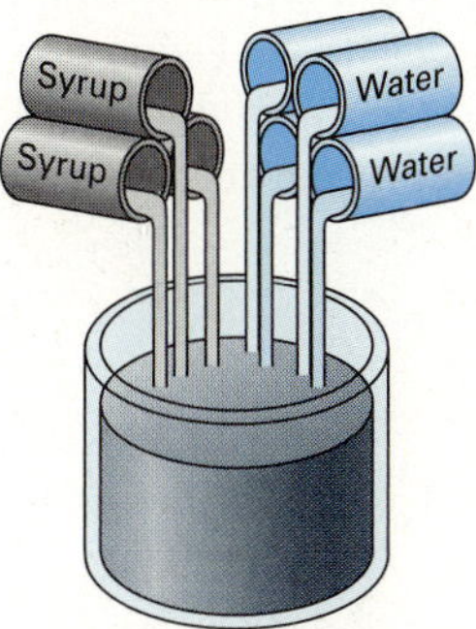

62. In a wildlife preserve, 16 deer are captured, tagged, and then released. Later, 30 deer are captured. Two of the 30 are found to have tags. Estimate the number of deer in the preserve.

63. In a lake, 50 fish are caught, tagged, and then released. Later 70 fish are caught. Five of the 70 are found to have tags. Estimate the number of fish in the lake.

64. A liquid plant food is prepared by using 1 gal of water for each 2 oz of plant food. At this rate, how many ounces of plant food are required for 4 gal of water?

65. The directions on a bag of lawn fertilizer recommend 2 lb of fertilizer for every 100 ft^2 of lawn. At this rate, how many pounds of fertilizer are used on a lawn that measures 2500 ft^2?

66. A painter estimates that 5 gal of paint will cover 1200 ft^2 of wall surface. How many additional gallons are required to cover 1680 ft^2?

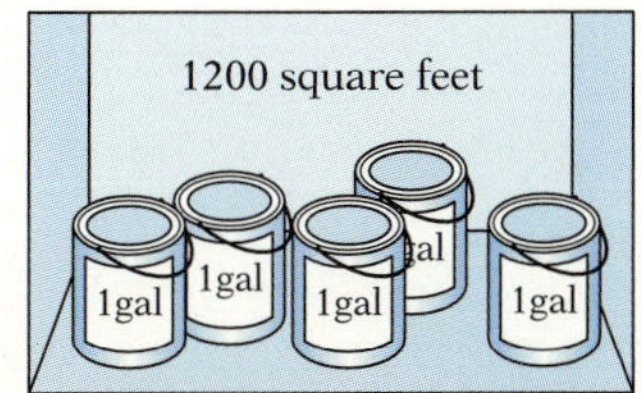

Solve.

67. The scale on a map is 1 in. equals 6 mi. What is the distance between two points that measure 7 in. on the map?

68. The scale on the plans for a new house is 1 in. equals 2 ft. How long is a room that measures $5\frac{1}{2}$ in. on the drawing?

69. A stock investment of 100 shares paid a dividend of \$124. At this rate, how many additional shares are required to earn a dividend of \$186?

70. A chef estimates that 50 lb of vegetables will serve 130 people. Using this estimate, how many additional pounds will be necessary to serve 156 people?

71. A caterer estimates that 3 gal of fruit punch will serve 40 people. How much additional punch is necessary to serve 60 people?

72. A farmer estimates that 5600 bushels of wheat can be harvested from 160 acres of land. Using this estimate, how many additional acres are needed to harvest 8120 bushels of wheat?

73. To conserve energy and still allow for as much natural lighting as possible, an architect suggests that the ratio of the area of a window to the area of the total wall surface be 5 to 12. Using this ratio, determine the recommended area of a window to be installed in a wall that measures 8 ft by 12 ft.

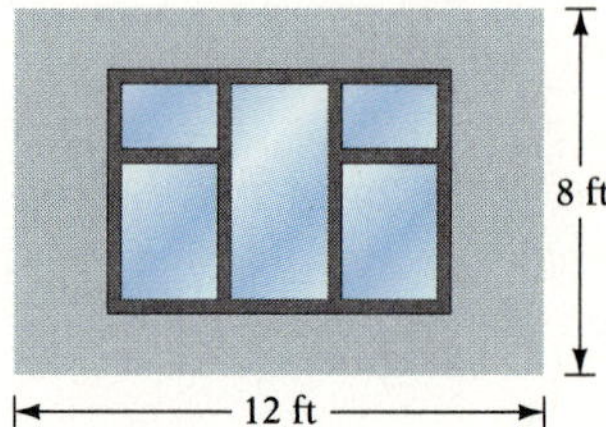

Objective D

Solve.

Triangles *ABC* and *DEF* are congruent.

74. Find the measure of $\angle E$.

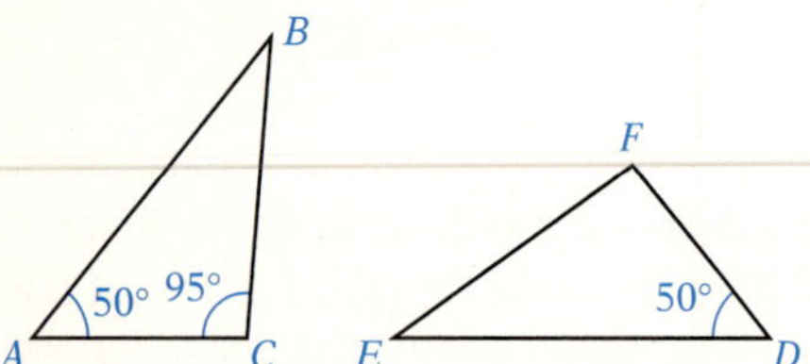

75. Find the measure of $\angle E$.

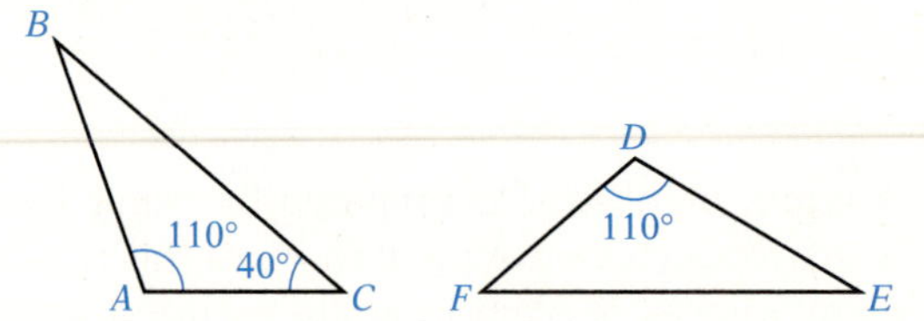

Triangles *ABC* and *DEF* are similar. Round to the nearest tenth.

76. Find side *DE*.

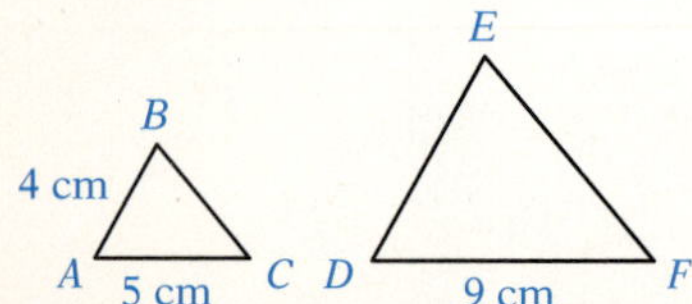

77. Find side *DE*.

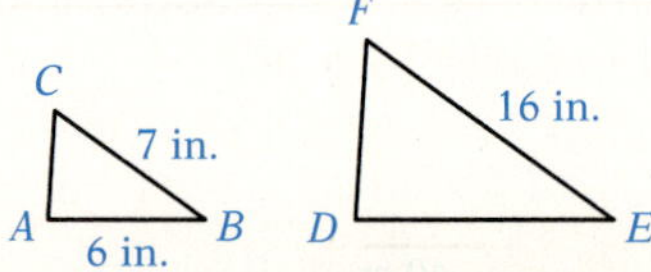

Triangles *ABC* and *DEF* are similar. Round to the nearest tenth.

78. Find the height of triangle *DEF*.

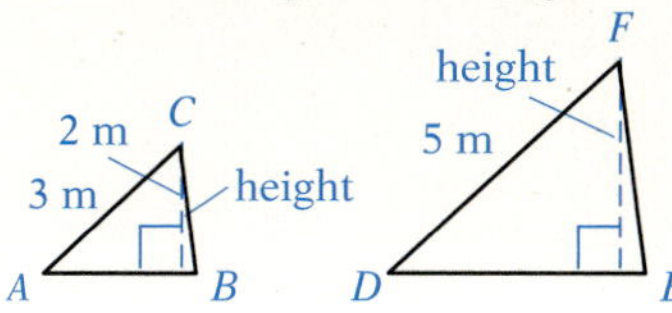

79. Find the height of triangle *ABC*.

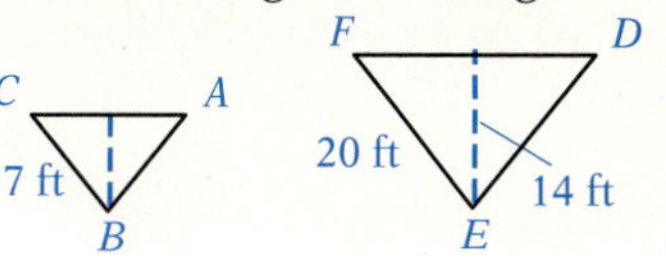

The sun's rays, objects on Earth, and the shadows cast by them form similar triangles.

80. Find the height of the flagpole.

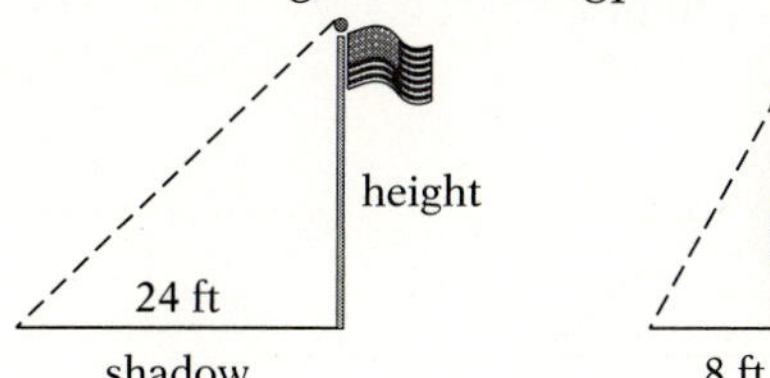

81. Find the height of the flagpole.

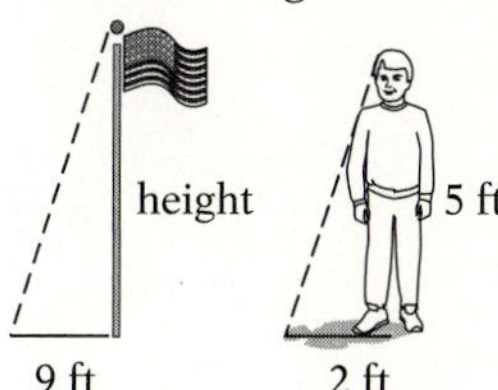

82. Find the height of the building.

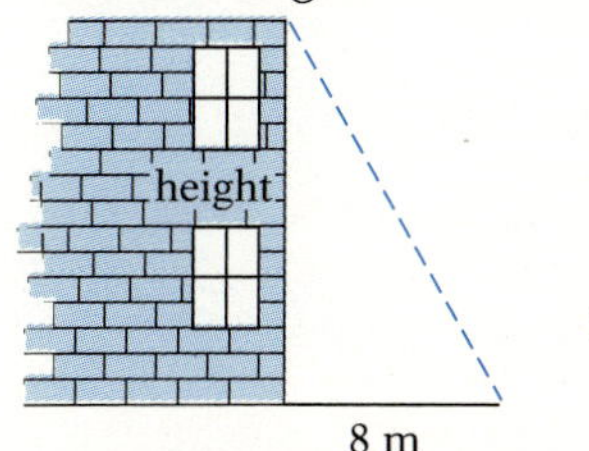

83. Find the height of the building.

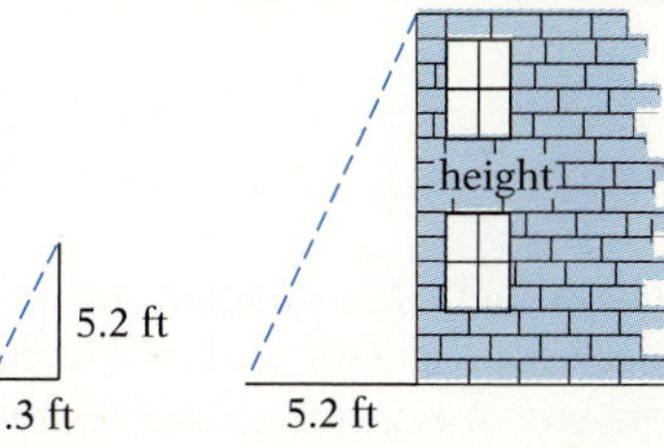

Solve. Triangles *ABC* and *DEF* are similar. Round to the nearest tenth.

84. Find the perimeter of triangle *ABC*.

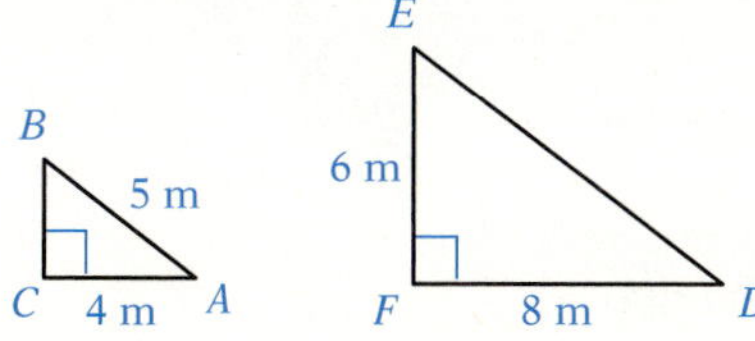

85. Find the perimeter of triangle *DEF*.

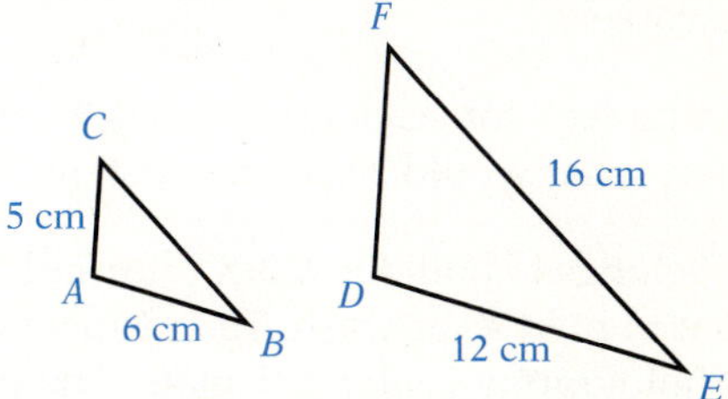

86. Find the perimeter of triangle *ABC*.

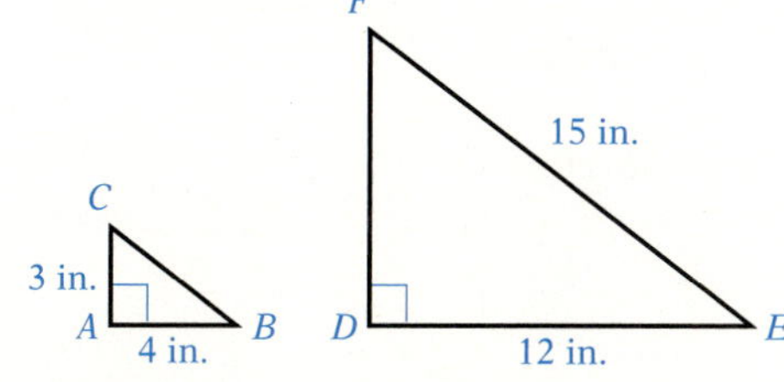

87. Find the area of triangle *DEF*.

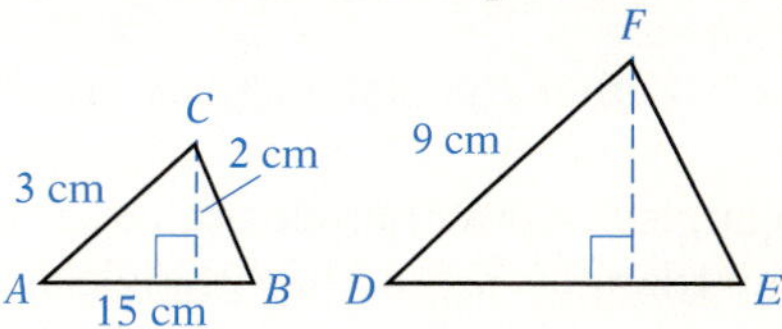

88. Find the area of triangle *ABC*.

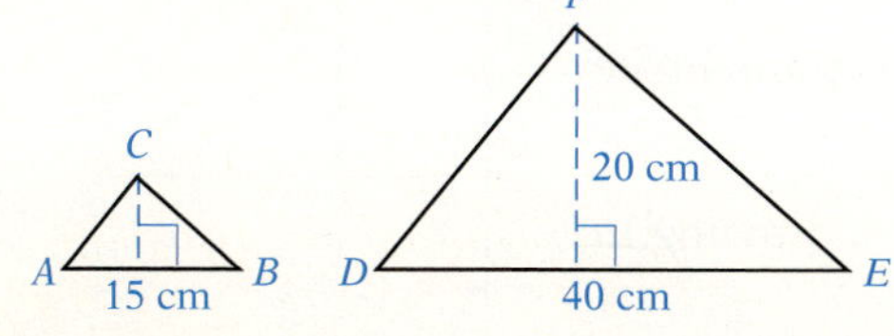

89. Find the area of triangle *DEF*.

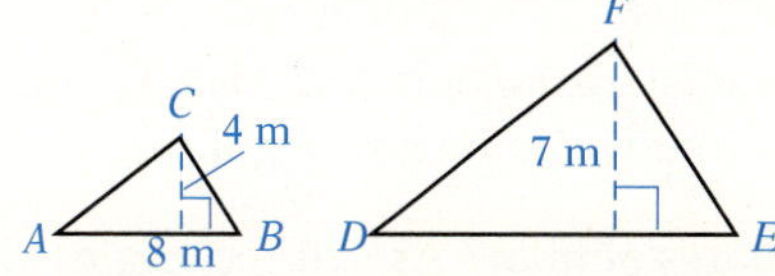

90. Given $BD \parallel AE$, $BD = 5$, $AE = 8$, and $AC = 10$, find the length of BC.

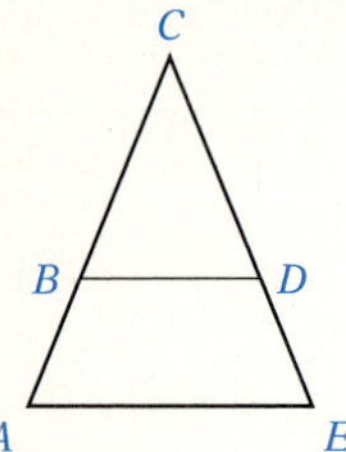

91. Given $AC \parallel DE$, $BD = 8$, $AD = 12$, and $BE = 6$, find the length of BC.

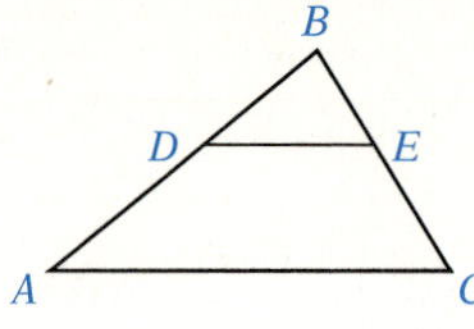

92. Given $DE \parallel AC$, $DE = 6$, $AC = 10$, and $AB = 15$, find the length of DA.

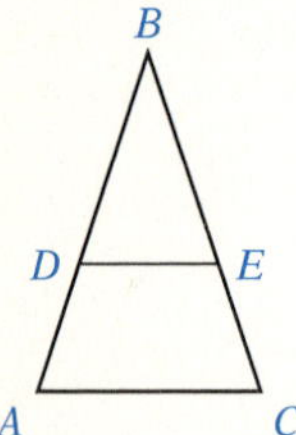

93. Given $AE \parallel BD$, $AB = 3$, $ED = 4$, and $BC = 3$, find the length of CE.

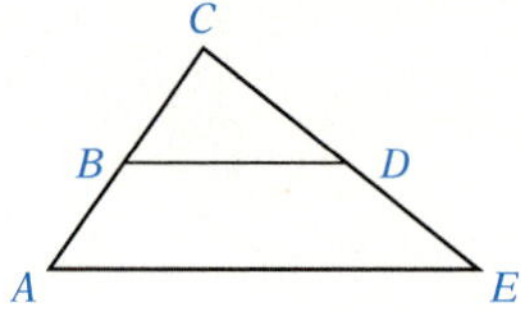

Critical Thinking

94. Three people put their money together to buy lottery tickets. The first person put in \$25, the second person put in \$30, and the third person put in \$35. One of their tickets was a winning ticket. If they won \$4.5 million, what was the first person's share of the winnings?

95. A basketball player has made 5 out of every 6 foul shots attempted. If 42 foul shots are missed in the player's career, how many foul shots were made in the player's career?

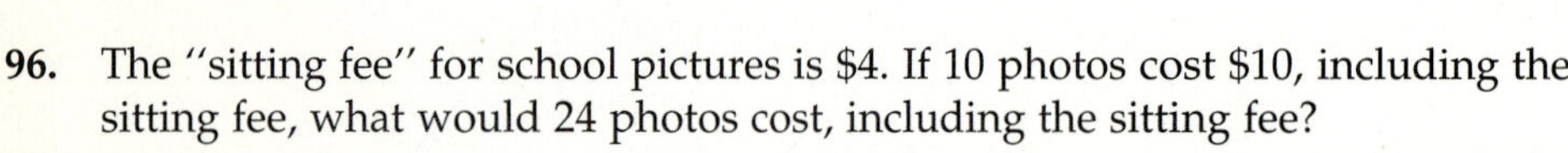

96. The "sitting fee" for school pictures is \$4. If 10 photos cost \$10, including the sitting fee, what would 24 photos cost, including the sitting fee?

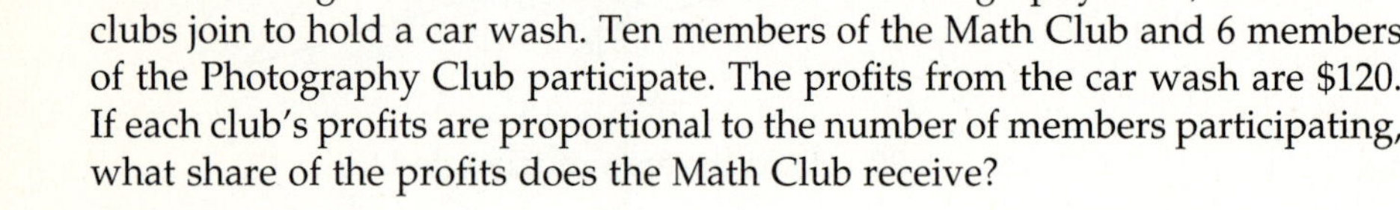

97. No one belongs to both the Math Club and the Photography Club, but the two clubs join to hold a car wash. Ten members of the Math Club and 6 members of the Photography Club participate. The profits from the car wash are \$120. If each club's profits are proportional to the number of members participating, what share of the profits does the Math Club receive?

98. Determine whether the statement is always true, sometimes true, or never true.

a. If two angles of one triangle are equal to two angles of a second triangle, then the triangles are similar triangles.

b. Two isosceles triangles are similar triangles.

c. Two equilateral triangles are similar triangles.

99. Figure ABC is a right triangle and DE is parallel to AB. What is the perimeter of the trapezoid $ABED$?

B, E, 10, 6, A, D, C, 5, 8

100. [W] Explain how, by using only a yardstick, you could determine the approximate height of a tree without climbing it.

101. [W] Are all squares similar? Are all rectangles similar? Explain. Use a drawing in your explanation.

SECTION 10.5 Work and Uniform Motion Problems

OBJECTIVE 

Work problems

If a painter can paint a room in 4 h, then in 1 h the painter can paint $\frac{1}{4}$ of the room. The painter's rate of work is $\frac{1}{4}$ of the room each hour. The **rate of work** is that part of a task that is completed in one unit of time.

A pipe can fill a tank in 30 min. This pipe can fill $\frac{1}{30}$ of the tank in 1 min. The rate of work is $\frac{1}{30}$ of the tank each minute. If a second pipe can fill the tank in x minutes, the rate of work for the second pipe is $\frac{1}{x}$ of the tank each minute.

CONSIDER THIS

Here is a challenge problem that you may want to try. A construction project must be completed in 15 days. Twenty-five workers did one-half of the job in 10 days. How many additional workers are necessary to complete the job in 5 days?

Answer: 50 additional workers

The basic equation that is used to solve work problems is:

Rate of work • time worked = part of task completed

For example, if a faucet can fill a sink in 6 min, then in 5 min the faucet will fill $\frac{1}{6} \cdot 5 = \frac{5}{6}$ of the sink. In 5 min the faucet completes $\frac{5}{6}$ of the task.

A painter can paint a wall in 20 min. The painter's apprentice can paint the same wall in 30 min. How long will it take to paint the wall when they work together?

Strategy for Solving a Work Problem

► For each person or machine, write a numerical or variable expression for the rate of work, the time worked, and the part of the task completed. The results can be recorded in a table.

CONSIDER THIS

Use the information given in the problem to fill in the rate and time columns of the table. Fill in the part-completed column by multiplying the two expressions you wrote in each row.

Time to paint the wall working together: t

	Rate of work	·	Time worked	=	Part of task completed
Painter	$\frac{1}{20}$	·	t	=	$\frac{t}{20}$
Apprentice	$\frac{1}{30}$	·	t	=	$\frac{t}{30}$

► Determine how the parts of the task completed are related. Use the fact that the sum of the parts of the task completed must equal 1, the complete task.

The sum of the part of the task completed by the painter and the part of the task completed by the apprentice is 1.

$$\frac{t}{20} + \frac{t}{30} = 1$$
$$60\left(\frac{t}{20} + \frac{t}{30}\right) = 60 \cdot 1$$
$$3t + 2t = 60$$
$$5t = 60$$
$$t = 12$$

Working together, they will paint the wall in 12 min.

Example 1

A small water pipe takes three times longer to fill a tank than does a larger water pipe. With both pipes open it takes 4 h to fill the tank. Find the time it would take the small pipe working alone to fill the tank.

Strategy

- Time for large pipe to fill the tank: t
  Time for small pipe to fill the tank: $3t$

	Rate	Time	Part
Small pipe	$\frac{1}{3t}$	4	$\frac{4}{3t}$
Large pipe	$\frac{1}{t}$	4	$\frac{4}{t}$

- The sum of the parts of the task completed by each pipe must equal one.

Solution

$$\frac{4}{3t} + \frac{4}{t} = 1$$
$$3t\left(\frac{4}{3t} + \frac{4}{t}\right) = 3t \cdot 1$$
$$4 + 12 = 3t$$
$$16 = 3t$$
$$\frac{16}{3} = t$$
$$3t = 3\left(\frac{16}{3}\right) = 16$$

The small pipe working alone takes 16 h to fill the tank.

You Try It 1

Two computer printers that work at the same rate are working together to print the payroll checks for a large corporation. After working together for 2 h, one of the printers quits. The second requires 3 more hours to complete the payroll checks. Find the time it would take one printer working alone to print the payroll.

Your Strategy

Your Solution

Solution on p. A35

OBJECTIVE B Uniform motion problems

A car that travels constantly in a straight line at 30 mph is in uniform motion. **Uniform motion** means that the speed of an object does not change.

The basic equation used to solve uniform motion problems is:

$$\textbf{Distance} = \textbf{rate} \cdot \textbf{time}$$

CONSIDER THIS

Here is a challenge problem that you may want to try. Two trains on the same track are heading toward each other. One is traveling at 40 mph; the other is traveling at 50 mph. How far apart are they one minute before they would collide?

Answer: 1.5 mi.

An alternate form of this formula can be written by solving the equation for time.

$$\frac{\textbf{Distance}}{\textbf{Rate}} = \textbf{time}$$

This form of the equation is useful when the total time of travel for two objects or the time of travel between two points is known.

The speed of a boat in still water is 20 mph. The boat traveled 75 mi down a river in the same time as it traveled 45 mi up the river. Find the rate of the river's current.

Strategy for Solving a Uniform Motion Problem

▶ For each object, write a numerical or variable expression for the distance, rate, and time. The results can be recorded in a table.

The rate of the river's current: r

CONSIDER THIS

Use the information given in the problem to fill in the distance and rate columns of the table. Fill in the time column by dividing the two expressions you wrote in each row.

	Distance	÷	Rate	=	Time
Down river	75	÷	$20 + r$	=	$\frac{75}{20 + r}$
Up river	45	÷	$20 - r$	=	$\frac{45}{20 - r}$

▶ Determine how the times traveled by each object are related. For example, it may be known that the times are equal or the total time may be known.

The time traveling down the river is equal to the time traveling up the river.

$$\frac{75}{20 + r} = \frac{45}{20 - r}$$

$$(20 + r)(20 - r)\frac{75}{20 + r} = (20 + r)(20 - r)\frac{45}{20 - r}$$

$$(20 - r)75 = (20 + r)45$$

$$1500 - 75r = 900 + 45r$$

$$-120r = -600$$

$$r = 5$$

The rate of the river's current is 5 mph.

Example 2

A jet can fly 600 mph in calm air. Traveling with the wind, the plane can fly 2100 mi in the same amount of time as it flies 1500 mi against the wind. Find the rate of the wind.

Strategy

- Rate of wind: r

	Distance	Rate	Time
With wind	2100	$600 + r$	$\frac{2100}{600 + r}$
Against wind	1500	$600 - r$	$\frac{1500}{600 - r}$

- The time spent flying with the wind equals the time spent flying against the wind.

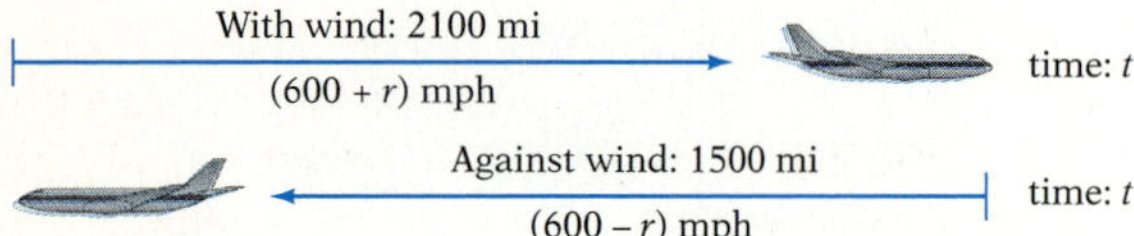

Solution

$$\frac{2100}{600 + r} = \frac{1500}{600 - r}$$

$$(600 + r)(600 - r)\frac{2100}{600 + r} = (600 + r)(600 - r)\frac{1500}{600 - r}$$

$$(600 - r)2100 = (600 + r)1500$$

$$1{,}260{,}000 - 2100r = 900{,}000 + 1500r$$

$$360{,}000 = 3600r$$

$$100 = r$$

The rate of the wind is 100 mph.

You Try It 2

The total time for a sailboat to sail back and forth across a lake 6 km wide was 2 h. The rate sailing back was three times the rate sailing across. Find the rate across the lake.

Your Strategy

Your Solution

Solution on p. A35

10.5 EXERCISES

Objective A

Solve.

1. A park has two sprinklers that are used to fill a fountain. One sprinkler can fill the fountain in 3 h, while the second sprinkler can fill the fountain in 6 h. How long will it take to fill the fountain with both sprinklers operating?

2. A new printing press can complete the weekly edition of a news magazine in 10 h. An older printing press requires 15 h to do the same task. How long would it take to print the weekly edition with both presses operating?

3. Two farmers are plowing a field. One farmer, using an old tractor and working alone, requires 12 h to plow the field. A second farmer, using a modern tractor, can plow the same field in 4 h. How long would it take to plow the field with both tractors working together?

4. A small air conditioner can cool a room 5° in 75 min. A larger air conditioner can cool the room 5° in 50 min. How long would it take to cool the room 5° with both air conditioners operating?

5. A new machine can fill soda bottles three times faster than an old machine. With both machines working together, they can complete the task in 9 h. How long would it take the new machine working alone to complete the task?

6. An experienced painter can paint a fence twice as fast as an inexperienced painter. Working together, the painters require 4 h to paint the fence. How long would it take the experienced painter working alone to paint the fence?

7. A plumber can install a garbage disposal in 45 min. With the plumber's assistant helping, the task would take 30 min. How long would it take the assistant working alone to complete the task?

8. A mason can construct a retaining wall in 10 h. With the mason's apprentice assisting, the task would take 6 h. How long would it take the apprentice working alone to construct the wall?

9. One welder requires 2 h to make the welds on a steel frame, while a second welder requires 4 h to do the same job. The first welder worked for 1 h and then quit. How long would it take the second welder to complete the welds?

10. One cement mason can lay a cement foundation in 8 h, while it takes a second mason 12 h to do the same task. After working alone for 4 h, the first mason quit. How long would it take the second mason to complete the task?

Objective B

Solve.

11. An express train travels 300 mi in the same amount of time that a freight train travels 180 mi. The rate of the express train is 20 mph faster than that of the freight train. Find the rate of each train.

12. A postal clerk on vacation took an 8-mile cruise in a sailboat in the same time as she took a 20-mile cruise on a power boat. The speed of the power boat is 12 mph faster than the speed of the sailboat. Find the speed of each boat.

Solve.

13. A twin-engine plane can fly 660 mi in the same amount of time as it takes a single-engine plane to fly 330 mi. The rate of the twin-engine plane is 100 mph faster than that of the single-engine plane. Find the speed of the twin-engine plane.

14. The rate of a motorcycle is 36 mph faster than the rate of a bicycle. The motorcycle travels 192 mi in the same amount of time as the bicycle travels 48 mi. Find the rate of the motorcycle.

15. A sales accountant traveled 1800 mi by jet and 300 mi on a prop plane. The rate of the jet is four times the rate of the prop plane. The entire trip took 5 h. Find the rate of each plane.

16. A motorist drove 90 mi before running out of gas and then walking 5 mi to a gas station. The rate of the motorist in the car was nine times the rate walking. The time spent walking and driving was 3 h. Find the rate at which the motorist walks.

17. A computer representative traveled 135 mi by train and then an additional 855 mi by plane. The rate of the plane was three times the rate of the train and the total time for the trip was 6 h. Find the rate of the plane.

18. A marketing manager traveled 1080 mi on a corporate jet and then an additional 180 mi by helicopter. The rate of the jet is four times the rate of the helicopter. The entire trip took 5 h. Find the rate of the jet.

19. A light plane can fly at a rate of 100 mph in calm air. Traveling with the wind, the plane flew 360 mi in the same amount of time as it flew 240 mi against the wind. Find the rate of the wind.

20. A tour boat used for river excursions can travel 6 mph in calm water. The amount of time it takes to travel 12 mi against the river's current is the same as the amount of time it takes to travel 24 mi with the current. Find the rate of the current.

Critical Thinking

Solve.

21. One pipe can fill a tank in 2 h, a second pipe can fill the tank in 4 h, and a third pipe can fill the tank in 5 h. How long will it take to fill the tank with all three pipes operating?

22. A mason can construct a retaining wall in 10 h. The mason's more experienced apprentice can do the same job in 15 h. How long would it take the mason's less experienced apprentice to do the job if, working together, all three can complete the wall in 5 h?

23. The Outing Club traveled 32 mi by canoe and then hiked 4 mi. The rate of travel by boat was four times the rate on foot. If the time spent walking was 1 h less than the time spent canoeing, find the amount of time spent traveling by canoe.

24. A motorist drove 120 mi before running out of gas and walking 4 mi to a gas station. The rate of the motorist in the car was ten times the rate walking. The time spent walking was 2 h less than the time spent driving. How long did it take for the motorist to drive the 120 mi?

25. Because of bad weather, a bus driver reduced the usual speed along a 150-mile bus route by 10 mph. The bus arrived only 30 min later than its usual arrival time. How fast does the bus usually travel?

Projects in Mathematics

Joint and Combined Variation

A variation may involve more than two variables. If a quantity varies directly as the product of two or more variables, it is known as a **joint variation**.

The weight of a rectangular metal box is directly proportional to the volume of the box, given by length · width · height.

Thus, Weight $= kLWH$.

The weight of a box with $L = 24$ in., $W = 12$ in., and $H = 12$ in. is 72 lb. Find the weight of another box with $L = 18$ in., $W = 9$ in., and $H = 18$ in.

$$\text{Weight} = kLWH$$

$$72 = k(24)(12)(12)$$

$$\frac{72}{(24)(12)(12)} = k$$

$$\frac{1}{48} = k$$

$$\text{Weight} = kLWH$$

$$\text{Weight} = \frac{1}{48}(18)(9)(18)$$ Substitute $\frac{1}{48}$ for k.

$$\text{Weight} = 60.75$$ Substitute the dimensions of the other box into the equation.

The weight of the other box is 60.75 lb.

Direct and inverse variation can occur in the same problem. When this occurs, it is called a **combined variation**.

The electrical resistance of a wire is directly proportional to the length and inversely proportional to the square of the diameter of the wire. This is written as:

$$R = \frac{kL}{d^2}$$

The weight that a horizontal beam with a rectangular cross section can safely support varies jointly as the width and square of the depth of the cross section and inversely as the length of the beam.

a. Write the joint variation.

b. If a 2-inch by 4-inch beam 8 ft long can safely support a load of 300 lb, what load can be safely supported by a beam made of the same material which is 4 in. wide, 6 in. deep, and 12 ft long?

Chapter Summary

Key Words

An *algebraic fraction* is a fraction in which the numerator or denominator is a variable expression.

An algebraic fraction is in *simplest form* when the numerator and denominator have no common factors.

The *reciprocal* of a fraction is the fraction with the numerator and denominator interchanged.

The *least common multiple* (LCM) of two or more numbers is the smallest number that contains the prime factorization of each number.

The equation $y = kx$, where k is a constant value, is an example of a variation in which y varies directly as x.

The equation $y = \frac{k}{x}$, where k is a constant value, is an example of a variation in which y varies inversely as x.

A *ratio* is the quotient of two quantities that have the same unit.

A *rate* is the quotient of two quantities that have different units.

A *unit rate* is a rate in which the number in the denominator is 1.

A *proportion* is an equation that states the equality of two ratios or rates.

A *literal equation* is an equation that contains more than one variable.

A *formula* is a literal equation that states rules about measurements.

Essential Rules

To multiply fractions $\frac{a}{b} \cdot \frac{c}{d} = \frac{ac}{bd}$

To divide fractions $\frac{a}{b} \div \frac{c}{d} = \frac{a}{b} \cdot \frac{d}{c}$

To add fractions $\frac{a}{b} + \frac{c}{b} = \frac{a+c}{b}$

To subtract fractions $\frac{a}{b} - \frac{c}{b} = \frac{a-c}{b}$

Equation for Work Problems Rate of work · Time worked = Part of task completed

Uniform Motion Equation $\frac{\text{Distance}}{\text{Rate}} = \text{Time}$

Chapter Review Exercises

1. Simplify: $\dfrac{16x^5y}{24x^2y^4}$

2. Simplify: $\dfrac{x^2 + 4x - 5}{1 - x^2}$

3. Write "$27,000 earned in 12 months" as a unit rate.

4. Write each fraction in terms of the LCM of the denominators.
$\dfrac{3}{x^2 - 2x}, \dfrac{x}{x^2 - 4}$

5. Subtract: $\dfrac{2x}{x^2 + 3x - 10} - \dfrac{4}{x^2 + 3x - 10}$

6. Subtract: $\dfrac{3x}{x^2 + 5x - 24} - \dfrac{9}{x^2 + 5x - 24}$

7. Simplify: $\dfrac{y}{x} - \dfrac{5y}{2x} + \dfrac{3y}{4x}$

8. Subtract: $\dfrac{2}{2x - 1} - \dfrac{1}{x + 1}$

9. Multiply: $\dfrac{4x - 8}{6x + 18} \cdot \dfrac{2x + 6}{x^2 - 2x}$

10. Multiply: $\dfrac{x^5y^3}{x^2 - x - 6} \cdot \dfrac{x^2 - 9}{x^2y^4}$

11. Divide: $\dfrac{ab^2 - 2a^2b}{c^2} \div \dfrac{4a^2 - 2ab}{c^3}$

12. Divide: $\dfrac{x^2 - x - 56}{x^2 + 8x + 7} \div \dfrac{x^2 - 13x + 40}{x^2 - 4x - 5}$

13. A small plane used 2 qt of oil on a 1200-mile trip. At this rate, how many quarts of oil would be used on a trip of 2000 mi?

14. Triangles ABC and DEF are similar. Find the perimeter of triangle ABC.

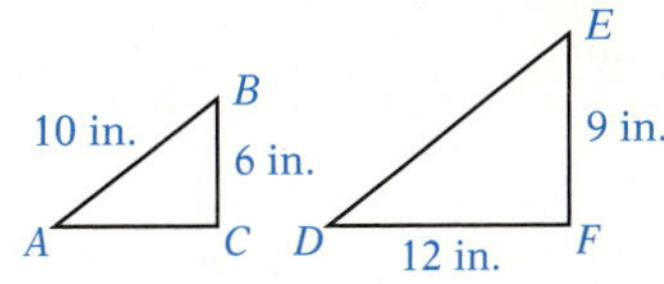

15. Solve: $\dfrac{6}{x} - 2 = 1$

16. Solve: $\dfrac{2}{x + 1} - 3 = \dfrac{-2}{x + 1}$

17. P varies directly as x and $P = 14$ when $x = 2$. Find P when $x = 5$.

18. W is inversely proportional to V and $W = 6$ when $V = 2$. Find W when $V = 3$.

19. Solve $3x - 8y = 16$ for x.

20. Solve $d = s + rt$ for t.

21. One water pipe can fill a tank in 9 min while a second pipe requires 18 min to fill the tank. How long would it take both pipes working together to fill the tank?

22. The rower of a boat can row at a rate of 5 mph in calm water. Rowing with the current, the boat travels 14 mi in the same amount of time as it travels 6 mi against the current. Find the rate of the current.

23. With cold and hot water running, a bathtub can be filled in 6 min. The hot water faucet working alone requires 15 min to fill the tub. How long would it take the cold water faucet working alone to fill the tub?

24. A freight train and a passenger train leave a town at 10 a.m. and head for a town 300 mi away. The rate of the passenger train is twice the rate of the freight train. The passenger train arrives 5 h ahead of the freight train. Find the rate of each train.

Cumulative Review Exercises

1. Multiply: $(a^2b^5)(ab^2)$

2. Multiply: $(a - 3b)(a + 4b)$

3. Solve: $4 - \frac{2}{3}x = 7$

4. Factor: $2a^3 + 7a^2 - 15a$

5. Find $16\frac{2}{3}\%$ of 60.

6. Multiply: $\frac{3x^2 - 6x}{4x - 6} \cdot \frac{2x^2 + x - 6}{6x^3 - 24x}$

7. Find the GCF of $12x^3y^2$ and $42xy^6$.

8. Factor: $9x^2 + 15x - 14$

9. Solve by the addition method: $2x - 3y = -4$
   $5x + y = 7$

10. Simplify: $(a^3b^5)^3$

11. Find the slope of the line containing points $(2, -3)$ and $(-1, 4)$.

12. Solve: $-\frac{3}{5}x = -\frac{9}{10}$

13. Factor: $4x^2 + 28xy + 49y^2$

14. Find the equation of the line that contains the points $(2, -1)$ and has slope $\frac{1}{2}$.

15. Graph: $y = \frac{3}{4}x - 2$

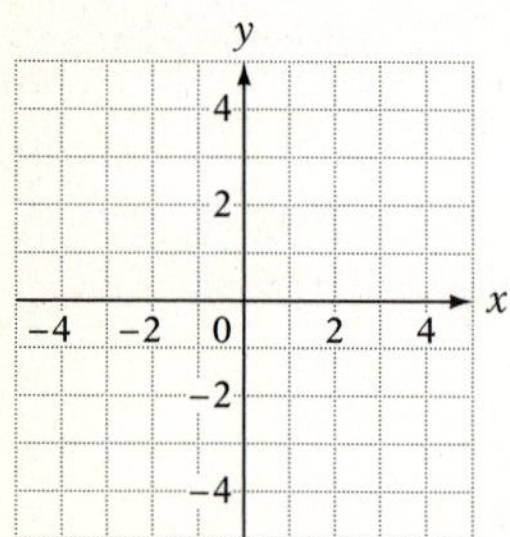

16. Given that $\angle a = 35°$ and $\angle x = 115°$, find the measures of angles b, c, and y.

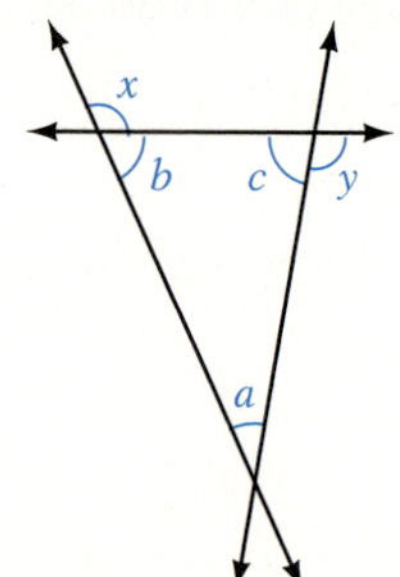

17. How long does it take light to travel to the earth from the moon when the moon is 232,500 mi from the earth? Light travels at 1.86×10^5 mi/s.

18. Find the perimeter of a rectangle that has a width of $2\frac{1}{3}$ ft and a length of $5\frac{1}{4}$ ft.

19. How many ounces of pure water must be added to 40 oz of a 12% salt solution to make a salt solution that is 5% salt?

20. A plane can fly 160 mi in calm air. Flying with the wind, the plane can fly 570 mi in the same amount of time as it takes to fly 390 mi against the wind. Find the rate of the wind.

21. Find the cost per pound of a mixture made from 20 lb of cashews that cost \$3.50 per pound and 50 lb of peanuts that cost \$1.75 per pound.

CHAPTER

11

Radical Expressions

Focus on Problem Solving

In solving application problems, it may be useful to include the units in order to organize the problem so that the answer is in the proper units. Using units to organize and check the correctness of an application is called **dimensional analysis**. We multiply and divide units in applying dimensional analysis to application problems.

Dimensional analysis is used to convert units of measure. The following example converts miles to feet. The equivalent measures 1 mi and 5280 ft are used to form the following rates, which are called conversion factors: $\frac{1\text{ mi}}{5280\text{ ft}}$ or $\frac{5280\text{ ft}}{1\text{ mi}}$. Because 1 mi = 5280 ft, both of the conversion factors $\frac{1\text{ mi}}{5280\text{ ft}}$ and $\frac{5280\text{ ft}}{1\text{ mi}}$ are equal to 1

To convert 3 mi to feet, multiply 3 mi by the conversion factor $\frac{5280\text{ ft}}{1\text{ mi}}$.

$$3\text{ mi} = 3\text{ mi} \cdot 1 = \frac{3\text{ mi}}{1} \cdot \frac{5280\text{ ft}}{1\text{ mi}} = \frac{3\text{ }\cancel{\text{mi}} \cdot 5280\text{ ft}}{1\text{ }\cancel{\text{mi}}} = 3 \cdot 5280\text{ ft} = 15{,}840\text{ ft}$$

Note that you can think of dividing the numerator and denominator by the common unit "mile" just as you would divide the numerator and denominator of a fraction by a common factor. The conversion factor $\frac{5280\text{ ft}}{1\text{ mi}}$ is equal to 1, and multiplying an expression by one does not change the value of the expression.

In the application problem below, the units are kept in the problem while working through the arithmetic.

In 1980, a horse named Fiddle Isle ran a 1.5-mile race in 2 min 23 s. Find Fiddle Isle's average speed in miles per hour for that race. Round to the nearest tenth.

To find the average speed, first convert 2 min 23 s to minutes. Then use the formula $r = \frac{d}{t}$, where r is the speed, d is the distance, and t is the time.

Convert 2 min 23 s to decimal minutes.

$$\begin{aligned} 2\text{ min } 23\text{ s} &= 2\text{ min} + 23\text{ s}\left(\frac{1\text{ min}}{60\text{ s}}\right) \\ &= 2\text{ min} + \frac{23}{60}\text{ min} \\ &= 2\frac{23}{60}\text{ min} \\ &\approx 2.383\text{ min} \end{aligned}$$

Divide the distance by the time.

$$r = \frac{d}{t} = \frac{1.5\text{ mi}}{2.383\text{ min}}$$

Use the conversion factor $\frac{60\text{ min}}{1\text{ h}}$.

$$= \frac{1.5\text{ mi}}{2.383\text{ min}} \cdot \frac{60\text{ min}}{1\text{ h}}$$

$$= \frac{90\text{ mi}}{2.383\text{ h}} \approx 37.8\text{ mph}$$

The horse's average speed was 37.8 mph.

SECTION 11.1 Addition and Subtraction of Radical Expressions

OBJECTIVE A Simplify numerical radical expressions

A **square root** of a positive number x is a number whose square is x.

A square root of 16 is 4 since $4^2 = 16$.
A square root of 16 is -4 since $(-4)^2 = 16$.

Every positive number has two square roots, one positive and one negative. The symbol $\sqrt{\ }$, called a **radical sign**, indicates the positive or **principal square root** of a number. For example, $\sqrt{16} = 4$ and $\sqrt{25} = 5$. The number under the radical sign is called the **radicand**.

POINT OF INTEREST

The radical symbol was first used in 1525 but was written as √. Some historians suggest that the radical symbol also developed into the symbols for "less than" and "greater than." Because typesetters of that time did not want to make additional symbols, the radical was rotated to the position ↘ and used as a "greater than" symbol and rotated to ↖ and used for the "less than" symbol. Other evidence, however, suggests that the "less than" and "greater than" symbols were developed independently of the radical symbol.

To indicate the negative square root of a number, place a negative sign in front of the radical. For example, $-\sqrt{16} = -4$ and $-\sqrt{25} = -5$.

The square of an integer is a **perfect square**.

$$7^2 = 49$$
$$9^2 = 81$$
$$12^2 = 144$$

An integer that is a perfect square can be written as the product of prime factors, each of which has an even exponent when expressed in exponential form.

$$49 = 7 \cdot 7 = 7^2$$
$$81 = 3 \cdot 3 \cdot 3 \cdot 3 = 3^4$$
$$144 = 2 \cdot 2 \cdot 2 \cdot 2 \cdot 3 \cdot 3 = 2^4 3^2$$

To find the square root of a perfect square written in exponential form, remove the radical sign and divide the exponent by 2.

Simplify: $\sqrt{625}$

Write the prime factorization of the radicand in exponential form. $\sqrt{625} = \sqrt{5^4}$

Remove the radical sign and divide the exponent by 2. $= 5^2$

Simplify. $= 25$

If a number is not a perfect square, its square root can only be approximated, for example, $\sqrt{2}$ and $\sqrt{7}$.

$$\sqrt{2} \approx 1.4142135\ldots$$
$$\sqrt{7} \approx 2.6457513\ldots$$

The numbers $\sqrt{2}$ and $\sqrt{7}$ are **irrational numbers**. Their decimal representations do not terminate or repeat. The rational numbers and the irrational numbers taken together are called the **real numbers**.

A radical expression is in simplest form when the radicand contains no factor that is a perfect square. The Product Property of Square Roots is used to simplify radical expressions.

The Product Property of Square Roots

If a and b are positive real numbers, then $\sqrt{ab} = \sqrt{a} \cdot \sqrt{b}$ and $\sqrt{a} \cdot \sqrt{b} = \sqrt{ab}$.

Simplify: $\sqrt{96}$

Write the prime factorization of the radicand in exponential form. $\sqrt{96} = \sqrt{2^5 \cdot 3}$

Write the radicand as a product of a perfect square and factors that do not contain a perfect square. $= \sqrt{2^4(2 \cdot 3)}$

Use the Product Property of Square Roots. $= \sqrt{2^4}\sqrt{2 \cdot 3}$

Simplify. $= 2^2\sqrt{2 \cdot 3}$

$= 4\sqrt{6}$

Simplify: $\sqrt{-4}$

The square root of a negative number is not a real number since the square of a real number is never negative. $\sqrt{-4}$ is not a real number.

Simplify: $\sqrt{125}$

Write the prime factorization of the radicand in exponential form. $\sqrt{125} = \sqrt{5^3}$

Write the radicand as a product of a perfect square and factors that do not contain a perfect square. $= \sqrt{5^2 \cdot 5}$

Use the Product Property of Square Roots. $= \sqrt{5^2}\sqrt{5}$

Simplify. The two expressions $\sqrt{125}$ and $5\sqrt{5}$ are different representations of the same number. $= 5\sqrt{5}$

Using a calculator, find the decimal approximation of $\sqrt{125}$ and $5\sqrt{5}$ to the nearest thousandth. Note that they are both 11.180. ≈ 11.180

Example 1

Simplify: $\sqrt{108}$

Solution

$$\sqrt{108} = \sqrt{2^2 \cdot 3^3} = \sqrt{2^2 \cdot 3^2 \cdot 3} = \sqrt{2^2 \cdot 3^2}\sqrt{3} = 2 \cdot 3\sqrt{3} = 6\sqrt{3}$$

You Try It 1

Simplify: $\sqrt{28}$

Your Solution

Solution on p. A35

Example 2

Simplify: $3\sqrt{90}$

Solution

$3\sqrt{90} = 3\sqrt{2 \cdot 3^2 \cdot 5} = 3\sqrt{3^2(2 \cdot 5)}$
$= 3\sqrt{3^2}\sqrt{2 \cdot 5} = 3 \cdot 3\sqrt{10} = 9\sqrt{10}$

You Try It 2

Simplify: $-5\sqrt{32}$

Your Solution

Example 3

Find the decimal approximation of $\sqrt{252}$. Round to the nearest thousandth.

Solution

$\sqrt{252} \approx 15.875$

You Try It 3

Find the decimal approximation of $\sqrt{216}$. Round to the nearest thousandth.

Your Solution

Solutions on p. A35

OBJECTIVE B Simplify variable radical expressions

Variable expressions that contain radicals do not always represent real numbers.

The variable expression at the right does not represent a real number when x is a negative number, for example, when x is -4

$\sqrt{x^3}$

$\sqrt{(-4)^3} = \sqrt{-64}$ Not a real number

For this reason, the variables in this chapter will represent nonnegative numbers unless otherwise stated.

A variable or a product of variables written in exponential form is a **perfect square** if each exponent is an even number.

To find the square root of a perfect square, remove the radical sign and divide each exponent by 2.

Simplify: $\sqrt{a^6}$

Remove the radical sign and divide the exponent by 2. $\sqrt{a^6} = a^3$

A variable radical expression is in simplest form when the radicand contains no factor that is a perfect square.

Simplify: $\sqrt{x^7}$

Write x^7 as the product of x and a perfect square. $\sqrt{x^7} = \sqrt{x^6 \cdot x}$

Use the Product Property of Square Roots. $= \sqrt{x^6}\sqrt{x}$

Simplify the perfect square. $= x^3\sqrt{x}$

Simplify: $3x\sqrt{8x^3y^{13}}$

Write the prime factorization of the coefficient of the radicand in exponential form. $3x\sqrt{8x^3y^{13}} = 3x\sqrt{2^3x^3y^{13}}$

Write the radicand as a product of a perfect square and factors that do not contain a perfect square. $= 3x\sqrt{2^2x^2y^{12}(2xy)}$

Use the Product Property of Square Roots. $= 3x\sqrt{2^2x^2y^{12}}\sqrt{2xy}$

Simplify. $= 3x \cdot 2xy^6\sqrt{2xy}$
$= 6x^2y^6\sqrt{2xy}$

Simplify: $\sqrt{25(x + 2)^2}$

Write the prime factorization of the coefficient in exponential form. $\sqrt{25(x + 2)^2} = \sqrt{5^2(x + 2)^2}$

Simplify. $= 5(x + 2)$
$= 5x + 10$

Example 4

Simplify: $\sqrt{b^{15}}$

Solution

$\sqrt{b^{15}} = \sqrt{b^{14} \cdot b} = \sqrt{b^{14}} \cdot \sqrt{b} = b^7\sqrt{b}$

You Try It 4

Simplify: $\sqrt{y^{19}}$

Your Solution

Example 5

Simplify: $\sqrt{24x^5}$

Solution

$\sqrt{24x^5} = \sqrt{2^3 \cdot 3 \cdot x^5} = \sqrt{2^2x^4(2 \cdot 3x)}$
$= \sqrt{2^2 \cdot x^4}\sqrt{2 \cdot 3x} = 2x^2\sqrt{6x}$

You Try It 5

Simplify: $\sqrt{45b^7}$

Your Solution

Example 6

Simplify: $2a\sqrt{18a^3b^{10}}$

Solution

$2a\sqrt{18a^3b^{10}} = 2a\sqrt{2 \cdot 3^2 \cdot a^3 \cdot b^{10}}$
$= 2a\sqrt{3^2a^2b^{10}(2a)} = 2a\sqrt{3^2a^2b^{10}}\sqrt{2a}$
$= 2a \cdot 3 \cdot a \cdot b^5\sqrt{2a} = 6a^2b^5\sqrt{2a}$

You Try It 6

Simplify: $3a\sqrt{28a^9b^{18}}$

Your Solution

Example 7

Simplify: $\sqrt{16(x + 5)^2}$

Solution

$\sqrt{16(x + 5)^2} = \sqrt{2^4(x + 5)^2} = 2^2(x + 5)$
$= 4(x + 5) = 4x + 20$

You Try It 7

Simplify: $\sqrt{25(a + 3)^2}$

Your Solution

Solutions on p. A35

Example 8

Simplify: $\sqrt{x^2 + 10x + 25}$

Solution

$\sqrt{x^2 + 10x + 25} = \sqrt{(x + 5)^2} = x + 5$

You Try It 8

Simplify: $\sqrt{x^2 + 14x + 49}$

Your Solution

Solution on p. A35

OBJECTIVE C Addition and subtraction of radical expressions

The Distributive Property is used to simplify the sum or difference of radical expressions with like radicands.

$5\sqrt{2} + 3\sqrt{2} = (5 + 3)\sqrt{2} = 8\sqrt{2}$

$6\sqrt{2x} - 4\sqrt{2x} = (6 - 4)\sqrt{2x} = 2\sqrt{2x}$

Radical expressions that are in simplest form and have unlike radicands cannot be simplified by the Distributive Property.

$2\sqrt{3} + 4\sqrt{2}$ cannot be simplified by the Distributive Property.

CONSIDER THIS

Note that doing operations with radical expressions is like doing operations with variable expressions.

Simplify: $4\sqrt{8} - 10\sqrt{2}$

Simplify each term.

$$4\sqrt{8} - 10\sqrt{2} = 4\sqrt{2^3} - 10\sqrt{2}$$
$$= 4\sqrt{2^2 \cdot 2} - 10\sqrt{2}$$
$$= 4\sqrt{2^2}\sqrt{2} - 10\sqrt{2}$$
$$= 4 \cdot 2\sqrt{2} - 10\sqrt{2}$$

Simplify the expression by using the Distributive Property.

$$= 8\sqrt{2} - 10\sqrt{2}$$
$$= (8 - 10)\sqrt{2}$$
$$= -2\sqrt{2}$$

Simplify: $8\sqrt{18x} - 2\sqrt{32x}$

Simplify each term.

$$8\sqrt{18x} - 2\sqrt{32x} = 8\sqrt{2 \cdot 3^2x} - 2\sqrt{2^5x}$$
$$= 8\sqrt{3^2 \cdot 2x} - 2\sqrt{2^4 \cdot 2x}$$
$$= 8\sqrt{3^2}\sqrt{2x} - 2\sqrt{2^4}\sqrt{2x}$$
$$= 8 \cdot 3\sqrt{2x} - 2 \cdot 2^2\sqrt{2x}$$
$$= 24\sqrt{2x} - 8\sqrt{2x}$$

Simplify the expression by using the Distributive Property.

$$= (24 - 8)\sqrt{2x}$$
$$= 16\sqrt{2x}$$

Example 9

Simplify: $5\sqrt{2} - 3\sqrt{2} + 12\sqrt{2}$

Solution

$5\sqrt{2} - 3\sqrt{2} + 12\sqrt{2} = 14\sqrt{2}$

You Try It 9

Simplify: $9\sqrt{3} + 3\sqrt{3} - 18\sqrt{3}$

Your Solution

Example 10

Subtract: $3\sqrt{12} - 5\sqrt{27}$

Solution

$$\begin{aligned} 3\sqrt{12} - 5\sqrt{27} &= 3\sqrt{2^2 \cdot 3} - 5\sqrt{3^3} \\ &= 3\sqrt{2^2}\sqrt{3} - 5\sqrt{3^2}\sqrt{3} \\ &= 3 \cdot 2\sqrt{3} - 5 \cdot 3\sqrt{3} = 6\sqrt{3} - 15\sqrt{3} \\ &= -9\sqrt{3} \end{aligned}$$

You Try It 10

Subtract: $2\sqrt{50} - 5\sqrt{32}$

Your Solution

Example 11

Simplify: $2\sqrt{8a} + 2\sqrt{18a} - 2\sqrt{32a}$

Solution

$$\begin{aligned} &2\sqrt{8a} + 2\sqrt{18a} - 2\sqrt{32a} \\ &= 2\sqrt{2^3a} + 2\sqrt{2 \cdot 3^2a} - 2\sqrt{2^5a} \\ &= 2\sqrt{2^2}\sqrt{2a} + 2\sqrt{3^2}\sqrt{2a} - 2\sqrt{2^4}\sqrt{2a} \\ &= 2 \cdot 2\sqrt{2a} + 2 \cdot 3\sqrt{2a} - 2 \cdot 2^2\sqrt{2a} \\ &= 4\sqrt{2a} + 6\sqrt{2a} - 8\sqrt{2a} = 2\sqrt{2a} \end{aligned}$$

You Try It 11

Simplify: $\sqrt{27b} - 2\sqrt{12b} + 7\sqrt{3b}$

Your Solution

Example 12

Subtract: $3\sqrt{12x^3} - 2x\sqrt{3x}$

Solution

$$\begin{aligned} 3\sqrt{12x^3} - 2x\sqrt{3x} &= 3\sqrt{2^2 \cdot 3 \cdot x^3} - 2x\sqrt{3x} \\ &= 3\sqrt{2^2 \cdot x^2}\sqrt{3x} - 2x\sqrt{3x} \\ &= 3 \cdot 2 \cdot x\sqrt{3x} - 2x\sqrt{3x} \\ &= 6x\sqrt{3x} - 2x\sqrt{3x} \\ &= 4x\sqrt{3x} \end{aligned}$$

You Try It 12

Add: $y\sqrt{28y} + 7\sqrt{63y^3}$

Your Solution

Example 13

Simplify: $2x\sqrt{8y} - 3\sqrt{2x^2y} + 2\sqrt{32x^2y}$

Solution

$$\begin{aligned} &2x\sqrt{8y} - 3\sqrt{2x^2y} + 2\sqrt{32x^2y} \\ &= 2x\sqrt{2^3y} - 3\sqrt{2x^2y} + 2\sqrt{2^5x^2y} \\ &= 2x\sqrt{2^2}\sqrt{2y} - 3\sqrt{x^2}\sqrt{2y} + 2\sqrt{2^4x^2}\sqrt{2y} \\ &= 2x \cdot 2\sqrt{2y} - 3 \cdot x\sqrt{2y} + 2 \cdot 2^2 \cdot x\sqrt{2y} \\ &= 4x\sqrt{2y} - 3x\sqrt{2y} + 8x\sqrt{2y} = 9x\sqrt{2y} \end{aligned}$$

You Try It 13

Simplify: $2\sqrt{27a^5} - 4a\sqrt{12a^3} + a^2\sqrt{75a}$

Your Solution

Solutions on pp. A35–A36

11.1 EXERCISES

Objective A

Simplify.

1. $\sqrt{16}$
2. $\sqrt{64}$
3. $\sqrt{49}$
4. $\sqrt{144}$
5. $\sqrt{32}$
6. $\sqrt{50}$
7. $\sqrt{8}$
8. $\sqrt{12}$
9. $6\sqrt{18}$
10. $-3\sqrt{48}$
11. $5\sqrt{40}$
12. $2\sqrt{28}$
13. $\sqrt{15}$
14. $\sqrt{21}$
15. $\sqrt{29}$
16. $\sqrt{13}$
17. $-9\sqrt{72}$
18. $11\sqrt{80}$
19. $\sqrt{45}$
20. $\sqrt{225}$
21. $\sqrt{0}$
22. $\sqrt{210}$
23. $6\sqrt{128}$
24. $9\sqrt{288}$
25. $\sqrt{105}$
26. $\sqrt{55}$
27. $\sqrt{900}$
28. $\sqrt{300}$

Find the decimal approximation. Round to the nearest thousandth.

29. $\sqrt{240}$
30. $\sqrt{300}$
31. $\sqrt{288}$
32. $\sqrt{600}$
33. $\sqrt{256}$
34. $\sqrt{729}$
35. $\sqrt{275}$
36. $\sqrt{450}$
37. $\sqrt{245}$
38. $\sqrt{525}$
39. $\sqrt{352}$
40. $\sqrt{363}$

Objective B

Simplify.

41. $\sqrt{x^6}$
42. $\sqrt{x^{12}}$
43. $\sqrt{y^{15}}$
44. $\sqrt{y^{11}}$
45. $\sqrt{a^{20}}$
46. $\sqrt{a^{16}}$
47. $\sqrt{x^4y^4}$
48. $\sqrt{x^{12}y^8}$

Simplify.

49. $\sqrt{4x^4}$

50. $\sqrt{25y^8}$

51. $\sqrt{24x^2}$

52. $\sqrt{x^3y^{15}}$

53. $\sqrt{x^3y^7}$

54. $\sqrt{a^{15}b^5}$

55. $\sqrt{a^3b^{11}}$

56. $\sqrt{24y^7}$

57. $\sqrt{60x^5}$

58. $\sqrt{72y^7}$

59. $\sqrt{49a^4b^8}$

60. $\sqrt{144x^2y^8}$

61. $x\sqrt{x^4y^2}$

62. $y\sqrt{x^3y^6}$

63. $4\sqrt{20a^4b^7}$

64. $5\sqrt{12a^3b^4}$

65. $3x\sqrt{12x^2y^7}$

66. $4y\sqrt{18x^5y^4}$

67. $2x^2\sqrt{8x^2y^3}$

68. $3y^2\sqrt{27x^4y^3}$

69. $\sqrt{25(a+4)^2}$

70. $\sqrt{81(x+y)^4}$

71. $\sqrt{4(x+2)^4}$

72. $\sqrt{9(x+2)^8}$

73. $\sqrt{x^2+4x+4}$

74. $\sqrt{b^2+8b+16}$

75. $\sqrt{y^2+2y+1}$

76. $\sqrt{a^2+6a+9}$

Objective C

Add or subtract.

77. $2\sqrt{2}+\sqrt{2}$

78. $3\sqrt{5}+8\sqrt{5}$

79. $-3\sqrt{7}+2\sqrt{7}$

80. $4\sqrt{5}-10\sqrt{5}$

81. $-3\sqrt{11}-8\sqrt{11}$

82. $-3\sqrt{3}-5\sqrt{3}$

83. $2\sqrt{x}+8\sqrt{x}$

84. $3\sqrt{y}+2\sqrt{y}$

85. $8\sqrt{y}-10\sqrt{y}$

86. $-5\sqrt{2a}+2\sqrt{2a}$

87. $-2\sqrt{3b}-9\sqrt{3b}$

88. $-7\sqrt{5a}-5\sqrt{5a}$

89. $3x\sqrt{2}-x\sqrt{2}$

90. $2y\sqrt{3}-9y\sqrt{3}$

91. $2a\sqrt{3a}-5a\sqrt{3a}$

92. $-5b\sqrt{3x}-2b\sqrt{3x}$

93. $3\sqrt{xy}-8\sqrt{xy}$

94. $-4\sqrt{xy}+6\sqrt{xy}$

Simplify.

95. $\sqrt{45} + \sqrt{125}$

96. $\sqrt{32} - \sqrt{98}$

97. $2\sqrt{2} + 3\sqrt{8}$

98. $4\sqrt{128} - 3\sqrt{32}$

99. $5\sqrt{18} - 2\sqrt{75}$

100. $5\sqrt{75} - 2\sqrt{18}$

101. $5\sqrt{4x} - 3\sqrt{9x}$

102. $-3\sqrt{25y} + 8\sqrt{49y}$

103. $3\sqrt{3x^2} - 5\sqrt{27x^2}$

104. $-2\sqrt{8y^2} + 5\sqrt{32y^2}$

105. $2x\sqrt{xy^2} - 3y\sqrt{x^2y}$

106. $4a\sqrt{b^2a} - 3b\sqrt{a^2b}$

107. $3x\sqrt{12x} - 5\sqrt{27x^3}$

108. $2a\sqrt{50a} + 7\sqrt{32a^3}$

109. $4y\sqrt{8y^3} - 7\sqrt{18y^5}$

110. $2a\sqrt{8ab^2} - 2b\sqrt{2a^3}$

111. $b^2\sqrt{a^5b} + 3a^2\sqrt{ab^5}$

112. $y^2\sqrt{x^5y} + x\sqrt{x^3y^5}$

113. $4\sqrt{2} - 5\sqrt{2} + 8\sqrt{2}$

114. $3\sqrt{3} + 8\sqrt{3} - 16\sqrt{3}$

115. $5\sqrt{x} - 8\sqrt{x} + 9\sqrt{x}$

116. $\sqrt{x} - 7\sqrt{x} + 6\sqrt{x}$

117. $8\sqrt{2} - 3\sqrt{y} - 8\sqrt{2}$

118. $8\sqrt{3} - 5\sqrt{2} - 5\sqrt{3}$

119. $8\sqrt{8} - 4\sqrt{32} - 9\sqrt{50}$

120. $2\sqrt{12} - 4\sqrt{27} + \sqrt{75}$

121. $-2\sqrt{3} + 5\sqrt{27} - 4\sqrt{45}$

122. $-2\sqrt{8} - 3\sqrt{27} + 3\sqrt{50}$

123. $4\sqrt{75} + 3\sqrt{48} - \sqrt{99}$

124. $2\sqrt{75} - 5\sqrt{20} + 2\sqrt{45}$

125. $\sqrt{25x} - \sqrt{9x} + \sqrt{16x}$

126. $\sqrt{4x} - \sqrt{100x} - \sqrt{49x}$

Simplify.

127. $2a\sqrt{75b} - a\sqrt{20b} + 4a\sqrt{45b}$

128. $2b\sqrt{75a} - 5b\sqrt{27a} + 2b\sqrt{20a}$

129. $x\sqrt{3y^2} - 2y\sqrt{12x^2} + xy\sqrt{3}$

130. $a\sqrt{27b^2} + 3b\sqrt{147a^2} - ab\sqrt{3}$

131. $3\sqrt{ab^3} + 4a\sqrt{ab} - 5b\sqrt{4ab}$

132. $5\sqrt{a^3b} + a\sqrt{4ab} - 3\sqrt{49a^3b}$

133. $3a\sqrt{2ab^2} - \sqrt{a^2b^2} + 4b\sqrt{3a^2b}$

134. $2\sqrt{4a^2b^2} - 3a\sqrt{9ab^2} + 4b\sqrt{a^2b}$

Critical Thinking

Simplify.

135. $\sqrt{0.0025a^3b^5}$

136. $-\frac{3y}{4}\sqrt{64x^4y^2}$

137. $\sqrt{x^2y^3 + x^3y^2}$

138. $\sqrt{4a^5b^4 - 4a^4b^5}$

139. If a and b are positive real numbers, does $\sqrt{a+b} = \sqrt{a} + \sqrt{b}$? If not, give an example when the expressions are not equal.

140. Grade the following solution to the problem "Write $\sqrt{72}$ in simplest form."

$$\begin{aligned}\sqrt{72} &= \sqrt{4}\sqrt{18} \\ &= 2\sqrt{18}\end{aligned}$$

Is the solution correct? If not, what error was made? What is the correct solution?

Simplify.

141. $\sqrt{\sqrt{16}}$

142. $\sqrt{\sqrt{81}}$

143. $2\sqrt{8x + 4y} - 5\sqrt{18x + 9y}$

144. $6\sqrt{16x - 16} + \sqrt{25x - 25}$

145. $3\sqrt{a^3 + a^2} + 5\sqrt{4a^3 + 4a^2}$

146. $3\sqrt{x^3y^2 + x^2y^3} + xy\sqrt{4x + 4y}$

147. [W] Describe in your own words how to simplify a radical expression.

148. [W] Explain why $2\sqrt{2}$ is in simplest form and $\sqrt{8}$ is not in simplest form.

SECTION 11.2 Multiplication and Division of Radical Expressions

OBJECTIVE A Multiplication of radical expressions

The Product Property of Square Roots can also be used to multiply variable radical expressions.

$$\sqrt{2x}\sqrt{3y} = \sqrt{2x \cdot 3y} = \sqrt{6xy}$$

Simplify: $(\sqrt{x})^2$

Multiply the radicands. $(\sqrt{x})^2 = \sqrt{x}\sqrt{x} = \sqrt{x \cdot x}$

Simplify. $= \sqrt{x^2}$

$= x$

Note: For $a > 0$, $(\sqrt{a})^2 = \sqrt{a^2} = a$

Multiply: $\sqrt{2x^2}\sqrt{32x^5}$

Use the Product Property of Square Roots. $\sqrt{2x^2}\sqrt{32x^5} = \sqrt{2x^2 \cdot 32x^5}$

Multiply the radicands. $= \sqrt{64x^7}$

Simplify. $= \sqrt{2^6x^7}$

$= \sqrt{2^6x^6}\sqrt{x}$

$= 2^3x^3\sqrt{x}$

$= 8x^3\sqrt{x}$

Multiply: $\sqrt{2x}(x + \sqrt{2x})$

Use the Distributive Property to remove parentheses. $\sqrt{2x}(x + \sqrt{2x}) = \sqrt{2x}(x) + (\sqrt{2x})^2$

$= x\sqrt{2x} + 2x$

Multiply: $(\sqrt{2} - 3x)(\sqrt{2} + x)$

Use the FOIL method to remove parentheses. $(\sqrt{2} - 3x)(\sqrt{2} + x) = (\sqrt{2})^2 + x\sqrt{2} - 3x\sqrt{2} - 3x^2$

$= 2 + (x - 3x)\sqrt{2} - 3x^2$

$= 2 - 2x\sqrt{2} - 3x^2$

The expressions $a + b$ and $a - b$, which are the sum and difference of two terms, are called **conjugates** of each other. Conjugates differ only in the sign of one of the terms.

Multiply: $(2 + \sqrt{7})(2 - \sqrt{7})$

The product of conjugates of the form $(a + b)(a - b) = a^2 - b^2$. $(2 + \sqrt{7})(2 - \sqrt{7}) = 2^2 - (\sqrt{7})^2$

$= 4 - 7$

$= -3$

Multiply: $(3 + \sqrt{y})(3 - \sqrt{y})$

The product of conjugates of the form $(a + b)(a - b) = a^2 - b^2$. $(3 + \sqrt{y})(3 - \sqrt{y}) = 3^2 - (\sqrt{y})^2$

$= 9 - y$

Example 1

Multiply: $\sqrt{3x^4}\sqrt{2x^2y}\sqrt{6xy^2}$

Solution

$\sqrt{3x^4}\sqrt{2x^2y}\sqrt{6xy^2} = \sqrt{36x^7y^3} = \sqrt{2^2 3^2 x^7 y^3}$

$= \sqrt{2^2 3^2 x^6 y^2}\sqrt{xy} = 2 \cdot 3x^3y\sqrt{xy} = 6x^3y\sqrt{xy}$

You Try It 1

Multiply: $\sqrt{5a}\sqrt{15a^3b^4}\sqrt{3b^5}$

Your Solution

Example 2

Multiply: $\sqrt{3ab}(\sqrt{3a} + \sqrt{9b})$

Solution

$\sqrt{3ab}(\sqrt{3a} + \sqrt{9b}) = \sqrt{3^2a^2b} + \sqrt{3^3ab^2}$

$= \sqrt{3^2a^2}\sqrt{b} + \sqrt{3^2b^2}\sqrt{3a} = 3a\sqrt{b} + 3b\sqrt{3a}$

You Try It 2

Multiply: $\sqrt{5x}(\sqrt{5x} - \sqrt{25y})$

Your Solution

Example 3

Multiply: $(\sqrt{a} - \sqrt{b})(\sqrt{a} + \sqrt{b})$

Solution

$(\sqrt{a} - \sqrt{b})(\sqrt{a} + \sqrt{b}) = (\sqrt{a})^2 - (\sqrt{b})^2$

$= a - b$

You Try It 3

Multiply: $(2\sqrt{x} + 7)(2\sqrt{x} - 7)$

Your Solution

Example 4

Multiply: $(2\sqrt{x} - \sqrt{y})(5\sqrt{x} - 2\sqrt{y})$

Solution

$(2\sqrt{x} - \sqrt{y})(5\sqrt{x} - 2\sqrt{y})$

$= 10(\sqrt{x})^2 - 4\sqrt{xy} - 5\sqrt{xy} + 2(\sqrt{y})^2$

$= 10x - 9\sqrt{xy} + 2y$

You Try It 4

Multiply: $(3\sqrt{x} - \sqrt{y})(5\sqrt{x} - 2\sqrt{y})$

Your Solution

Solutions on p. A36

OBJECTIVE B Division of radical expressions

POINT OF INTEREST

A radical expression that occurs in Einstein's theory of relativity is

$$\frac{1}{\sqrt{1 - \frac{v^2}{c^2}}}$$

where v is the velocity of an object and c is the speed of light.

The Quotient Property of Square Roots

The square root of a quotient equals the quotient of the square roots. If a and b are positive real numbers, then

$$\sqrt{\frac{a}{b}} = \frac{\sqrt{a}}{\sqrt{b}} \text{ and } \frac{\sqrt{a}}{\sqrt{b}} = \sqrt{\frac{a}{b}}.$$

Simplify: $\sqrt{\frac{4x^2}{z^6}}$

Rewrite the radical expression as the quotient of the square roots.

$$\sqrt{\frac{4x^2}{z^6}} = \frac{\sqrt{4x^2}}{\sqrt{z^6}}$$

Simplify.

$$= \frac{\sqrt{2^2x^2}}{\sqrt{z^6}} = \frac{2x}{z^3}$$

Simplify: $\sqrt{\dfrac{24x^3y^7}{3x^7y^2}}$

Simplify the radicand. $\sqrt{\dfrac{24x^3y^7}{3x^7y^2}} = \sqrt{\dfrac{8y^5}{x^4}}$

Rewrite the radical expression as the quotient of the square roots. $= \dfrac{\sqrt{8y^5}}{\sqrt{x^4}}$

Simplify. $= \dfrac{\sqrt{2^3y^5}}{\sqrt{x^4}}$

$= \dfrac{\sqrt{2^2y^4}\sqrt{2y}}{\sqrt{x^4}}$

$= \dfrac{2y^2\sqrt{2y}}{x^2}$

Simplify: $\dfrac{\sqrt{4x^2y}}{\sqrt{xy}}$

Use the Quotient Property of Square Roots. $\dfrac{\sqrt{4x^2y}}{\sqrt{xy}} = \sqrt{\dfrac{4x^2y}{xy}}$

Simplify the radicand. $= \sqrt{4x}$

Simplify the radical expression. $= \sqrt{2^2}\sqrt{x}$

$= 2\sqrt{x}$

CONSIDER THIS

Rationalizing the denominator is another case where the Multiplication Property of One is used. Use a calculator to verify that $\dfrac{2}{\sqrt{3}}$ and $\dfrac{2\sqrt{3}}{3}$ are equal (to the limits of the calculator).

A radical expression is not in simplest form if a radical remains in the denominator. Removing a radical from the denominator is called **rationalizing the denominator**.

Simplify: $\dfrac{2}{\sqrt{3}}$

Multiply the expression by 1 in the form $\dfrac{\sqrt{3}}{\sqrt{3}}$. $\dfrac{2}{\sqrt{3}} = \dfrac{2}{\sqrt{3}} \cdot \dfrac{\sqrt{3}}{\sqrt{3}}$

The radicand in the denominator is a perfect square. $= \dfrac{2\sqrt{3}}{\sqrt{3^2}}$

Simplify.
The radical has been removed from the denominator. $= \dfrac{2\sqrt{3}}{3}$

When the denominator is a binomial expression with a radical, multiply the numerator and denominator by the conjugate of the denominator.

Simplify: $\dfrac{1}{\sqrt{y}+3}$

Multiply the numerator and denominator by $\sqrt{y}-3$, the conjugate of $\sqrt{y}+3$. $\dfrac{1}{\sqrt{y}+3} = \dfrac{1}{\sqrt{y}+3} \cdot \dfrac{\sqrt{y}-3}{\sqrt{y}-3}$

Simplify. $= \dfrac{\sqrt{y}-3}{(\sqrt{y})^2 - 3^2}$

$= \dfrac{\sqrt{y}-3}{y-9}$

Simplify: $\dfrac{\sqrt{2} + \sqrt{18y^2}}{\sqrt{2}}$

Divide each term in the numerator by the denominator.

$$\frac{\sqrt{2} + \sqrt{18y^2}}{\sqrt{2}} = \frac{\sqrt{2}}{\sqrt{2}} + \frac{\sqrt{18y^2}}{\sqrt{2}}$$

Use the Quotient Property of Square Roots.

$$= 1 + \sqrt{\frac{18y^2}{2}}$$

Simplify.

$$= 1 + \sqrt{9y^2}$$
$$= 1 + \sqrt{3^2y^2}$$
$$= 1 + 3y$$

Example 5

Simplify: $\dfrac{\sqrt{4x^2y^5}}{\sqrt{3x^4y}}$

Solution

$$\frac{\sqrt{4x^2y^5}}{\sqrt{3x^4y}} = \sqrt{\frac{2^2x^2y^5}{3x^4y}} = \sqrt{\frac{2^2y^4}{3x^2}} = \frac{2y^2}{x\sqrt{3}}$$
$$= \frac{2y^2}{x\sqrt{3}} \cdot \frac{\sqrt{3}}{\sqrt{3}} = \frac{2y^2\sqrt{3}}{3x}$$

You Try It 5

Simplify: $\dfrac{\sqrt{15x^6y^7}}{\sqrt{3x^7y^9}}$

Your Solution

Example 6

Simplify: $\dfrac{\sqrt{2}}{\sqrt{2} - \sqrt{x}}$

Solution

$$\frac{\sqrt{2}}{\sqrt{2} - \sqrt{x}} = \frac{\sqrt{2}}{\sqrt{2} - \sqrt{x}} \cdot \frac{\sqrt{2} + \sqrt{x}}{\sqrt{2} + \sqrt{x}}$$
$$= \frac{2 + \sqrt{2x}}{2 - x}$$

You Try It 6

Simplify: $\dfrac{\sqrt{y}}{\sqrt{y} + 3}$

Your Solution

Example 7

Simplify: $\dfrac{\sqrt{20} - 2\sqrt{125}}{\sqrt{5}}$

Solution

$$\frac{\sqrt{20} - 2\sqrt{125}}{\sqrt{5}} = \frac{\sqrt{20}}{\sqrt{5}} - \frac{2\sqrt{125}}{\sqrt{5}}$$
$$= \sqrt{\frac{20}{5}} - 2\sqrt{\frac{125}{5}} = \sqrt{4} - 2\sqrt{25}$$
$$= \sqrt{2^2} - 2\sqrt{5^2} = 2 - 2 \cdot 5 = 2 - 10$$
$$= -8$$

You Try It 7

Simplify: $\dfrac{\sqrt{27x^3} - 3\sqrt{12x}}{\sqrt{3x}}$

Your Solution

Solutions on p. A36

11.2 EXERCISES

Objective A

Multiply.

1. $\sqrt{5} \cdot \sqrt{5}$

2. $\sqrt{11} \cdot \sqrt{11}$

3. $\sqrt{3} \cdot \sqrt{12}$

4. $\sqrt{2} \cdot \sqrt{8}$

5. $\sqrt{x} \cdot \sqrt{x}$

6. $\sqrt{y} \cdot \sqrt{y}$

7. $\sqrt{xy^3} \cdot \sqrt{x^5y}$

8. $\sqrt{a^3b^5} \cdot \sqrt{ab^5}$

9. $\sqrt{3a^2b^5} \cdot \sqrt{6ab^7}$

10. $\sqrt{5x^3y} \cdot \sqrt{10x^2y}$

11. $\sqrt{6a^3b^2} \cdot \sqrt{24a^5b}$

12. $\sqrt{8ab^5} \cdot \sqrt{12a^7b}$

13. $\sqrt{2}(\sqrt{2} - \sqrt{3})$

14. $3(\sqrt{12} - \sqrt{3})$

15. $\sqrt{x}(\sqrt{x} - \sqrt{y})$

16. $\sqrt{b}(\sqrt{a} - \sqrt{b})$

17. $\sqrt{5}(\sqrt{10} - \sqrt{x})$

18. $\sqrt{6}(\sqrt{y} - \sqrt{18})$

19. $\sqrt{8}(\sqrt{2} - \sqrt{5})$

20. $\sqrt{10}(\sqrt{20} - \sqrt{a})$

21. $(\sqrt{x} - 3)^2$

22. $(2\sqrt{a} - y)^2$

23. $\sqrt{3a}(\sqrt{3a} - \sqrt{3b})$

24. $\sqrt{5x}(\sqrt{10x} - \sqrt{x})$

25. $\sqrt{2ac} \cdot \sqrt{5ab} \cdot \sqrt{10cb}$

26. $\sqrt{3xy} \cdot \sqrt{6x^3y} \cdot \sqrt{2y^2}$

27. $(3\sqrt{x} - 2y)(5\sqrt{x} - 4y)$

28. $(5\sqrt{x} + 2\sqrt{y})(3\sqrt{x} - \sqrt{y})$

29. $(\sqrt{x} - \sqrt{y})(\sqrt{x} + \sqrt{y})$

30. $(\sqrt{3x} + y)(\sqrt{3x} - y)$

Objective B

Simplify.

31. $\dfrac{\sqrt{32}}{\sqrt{2}}$

32. $\dfrac{\sqrt{45}}{\sqrt{5}}$

33. $\dfrac{\sqrt{98}}{\sqrt{2}}$

Simplify.

34. $\dfrac{\sqrt{15x^3y}}{\sqrt{3xy}}$

35. $\dfrac{\sqrt{40x^5y^2}}{\sqrt{5xy}}$

36. $\dfrac{\sqrt{2a^5b^4}}{\sqrt{98ab^4}}$

37. $\dfrac{\sqrt{48x^5y^2}}{\sqrt{3x^3y}}$

38. $\dfrac{1}{\sqrt{3}}$

39. $\dfrac{1}{\sqrt{8}}$

40. $\dfrac{3}{\sqrt{x}}$

41. $\dfrac{4}{\sqrt{2x}}$

42. $\dfrac{\sqrt{8x^2y}}{\sqrt{2x^4y^2}}$

43. $\dfrac{\sqrt{9xy^2}}{\sqrt{27x}}$

44. $\dfrac{\sqrt{4x^2y}}{\sqrt{3xy^3}}$

45. $\dfrac{\sqrt{16x^3y^2}}{\sqrt{8x^3y}}$

46. $\dfrac{1}{\sqrt{2} - 3}$

47. $\dfrac{5}{\sqrt{7} - 3}$

48. $\dfrac{3}{5 + \sqrt{5}}$

49. $\dfrac{7}{\sqrt{2} - 7}$

50. $\dfrac{\sqrt{xy}}{\sqrt{x} - \sqrt{y}}$

51. $\dfrac{\sqrt{x}}{\sqrt{x} - \sqrt{y}}$

52. $\dfrac{3\sqrt{2} - 8\sqrt{2}}{\sqrt{2}}$

53. $\dfrac{5\sqrt{3} - 2\sqrt{3}}{2\sqrt{3}}$

54. $\dfrac{2\sqrt{8} + 3\sqrt{2}}{\sqrt{32}}$

Critical Thinking

Simplify.

55. $-\sqrt{1.3} \cdot \sqrt{1.3}$

56. $\sqrt{\dfrac{5}{8}} \cdot \sqrt{\dfrac{5}{8}}$

57. $-\sqrt{\dfrac{16}{81}}$

58. $\sqrt{1\dfrac{9}{16}}$

59. $\sqrt{2\dfrac{1}{4}}$

60. $-\sqrt{6\dfrac{1}{4}}$

61. Answer true or false. If the answer is false, write the correct answer.

a. $(\sqrt{y})^4 = y^2$

b. $(2\sqrt{x})^3 = 8x\sqrt{x}$

c. $(\sqrt{x} + 1)^2 = x + 1$

d. $\dfrac{1}{2 - \sqrt{3}} = 2 + \sqrt{3}$

62. [W] In your own words, describe the process of rationalizing the denominator.

63. [W] The number $\dfrac{\sqrt{5} + 1}{2}$ is called the golden ratio. Research the golden ratio and write a few paragraphs about this number and its applications.

SECTION 11.3 Radical Equations

OBJECTIVE A Equations with radical expressions

An equation that contains a variable in a radicand is a **radical equation.**

$$\left.\begin{array}{r}\sqrt{x} = 4 \\ \sqrt{x+2} = \sqrt{x-7}\end{array}\right\} \text{Radical Equations}$$

The following property of equality is used to solve radical equations.

CONSIDER THIS

Does $a^2 = b^2$ mean that $a = b$? NO! and that is why you need to check solutions of radical equations.

> ***Property of Squaring Both Sides of an Equation***
>
> If two numbers are equal, then the squares of the number are equal. If a and b are real numbers and $a = b$, then $a^2 = b^2$.

Solve: $\sqrt{x-2} - 7 = 0$

Rewrite the equation with the radical on one side of the equation and the constant on the other side.

$$\sqrt{x-2} - 7 = 0$$
$$\sqrt{x-2} = 7$$

Square each side of the equation.

$$(\sqrt{x-2})^2 = 7^2$$

Solve the resulting equation.

$$x - 2 = 49$$
$$x = 51$$

Check the solution.
When squaring each side of an equation, the resulting equation may have a solution that is not a solution of the original equation.

Check: $\sqrt{x-2} - 7 = 0$

$\sqrt{51-2} - 7$	0
$\sqrt{49} - 7$	0
$\sqrt{7^2} - 7$	0
$7 - 7$	0

$0 = 0$ True

The solution is 51.

Example 1

Solve: $\sqrt{3x} + 2 = 5$

Solution

$$\sqrt{3x} + 2 = 5$$
$$\sqrt{3x} = 3$$
$$(\sqrt{3x})^2 = 3^2$$
$$3x = 9$$
$$x = 3$$

Check: $\sqrt{3x} + 2 = 5$

$\sqrt{3 \cdot 3} + 2$	5
$\sqrt{3^2} + 2$	5
$3 + 2$	5

$5 = 5$

The solution is 3.

You Try It 1

Solve: $\sqrt{4x} + 3 = 7$

Your Solution

Solution on p. A36

Example 2

Solve: $0 = 3 - \sqrt{2x - 3}$

Solution

$$\begin{aligned} 0 &= 3 - \sqrt{2x - 3} \\ -3 &= -\sqrt{2x - 3} \\ (-3)^2 &= (-\sqrt{2x - 3})^2 \\ 9 &= 2x - 3 \\ 12 &= 2x \\ 6 &= x \end{aligned}$$

Check:

$$\begin{array}{c|l} \multicolumn{2}{l}{0 = 3 - \sqrt{2x - 3}} \\ \hline 0 & 3 - \sqrt{2 \cdot 6 - 3} \\ 0 & 3 - \sqrt{12 - 3} \\ 0 & 3 - \sqrt{9} \\ 0 & 3 - \sqrt{3^2} \\ 0 & 3 - 3 \\ \multicolumn{2}{l}{0 = 0} \end{array}$$

The solution is 6.

You Try It 2

Solve: $\sqrt{3x - 2} - 5 = 0$

Your Solution

Example 3

Solve: $\sqrt{2x + 1} = \sqrt{3x - 4}$

Solution

$$\begin{aligned} \sqrt{2x + 1} &= \sqrt{3x - 4} \\ (\sqrt{2x + 1})^2 &= (\sqrt{3x - 4})^2 \\ 2x + 1 &= 3x - 4 \\ 2x &= 3x - 5 \\ -x &= -5 \\ x &= 5 \end{aligned}$$

Check:

$$\begin{array}{c|c} \sqrt{2x + 1} & \sqrt{3x - 4} \\ \hline \sqrt{2 \cdot 5 + 1} & \sqrt{3 \cdot 5 - 4} \\ \sqrt{10 + 1} & \sqrt{15 - 4} \\ \multicolumn{2}{c}{\sqrt{11} = \sqrt{11}} \end{array}$$

The solution is 5.

You Try It 3

Solve: $\sqrt{4x + 3} = \sqrt{x + 12}$

Your Solution

Solutions on pp. A36–A37

OBJECTIVE B Right triangles

A right triangle contains one 90° angle. The side opposite the 90° angle is called the **hypotenuse**. The other two sides are called the **legs**.

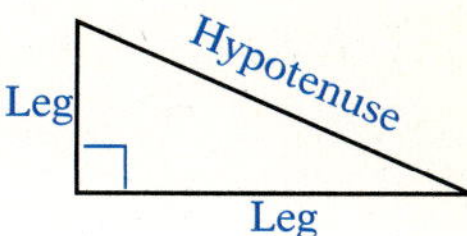

The angles in a right triangle are usually labeled with the capital letters A, B, and C, with C reserved for the right angle. The side opposite angle A is side a, the side opposite angle B is side b, and c is the hypotenuse.

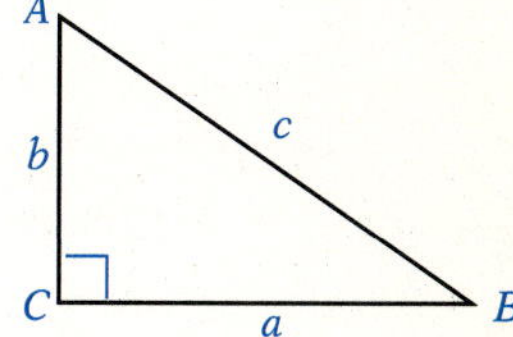

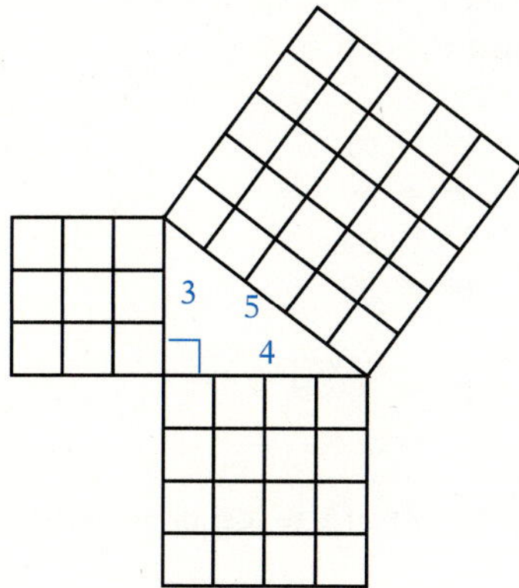

The Greek mathematician Pythagoras is generally credited with the discovery that the square of the hypotenuse of a right triangle is equal to the sum of the squares of the two legs. This is called the **Pythagorean Theorem.** However, the Babylonians used this theorem more than 1000 years before Pythagoras's time.

POINT OF INTEREST

The first known proof of this theorem is in a Chinese text, *Arithmetic Classic,* which was written about 600 B.C. (but there are no existing copies) and revised over a period of 500 years. The earliest known copy of this text dates from approximately 100 B.C.

The Pythagorean Theorem

If a and b are the lengths of the legs of a right triangle and c is the length of the hypotenuse, then $c^2 = a^2 + b^2$.

If the lengths of two sides of a right triangle are known, the Pythagorean Theorem can be used to find the length of the third side.

Consider a right triangle with legs that measure 5 cm and 12 cm. Use the Pythagorean Theorem, with $a = 5$ and $b = 12$, to find the length of the hypotenuse. (If you let $a = 12$ and $b = 5$, the result is the same.)

$$c^2 = a^2 + b^2$$
$$c^2 = 5^2 + 12^2$$
$$c^2 = 25 + 144$$
$$c^2 = 169$$

This equation states that the square of c is 169. Since $13^2 = 169$, $c = 13$, and the length of the hypotenuse is 13 cm. We can find c by taking the square root of 169: $\sqrt{169} = 13$. This suggests the following property.

The Principal Square Root Property

If $r^2 = s$, then $r = \sqrt{s}$, and r is called the square root of s.

The Principal Square Root Property and its application can be illustrated as follows:

$$\text{Because } 5^2 = 25,\ 5 = \sqrt{25}. \text{ Therefore, if } c^2 = 25,\ c = \sqrt{25} = 5.$$

Recall that numbers whose square roots are integers, such as 25, are perfect squares. If a number is not a perfect square, a calculator can be used to find an approximate square root when a decimal approximation is required. For example, $\sqrt{35} \approx 5.916$.

The length of one leg of a right triangle is 5 cm. The hypotenuse is 8 cm. Find the length of the other leg. Round to the nearest hundredth.

Use the Pythagorean Theorem.
$a = 5, c = 8$

$$a^2 + b^2 = c^2$$
$$5^2 + b^2 = 8^2$$
$$25 + b^2 = 64$$
$$b^2 = 39$$

Take the principal square root of each side of the equation.

$$\sqrt{b^2} = \sqrt{39}$$
$$b \approx 6.24$$

The length of the leg is 6.24 cm.

The Pythagorean Theorem can be used to find a relationship among the sides of two special right triangles.

Recall that an isosceles triangle has two equal angles. In an **isosceles right triangle,** $\angle A$ and $\angle B$ both measure 45°. For this reason, an isosceles right triangle is also called a **45°-45°-90° triangle**.

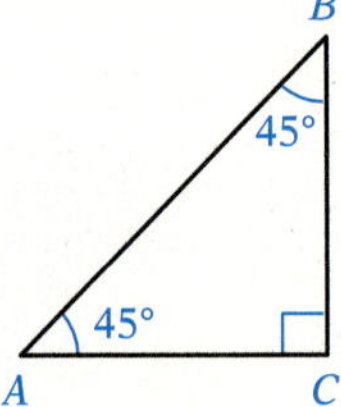

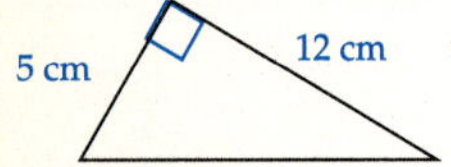

Find the hypotenuse of an isosceles right triangle if the length of one of the legs measures 1 unit.

Use the Pythagorean Theorem. In an isosceles right triangle, both legs are of equal length.
$a = 1, b = 1$

$$c^2 = a^2 + b^2$$
$$c^2 = 1^2 + 1^2$$
$$c^2 = 1 + 1$$
$$c^2 = 2$$

Take the principal square root of each side of the equation.

$$\sqrt{c^2} = \sqrt{2}$$
$$c = \sqrt{2}$$

The hypotenuse is $\sqrt{2}$ units.

The figure corresponding to this example is shown at the right.

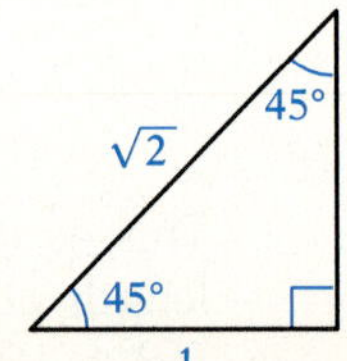

For any isosceles right triangle, the hypotenuse, c, equals $\sqrt{2}$ times the length of a leg.

$$c = \sqrt{2}\ \textbf{(length of a leg)}$$

For example, if the length of one of the legs in a 45°-45°-90° triangle is $4\sqrt{2}$ m, then the hypotenuse is 8 m.

$$c = \sqrt{2}(4\sqrt{2}) = 4 \cdot 2 = 8 \text{ m}$$

A leg of an isosceles right triangle is 4 m long. Find the perimeter of the triangle. Round to the nearest hundredth.

Find the hypotenuse.

$$c = \sqrt{2}(4) = 4\sqrt{2}$$

Find the perimeter.
An isosceles right triangle has 2 equal legs.
$a = 4, b = 4, c = 4\sqrt{2}$

$$P = a + b + c$$
$$P = 4 + 4 + 4\sqrt{2}$$
$$P = 8 + 4\sqrt{2}$$
$$P \approx 8 + 4(1.414)$$
$$P \approx 8 + 5.66$$
$$P \approx 13.66$$

The perimeter is 13.66 m.

The second special right triangle is the **30°-60°-90° triangle.** The acute angles in a 30°-60°-90° triangle measure 30° and 60°.

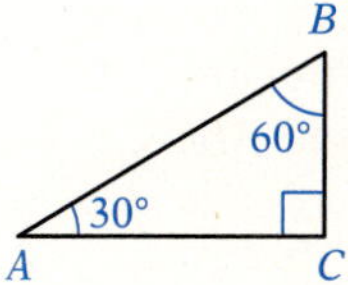

If two 30°-60°-90° triangles, each with a hypotenuse of 2 units, are positioned so that the longer legs of each triangle lie on the same line segment, then an equilateral triangle is formed and the shorter leg of each triangle is 1 unit.

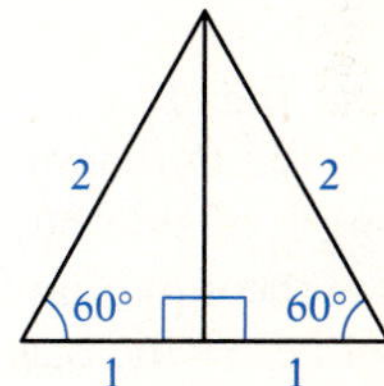

Using the Pythagorean Theorem, the length of the longer leg can be found. $a = 1, c = 2.$

$$a^2 + b^2 = c^2$$
$$1^2 + b^2 = 2^2$$
$$1 + b^2 = 4$$
$$b^2 = 3$$

Take the principal square root of each side of the equation.

$$\sqrt{b^2} = \sqrt{3}$$
$$b = \sqrt{3}$$

The relationship among the sides of a 30°-60°-90° triangle is shown in the figure at the right.

For any 30°-60°-90° triangle, the hypotenuse, c, equals twice the length of the shorter leg (the leg opposite the 30° angle).

$$c = \textbf{2(length of shorter leg)}$$

Example 4

The lengths of the legs of a 30°-60°-90° triangle measure 4 cm and $4\sqrt{3}$ cm. Find the perimeter of the triangle. Round to the nearest tenth.

Strategy

To find the perimeter:

- ▶ Find the hypotenuse.
- ▶ Use the equation for the perimeter of a triangle. $4\sqrt{3} \approx 6.9$

Solution

$c = 2(\text{length of the shorter leg}) = 2(4) = 8$

$P = a + b + c$

$P \approx 4 + 6.9 + 8 = 18.9$

The perimeter is 18.9 cm.

You Try It 4

Find the perimeter of a right triangle with legs that measure 4 in. and 8 in. Round to the nearest tenth.

Your Strategy

Your Solution

Example 5

Find the area of triangle ABC.

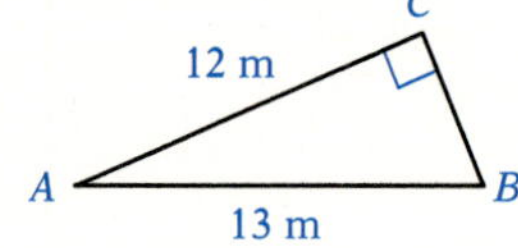

Strategy

To find the area:

- ▶ Use the Pythagorean Theorem to find the length of side BC.
- ▶ Use the equation for the area of a triangle. Let BC = the base and AC = the height.

Solution

$$c^2 = a^2 + b^2$$
$$13^2 = a^2 + 12^2$$
$$169 = a^2 + 144$$
$$25 = a^2$$
$$\sqrt{25} = \sqrt{a^2}$$
$$5 = a \qquad \text{Side } BC = 5 \text{ m}$$

$$A = \frac{1}{2}bh = \frac{1}{2}(5)(12) = 30$$

The area is 30 m^2.

You Try It 5

The length of a side of a square is 5 cm. Find the length of the diagonal of the square. Round to the nearest tenth.

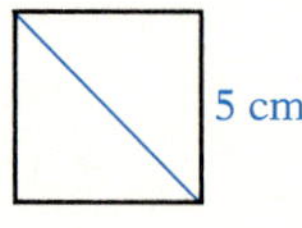

Your Strategy

Your Solution

Solutions on p. A37

11.3 EXERCISES

Objective A

Solve and check.

1. $\sqrt{x} = 5$
2. $\sqrt{y} = 7$
3. $\sqrt{a} = 12$
4. $\sqrt{a} = 9$
5. $\sqrt{b} = -8$
6. $\sqrt{y} = -6$
7. $\sqrt{5x} = 5$
8. $\sqrt{3x} = 4$
9. $\sqrt{4x} = 8$
10. $\sqrt{6x} = 3$
11. $\sqrt{2x} - 4 = 0$
12. $3 - \sqrt{5x} = 0$
13. $\sqrt{4x} + 5 = 2$
14. $\sqrt{3x} + 9 = 4$
15. $\sqrt{3x} - 6 = 0$
16. $8 - \sqrt{4x} = 0$
17. $\sqrt{2x} + 7 = 8$
18. $6 + \sqrt{3x} = 18$
19. $\sqrt{5x + 4} = 3$
20. $\sqrt{3x + 3} = 3$
21. $3 = \sqrt{5 - 2x}$
22. $\sqrt{12 - x} = 4$
23. $\sqrt{4x + 8} = 2$
24. $0 = 2 - \sqrt{3 - x}$
25. $0 = 5 - \sqrt{10 + x}$
26. $0 = 4 - \sqrt{7x + 2}$
27. $0 = 3 - \sqrt{4x + 1}$
28. $\sqrt{5x + 2} = 0$
29. $\sqrt{3x - 7} = 0$
30. $\sqrt{4x + 9} = 0$
31. $\sqrt{6x - 8} = 0$
32. $\sqrt{3x} - 6 = -4$
33. $\sqrt{5x} + 8 = 23$
34. $\sqrt{4x} - 9 = 1$
35. $\sqrt{2x} + 5 = 11$
36. $\sqrt{6x} + 7 = 2$

Solve and check.

37. $0 = \sqrt{3x - 9} - 6$

38. $0 = \sqrt{2x + 7} - 3$

39. $\sqrt{2x} = \sqrt{3x - 4}$

40. $\sqrt{5x - 3} = \sqrt{4x - 2}$

41. $\sqrt{5x - 9} = \sqrt{2x - 3}$

42. $\sqrt{3x + 2} = \sqrt{5x - 8}$

43. $\sqrt{4 - 3x} = \sqrt{8 - 2x}$

44. $\sqrt{5 - 2x} = \sqrt{6 - 5x}$

45. $\sqrt{3 - 7x} = \sqrt{6 - 4x}$

46. $\sqrt{3x - 5} - \sqrt{x + 7} = 0$

47. $\sqrt{2x + 8} - \sqrt{6x + 8} = 0$

48. $\sqrt{4x - 3} - \sqrt{2x + 5} = 0$

49. $\sqrt{5x + 1} - \sqrt{3x + 7} = 0$

50. $\sqrt{x^2 - 5x + 6} = \sqrt{x^2 - 8x + 9}$

51. $\sqrt{x^2 - 4x + 4} = \sqrt{x^2 - 6x + 8}$

Objective B

Find the unknown side of the triangle. Round to the nearest tenth.

52.

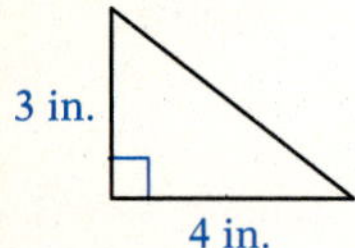

53.

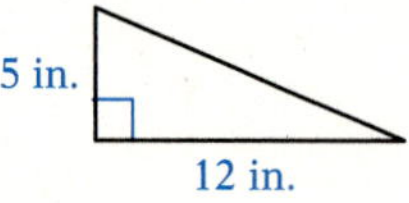

54.

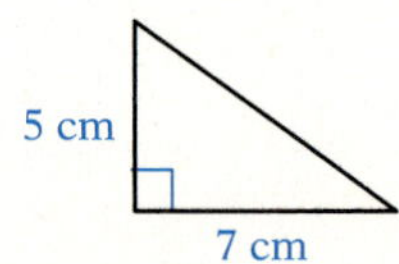

55.

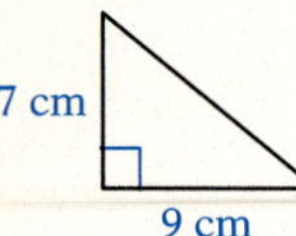

56.

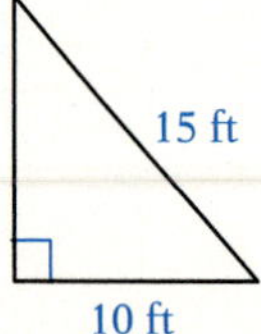

57.

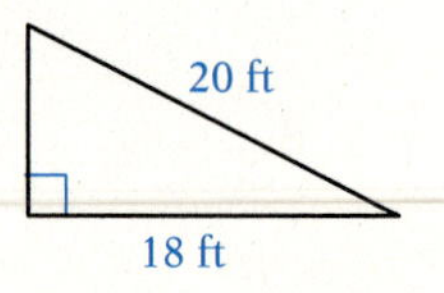

58.

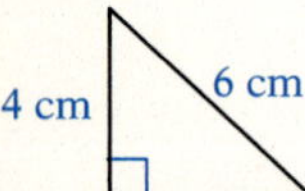

59.

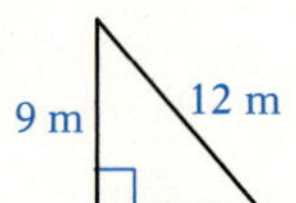

60.

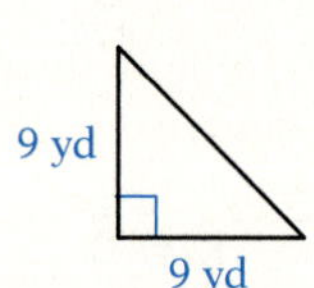

Find the lengths of the two legs. Round to the nearest tenth.

61.

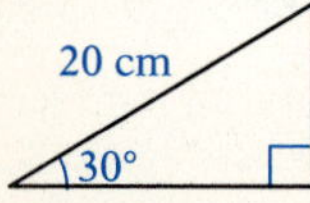

62.

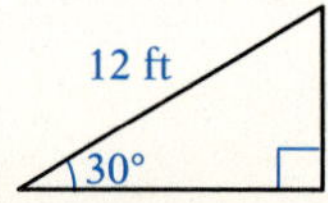

63.

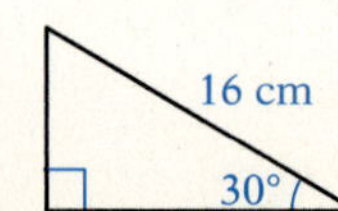

Find the hypotenuse of the right triangle. Round to the nearest tenth.

64.

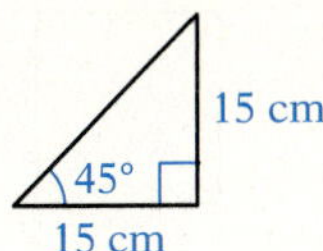

65.

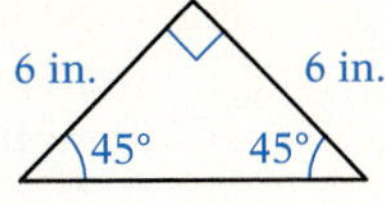

66.

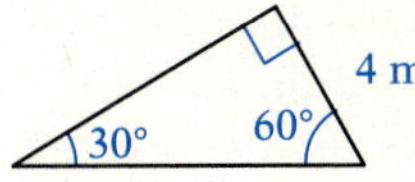

67.

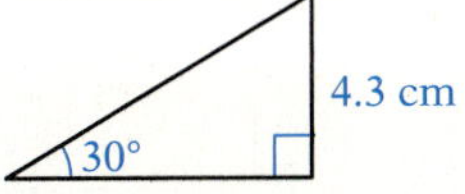

68.

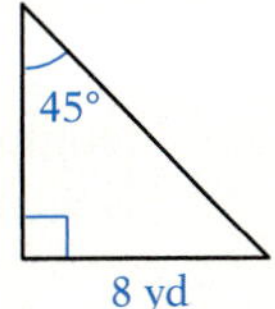

69.

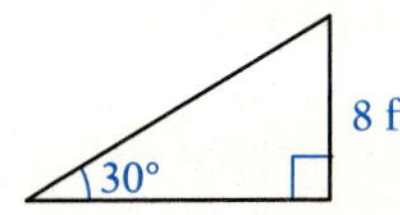

Solve. Round to the nearest tenth.

70. A ladder 8 m long is leaning against a building. How high on the building will the ladder reach when the bottom of the ladder is 3 m from the building?

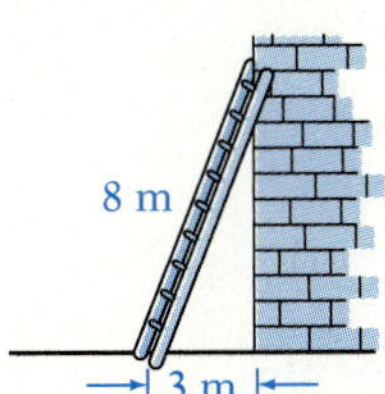

71. Find the distance between the centers of the holes in the metal plate.

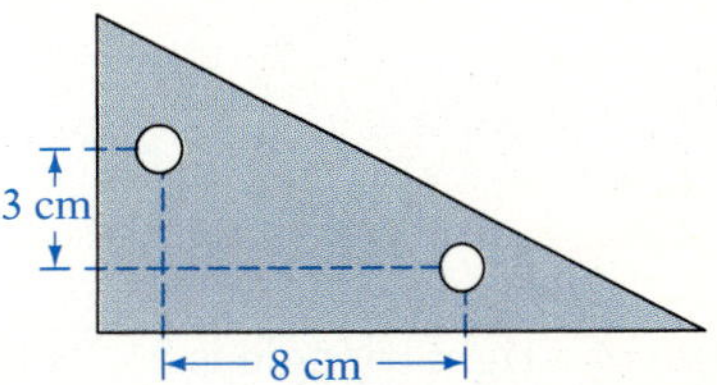

72. Find the perimeter of a right triangle with legs that measure 5 cm and 9 cm.

73. Find the perimeter of a right triangle whose legs measure 6 in. and 8 in.

74. The lengths of the legs of a 30°-60°-90° triangle measure 6 m and $6\sqrt{3}$ m. Find the perimeter of the triangle.

75. The lengths of the legs of a 30°-60°-90° triangle measure 3 ft and $3\sqrt{3}$ ft. Find the perimeter of the triangle.

76. The length of a leg of an isosceles right triangle is 3 cm. Find the perimeter of the triangle.

77. The length of a leg of an isosceles right triangle is 5 in. Find the perimeter of the triangle.

78. Find the area of triangle *ABC*.

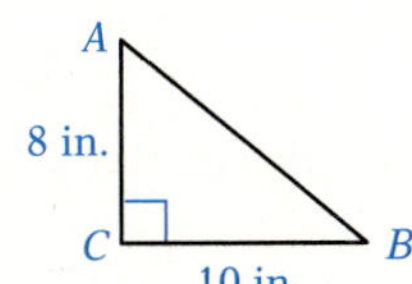

79. Find the area of triangle *ABC*.

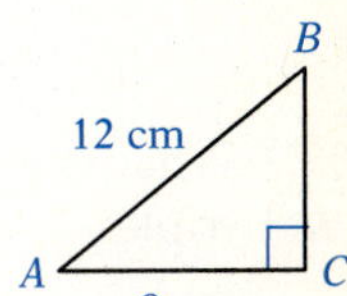

80. Find the area of triangle *ABC*.

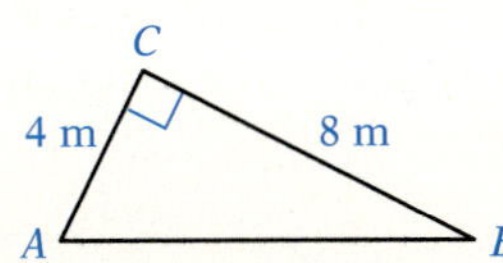

81. Find the area of triangle *ABC*.

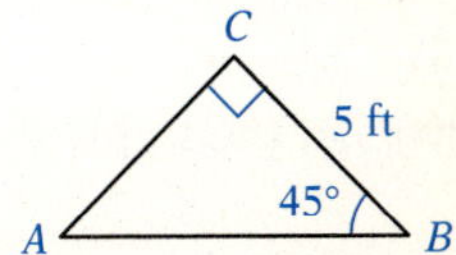

Solve. Round to the nearest tenth.

82. The length of a side of a square is 8 m. Find the length of its diagonal.

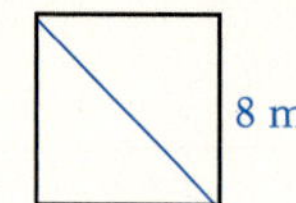

83. The length of a side of a square is 6 in. Find the length of its diagonal.

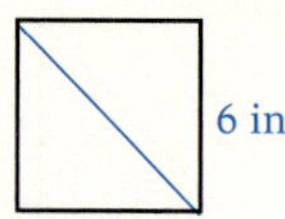

84. Find the perimeter of rectangle *ABCD* if the length of diagonal *AC* is 6 m.

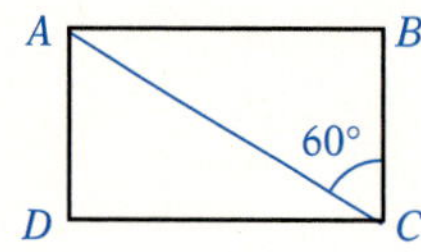

85. Find the perimeter of rectangle *ABCD* if the length of diagonal *AC* is 4 cm.

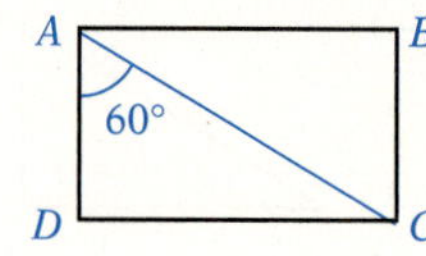

Critical Thinking

Solve.

86. $\sqrt{\dfrac{3x - 2}{4}} = 2$

87. $\sqrt{\dfrac{4x}{3}} - 1 = 1$

88. $\sqrt{\dfrac{3y}{5}} - 1 = 2$

89. The hypotenuse of a right triangle is $5\sqrt{2}$ cm, and one leg is $4\sqrt{2}$ cm.
a. Find the perimeter of the triangle.
b. Find the area of the triangle.

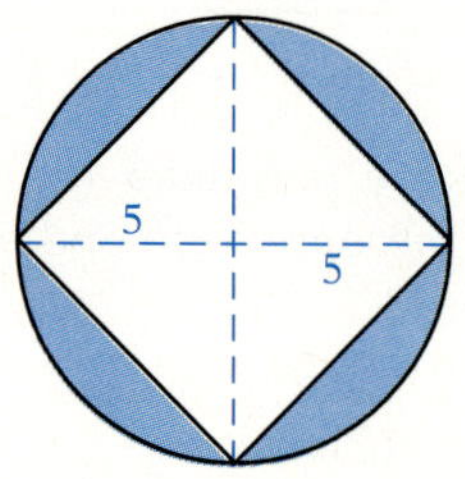

90. Find the area of the shaded region in the diagram at the right. Round to the nearest tenth.

91. A circular fountain is being designed for a triangular plaza in a cultural center. The fountain is placed so that each side of the triangle touches the fountain as shown in the diagram at the right. Find the area of the fountain to the nearest hundredth. The formula for the radius of the circle is given by

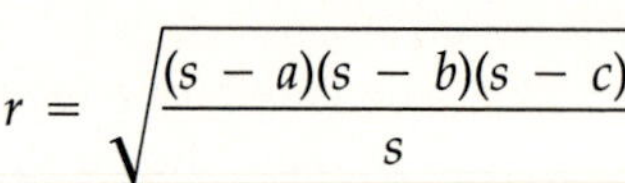

$$r = \sqrt{\frac{(s - a)(s - b)(s - c)}{s}}$$

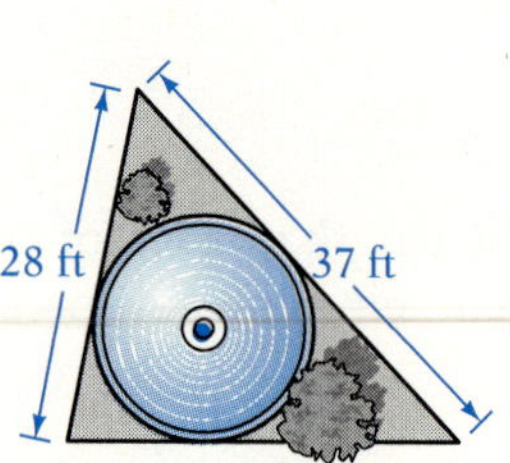

where $s = \frac{1}{2}(a + b + c)$ and a, b, and c are the lengths of the sides of the triangle.

92. [W] What is a Pythagorean triple? Provide at least three examples of Pythagorean triples.

93. [W] Can the Pythagorean Theorem be used to find the length of side c of the triangle at the right? If so, determine c. If not, explain why.

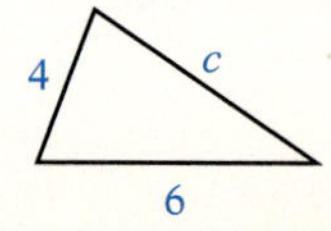

Projects in Mathematics

Distance to the Horizon

The formula $d = \sqrt{1.5h}$ is used to calculate the approximate distance d (in miles) that a person could see who used a periscope h feet above the water. This formula is derived by using the Pythagorean Theorem.

Consider the diagram (not to scale) at the right, which shows the earth as a sphere and the periscope extending h feet above the surface. Because AB is tangent to the circle and OA is a radius, triangle AOB is a right triangle. Therefore,

$$(OA)^2 + (AB)^2 = (OB)^2$$

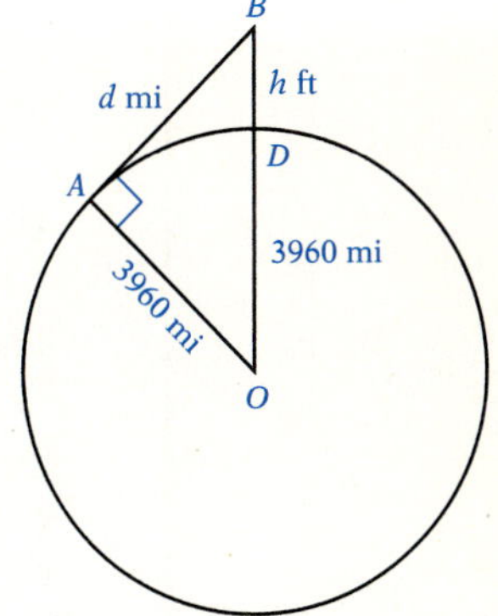

The radius of the earth is 3960 mi.

Because h is in feet and 1 mi = 5280 ft, $\frac{h}{5280}$ is in miles.
$$3960^2 + d^2 = \left(3960 + \frac{h}{5280}\right)^2$$

Square the binomial.
$$3960^2 + d^2 = 3960^2 + \frac{2 \cdot 3960}{5280}h + \left(\frac{h}{5280}\right)^2$$

Subtract 3960^2 from each side and simplify.
$$d^2 = \frac{3}{2}h + \left(\frac{h}{5280}\right)^2$$

Use the Principal Square Root Property.
$$d = \sqrt{\frac{3}{2}h + \left(\frac{h}{5280}\right)^2}$$

Now $\left(\frac{h}{5280}\right)^2$ is very small, so we can assume that $\sqrt{\frac{3}{2}h + \left(\frac{h}{5280}\right)^2} \approx \sqrt{1.5h}$, where we have written $\frac{3}{2}$ as 1.5. Thus $d = \sqrt{1.5h}$ approximates the distance that can be seen using a periscope h feet above the water.

1. Write a paragraph that justifies the assumption that

$$\sqrt{\frac{3}{2}h + \left(\frac{h}{5280}\right)^2} \approx \sqrt{1.5h}.$$

(*Suggestion:* Evaluate each expression for various values of h. Because h is the height of a periscope above water, it is unlikely that $h > 25$ ft.)

2. The distance d is the distance from the top of the periscope to A. The distance along the surface of the water is the length of arc AD. This distance can be approximated by the equation

$$d \approx \sqrt{1.5h} + 0.306186(\sqrt{h})^3$$

Using this formula, calculate d when $h = 10$.

Chapter Summary

Key Words A *square root* of a positive number x is a number whose square is x.

The *principal square root* of a number is its positive square root.

The symbol $\sqrt{\ }$ is called a *radical sign* and is used to indicate the principal square root of a number.

The *radicand* is the expression under the radical sign.

The square of an integer is a *perfect square.*

If a number is not a perfect square, its square root can only be approximated. Such numbers are *irrational numbers.* Their decimal representations do not terminate or repeat. The rational numbers and the irrational numbers taken together are called the *real numbers.*

Conjugates are binomial expressions that differ only in the sign of the second term. (The expressions $a + b$ and $a - b$ are conjugates.)

Rationalizing the denominator removes a radical from the denominator of a fraction.

A *radical equation* is an equation that contains a variable in a radicand.

A *right triangle* contains one right angle. The side opposite the right angle is the *hypotenuse.* The other two sides are called *legs.* 45°-45°-90° triangles and 30°-60°-90° triangles are special right triangles.

Essential Rules

The Product Property of Square Roots If a and b are positive real numbers, then $\sqrt{ab} = \sqrt{a}\sqrt{b}$.

The Quotient Property of Square Roots If a and b are positive real numbers, then $\sqrt{\frac{a}{b}} = \frac{\sqrt{a}}{\sqrt{b}}$.

Property of Squaring Both Sides of an Equation If a and b are real numbers and $a = b$, then $a^2 = b^2$.

Pythagorean Theorem $c^2 = a^2 + b^2$, where a and b are the legs of a right triangle and c is the hypotenuse.

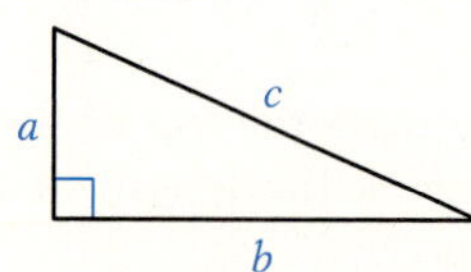

The Principal Square Root Property If $r^2 = s$, then $r = \sqrt{s}$, and r is called the square root of s.

Chapter Review Exercises

1. Simplify: $\sqrt{45}$

2. Simplify: $\sqrt{75}$

3. Simplify: $-2\sqrt{80}$

4. Find the decimal representation of $\sqrt{175}$. Round to the nearest thousandth.

5. Simplify: $\sqrt{121x^8y^2}$

6. Find the perimeter of a right triangle with legs that measure 7 cm and 10 cm. Round to the nearest tenth.

7. Simplify: $\sqrt{32a^5b^{11}}$

8. Simplify: $\sqrt{x^2 + 4x + 4}$

9. Subtract: $8\sqrt{y} - 3\sqrt{y}$

10. Subtract: $5\sqrt{8} - 3\sqrt{50}$

11. Simplify: $3\sqrt{8y} - 2\sqrt{72x} + 5\sqrt{18y}$

12. Subtract: $2x\sqrt{3xy^3} - 2y\sqrt{12x^3y} - 3xy\sqrt{xy}$

13. Multiply: $\sqrt{8x^3y}\sqrt{10xy^4}$

14. Multiply: $\sqrt{3x^2y}\sqrt{6xy^2}\sqrt{2x}$

15. Multiply: $\sqrt{a}(\sqrt{a} - \sqrt{b})$

16. Multiply: $(\sqrt{y} - 3)(\sqrt{y} + 5)$

17. Simplify: $\dfrac{\sqrt{162}}{\sqrt{2}}$

18. Simplify: $\dfrac{\sqrt{98a^6b^4}}{\sqrt{2a^3b^2}}$

19. Simplify: $\dfrac{2}{\sqrt{3} - 1}$

20. Simplify: $\dfrac{3\sqrt{x^3} - 4\sqrt{9x}}{3\sqrt{x}}$

21. Solve: $\sqrt{5x - 6} = 7$

22. Solve: $\sqrt{9x} + 3 = 18$

23. Find the unknown sides of the triangle.

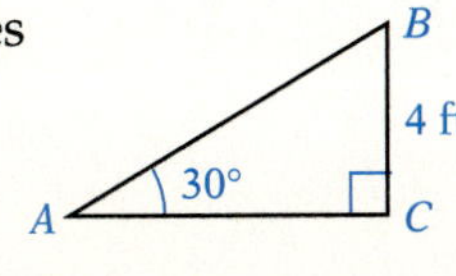

24. Solve: $\sqrt{8x - 3} = \sqrt{4x + 5}$

Cumulative Review Exercises

1. 28% of what number is 210?

2. Factor: $2a^3 - 16a^2 + 30a$

3. Simplify: $\dfrac{x^2 - 7x + 10}{25 - x^2}$

4. Solve: $\dfrac{2}{x + 3} = \dfrac{12}{x - 2}$

5. Solve: $\begin{aligned} 5x - 3y &= 29 \\ 4x + 7y &= -5 \end{aligned}$

6. Multiply: $(4xy^3)(-2x^2y^3)$

7. Add: $(4b^3 - 7b^2 - 7) + (3b^2 - 8b + 3)$

8. Divide: $\dfrac{-18a^3 + 12a^2 - 6}{-3a^2}$

9. Factor: $5xy^2 - 20xy^4$

10. Multiply: $\dfrac{3x - 6}{9x + 3} \cdot \dfrac{5x + 10}{x^2 - 4}$

11. Solve: $\dfrac{3x}{x - 3} - 2 = \dfrac{10}{x - 3}$

12. Subtract: $\dfrac{2}{2x - 1} - \dfrac{1}{x + 1}$

13. Divide: $\dfrac{x^2 - 3x - 18}{x^2 + 8x + 15} \div \dfrac{x^2 - 10x + 24}{x^2 + 4x - 5}$

14. Simplify: $\dfrac{\sqrt{320}}{\sqrt{5}}$

15. Solve: $\sqrt{3x - 2} - 4 = 0$

16. Subtract: $\dfrac{x + 2}{x - 4} - \dfrac{6}{x^2 - 7x + 12}$

17. Solve and graph the solution:
$x + 5 \geq 7$

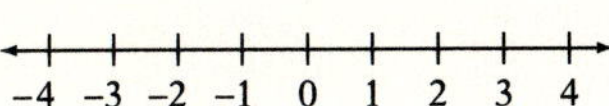

18. Multiply: $\sqrt{3}(\sqrt{6} - \sqrt{x^2})$

19. Factor: $12x^3y^2 - 9x^2y^3$

20. Multiply: $\sqrt{2a^9b}\sqrt{98ab^3}\sqrt{2a}$

21. Find the area of the figure.

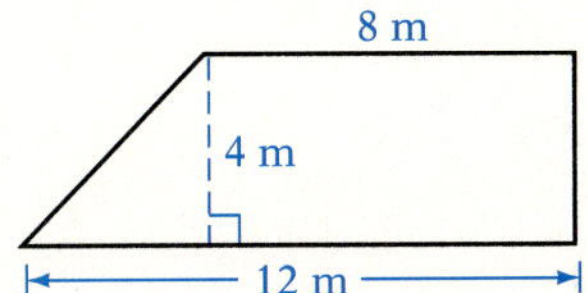

22. Triangles *ABC* and *DEF* are similar. Find side *DE*.

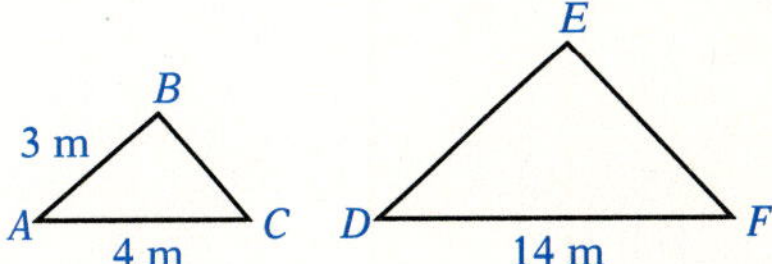

23. Two solar heating panels will raise the temperature of water 1° in 60 min. One panel, working alone, requires 90 min to raise the temperature of the water 1°. How long would it take the second panel working alone to heat the water?

24. In a lake 100 fish are caught, tagged, and then released. Later 150 fish are caught. Three of the fish are found to have tags. Estimate the number of fish in the lake.

25. A silo, which is in the shape of a cylinder, is 14 ft in diameter and 25 ft high. The silo is 60% full. Find the volume of the portion of the silo that is being used for storage. Round to the nearest hundredth.

CHAPTER

12

Quadratic Equations

Focus on Problem Solving

Sometimes the solution to a problem can be found by *working backwards.* This problem-solving technique can be used to find a winning strategy for a game called Nim.

There are many variations of Nim. For our game, there are two players, Player A and Player B, who alternately place 1, 2, or 3 matchsticks in a pile. The object of the game is to place the 32nd matchstick in the pile. Is there a strategy that Player A can use to guarantee winning the game?

Working backward, if there are 29, 30, or 31 matchsticks in the pile when it's A's turn to play, A can win by placing 3 ($29 + 3 = 32$), 2 ($30 + 2 = 32$), or 1 ($31 + 1 = 32$) matchsticks on the pile. If there are to be 29, 30, or 31 matchsticks in the pile when it's A's turn, there must be 28 matchsticks in the pile when it's B's turn.

Working backward from 28, if there are to be 28 matches in the pile at B's turn, there must be 25, 26, or 27 at A's turn. Player A can then add 3, 2, or 1 matchsticks to the pile to bring the number to 28. For there to be 25, 26, or 27 matchsticks in the pile at A's turn, there must be 24 matchsticks at B's turn.

Now working backward from 24, if there are to be 24 matches in the pile at B's turn, there must be 21, 22, or 23 at A's turn. Player A can then add 3, 2, or 1 matchsticks to the pile to bring the number to 24. For there to be 21, 22, or 23 matchsticks in the pile at A's turn, there must be 20 matchsticks at B's turn.

So far, we have found that for Player A to win, there must be 28, 24, or 20 matchsticks in the pile when it's B's turn to play. Note that each time, the number is decreasing by 4. Continuing this pattern, Player A will win if there are 16, 12, 8, or 4 matchsticks in the pile when it's B's turn.

Player A can guarantee winning by making sure the number of matchsticks in the pile is a multiple of 4. To ensure this, at the beginning of the game, Player A allows Player B to go first and then adds exactly enough matchsticks to the pile to bring the total to a multiple of 4.

For example, suppose B places 3 matchsticks in the pile; then A places 1 matchstick ($3 + 1 = 4$). Now B places 2 matchsticks in the pile. The total is now 6 matchsticks. Player A then places 2 matchsticks in the pile to bring the total to 8, a multiple of 4. If play continues in this way, Player A will win.

Here are some variations of Nim. See if you can develop a winning strategy for Player A. *Hint:* It may not be possible.

1. Suppose the goal is to place the last matchstick in a pile of 30 matchsticks.

2. Suppose the players make two piles of matchsticks, with the maximum number of matchsticks in each pile to be 20.

3. Suppose there are 40 matchsticks in a pile. Each player alternately takes 1, 2, or 3 matchsticks from the pile. The player who takes the last matchstick wins.

SECTION 12.1 Solving Quadratic Equations by Factoring or by Taking Square Roots

OBJECTIVE A Solve a quadratic equation by factoring

CONSIDER THIS
Why is the condition $a \neq 0$ placed on the quadratic function?

An equation of the form $ax^2 + bx + c = 0$, where a, b, and c are constants and $a \neq 0$, is a **quadratic equation**.

$4x^2 - 3x + 1 = 0, a = 4, b = -3, c = 1$
$3x^2 - 4 = 0, a = 3, b = 0, c = -4$
$\frac{x^2}{2} - 2x + 4 = 0, a = \frac{1}{2}, b = -2, c = 4$

A quadratic equation is also called a **second-degree equation**.

A quadratic equation is in **standard form** when the polynomial is in descending order and equal to zero.

Recall that the Principle of Zero Products states that if the product of two factors is zero, then at least one of the factors must be zero.

If $a \cdot b = 0$,
then $a = 0$ or $b = 0$.

The Principle of Zero Products can be used to solve quadratic equations.

Solve by factoring: $2x^2 - x = 1$

$2x^2 - x = 1$

Write the equation in standard form. $2x^2 - x - 1 = 0$

Factor. $(2x + 1)(x - 1) = 0$

Use the Principle of Zero Products to set each factor equal to zero. $2x + 1 = 0 \quad x - 1 = 0$

Solve each equation for x. $2x = -1 \quad x = 1$

$x = -\frac{1}{2}$

CONSIDER THIS
You should always check your solutions by substituting the proposed solutions back into the *original* equation.

Check:

$2x^2 - x = 1$		$2x^2 - x = 1$	
$2\left(-\frac{1}{2}\right)^2 - \left(-\frac{1}{2}\right)$	1	$2(1)^2 - 1$	1
$2 \cdot \frac{1}{4} + \frac{1}{2}$	1	$2 \cdot 1 - 1$	1
$\frac{1}{2} + \frac{1}{2}$	1	$2 - 1$	1
$1 = 1$		$1 = 1$	

The solutions are $-\frac{1}{2}$ and 1.

Solve by factoring: $3x^2 - 4x + 8 = (4x + 1)(x - 2)$

$$3x^2 - 4x + 8 = (4x + 1)(x - 2)$$

Multiply the factors on the right side of the equation. $3x^2 - 4x + 8 = 4x^2 - 7x - 2$

Write the equation in standard form. $0 = x^2 - 3x - 10$

Factor. $0 = (x - 5)(x + 2)$

Use the Principle of Zero Products to set each factor equal to zero. $x - 5 = 0 \quad x + 2 = 0$

Solve each equation for x. $x = 5 \quad x = -2$

Check:

$3x^2 - 4x + 8$	$= (4x + 1)(x - 2)$
$3(5)^2 - 4(5) + 8$	$(4[5] + 1)(5 - 2)$
$3(25) - 20 + 8$	$(20 + 1)(3)$
$75 - 12$	$(21)(3)$
63	$= 63$

$3x^2 - 4x + 8$	$= (4x + 1)(x - 2)$
$3(-2)^2 - 4(-2) + 8$	$(4[-2] + 1)(-2 - 2)$
$3(4) + 8 + 8$	$(-8 + 1)(-4)$
$12 + 16$	$(-7)(-4)$
28	$= 28$

The solutions are 5 and -2.

Solve by factoring: $x^2 - 10x + 25 = 0$

$$x^2 - 10x + 25 = 0$$

Factor. $(x - 5)(x - 5) = 0$

Use the Principle of Zero Products. $x - 5 = 0 \quad x - 5 = 0$

Solve each equation for x. $x = 5 \quad x = 5$

The solution is 5.

In this last example, 5 is called a **double root** of the quadratic equation.

Example 1

Solve by factoring: $\frac{z^2}{2} - \frac{z}{4} - \frac{1}{4} = 0$

Solution

$$\frac{z^2}{2} - \frac{z}{4} - \frac{1}{4} = 0$$

$$4\left(\frac{z^2}{2} - \frac{z}{4} - \frac{1}{4}\right) = 4(0)$$ Multiply each side by 4.

$$2z^2 - z - 1 = 0$$

$$(2z + 1)(z - 1) = 0$$

$$2z + 1 = 0 \qquad z - 1 = 0$$

$$2z = -1 \qquad z = 1$$

$$z = -\frac{1}{2}$$

The solutions are $-\frac{1}{2}$ and 1.

You Try It 1

Solve by factoring: $\frac{3y^2}{2} + y - \frac{1}{2} = 0$

Your Solution

Solution on p. A37

OBJECTIVE B Solve a quadratic equation by taking square roots

Consider a quadratic equation of the form $x^2 = a$. This equation can be solved by factoring.

$$x^2 = 25$$
$$x^2 - 25 = 0$$
$$(x - 5)(x + 5) = 0$$
$$x = 5 \qquad x = -5$$

The solutions are 5 and -5. The fact that the solutions are plus or minus the same number is frequently written by using $\pm$; for example, "the solutions are ± 5." Because ± 5 can be written as $\pm\sqrt{25}$, an alternative method of solving this equation is suggested.

The Square Root Property

If $x^2 = a$, then $x = \pm\sqrt{a}$.

Solve by taking square roots: $x^2 = 121$

Take the square root of each side of the equation. Then simplify.

$$x^2 = 121$$
$$\sqrt{x^2} = \pm\sqrt{121}$$
$$x = \pm 11$$

The solutions are 11 and -11.

Solve by taking square roots: $3x^2 = 36$

$$3x^2 = 36$$

Solve for x^2. $\quad x^2 = 12$

Take the square root of each side. $\quad \sqrt{x^2} = \pm\sqrt{12}$

Simplify. $\quad x = \pm 2\sqrt{3}$

The solutions are $2\sqrt{3}$ and $-2\sqrt{3}$.

Solve by taking square roots: $49y^2 - 25 = 0$

$$49y^2 - 25 = 0$$

Solve for y^2.

$$49y^2 = 25$$
$$y^2 = \frac{25}{49}$$

Take the square root of each side. $\quad \sqrt{y^2} = \pm\sqrt{\frac{25}{49}}$

Simplify. $\quad y = \pm\frac{5}{7}$

The solutions are $\frac{5}{7}$ and $-\frac{5}{7}$.

An equation that contains the square of a binomial can be solved by taking square roots.

Solve by taking square roots: $2(x - 1)^2 - 36 = 0$

$$2(x - 1)^2 - 36 = 0$$

Solve for $(x - 1)^2$.

$$2(x - 1)^2 = 36$$
$$(x - 1)^2 = 18$$

Take the square root of each side of the equation.

$$\sqrt{(x - 1)^2} = \pm\sqrt{18}$$

Simplify.

$$x - 1 = \pm 3\sqrt{2}$$

Solve for x.

$$x = 1 \pm 3\sqrt{2}$$

The solutions are $1 + 3\sqrt{2}$ and $1 - 3\sqrt{2}$.

Example 2

Solve by taking square roots: $x^2 + 16 = 0$

Solution

$x^2 + 16 = 0$

$x^2 = -16$

$\sqrt{x^2} = \pm\sqrt{-16}$

$\sqrt{-16}$ is not a real number.

The equation has no real number solution.

You Try It 2

Solve by taking square roots: $x^2 + 81 = 0$

Your Solution

Example 3

Solve by taking square roots: $2(x - 4)^2 = 8$

Solution

$2(x - 4)^2 = 8$

$(x - 4)^2 = 4$

$\sqrt{(x - 4)^2} = \pm\sqrt{4}$

$x - 4 = \pm 2$

$x - 4 = 2 \qquad x - 4 = -2$

$x = 6 \qquad x = 2$

The solutions are 2 and 6.

You Try It 3

Solve by taking square roots: $4(x + 2)^2 = 9$

Your Solution

Example 4

Solve by taking square roots: $5(y - 4)^2 = 25$

Solution

$5(y - 4)^2 = 25$

$(y - 4)^2 = 5$

$\sqrt{(y - 4)^2} = \pm\sqrt{5}$

$y - 4 = \pm\sqrt{5}$

$y = 4 \pm \sqrt{5}$

The solutions are $4 + \sqrt{5}$ and $4 - \sqrt{5}$.

You Try It 4

Solve by taking square roots: $7(z + 2)^2 = 21$

Your Solution

Solutions on pp. A37–A38

12.1 EXERCISES

Objective A

Solve.

1. $(x - 5)(x + 3) = 0$
2. $(x - 2)(x + 6) = 0$
3. $(x + 7)(x - 8) = 0$
4. $(2x - 3)(x + 7) = 0$
5. $(3x + 5)(x - 4) = 0$
6. $(4x - 1)(3x + 5) = 0$

Solve by factoring.

7. $x^2 + 2x - 15 = 0$
8. $t^2 + 3t - 10 = 0$
9. $z^2 - 4z + 3 = 0$
10. $s^2 - 5s + 4 = 0$
11. $p^2 + 3p + 2 = 0$
12. $v^2 + 6v + 5 = 0$
13. $x^2 - 6x + 9 = 0$
14. $y^2 - 8y + 16 = 0$
15. $12y^2 + 8y = 0$
16. $6x^2 - 9x = 0$
17. $r^2 - 10 = 3r$
18. $t^2 - 12 = 4t$
19. $3v^2 - 5v + 2 = 0$
20. $2p^2 - 3p - 2 = 0$
21. $3s^2 + 8s = 3$
22. $3x^2 + 5x = 12$
23. $9z^2 = 12z - 4$
24. $6r^2 = 12 - r$
25. $4t^2 = 4t + 3$
26. $5y^2 + 11y = 12$
27. $4v^2 - 4v + 1 = 0$
28. $9s^2 - 6s + 1 = 0$
29. $x^2 - 9 = 0$
30. $t^2 - 16 = 0$
31. $4y^2 - 1 = 0$
32. $9z^2 - 4 = 0$
33. $x + 15 = x(x - 1)$
34. $p + 18 = p(p - 2)$
35. $r^2 - r - 2 = (2r - 1)(r - 3)$
36. $s^2 + 5s - 4 = (2s + 1)(s - 4)$
37. $x^2 + x + 5 = (3x + 2)(x - 4)$

Objective B

Solve by taking square roots.

38. $x^2 = 36$

39. $y^2 = 49$

40. $v^2 - 1 = 0$

41. $z^2 - 64 = 0$

42. $4x^2 - 49 = 0$

43. $9w^2 - 64 = 0$

44. $9y^2 = 4$

45. $4z^2 = 25$

46. $16y^2 - 9 = 0$

47. $25x^2 - 64 = 0$

48. $y^2 + 81 = 0$

49. $z^2 + 49 = 0$

50. $2(x + 5)^2 = 8$

51. $4(z - 3)^2 = 100$

52. $9(x - 1)^2 - 16 = 0$

53. $4(y + 3)^2 - 81 = 0$

54. $49(v + 1)^2 - 25 = 0$

55. $81(y - 2)^2 - 64 = 0$

56. $(x - 4)^2 - 20 = 0$

57. $(y + 5)^2 - 50 = 0$

58. $(x + 1)^2 + 36 = 0$

59. $(y - 7)^2 + 49 = 0$

60. $2\left(z - \frac{1}{2}\right)^2 = 12$

61. $3\left(v + \frac{3}{4}\right)^2 = 36$

Critical Thinking

Solve for x.

62. $(x^2 - 1)^2 = 9$

63. $(x^2 + 3)^2 = 25$

64. $(6x^2 - 5)^2 = 1$

65. $ax^2 - bx = 0$, $a > 0$ and $b > 0$

66. $ax^2 - b = 0$, $a > 0$ and $b > 0$

67. $x^2 = x$

68. The value P of an initial investment of A dollars after 2 years is given by $P = A(1 + r)^2$, where r is the annual percentage rate earned by the investment. If an initial investment of \$1500 grew to a value of \$1782.15 in 2 years, what was the annual percentage rate?

69. On a certain street surface, the equation $d = 0.0074v^2$ can be used to approximate the distance d a car traveling v miles per hour will slide when its brakes are applied. After applying the brakes, the owner of a car involved in an accident skidded 40 ft. Did the traffic officer investigating the accident issue the car owner a ticket for speeding if the speed limit is 65 mph?

SECTION 12.2 Solving Quadratic Equations by Completing the Square

OBJECTIVE A Solve a quadratic equation by completing the square

Recall that a perfect square trinomial is the square of a binomial.

Perfect Square Trinomial		Square of a Binomial
$x^2 + 6x + 9$	=	$(x + 3)^2$
$x^2 - 10x + 25$	=	$(x - 5)^2$
$x^2 + 8x + 16$	=	$(x + 4)^2$

For each perfect square trinomial, the square of $\frac{1}{2}$ of the coefficient of x equals the constant term.

$$x^2 + 6x + 9, \quad \left(\frac{1}{2} \cdot 6\right)^2 = 9$$

$$x^2 - 10x + 25, \quad \left[\frac{1}{2}(-10)\right]^2 = 25$$

$$x^2 + 8x + 16, \quad \left(\frac{1}{2} \cdot 8\right)^2 = 16$$

$$\left(\frac{1}{2} \textbf{ coefficient of } x\right)^2 = \textbf{constant term}$$

This relationship can be used to write the constant term for a perfect square trinomial. Adding to a binomial the constant term that makes it a perfect square trinomial is called **completing the square**.

POINT OF INTEREST

Early mathematicians solved quadratic equations by literally *completing the square.* For these mathematicians, all equations had geometric interpretations. They found that a quadratic equation could be solved by making certain figures into squares.

Complete the square on $x^2 - 8x$. Write the resulting perfect square trinomial as the square of a binomial.

Find the constant term. $\left[\frac{1}{2}(-8)\right]^2 = 16$

Complete the square on $x^2 - 8x$ by adding the constant term. $x^2 - 8x + 16$

Write the resulting perfect square trinomial as the square of a binomial. $x^2 - 8x + 16 = (x - 4)^2$

Complete the square on $y^2 + 5y$. Write the resulting perfect square trinomial as the square of a binomial.

Find the constant term. $\left(\frac{1}{2} \cdot 5\right)^2 = \left(\frac{5}{2}\right)^2 = \frac{25}{4}$

Complete the square on $y^2 + 5y$ by adding the constant term. $y^2 + 5y + \frac{25}{4}$

Write the resulting perfect square trinomial as the square of a binomial. $y^2 + 5y + \frac{25}{4} = \left(y + \frac{5}{2}\right)^2$

A quadratic equation that cannot be solved by factoring can be solved by completing the square. Add to each side of the equation the term that completes the square. Then rewrite the quadratic equation in the form $(x + a)^2 = b$. Take the square root of each side of the equation and then solve for x.

Solve by completing the square: $x^2 - 6x - 3 = 0$

$$x^2 - 6x - 3 = 0$$

Add the opposite of the constant term to each side of the equation.

$$x^2 - 6x = 3$$

Find the constant term that completes the square on $x^2 - 6x$.

$$\left[\frac{1}{2}(-6)\right]^2 = 9$$

Add this term to each side of the equation.

$$x^2 - 6x + 9 = 3 + 9$$

Factor the perfect square trinomial.

$$(x - 3)^2 = 12$$

Take the square root of each side of the equation.

$$\sqrt{(x - 3)^2} = \pm\sqrt{12}$$

Simplify.

$$x - 3 = \pm 2\sqrt{3}$$

Solve for x.

$$x - 3 = 2\sqrt{3} \qquad x - 3 = -2\sqrt{3}$$
$$x = 3 + 2\sqrt{3} \qquad x = 3 - 2\sqrt{3}$$

Check:

$x^2 - 6x - 3 = 0$	
$(3 + 2\sqrt{3})^2 - 6(3 + 2\sqrt{3}) - 3$	0
$9 + 12\sqrt{3} + 12 - 18 - 12\sqrt{3} - 3$	0
$0 = 0$	

$x^2 - 6x - 3 = 0$	
$(3 - 2\sqrt{3})^2 - 6(3 - 2\sqrt{3}) - 3$	0
$9 - 12\sqrt{3} + 12 - 18 + 12\sqrt{3} - 3$	0
$0 = 0$	

Write the solution.

The solutions are $3 + 2\sqrt{3}$ and $3 - 2\sqrt{3}$.

In the example above, the coefficient of the x^2 term is 1. In order to complete the square, the coefficient of the x^2 term must be 1. The example on the next page illustrates the procedure for solving an equation by completing the square when the coefficient of the x^2 term is not 1.

Solve by completing the square: $2x^2 - x - 1 = 0$

$$2x^2 - x - 1 = 0$$

Add the opposite of the constant term to each side of the equation.

$$2x^2 - x = 1$$

To complete the square, the coefficient of the x^2 term must be 1. Multiply each term by the reciprocal of the coefficient of x^2.

$$\frac{1}{2}(2x^2 - x) = \frac{1}{2} \cdot 1$$

$$x^2 - \frac{1}{2}x = \frac{1}{2}$$

Complete the square on $x^2 - \frac{1}{2}x$.

$$x^2 - \frac{1}{2}x + \frac{1}{16} = \frac{1}{2} + \frac{1}{16}$$

Factor the perfect square trinomial.

$$\left(x - \frac{1}{4}\right)^2 = \frac{9}{16}$$

Take the square root of each side of the equation.

$$\sqrt{\left(x - \frac{1}{4}\right)^2} = \pm\sqrt{\frac{9}{16}}$$

Simplify.

$$x - \frac{1}{4} = \pm\frac{3}{4}$$

Solve for x.

$$x - \frac{1}{4} = \frac{3}{4} \qquad x - \frac{1}{4} = -\frac{3}{4}$$

$$x = 1 \qquad x = -\frac{1}{2}$$

1 and $-\frac{1}{2}$ check as solutions.

The solutions are 1 and $-\frac{1}{2}$.

Example 1 Solve by completing the square: $2x^2 - 4x - 1 = 0$

Solution

$$2x^2 - 4x - 1 = 0$$

$$2x^2 - 4x = 1$$

$$\frac{1}{2}(2x^2 - 4x) = \frac{1}{2} \cdot 1$$

$$x^2 - 2x = \frac{1}{2}$$

$$x^2 - 2x + 1 = \frac{1}{2} + 1 \quad \text{Complete the square.}$$

$$(x - 1)^2 = \frac{3}{2}$$

$$\sqrt{(x - 1)^2} = \pm\sqrt{\frac{3}{2}}$$

$$x - 1 = \pm\frac{\sqrt{6}}{2}$$

$$x = 1 \pm \frac{\sqrt{6}}{2} = \frac{2 \pm \sqrt{6}}{2}$$

The solutions are $\frac{2 + \sqrt{6}}{2}$ and $\frac{2 - \sqrt{6}}{2}$.

You Try It 1 Solve by completing the square: $3x^2 - 6x - 2 = 0$

Your Solution

Solution on p. A38

Example 2

Solve by completing the square:
$x^2 + 4x + 5 = 0$

Solution

$$x^2 + 4x + 5 = 0$$
$$x^2 + 4x = -5$$

Complete the square.

$$x^2 + 4x + 4 = -5 + 4$$
$$(x + 2)^2 = -1$$
$$\sqrt{(x + 2)^2} = \pm\sqrt{-1}$$

$\sqrt{-1}$ is not a real number.

The quadratic equation has no real number solution.

You Try It 2

Solve by completing the square:
$x^2 + 6x + 12 = 0$

Your Solution

Example 3

Solve by completing the square:
$x^2 + 6x + 4 = 0$

Approximate the solutions. Round to the nearest thousandth.

Solution

$$x^2 + 6x + 4 = 0$$
$$x^2 + 6x = -4$$

Complete the square.

$$x^2 + 6x + 9 = -4 + 9$$
$$(x + 3)^2 = 5$$
$$\sqrt{(x + 3)^2} = \pm\sqrt{5}$$
$$x + 3 = \pm\sqrt{5}$$
$$x = -3 \pm \sqrt{5}$$

$$x = -3 + \sqrt{5} \approx -3 + 2.236 \approx -0.764$$
$$x = -3 - \sqrt{5} \approx -3 - 2.236 \approx -5.236$$

The solutions are approximately -0.764 and -5.236.

You Try It 3

Solve by completing the square:
$x^2 + 8x + 8 = 0$

Approximate the solutions. Round to the nearest thousandth.

Your Solution

Solutions on pp. A38–A39

12.2 EXERCISES

Objective A

Solve by completing the square.

1. $x^2 + 2x - 3 = 0$
2. $y^2 + 4y - 5 = 0$
3. $z^2 - 6z - 16 = 0$
4. $w^2 + 8w - 9 = 0$
5. $x^2 = 4x - 4$
6. $z^2 = 8z - 16$
7. $v^2 - 6v + 13 = 0$
8. $x^2 + 4x + 13 = 0$
9. $y^2 + 5y + 4 = 0$
10. $v^2 - 5v - 6 = 0$
11. $w^2 + 7w = 8$
12. $y^2 + 5y = -4$
13. $v^2 + 4v + 1 = 0$
14. $y^2 - 2y - 5 = 0$
15. $x^2 + 6x = 5$
16. $w^2 - 8w = 3$
17. $z^2 = 2z + 1$
18. $y^2 = 10y - 20$
19. $p^2 + 3p = 1$
20. $r^2 + 5r = 2$
21. $t^2 - 3t = -2$
22. $z^2 - 5z = -3$
23. $v^2 + v - 3 = 0$
24. $x^2 - x = 1$
25. $y^2 = 7 - 10y$
26. $v^2 = 14 + 16v$
27. $r^2 - 3r = 5$
28. $s^2 + 3s = -1$
29. $t^2 - t = 4$
30. $y^2 + y - 4 = 0$
31. $x^2 - 3x + 5 = 0$
32. $z^2 + 5z + 7 = 0$
33. $2t^2 - 3t + 1 = 0$
34. $2x^2 - 7x + 3 = 0$
35. $2r^2 + 5r = 3$
36. $2y^2 - 3y = 9$
37. $2s^2 = 7s - 6$
38. $2x^2 = 3x + 20$
39. $2v^2 = v + 1$

Solve by completing the square.

40. $2z^2 = z + 3$

41. $3r^2 + 5r = 2$

42. $3t^2 - 8t = 3$

43. $3y^2 + 8y + 4 = 0$

44. $3z^2 - 10z - 8 = 0$

45. $4x^2 + 4x - 3 = 0$

46. $4v^2 + 4v - 15 = 0$

47. $6s^2 + 7s = 3$

48. $6z^2 = z + 2$

49. $6p^2 = 5p + 4$

50. $6t^2 = t - 2$

51. $4v^2 - 4v - 1 = 0$

52. $2s^2 - 4s - 1 = 0$

53. $4z^2 - 8z = 1$

54. $3r^2 - 2r = 2$

55. $3y - 6 = (y - 1)(y - 2)$

56. $7s + 55 = (s + 5)(s + 4)$

57. $4p + 2 = (p - 1)(p + 3)$

58. $v - 10 = (v + 3)(v - 4)$

Critical Thinking

Solve.

59. $\frac{x^2}{6} - \frac{x}{3} = 1$

60. $\sqrt{x + 2} = x - 4$

61. $\sqrt{3x + 4} - x = 2$

62. $\frac{x}{3} + \frac{3}{x} = \frac{8}{3}$

63. $\frac{x + 1}{2} + \frac{3}{x - 1} = 4$

64. $\frac{x - 2}{3} + \frac{2}{x + 2} = 4$

65. A basketball player shoots at a basket 25 ft away. The height of the ball above the ground at time t is given by $h = -16t^2 + 32t + 6.5$. How many seconds after the ball is released does it hit the basket? Round to the nearest hundredth. *Hint:* When it hits the basket, $h = 10$ ft.

66. A ball player hits a ball. The height of the ball above the ground can be approximated by the equation $h = -16t^2 + 76t + 5$. When will the ball hit the ground? Round to the nearest hundredth. *Hint:* The ball strikes the ground when $h = 0$ ft.

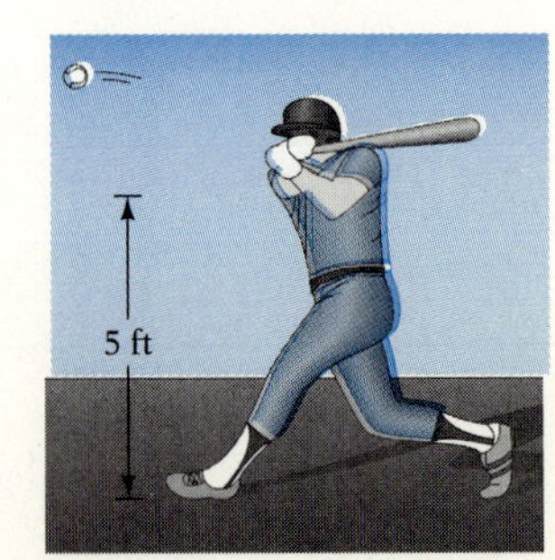

67. **[W]** Explain why the equation $(x - 2)^2 = -4$ does not have a real number solution.

SECTION 12.3 Solving Quadratic Equations by Using the Quadratic Formula

OBJECTIVE 

Solve a quadratic equation by using the quadratic formula

Any quadratic equation can be solved by completing the square. Applying this method to the standard form of a quadratic equation produces a formula that can be used to solve any quadratic equation.

CONSIDER THIS

Factoring, when it is easy, is the quickest way to solve a quadratic equation. Completing the square has other uses besides deriving the quadratic formula. It is rarely used to solve a quadratic equation. The quadratic formula is used for those equations that do not factor.

Solve $ax^2 + bx + c = 0$ by completing the square.

$$ax^2 + bx + c = 0$$

Subtract the constant term from each side of the equation.

$$ax^2 + bx + c - c = 0 - c$$
$$ax^2 + bx = -c$$

Multiply each side of the equation by the reciprocal of a, the coefficient of x^2.

$$\frac{1}{a}(ax^2 + bx) = \frac{1}{a}(-c)$$
$$x^2 + \frac{b}{a}x = -\frac{c}{a}$$

Complete the square by adding $\left(\frac{1}{2} \cdot \frac{b}{a}\right)^2$ to each side of the equation.

$$x^2 + \frac{b}{a}x + \left(\frac{1}{2} \cdot \frac{b}{a}\right)^2 = \left(\frac{1}{2} \cdot \frac{b}{a}\right)^2 - \frac{c}{a}$$
$$x^2 + \frac{b}{a}x + \frac{b^2}{4a^2} = \frac{b^2}{4a^2} - \frac{c}{a}$$

Simplify the right side of the equation.

$$x^2 + \frac{b}{a}x + \frac{b^2}{4a^2} = \frac{b^2}{4a^2} - \left(\frac{c}{a} \cdot \frac{4a}{4a}\right)$$
$$x^2 + \frac{b}{a}x + \frac{b^2}{4a^2} = \frac{b^2}{4a^2} - \frac{4ac}{4a^2}$$
$$x^2 + \frac{b}{a}x + \frac{b^2}{4a^2} = \frac{b^2 - 4ac}{4a^2}$$

Factor the perfect square trinomial on the left side of the equation.

$$\left(x + \frac{b}{2a}\right)^2 = \frac{b^2 - 4ac}{4a^2}$$

Take the square root of each side of the equation.

$$\sqrt{\left(x + \frac{b}{2a}\right)^2} = \pm\sqrt{\frac{b^2 - 4ac}{4a^2}}$$
$$x + \frac{b}{2a} = \pm\frac{\sqrt{b^2 - 4ac}}{2a}$$

Solve for x.

$$x + \frac{b}{2a} = \frac{\sqrt{b^2 - 4ac}}{2a} \qquad\qquad x + \frac{b}{2a} = -\frac{\sqrt{b^2 - 4ac}}{2a}$$
$$x = -\frac{b}{2a} + \frac{\sqrt{b^2 - 4ac}}{2a} \qquad\qquad x = -\frac{b}{2a} - \frac{\sqrt{b^2 - 4ac}}{2a}$$
$$= \frac{-b + \sqrt{b^2 - 4ac}}{2a} \qquad\qquad = \frac{-b - \sqrt{b^2 - 4ac}}{2a}$$

The Quadratic Formula

The solution of $ax^2 + bx + c = 0$ is $x = \frac{-b + \sqrt{b^2 - 4ac}}{2a}$ or $x = \frac{-b - \sqrt{b^2 - 4ac}}{2a}$. The quadratic formula is frequently written in the form $x = \frac{-b \pm \sqrt{b^2 - 4ac}}{2a}$.

Solve by using the quadratic formula: $2x^2 = 4x - 1$

Write the equation in standard form.

$$2x^2 = 4x - 1$$
$$2x^2 - 4x + 1 = 0$$

$a = 2, b = -4$, and $c = 1$.

$$x = \frac{-b \pm \sqrt{b^2 - 4ac}}{2a}$$

Replace a, b, and c in the quadratic formula by their values.

$$= \frac{-(-4) \pm \sqrt{(-4)^2 - 4(2)(1)}}{2 \cdot 2}$$

Simplify.

$$= \frac{4 \pm \sqrt{16 - 8}}{4} = \frac{4 \pm \sqrt{8}}{4}$$
$$= \frac{4 \pm 2\sqrt{2}}{4} = \frac{2 \pm \sqrt{2}}{2}$$

The solutions are $\frac{2 + \sqrt{2}}{2}$ and $\frac{2 - \sqrt{2}}{2}$.

Example 1 Solve by using the quadratic formula: $2x^2 - 3x + 1 = 0$

Solution

$2x^2 - 3x + 1 = 0$

$a = 2, b = -3, c = 1$

$$x = \frac{-(-3) \pm \sqrt{(-3)^2 - 4(2)(1)}}{2 \cdot 2}$$
$$= \frac{3 \pm \sqrt{9 - 8}}{4} = \frac{3 \pm \sqrt{1}}{4} = \frac{3 \pm 1}{4}$$
$$x = \frac{3 + 1}{4} = \frac{4}{4} = 1 \qquad x = \frac{3 - 1}{4} = \frac{2}{4} = \frac{1}{2}$$

The solutions are 1 and $\frac{1}{2}$.

You Try It 1 Solve by using the quadratic formula: $3x^2 + 4x - 4 = 0$

Your Solution

Example 2 Solve by using the quadratic formula: $2x^2 = 8x - 5$

Solution

$$2x^2 = 8x - 5$$
$$2x^2 - 8x + 5 = 0$$

$a = 2, b = -8, c = 5$

$$x = \frac{-(-8) \pm \sqrt{(-8)^2 - 4(2)(5)}}{2 \cdot 2}$$
$$= \frac{8 \pm \sqrt{64 - 40}}{4} = \frac{8 \pm \sqrt{24}}{4}$$
$$= \frac{8 \pm 2\sqrt{6}}{4} = \frac{4 \pm \sqrt{6}}{2}$$

The solutions are $\frac{4 + \sqrt{6}}{2}$ and $\frac{4 - \sqrt{6}}{2}$.

You Try It 2 Solve by using the quadratic formula: $x^2 + 2x = 1$

Your Solution

Solutions on p. A39

12.3 EXERCISES

Objective A

Solve by using the quadratic formula.

1. $x^2 - 4x - 5 = 0$
2. $y^2 + 3y + 2 = 0$
3. $z^2 - 2z - 15 = 0$
4. $v^2 + 5v + 4 = 0$
5. $z^2 + 6z - 7 = 0$
6. $s^2 + 3s - 10 = 0$
7. $t^2 + t - 6 = 0$
8. $x^2 - x - 2 = 0$
9. $y^2 = 2y + 3$
10. $w^2 = 3w + 18$
11. $r^2 = 5 - 4r$
12. $z^2 = 3 - 2z$
13. $2y^2 - y - 1 = 0$
14. $2t^2 - 5t + 3 = 0$
15. $w^2 + 3w + 5 = 0$
16. $x^2 - 2x + 6 = 0$
17. $p^2 - p = 0$
18. $2v^2 + v = 0$
19. $4t^2 - 9 = 0$
20. $4s^2 - 25 = 0$
21. $4y^2 + 4y = 15$
22. $4r^2 + 4r = 3$
23. $3t^2 = 7t + 6$
24. $3x^2 = 10x + 8$
25. $5z^2 + 11z = 12$
26. $4v^2 = v + 3$
27. $6s^2 - s - 2 = 0$
28. $6y^2 + 5y - 4 = 0$
29. $2x^2 + x + 1 = 0$
30. $3r^2 - r + 2 = 0$
31. $t^2 - 2t = 5$
32. $y^2 - 4y = 6$
33. $t^2 + 6t - 1 = 0$
34. $z^2 + 4z + 1 = 0$
35. $w^2 = 4w + 9$
36. $y^2 = 8y + 3$

Solve by using the quadratic formula.

37. $9y^2 + 6y - 1 = 0$

38. $9s^2 - 6s - 2 = 0$

39. $4p^2 + 4p + 1 = 0$

40. $9z^2 + 12z + 4 = 0$

41. $2x^2 = 4x - 5$

42. $3r^2 = 5r - 6$

43. $4p^2 + 16p = -11$

44. $4y^2 - 12y = -1$

45. $4x^2 = 4x + 11$

46. $4s^2 + 12s = 3$

47. $9v^2 = -30v - 23$

48. $9t^2 = 30t + 17$

Critical Thinking

49. Solve $x^2 + ax + b = 0$ for x.

50. True or false?

a. The equations $x = \sqrt{12 - x}$ and $x^2 = 12 - x$ have the same solution.

b. If $\sqrt{a} + \sqrt{b} = c$, then $a + b = c^2$.

c. $\sqrt{9} = \pm 3$

d. $\sqrt{x^2} = |x|$.

Solve.

51. $\sqrt{x + 3} = x - 3$

52. $\sqrt{x + 4} = x + 4$

53. $\sqrt{x + 1} = x - 1$

54. $\sqrt{x^2 + 2x + 1} = x - 1$

55. $\frac{x}{4} + \frac{3}{x} = \frac{5}{2}$

56. $\frac{x + 1}{5} - \frac{4}{x - 1} = 2$

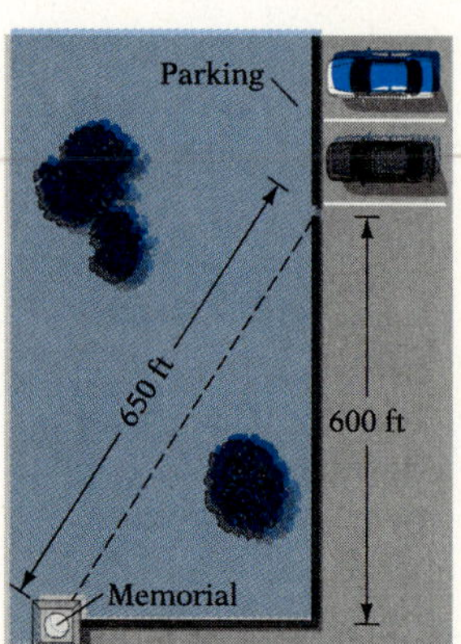

57. An L-shaped sidewalk from the parking lot to a memorial is shown in the figure at the right. The distance directly across the grass to the memorial is 650 ft. The distance to the corner is 600 ft. Find the distance from the corner to the memorial.

58. A commuter plane leaves an airport traveling due south at 400 mph. Another plane leaving at the same time travels due east at 300 mph. Find the distance between the two planes after two hours.

59. [W] Factoring, completing the square, and using the quadratic formula are three methods of solving quadratic equations. Describe each method, and cite the advantages and disadvantages of each.

60. [W] Explain why the equation $0x^2 + 3x + 4 = 0$ cannot be solved by the quadratic formula.

SECTION 12.4 Graphing Quadratic Equations in Two Variables

OBJECTIVE A Graph a quadratic equation of the form $y = ax^2 + bx + c$

An equation of the form $y = ax^2 + bx + c$, where $a \neq 0$, is a **quadratic equation in two variables.** Examples of quadratic equations in two variables are shown at the right.

$$y = 3x^2 - x + 1$$
$$y = -x^2 - 3$$
$$y = 2x^2 - 5x$$

The graph of a quadratic equation in two variables is a **parabola**. The graph is cup-shaped and opens either up or down. The graphs of two parabolas are shown below.

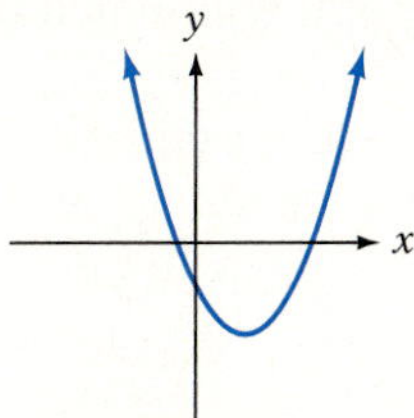

Parabola that opens up

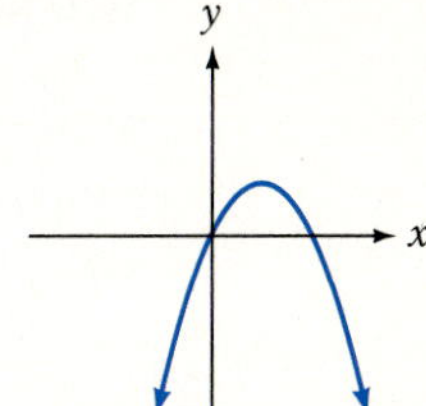

Parabola that opens down

Graph: $y = x^2 - 2x - 3$

Find several solutions of the equation. Because the graph is not a straight line, several solutions must be found in order to determine the cup shape.

Display the ordered-pair solutions in a table.

x	y
0	−3
1	−4
−1	0
2	−3
3	0

Graph the ordered-pair solutions on a rectangular coordinate system.

Draw a parabola through the points.

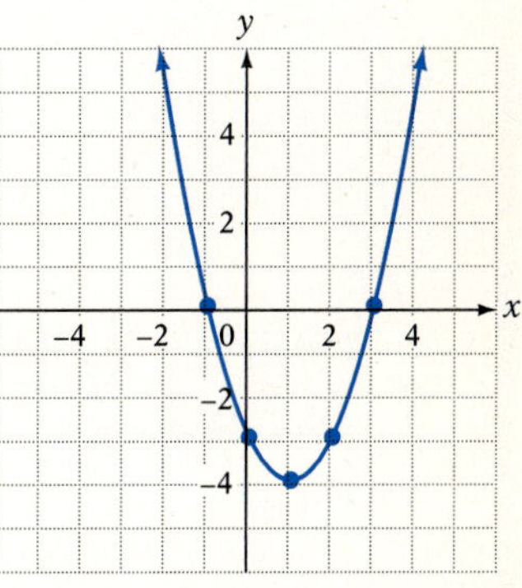

The graph of $y = -2x^2 + 1$ is shown at the right.

x	y
0	1
1	−1
−1	−1
2	−7
−2	−7

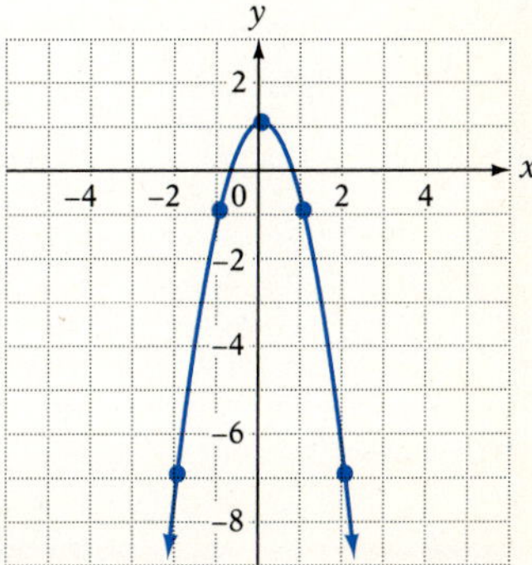

POINT OF INTEREST

Mirrors in some telescopes are ground into the shape of a parabola. The mirror at the Palomar Mountain Observatory is 2 ft thick at the ends and weighs 14.75 tons. The mirror has been ground to a true paraboloid (the three-dimensional version of a parabola) to within 0.0000015 in. An equation of the mirror is $y = 2640x^2$.

Note that the graph of $y = x^2 - 2x - 3$, shown on the previous page, opens up and that the coefficient of x^2 is positive. The graph of $y = -2x^2 + 1$ opens down, and the coefficient of x^2 is negative. For any quadratic equation in two variables, the coefficient of x^2 determines whether the parabola opens up or down. When a is positive, the parabola opens up. When a is negative, the parabola opens down.

Every parabola has an axis of symmetry and a vertex that is on the axis of symmetry. If the parabola opens up, the vertex is the lowest point on the graph. If the parabola opens down, the vertex is the highest point on the parabola.

y
x
Vertex
Axis of symmetry

To understand the axis of symmetry, think of folding the paper along that axis. The two halves of the graph will match up.

When graphing a quadratic equation in two variables, use the value of a to determine whether the parabola opens up or down. After graphing ordered-pair solutions of the equation, use symmetry to help you draw the parabola.

Example 1 Graph: $y = x^2 - 2x$

Solution

x	y
0	0
1	−1
−1	3
2	0
3	3

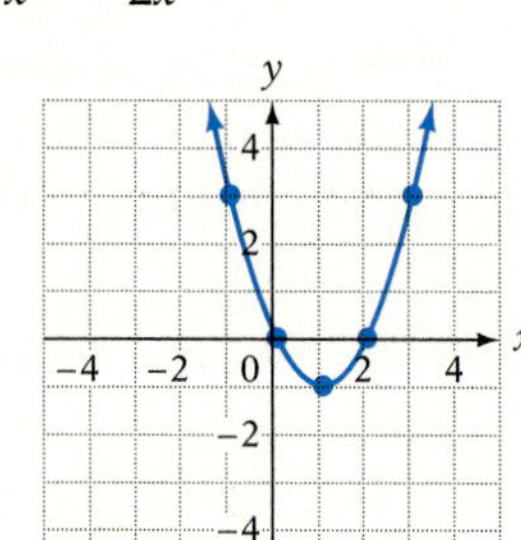

You Try It 1 Graph: $y = x^2 + 2$

Your Solution

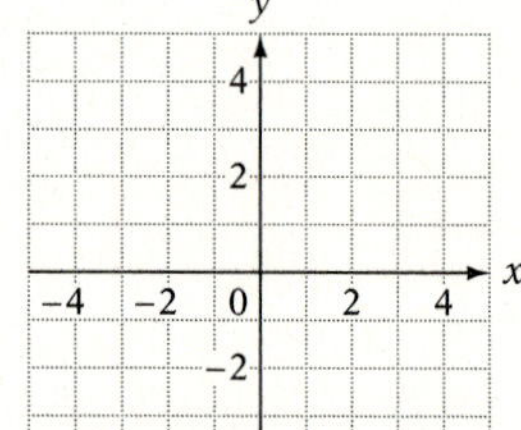

Example 2 Graph: $y = -x^2 + 4x - 4$

Solution

x	y
0	−4
1	−1
2	0
3	−1
4	−4

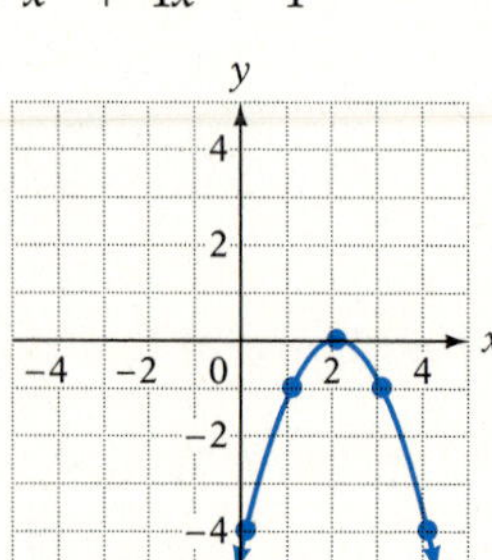

You Try It 2 Graph: $y = -x^2 - 2x - 1$

Your Solution

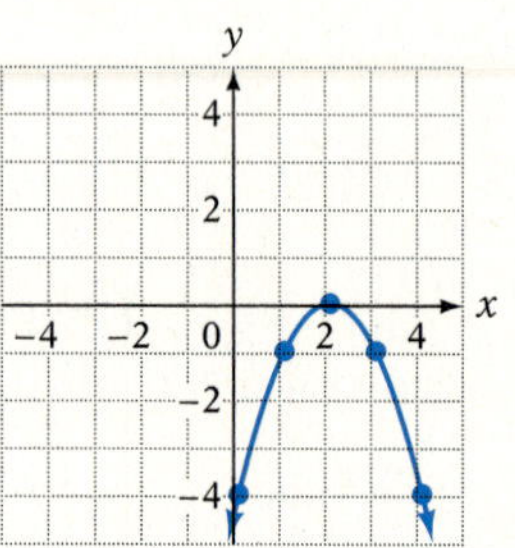

Solutions on p. A39

12.4 EXERCISES

Objective A

Determine whether the graph of the equation opens up or down. Is the vertex the highest or lowest point on the graph?

1. $y = x^2 - 3$

2. $y = -x^2 + 4$

3. $y = \frac{1}{2}x^2 - 2$

4. $y = -\frac{1}{3}x^2 + 5$

5. $y = x^2 - 2x + 3$

6. $y = -x^2 + 4x - 1$

Graph.

7. $y = x^2$

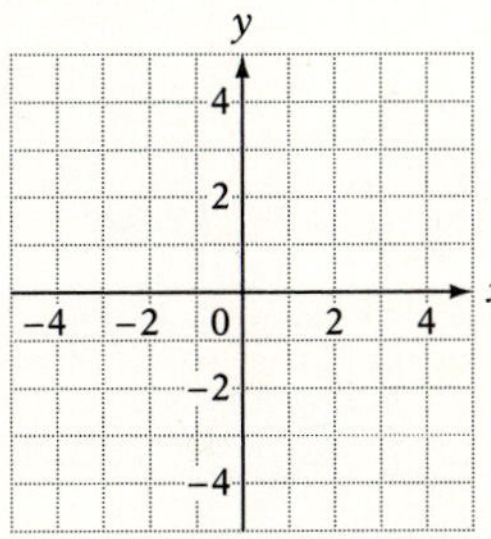

8. $y = -x^2$

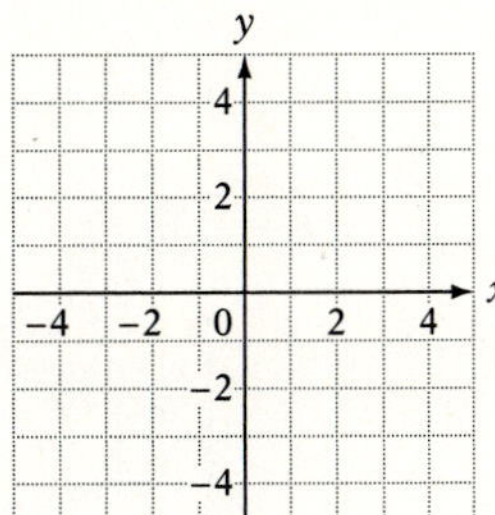

9. $y = -x^2 + 1$

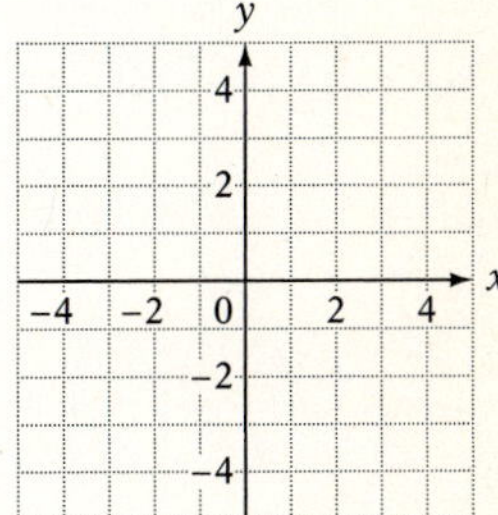

10. $y = x^2 - 1$

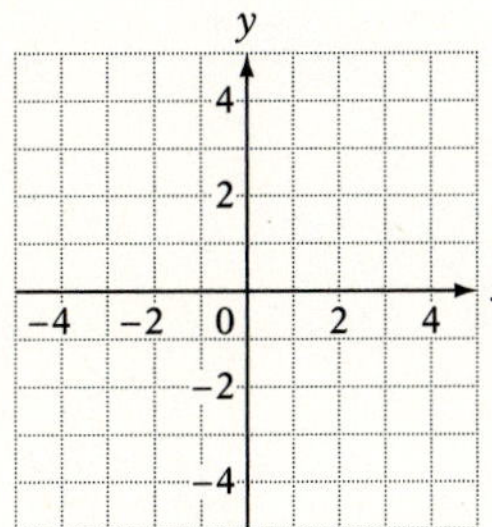

11. $y = 2x^2$

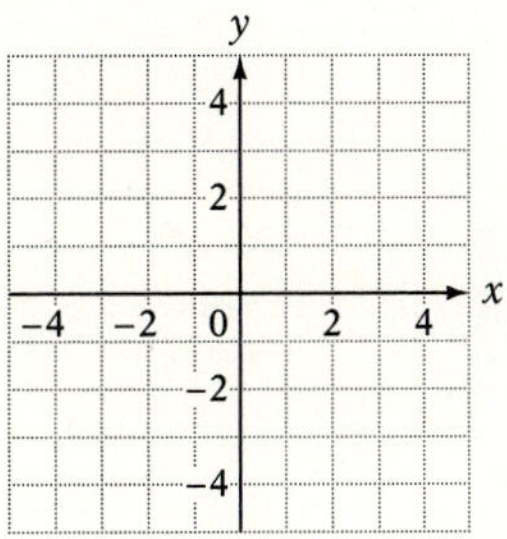

12. $y = \frac{1}{2}x^2$

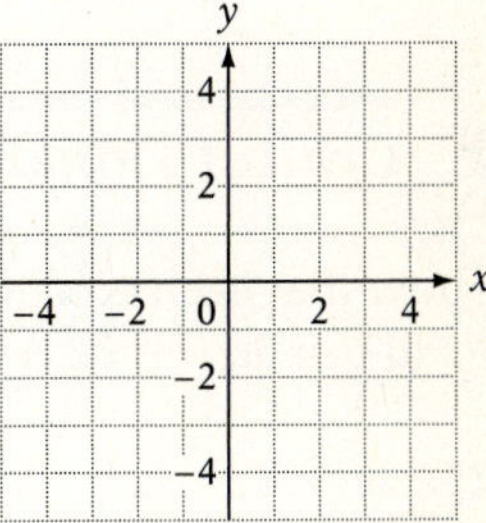

13. $y = -\frac{1}{2}x^2 + 1$

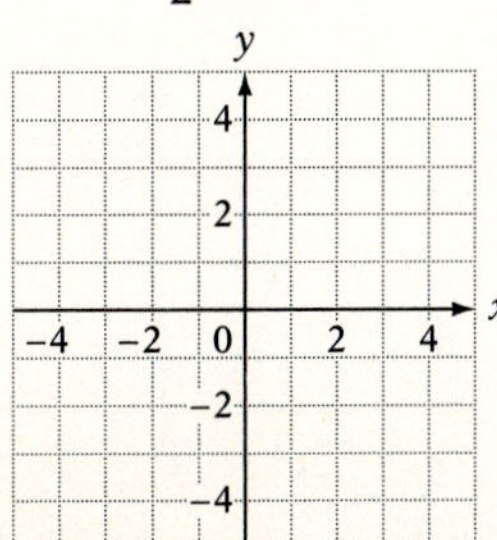

14. $y = 2x^2 - 1$

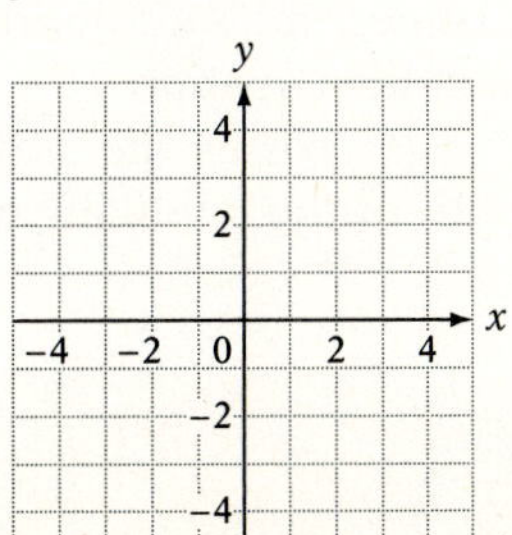

15. $y = x^2 - 4x$

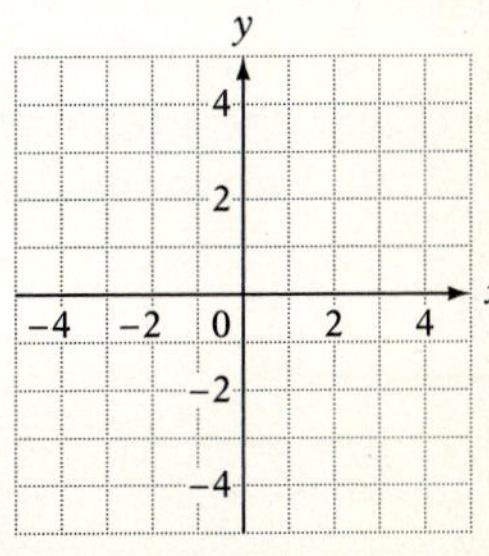

Graph.

16. $y = x^2 + 4x$

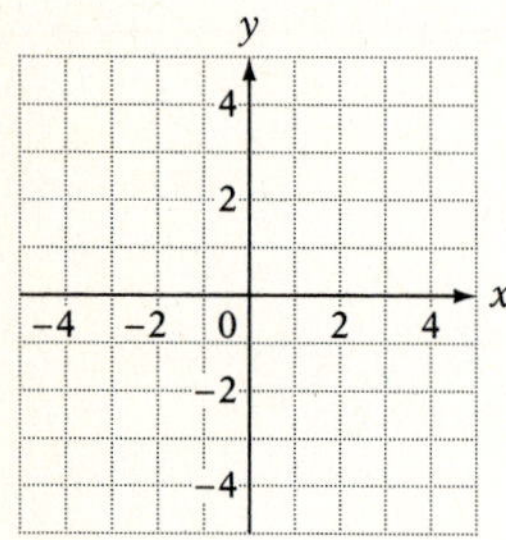

17. $y = x^2 - 2x + 3$

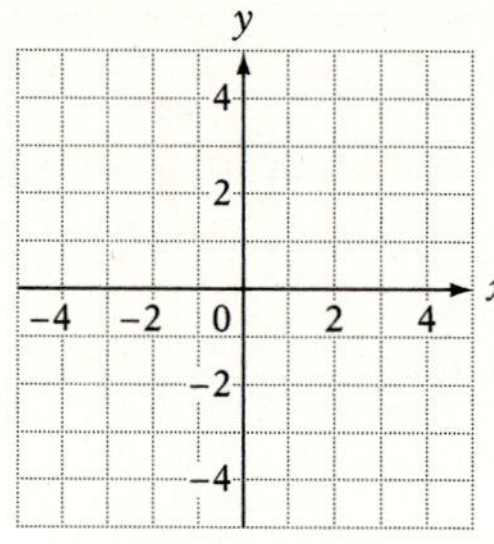

18. $y = x^2 - 4x + 2$

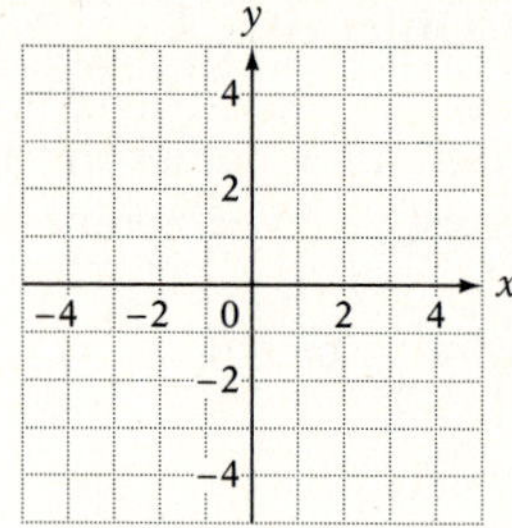

19. $y = -x^2 + 2x + 3$

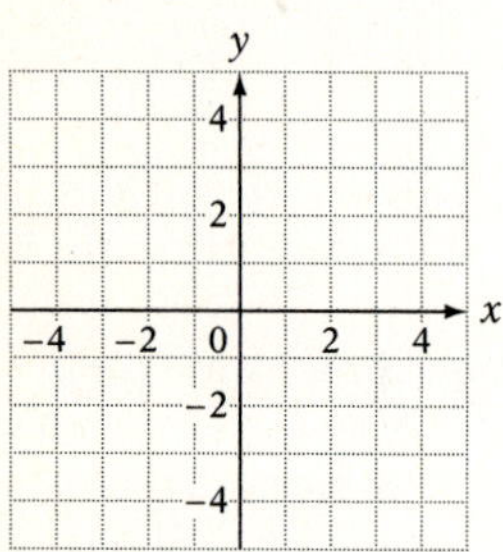

20. $y = -x^2 - 2x + 3$

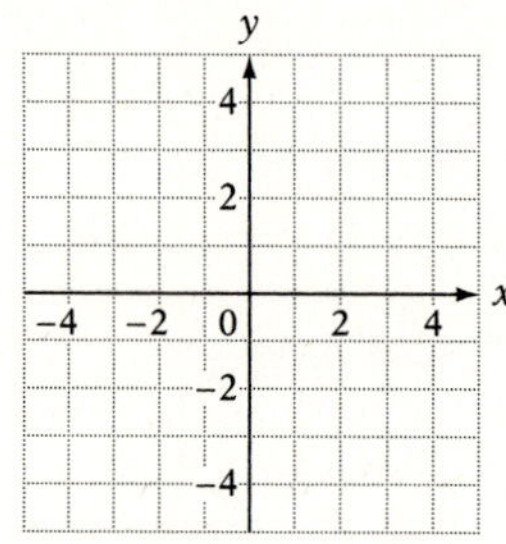

21. $y = -x^2 + 4x - 4$

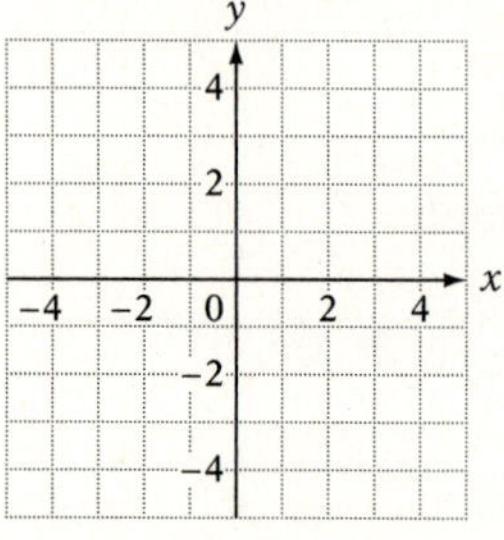

22. $y = -x^2 + 6x - 9$

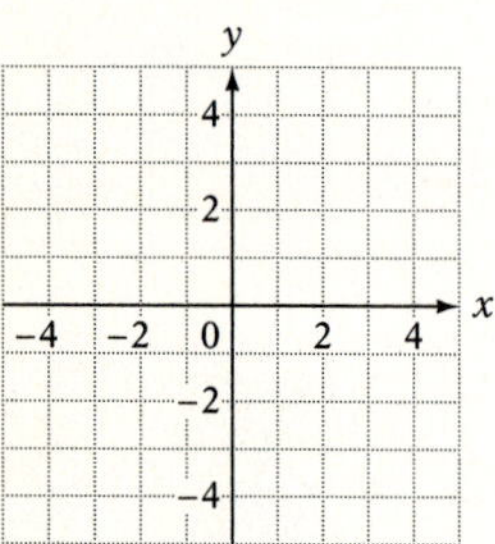

23. $y = (x - 2)^2$

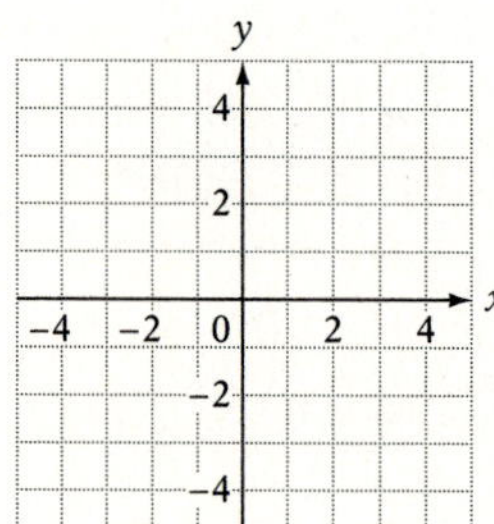

24. $y = -(x + 1)^2$

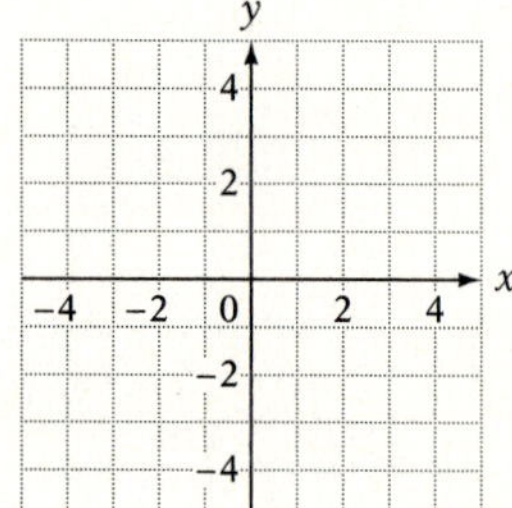

Critical Thinking

Show that each equation is a quadratic equation in two variables by writing it in the form $y = ax^2 + bx + c$.

25. $y + 1 = (x - 4)^2$

26. $y - 2 = 3(x + 1)^2$

27. $y - 4 = 2(x - 3)^2$

The x-intercepts of a parabola are the points where the graph crosses the x-axis. Since any point on the x-axis has y-coordinate 0, the x-intercepts of a parabola occur when $y = 0$. Therefore, the x-coordinate of an x-intercept is a solution of the equation $ax^2 + bx + c = 0$. For example, the solutions of the equation $x^2 - 1 = 0$ are 1 and -1, and the graph of the equation $y = x^2 - 1$ crosses the x-axis at (1, 0) and $(-1, 0)$. Determine the x-intercepts of the graphs of the following equations.

28. $y = x^2 - 4$

29. $y = -x^2 + 4$

30. $y = x^2 - x$

31. $y = x^2 - 4x$

32. $y = x^2 - 4x + 3$

33. $y = 2x^2 - x - 1$

Projects in Mathematics

Sun	Mon	Tue	Wed	Thu	Fri	Sat
					1	2
3	4	5	6	7	8	9
10	11	12	13	14	15	16
17	18	19	20	21	22	23
24	25	26	27	28	29	30

For the calendar at the left, note that if each of the dates under Friday were divided by 7, the remainder would be the same. For example,

$15 \div 7 = 2$ with remainder 1
$29 \div 7 = 4$ with remainder 1

Dividing each of the dates under Tuesday by 7 results in a remainder of 5.

$19 \div 7 = 2$ with remainder 5

The same idea can be applied to each of the seven days of the week. Numbers that have the same remainder when divided by a given number n are said to be **congruent modulo n**. For example, 5, 12, 19, and 26 are congruent modulo 7.

The reason the remainders are the same is that there are 7 days in one week. On this calendar, Tuesdays are on the 5th, 12th $(5 + 7)$, 19th $(12 + 7)$, and 26th $(19 + 7)$.

The notation $a \equiv b \pmod{n}$ indicates that a and b have the same remainder when divided by n. For example, $19 \equiv 26 \pmod{7}$ because 19 and 26 have the same remainder when divided by 7. The remainder is 5.

For each of the problems below, mark the statement true or false.

1. $34 \equiv 9 \pmod{5}$
2. $78 \equiv 23 \pmod{9}$
3. $16 \equiv 52 \pmod{12}$
4. $20 \equiv 0 \pmod{10}$

ISBN 0-395-71069-3

There are many applications of congruence. The Universal Product Code (UPC) that is used by many grocery stores is one application. The UPC identification number consists of 12 digits.

To be a valid UPC number, the following modular equation must be true.

$$3a_1 + a_2 + 3a_3 + a_4 + 3a_5 + a_6 + 3a_7 + a_8 + 3a_9 + a_{10} + 3a_{11} + a_{12} \equiv 0 \pmod{10}$$

Each a in this equation is one of the numbers of the UPC identification number. The first 11 numbers identify the country in which the product was made, the manufacturer, and the type of product. The twelfth digit, a_{12}, is called the **check digit** and is chosen so that the equation is true.

For example, the first 11 numbers of the UPC at the left identify *The 1995 Information Please Almanac* published by Houghton Mifflin Company. Substituting the first 11 numbers into the equation gives

$$3(0) + 4 + 3(6) + 4 + 3(4) + 2 + 3(7) + 7 + 3(0) + 1 + 3(8) + a_{12} \equiv 0 \pmod{10}$$
$$0 + 4 + 18 + 4 + 12 + 2 + 21 + 7 + 0 + 1 + 24 + a_{12} \equiv 0 \pmod{10}$$
$$93 + a_{12} \equiv 0 \pmod{10}$$

To have a valid UPC number, a_{12} is chosen so that the result is congruent to 0 (mod 10). For $93 + a_{12} \equiv 0 \pmod{10}$, $93 + a_{12}$ must be divisible by 10. The single digit that can be added to 93 so that the sum is divisible by 10 is 7. Therefore, $a_{12} = 7$, which is the check digit shown in the UPC number.

If a bookstore ordered this book and incorrectly wrote the number 046443770187, a computer processing the order would be able to determine that there had been a mistake because the number is not congruent to 0 (mod 10) and therefore does not belong to any product.

Another number shown above the bar coding is 0-395-71069-3, which is the International Standard Book Number (ISBN). The first number identifies the book as being published in an English-speaking country. The next group of numbers is the publisher, the next group of five digits identifies the title, and the last digit is the check digit. In this case, a certain sum must be congruent to 0 (mod 11). The formula for an ISBN is

$$10a_1 + 9a_2 + 8a_3 + 7a_4 + 6a_5 + 5a_6 + 4a_7 + 3a_8 + 2a_9 + a_{10} \equiv 0 \pmod{11}$$

Use this formula to verify the ISBN for *The 1995 Information Please Almanac.*

Chapter Summary

Key Words

A *quadratic equation* is an equation of the form $ax^2 + bx + c = 0$, where $a \neq 0$. A quadratic equation is also called a *second-degree* equation.

A quadratic equation is in *standard form* when the polynomial is in descending order and equal to zero.

Adding to a binomial the constant term that makes it a perfect square trinomial is called *completing the square.*

Essential Rules

Some quadratic equations can be solved by factoring or by taking square roots.

Any quadratic equation can be solved by completing the square or by using the quadratic formula.

For a perfect square trinomial, $\left(\frac{1}{2} \textbf{ the coefficient of } x\right)^2$ **= the constant term.**

The Principle of Zero Products If $ab = 0$, then $a = 0$ or $b = 0$.

The Quadratic Formula $x = \dfrac{-b \pm \sqrt{b^2 - 4ac}}{2a}$

The Square Root Property If $x^2 = a$, then $x = \pm a$.

Chapter Review Exercises

1. Solve: $(x - 5)(x + 6) = 0$

2. Solve: $(3x + 1)(x - 8) = 0$

3. Graph: $y = -3x^2$

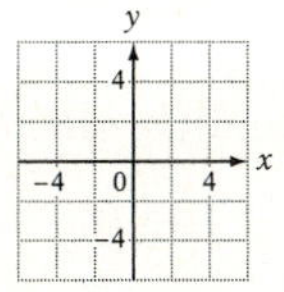

4. Solve by factoring: $3x^2 + 7x = 20$

5. Solve by factoring: $6x^2 - 17x = -5$

6. Solve by factoring: $2x = (x + 5)(x - 4)$

7. Solve by taking square roots: $x^2 - 81 = 0$

8. Solve by taking square roots: $16x^2 = 25$

9. Solve by taking square roots: $9x^2 - 36 = 0$

10. Graph: $y = \frac{1}{2}x^2 - 1$

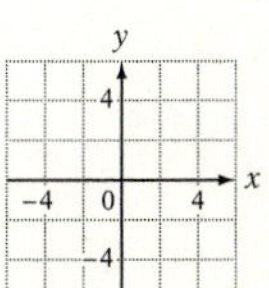

11. Solve by taking square roots:
$2(x - 5)^2 = 36$

12. Solve by taking square roots:
$3(x + 4)^2 - 60 = 0$

13. Solve by completing the square:
$x^2 + 4x - 16 = 0$

14. Graph: $y = 2x^2 + 1$

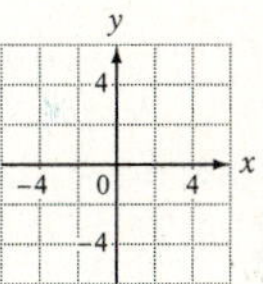

15. Solve by completing the square: $x^2 - 5x = 2$

16. Solve by completing the square: $x^2 + 3x = 8$

17. Solve by completing the square:
$2x^2 - 6x + 1 = 0$

18. Solve by completing the square:
$2x^2 + 8x = 3$

19. Solve by using the quadratic formula:
$x^2 + 4x + 2 = 0$

20. Solve by using the quadratic formula:
$x^2 + 3x - 7 = 0$

21. Solve by using the quadratic formula:
$x^2 - 3x = 6$

22. Graph: $y = -\frac{1}{4}x^2$

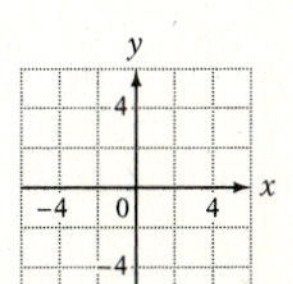

23. Solve by using the quadratic formula:
$2x^2 - 5x - 3 = 0$

24. Solve by using the quadratic formula:
$3x^2 - x = 1$

Cumulative Review Exercises

1. Solve: $\sqrt{2x - 3} - 5 = 0$

2. Simplify: $\dfrac{\sqrt{108a^7b^3}}{\sqrt{3a^4b}}$

3. Simplify:
$2x - 3[2x - 4(3 - 2x) + 2] - 3$

4. Find the equation of the line that contains the point $(-3, 2)$ and has slope $-\frac{4}{3}$.

5. Solve: $3x + 2y = 2$
$5x - 2y = 14$

6. Multiply: $\dfrac{3x^3 - 6x^2}{4x^2 + 4x} \cdot \dfrac{3x - 9}{9x^3 - 45x^2 + 54x}$

7. Subtract: $3\sqrt{32} - 2\sqrt{128}$

8. Solve: $\dfrac{x}{2x - 5} - 2 = \dfrac{3x}{2x - 5}$

9. Simplify: $\dfrac{3}{2 - \sqrt{3}}$

10. Simplify: $2a\sqrt{2ab^3} + b\sqrt{8a^3b} - 5ab\sqrt{ab}$

11. Solve by taking square roots:
$2(x - 5)^2 = 72$

12. Simplify: $\dfrac{\sqrt{12x^2} - \sqrt{27}}{\sqrt{3}}$

13. Factor: $18a^3 + 57a^2 + 30a$

14. Simplify: $(3 - 7)^2 \div (-2) - 3 \cdot (-4)$

15. Simplify: $(3b + 2)^2$

16. Solve: $2x^2 - 4x - 5 = 0$

17. Divide: $(x^2 - 8) \div (x - 2)$

18. Solve $R = \dfrac{C - S}{t}$ for C.

19. Solve: $\frac{5y}{6} - \frac{5}{9} = \frac{y}{3} - \frac{5}{6}$

20. Simplify: $\dfrac{27a^3b^2}{(-3ab^2)^3}$

21. Find the measure of x.

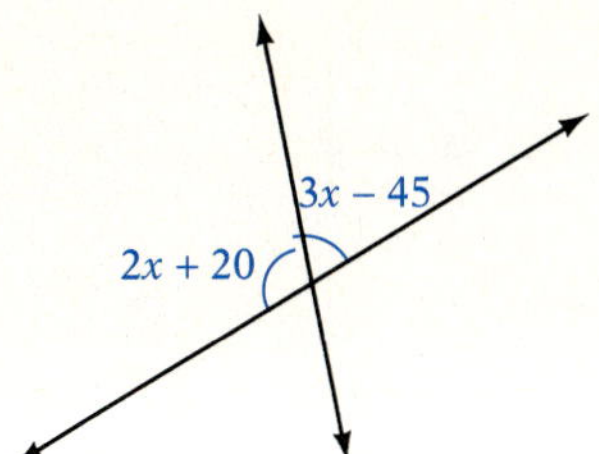

22. Graph: $x + 2y = 4$

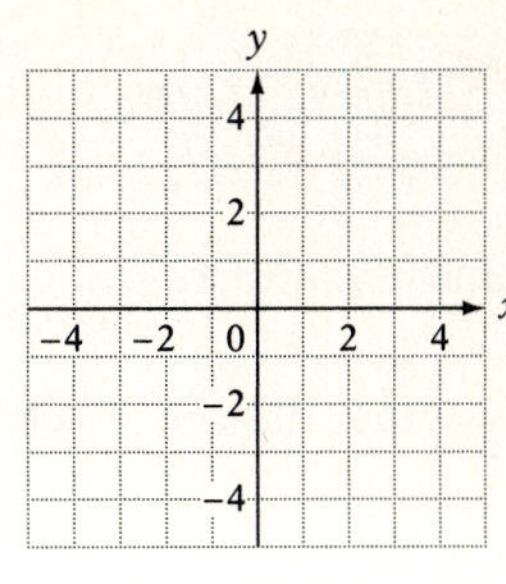

23. A small water pipe takes twice as long to fill a tank as does a larger water pipe. With both pipes open it takes 16 h to fill the tank. Find the time it would take the small pipe working alone to fill the tank.

24. A stock dividend of 100 shares paid \$215. At this rate, how many additional shares are required to earn a dividend of \$752.50?

25. Find the perimeter of a right triangle with legs that measure 5 in. and 7 in. Round to the nearest tenth.

26. The length of each side of a cube is $(x + 4)$ cm. Express the volume of the cube in terms of the variable x.

Final Exam

1. Simplify: $4^2(3^2)$

2. Simplify: $12 \div (6 - 3) \cdot 4 + 3^2$

3. Find the difference between $6\frac{1}{4}$ and $3\frac{5}{8}$.

4. Subtract: $90.001 - 29.796$

5. Find the quotient of 0.0426 and 0.062. Round to the nearest hundredth.

6. Evaluate: $-|-3.8|$

7. Multiply: $\left(2\frac{1}{2}\right)\left(-\frac{1}{5}\right)$

8. Evaluate $\frac{-a^2 - b}{a - b}$ when $a = 2$ and $b = -3$.

9. Simplify: $3x - 3(2x - 4) - 1$

10. Solve: $3x - 5 = 10$

11. Solve: $3x - 2(4 - 3x) = -3(2 - x)$

12. Solve: $-\frac{2}{3}x < -\frac{4}{9}$

13. Evaluate x^2y^4 when $x = 3$ and $y = 2$.

14. Evaluate $x + y + z$ when $x = \frac{1}{6}$, $y = \frac{2}{9}$, and $z = \frac{2}{3}$.

15. Solve: $3(3y - 2) + 2 < -3(1 - 2y)$

16. Write "323.4 mi on 13.2 gal of gasoline" as a unit rate.

17. Solve the proportion $\frac{5}{4} = \frac{n}{18}$.

18. Write $\frac{3}{8}$ as a percent.

19. 22% of what number is 9.9?

20. Subtract: $(3x^2 - 2x + 4) - (5x^2 - 4x - 8)$

21. Simplify: $(-2x^2y)^5$

22. Multiply: $(-2x^2 + 2x - 1)(x - 3)$

23. Simplify: $\frac{(3xy^4)^2}{12x^3y^3}$

24. Divide: $(x^3 + 5x^2 + 2x - 8) \div (x + 2)$

25. Simplify: $(-2x^{-2}y^4)(3xy^{-1})^2$

26. Write 0.00000473 in scientific notation.

27. Factor: $-3y^4 + 6y^3 - 21y^2$

28. Factor: $x^2 - 8x - 9$

29. Factor: $6x^2 - x - 12$

30. Factor: $-6x^3 - 21x^2 - 18x$

31. Factor: $49x^2 - 1$

32. Factor: $x(y - 1) - 2(y - 1)$

33. Factor: $12x^2 - 27x^2y^2$

34. Simplify: $\frac{6 - 9x}{3x^2 - 2x}$

35. Divide: $\frac{x^2 - 7x + 12}{x^2 - 4x} \div \frac{2x^2 - 5x - 3}{2x^2 + x}$

36. Subtract: $\frac{3y}{y - 2} - \frac{6}{y - 2}$

37. Add: $\frac{3}{x + 5} + \frac{7}{x - 1}$

38. A triangle has a 32° angle and a 106° angle. What is the measure of the third angle?

39. Solve: $\frac{3x}{2x - 3} - 4 = \frac{2}{2x - 3}$

40. W is inversely proportional to V and $W = 8$ when $V = 2$. Find W when $V = 4$.

41. Given that angle $x = 35°$, find the measures of angles a and b.

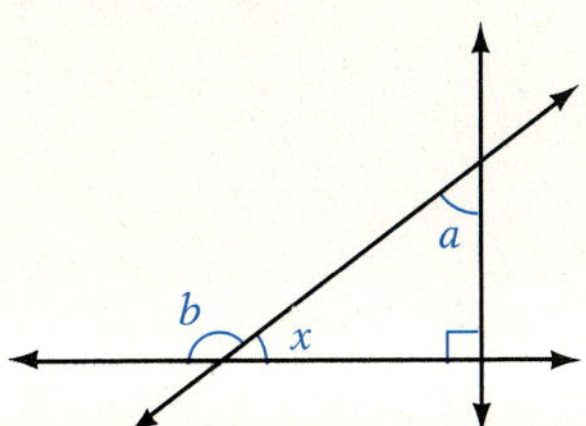

42. Graph: $2x - 3y = 3$

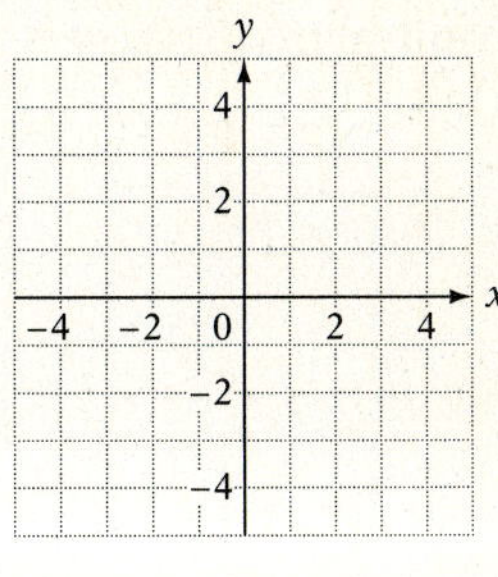

43. Solve $L = a(1 + ct)$ for t.

44. The sum of three times a number and 12 is 27. Find the number.

45. Find the slope of the line containing the points $(-2, 4)$ and $(-1, 1)$.

46. Find the equation of the line that contains the point $(4, -4)$ and has slope $-\frac{3}{4}$.

47. Solve:
$$\begin{aligned} 2x - 3y &= 7 \\ 3x + y &= 5 \end{aligned}$$

48. Simplify: $4\sqrt{32}$

49. Simplify: $a\sqrt{81a^4}$

50. Add: $2\sqrt{27y} + 8\sqrt{48y}$

51. Multiply: $\sqrt{3x}(\sqrt{6x} - \sqrt{x})$

52. Simplify: $\dfrac{3}{\sqrt{2} - 1}$

53. Solve: $\sqrt{3x} - 2 = 1$

54. Solve: $2x^2 - x = 3$

55. Solve: $4(x - 1)^2 = 12$

56. Solve: $3x^2 - 2x = 3$

57. Given that $\ell_1 \parallel \ell_2$, find x.

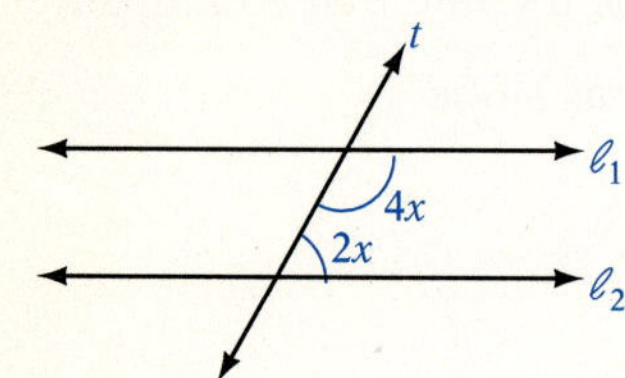

58. Triangles ABC and BDE are similar triangles. If $BE = 8$, $CE = 4$, and $BD = 10$, find the length of AB.

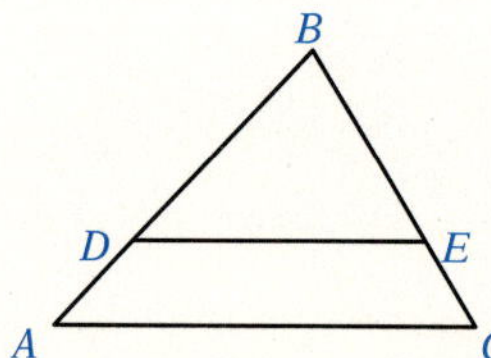

59. The height of a triangle is 4.8 cm and the base is 6.4 cm. Find the area of the triangle.

60. Find the perimeter of a right triangle with legs that measure 5 in. and 8 in. Round to the nearest tenth.

61. Given that $\angle a = 40°$ and $\angle x = 105°$, find the measures of angles b, c, and y.

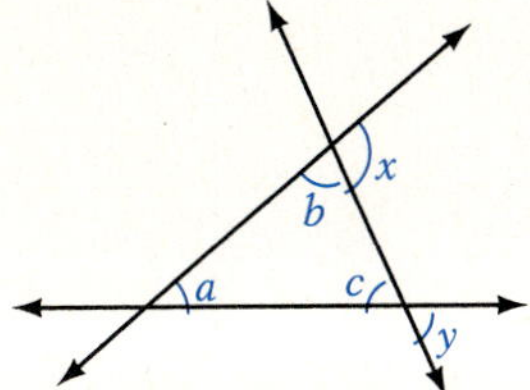

62. A compact car gets 30 mi on each gallon of gasoline. How many miles can the car travel on $4\frac{3}{10}$ gal of gasoline?

63. The diameter of the base of a cylinder is 9 in. and the height is 15 in. Find the volume of the cylinder. Round to the nearest hundredth.

64. The length of a rectangular solid is 8 ft, the width is 5 ft, and the height is 4 ft. Find the surface area of the solid.

65. A pre-election survey showed that 4 out of 7 voters would vote in an election. At this rate, how many people would be expected to vote in a city of 133,000?

66. A city has a population of 72,000. This is 180% of the city's population 5 years ago. What was the city's population 5 years ago?

67. An investment of $6000 is made at an annual simple interest rate of 7.5%. How much additional money must be invested at 9.5% so that the total interest earned is $735?

68. Find the cost per pound of a mixture of coffee made from 60 lb of coffee that cost $3.20 per pound and 20 lb of coffee that cost $8.40 per lb.

69. Eight grams of sugar are added to a 60-gram serving of a breakfast cereal that is 15% sugar. What is the percent concentration of the resulting mixture?

70. The rate of a motorcycle is 38 mph faster than the rate of a bicycle. The motorcycle travels 165 mi in the same amount of time as the bicycle travels 51 mi. Find the rate of the bicycle.

Solutions to Chapter 1 You Try It

Section 1.1 *pages 3–8*

You Try It 1 500,000 + 7000 + 200 + 4

You Try It 2

$$\begin{array}{r} {}^{1} \\ 47 \\ +\,94 \\ \hline 141 \end{array}$$

You Try It 3 $x + y + z$

1692 + 4783 + 5046

$$\begin{array}{r} {}^{1\,2\,1} \\ 1692 \\ 4783 \\ +\,5046 \\ \hline 11{,}521 \end{array}$$

You Try It 4

$$\begin{array}{r} {}^{1\;16\;6\;13} \\ \not{2}\not{6}\not{7}\not{3} \\ -\,814 \\ \hline 1859 \end{array}$$

You Try It 5 $x - y$

7061 − 3229

$$\begin{array}{r} {}^{6\;10\;5\;11} \\ \not{7}\not{0}\not{6}\not{1} \\ -\,3229 \\ \hline 3832 \end{array}$$

You Try It 6

Strategy

To find the price, replace C by 148 and M by 74 in the given formula and solve for P.

Solution

$P = C + M$

$P = 148 + 74$

$P = 222$

The price of the leather jacket is $222.

Section 1.2 *pages 13–20*

You Try It 1

$$\begin{array}{r} 657 \\ \times\,408 \\ \hline 5256 \\ 26280 \\ \hline 268{,}056 \end{array}$$

You Try It 2 $5xy$

$$\begin{aligned} 5(20)(60) &= 100(60) \\ &= 6000 \end{aligned}$$

You Try It 3 $2 \cdot 2 \cdot 2 \cdot 3 \cdot 3 \cdot 3 \cdot 3 = 2^3 \cdot 3^4$

You Try It 4 x^4y^2

$$\begin{aligned} 1^4 \cdot 3^2 &= (1 \cdot 1 \cdot 1 \cdot 1) \cdot (3 \cdot 3) \\ &= 1 \cdot 9 \\ &= 9 \end{aligned}$$

You Try It 5 $\dfrac{x}{y}$

$\dfrac{672}{8}$

$$\begin{array}{r} 84 \\ 8\overline{)672} \\ -64 \\ \hline 32 \\ -32 \\ \hline 0 \end{array}$$

You Try It 6

$30 \div 1 = 30$

$30 \div 2 = 15$

$30 \div 3 = 10$

$30 \div 4$ Does not divide evenly.

$30 \div 5 = 6$

$30 \div 6 = 5$ The factors are repeating.

The factors of 30 are 1, 2, 3, 5, 6, 10, 15, and 30.

You Try It 7

Strategy

To find the speed, replace d by 486 and t by 9 in the given formula and solve for r.

Solution

$r = \dfrac{d}{t}$

$r = \dfrac{486}{9}$

$r = 54$

You would need to travel at 54 mph.

Section 1.3 *pages 25–26*

You Try It 1

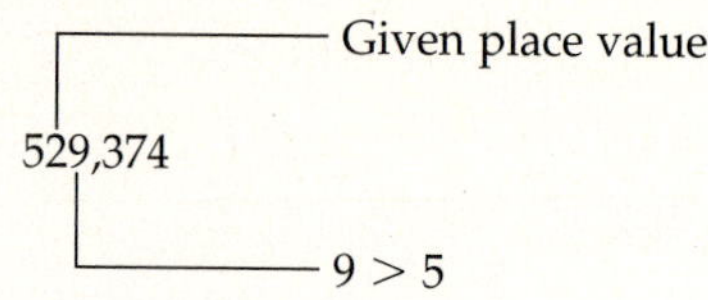

529,374 rounded to the nearest ten-thousand is 530,000.

You Try It 2

216,936 ⟶ 200,000
207 ⟶ 200

$200{,}000 \div 200 = 1000$

Section 1.4 *pages 29–32*

You Try It 1

$$\begin{array}{r} 11 \\ 2\overline{)22} \\ 2\overline{)44} \\ 2\overline{)88} \end{array}$$

$88 = 2 \cdot 2 \cdot 2 \cdot 11 = 2^3 \cdot 11$

You Try It 2

$16 = \textcircled{2^4}$

$24 = 2^3 \cdot \textcircled{3}$

$28 = 2^2 \cdot \textcircled{7}$

The LCM $= 2^4 \cdot 3 \cdot 7 = 16 \cdot 3 \cdot 7 = 336.$

You Try It 3

$25 = 5^2$

$52 = 2^2 \cdot 13$

No prime factor occurs in both factorizations. The GCF is 1.

You Try It 4

$32 = 2^5$

$40 = \textcircled{2^3} \cdot 5$

$56 = 2^3 \cdot 7$

The GCF $= 2^3 = 8.$

You Try It 5

$$\begin{aligned} &(a - b)^2 + 5c \\ &(7 - 2)^2 + 5(4) = 5^2 + 5(4) \\ &\qquad = 25 + 5(4) \\ &\qquad = 25 + 20 \\ &\qquad = 45 \end{aligned}$$

Solutions to Chapter 2 You Try It

Section 2.1 *pages 43–48*

You Try It 1

−3 is 4 units to the left of 1.

You Try It 2

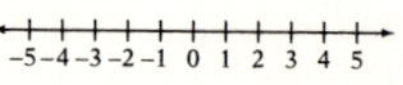

A is −5, and C is −3.

You Try It 3

a. 2 is the right of −5 on the number line.

$2 > -5$

b. −4 is to the left of 3 on the number line.

$-4 < 3$

You Try It 4

−7, −1, 0, 4, 8

You Try It 5 **a.** −24 **b.** 13 **c.** b

You Try It 6

a. negative three minus twelve

b. eight plus negative five

You Try It 7 **a.** $-(-59) = 59$ **b.** $-(y) = -y$

You Try It 8 **a.** $|-8| = 8$ **b.** $|12| = 12$

You Try It 9 **a.** $|0| = 0$ **b.** $-|35| = -35$

You Try It 10 $|-y| = |-2| = 2$

You Try It 11

$|-3| = 3, |5| = 5$

$3 < 5$

$|-3| < |5|$

You Try It 12

$|6| = 6, |-2| = 2, -(-1) = 1, -|-8| = -8$

$-8, -4, 1, 2, 6$

$-|-8|, -4, -(-1), |-2|, |6|$

You Try It 13

Strategy

To determine which is closer to blast-off, find the absolute value of each number. The number with the smaller absolute value is closer to zero and, therefore, closer to blast-off.

Solution

$|-9| = 9, |-7| = 7$

$7 < 9$

-7 s and counting is closer to blast-off than -9 s and counting.

Section 2.2 *pages 53–58*

You Try It 1

$$\begin{aligned} -36 + 17 + (-21) &= -19 + (-21) \\ &= -40 \end{aligned}$$

You Try It 2

$$-154 + (-37) = -191$$

You Try It 3

$$\begin{aligned} &-x + y \\ -(-3) + (-10) &= 3 + (-10) \\ &= -7 \end{aligned}$$

You Try It 4

$$\begin{aligned} -8 - 14 &= -8 + (-14) \\ &= -22 \end{aligned}$$

You Try It 5

$$\begin{aligned} &-4 - (-3) + 12 - (-7) - 20 \\ &= -4 + 3 + 12 + 7 + (-20) \\ &= -1 + 12 + 7 + (-20) \\ &= 11 + 7 + (-20) \\ &= 18 + (-20) \\ &= -2 \end{aligned}$$

You Try It 6

$$\begin{aligned} &x - y \\ -9 - 7 &= -9 + (-7) \\ &= -16 \end{aligned}$$

You Try It 7

Strategy

To find the temperature, add the increase (10) to the previous temperature (-3).

Solution

$-3 + 10 = 7$

The temperature is 7°C.

You Try It 8

Strategy

To find the difference, subtract the lower temperature (-70) from the higher temperature (57).

Solution

$$\begin{aligned} 57 - (-70) &= 57 + 70 \\ &= 127 \end{aligned}$$

The difference between the average temperatures is 127°F.

You Try It 9

Strategy

To find d, replace a by -6 and b by 5 in the given formula and solve for d.

Solution

$$\begin{aligned} d &= |a - b| \\ d &= |-6 - 5| \\ d &= |-11| \\ d &= 11 \end{aligned}$$

The distance between the two points is 11 units.

Section 2.3 *pages 65–68*

You Try It 1 $-3 \cdot 12 = -36$

You Try It 2 $-38 \cdot 51 = -1938$

You Try It 3 $-7(-8)(9)(-2) = 56(9)(-2) = 504(-2) = -1008$

You Try It 4 $-9y$
$-9(20) = -180$

You Try It 5 $(-135) \div (-9) = 15$

You Try It 6 $0 \div (-17) = 0$

You Try It 7 $\frac{84}{-6} = -14$

You Try It 8 Any number divided by one equals the number.

$x \div 1 = x$

You Try It 9 $\frac{a}{-b}$

$\frac{-14}{-(-7)} = \frac{-14}{7} = -2$

You Try It 10

Strategy

To find the melting point of argon, multiply the melting point of mercury (-38) by 5.

Solution

$5(-38) = -190$

The melting point of argon is -190°C.

Section 2.4 *pages 73–74*

You Try It 1 $(-5)^2 = (-5)(-5) = 25$

$-5^2 = -(5 \cdot 5) = -25$

You Try It 2 $8 \div 4 \cdot 4 - (-2)^2 = 8 \div 4 \cdot 4 - 4 = 2 \cdot 4 - 4 = 8 - 4 = 4$

You Try It 3 $(-2)^2(3 - 7)^2 - (-16) \div (-4) = (-2)^2(-4)^2 - (-16) \div (-4) = (4)(16) - (-16) \div (-4) = 64 - (-16) \div (-4) = 64 - 4 = 60$

You Try It 4 $3a - 4b$

$3(-2) - 4(5) = -6 - 4(5) = -6 - 20 = -6 + (-20) = -26$

Solutions to Chapter 3 You Try It

Section 3.1 *pages 83–88*

You Try It 1 $3\overline{)26}$ quotient 8, -24, remainder 2

$\frac{26}{3} = 8\frac{2}{3}$

You Try It 2 $9\frac{4}{7} = \frac{(7 \cdot 9) + 4}{7} = \frac{63 + 4}{7} = \frac{67}{7}$

You Try It 3 $48 \div 8 = 6$

$\frac{5}{8} = \frac{5 \cdot 6}{8 \cdot 6} = \frac{30}{48}$

$\frac{30}{48}$ is equivalent to $\frac{5}{8}$.

You Try It 4 $8 = \frac{8}{1}$ $\quad 12 \div 1 = 12$

$8 = \frac{8}{1} = \frac{8 \cdot 12}{1 \cdot 12} = \frac{96}{12}$

$\frac{96}{12}$ is equivalent to 8.

You Try It 5 $\frac{21}{84} = \frac{\overset{1}{\cancel{3}} \cdot \overset{1}{\cancel{7}}}{2 \cdot 2 \cdot \underset{1}{\cancel{3}} \cdot \underset{1}{\cancel{7}}} = \frac{1}{4}$

You Try It 6 $\frac{4}{9} = \frac{28}{63}$ $\quad \frac{8}{21} = \frac{24}{63}$

$\frac{28}{63} > \frac{24}{63}$

$\frac{4}{9} > \frac{8}{21}$

Section 3.2 *pages 91–104*

You Try It 1 $\frac{3}{5}+\frac{2}{3}+\frac{5}{6}=\frac{18}{30}+\frac{20}{30}+\frac{25}{30}=\frac{63}{30}$

$=2\frac{3}{30}=2\frac{1}{10}$

You Try It 2 $x+y+z$

$-\frac{5}{12}+\frac{5}{8}+\left(-\frac{1}{6}\right)=\frac{-5}{12}+\frac{5}{8}+\frac{-1}{6}$

$=\frac{-10}{24}+\frac{15}{24}+\frac{-4}{24}$

$=\frac{-10+15+(-4)}{24}$

$=\frac{1}{24}$

You Try It 3 $\frac{7}{12}-\frac{4}{9}=\frac{21}{36}-\frac{16}{36}=\frac{21-16}{36}=\frac{5}{36}$

You Try It 4 $x-y$

$-\frac{5}{6}-\left(\frac{-7}{9}\right)=\frac{-5}{6}-\left(\frac{-7}{9}\right)$

$=\frac{-15}{18}-\frac{-14}{18}$

$=\frac{-15-(-14)}{18}$

$=\frac{-15+14}{18}$

$=\frac{-1}{18}=-\frac{1}{18}$

You Try It 5 $-\frac{1}{3}\left(-\frac{5}{12}\right)\left(\frac{8}{15}\right)=\frac{1}{3}\cdot\frac{5}{12}\cdot\frac{8}{15}$

$=\frac{1\cdot 5\cdot 8}{3\cdot 12\cdot 15}$

$=\frac{1\cdot 5\cdot 2\cdot 2\cdot 2}{3\cdot 2\cdot 2\cdot 3\cdot 3\cdot 5}$

$=\frac{2}{27}$

You Try It 6 $\frac{8}{9}\cdot 6=\frac{8}{9}\cdot\frac{6}{1}=\frac{8\cdot 6}{9\cdot 1}$

$=\frac{2\cdot 2\cdot 2\cdot 2\cdot 3}{3\cdot 3\cdot 1}=\frac{16}{3}=5\frac{1}{3}$

You Try It 7 xy

$5\frac{1}{8}\cdot\frac{2}{3}=\frac{41}{8}\cdot\frac{2}{3}$

$=\frac{41\cdot 2}{8\cdot 3}$

$=\frac{41\cdot 2}{2\cdot 2\cdot 2\cdot 3}$

$=\frac{41}{12}=3\frac{5}{12}$

You Try It 8 $4\div\left(-\frac{6}{7}\right)=-\left(\frac{4}{1}\div\frac{6}{7}\right)$

$=-\left(\frac{4}{1}\cdot\frac{7}{6}\right)$

$=-\frac{4\cdot 7}{1\cdot 6}$

$=-\frac{2\cdot 2\cdot 7}{1\cdot 2\cdot 3}=-\frac{14}{3}=-4\frac{2}{3}$

You Try It 9 $\frac{x}{8}\div\frac{y}{6}=\frac{x}{8}\cdot\frac{6}{y}$

$=\frac{x\cdot 6}{8\cdot y}=\frac{x\cdot 2\cdot 3}{2\cdot 2\cdot 2\cdot y}=\frac{3x}{4y}$

You Try It 10 $x\div y$

$2\frac{1}{4}\div 9=\frac{9}{4}\div\frac{9}{1}=\frac{9}{4}\cdot\frac{1}{9}=\frac{9\cdot 1}{4\cdot 9}$

$=\frac{3\cdot 3\cdot 1}{2\cdot 2\cdot 3\cdot 3}=\frac{1}{4}$

You Try It 11

Strategy

To find the total cost of the material:

- Multiply the amount of material per sash $\left(1\frac{3}{8}\right)$ by the number of sashes (22) to find the total number of yards of material needed.
- Multiply the total number of yards of material needed by the cost per yard (8).

Solution

$$1\frac{3}{8} \cdot 22 = \frac{11}{8} \cdot \frac{22}{1} = \frac{11 \cdot 22}{8 \cdot 1} = \frac{11 \cdot 2 \cdot 11}{2 \cdot 2 \cdot 2 \cdot 1}$$
$$= \frac{121}{4} = 30\frac{1}{4}$$

$$30\frac{1}{4} \cdot 8 = \frac{121}{4} \cdot \frac{8}{1} = \frac{121 \cdot 8}{4 \cdot 1}$$
$$= \frac{11 \cdot 11 \cdot 2 \cdot 2 \cdot 2}{2 \cdot 2 \cdot 1} = 242$$

The total cost of the material is $242.

You Try It 12

Strategy

To find the Celsius temperature, replace F by 68 in the given formula and solve for C.

Solution

$$C = \frac{5}{9}(F - 32)$$

$$C = \frac{5}{9}(68 - 32) = \frac{5}{9}(36) = \frac{5}{9} \cdot \frac{36}{1} = \frac{5 \cdot 36}{9 \cdot 1} = 20$$

The Celsius temperature is 20°.

You Try It 13

Strategy

- To express $\frac{1}{3}$ as the sum of two other unit fractions, replace n with 3 in the given formula and simplify.
- To verify the results, find the sum of the two unit fractions found.

Solution

$$\frac{1}{n} = \frac{1}{n + 1} + \frac{1}{n(n + 1)}$$

$$\frac{1}{3} = \frac{1}{3 + 1} + \frac{1}{3(3 + 1)}$$

$$\frac{1}{3} = \frac{1}{4} + \frac{1}{12}$$

$$\frac{1}{4} + \frac{1}{12} = \frac{3}{12} + \frac{1}{12} = \frac{4}{12} = \frac{1}{3}$$

The unit fraction $\frac{1}{3}$ is equal to the sum of the unit fractions $\frac{1}{4}$ and $\frac{1}{12}$.

Section 3.3 *pages 113–116*

You Try It 1

$$x^4y^3$$
$$\left(2\frac{1}{3}\right)^4 \cdot \left(\frac{3}{7}\right)^3 = \left(\frac{7}{3}\right)^4 \cdot \left(\frac{3}{7}\right)^3$$
$$= \frac{7}{3} \cdot \frac{7}{3} \cdot \frac{7}{3} \cdot \frac{7}{3} \cdot \frac{3}{7} \cdot \frac{3}{7} \cdot \frac{3}{7}$$
$$= \frac{7 \cdot 7 \cdot 7 \cdot 7 \cdot 3 \cdot 3 \cdot 3}{3 \cdot 3 \cdot 3 \cdot 3 \cdot 7 \cdot 7 \cdot 7} = \frac{7}{3} = 2\frac{1}{3}$$

You Try It 2

$$\frac{x}{y - z}$$
$$\frac{-\frac{4}{9}}{3 - \frac{1}{3}} = \frac{-\frac{4}{9}}{\frac{8}{3}} = -\frac{4}{9} \div \frac{8}{3} = -\frac{4}{9} \cdot \frac{3}{8}$$
$$= -\frac{1}{6}$$

You Try It 3 $\left(-\frac{1}{2}\right)^3 \cdot \frac{7-3}{4-9} + \frac{4}{5}$

$$= \left(-\frac{1}{2}\right)^3 \cdot \frac{4}{-5} + \frac{4}{5}$$

$$= -\frac{1}{8} \cdot \frac{4}{-5} + \frac{4}{5}$$

$$= \frac{1}{10} + \frac{4}{5} = \frac{9}{10}$$

Solutions to Chapter 4 You Try It

Section 4.1 *pages 127–132*

You Try It 1 The digit 4 is in the thousandths place.

You Try It 2 $\frac{501}{1000} = 0.501$ [five hundred one thousandths]

You Try It 3 $0.67 = \frac{67}{100}$ [sixty-seven hundredths]

You Try It 4 fifty-five and six thousand eighty-three ten-thousandths

You Try It 5 806.00491

You Try It 6 $0.065 = 0.0650$

650 ten-thousandths < 802 ten-thousandths

$0.0650 < 0.0802$

$0.065 < 0.0802$

You Try It 7 3.03, 0.33, 0.30, 3.30, 0.03

0.03, 0.30, 0.33, 3.03, 3.30

0.03, 0.3, 0.33, 3.03, 3.3

You Try It 8

3.675849 — Given place value: 5; $4 < 5$

3.675849 rounded to the nearest ten-thousandth is 3.6758.

You Try It 9

48.907 — Given place value: 9; $0 < 5$

48.907 rounded to the nearest tenth is 48.9.

You Try It 10

31.8652 — Given place value: 1; $8 > 5$

31.8652 rounded to the nearest whole number is 32.

You Try It 11

Strategy

To determine who had the higher score, compare the numbers 9.675 and 9.725.

Solution

$9.675 < 9.725$

Patrick Kirkey had the higher score.

You Try It 12

Strategy

To find the time, round 4.74 to the nearest tenth.

Solution

4.74 rounded to the nearest tenth is 4.7.

To the nearest tenth of a second, it takes an object 4.7 s to descend the length of the pipe.

Section 4.2 *pages 137–148*

You Try It 1

$$\begin{array}{r} {}^{1\,1\,1} \\ 8.64 \\ 52.7 \\ +\ \ 0.39105 \\ \hline 61.73105 \end{array}$$

You Try It 2

$$\begin{array}{r} {}^{4\ 9\ 10} \\ 2\not{5}.\not{0}\not{0} \\ -\ \ 4.91 \\ \hline 20.09 \end{array} \qquad \text{Check: } \begin{array}{r} 4.91 \\ +\ 20.09 \\ \hline 25.00 \end{array}$$

You Try It 3 $-41.65 + 29.303 = -12.347$

You Try It 4 $4.002 - 9.378 = 4.002 + (-9.378)$
$= -5.376$

You Try It 5

$x + y + z$
$-7.84 + (-3.05) + 2.19$
$= -10.89 + 2.19$
$= -8.7$

You Try It 6

$$\begin{array}{lr} 6.514 \longrightarrow & 7 \\ 8.903 \longrightarrow & 9 \\ 2.275 \longrightarrow & +\ 2 \\ \hline & 18 \end{array}$$

You Try It 7

$$\begin{array}{lr} 487.52 \longrightarrow & 500 \\ 61.903 \longrightarrow & -\ 60 \\ \hline & 440 \end{array}$$

You Try It 8

$$\begin{array}{r} 44.356 \\ \times\ \ 0.72 \\ \hline 88712 \\ 310492 \\ \hline 31.93632 \end{array}$$

You Try It 9

$$\begin{array}{lr} 6.407 \longrightarrow & 6 \\ 0.959 \longrightarrow & \times\ 1 \\ \hline & 6 \end{array}$$

You Try It 10 $835.294 \cdot 1000 = 835{,}294$

You Try It 11

$$\begin{array}{r} 6.0391 \approx 6.039 \\ 86\overline{)519.3700} \\ -\,516 \\ \hline 3\,3 \\ -\ \ \ 0 \\ \hline 337 \\ -\,258 \\ \hline 790 \\ -\,774 \\ \hline 160 \\ -\ \ 86 \\ \hline 74 \end{array}$$

You Try It 12 $592.4 \div 10^4 = 0.05924$

You Try It 13 The quotient is negative.

$-25.7 \div 0.31 \approx -82.9$

You Try It 14 $(-0.7)(-5.8) = 4.06$

You Try It 15

$25xy$
$25(-0.8)(0.6) = -20(0.6) = -12$

You Try It 16

$\frac{x}{y}$

$\frac{-40.6}{-0.7} = -40.6 \div (-0.7) = 58$

You Try It 17

Strategy

To find the change you receive:

- ▶ Multiply the number of stamps (12) by the cost of each stamp (32¢) to find the total cost of the stamps.
- ▶ Convert the total cost of the stamps to dollars and cents.
- ▶ Subtract the total cost of the stamps from \$5.

Solution

$12(32) = 384$ The stamps cost 384¢.

$384¢ = \$3.84$ The stamps cost \$3.84.

$$\begin{array}{r} 5.00 \\ -3.84 \\ \hline 1.16 \end{array}$$

You receive \$1.16 in change.

You Try It 18

Strategy

To find the equity, replace V by 240,000 and L by 142,976.80 in the given formula and solve for E.

Solution

$E = V - L$
$E = 240{,}000 - 142{,}976.80$
$E = 97{,}023.20$

The equity on the home is \$97,023.20.

You Try It 19

Strategy

To find the profit:

- ▶ Divide the number of pounds per 100-pound container (100) by the number of pounds packaged in each bag (2) to find the number of bags sold.
- ▶ Multiply the number of bags sold by the selling price per bag (8.50) to find the income from selling the nuts.
- ▶ Substract the cost (325) from the income.

Solution

$100 \div 2 = 50$ Each container makes 50 bags of nuts.

$50(8.50) = 425$ The income from 50 bags is \$425.

$425 - 325 = 100$

The profit is \$100.

You Try It 20

Strategy

To find the insurance premium due, replace B by 276.25 and F by 1.8 in the given formula and solve for P.

Solution

$P = BF$
$P = 276.25(1.8)$
$P = 497.25$

The insurance premium due is \$497.25.

Section 4.3 *pages 157–160*

You Try It 1 $\begin{array}{r} 0.8 \\ 5\overline{)4.0} \end{array}$ $\quad \frac{4}{5} = 0.8$

You Try It 2 $\begin{array}{r} 0.666 \\ 3\overline{)2.000} \end{array} = 0.\overline{6}$ $\left[\text{Write } \frac{2}{3} \text{ as a decimal.}\right]$

$1\frac{2}{3} = 1.\overline{6}$

You Try It 3 $6.2 = 6\frac{2}{10} = 6\frac{1}{5}$

You Try It 4 $\frac{7}{12} \approx 0.5833$
$0.5880 > 0.5833$
$0.588 > \frac{7}{12}$

You Try It 5 $110\% = 110\left(\frac{1}{100}\right) = \frac{110}{100} = 1\frac{1}{10}$
$110\% = 110(0.01) = 1.10$

You Try It 6 $0.8\% = 0.8(0.01) = 0.008$

You Try It 7 $0.038 = 0.038(100\%) = 3.8\%$

You Try It 8 $1.05 = 1.05(100\%) = 105\%$

You Try It 9 $\frac{9}{7} = \frac{9}{7}(100\%) = \frac{900}{7}\% = 128\frac{4}{7}\%$

You Try It 10 $1\frac{5}{9} = \frac{14}{9} = \frac{14}{9}(100\%) = \frac{1400}{9}\% \approx 155.6\%$

Section 4.4 *pages 165–170*

You Try It 1 −5 −4 −3 −2 −1 0 1 2 3 4 5

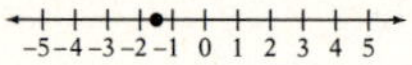

You Try It 2 −5 −4 −3 −2 −1 0 1 2 3 4 5

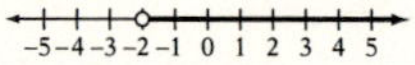

You Try It 3 −5 −4 −3 −2 −1 0 1 2 3 4 5

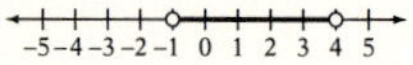

You Try It 4

a. $x \geq 4$
$-1 \geq 4$ False

b. $x \geq 4$
$0 \geq 4$ False

c. $x \geq 4$
$4 \geq 4$ True

d. $x \geq 4$
$6 \geq 4$ True

The numbers 4 and 6 make the inequality true.

You Try It 5 All real numbers greater than -7 make the inequality $x > -7$ true.

You Try It 6 

You Try It 7

Strategy

- ▶ To write the inequality, let s represent the minimum speed at which a motorist is ticketed. Motorists are ticketed at speeds greater than 55.
- ▶ To determine if a motorist traveling at 58 mph will be ticketed, replace s in the inequality by 58. If the inequality is true, the motorist will be ticketed. If the inequality is false, the motorist will not be ticketed.

Solution

$s > 55$

$58 > 55$ True

Yes, a motorist traveling at 58 mph will be ticketed.

Solutions to Chapter 5 You Try It

Section 5.1 *pages 181–190*

You Try It 1

$$\begin{aligned} 12a^2 - 8a + 3 - 16a^2 + 8a &= 12a^2 - 16a^2 - 8a + 8a + 3 \\ &= -4a^2 + 0a + 3 \\ &= -4a^2 + 3 \end{aligned}$$

You Try It 2

$$\begin{aligned} &8x^2y - 15xy^2 + 12xy^2 - 7x^2y \\ &\quad = 8x^2y - 7x^2y - 15xy^2 + 12xy^2 \\ &\quad = x^2y - 3xy^2 \end{aligned}$$

You Try It 3

$$\begin{aligned} -6(-3p) &= [-6(-3)]p \\ &= 18p \end{aligned}$$

You Try It 4

$$\begin{aligned} (-2m)(-8n) &= [(-2)(-8)](m \cdot n) \\ &= 16mn \end{aligned}$$

You Try It 5

$$\begin{aligned} \left(\frac{2}{5}a\right)\left(-\frac{3}{4}a\right) &= \left[\frac{2}{5}\left(-\frac{3}{4}\right)\right](a \cdot a) \\ &= -\frac{3}{10}a^2 \end{aligned}$$

You Try It 6

$$\begin{aligned} -7(2k - 5) &= -7(2k) - (-7)5 \\ &= -14k + 35 \end{aligned}$$

You Try It 7

$$-4(x - 2y) = (-4)(x) - (-4)(2y)$$
$$= -4x + 8y$$

You Try It 8

$$3(-2v + 3w - 7) = 3(-2v) + 3(3w) - 3(7)$$
$$= -6v + 9w - 21$$

You Try It 9

$$-4z(2z - 7) = -4z(2z) - (-4z)(7)$$
$$= (-4 \cdot 2)(z \cdot z) - (-4 \cdot 7)z$$
$$= -8z^2 - (-28)z$$
$$= -8z^2 + 28z$$

You Try It 10

$$-(c - 9d + 1) = -c + 9d - 1$$

You Try It 11

$$6 - 4(2x - y) + 3(x - 4y)$$
$$= 6 - 8x + 4y + 3x - 12y$$
$$= -5x - 8y + 6$$

You Try It 12

$$8c - 4(3c - 8) - 5(c + 4)$$
$$= 8c - 12c + 32 - 5c - 20$$
$$= -9c + 12$$

You Try It 13

$$6p + 5[3(2 - 3p) - 2(5 - 4p)]$$
$$= 6p + 5[6 - 9p - 10 + 8p]$$
$$= 6p + 5[-p - 4]$$
$$= 6p - 5p - 20$$
$$= p - 20$$

Section 5.2 *pages 197–208*

You Try It 1

Strategy

To find the perimeter, use the formula for the perimeter of a rectangle. Substitute 11 for L and $8\frac{1}{2}$ for W and solve for P.

Solution

$$P = 2L + 2W$$
$$P = 2(11) + 2\left(8\frac{1}{2}\right)$$
$$P = 2(11) + 2\left(\frac{17}{2}\right)$$
$$P = 22 + 17$$
$$P = 39$$

The perimeter of a standard sheet of typing paper is 39 in.

You Try It 2

Strategy

To find the circumference, use the formula that gives the circumference in terms of the diameter. Leave the answer in terms of π.

Solution

$$C = \pi d$$
$$C = \pi(9)$$
$$C = 9\pi$$

The circumference is 9π in.

You Try It 3

Strategy

To find the number of rolls of wallpaper to be purchased:

- ▶ Use the formula for the area of a rectangle to find the area of one wall.
- ▶ Multiply the area of one wall by the number of walls to be covered (2).
- ▶ Divide the area of wall to be covered by the area one roll of wallpaper will cover (30).

Solution

$A = LW$

$A = 12 \cdot 8 = 96$ The area of one wall is 96 ft^2.

$2(96) = 192$ The area of the two walls is 192 ft^2.

$192 \div 30 = 6.4$

Because a portion of a seventh roll is needed, 7 rolls of wallpaper should be purchased.

You Try It 4

Strategy

To find the area, use the formula for the area of a circle. An approximation is asked for; use the π key on a calculator. $r = 11$.

Solution

$$A = \pi r^2$$
$$A = \pi(11)^2$$
$$A = 121\pi$$
$$A \approx 380.13$$

The area is approximately 380.13 cm^2.

Section 5.3 *pages 215–220*

You Try It 1

Strategy

To find the volume, use the formula for the volume of a cube. $s = 2.5$.

Solution

$V = s^3$
$V = (2.5)^3 = 15.625$

The volume of the cube is 15.625 m^3.

You Try It 2

Strategy

To find the volume:

- Find the radius of the base of the cylinder. $d = 8$.
- Use the formula for the volume of a cylinder. Leave the answer in terms of π.

Solution

$r = \frac{1}{2}d = \frac{1}{2}(8) = 4$
$V = \pi r^2 h = \pi(4)^2(22) = \pi(16)(22) = 352\pi$

The volume of the cylinder is 352π ft^3.

You Try It 3

Strategy

To find the surface area of the cylinder:

- Find the radius of the base of the cylinder. $d = 6$.
- Use the formula for the surface area of a cylinder. An approximation is asked for; use the π key on a calculator.

Solution

$r = \frac{1}{2}d = \frac{1}{2}(6) = 3$

$$\begin{aligned} S &= 2\pi r^2 + 2\pi rh \\ S &= 2\pi(3)^2 + 2\pi(3)(8) \\ &= 2\pi(9) + 2\pi(3)(8) \\ &= 18\pi + 48\pi \\ &= 66\pi \\ &\approx 207.35 \end{aligned}$$

The surface area of the cylinder is approximately 207.35 ft^2.

You Try It 4

Strategy

To find which solid has the larger surface area:

- Use the formula for the surface area of a cube. $s = 10$.
- Find the radius of the sphere. $d = 8$.
- Use the formula for the surface area of a sphere. Since this number is to be compared to another number, use the π key on a calculator to approximate the surface area.
- Compare the two areas.

Solution

$S = 6s^2$
$S = 6(10)^2 = 6(100) = 600$
The surface area of the cube is 600 cm^2.

$r = \frac{1}{2}d = \frac{1}{2}(8) = 4$
$S = 4\pi r^2$
$S = 4\pi(4)^2 = 4\pi(16) = 64\pi \approx 201.06$
The surface area of the sphere is 201.06 cm^2.

$600 > 201.06$

The cube has a larger surface area than the sphere.

Section 5.4 *pages 225–228*

You Try It 1

twice x divided by the difference between x and 7

$\frac{2x}{x - 7}$

You Try It 2

the product of negative three and the square of d

$-3d^2$

You Try It 3

Let the smaller number be x.
The larger number is $16 - x$.

the difference between the larger number and twice the smaller number

$16 - x - 2x$
$16 - 3x$

You Try It 4

the difference between fourteen and the sum of a number and seven

Let the unknown number be n.

$14 - (n + 7)$
$14 - n - 7$
$-n + 7$

You Try It 5

pounds of caramel: x
pounds of milk chocolate: $x + 3$

Solutions to Chapter 6 You Try It

Section 6.1 *pages 239–244*

You Try It 1

$$4 - 3x = 6x + 10$$

$$\begin{array}{c|c} 4 - 3\left(-\frac{2}{3}\right) & 6\left(-\frac{2}{3}\right) + 10 \\ 4 - (-2) & (-4) + 10 \end{array}$$

$$6 = 6$$

Yes, $-\frac{2}{3}$ is a solution of $4 - 3x = 6x + 10$.

You Try It 2

$$\begin{aligned} 7 + y &= 12 \\ 7 - 7 + y &= 12 - 7 \\ y &= 5 \end{aligned}$$

The solution is 5.

You Try It 3

$$\begin{aligned} -5r + 3 + 6r &= 1 \\ r + 3 &= 1 \\ r + 3 - 3 &= 1 - 3 \\ r &= -2 \end{aligned}$$

The solution is -2.

You Try It 4

$$\begin{aligned} 10 &= \frac{-2x}{5} \\ \left(-\frac{5}{2}\right)10 &= \left(-\frac{5}{2}\right)\left(-\frac{2}{5}x\right) \\ -25 &= x \end{aligned}$$

• $-\frac{2x}{5} = -\frac{2}{5}x$

The solution is -25.

You Try It 5

$$\begin{aligned} \frac{1}{3}x - \frac{5}{6}x &= 4 \\ \frac{2}{6}x - \frac{5}{6}x &= 4 \\ -\frac{1}{2}x &= 4 \\ -2\left(-\frac{1}{2}x\right) &= -2(4) \\ x &= -8 \end{aligned}$$

• $\frac{2}{6} - \frac{5}{6} = -\frac{3}{6} = -\frac{1}{2}$

Check:

$$\frac{1}{3}x - \frac{5}{6}x = 4$$

$$\begin{array}{c|c} \frac{1}{3}(-8) - \frac{5}{6}(-8) & 4 \\ -\frac{8}{3} - \left(-\frac{20}{3}\right) & 4 \\ -\frac{8}{3} + \frac{20}{3} & 4 \end{array}$$

$$4 = 4$$

-8 checks as the solution. The solution is -8.

Section 6.2 *pages 249–256*

You Try It 1

Strategy

To find the amount, solve the basic percent equation.

Percent $= 33\frac{1}{3}\% = \frac{1}{3}$, base $= 45$, amount $= A$

Solution

Percent · base = amount

$$\begin{aligned} \frac{1}{3}(45) &= A \\ 15 &= A \end{aligned}$$

15 is $33\frac{1}{3}\%$ of 45.

You Try It 2

Strategy

To find the percent, solve the basic percent equation.

Percent $= P$, base $= 40$, amount $= 25$

Solution

Percent · base = amount

$$\begin{aligned} P \cdot 40 &= 25 \\ P &= \frac{25}{40} = 0.625 \\ P &= 62.5\% \end{aligned}$$

25 is 62.5% of 40.

You Try It 3

Strategy

To find the base, solve the basic percent equation.

Percent = $16\frac{2}{3}\% = \frac{1}{6}$, base = B, amount = 15

Solution

Percent · base = amount

$$\frac{1}{6} \cdot B = 15$$
$$B = 15 \cdot 6$$
$$B = 90$$

$16\frac{2}{3}\%$ of 90 is 15.

You Try It 4

Strategy

To find the amount budgeted for food, solve the basic percent equation for the amount. Percent = 8% = 0.08, base = 26,000, amount = A

Solution

Percent · base = amount

$$0.08 \cdot 26{,}000 = A$$
$$2080 = A$$

The amount budgeted for food is $2080.

You Try It 5

Strategy

To find the percent, use the basic percent equation.
Percent = P, base = 2165, amount = 324.75

Solution

Percent · base = amount

$$P \cdot 2165 = 324.75$$
$$P = \frac{324.75}{2165}$$
$$P = 0.15$$

15% of the instructor's salary is deducted for income tax.

You Try It 6

Strategy

To find the number polled, solve the basic percent equation.
Percent = 70% = 0.70, base = B, amount = 210

Solution

Percent · base = amount

$$0.70 \cdot B = 210$$
$$B = \frac{210}{0.70}$$
$$B = 300$$

300 people were polled.

You Try It 7

Strategy

To find the increase in the hourly wage:

- ▶ Find last year's wage. Solve the basic percent equation. Percent = 115% = 1.15, base = B, amount = 20.01
- ▶ Subtract last year's wage from this year's wage.

Solution

Percent · base = amount

$$1.15 \cdot B = 20.01$$
$$B = \frac{20.01}{1.15}$$
$$B = 17.40$$

20.01 − 17.40 = 2.61

The increase in the hourly wage was $2.61.

You Try It 8

Strategy

To find the percent increase in mileage:

- ▶ Find the amount of increase in mileage.
- ▶ Solve the basic percent equation. Percent = P, base = 16.4, amount = amount of increase

Solution

17.2 − 16.4 = 0.8

Percent · base = amount

$$P \cdot 16.4 = 0.8$$
$$P = \frac{0.8}{16.4}$$
$$P \approx 0.049$$

The percent increase in mileage is 4.9%.

You Try It 9

Strategy

To find the value of the car:

- ▶ Solve the basic percent equation to find the amount of decrease in value. Percent = 24% = 0.24, base = 47,000, amount = A
- ▶ Subtract the amount of decrease from the cost.

Solution

Percent · base = amount

$$0.24 \cdot 47{,}000 = A$$
$$11{,}280 = A$$

47,000 − 11,280 = 35,720

The value of the car is $35,720.

Section 6.3 *pages 263–268*

You Try It 1

$$\begin{aligned} -5 - 4t &= 7 \\ -5 + 5 - 4t &= 7 + 5 \\ -4t &= 12 \\ \frac{-4t}{-4} &= \frac{12}{-4} \\ t &= -3 \end{aligned}$$

The solution is -3.

You Try It 2

$$\begin{aligned} 4m - 7 + m &= 8 \\ 5m - 7 &= 8 \\ 5m - 7 + 7 &= 8 + 7 \\ 5m &= 15 \\ \frac{5m}{5} &= \frac{15}{5} \\ m &= 3 \end{aligned}$$

The solution is 3.

You Try It 3

$$\begin{aligned} r - 7 &= 5 - 3r \\ r + 3r - 7 &= 5 - 3r + 3r \\ 4r - 7 &= 5 \\ 4r - 7 + 7 &= 5 + 7 \\ 4r &= 12 \\ \frac{4r}{4} &= \frac{12}{4} \\ r &= 3 \end{aligned}$$

The solution is 3.

You Try It 4

$$\begin{aligned} 4a - 2 + 5a &= 2a - 2 + 3a \\ 9a - 2 &= 5a - 2 \\ 9a - 5a - 2 &= 5a - 5a - 2 \\ 4a - 2 &= -2 \\ 4a - 2 + 2 &= -2 + 2 \\ 4a &= 0 \\ \frac{4a}{4} &= \frac{0}{4} \\ a &= 0 \end{aligned}$$

The solution is 0.

You Try It 5

$$\begin{aligned} 2w - 7(3w + 1) &= 5(5 - 3w) \\ 2w - 21w - 7 &= 25 - 15w \\ -19w - 7 &= 25 - 15w \\ -19w + 15w - 7 &= 25 - 15w + 15w \\ -4w - 7 &= 25 \\ -4w - 7 + 7 &= 25 + 7 \\ -4w &= 32 \\ \frac{-4w}{-4} &= \frac{32}{-4} \\ w &= -8 \end{aligned}$$

The solution is -8.

You Try It 6

the unknown number: x

six more than one-half a number	is	the total of the number and nine

$$\begin{aligned} \frac{1}{2}x + 6 &= x + 9 \\ \frac{1}{2}x - x + 6 &= x - x + 9 \\ -\frac{1}{2}x + 6 &= 9 \\ -\frac{1}{2}x + 6 - 6 &= 9 - 6 \\ -\frac{1}{2}x &= 3 \\ (-2)\left(-\frac{1}{2}x\right) &= (-2)3 \\ x &= -6 \end{aligned}$$

-6 checks as the solution.

The solution is -6.

You Try It 7

the unknown number: x

seven less than a number	is equal to	five more than three times the number

$$x - 7 = 3x + 5$$
$$x - 3x - 7 = 3x - 3x + 5$$
$$-2x - 7 = 5$$
$$-2x - 7 + 7 = 5 + 7$$
$$-2x = 12$$
$$\frac{-2x}{-2} = \frac{12}{-2}$$
$$x = -6$$

-6 checks as the solution.

The solution is -6.

You Try It 8

the smaller number: n
the larger number: $14 - n$

one more than three times the smaller number	equals	the sum of the larger number and three

$$3n + 1 = (14 - n) + 3$$
$$3n + 1 = 17 - n$$
$$3n + n + 1 = 17 - n + n$$
$$4n + 1 = 17$$
$$4n + 1 - 1 = 17 - 1$$
$$4n = 16$$
$$\frac{4n}{4} = \frac{16}{4}$$
$$n = 4$$

$14 - n = 14 - 4 = 10$

These numbers check as solutions.
The smaller number is 4.
The larger number is 10.

Section 6.4 *pages 275–282*

You Try It 1

$$AC = AB + BC$$
$$AC = \frac{1}{4}(BC) + BC$$
$$AC = \frac{1}{4}(16) + 16$$
$$AC = 4 + 16$$
$$AC = 20$$
$$AC = 20 \text{ ft}$$

You Try It 2

$$QR + RS + ST = QT$$
$$24 + RS + 17 = 62$$
$$41 + RS = 62$$
$$RS = 21$$
$$RS = 21 \text{ cm}$$

You Try It 3

Strategy

Supplementary angles are two angles whose sum is 180°. To find the supplement, let x represent the supplement of a 129° angle. Write an equation and solve for x.

Solution

$$x + 129° = 180°$$
$$x = 51°$$

The supplement of a 129° angle is a 51° angle.

You Try It 4

Strategy

To find the measure of $\angle a$, write an equation using the fact that the sum of the measure of $\angle a$ and 68° is 118°. Solve for $\angle a$.

Solution

$$\angle a + 68° = 118°$$
$$\angle a = 50°$$

The measure of $\angle a$ is 50°.

You Try It 5

Strategy

The angles labeled are adjacent angles of intersecting lines and are, therefore, supplementary angles. To find x, write an equation and solve for x.

Solution

$$(x + 16°) + 3x = 180°$$
$$4x + 16° = 180°$$
$$4x = 164°$$
$$x = 41°$$

You Try It 6

Strategy

$3x = y$ because corresponding angles have the same measure. $y + (x + 40°) = 180°$ because adjacent angles of intersecting lines are supplementary angles. Substitute $3x$ for y and solve for x.

Solution

$$3x + (x + 40°) = 180°$$
$$4x + 40° = 180°$$
$$4x = 140°$$
$$x = 35°$$

You Try It 7

Strategy

- To find the measure of angle b, use the fact that $\angle b$ and $\angle x$ are supplementary angles.
- To find the measure of angle c, use the fact that the sum of the interior angles of a triangle is 180°.
- To find the measure of angle y, use the fact that $\angle c$ and $\angle y$ are vertical angles.

Solution

$$\angle b + \angle x = 180^\circ$$
$$\angle b + 100^\circ = 180$$
$$\angle b = 80^\circ$$

$$\angle a + \angle b + \angle c = 180^\circ$$
$$45^\circ + 80^\circ + \angle c = 180^\circ$$
$$125^\circ + \angle c = 180^\circ$$
$$\angle c = 55^\circ$$

$$\angle y = \angle c = 55^\circ$$

You Try It 8

Strategy

To find the measure of the third angle, use the facts that the measure of a right angle is 90° and the sum of the measures of the interior angles of a triangle is 180°. Write an equation using x to represent the measure of the third angle. Solve the equation for x.

Solution

$$x + 90^\circ + 34^\circ = 180^\circ$$
$$x + 124^\circ = 180^\circ$$
$$x = 56^\circ$$

The measure of the third angle is 56°.

Section 6.5 *pages 289–296*

You Try It 1

Strategy

- Pounds of \$.55 fertilizer: x

	Amount	Cost	Value
\$.80 fertilizer	20	\$.80	0.80(20)
\$.55 fertilizer	x	\$.55	$0.55x$
\$.75 fertilizer	$20 + x$	\$.75	$0.75(20 + x)$

- The sum of the values before mixing equals the value after mixing.

Solution

$$0.80(20) + 0.55x = 0.75(20 + x)$$
$$16 + 0.55x = 15 + 0.75x$$
$$16 - 0.20x = 15$$
$$-0.20x = -1$$
$$x = 5$$

5 lb of the \$.55 fertilizer must be added.

You Try It 2

Strategy

- Liters of water: x

	Amount	Percent	Quantity
Water	x	0	$0x$
12%	5	0.12	5(0.12)
8%	$x + 5$	0.08	$0.08(x + 5)$

- The sum of the quantities before mixing is equal to the quantity after mixing.

Solution

$$0x + 5(0.12) = 0.08(x + 5)$$
$$0.60 = 0.08x + 0.40$$
$$0.20 = 0.08x$$
$$2.5 = x$$

The pharmacist adds 2.5 L of water to the 12% solution to make an 8% solution.

You Try It 3

Strategy

▶ Additional amount: x

Principal	Rate	Interest
5000	0.08	0.08(5000)
x	0.11	$0.11x$
$5000 + x$	0.09	$0.09(5000 + x)$

▶ The sum of the interest earned by the two investments equals 9% of the total investment.

Solution

$$\begin{aligned} 0.08(5000) + 0.11x &= 0.09(5000 + x) \\ 400 + 0.11x &= 450 + 0.09x \\ 400 + 0.02x &= 450 \\ 0.02x &= 50 \\ x &= 2500 \end{aligned}$$

$2500 more must be invested at 11%.

You Try It 4

Strategy

▶ Rate of the first train: r
Rate of the second train: $2r$

	Rate	Time	Distance
1st train	r	3	$3r$
2nd train	$2r$	3	$3(2r)$

▶ The sum of the distances traveled by each train equals 288 mi.

Solution

$$\begin{aligned} 3r + 3(2r) &= 288 \\ 3r + 6r &= 288 \\ 9r &= 288 \\ r &= 32 \\ 2r = 2(32) &= 64 \end{aligned}$$

The first train is traveling at 32 mph.
The second train is traveling at 64 mph.

You Try It 5

Strategy

▶ Time spent flying out: t
Time spent flying back: $5 - t$

	Rate	Time	Distance
Out	150	t	$150t$
Back	100	$5 - t$	$100(5 - t)$

▶ The distance out equals the distance back.

Solution

$$\begin{aligned} 150t &= 100(5 - t) \\ 150t &= 500 - 100t \\ 250t &= 500 \\ t &= 2 \text{ (The time out was 2 h.)} \end{aligned}$$

The distance $= 150t = 150(2) = 300$ mi.
The parcel of land was 300 mi away.

Section 6.6 *pages 305–310*

You Try It 1

$$\begin{aligned} x + 2 &< -2 \\ x + 2 - 2 &< -2 - 2 \\ x &< -4 \end{aligned}$$

−5 −4 −3 −2 −1 0 1 2 3 4 5

You Try It 2

$$\begin{aligned} 5x + 3 &> 4x + 5 \\ 5x - 4x + 3 &> 4x - 4x + 5 \\ x + 3 &> 5 \\ x + 3 - 3 &> 5 - 3 \\ x &> 2 \end{aligned}$$

You Try It 3

$$-3x > -9$$
$$\frac{-3x}{-3} < \frac{-9}{-3}$$
$$x < 3$$

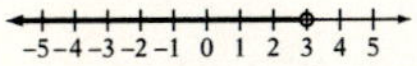

You Try It 4

$$-\frac{3}{4}x \geq 18$$
$$-\frac{4}{3}\left(-\frac{3}{4}x\right) \leq -\frac{4}{3}(18)$$
$$x \leq -24$$

You Try It 5

$$5 - 4x > 9 - 8x$$
$$5 - 4x + 8x > 9 - 8x + 8x$$
$$5 + 4x > 9$$
$$5 - 5 + 4x > 9 - 5$$
$$4x > 4$$
$$\frac{4x}{4} > \frac{4}{4}$$
$$x > 1$$

You Try It 6

$$8 - 4(3x + 5) \leq 6(x - 8)$$
$$8 - 12x - 20 \leq 6x - 48$$
$$-12 - 12x \leq 6x - 48$$
$$-12 - 12x - 6x \leq 6x - 6x - 48$$
$$-12 - 18x \leq -48$$
$$-12 + 12 - 18x \leq -48 + 12$$
$$-18x \leq -36$$
$$\frac{-18x}{-18} \geq \frac{-36}{-18}$$
$$x \geq 2$$

You Try It 7

Strategy

To find the maximum number of miles:

- ▶ Write an expression for the cost of each car, using x to represent the number of miles driven during the week.
- ▶ Write and solve an inequality.

Solution

Cost of a Company A car	is less than	Cost of a Company B car

$$8(7) + 0.10x < 10(7) + 0.08x$$
$$56 + 0.10x < 70 + 0.08x$$
$$56 + 0.10x - 0.08x < 70 + 0.08x - 0.08x$$
$$56 + 0.02x < 70$$
$$56 - 56 + 0.02x < 70 - 56$$
$$0.02x < 14$$
$$\frac{0.02x}{0.02} < \frac{14}{0.02}$$
$$x < 700$$

The maximum number of miles is 699.

Solutions to Chapter 7 You Try It

Section 7.1 *pages 321–330*

You Try It 1

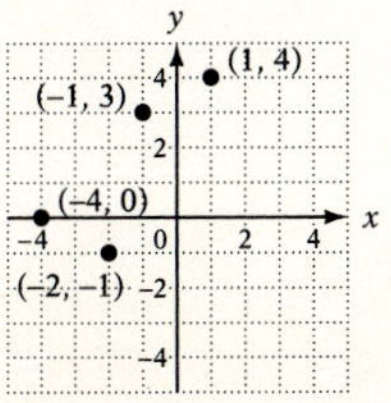

You Try It 2 $A\ (4, 2)$
$B\ (-3, 4)$
$C\ (-3, 0)$
$D\ (0, 0)$

You Try It 3

$$\begin{array}{c|c} \multicolumn{2}{c}{y = -\frac{1}{2}x - 3} \\ \hline -4 & -\frac{1}{2}(2) - 3 \\ & -1 - 3 \\ & -4 \end{array}$$

$-4 = -4$

Yes, $(2, -4)$ is a solution of $y = -\frac{1}{2}x - 3$.

You Try It 4

x	y
-3	-3
-1	1
0	3
1	5

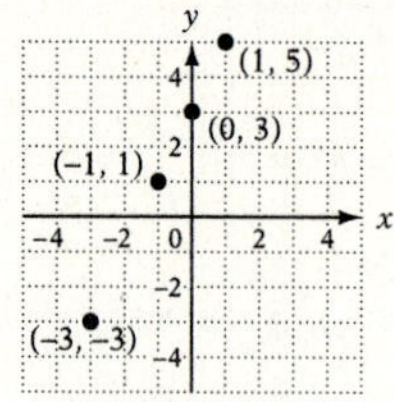

You Try It 5

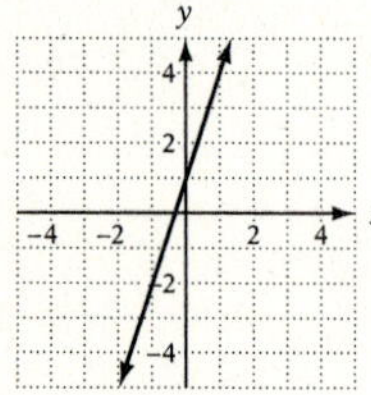

You Try It 6

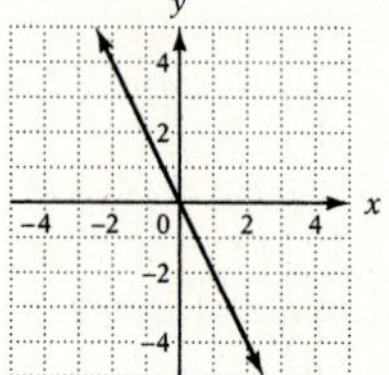

You Try It 7

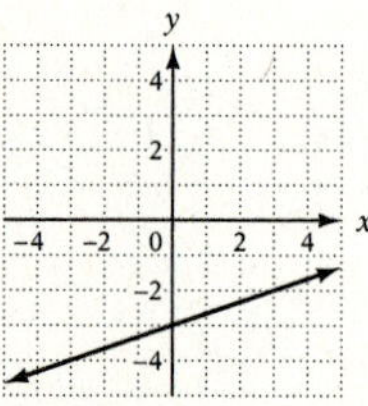

You Try It 8

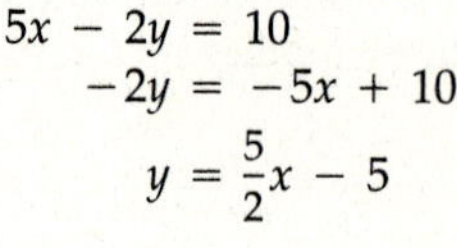

$$\begin{aligned} 5x - 2y &= 10 \\ -2y &= -5x + 10 \\ y &= \frac{5}{2}x - 5 \end{aligned}$$

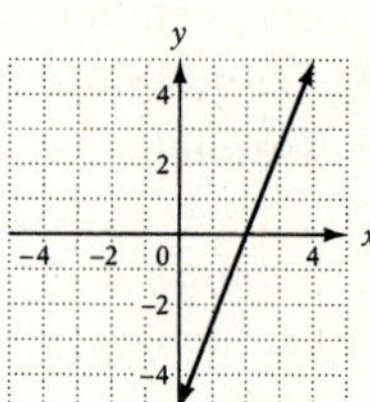

You Try It 9

$$\begin{aligned} x - 3y &= 9 \\ -3y &= -x + 9 \\ y &= \frac{1}{3}x - 3 \end{aligned}$$

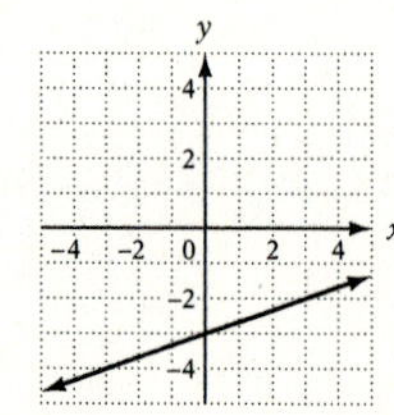

You Try It 10

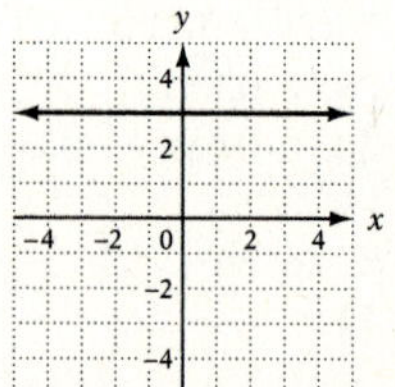

You Try It 11

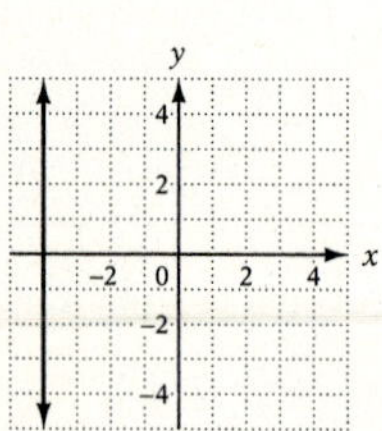

You Try It 12

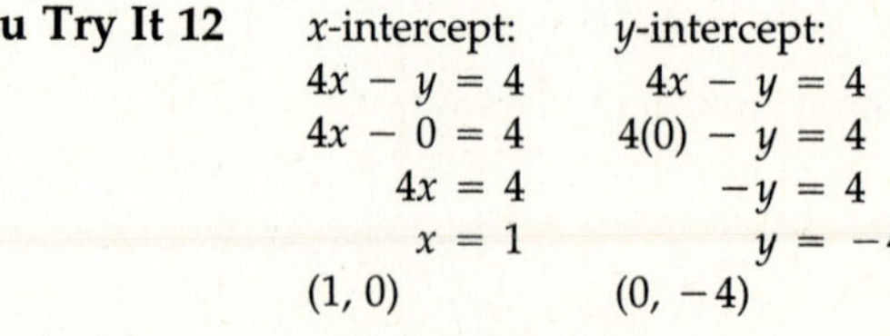

x-intercept:	y-intercept:
$4x - y = 4$	$4x - y = 4$
$4x - 0 = 4$	$4(0) - y = 4$
$4x = 4$	$-y = 4$
$x = 1$	$y = -4$
$(1, 0)$	$(0, -4)$

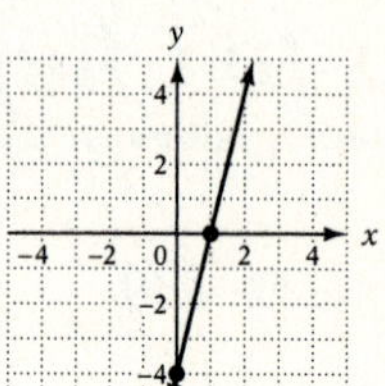

Section 7.2 *pages 337–342*

You Try It 1 Let $P_1 = (-1, 2)$ and $P_2 = (1, 3)$.

$$m = \frac{y_2 - y_1}{x_2 - x_1} = \frac{3 - 2}{1 - (-1)} = \frac{1}{2}$$

The slope is $\frac{1}{2}$.

You Try It 2 Let $P_1 = (1, 2)$ and $P_2 = (4, -5)$.

$$m = \frac{y_2 - y_1}{x_2 - x_1} = \frac{-5 - 2}{4 - 1} = \frac{-7}{3}$$

The slope is $-\frac{7}{3}$.

You Try It 3 Let $P_1 = (2, 3)$ and $P_2 = (2, 7)$.

$$m = \frac{y_2 - y_1}{x_2 - x_1} = \frac{7 - 3}{2 - 2} = \frac{4}{0}$$

The slope of the line is undefined.

You Try It 4 Let $P_1 = (1, -3)$ and $P_2 = (-5, -3)$.

$$m = \frac{y_2 - y_1}{x_2 - x_1} = \frac{-3 - (-3)}{-5 - 1} = \frac{0}{-6} = 0$$

The line has zero slope.

You Try It 5

$$m = \frac{8650 - 6100}{1 - 4} = \frac{2550}{-3}$$
$$= -850$$

A slope of -850 means that the value of the car is decreasing at a rate of \$850 per year.

You Try It 6 y-intercept $= (0, b) = (0, -1)$

$$m = -\frac{1}{4}$$

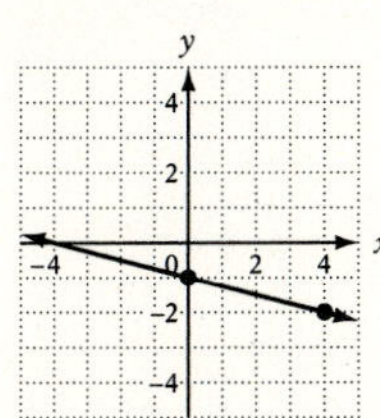

You Try It 7 y-intercept $= (0, b) = (0, 0)$

$$m = -\frac{3}{5}$$

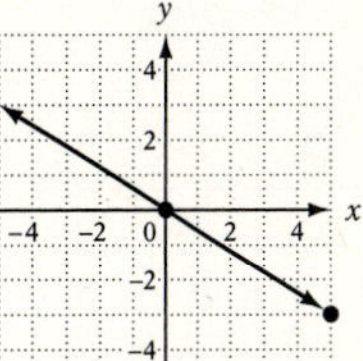

You Try It 8 Solve the equation for y.

$$x - 2y = 4$$
$$-2y = -x + 4$$
$$y = \frac{1}{2}x - 2$$

y-intercept $= (0, b) = (0, -2)$

$$m = \frac{1}{2}$$

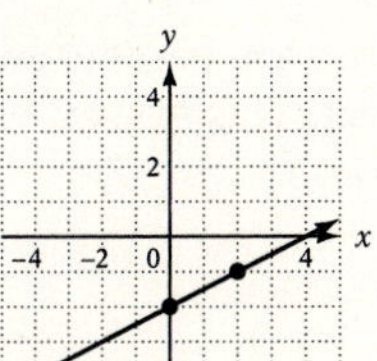

Section 7.3 *pages 347–350*

You Try It 1 Because the slope and y-intercept are known, use the slope–intercept formula, $y = mx + b$.

$$m = \frac{5}{3}, b = 2$$
$$y = \frac{5}{3}x + 2$$

You Try It 2 $m = \frac{3}{4}$ $\quad (x_1, y_1) = (4, -2)$

$$y - y_1 = m(x - x_1)$$
$$y - (-2) = \frac{3}{4}(x - 4)$$
$$y + 2 = \frac{3}{4}x - 3$$
$$y = \frac{3}{4}x - 5$$

The equation of the line is $y = \frac{3}{4}x - 5$.

You Try It 3 Find the slope of the line between the two points.

$(x_1, y_1) = (-6, -1), (x_2, y_2) = (3, 1)$

$$m = \frac{y_2 - y_1}{x_2 - x_1} = \frac{1 - (-1)}{3 - (-6)} = \frac{2}{9}$$

Use the point–slope formula.

$$y - y_1 = m(x - x_1)$$
$$y - (-1) = \frac{2}{9}[x - (-6)]$$
$$y + 1 = \frac{2}{9}x + \frac{4}{3}$$
$$y = \frac{2}{9}x + \frac{1}{3}$$

You Try It 4

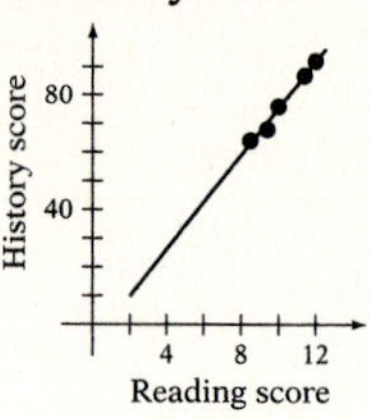

The slope of the line means that the grade on the history test increases 8.3 points for each 1 point increase in the grade on the reading test.

Section 7.4 *pages 355–362*

You Try It 1

$2x - 5y = 8$	
$2(-1) - 5(-2)$	8
$-2 + 10$	8
$8 = 8$	

$-x + 3y = -5$	
$-(-1) + 3(-2)$	-5
$1 + (-6)$	-5
$-5 = -5$	

Yes, $(-1, -2)$ is a solution of the system of equations.

You Try It 2

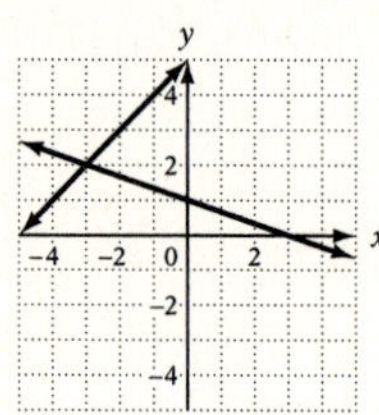

The solution is $(-3, 2)$.

You Try It 3

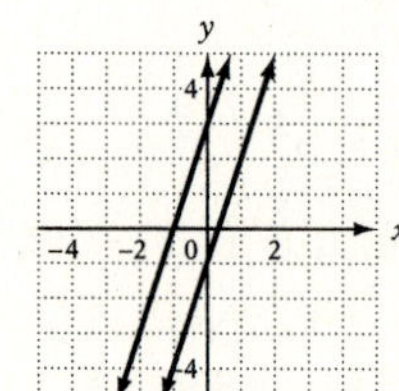

The lines are parallel and therefore do not intersect. This system of equations has no solution.

You Try It 4

(1) $7x - y = 4$
(2) $3x + 2y = 9$

Solve equation (1) for y.

$$7x - y = 4$$
$$-y = -7x + 4$$
$$y = 7x - 4$$

Substitute in equation (2).

$$3x + 2y = 9$$
$$3x + 2(7x - 4) = 9$$
$$3x + 14x - 8 = 9$$
$$17x - 8 = 9$$
$$17x = 17$$
$$x = 1$$

Substitute in equation (1).

$$7x - y = 4$$
$$7(1) - y = 4$$
$$7 - y = 4$$
$$-y = -3$$
$$y = 3$$

The solution is (1, 3).

You Try It 5

(1) $3x - y = 4$
(2) $y = 3x + 2$

$$3x - y = 4$$
$$3x - (3x + 2) = 4$$
$$3x - 3x - 2 = 4$$
$$-2 = 4$$

The lines are parallel. This system has no solution.

You Try It 6

$$\begin{aligned} (1)\qquad y &= -2x + 1 \\ (2)\qquad 6x + 3y &= 3 \end{aligned}$$

$$\begin{aligned} 6x + 3y &= 3 \\ 6x + 3(-2x + 1) &= 3 \\ 6x - 6x + 3 &= 3 \\ 3 &= 3 \end{aligned}$$

The two equations represent the same line. Any ordered pair that is a solution of one equation is also a solution of the other equation.

You Try It 7

$$\begin{aligned} (1)\qquad x - 2y &= 1 \\ (2)\qquad 2x + 4y &= 0 \end{aligned}$$

Eliminate y.

$$\begin{aligned} 2(x - 2y) &= 2 \cdot 1 \\ 2x + 4y &= 0 \end{aligned}$$

$$\begin{aligned} 2x - 4y &= 2 \\ 2x + 4y &= 0 \end{aligned}$$

Add the equations.

$$\begin{aligned} 4x &= 2 \\ x &= \frac{2}{4} = \frac{1}{2} \end{aligned}$$

Replace x in equation (2).

$$\begin{aligned} 2\left(\frac{1}{2}\right) + 4y &= 0 \\ 1 + 4y &= 0 \\ 4y &= -1 \\ y &= -\frac{1}{4} \end{aligned}$$

The solution is $\left(\frac{1}{2}, -\frac{1}{4}\right)$.

You Try It 8

$$\begin{aligned} (1)\qquad 2x - 3y &= 4 \\ (2)\qquad -4x + 6y &= -8 \end{aligned}$$

Eliminate y.

$$\begin{aligned} 2(2x - 3y) &= 2 \cdot 4 \\ -4x + 6y &= -8 \end{aligned}$$

$$\begin{aligned} 4x - 6y &= 8 \\ -4x + 6y &= -8 \end{aligned}$$

Add the equations.

$$\begin{aligned} 0 + 0 &= 0 \\ 0 &= 0 \end{aligned}$$

The two equations represent the same line. Any ordered pair that is a solution of one equation is also a solution of the other equation.

You Try It 9

$$\begin{aligned} (1)\qquad 4x + 5y &= 11 \\ (2)\qquad 3y &= x + 10 \end{aligned}$$

Write equation (2) in the form $Ax + By = C$.

$$\begin{aligned} 3y &= x + 10 \\ -x + 3y &= 10 \end{aligned}$$

Eliminate x.

$$\begin{aligned} 4x + 5y &= 11 \\ 4(-x + 3y) &= 4 \cdot 10 \end{aligned}$$

$$\begin{aligned} 4x + 5y &= 11 \\ -4x + 12y &= 40 \end{aligned}$$

Add the equations.

$$\begin{aligned} 17y &= 51 \\ y &= 3 \end{aligned}$$

Replace y in equation (1).

$$\begin{aligned} 4x + 5y &= 11 \\ 4x + 5 \cdot 3 &= 11 \\ 4x + 15 &= 11 \\ 4x &= -4 \\ x &= -1 \end{aligned}$$

The solution is $(-1, 3)$.

Section 7.5 *pages 369–372*

You Try It 1

Strategy

▶ Rate of the current: c
Rate of the canoeist in calm water: r

	Rate	Time	Distance
With current	$r + c$	3	$3(r + c)$
Against current	$r - c$	5	$5(r - c)$

▶ The distance traveled with the current is 15 mi.
The distance traveled against the current is 15 mi.

Solution

$$3(r + c) = 15 \qquad \frac{1}{3} \cdot 3(r + c) = \frac{1}{3} \cdot 15$$

$$5(r - c) = 15 \qquad \frac{1}{5} \cdot 5(r - c) = \frac{1}{5} \cdot 15$$

$$\begin{aligned} r + c &= 5 \\ r - c &= 3 \\ 2r &= 8 \\ r &= 4 \end{aligned}$$

$$\begin{aligned} r + c &= 5 \\ 4 + c &= 5 \\ c &= 1 \end{aligned}$$

The rate of the current is 1 mph.
The rate of the canoeist in calm water is 4 mph.

You Try It 2

Strategy

▶ Cost of an orange tree: x
Cost of a grapefruit tree: y

First purchase

	Amount	Unit Cost	Value
Orange trees	25	x	$25x$
Grapefruit trees	20	y	$20y$

Second purchase

	Amount	Unit Cost	Value
Orange trees	20	x	$20x$
Grapefruit trees	30	y	$30y$

▶ The total of the first purchase was \$290.
The total of the second purchase was \$330.

Solution

$$25x + 20y = 290 \qquad 4(25x + 20y) = 4 \cdot 290$$

$$20x + 30y = 330 \qquad -5(20x + 30y) = -5 \cdot 330$$

$$\begin{aligned} 100x + 80y &= 1160 \\ -100x - 150y &= -1650 \\ -70y &= -490 \\ y &= 7 \end{aligned}$$

$$\begin{aligned} 25x + 20y &= 290 \\ 25x + 20(7) &= 290 \\ 25x + 140 &= 290 \\ 25x &= 150 \\ x &= 6 \end{aligned}$$

The cost of an orange tree is \$6.
The cost of a grapefruit tree is \$7.

Solutions to Chapter 8 You Try It

Section 8.1 *pages 381–382*

You Try It 1

$$\begin{aligned} &(-4x^3 + 2x^2 - 8) + (4x^3 + 6x^2 - 7x + 5) \\ &= (-4x^3 + 4x^3) + (2x^2 + 6x^2) + (-7x) + (-8 + 5) \\ &= 8x^2 - 7x - 3 \end{aligned}$$

You Try It 2

$$\begin{array}{r} 6x^3 \qquad\quad + \;\; 2x + 8 \\ -9x^3 + 2x^2 - 12x - 8 \\ \hline -3x^3 + 2x^2 - 10x \quad\;\; \end{array}$$

You Try It 3

$(-4w^3 + 8w - 8) - (3w^3 - 4w^2 - 2w - 1)$
$= (-4w^3 + 8w - 8) + (-3w^3 + 4w^2 + 2w + 1)$
$= -7w^3 + 4w^2 + 10w - 7$

You Try It 4

$$\begin{array}{r} 13y^3 \qquad\quad - 6y - 7 \\ -\ 4y^2 + 6y + 9 \\ \hline 13y^3 - 4y^2 \qquad\quad + 2 \end{array}$$

Section 8.2 *pages 385–390*

You Try It 1 $(3x^2)(6x^3) =$
$(3 \cdot 6)(x^2 \cdot x^3) = 18x^5$

You Try It 2 $(-3xy^2)(-4x^2y^3) =$
$[(-3)(-4)](x \cdot x^2)(y^2 \cdot y^3) =$
$12x^3y^5$

You Try It 3 $(3x)(2x^2y)^3 =$
$(3x)(2^3x^6y^3) = (3x)(8x^6y^3) =$
$(3 \cdot 8)(x \cdot x^6)y^3 = 24x^7y^3$

You Try It 4 $(3x^2)^2(-2xy^2)^3 =$
$(3^2x^4)[(-2)^3x^3y^6] =$
$(9x^4)(-8x^3y^6) =$
$[9(-8)](x^4 \cdot x^3)y^6 =$
$-72x^7y^6$

You Try It 5 $(-2y + 3)(-4y) = 8y^2 - 12y$

You Try It 6 $-a^2(3a^2 + 2a - 7) =$
$-3a^4 - 2a^3 + 7a^2$

You Try It 7

$$\begin{array}{r} 2y^3 + 2y^2 - 3 \\ \times \qquad\qquad 3y - 1 \\ \hline -2y^3 - 2y^2 \qquad + 3 \\ 6y^4 + 6y^3 \qquad - 9y \qquad \\ \hline 6y^4 + 4y^3 - 2y^2 - 9y + 3 \end{array}$$

You Try It 8

$$\begin{array}{r} a^3 - 2 \\ \times \qquad a + 7 \\ \hline 7a^3 \qquad - 14 \\ a^4 \qquad - 2a \qquad \\ \hline a^4 + 7a^3 - 2a - 14 \end{array}$$

You Try It 9 $(b - 5)(b + 8)$
$= b^2 + 8b - 5b - 40$
$= b^2 + 3b - 40$

You Try It 10 $(4y - 5)(2y - 3)$
$= 8y^2 - 12y - 10y + 15$
$= 8y^2 - 22y + 15$

You Try It 11 $(3b + 2)(3b - 5)$
$= 9b^2 - 15b + 6b - 10$
$= 9b^2 - 9b - 10$

You Try It 12 $(4x - y)(2x + 3y)$
$= 8x^2 + 12xy - 2xy - 3y^2$
$= 8x^2 + 10xy - 3y^2$

You Try It 13 $(2a + 5c)(2a - 5c)$
$= 4a^2 - 25c^2$

You Try It 14 $(3x + 2y)^2 = 9x^2 + 12xy + 4y^2$

Section 8.3 *pages 395–398*

You Try It 1 $\dfrac{42y^{12}}{-14y^{17}} = -\dfrac{\overset{1}{\cancel{2}} \cdot 3 \cdot \overset{1}{\cancel{7}}y^{12}}{\underset{1}{\cancel{2}} \cdot \underset{1}{\cancel{7}}y^{17}} = -\dfrac{3}{y^5}$

You Try It 2 $\dfrac{12r^4s^2}{-8r^3s} = -\dfrac{\overset{1}{\cancel{2}} \cdot \overset{1}{\cancel{2}} \cdot 3r^4s^2}{\underset{1}{\cancel{2}} \cdot \underset{1}{\cancel{2}} \cdot 2r^3s} = -\dfrac{3rs}{2}$

You Try It 3 $\dfrac{(2x^2y)^3}{-4xy^5} = -\dfrac{2^3x^6y^3}{4xy^5} =$

$-\dfrac{\overset{1}{\cancel{2}} \cdot \overset{1}{\cancel{2}} \cdot 2x^6y^3}{\underset{1}{\cancel{2}} \cdot \underset{1}{\cancel{2}}xy^5} = -\dfrac{2x^5}{y^2}$

You Try It 4 $\dfrac{4x^3y + 8x^2y^2 - 4xy^3}{2xy} =$

$\dfrac{4x^3y}{2xy} + \dfrac{8x^2y^2}{2xy} - \dfrac{4xy^3}{2xy} =$

$2x^2 + 4xy - 2y^2$

You Try It 5
$$\frac{24x^2y^2 - 18xy + 6y}{6xy} =$$
$$\frac{24x^2y^2}{6xy} - \frac{18xy}{6xy} + \frac{6y}{6xy} =$$
$$4xy - 3 + \frac{1}{x}$$

You Try It 6
$$\begin{array}{r} x^2 + 2x - 1 \\ 2x - 3\overline{)2x^3 + x^2 - 8x - 3} \\ \underline{2x^3 - 3x^2} \\ 4x^2 - 8x \\ \underline{4x^2 - 6x} \\ -2x - 3 \\ \underline{-2x + 3} \\ -6 \end{array}$$
$$(2x^3 + x^2 - 8x - 3) \div (2x - 3) =$$
$$x^2 + 2x - 1 - \frac{6}{2x - 3}$$

Section 8.4 *pages 401–404*

You Try It 1 $\frac{2^{-2}}{2^3} = 2^{-5} = \frac{1}{2^5} = \frac{1}{32}$

You Try It 2 $(-2x^2)(x^{-3}y^{-4})^{-2} =$
$(-2x^2)(x^6y^8) = -2x^8y^8$

You Try It 3
$$\frac{(3x^{-2}y)^3}{9xy^0} = \frac{3^3x^{-6}y^3}{9xy^0} =$$
$$\frac{\overset{1}{\cancel{3}} \cdot \overset{1}{\cancel{3}} \cdot 3x^{-6}y^3}{\underset{1}{\cancel{3}} \cdot \underset{1}{\cancel{3}}xy^0} = 3x^{-7}y^3 = \frac{3y^3}{x^7}$$

You Try It 4
$942{,}000{,}000 = 9.42 \cdot 10^8$

You Try It 5
$2.7 \cdot 10^{-5} = 0.000027$

You Try It 6
$$\frac{5{,}600{,}000 \cdot 0.000000081}{900 \cdot 0.000000028} =$$
$$\frac{5.6 \cdot 10^6 \cdot 8.1 \cdot 10^{-8}}{9 \cdot 10^2 \cdot 2.8 \cdot 10^{-8}} =$$
$$\frac{(5.6)(8.1) \cdot 10^{6+(-8)-(2)-(-8)}}{(9)(2.8)} =$$
$$1.8 \cdot 10^4 = 18{,}000$$

You Try It 7

Strategy

To find the time, use the equation $d = rt$, where r is the speed of light and d is the distance of the sun from the earth.

Solution
$$d = rt$$
$$9.3 \cdot 10^7 = (1.86 \cdot 10^5)t$$
$$\frac{9.3 \cdot 10^7}{1.86 \cdot 10^5} = t$$
$$5 \cdot 10^2 = t$$

Light travels to the earth from the sun in $5 \cdot 10^2$ s.

Solutions to Chapter 9 You Try It

Section 9.1 *pages 413–414*

You Try It 1

$8a^4bc = 2 \cdot 2 \cdot 2 \cdot a^4 \cdot b \cdot c$
$12ab^3 = 2 \cdot 2 \cdot 3 \cdot a \cdot b^3$
$20abc^2 = 2 \cdot 2 \cdot 5 \cdot a \cdot b \cdot c^2$

The GCF is $4ab$.

You Try It 2

$18a^4 = 2 \cdot 3 \cdot 3 \cdot a^4$
$27a^3 = 3 \cdot 3 \cdot 3 \cdot a^3$
$9a^2 = 3 \cdot 3 \cdot a^2$

The GCF is $9a^2$.

$18a^4 + 27a^3 - 9a^2$
$= 9a^2(2a^2) + 9a^2(3a) + 9a^2(-1)$
$= 9a^2(2a^2 + 3a - 1)$

Section 9.2 *pages 417–420*

You Try It 1

$(x - ■)(x - ■)$

Factors	Sums
−1, −20	−21
−2, −10	−12
−4, −5	−9

$(x - 4)(x - 5)$

$x^2 - 9x + 20 = (x - 4)(x - 5)$

You Try It 2

$(x + ■)(x - ■)$

Factors	Sums
+1, −18	−17
−1, +18	17
+2, −9	−7
−2, +9	7
+3, −6	−3
−3, +6	3

$(x + 6)(x - 3)$

$x^2 + 3x - 18 = (x + 6)(x - 3)$

You Try It 3

$(■x + ■)(■x - ■)$ or
$(■x - ■)(■x + ■)$

Factors of 2: 1, 2   Factors of −3: +1, −3
−1, +3

Trial Factors	Middle Term
$(1x + 1)(2x - 3)$	$-3x + 2x = -x$
$(1x - 3)(2x + 1)$	$x - 6x = -5x$
$(1x - 1)(2x + 3)$	$3x - 2x = x$
$(1x + 3)(2x - 1)$	$-x + 6x = 5x$

$(x + 1)(2x - 3)$

$2x^2 - x - 3 = (x + 1)(2x - 3)$

Section 9.3 *pages 423–428*

You Try It 1

$25a^2 - b^2 = (5a)^2 - b^2 = (5a + b)(5a - b)$

You Try It 2

$n^8 - 36 = (n^4)^2 - 6^2 = (n^4 + 6)(n^4 - 6)$

You Try It 3

$\sqrt{a^2} = a$
$\sqrt{100} = 10$ $2(10a) = 20a$

The trinomial is a perfect square.

$a^2 + 20a + 100 = (a + 10)^2$

You Try It 4

$\sqrt{25a^2} = 5a$
$\sqrt{9b^2} = 3b$ $2(5a \cdot 3b) = 30ab$

The trinomial is a perfect square.

$25a^2 - 30ab + 9b^2 = (5a - 3b)^2$

You Try It 5

$5x(2x + 3) - 4(2x + 3) = (2x + 3)(5x - 4)$

You Try It 6

$2y(5x - 2) - 3(2 - 5x) =$
$2y(5x - 2) + 3(5x - 2) = (5x - 2)(2y + 3)$

You Try It 7

The GCF is 3.

$3x^2 - 9xy - 12y^2 = 3(x^2 - 3xy - 4y^2)$

Factor the trinomial.

$3(x + \blacksquare y)(x - \blacksquare y)$

Factors	Sums
+1, −4	−3
−1, +4	3
+2, −2	0

$3(x + y)(x - 4y)$

$3x^2 - 9xy - 12y^2 = 3(x + y)(x - 4y)$

You Try It 8

The GCF is $3x$.

$12x^3 - 75x = 3x(4x^2 - 25)$

Factor the difference of two squares.

$3x(2x + 5)(2x - 5)$

You Try It 9

The common binomial factor is $b - 7$.

$a^2(b - 7) + (7 - b) =$
$a^2(b - 7) - (b - 7) = (b - 7)(a^2 - 1)$

Factor the difference of two squares.

$(b - 7)(a + 1)(a - 1)$

You Try It 10

The GCF is $2a^2$.

$4a^2b^2 + 26a^2b - 14a^2 = 2a^2(2b^2 + 13b - 7)$

Factor the trinomial.

$2a^2(\blacksquare b + \blacksquare)(\blacksquare b - \blacksquare)$ or
$2a^2(\blacksquare b - \blacksquare)(\blacksquare b + \blacksquare)$

Factors of 2: 1, 2 Factors of −7: +1, −7
−1, +7

Trial Factors	Middle Term
$(1b + 1)(2b - 7)$	$-7b + 2b = -5b$
$(1b - 7)(2b + 1)$	$b - 14b = -13b$
$(1b - 1)(2b + 7)$	$7b - 2b = 5b$
$(1b + 7)(2b - 1)$	$-b + 14b = 13b$

$2a^2(b + 7)(2b - 1)$
$4a^2b^2 + 26a^2b - 14a^2 = 2a^2(b + 7)(2b - 1)$

You Try It 11

The GCF is $3y$.

$12y + 12y^2 - 45y^3 = 3y(4 + 4y - 15y^2)$

Factor the trinomial.

$3y(\blacksquare + \blacksquare y)(\blacksquare - \blacksquare y)$ or $3y(\blacksquare - \blacksquare y)(\blacksquare + \blacksquare y)$

Factors of 4: 1, 4 Factors of −15: +1, −15
2, 2 −1, +15
+3, −5
−3, +5

Trial Factors	Middle Term
$(1 + 1y)(4 - 15y)$	$-15y + 4y = -11y$
$(1 - 15y)(4 + 1y)$	$y - 60y = -59y$
$(1 - 1y)(4 + 15y)$	$15y - 4y = 11y$
$(1 + 15y)(4 - 1y)$	$-y + 60y = 59y$
$(1 + 3y)(4 - 5y)$	$-5y + 12y = 7y$
$(1 - 5y)(4 + 3y)$	$3y - 20y = -17y$
$(1 - 3y)(4 + 5y)$	$5y - 12y = -7y$
$(1 + 5y)(4 - 3y)$	$-3y + 20y = 17y$
$(2 + 1y)(2 - 15y)$	$-30y + 2y = -28y$
$(2 - 1y)(2 + 15y)$	$30y - 2y = 28y$
$(2 + 3y)(2 - 5y)$	$-10y + 6y = -4y$
$(2 - 3y)(2 + 5y)$	$10y - 6y = 4y$

$3y(2 - 3y)(2 + 5y)$

$12y + 12y^2 - 45y^3 = 3y(2 - 3y)(2 + 5y)$

Section 9.4 *pages 433–436*

You Try It 1

$2x(x + 7) = 0$

$2x = 0 \qquad x + 7 = 0$
$x = 0 \qquad x = -7$

The solutions are 0 and -7.

You Try It 2

$4x^2 - 9 = 0$
$(2x - 3)(2x + 3) = 0$

$2x - 3 = 0 \qquad 2x + 3 = 0$
$2x = 3 \qquad 2x = -3$
$x = \frac{3}{2} \qquad x = -\frac{3}{2}$

The solutions are $\frac{3}{2}$ and $-\frac{3}{2}$.

You Try It 3

$(x + 2)(x - 7) = 52$
$x^2 - 5x - 14 = 52$
$x^2 - 5x - 66 = 0$
$(x + 6)(x - 11) = 0$

$x + 6 = 0 \qquad x - 11 = 0$
$x = -6 \qquad x = 11$

The solutions are -6 and 11.

You Try It 4

Strategy

First positive consecutive integer: n
Second positive consecutive integer: $n + 1$
Square of the first integer: n^2
Square of the second integer: $(n + 1)^2$

The sum of the squares of two positive consecutive integers is 61.

Solution

$n^2 + (n + 1)^2 = 61$
$n^2 + n^2 + 2n + 1 = 61$
$2n^2 + 2n + 1 = 61$
$2n^2 + 2n - 60 = 0$
$2(n^2 + n - 30) = 0$
$2(n - 5)(n + 6) = 0$

$n - 5 = 0 \qquad n + 6 = 0$
$n = 5 \qquad n = -6$

Since -6 is not a positive integer, it is not a solution.

$n = 5$
$n + 1 = 5 + 1 = 6$

The two integers are 5 and 6.

You Try It 5

Strategy

Width $= x$
Length $= 2x + 4$

The area of a rectangle is 96 in^2.
Use the equation $A = LW$.

Solution

$A = LW$
$96 = (2x + 4)x$
$96 = 2x^2 + 4x$
$0 = 2x^2 + 4x - 96$
$0 = 2(x^2 + 2x - 48)$
$0 = 2(x + 8)(x - 6)$

$x + 8 = 0 \qquad x - 6 = 0$
$x = -8 \qquad x = 6$

Since the width cannot be negative, -8 is not a solution.

$x = 6$
$2x + 4 = 2(6) + 4 = 12 + 4 = 16$

The width is 6 in.
The length is 16 in.

Solutions to Chapter 10 You Try It

Section 10.1 *pages 447–450*

You Try It 1

$$\frac{6x^5y}{12x^2y^3} = \frac{\overset{1}{\cancel{2}} \cdot \overset{1}{\cancel{3}} \cdot x^5y}{\underset{1}{\cancel{2}} \cdot 2 \cdot \underset{1}{\cancel{3}} \cdot x^2y^3} = \frac{x^3}{2y^2}$$

You Try It 2

$$\frac{x^2 + 2x - 24}{16 - x^2} = \frac{\overset{-1}{\cancel{(x - 4)}}(x + 6)}{\underset{1}{\cancel{(4 - x)}}(4 + x)} = -\frac{x + 6}{x + 4}$$

You Try It 3

$$\frac{12x^2+3x}{10x-15}\cdot\frac{8x-12}{9x+18}$$
$$=\frac{3x(4x+1)}{5(2x-3)}\cdot\frac{4(2x-3)}{9(x+2)}$$
$$=\frac{\overset{1}{\cancel{3}}x(4x+1)\cdot 2\cdot 2\overset{1}{\cancel{(2x-3)}}}{5\underset{1}{\cancel{(2x-3)}}\cdot\underset{1}{\cancel{3}}\cdot 3(x+2)}=\frac{4x(4x+1)}{15(x+2)}$$

You Try It 4

$$\frac{x^2+2x-15}{9-x^2}\cdot\frac{x^2-3x-18}{x^2-7x+6}$$
$$=\frac{(x-3)(x+5)}{(3-x)(3+x)}\cdot\frac{(x+3)(x-6)}{(x-1)(x-6)}$$
$$=\frac{\overset{-1}{\cancel{(x-3)}}(x+5)\cdot\overset{1}{\cancel{(x+3)}}\overset{1}{\cancel{(x-6)}}}{\underset{1}{\cancel{(3-x)}}\underset{1}{\cancel{(3+x)}}\cdot(x-1)\underset{1}{\cancel{(x-6)}}}=-\frac{x+5}{x-1}$$

You Try It 5

$$\frac{a^2}{4bc^2-2b^2c}\div\frac{a}{6bc-3b^2}$$
$$=\frac{a^2}{4bc^2-2b^2c}\cdot\frac{6bc-3b^2}{a}$$
$$=\frac{a^2\cdot 3b\overset{1}{\cancel{(2c-b)}}}{2bc\underset{1}{\cancel{(2c-b)}}\cdot a}=\frac{3a}{2c}$$

You Try It 6

$$\frac{3x^2+26x+16}{3x^2-7x-6}\div\frac{2x^2+9x-5}{x^2+2x-15}$$
$$=\frac{3x^2+26x+16}{3x^2-7x-6}\cdot\frac{x^2+2x-15}{2x^2+9x-5}$$
$$=\frac{\overset{1}{\cancel{(3x+2)}}(x+8)}{\underset{1}{\cancel{(3x+2)}}\underset{1}{\cancel{(x-3)}}}\cdot\frac{\overset{1}{\cancel{(x+5)}}\overset{1}{\cancel{(x-3)}}}{(2x-1)\underset{1}{\cancel{(x+5)}}}=\frac{x+8}{2x-1}$$

Section 10.2 *pages 455–460*

You Try It 1

$8uv^2=2\cdot2\cdot2\cdot u\cdot v\cdot v$ $\quad 12uw=2\cdot2\cdot3\cdot u\cdot w$
LCM $=2\cdot2\cdot2\cdot3\cdot u\cdot v\cdot v\cdot w=24uv^2w$

You Try It 2

$m^2-6m+9=(m-3)(m-3)$
$m^2-2m-3=(m+1)(m-3)$
LCM $=(m-3)(m-3)(m+1)$

You Try It 3

The LCM is $36xy^2z$.

$$\frac{x-3}{4xy^2}=\frac{x-3}{4xy^2}\cdot\frac{9z}{9z}=\frac{9xz-27z}{36xy^2z}$$
$$\frac{2x+1}{9y^2z}=\frac{2x+1}{9y^2z}\cdot\frac{4x}{4x}=\frac{8x^2+4x}{36xy^2z}$$

You Try It 4

$$\frac{2x}{25-x^2}=\frac{2x}{-(x^2-25)}=-\frac{2x}{x^2-25}$$

The LCM is $(x+2)(x-5)(x+5)$.

$$\frac{x+4}{x^2-3x-10}=\frac{x+4}{(x+2)(x-5)}\cdot\frac{x+5}{x+5}$$
$$=\frac{x^2+9x+20}{(x+2)(x-5)(x+5)}$$
$$\frac{2x}{25-x^2}=-\frac{2x}{(x-5)(x+5)}\cdot\frac{x+2}{x+2}$$
$$=-\frac{2x^2+4x}{(x+2)(x-5)(x+5)}$$

You Try It 5

$$\frac{3}{xy}+\frac{12}{xy}=\frac{3+12}{xy}=\frac{15}{xy}$$

You Try It 6

$$\frac{2x^2}{x^2-x-12}-\frac{7x+4}{x^2-x-12}=\frac{2x^2-(7x+4)}{x^2-x-12}$$
$$=\frac{2x^2-7x-4}{x^2-x-12}=\frac{(2x+1)\overset{1}{\cancel{(x-4)}}}{(x+3)\underset{1}{\cancel{(x-4)}}}=\frac{2x+1}{x+3}$$

You Try It 7

$$\frac{x^2-1}{x^2-8x+12}-\frac{2x+1}{x^2-8x+12}+\frac{x}{x^2-8x+12}$$
$$=\frac{(x^2-1)-(2x+1)+x}{x^2-8x+12}=\frac{x^2-1-2x-1+x}{x^2-8x+12}$$
$$=\frac{x^2-x-2}{x^2-8x+12}=\frac{(x+1)\overset{1}{\cancel{(x-2)}}}{\underset{1}{\cancel{(x-2)}}(x-6)}=\frac{x+1}{x-6}$$

You Try It 8

The LCM of the denominators is $24y$.

$$\frac{z}{8y}=\frac{z}{8y}\cdot\frac{3}{3}=\frac{3z}{24y} \qquad \frac{4z}{3y}=\frac{4z}{3y}\cdot\frac{8}{8}=\frac{32z}{24y}$$
$$\frac{5z}{4y}=\frac{5z}{4y}\cdot\frac{6}{6}=\frac{30z}{24y}$$
$$\frac{z}{8y}-\frac{4z}{3y}+\frac{5z}{4y}=\frac{3z}{24y}-\frac{32z}{24y}+\frac{30z}{24y}$$
$$=\frac{3z-32z+30z}{24y}=\frac{z}{24y}$$

You Try It 9

The LCM is $x-2$.

$$\frac{5x}{x-2}=\frac{5x}{x-2}\cdot\frac{1}{1}=\frac{5x}{x-2}$$
$$\frac{3}{2-x}=\frac{3}{-(x-2)}\cdot\frac{-1}{-1}=\frac{-3}{x-2}$$
$$\frac{5x}{x-2}-\frac{3}{2-x}=\frac{5x}{x-2}-\frac{-3}{x-2}$$
$$=\frac{5x-(-3)}{x-2}=\frac{5x+3}{x-2}$$

You Try It 10

The LCM is $(3x-1)(x+4)$.

$$\frac{4x}{3x-1}=\frac{4x}{3x-1}\cdot\frac{x+4}{x+4}=\frac{4x^2+16x}{(3x-1)(x+4)}$$
$$\frac{9}{x+4}=\frac{9}{x+4}\cdot\frac{3x-1}{3x-1}=\frac{27x-9}{(3x-1)(x+4)}$$
$$\frac{4x}{3x-1}-\frac{9}{x+4}$$
$$=\frac{4x^2+16x}{(3x-1)(x+4)}-\frac{27x-9}{(3x-1)(x+4)}$$
$$=\frac{4x^2+16x-(27x-9)}{(3x-1)(x+4)}=\frac{4x^2+16x-27x+9}{(3x-1)(x+4)}$$
$$=\frac{4x^2-11x+9}{(3x-1)(x+4)}$$

You Try It 11

The LCM is $(x+5)(x-5)$.

$$\frac{2x-1}{x^2-25}=\frac{2x-1}{(x+5)(x-5)}$$
$$\frac{2}{5-x}=\frac{2}{-(x-5)}\cdot\frac{-1\cdot(x+5)}{-1\cdot(x+5)}=\frac{-2(x+5)}{(x+5)(x-5)}$$
$$\frac{2x-1}{x^2-25}+\frac{2}{5-x}$$
$$=\frac{2x-1}{(x+5)(x-5)}+\frac{-2(x+5)}{(x+5)(x-5)}$$
$$=\frac{2x-1+(-2)(x+5)}{(x+5)(x-5)}=\frac{2x-1-2x-10}{(x+5)(x-5)}$$
$$=\frac{-11}{(x+5)(x-5)}=-\frac{11}{(x+5)(x-5)}$$

You Try It 12

The LCM is $(3x+2)(x-1)$.

$$\frac{2x-3}{3x^2-x-2}=\frac{2x-3}{(3x+2)(x-1)}$$
$$\frac{5}{3x+2}=\frac{5}{3x+2}\cdot\frac{x-1}{x-1}=\frac{5x-5}{(3x+2)(x-1)}$$
$$\frac{1}{x-1}=\frac{1}{x-1}\cdot\frac{3x+2}{3x+2}=\frac{3x+2}{(3x+2)(x-1)}$$
$$\frac{2x-3}{3x^2-x-2}+\frac{5}{3x+2}-\frac{1}{x-1}$$
$$=\frac{2x-3}{(3x+2)(x-1)}+\frac{5x-5}{(3x+2)(x-1)}-\frac{3x+2}{(3x+2)(x-1)}$$
$$=\frac{(2x-3)+(5x-5)-(3x+2)}{(3x+2)(x-1)}$$
$$=\frac{2x-3+5x-5-3x-2}{(3x+2)(x-1)}$$
$$=\frac{4x-10}{(3x+2)(x-1)}=\frac{2(2x-5)}{(3x+2)(x-1)}$$

Section 10.3 *pages 467–472*

You Try It 1

$$\frac{2}{x+6} = \frac{3}{x} \quad \text{The LCM is } x(x+6).$$

$$\frac{x(\cancel{x+6})}{1} \cdot \frac{2}{\cancel{x+6}} = \frac{x(x+6)}{1} \cdot \frac{3}{x}$$

$$2x = (x+6)3$$
$$2x = 3x + 18$$
$$-x = 18$$
$$x = -18$$

-18 checks as a solution.
The solution is -18.

You Try It 2

$$\frac{5x}{x+2} = 3 - \frac{10}{x+2} \quad \text{The LCM is } x + 2.$$

$$\frac{(x+2)}{1} \cdot \frac{5x}{x+2} = \frac{(x+2)}{1}\left(3 - \frac{10}{x+2}\right)$$

$$\frac{\cancel{x+2}}{1} \cdot \frac{5x}{\cancel{x+2}} = \frac{x+2}{1} \cdot 3 - \frac{\cancel{x+2}}{1} \cdot \frac{10}{\cancel{x+2}}$$

$$5x = (x+2)3 - 10$$
$$5x = 3x + 6 - 10$$
$$5x = 3x - 4$$
$$2x = -4$$
$$x = -2$$

-2 does not check as a solution.
This equation has no solution.

You Try It 3

Strategy

To find the number of computers:

- ▶ Write the basic inverse variation equation, replace the variables by the given values, and solve for k.
- ▶ Write the inverse variation equation, replacing k by its value. Substitute 2000 for P and solve for s.

Solution

$$s = \frac{k}{P}$$

$$2000 = \frac{k}{2500}$$

$$5{,}000{,}000 = k$$

$$s = \frac{5{,}000{,}000}{P} = \frac{5{,}000{,}000}{2000} = 2500$$

At a price of $2000, 2500 computers can be sold.

You Try It 4

$$5x - 2y = 10$$
$$5x - 2y + 2y = 2y + 10$$
$$5x = 2y + 10$$
$$\frac{5x}{5} = \frac{2y + 10}{5}$$
$$x = \frac{2}{5}y + 2$$

You Try It 5

$$s = \frac{A + L}{2}$$
$$2 \cdot s = 2\left(\frac{A + L}{2}\right)$$
$$2s = A + L$$
$$2s - A = A - A + L$$
$$2s - A = L$$

You Try It 6

$$S = a + (n - 1)d$$
$$S = a + nd - d$$
$$S - a = a - a + nd - d$$
$$S - a = nd - d$$
$$S - a + d = nd - d + d$$
$$S - a + d = nd$$
$$\frac{S - a + d}{d} = \frac{nd}{d}$$
$$\frac{S - a + d}{d} = n$$

You Try It 7

$$S = C + rC$$
$$S = (1 + r)C$$
$$\frac{S}{1 + r} = \frac{(1 + r)C}{1 + r}$$
$$\frac{S}{1 + r} = C$$

Section 10.4 *pages 479–484*

You Try It 1

$\frac{260 \text{ mi}}{8 \text{ h}}$ $8\overline{)260.0}$ = 32.5

32.5 mi/h

You Try It 2

$$\frac{x}{14} = \frac{3}{7}$$
$$x \cdot 7 = 14 \cdot 3$$
$$7x = 42$$
$$x = 6$$

Check:

$\frac{6}{14} = \frac{3}{7}$ ⟹ $14 \times 3 = 42$, $6 \times 7 = 42$

The solution is 6.

You Try It 3

$$\frac{5}{8} = \frac{x}{20}$$
$$5 \cdot 20 = 8 \cdot x$$
$$100 = 8x$$
$$12.5 = x$$

The solution is 12.5.

You Try It 4

$$\frac{2}{x + 3} = \frac{6}{5x + 5}$$
$$2(5x + 5) = (x + 3)6$$
$$10x + 10 = 6x + 18$$
$$4x + 10 = 18$$
$$4x = 8$$
$$x = 2$$

The solution is 2.

You Try It 5

Strategy

To find the rate in miles per hour, divide the number of miles (47) by the number of hours (3).

Solution

$3\overline{)47.00}$ = 15.66 ≈ 15.7

The rate is 15.7 mph.

You Try It 6

Strategy

To find the number of tablespoons of fertilizer needed, write and solve a proportion using n to represent the number of tablespoons of fertilizer.

Solution

$$\frac{3 \text{ tablespoons}}{4 \text{ gal}} = \frac{n \text{ tablespoons}}{10 \text{ gal}}$$

$$\frac{3}{4} = \frac{n}{10}$$

$$3 \cdot 10 = 4 \cdot n$$

$$30 = 4n$$

$$7.5 = n$$

For 10 gal of water, 7.5 tablespoons of fertilizer are required.

You Try It 7

Strategy

To find the additional amount of medication required for a 200-pound adult, write and solve a proportion using x to represent the additional medication. Then $3 + x$ is the total amount required for a 200-pound adult.

Solution

$$\frac{150}{3} = \frac{200}{3 + x}$$

$$150(3 + x) = 3 \cdot 200$$

$$450 + 150x = 600$$

$$150x = 150$$

$$x = 1$$

One additional ounce is required for a 200-pound adult.

You Try It 8

Strategy

To find FG, solve a proportion to find the height.

Solution

$$\frac{AC}{DF} = \frac{CH}{FG}$$

$$\frac{10}{15} = \frac{7}{FG}$$

$$10(FG) = 15(7)$$

$$10(FG) = 105$$

$$FG = 10.5$$

The height FG of triangle DEF is 10.5 m.

You Try It 9

Strategy

To find the perimeter of triangle ABC:

- ▶ Solve a proportion to find the length of side AC.
- ▶ Solve a proportion to find the length of side BC.
- ▶ Use the equation for the perimeter of a triangle.

Solution

$$\frac{AC}{DF} = \frac{AB}{DE}$$

$$\frac{AC}{6} = \frac{4}{8}$$

$$(AC)8 = 6(4)$$

$$8(AC) = 24$$

$$AC = 3$$

$$\frac{BC}{EF} = \frac{AB}{DE}$$

$$\frac{BC}{10} = \frac{4}{8}$$

$$(BC)8 = 10(4)$$

$$8(BC) = 40$$

$$BC = 5$$

$$P = a + b + c$$

$$P = 3 + 4 + 5$$

$$P = 12$$

The perimeter of triangle ABC is 12 in.

Section 10.5 *pages 491–494*

You Try It 1

Strategy

▶ Time for one printer to complete the job: t

	Rate	Time	Part
1st printer	$\frac{1}{t}$	2	$\frac{2}{t}$
2nd printer	$\frac{1}{t}$	5	$\frac{5}{t}$

▶ The sum of the parts of the task completed must equal 1.

Solution

$$\frac{2}{t} + \frac{5}{t} = 1$$
$$t\left(\frac{2}{t} + \frac{5}{t}\right) = t \cdot 1$$
$$2 + 5 = t$$
$$7 = t$$

Working alone, one printer takes 7 h to print the payroll.

You Try It 2

Strategy

▶ Rate sailing across the lake: r
Rate sailing back: $3r$

	Distance	Rate	Time
Across	6	r	$\frac{6}{r}$
Back	6	$3r$	$\frac{6}{3r}$

▶ The total time for the trip was 2 h.

Solution

$$\frac{6}{r} + \frac{6}{3r} = 2$$
$$3r\left(\frac{6}{r} + \frac{6}{3r}\right) = 3r(2)$$
$$3r \cdot \frac{6}{r} + 3r \cdot \frac{6}{3r} = 6r$$
$$18 + 6 = 6r$$
$$24 = 6r$$
$$4 = r$$

The rate across the lake was 4 km/h.

Solutions to Chapter 11 You Try It

Section 11.1 *pages 503–508*

You Try It 1

$$\sqrt{28} = \sqrt{2^2 \cdot 7} = \sqrt{2^2}\sqrt{7} = 2\sqrt{7}$$

You Try It 2

$$-5\sqrt{32} = -5\sqrt{2^5} = -5\sqrt{2^4 \cdot 2} = -5\sqrt{2^4}\sqrt{2}$$
$$= -5 \cdot 2^2\sqrt{2} = -20\sqrt{2}$$

You Try It 3

$$\sqrt{216} \approx 14.697$$

You Try It 4

$$\sqrt{y^{19}} = \sqrt{y^{18} \cdot y} = \sqrt{y^{18}}\sqrt{y} = y^9\sqrt{y}$$

You Try It 5

$$\sqrt{45b^7} = \sqrt{3^2 \cdot 5 \cdot b^7} = \sqrt{3^2b^6(5 \cdot b)} = \sqrt{3^2b^6}\sqrt{5b}$$
$$= 3b^3\sqrt{5b}$$

You Try It 6

$$3a\sqrt{28a^9b^{18}} = 3a\sqrt{2^2 \cdot 7 \cdot a^9 \cdot b^{18}}$$
$$= 3a\sqrt{2^2a^8b^{18}(7a)} = 3a\sqrt{2^2a^8b^{18}}\sqrt{7a}$$
$$= 3a \cdot 2 \cdot a^4b^9\sqrt{7a} = 6a^5b^9\sqrt{7a}$$

You Try It 7

$$\sqrt{25(a+3)^2} = \sqrt{5^2(a+3)^2} = 5(a+3)$$
$$= 5a + 15$$

You Try It 8

$$\sqrt{x^2 + 14x + 49} = \sqrt{(x+7)^2} = x + 7$$

You Try It 9

$$9\sqrt{3} + 3\sqrt{3} - 18\sqrt{3} = -6\sqrt{3}$$

You Try It 10

$$2\sqrt{50} - 5\sqrt{32} = 2\sqrt{2 \cdot 5^2} - 5\sqrt{2^5}$$
$$= 2\sqrt{5^2}\sqrt{2} - 5\sqrt{2^4}\sqrt{2} = 2 \cdot 5\sqrt{2} - 5 \cdot 2^2\sqrt{2}$$
$$= 10\sqrt{2} - 20\sqrt{2} = -10\sqrt{2}$$

You Try It 11

$$\sqrt{27b} - 2\sqrt{12b} + 7\sqrt{3b}$$
$$= \sqrt{3^3 b} - 2\sqrt{2^2 \cdot 3b} + 7\sqrt{3b}$$
$$= \sqrt{3^2}\sqrt{3b} - 2\sqrt{2^2}\sqrt{3b} + 7\sqrt{3b}$$
$$= 3\sqrt{3b} - 2 \cdot 2\sqrt{3b} + 7\sqrt{3b}$$
$$= 3\sqrt{3b} - 4\sqrt{3b} + 7\sqrt{3b} = 6\sqrt{3b}$$

You Try It 12

$$y\sqrt{28y} + 7\sqrt{63y^3} = y\sqrt{2^2 \cdot 7y} + 7\sqrt{3^2 \cdot 7 \cdot y^3}$$
$$= y\sqrt{2^2}\sqrt{7y} + 7\sqrt{3^2 \cdot y^2}\sqrt{7y}$$
$$= y \cdot 2\sqrt{7y} + 7 \cdot 3 \cdot y\sqrt{7y} = 2y\sqrt{7y} + 21y\sqrt{7y}$$
$$= 23y\sqrt{7y}$$

You Try It 13

$$2\sqrt{27a^5} - 4a\sqrt{12a^3} + a^2\sqrt{75a}$$
$$= 2\sqrt{3^3 \cdot a^5} - 4a\sqrt{2^2 \cdot 3 \cdot a^3} + a^2\sqrt{3 \cdot 5^2 \cdot a}$$
$$= 2\sqrt{3^2 \cdot a^4}\sqrt{3a} - 4a\sqrt{2^2 \cdot a^2}\sqrt{3a} + a^2\sqrt{5^2}\sqrt{3a}$$
$$= 2 \cdot 3 \cdot a^2\sqrt{3a} - 4a \cdot 2 \cdot a\sqrt{3a} + a^2 \cdot 5\sqrt{3a}$$
$$= 6a^2\sqrt{3a} - 8a^2\sqrt{3a} + 5a^2\sqrt{3a} = 3a^2\sqrt{3a}$$

Section 11.2 *pages 513–516*

You Try It 1

$$\sqrt{5a}\sqrt{15a^3b^4}\sqrt{3b^5} = \sqrt{225a^4b^9} = \sqrt{3^2 5^2 a^4 b^9}$$
$$= \sqrt{3^2 5^2 a^4 b^8}\sqrt{b} = 3 \cdot 5a^2b^4\sqrt{b} = 15a^2b^4\sqrt{b}$$

You Try It 2

$$\sqrt{5x}(\sqrt{5x} - \sqrt{25y}) = (\sqrt{5x})^2 - \sqrt{5^3 xy}$$
$$= \sqrt{5^2x^2} - \sqrt{5^2}\sqrt{5xy} = 5x - 5\sqrt{5xy}$$

You Try It 3

$$(2\sqrt{x} + 7)(2\sqrt{x} - 7) = 4(\sqrt{x})^2 - 7^2 = 4x - 49$$

You Try It 4

$$(3\sqrt{x} - \sqrt{y})(5\sqrt{x} - 2\sqrt{y})$$
$$= 15(\sqrt{x})^2 - 6\sqrt{xy} - 5\sqrt{xy} + 2(\sqrt{y})^2$$
$$= 15\sqrt{x^2} - 11\sqrt{xy} + 2\sqrt{y^2}$$
$$= 15x - 11\sqrt{xy} + 2y$$

You Try It 5

$$\frac{\sqrt{15x^6y^7}}{\sqrt{3x^7y^9}} = \sqrt{\frac{15x^6y^7}{3x^7y^9}} = \sqrt{\frac{5}{xy^2}}$$
$$= \frac{\sqrt{5}}{y\sqrt{x}} = \frac{\sqrt{5}}{y\sqrt{x}} \cdot \frac{\sqrt{x}}{\sqrt{x}} = \frac{\sqrt{5x}}{xy}$$

You Try It 6

$$\frac{\sqrt{y}}{\sqrt{y} + 3} = \frac{\sqrt{y}}{\sqrt{y} + 3} \cdot \frac{\sqrt{y} - 3}{\sqrt{y} - 3} = \frac{y - 3\sqrt{y}}{y - 9}$$

You Try It 7

$$\frac{\sqrt{27x^3} - 3\sqrt{12x}}{\sqrt{3x}} = \frac{\sqrt{27x^3}}{\sqrt{3x}} - \frac{3\sqrt{12x}}{\sqrt{3x}} = \sqrt{\frac{27x^3}{3x}} - 3\sqrt{\frac{12x}{3x}}$$
$$= \sqrt{9x^2} - 3\sqrt{4} = \sqrt{3^2x^2} - 3\sqrt{2^2}$$
$$= 3x - 3 \cdot 2 = 3x - 6$$

Section 11.3 *pages 519–524*

You Try It 1

$$\sqrt{4x} + 3 = 7$$
$$\sqrt{4x} = 4$$
$$(\sqrt{4x})^2 = 4^2$$
$$4x = 16$$
$$x = 4$$

Check: $\sqrt{4x} + 3 = 7$

$\sqrt{4 \cdot 4} + 3$	7
$\sqrt{4^2} + 3$	7
$4 + 3$	7
$7 = 7$	

The solution is 4.

You Try It 2

$$\sqrt{3x - 2} - 5 = 0$$
$$\sqrt{3x - 2} = 5$$
$$(\sqrt{3x - 2})^2 = 5^2$$
$$3x - 2 = 25$$
$$3x = 27$$
$$x = 9$$

Check: $\sqrt{3x - 2} - 5 = 0$

$\sqrt{3 \cdot 9 - 2} - 5$	0
$\sqrt{27 - 2} - 5$	0
$\sqrt{25} - 5$	0
$\sqrt{5^2} - 5$	0
$5 - 5$	0
$0 = 0$	

The solution is 9.

You Try It 3

$$\begin{aligned} \sqrt{4x+3} &= \sqrt{x+12} \\ (\sqrt{4x+3})^2 &= (\sqrt{x+12})^2 \\ 4x+3 &= x+12 \\ 4x &= x+9 \\ 3x &= 9 \\ x &= 3 \end{aligned}$$

Check:
$$\begin{array}{c|c} \sqrt{4+3} & = \sqrt{x+12} \\ \hline \sqrt{4\cdot 3+3} & \sqrt{3+12} \\ \sqrt{12+3} & \sqrt{15} \\ \sqrt{15} & = \sqrt{15} \end{array}$$

The solution is 3.

You Try It 4

Strategy

To find the perimeter:
- ▶ Use the Pythagorean Theorem to find the length of the hypotenuse.
- ▶ Use the equation for the perimeter of a triangle.

Solution

$$\begin{aligned} c^2 &= a^2 + b^2 \\ c^2 &= 4^2 + 8^2 \\ c^2 &= 16 + 64 \\ c^2 &= 80 \\ \sqrt{c^2} &= \sqrt{80} \\ c &= 4\sqrt{5} \approx 8.9 \end{aligned}$$

$$\begin{aligned} P &= a + b + c \\ P &= 4 + 8 + 8.9 \\ P &= 20.9 \end{aligned}$$

The perimeter is 20.9 in.

You Try It 5

Strategy

The diagonal of a square separates the square into two isosceles right triangles. To find the length of the diagonal of the square, find the hypotenuse of an isosceles right triangle with a leg of length 5 cm.

Solution

$c = \sqrt{2}(\text{length of a leg})$
$c = \sqrt{2}(5) \approx (1.414)5 = 7.07$

The length of a diagonal is 7.1 cm.

Solutions to Chapter 12 You Try It

Section 12.1 *pages 537–540*

You Try It 1

$$\begin{aligned} \frac{3y^2}{2} + y - \frac{1}{2} &= 0 \\ 2\left(\frac{3y^2}{2} + y - \frac{1}{2}\right) &= 2(0) \\ 3y^2 + 2y - 1 &= 0 \\ (3y-1)(y+1) &= 0 \end{aligned}$$

$$\begin{aligned} 3y - 1 &= 0 & \quad y + 1 &= 0 \\ 3y &= 1 & y &= -1 \\ y &= \frac{1}{3} \end{aligned}$$

The solutions are $\frac{1}{3}$ and -1.

You Try It 2

$$\begin{aligned} x^2 + 81 &= 0 \\ x^2 &= -81 \\ \sqrt{x^2} &= \pm\sqrt{-81} \end{aligned}$$

$\sqrt{-81}$ is not a real number.

The equation has no real number solution.

You Try It 3

$$4(x+2)^2 = 9$$
$$(x+2)^2 = \frac{9}{4}$$
$$\sqrt{(x+2)^2} = \pm\sqrt{\frac{9}{4}}$$
$$x+2 = \pm\frac{3}{2}$$
$$x+2 = \frac{3}{2} \qquad x+2 = -\frac{3}{2}$$
$$x = -\frac{1}{2} \qquad x = -\frac{7}{2}$$

The solutions are $-\frac{1}{2}$ and $-\frac{7}{2}$.

You Try It 4

$$7(z+2)^2 = 21$$
$$(z+2)^2 = 3$$
$$\sqrt{(z+2)^2} = \pm\sqrt{3}$$
$$z+2 = \pm\sqrt{3}$$
$$z = -2 \pm \sqrt{3}$$

The solutions are $-2+\sqrt{3}$ and $-2-\sqrt{3}$.

Section 12.2 *pages 543–546*

You Try It 1

$$3x^2 - 6x - 2 = 0$$
$$3x^2 - 6x = 2$$
$$\frac{1}{3}(3x^2 - 6x) = \frac{1}{3}\cdot 2$$
$$x^2 - 2x = \frac{2}{3}$$

Complete the square.

$$x^2 - 2x + 1 = \frac{2}{3} + 1$$
$$(x-1)^2 = \frac{5}{3}$$
$$\sqrt{(x-1)^2} = \pm\sqrt{\frac{5}{3}}$$
$$x - 1 = \pm\frac{\sqrt{15}}{3}$$
$$x-1 = \frac{\sqrt{15}}{3} \qquad x-1 = -\frac{\sqrt{15}}{3}$$
$$x = 1 + \frac{\sqrt{15}}{3} \qquad x = 1 - \frac{\sqrt{15}}{3}$$
$$= \frac{3+\sqrt{15}}{3} \qquad = \frac{3-\sqrt{15}}{3}$$

The solutions are $\frac{3+\sqrt{15}}{3}$ and $\frac{3-\sqrt{15}}{3}$.

You Try It 2

$$x^2 + 6x + 12 = 0$$
$$x^2 + 6x = -12$$
$$x^2 + 6x + 9 = -12 + 9$$
$$(x+3)^2 = -3$$
$$\sqrt{(x+3)^2} = \pm\sqrt{-3}$$

$\sqrt{-3}$ is not a real number.

The quadratic equation has no real number solution.

You Try It 3

$$x^2 + 8x + 8 = 0$$
$$x^2 + 8x = -8$$
$$x^2 + 8x + 16 = -8 + 16$$
$$(x + 4)^2 = 8$$
$$\sqrt{(x + 4)^2} = \pm\sqrt{8}$$
$$x + 4 = \pm 2\sqrt{2}$$

$$x + 4 = 2\sqrt{2} \qquad\qquad x + 4 = -2\sqrt{2}$$
$$x = -4 + 2\sqrt{2} \qquad\qquad x = -4 - 2\sqrt{2}$$
$$\approx -4 + 2.828 \qquad\qquad \approx -4 - 2.828$$
$$\approx -1.172 \qquad\qquad \approx -6.828$$

The solutions are approximately -1.172 and -6.828.

Section 12.3 *pages 549–550*

You Try It 1

$$3x^2 + 4x - 4 = 0$$
$$a = 3, b = 4, c = -4$$
$$x = \frac{-(4) \pm \sqrt{(4)^2 - 4(3)(-4)}}{2 \cdot 3}$$
$$= \frac{-4 \pm \sqrt{16 + 48}}{6}$$
$$= \frac{-4 \pm \sqrt{64}}{6} = \frac{-4 \pm 8}{6}$$

$$x = \frac{-4 + 8}{6} = \frac{4}{6} = \frac{2}{3} \qquad\qquad x = \frac{-4 - 8}{6} = \frac{-12}{6} = -2$$

The solutions are $\frac{2}{3}$ and -2.

You Try It 2

$$x^2 + 2x = 1$$
$$x^2 + 2x - 1 = 0$$
$$a = 1, b = 2, c = -1$$
$$x = \frac{-(2) \pm \sqrt{(2)^2 - 4(1)(-1)}}{2 \cdot 1}$$
$$= \frac{-2 \pm \sqrt{4 + 4}}{2} = \frac{-2 \pm \sqrt{8}}{2}$$
$$= \frac{-2 \pm 2\sqrt{2}}{2} = -1 \pm \sqrt{2}$$

The solutions are $-1 + \sqrt{2}$ and $-1 - \sqrt{2}$.

Section 12.4 *pages 553–554*

You Try It 1

x	y
−2	6
−1	3
0	2
1	3
2	6

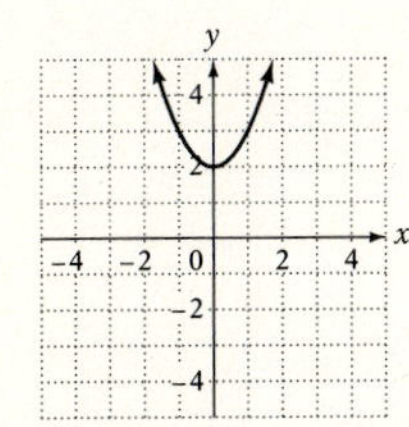

You Try It 2

x	y
−3	−4
−2	−1
−1	0
0	−1
1	−4

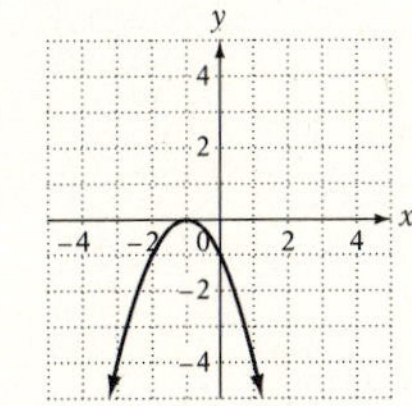

Answers to Chapter 1 Selected Exercises

Section 1.1 *pages 9–12*

1. 6000 + 300 + 90 + 8 **3.** 40,000 + 6000 + 100 + 80 + 2 **5.** 300,000 + 20,000 + 8000 + 400 + 70 + 6
7. 90,000 + 800 + 30 + 4 **9.** 400,000 + 600 + 30 + 5 **11.** 500,000 + 4000 + 600 + 3 **13.** 168,574
15. 7947 **17.** 99,637 **19.** 799 **21.** 12,150 **23.** 89,900 **25.** 1572 **27.** 14,591 **29.** 97,413
31. The Commutative Property of Addition **33.** The Associative Property of Addition **35.** 28 **37.** 4
39. 1618 **41.** 7378 **43.** 17,548 **45.** 13 **47.** 643 **49.** 355 **51.** 5211 **53.** 766 **55.** 18,231
57. The difference is 901. **59a.** $29 billion more is spent on fast food. **b.** $13 billion less is spent on cable TV and video.
61. The total cost is $1645. **63.** The value of the investment is $9284. **65.** The mortgage loan amount is $94,400.
67. The ground speed is 410 mph. **69.** There are 90 two-digit numbers. There are 900 three-digit numbers.
71a. Always true **b.** Always true

Section 1.2 *pages 21–24*

1. 1500 **3.** 2000 **5.** fg **7.** 2292 **9.** 4000 **11.** 12,540 **13.** 265,684
15. The Associative Property of Multiplication **17.** The Commutative Property of Multiplication **19.** 5 **21.** 1
23. $3^6 \cdot 5^3$ **25.** $7^2 \cdot 11^3 \cdot 19^4$ **27.** a^2b^4 **29.** 675 **31.** 17,150 **33.** 36 **35.** 448 **37.** 90,000 **39.** 907
41. 881 r1 **43.** 48 **45.** undefined **47.** 29 r5 **49.** 9800 **51.** 297 **53.** 1, 2, 3, 4, 6, 12
55. 1, 2, 3, 6, 9, 18 **57.** 1, 17 **59.** 1, 2, 3, 4, 6, 9, 12, 18, 36 **61.** 1, 3, 5, 9, 15, 45 **63.** 1, 2, 4, 8, 16, 32
65. 1, 2, 4, 13, 26, 52 **67.** 1, 2, 3, 4, 6, 8, 12, 16, 24, 48 **69.** 1, 2, 3, 6, 9, 18, 27, 54 **71.** 1, 2, 4, 5, 8, 10, 16, 20, 40, 80
73. 1 (the number 60)
75a. There are 4 times more grams of sugar than of protein.
b. There are 11 times more grams of total carbohydrate than of dietary fiber.
77. The total amount paid is $6732. **79.** It would take 8 h to drive 432 mi.
81. The value per share of the fund is $18. **83.** The number 1 million is in column A.

Section 1.3 *pages 27–28*

1. 840 **3.** 3050 **5.** 1600 **7.** 17,600 **9.** 5000 **11.** 85,000 **13.** 390,000 **15.** 750,000
17. 37,000,000 **19.** 15,000; 15,040 **21.** 2000; 2136 **23.** 1,200,000; 1,244,653 **25.** 800; 776
27. 1,400,000; 1,388,917 **29.** 40,000; 38,283 **31.** 1,200,000; 1,138,134 **33.** 5000; 5129 **35.** 6,300,000; 6,491,166
37. 1500; 1516 **39.** 2000; 1998 **41.** 100,000; 125,665 **43.** It is an approximation.

Section 1.4 *pages 33–36*

1. 2^4 **3.** $2^2 \cdot 3$ **5.** $3 \cdot 5$ **7.** $2^3 \cdot 5$ **9.** 37 **11.** $5 \cdot 13$ **13.** $2^4 \cdot 5$ **15.** $2^2 \cdot 7$ **17.** $2 \cdot 3 \cdot 7$
19. 3^4 **21.** 89 **23.** $2 \cdot 3 \cdot 11$ **25.** $2^3 \cdot 3 \cdot 5$ **27.** 9 **29.** 55 **31.** 24 **33.** 42 **35.** 56
37. 90 **39.** 48 **41.** 112 **43.** 140 **45.** 216 **47.** 30 **49.** 24 **51.** 180 **53.** 210 **55.** 3
57. 5 **59.** 25 **61.** 17 **63.** 1 **65.** 4 **67.** 6 **69.** 12 **71.** 15 **73.** 12 **75.** 4 **77.** 5
79. 21 **81.** 8 **83.** 9 **85.** 38 **87.** 22 **89.** 10 **91.** 7 **93.** 15 **95.** 63 **97.** 30 **99.** 29
101. 42 **103.** 2 **105.** 19 **107.** 13 **109.** 9 **111.** 3 **113.** 29 **115.** 21 **117.** 20 **119.** 69
121. 66 **123.** The LCM of x and $2x$ is $2x$. The GCF of x and $2x$ is x. **125.** Answers will vary. For example, 5 + 3(7).

Chapter Review Exercises *pages 39–40*

1. $2 \cdot 3^2 \cdot 5$ [1.4A] **2.** 90,000 [1.2B] **3.** 2583 [1.1C] **4.** $3^2 \cdot 5^4$ [1.2B] **5.** 80 [1.2C] **6.** 38,700 [1.3A]
7. 1 [1.2A] **8.** The Commutative Property of Addition [1.1B] **9.** 700 [1.2A] **10.** 2607 [1.2C] **11.** 675 [1.2B]
12. 1500 [1.3B] **13.** 1, 2, 5, 10, 25, 50 [1.2D] **14.** 932 [1.1B] **15.** 18 [1.4D] **16.** 20 [1.4C] **17.** 2636 [1.2A]
18. 137 [1.1C] **19.** 210 [1.4B] **20.** 900,000 + 6000 + 300 + 70 + 8 [1.1A] **21.** $2 \cdot 3 \cdot 17$ [1.4A]
22. 4,900,000 [1.2B] **23.** 428 [1.1C] **24.** x^4y^2 [1.2B] **25.** 21 [1.2C] **26.** 75,000 [1.3A] **27.** 8 [1.1B]
28. The Commutative Property of Multiplication [1.2A] **29.** 800 [1.2A] **30.** 6971 r23 [1.2C] **31.** 90,000 [1.2B]
32. 2500 [1.3B] **33.** 1, 2, 4, 8, 13, 26, 52, 104 [1.2D] **34.** 7221 [1.1B] **35.** 38 [1.4D] **36.** 6 [1.4C]
37. 3784 [1.2A] **38.** 288 [1.1C] **39.** 90 [1.4B] **40.** 60,000 + 700 + 5 [1.1A] **41.** 116 [1.4D] **42.** 56 [1.4D]
43. Kareem Abdul-Jabbar had 1161 more rebounds. [1.1D] **44.** Each partner's share is $32,000. [1.2E]
45. The amount financed is $379. [1.1D] **46.** The annual dividend is $1275. [1.2E]
47. The odometer should read 407 mi. [1.1D] **48.** The total cost will be about $210,000. [1.2E]
49. The distance traveled is 42 mi. [1.2E] **50.** The markup is $449. [1.1D]

Answers to Chapter 2 Selected Exercises

Section 2.1 *pages 49–52*

1. 1 **3.** −1 **5.** 3 **7.** A is −4. C is −2. **9.** A is −7. D is −4. **11.** $-36 < 49$ **13.** $53 > -46$
15. $-51 < -20$ **17.** $-131 < 101$ **19.** −16, −13, −8, 21 **21.** −6, −3, −1, 2, 8 **23.** −8, −6, −1, 7, 11
25. −4 **27.** 2 **29.** 31 **31.** $-c$ **33.** w **35.** the opposite of negative 11
37. the opposite of the opposite of d **39.** negative 2 plus negative 5 **41.** six minus negative 7 **43.** nine minus 12
45. the opposite of a minus b **47.** 38 **49.** −29 **51.** 52 **53.** m **55.** $-b$ **57.** 4 **59.** 7 **61.** 11
63. 12 **65.** 65 **67.** 15 **69.** −33 **71.** 32 **73.** −36 **75.** −81 **77.** 7 **79.** 2 **81.** 3
83. $|7| < |-9|$ **85.** $|-5| > |-2|$ **87.** $|-8| > |3|$ **89.** $|-14| = |14|$ **91.** $-|-5|, |2|, -(-3), |-8|$
93. $-|3|, |0|, -(-1), |-6|$ **95.** $-7, -|2|, |1|, 6, -(-8)$ **97.** The windchill factor is −25°F.
99. A temperature of 0°F with a 15 mph wind feels colder. **101.** −12 min and counting is closer to blast-off.
103. Stock B showed the least net change. **105.** The loss was greater during the third quarter.
107. The value of x can be −6, −5, −4, −3, −2, −1, 0, 1, 2, 3, 4, 5, 6. **109a.** sometimes true **b.** sometimes true
c. sometimes true **d.** sometimes true **e.** always true **f.** always true

Section 2.2 *pages 59–64*

1. −2 **3.** −11 **5.** −9 **7.** −3 **9.** 1 **11.** −15 **13.** 20 **15.** 0 **17.** 3 **19.** −17 **21.** 11
23. −19 **25.** −21 **27.** −14 **29.** 19 **31.** 47 **33.** −5 **35.** −30 **37.** 9 **39.** −12 **41.** −28
43. −13 **45.** −18 **47.** 11 **49.** 1 **51.** −12 **53.** 2 **55.** 5 **57.** −7 **59.** 20
61. The Commutative Property of Addition **63.** The Inverse Property of Addition **65.** $16 + (-16) = 0$
67. $-11 + (6 + 9) = (-11 + 6) + 9$ **69.** $-2 + (-4) = -4 + (-2)$ **71.** −7 **73.** −9 **75.** 9 **77.** 0
79. 8 **81.** 18 **83.** −28 **85.** −6 **87.** −15 **89.** −15 **91.** 18 **93.** 12 **95.** 4 **97.** 27
99. −106 **101.** −29 **103.** 18 **105.** −6 **107.** −15 **109.** −9 **111.** 11 **113.** 0 **115.** 2
117. 18 **119.** 26 **121.** 11 **123.** −8 **125.** 17 **127.** −1 **129.** 12 **131.** −6 **133.** 12 **135.** 4
137. 7 **139.** 7 **141.** The difference in elevation is 7046 m. **143.** The difference in elevation is 9248 m.
145. After a rise of 7°C, the temperature is −11°C. **147.** The difference was 13°C. **149.** The golfer's score was −3.
151. The distance d is 19 units.

Section 2.3 *pages 69–72*

1. −24 **3.** 6 **5.** 18 **7.** −20 **9.** −16 **11.** 25 **13.** 0 **15.** 42 **17.** −128 **19.** 208
21. −243 **23.** −115 **25.** 238 **27.** −96 **29.** −210 **31.** −224 **33.** −40 **35.** 180 **37.** $-qr$
39. 24 **41.** −24 **43.** 192 **45.** 90 **47.** 800 **49.** The Multiplication Property of Zero
51. The Commutative Property of Multiplication **53.** $-3(-9) = -9(-3)$ **55.** $-81 \cdot 0 = 0$ **57.** −2 **59.** 8
61. 0 **63.** −9 **65.** −24 **67.** 31 **69.** 19 **71.** undefined **73.** −32 **75.** −25 **77.** −10
79. 67 **81.** 9 **83.** −9 **85.** 6 **87.** −6 **89.** The boiling point of argon is −186°C.
91. The average score of the four golfers was −3.
93. The next three numbers in the geometric sequence are 135, −405, and 1215.
95. The next three numbers in the geometric sequence are −192, −768, and −3072.
97. Answers will vary. For example, $(-3)(4) = (-3) + (-3) + (-3) + (-3) = -12$.
99a. sometimes true **b.** always true **c.** always true

Section 2.4 *pages 75–76*

1. −3 **3.** −6 **5.** −5 **7.** −12 **9.** −3 **11.** 19 **13.** 2 **15.** 1 **17.** 14 **19.** 42 **21.** −13
23. −12 **25.** 32 **27.** 30 **29.** 94 **31.** −6 **33.** 2 **35.** 8 **37.** 1 **39.** 15 **41.** 32 **43.** 1
45. 1 **47.** 5 **49.** 28 **51.** −4 **53.** $w = 0, x = 1, y = -1, z = 2$

Chapter Review Exercises *page 79*

1. eight minus negative one [2.1B] **2.** −36 [2.1C] **3.** 13 [2.2B] **4.** 14 [2.2B] **5.** −60 [2.3A] **6.** 3 [2.2A]
7. −9 [2.3B] **8.** −210 [2.3B] **9.** −2 [2.2B] **10.** −18 [2.3A] **11.** −1 [2.2A] **12.** −2 [2.2B]
13. $-7, |-2|, 4, |5|, -(-6)$ [2.1C] **14.** The Commutative Property of Multiplication [2.3A] **15.** −4 [2.4A]
16. 200 [2.3A] **17.** 0 [2.3B] **18.** 15 [2.1B] **19.** −12 [2.2A] **20.** 5 [2.4A] **21.** $-8 > -10$ [2.1A]
22. $-21 + 21 = 0$ [2.2A] **23.** The melting point of oxygen is −213°C. [2.3C]
24. After an increase of 5°C, the temperature is −3°C. [2.2C] **25.** The distance d is 12 units. [2.2C]

Cumulative Review Exercises *page 80*

1. 5 [2.2B] **2.** 12,000 [1.3B] **3.** 3209 [1.2C] **4.** 2 [1.4D] **5.** −82 [2.1C] **6.** 630,000 [1.3A] **7.** 2400 [1.2A] **8.** 21 [2.3B] **9.** 126 [2.3A] **10.** −40 [2.2A] **11.** 1, 2, 4, 11, 22, 44 [1.2D] **12.** 1936 [1.2B] **13.** −26 [2.2B] **14.** 1300 [1.3B] **15.** 9 [2.2B] **16.** −2500 [2.3A] **17.** $3 \cdot 23$ [1.4A] **18.** 3,000,000 + 40,000 + 7000 + 900 + 50 + 3 [1.1A] **19.** −32 [2.4A] **20.** −4 [2.3B] **21.** −3 [2.3B] **22.** 47 [1.4D] **23.** Einstein was 76 years old when he died. [1.1D] **24.** After an increase of 7°C, the temperature is −5°C. [2.2C] **25.** Your sales must be $24,900. [1.1D] **26.** The golfer's score is −8. [2.2C]

Answers to Chapter 3 Selected Exercises

Section 3.1 *pages 89–90*

1. $10\frac{1}{3}$ **3.** 3 **5.** $2\frac{2}{3}$ **7.** $2\frac{3}{8}$ **9.** 21 **11.** $2\frac{11}{14}$ **13.** 1 **15.** $8\frac{3}{5}$ **17.** $\frac{22}{5}$ **19.** $\frac{11}{3}$ **21.** $\frac{51}{8}$ **23.** $\frac{46}{5}$ **25.** $\frac{4}{1}$ **27.** $\frac{16}{9}$ **29.** $\frac{45}{7}$ **31.** $\frac{6}{12}$ **33.** $\frac{9}{24}$ **35.** $\frac{6}{51}$ **37.** $\frac{24}{32}$ **39.** $\frac{108}{18}$ **41.** $\frac{30}{90}$ **43.** $\frac{14}{21}$ **45.** $\frac{42}{49}$ **47.** $\frac{5}{11}$ **49.** $\frac{3}{7}$ **51.** $\frac{1}{3}$ **53.** $\frac{9}{25}$ **55.** 0 **57.** $\frac{5}{3} = 1\frac{2}{3}$ **59.** 1 **61.** $\frac{1}{5}$ **63.** $\frac{2}{21}$ **65.** $\frac{2}{3}$ **67.** $\frac{5}{7} > \frac{2}{3}$ **69.** $\frac{7}{12} < \frac{5}{8}$ **71.** $\frac{11}{14} > \frac{3}{4}$ **73.** $\frac{11}{12} > \frac{7}{9}$ **75.** $\frac{5}{8} > \frac{4}{7}$ **77.** $\frac{11}{30} > \frac{7}{24}$ **79.** $\frac{5}{7}$ is between $\frac{2}{3}$ and $\frac{3}{4}$.

Section 3.2 *pages 105–112*

1. 1 **3.** $\frac{16}{b}$ **5.** $\frac{a + b}{9}$ **7.** $1\frac{1}{6}$ **9.** $\frac{11}{x}$ **11.** $\frac{-a + b}{2}$ **13.** $\frac{11}{12}$ **15.** $\frac{11}{12}$ **17.** $1\frac{7}{12}$ **19.** $-\frac{1}{12}$ **21.** $-\frac{1}{3}$ **23.** $\frac{11}{24}$ **25.** $\frac{1}{12}$ **27.** $15\frac{2}{3}$ **29.** $-\frac{7}{18}$ **31.** $-1\frac{1}{2}$ **33.** $\frac{13}{24}$ **35.** $1\frac{2}{5}$ **37.** $-\frac{5}{12}$ **39.** $1\frac{13}{18}$ **41.** $\frac{29}{30}$ **43.** $-\frac{7}{18}$ **45.** $-\frac{19}{24}$ **47.** $1\frac{5}{24}$ **49.** $-\frac{1}{6}$ **51.** $\frac{1}{6}$ **53.** $\frac{5}{y}$ **55.** $\frac{y - z}{4}$ **57.** $\frac{m + n}{30}$ **59.** $\frac{1}{2}$ **61.** $\frac{9}{16}$ **63.** $\frac{13}{45}$ **65.** $\frac{1}{9}$ **67.** $\frac{1}{4}$ **69.** $\frac{1}{2}$ **71.** $-\frac{17}{18}$ **73.** $-\frac{23}{30}$ **75.** $\frac{8}{15}$ **77.** $-\frac{1}{24}$ **79.** $1\frac{3}{10}$ **81.** $-\frac{1}{6}$ **83.** $\frac{13}{18}$ **85.** $1\frac{3}{20}$ **87.** $-\frac{41}{45}$ **89.** $\frac{2}{3}$ **91.** $-1\frac{1}{4}$ **93.** $\frac{5}{36}$ **95.** $-\frac{7}{10}$ **97.** $\frac{13}{16}$ **99.** $-\frac{11}{18}$ **101.** $\frac{3}{10}$ **103.** $-\frac{1}{3}$ **105.** undefined **107.** $\frac{1}{6}$ **109.** $-\frac{2}{9}$ **111.** $\frac{32}{cd}$ **113.** $\frac{ab}{60}$ **115.** $\frac{2}{27}$ **117.** $-\frac{1}{15}$ **119.** 1 **121.** −9 **123.** $-4\frac{1}{5}$ **125.** −10 **127.** $6\frac{1}{3}$ **129.** $\frac{3}{20}$ **131.** $-\frac{2}{5}$ **133.** $\frac{1}{12}$ **135.** $\frac{9}{28}$ **137.** $\frac{1}{3}$ **139.** $10\frac{1}{2}$ **141.** −1 **143.** $\frac{2}{7}$ **145.** $\frac{1}{38}$ **147.** $-10\frac{1}{2}$ **149.** $\frac{9}{16}$ **151.** $\frac{6}{7}$ **153.** $-\frac{3}{7}$ **155.** $\frac{9}{10}$ **157.** $\frac{1}{6}$ **159.** 12 **161.** $-\frac{1}{12}$ **163.** 0 **165.** $-\frac{9}{10}$ **167.** $\frac{2}{27}$ **169.** $-\frac{63}{mn}$ **171.** $\frac{yz}{40}$ **173.** −8 **175.** $9\frac{5}{6}$ **177.** $1\frac{1}{4}$ **179.** $-1\frac{3}{4}$ **181.** $-\frac{25}{44}$ **183.** $\frac{9}{10}$ **185.** 6 **187.** $2\frac{1}{2}$ **189.** −24 **191.** undefined **193.** $\frac{6}{7}$

195. $\frac{5}{12}$ of the job remains to be done. Yes, they can complete the job in one more day.

197. There were 354 days in one year in the Assyrian calendar.

199. There are $16\frac{1}{2}$ ft in one rod. There are 198 in. in one rod.

201. The car can travel 22 mi on one gallon of gasoline. **203.** The dimensions are 14 in. by 7 in. by $1\frac{3}{4}$ in.

205. The worker can assemble 8 products in one hour. **207.** The pressure on the diver is $21\frac{1}{4}$ lb/in^2.

209. The necessary force is $28\frac{1}{8}$ lb. **211.** $\frac{2}{15} = \frac{1}{10} + \frac{1}{30}; \frac{1}{10} + \frac{1}{30} = \frac{3}{30} + \frac{1}{30} = \frac{4}{30} = \frac{2}{15}$

213a. sometimes true **b.** sometimes true **215a.** $1\frac{1}{12}$ **b.** $1\frac{17}{60}$

Section 3.3 *pages 117–120*

1. $\frac{9}{16}$ **3.** $-\frac{1}{216}$ **5.** $5\frac{1}{16}$ **7.** $\frac{5}{128}$ **9.** $\frac{4}{45}$ **11.** $-\frac{1}{10}$ **13.** $1\frac{1}{7}$ **15.** $-\frac{27}{49}$ **17.** $\frac{16}{81}$
19. $37\frac{1}{27}$ **21.** $\frac{25}{144}$ **23.** $\frac{2}{3}$ **25.** 36 **27.** $\frac{3}{4}$ **29.** $-\frac{8}{9}$ **31.** $\frac{1}{28}$ **33.** $\frac{1}{6}$ **35.** -40 **37.** 6
39. $\frac{18}{35}$ **41.** $-\frac{1}{2}$ **43.** -4 **45.** $-\frac{4}{7}$ **47.** 17 **49.** $-\frac{4}{5}$ **51.** 1 **53.** $\frac{1}{5}$ **55.** $1\frac{1}{5}$ **57.** $\frac{5}{36}$
59. $\frac{11}{32}$ **61.** 1 **63.** 3 **65.** 0 **67.** $1\frac{3}{10}$ **69.** $1\frac{1}{9}$ **71.** $1\frac{15}{16}$ **73.** $\frac{1}{2}$ **75.** $-\frac{1}{3}$ **77.** $\frac{5}{13}$
79. The computer can perform 10^8 operations in approximately 3 min.
81. The total capacity of the gasoline tank is 16 gal.

Chapter Review Exercises *page 123*

1. $9\frac{1}{2}$ [3.1A] **2.** $-\frac{1}{6}$ [3.2B] **3.** $\frac{5}{6}$ [3.2D] **4.** $\frac{2}{7}$ [3.3C] **5.** $3\frac{3}{7}$ [3.2D] **6.** $-2\frac{2}{3}$ [3.2C]
7. $2\frac{11}{12}$ [3.3B] **8.** $-\frac{2}{3}$ [3.2C] **9.** $-3\frac{1}{3}$ [3.2C] **10.** $-\frac{1}{30}$ [3.2A] **11.** $\frac{7}{8} > \frac{17}{20}$ [3.1C] **12.** $\frac{32}{72}$ [3.1B]
13. $\frac{1}{18}$ [3.2B] **14.** $-\frac{1}{3}$ [3.3A] **15.** $\frac{33}{14}$ [3.1A] **16.** $\frac{3}{8}$ [3.2A] **17.** $-\frac{5}{6}$ [3.2D] **18.** $1\frac{3}{40}$ [3.3C]
19. $-\frac{9}{10}$ [3.2B] **20.** $2\frac{1}{4}$ [3.3A] **21.** $\frac{1}{12}$ [3.2A] **22.** $\frac{2}{7}$ [3.1B]
23. $\frac{3}{20}$ of the tank remains to be filled. [3.2E]
24. The employee can assemble 192 products during an 8-hour day. [3.2E]
25. The final velocity is 496 ft/s. [3.2E]

Cumulative Review Exercises *page 124*

1. $\frac{1}{18}$ [3.2A] **2.** $-3\frac{1}{2}$ [3.2C] **3.** 36 [1.4C] **4.** -21 [2.4A] **5.** $-\frac{2}{7}$ [3.2C] **6.** $1\frac{13}{28}$ [3.2B]
7. -13 [2.2B] **8.** $1\frac{7}{12}$ [3.3C] **9.** $-1\frac{1}{9}$ [3.2D] **10.** $-\frac{1}{16}$ [3.2A] **11.** $-\frac{3}{5}$ [3.2D] **12.** 11,272 [1.1B]
13. $\frac{1}{7}$ [3.2B] **14.** $\frac{3}{28}$ [3.3A] **15.** 48 [1.4D] **16.** 75 [1.4B] **17.** 10,000 [1.3B] **18.** $1\frac{5}{8}$ [3.2B]
19. $\frac{31}{4}$ [3.1A] **20.** $2^2 \cdot 5 \cdot 7$ [1.4A] **21.** The projected increase in the population is 330,000 people. [1.1D]
22. 458,400 lb of the fish were caught by anglers. [3.2E] **23.** The cost is \$112.50. [3.2E]
24. The total wages are \$150. [3.2E] **25.** The pressure on the diver is $22\frac{3}{8}$ lb/in^2. [3.2E]

Answers to Chapter 4 Selected Exercises

Section 4.1 *pages 133–136*

1. thousandths **3.** ten-thousandths **5.** hundredths **7.** 0.3 **9.** 0.21 **11.** 0.461 **13.** 0.093 **15.** $\frac{1}{10}$
17. $\frac{47}{100}$ **19.** $\frac{289}{1000}$ **21.** $\frac{9}{100}$ **23.** thirty-seven hundredths **25.** nine and four tenths
27. fifty-three ten-thousandths **29.** sixteen and three thousand one hundred fifty-two ten-thousandths

31. three and one hundred fifty-seven thousandths **33.** nine and thirty-seven hundredths **35.** 0.672 **37.** 9.0407
39. 612.704 **41.** 8034.3003 **43.** 73.02684 **45.** $0.16 < 0.6$ **47.** $5.54 > 5.45$ **49.** $0.047 < 0.407$
51. $1.0008 < 1.008$ **53.** $7.6005 < 7.605$ **55.** $0.31502 < 0.3152$ **57.** 0.309, 0.39, 0.399
59. 0.0024, 0.024, 0.204, 0.24 **61.** 0.0061, 0.059, 0.06, 0.061 **63.** 6.2 **65.** 21.0 **67.** 18.41 **69.** 72.50
71. 936.291 **73.** 47 **75.** 7015 **77.** 2.97527 **79.** The total cost is $83.72.
81. The average life expectancy is longer in Italy.
83a. The minimum payment due is $20. **b.** The minimum payment due is $35.
c. The minimum payment due is $30. **d.** The minimum payment due is $16.99.
e. The minimum payment due is $35. **f.** The minimum payment due is $20.
g. The minimum payment due is $25.

Section 4.2 *pages 149–156*

1. 92.37 **3.** 34.6925 **5.** 42.558 **7.** 83.56 **9.** 25.2653 **11.** 21.26 **13.** 2.768 **15.** −50.7
17. −3.312 **19.** −5.905 **21.** 269.2 **23.** −12.25 **25.** −9.55 **27.** −19.189 **29.** 21.352 **31.** 56.361
33. 3.4716 **35.** −98.38 **37.** −1.714 **39.** −649.36 **41.** 31.09 **43.** 18.39 **45.** −25.665 **47.** 13.535
49. 28.3925 **51.** 23.36 **53.** 11.316 **55.** −27.553 **57.** −1.412 **59.** 120; 119.55 **61.** 35; 31.86
63. 1.6; 1.58 **65.** 12; 12.325 **67.** 30; 28.847 **69.** 70; 72.49 **71.** 0.3; 0.303 **73.** 350; 332.68 **75.** 0.324
77. 0.03316 **79.** 0.54708 **81.** 15.12 **83.** −5.46 **85.** −0.00786 **87.** −473 **89.** 0.141 **91.** 32.3
93. 1.95 **95.** −9.91 **97.** −49.8 **99.** 4.14 **101.** −3.8 **103.** −128.8 **105.** 37.0 **107.** 592
109. 37.942 **111.** 3587 **113.** 0.2481 **115.** 71,920 **117.** 0.38255 **119.** 3540 **121.** 0.009407
123. 3294.2 **125.** 8; 7.5537 **127.** 1.6; 1.9516 **129.** 70; 68.5936 **131.** 30; 32.1485 **133.** 1000; 954.93
135. 2; 2.18 **137.** 100; 103.14 **139.** 25; 28.94 **141.** 50.16 **143.** 174 **145.** −315 **147.** −870
149. −10.759 **151.** −0.08338 **153.** 23.0867 **155.** 132 **157.** −4.06 **159.** −0.24 **161.** 2.06
163. −1.7 **165.** An eighth grader spends 16.1 h more watching TV than doing homework.
167. You can travel 23.6 mi on one gallon of gasoline. **169.** The total cost is $840.06.
171. The bill would be approximately $50. **173.** The cost would be approximately $120.
175. A family of four discards 5256 lb of garbage per year. **177.** The difference is 35.438°C.
179. The temperatue fell 32.22°C. **181.** The markup is $57.07. **183.** The federal earnings are $483.99.
185. The cost per mile is $.46. **187.** The cost is $.06. **189.** The force is −66.15 newtons.
191a. The two zeros do not need to be entered. **b.** The first zero does not need to be entered.
c. Both zeros must be entered. **d.** The first zero does not need to be entered.
193a. never true **b.** always true **c.** always true **d.** always true

Section 4.3 *pages 161–164*

1. 0.375 **3.** $0.4\overline{6}$ **5.** 0.5625 **7.** $1.\overline{6}$ **9.** 0.12 **11.** 2.75 **13.** $3.\overline{2}$ **15.** 2.25 **17.** $4.208\overline{3}$
19. 4.55 **21.** $\frac{3}{5}$ **23.** $\frac{1}{4}$ **25.** $\frac{12}{25}$ **27.** $\frac{1}{8}$ **29.** $\frac{9}{200}$ **31.** $\frac{7}{250}$ **33.** $2\frac{1}{2}$ **35.** $4\frac{11}{20}$ **37.** $1\frac{18}{25}$
39. $7\frac{431}{1000}$ **41.** $\frac{9}{10} > 0.89$ **43.** $\frac{4}{5} < 0.803$ **45.** $0.444 < \frac{4}{9}$ **47.** $0.13 > \frac{3}{25}$ **49.** $\frac{5}{16} > 0.312$ **51.** $\frac{10}{11} > 0.909$
53. $7\frac{13}{20}$ in. of rain fell in West Palm Beach during Tropical Storm Gordon. **55.** $\frac{3}{5}$; 0.6 **57.** $\frac{9}{10}$; 0.9 **59.** $1\frac{2}{5}$; 1.4
61. $\frac{33}{50}$; 0.66 **63.** $\frac{17}{20}$; 0.85 **65.** $\frac{2}{25}$; 0.08 **67.** $1\frac{13}{20}$; 1.65 **69.** $\frac{9}{100}$; 0.09 **71.** $\frac{83}{100}$; 0.83 **73.** $\frac{16}{25}$; 0.64
75. $\frac{3}{70}$ **77.** $\frac{3}{8}$ **79.** $\frac{1}{32}$ **81.** $\frac{5}{11}$ **83.** $\frac{1}{15}$ **85.** $\frac{1}{400}$ **87.** $\frac{23}{400}$ **89.** $\frac{5}{6}$ **91.** $\frac{13}{150}$ **93.** $\frac{1}{30}$
95. 0.091 **97.** 0.167 **99.** 0.009 **101.** 0.1823 **103.** 0.0015 **105.** 0.0505 **107.** 0.0006
109. $\frac{1}{5}$ of those surveyed would use a $50,000 inheritance for retirement. **111.** 37% **113.** 2% **115.** 12.5%
117. 136% **119.** 96% **121.** 209% **123.** 20% **125.** 54% **127.** 18.5% **129.** 37.5% **131.** 44.4%
133. 250% **135.** 191.7% **137.** 40% **139.** 177.8% **141.** 34% **143.** $37\frac{1}{2}$% **145.** $35\frac{5}{7}$% **147.** $57\frac{1}{7}$%
149. $155\frac{5}{9}$% **151.** $23\frac{1}{3}$% **153.** $27\frac{3}{11}$% **155.** $\frac{1}{4}$; 0.25; 25%; $\frac{3}{4}$; 0.75; 75%
157a. This represents $33\frac{1}{3}$% off the regular price. **b.** This represents $\frac{1}{2}$ off the regular price.
159. The medication will have less than 10% of its original potency after 4 weeks.

Section 4.4 *pages 171–174*

1. −5 −4 −3 −2 −1 0 1 2 3 4 5 **3.** −5 −4 −3 −2 −1 0 1 2 3 4 5 **5.** −5 −4 −3 −2 −1 0 1 2 3 4 5 **7.** −5 −4 −3 −2 −1 0 1 2 3 4 5

9. −3 −2 −1 0 1 2 3 4 5 6 7 **11.** −5 −4 −3 −2 −1 0 1 2 3 4 5 **13.** −5 −4 −3 −2 −1 0 1 2 3 4 5 **15.** −7 −6 −5 −4 −3 −2 −1 0 1 2 3

17. −5 −4 −3 −2 −1 0 1 2 3 4 5 **19.** −5 −4 −3 −2 −1 0 1 2 3 4 5 **21.** −3 −2 −1 0 1 2 3 4 5 6 7 **23.** −7 −6 −5 −4 −3 −2 −1 0 1 2 3

25. 10.1 **27.** $-2, 0.4, 2.1$ **29.** the real numbers less than 3 **31.** the real numbers greater than or equal to -1

33. −5 −4 −3 −2 −1 0 1 2 3 4 5 **35.** −5 −4 −3 −2 −1 0 1 2 3 4 5 **37.** −5 −4 −3 −2 −1 0 1 2 3 4 5 **39.** −5 −4 −3 −2 −1 0 1 2 3 4 5

41. $s \geq 50{,}000$. No, a representative who sold 49,000 units has not met the sales goal.
43. $h \leq 9$. Yes, a student taking 8.5 credit hours does fulfill the requirement.
45. $b \leq 1200$. Yes, you have kept within the budget. **47.** $T > 50$. No, it is not safe to store a computer disk at 47.5°F.
49a. integer, negative integer, rational number, real number
b. whole number, integer, positive integer, rational number, real number **c.** rational number, real number
d. rational number, real number **e.** rational number, real number **f.** irrational number, real number
51a. $-2.5, 0$ **b.** $-6.3, -3, 0, 6.7$ **c.** 4, 13.6 **d.** $-4.9, 0, 2.1, 5$
53a. always true **b.** always true **c.** sometimes true

Chapter Review Exercises *page 177*

1. -8.301 [4.2A] **2.** 0.0142 [4.2C] **3.** 89.243 [4.2A] **4.** 5.034 [4.1A] **5.** $\frac{8}{25}$ [4.3B] **6.** 0.11 [4.2C]
7. 50.743 [4.2A] **8.** 3425 [4.2C] **9.** −5 −4 −3 −2 −1 0 1 2 3 4 5 [4.4B] **10.** −7 −6 −5 −4 −3 −2 −1 0 1 2 3 [4.4A]

11. $\frac{3}{7} < 0.429$ [4.3A] **12.** $-1, -0.5, 0.1$ [4.4B] **13.** 440 [4.2B] **14.** -0.1 [4.2C] **15.** 17.5% [4.3C]
16. $8.039 < 8.31$ [4.1B] **17.** -441.2 [4.2A] **18.** $\frac{7}{25}$ [4.3A] **19.** -1110 [4.2C] **20.** 1.25 [4.3B]
21. one hundred twenty-six and four hundred thirty-nine ten-thousandths [4.1A] **22.** The difference is 395.45°C. [4.2D]
23. $g \geq 3.5$. No, a student with a GPA of 3.48 does not qualify for the scholarship. [4.4C]
24. One cup of peas contains the greatest amount of thiamin. [4.1D] **25.** The price is $499.49. [4.2D]

Cumulative Review Exercises *page 178*

1. 0.03879 [4.2C] **2.** 11 [2.4A] **3.** 30 [4.2C] **4.** $\frac{1}{4}$ [4.3B] **5.** -4 [2.2B] **6.** 1900 [1.3B]
7. -18.42 [4.2A] **8.** $\frac{1}{7}$ [3.2C] **9.** −5 −4 −3 −2 −1 0 1 2 3 4 5 [4.4A] **10.** −5 −4 −3 −2 −1 0 1 2 3 4 5 [4.4B]

11. 0.76 [4.3A] **12.** $\frac{4}{9}$ [3.3A] **13.** undefined [2.3B] **14.** $-\frac{11}{21}$ [3.2A] **15.** $2^2 \cdot 5 \cdot 13$ [1.4A] **16.** 2.8 [4.2C]
17. 8,072,092 [1.1A] **18.** 128.6% [4.3C] **19.** 17 [2.4A] **20.** 1600 [1.2B] **21.** $\frac{1}{24}$ [3.2B]
22. The total cost was $5142.50. [4.2D] **23.** The temperature fell 46.62°C. [4.2D] **24.** The cost is $1.56 per visit. [4.2D]
25. The cost is $.64. [4.2D]

Answers to Chapter 5 Selected Exercises

Section 5.1 *pages 191–196*

1. $3x^2, 4x, -9$ **3.** $b, 5$ **5.** $9a^2, -12a, 4b^2$ **7.** $3x^2$ **9.** $1, -6$ **11.** $12, 4$ **13.** $16a$ **15.** $27x$ **17.** $3z$
19. $8x$ **21.** $-7z$ **23.** $-6w$ **25.** 0 **27.** $9s - 8t$ **29.** $6x - 3y$ **31.** $2r + 13p$ **33.** $-3w + 2v$
35. $-9p + 11$ **37.** $2p$ **39.** 6 **41.** $13y^2 + 1$ **43.** $12w^2 - 16$ **45.** $-14w$ **47.** $5a^2b + 8ab^2$ **49.** 5
51. $11x^2 - 2x$ **53.** $12x$ **55.** $-15x$ **57.** $21t$ **59.** $-21p$ **61.** $12q$ **63.** $2x$ **65.** $-15w$ **67.** x
69. $6x^2$ **71.** $-27x^2$ **73.** x^2 **75.** x **77.** c **79.** a **81.** $12w$ **83.** $16vw$ **85.** $10z + 4$
87. $12y + 30z$ **89.** $21x - 27$ **91.** $-2x + 7$ **93.** $4x + 9$ **95.** $-5y - 15$ **97.** $-12x + 18$
99. $-20n + 40$ **101.** $48z - 24$ **103.** $24p + 42$ **105.** $10a + 15b + 5$ **107.** $12x - 4y - 4$
109. $36m - 9n + 18$ **111.** $12v - 18w - 42$ **113.** $8m + 4n - 12$ **115.** $7x + 2$ **117.** $3n + 3$
119. $4a + 4$ **121.** $4a + 1$ **123.** $8x + 42$ **125.** $-12x + 28$ **127.** $-18m - 52$ **129.** $20c + 23$
131. $8a + 5b$ **133.** $15z - 12$ **135.** -19 **137.** $-13x - 2y$ **139.** $-2v + 13$ **141.** $-5c - 6$
143. $2a + 21$ **145.** $11n - 26$ **147.** $-9x + 6$ **149.** $111v - 246$ **151.** $27z^2 - 78z - 36$
153. **a.** False. $8 \div 4 \neq 4 \div 8$. **b.** False. $(8 \div 4) \div 2 \neq 8 \div (4 \div 2)$ **c.** False. $(7 - 5) - 1 \neq 7 - (5 - 1)$
d. False. $6 - 3 \neq 3 - 6$
155. No. Zero does not have a multiplicative inverse.

Section 5.2 *pages 209–214*

1. 56 in. **3.** 14 ft **5.** 47 mi **7.** 8π cm or approximately 25.13 cm **9.** 11π mi or approximately 34.56 mi
11. 17π ft or approximately 53.41 ft **13.** The perimeter is 17.4 cm. **15.** The perimeter is 8 cm.
17. The perimeter is 24 m. **19.** The perimeter is 48.8 cm. **21.** The perimeter is 17.5 in. **23.** The length is 8.4 cm.
25. The circumference is 1.5π in. **27.** The circumference is 226.19 cm. **29.** 60 ft of fencing should be purchased.
31. The carpet must be nailed down along 44 ft. **33.** The length is 13.19 ft. **35.** The bicycle travels 50.27 ft.
37. The distance is 39,935.93 km. **39.** 60 ft^2 **41.** 20.25 in^2 **43.** 546 ft^2 **45.** 16π cm^2 or approximately 50.27 cm^2
47. 30.25π mi^2 or approximately 95.03 mi^2 **49.** 72.25π ft^2 or approximately 226.98 ft^2 **51.** The area is 156.25 cm^2.
53. The area is 570 in^2. **55.** The area is 192 in^2. **57.** The area is 13.5 ft^2. **59.** The area is 330 cm^2.
61. The area is 25π in^2. **63.** The area is 9.08 ft^2. **65.** The area is $10{,}000\pi$ in^2. **67.** The area is 126 ft^2.
69. 7500 yd^2 of artificial turf must be purchased. **71.** You should buy 2 qt of stain. **73.** The cost is $74.
75. The increase in area is 113.10 in^2. **77.** The cost is $638. **79.** The area is 216 m^2.
81. The area of the resulting rectangle is four times larger.
83. **a.** sometimes true **b.** sometimes true **c.** always true **d.** always true **e.** always true **f.** always true

Section 5.3 *pages 221–224*

1. 840 in^3 **3.** 15 ft^3 **5.** 4.5π cm^3 or approximately 14.14 cm^3 **7.** The volume is 34 m^3.
9. The volume is 15.625 in^3. **11.** The volume is 36π ft^3. **13.** The volume is 8143.01 cm^3. **15.** The volume is 75π in^3.
17. The volume is 120 in^3. **19.** There are 75.40 m^3 of oil in the tank. **21.** 94 m^2 **23.** 56 m^2
25. 96π in^2 or approximately 301.59 in^2. **27.** The surface area is 184 ft^2. **29.** The surface area is 69.36 m^2.
31. The surface area is 225π cm^2. **33.** The surface area is 402.12 in^2. **35.** The surface area is 6π ft^2.
37. The surface area is 297 in^2. **39.** 11 cans of paint should be purchased. **41.** 456 in^2 of glass are needed.
43. The surface area of the pyramid is 22.53 cm^2 larger. **45.** **a.** always true **b.** never true **c.** sometimes true
47. S of sphere $= 4\pi r^2$; S of the side of the cylinder $= 2\pi rh = 2\pi r(2r) = 4\pi r^2$

Section 5.4 *pages 229–232*

1. $t + 3$ **3.** $6m - 5$ **5.** $3b - 7$ **7.** $7n$ **9.** $2(3 + w)$ **11.** $4(2r - 5)$ **13.** $\frac{v}{v - 4}$ **15.** $4t^2$
17. $m^2 + m^3$ **19.** $31 - s + 5$ **21.** $x - (x + 12); -12$ **23.** $\frac{2}{3}x - \frac{3}{8}x; \frac{7}{24}x$ **25.** $2(7x + 6); 14x + 12$
27. $11x + 3x; 14x$ **29.** $9(x + 7); 9x + 63$ **31.** $(x + 5) + 7; x + 12$ **33.** $7(x - 4); 7x - 28$ **35.** $10x - 3x; 7x$
37. $x + 2(x - 4); 3x - 8$ **39.** $7(x - 14); 7x - 98$ **41.** $8(x + 10); 8x + 80$ **43.** $x + (7x - 8); 8x - 8$
45. $5 + 2(x + 15); 2x + 35$ **47.** $14 - (x + 13); -x + 1$ **49.** $(8x)2; 16x$ **51.** $x^2 + \frac{x^2}{2}; \frac{3}{2}x^2$
53. $(x + 11) + (x - 17); 2x - 6$ **55.** $(x + 10) + (x - 11); 2x - 1$ **57.** $5(9 - y); 45 - 5y$
59. $3(17 - m) - 9; 42 - 3m$ **61.** $390d$ **63.** $H - 4430$ **65.** $2s$ **67.** $r + 6$ **69.** $3 - L$ **71.** $12 - L$
73. $2x$ **75.** $\frac{4}{7}x$

Chapter Review Exercises *page 235*

1. $2x - 16$ [5.1A] **2.** $-18z - 2$ [5.1C] **3.** x [5.1B] **4.** $-4s + 43t$ [5.1D] **5.** $6w$ [5.1B]
6. $-2m + 16$ [5.1A] **7.** $-4a - 2$ [5.1D] **8.** $-5a^2 + 3a + 9$ [5.1A] **9.** $12c - 4d$ [5.1D]
10. $14m - 42$ [5.1C] **11.** $20x^2 - 48x + 36$ [5.1D] **12.** $6z^2 - 6z$ [5.1A] **13.** $-12c + 32$ [5.1C]
14. $3x - 4y$ [5.1A] **15.** $-a + 13b$ [5.1D] **16.** $14v - 2$ [5.1A] **17.** $\frac{4x}{7} - 9$ [5.4A]
18. The volume is 39 ft^3. [5.3A] **19.** The volume is 288π mm^3. [5.3A] **20.** $3x + 2(x - 7)$; $5x - 14$ [5.4B]
21. Four cans of paint should be purchased. [5.3B] **22.** 208 yd of fencing are needed. [5.2A] **23.** $30 - m$ [5.4C]
24. The area is 90.25 m^2. [5.2B] **25.** The area is 276 m^2. [5.2B]

Cumulative Review Exercises *page 236*

1. $\frac{5}{6}$ [3.2D] **2.** 1, 2, 3, 6, 13, 26, 39, 78 [1.2D] **3.** $3x^2 + 3x + 3$ [5.1A] **4.** $\frac{4}{5}$ [3.3C] **5.** $2\frac{2}{5}$ [3.2C]
6. 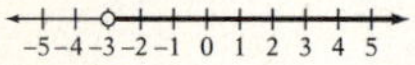

[4.4B] **7.** $7x + 18$ [5.1D] **8.** -2 [2.4A] **9.** $\frac{9}{16}$ [4.3A] **10.** 4 [2.4A] **11.** undefined [2.3B]
12. 1 [3.2C] **13.** 90 [2.4A] **14.** 12.8 [4.2C] **15.** $12x + 21y - 3$ [5.1D] **16.** 400,000 + 70,000 + 300 + 50 + 1 [4.1A]
17. $\frac{10}{x - 9}$ [5.4A] **18.** The temperature is -15°C. [3.2E] **19.** $30d$ [5.4C] **20.** $2(x + 4) - 2$; $2x + 6$ [5.4B]
21. The cost is \$3075. [3.2E] **22.** 23 gal of coffee should be prepared. [4.2D] **23.** The area is 20.25π cm^2. [5.2A]

Answers to Chapter 6 Selected Exercises

Section 6.1 *pages 245–248*

1. Yes **3.** No **5.** Yes **7.** Yes **9.** No **11.** Yes **13.** No **15.** Yes **17.** Yes **19.** Yes
21. No **23.** No **25.** Yes **27.** No **29.** 2 **31.** 9 **33.** 5 **35.** 12 **37.** 5 **39.** -6 **41.** -7
43. -4 **45.** 2 **47.** 7 **49.** 0 **51.** 0 **53.** -3 **55.** -1 **57.** -7 **59.** 9 **61.** 0 **63.** 7
65. $\frac{1}{5}$ **67.** 1 **69.** $\frac{1}{15}$ **71.** $\frac{5}{6}$ **73.** 2 **75.** -5 **77.** 0 **79.** -12 **81.** 3 **83.** 0 **85.** 4
87. -4 **89.** -7 **91.** 4 **93.** $\frac{5}{3}$ **95.** $-\frac{5}{2}$ **97.** $-\frac{7}{3}$ **99.** $\frac{3}{2}$ **101.** $-\frac{9}{4}$ **103.** $\frac{4}{3}$ **105.** 12
107. -25 **109.** -32 **111.** 40 **113.** 24 **115.** -24 **117.** $\frac{7}{10}$ **119.** $\frac{10}{9}$ **121.** 3 **123.** $-\frac{5}{2}$
125. $\frac{b}{a}$; No, $a \neq 0$. **127a.** 4 **b.** 4

Section 6.2 *pages 257–262*

1. 8 **3.** 0.075 **5.** $16\frac{2}{3}$% **7.** 37.5% **9.** 100 **11.** 1200 **13.** 51.895 **15.** 13 **17.** 2.7%
19. 400% **21.** 7.5 **23.** 200 **25.** 2.5% **27.** 37.5% **29.** 80 **31.** 9 **33.** The salary increased by 8%.
35. The estimated safe-life of the brakes is 50,000 mi. **37.** \$403.20 is deducted for income tax.
39. The rent is 32% of last month's income. **41.** The cost was \$71.25. **43.** 20.9% of the injuries happen on slides.
45. The previous year's snowfall was 165 in. **47.** The price was \$49.82. **49.** The tax credit would be \$12,750.
51. 9.4% of the units required are taken in mathematics.
53a. 8000 computer boards were tested. **b.** 7944 computer boards were not defective.
55. Yes, the account executive passed the test. **57.** The increase in price was approximately 6.7%.
59. The addition will increase the home by approximately 18.5%. **61.** The new weight of the chocolate bar is 4.4 oz.
63. The batting average decreased by approximately 11.4%. **65.** The monthly output increased by 300 machines.
67. The percent increase is 120%. **69.** The government order requires a 19.68 mi/gal average.
71. The year's income is \$756. **73.** The loss in value after one year is \$3300.
75. The fuel economy is increased by 20%. **77.** The dividend increase is approximately 8.6%.
79. Borden had the smallest percent decrease in stock price.
81. The monthly expense for gasoline decreased by \$20.24. The average bill for gas is now \$71.76.
83. The time waiting decreased approximately 34.2%.
85. No, the results of taking two consecutive 10% discounts or one 20% discount is not the same. The 20% discount was on the total of 100. The second 10% discount applied only to 90, not 100; thus the difference in results.

87. Yield $= \dfrac{\text{income}}{\text{quoted price}}$; income $= 0.08 \cdot 10{,}000 = 800$; yield $= \dfrac{800}{10{,}500} \approx 0.07619 \approx 7.62\%$
89. The answers are the same. Either method can be used to find the new wage.

Section 6.3 *pages 269–274*

1. 2 **3.** 10 **5.** 2 **7.** -1 **9.** -2 **11.** -4 **13.** 3 **15.** 9 **17.** 4 **19.** $\frac{3}{5}$ **21.** $\frac{7}{2}$ **23.** $\frac{3}{4}$
25. $-\frac{7}{4}$ **27.** 1 **29.** $\frac{2}{5}$ **31.** $-\frac{7}{6}$ **33.** 4 **35.** -6 **37.** $\frac{27}{2}$ **39.** $-\frac{8}{3}$ **41.** $\frac{5}{6}$ **43.** 1.2 **45.** 1.1
47. 5.8 **49.** 4 **51.** 1 **53.** $\frac{11}{3}$ **55.** 3 **57.** 3 **59.** -1 **61.** -4 **63.** 5 **65.** 2 **67.** 1 **69.** $\frac{7}{6}$
71. 3 **73.** $\frac{8}{5}$ **75.** $\frac{7}{2}$ **77.** $\frac{3}{2}$ **79.** $-\frac{3}{2}$ **81.** $-\frac{7}{2}$ **83.** $\frac{7}{2}$ **85.** $\frac{10}{3}$ **87.** 3 **89.** -1 **91.** 0
93. $\frac{5}{6}$ **95.** -3 **97.** 3 **99.** 2 **101.** 2 **103.** 2 **105.** 3 **107.** $\frac{1}{2}$ **109.** 8 **111.** -2
113. $-\frac{6}{5}$ **115.** $-\frac{1}{3}$ **117.** $x + 12 = 20; 8$ **119.** $\frac{3}{5}x = -30; -50$ **121.** $3x + 4 = 13; 3$ **123.** $9x - 6 = 12; 2$
125. $x + 2x = 9; 3$ **127.** $5x - 17 = 2; \frac{19}{5}$ **129.** $6x + 7 = 3x - 8; -5$ **131.** $30 = 7x - 9; \frac{39}{7}$
133. $2x = (21 - x) + 3; 8, 13$ **135.** $23 - x = 2x + 5; 6, 17$ **137.** 41,493
139. No, it does not make sense. The expression $2x - 3(4x + 1)$ is not an equation. Therefore, it cannot be solved.

Section 6.4 *pages 283–288*

1. 40°; acute **3.** 115°; obtuse **5.** 90°; right **7.** 28° **9.** 18° **11.** 14 cm **13.** 28 ft **15.** 30 m
17. 86° **19.** 71° **21.** 30° **23.** 36° **25.** 127° **27.** 116° **29.** 20° **31.** 20° **33.** 20° **35.** 141°
37. 106° **39.** 11° **41.** $\angle a = 38°, \angle b = 142°$ **43.** $\angle a = 47°, \angle b = 133°$ **45.** 20° **47.** 47°
49. $\angle x = 155°, \angle y = 70°$ **51.** $\angle a = 45°, \angle b = 135°$ **53.** $90° - x$ **55.** 60° **57.** 35° **59.** 102°
61a. 1° **b.** 179° **65.** 360°

Section 6.5 *pages 297–304*

1. The mixture contains 2 lb of diet supplement and 3 lb of vitamin supplement. **3.** The cost is \$6.98 per pound.
5. The combination contained 56 oz of the \$4.30 alloy and 144 oz of the \$1.80 alloy. **7.** The cost is \$2.90 per pound.
9. There must be 10 kg of hard candy. **11.** The mixture contains 30 lb of the \$2.20 meat and 20 lb of the \$4.20 meat.
13. 25 gal of ice cream and 75 gal of fruit juice should be used. **15.** The cost is \$2.75 per gallon.
17. The solution must contain 20 ml of the 13% solution and 30 ml of the 18% solution. **19.** The mixture is 50% silver.
21. 30 lb of the 60% mixture is used. **23.** The 150-gram cream contains 0.74% hydrocortisone.
25. 100 ml of the 7% solution and 200 ml of the 4% solution should be mixed. **27.** 25 oz of pure water must be added.
29. The resulting alloy is 27% gold. **31.** 10 oz of pure bran flakes must be added.
33. \$5000 more must be invested at 9%. **35.** There was \$9000 invested at 7% and \$6000 invested at 6.5%.
37. \$2500 was deposited in the mutual fund.
39. \$200,000 should be deposited in the 10% account and \$100,000 deposited in the 8.5% account.
41. \$3000 must be invested in additional bonds. **43.** \$40,500 was invested at 8%, and \$13,500 was invested at 12%.
45. The total amount invested was \$650,000. **47.** The total amount invested was \$500,000.
49. The plane flew 2 h at 105 mph and 3 h at 115 mph. **51.** The sailboat traveled 36 mi.
53. The rate of the passenger train is 50 mph. The rate of the freight train is 30 mph. **55.** The rate of the cyclist is 16 mph.
57. The rate of the first plane is 95 mph. The rate of the second plane is 120 mph. **59.** They will meet after 1 h.
61. The second runner will overtake the first runner after 3 h. **63.** It took the campers 20 min to canoe downstream.
65. The cost is \$3.65 per ounce. **67.** 10 oz of water must be evaporated.
69. 3.75 gal must be drained and replaced by pure antifreeze.
71. \$12,000 was invested in 9% bonds, \$21,000 was invested in the 8% account, and \$27,000 was invested in 9.5% bonds.
73. The campers turned around at 10:15 A.M. **75.** The cyclist's average speed is $13\frac{1}{3}$ mph.

Section 6.6 *pages 311–314*

1. $x < 2$

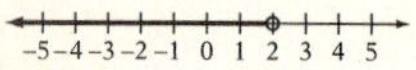

3. $x > 3$

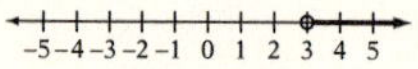

5. $n \geq 3$

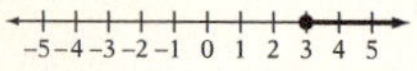

7. $x < 4$

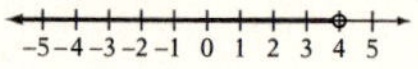

9. $y \geq 3$

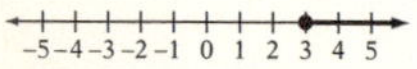

11. $x \leq 1$

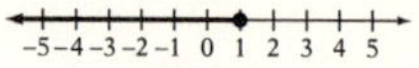

13. $y \geq -9$ **15.** $x < 12$ **17.** $x \geq 5$ **19.** $x < -11$ **21.** $x \leq 10$ **23.** $x \geq -\frac{31}{24}$ **25.** $x \leq -\frac{1}{6}$
27. $b \geq \frac{1}{48}$ **29.** $x < -7.3$ **31.** $n \leq 7.77$ **33.** $x \leq 0.70$ **35.** $z > 0$ **37.** $x \leq -\frac{1}{6}$ **39.** $n < 18$
41. $y \geq 6$ **43.** $x \leq 6$ **45.** $b \leq 33$ **47.** $x > \frac{10}{7}$ **49.** $y \leq \frac{5}{6}$ **51.** $x \leq 4.2$ **53.** $d < -2.1$ **55.** $x < 1$
57. $y < \frac{1}{2}$ **59.** $x \geq 1$ **61.** $n \leq -\frac{13}{3}$ **63.** $x > 0$ **65.** $x > 12.5$ **67.** $x \leq -\frac{5}{3}$ **69.** $d > \frac{10}{27}$ **71.** $x < \frac{14}{11}$
73. $x \leq -2$ **75.** $y \geq 3$ **77.** The team must win 11 or more games.
79. The student must receive a grade of 78 or better. **81.** The agent expects to sell $20,000 or less in one month.
83. A company must have more than 7 computers. **85.** The butcher can mix 75 lb of fat or less with the lean meat.
87. The maximum distance you can drive Company B's car is 166 mi. **89.** 1, 2 **91.** There are no solutions.

Chapter Review Exercises *page 317*

1. -4 [6.1C] **2.** $\frac{3}{2}$ [6.3A] **3.** $x > 2$ 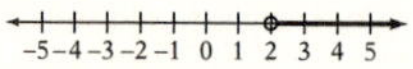 [6.6A] **4.** 44 cm [6.4A]
5. 3 [6.3C] **6.** 5.625% [6.2A] **7.** $x \geq -4$ [6.6B] **8.** -3 [6.1B] **9.** $\angle x = 22°$, $\angle y = 158°$ [6.4C]
10. $\angle a = 138°$, $\angle b = 42°$ [6.4B] **11.** $\frac{1}{2}$ [6.3B] **12.** Yes [6.1A] **13.** 562.50 [6.2A] **14.** $x \geq -3$ [6.6A]
15. $x \geq 4$ [6.6B] **16.** -5 [6.3C] **17.** $7 - 5x = 37; x = -6$ [6.3D] **18.** The airline would sell 196 tickets. [6.2B]
19. The percent decrease is 25%. [6.2C] **20.** $9600 is invested at 4%, and $14,400 is invested at 9%. [6.5C]
21. The resulting mixture is 14% butterfat. [6.5B] **22.** 7 qt of cranberry juice and 3 qt of apple juice were used. [6.5A]
23. The jet overtakes the propeller-driven plane 600 mi from the starting point. [6.5D]
24. The nursing home has more than 4 residents. [6.6C]

Cumulative Review Exercises *page 318*

1. 92 [2.4A] **2.** 9 [1.4C] **3.** -23 [2.1C] **4.** 11.09 [4.1A] **5.** $-2y^2 - 4$ [5.1A] **6.** $\frac{1}{6}$ [3.2C]
7. 131° [6.4B] **8.** $-\frac{1}{2} > -\frac{2}{3}$ [3.1C] **9.** The Commutative Property of Multiplication [5.1B]
10. -8 and -6.1 [4.4B] **11.** 5 [6.3B] **12.** 56 [2.4A] **13.** $-21b - 36$ [5.1D] **14.** 20% [6.2A] **15.** 1 [3.3B]
16. $\frac{1}{2}$ [6.3C] **17.** The pressure is 25 lb/in^2. [3.2E] **18.** The rent increased 9%. [6.2C]
19. 80% of the students went on to college. [6.2B] **20.** The pool holds 519.54 ft^3 of water. [5.3A]
21. The perimeter is 15.5 m. [5.2A] **22.** The smallest integer that satisfies the inequality is 32. [6.6C]
23. 1 L of pure water should be added. [6.5B]

Answers to Chapter 7 Selected Exercises

Section 7.1 *pages 331–336*

1.
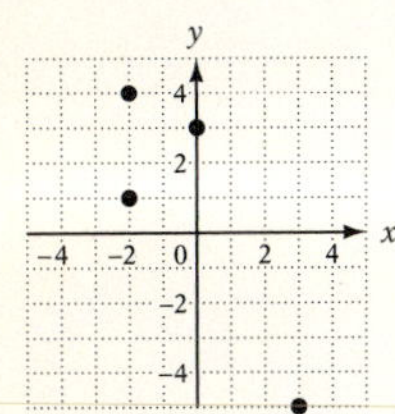

3.
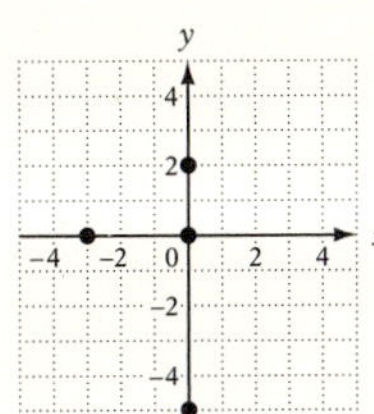

5.
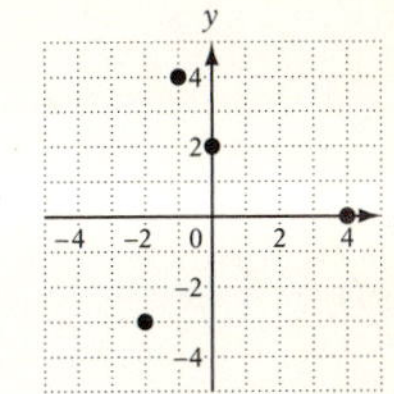

7. A is (2, 3), B is (4, 0), C is $(-4, 1)$, and D is $(-2, -2)$. **9.** A is $(-2, 5)$, B is (3, 4), C is (0, 0), and D is $(-3, -2)$. **11. a.** The abscissa of point A is 2. The abscissa of point C is -4. **b.** The ordinate of point B is 1. The ordinate of point D is -3. **13.** Yes **15.** No **17.** No **19.** Yes **21.** No **23.** (3, 7) **25.** (6, 3) **27.** (0, 1) **29.** $(-5, 0)$

31.
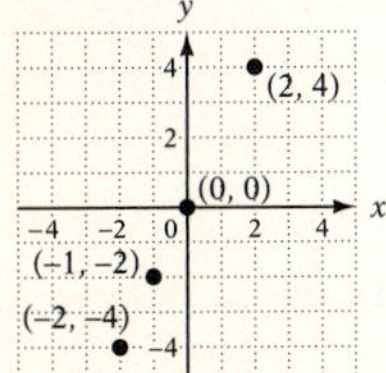

33.
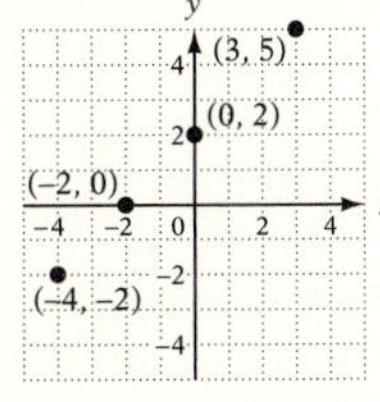

35.
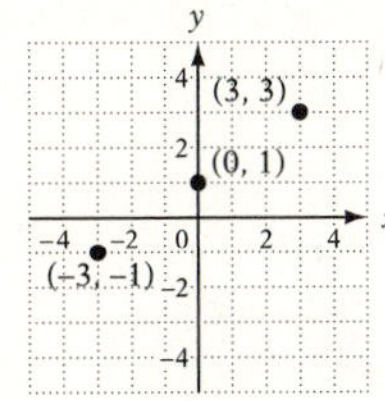

37.
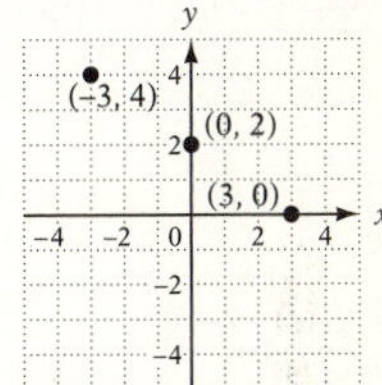

39.
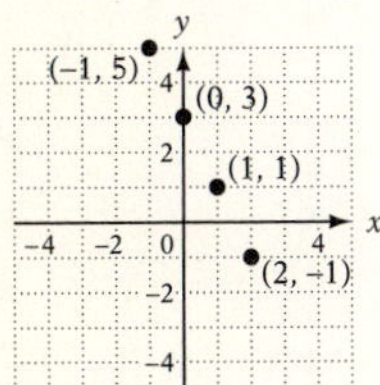

41.
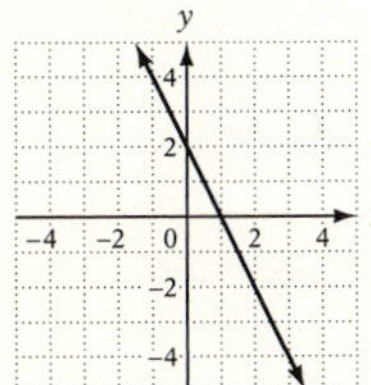

43.
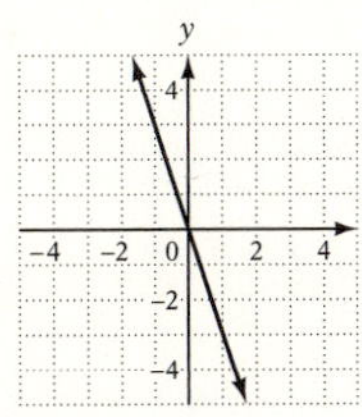

45.
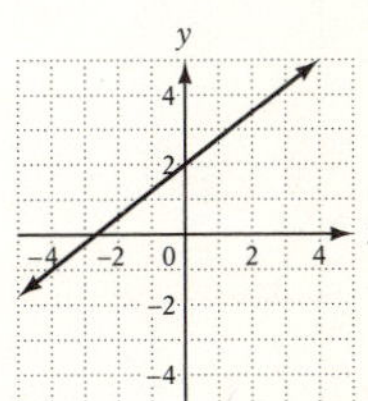

47.
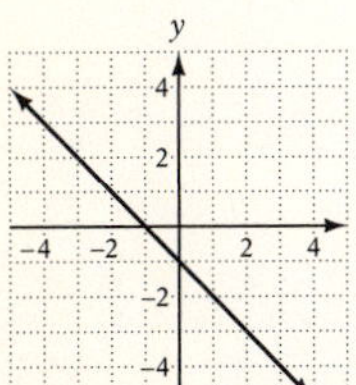

49.
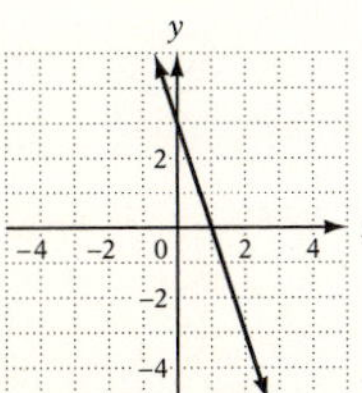

51.
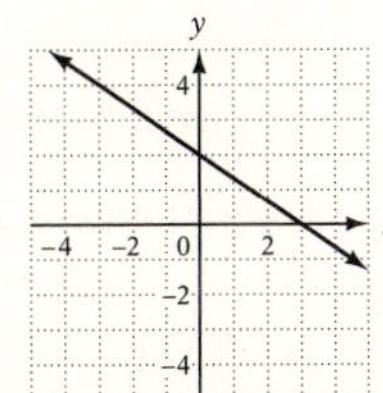

53.
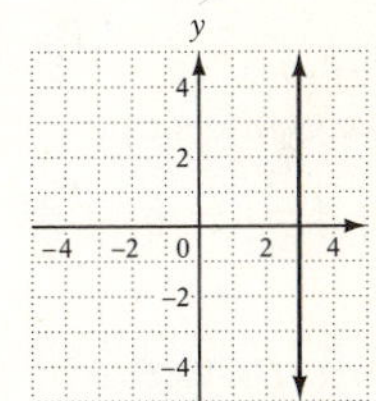

55.
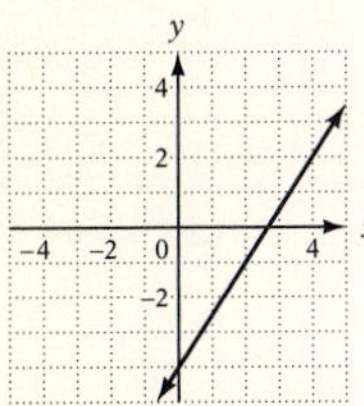

57.
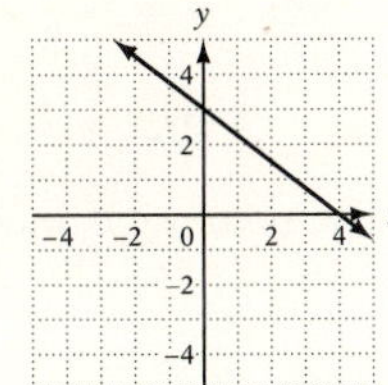

59.
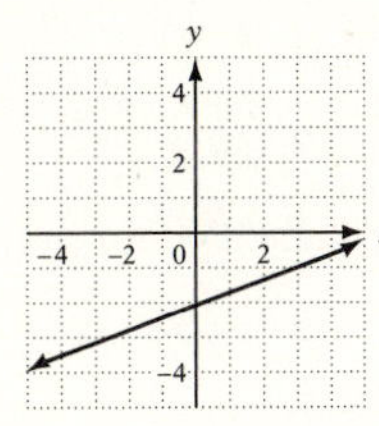

61. Quadrant IV **63.** Quadrant III **65.** 4 units

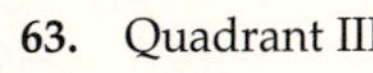

67. 2 units **69.** 5 units

71. a. The x-coordinate is positive.
The y-coordinate is positive.

b. The x-coordinate is negative.
The y-coordinate is positive.

c. The x-coordinate is negative.
The y-coordinate is negative.

d. The x-coordinate is positive.
The y-coordinate is negative.

73. a. $y = -\frac{1}{2}x - 5$
$(-2, -4)$

75. one unit

77.

Section 7.2 *pages 343–346*

1. -2 3. $\frac{1}{3}$ 5. $-\frac{5}{2}$ 7. $-\frac{1}{2}$ 9. -1 11. undefined 13. 0 15. $-\frac{1}{3}$ 17. 0 19. -5
21. undefined 23. $-\frac{2}{3}$
25. The slope is 0.8. After being connected, each minute of a transatlantic phone call costs an additional $.80.
27. The slope is -0.1. The price of Merck stock decreased by $.10 each day.

29.

31.

33. 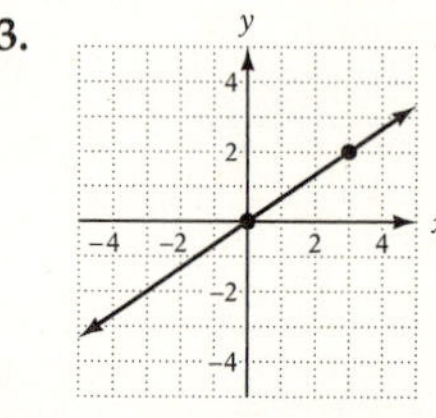

35. 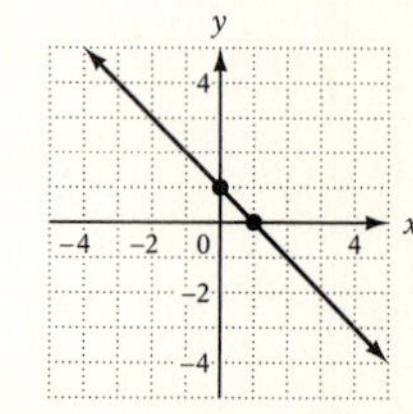

37.

39.

41. 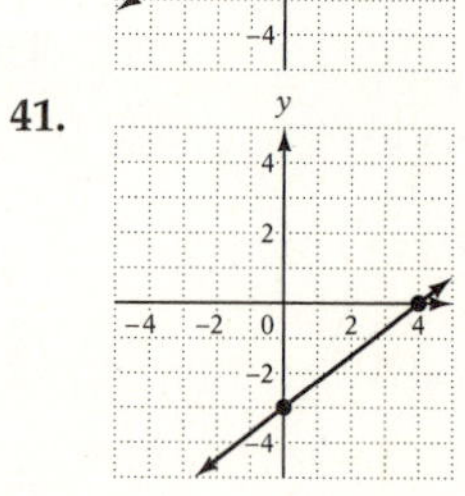

43. 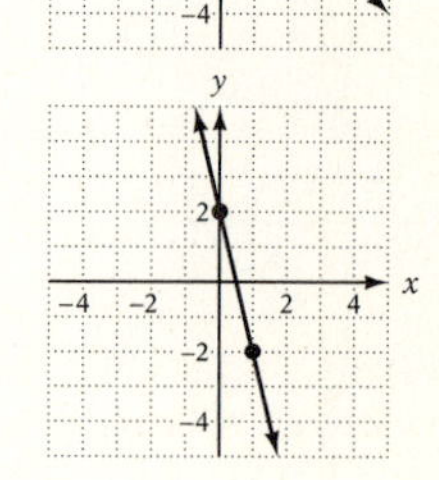

45. It increases the slope. 47. It moves the graph higher on the coordinate axis.
49. a. $y = -\frac{3}{4}x + 3$ b. x-intercept: (4.0); y-intercept: (0, 3) 51. yes

Section 7.3 *pages 351–354*

1. $y = 2x + 2$ 3. $y = -3x - 1$ 5. $y = \frac{1}{3}x$ 7. $y = \frac{3}{4}x - 5$ 9. $y = -\frac{3}{5}x$ 11. $y = \frac{1}{4}x + \frac{5}{2}$
13. $y = 2x - 3$ 15. $y = -2x - 3$ 17. $y = \frac{2}{3}x$ 19. $y = \frac{1}{2}x + 2$ 21. $y = -\frac{3}{4}x - 2$ 23. $y = \frac{3}{4}x + \frac{5}{2}$
25. The tread depth decreases 0.2 mm for each 1000 mi driven. 27. The revenue is increasing $1.6 billion per year.
29. No 31. Yes 33. $-\frac{3}{2}$ 35. -5 37. $y = -\frac{2}{3}x + \frac{5}{3}$

Section 7.4 *pages 363–368*

1. Yes 3. Yes 5. No 7. No 9. No

11. The solution is (4, 1).

13. The solution is (4, 1).

15. The solution is (4, 3).

17. The solution is (3, -2).

19. 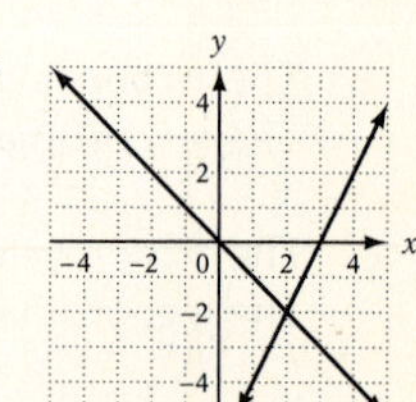
The solution is (2, -2).

21. 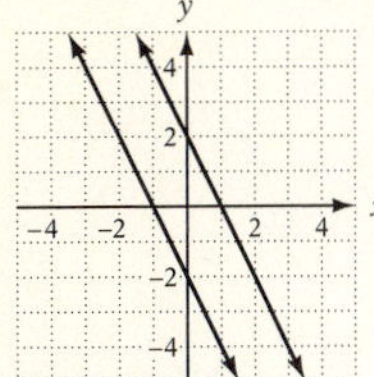
The system of equations has no solution.

23. 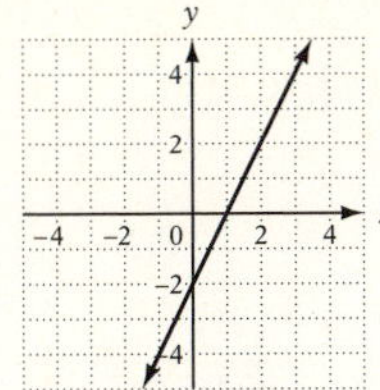
Any solution of one equation is the solution of the other equation.

25. 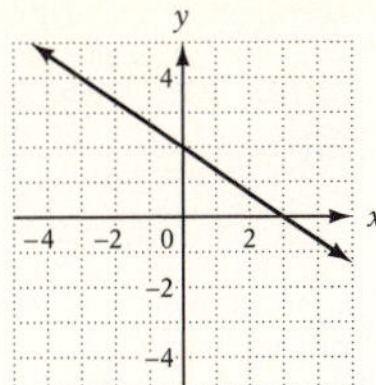
Any solution of one equation is the solution of the other equation.

27. 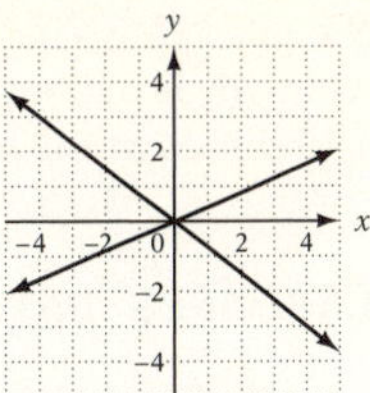
The solution is (0, 0).

29. (2, 1) **31.** (4, 1) **33.** $(-1, 1)$ **35.** (3, 1) **37.** (1, 1) **39.** $(-1, 1)$ **41.** No solution
43. No solution **45.** $\left(-\frac{3}{4}, -\frac{3}{4}\right)$ **47.** (5, 7) **49.** (1, 7) **51.** (0, 0) **53.** (10, 31) **55.** $(3, -10)$
57. $(-22, -5)$ **59.** $(-6, -19)$ **61.** $(-2, 1)$ **63.** $(2, -1)$ **65.** $(-1, -1)$ **67.** $(2, -1)$ **69.** (4, 1)
71. Any solution of one equation is a solution of the other equation. **73.** $(2, -1)$
75. Any solution of one equation is a solution of the other equation. **77.** (0, 0) **79.** $(5, -2)$ **81.** $\left(\frac{32}{19}, -\frac{9}{19}\right)$
83. (3, 4) **85.** $(1, -1)$ **87.** Any solution of one equation is a solution of the other equation. **89.** (3, 1)
91. $(-1, 2)$ **93.** (1, 1) **95.** $\left(\frac{1}{2}, -\frac{1}{2}\right)$ **97.** $\left(\frac{2}{3}, \frac{1}{9}\right)$ **99.** $\left(\frac{7}{25}, -\frac{1}{25}\right)$ **101.** 2 **103.** 2 **105.** $\frac{2}{3}$
107. $k \neq 1$ **109.** $k \neq 4$ **111.** $y = x - 1; x = 3$

Section 7.5 *pages 373–374*

1. The rate of the plane in calm air is 400 mph. The rate of the wind is 50 mph.
3. The rate of the boat in calm water is 7 mph. The rate of the current is 3 mph.
5. The rate of the plane in calm air is 125 mph. The rate of the wind is 25 mph.
7. The rate of the plane in calm air is 105 mph. The rate of the wind is 15 mph.
9. The cost per copy of a word processing program is \$245. The cost per copy of a spreadsheet program is \$325.
11. The dividend per share of oil stock is \$.25. The dividend per share of movie stock is \$.45.
13. The number of touchdowns is 3. The number of field goals is 4. **15.** The measures of the angles are 55° and 125°.
17. There is no solution to this problem. **19.** There are 7 different combinations of nickels and dimes.

Chapter Review Exercises *page 377*

1. 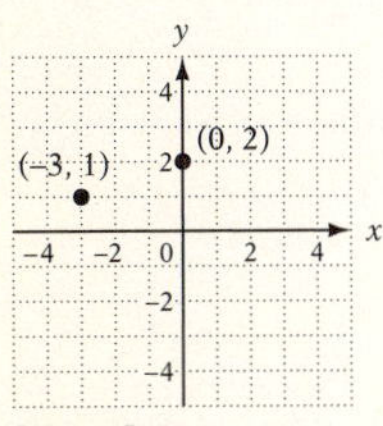

[7.1A]

2. 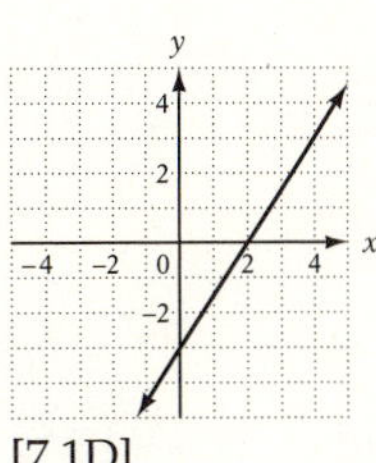
[7.1D]

3. $(0, -3)$ and (2, 0) [7.1D]

4. 2 [7.2A]

5. $y = 3x - 1$ [7.3A]

6. $y = \frac{1}{2}x + 2$ [7.3A]

7. 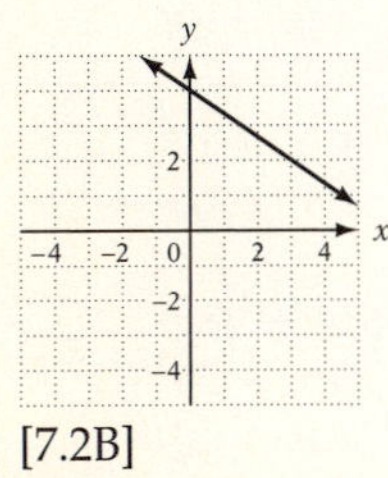
[7.2B]

8. 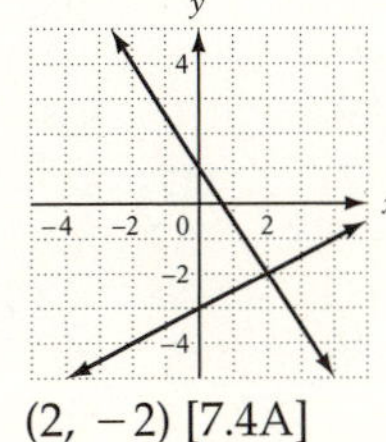
$(2, -2)$ [7.4A]

9. Yes [7.4A] **10.** $(2, -1)$ [7.4B] **11.** $\left(\frac{1}{2}, -1\right)$ [7.4C] **12.** $(1, -2)$ [7.4C]

13. Tuition is increasing \$1500 per year. [7.2A]
14. The rate of the plane in calm air is 110 mph. The rate of the wind is 30 mph. [7.5A]
15. There are 30 nickels and 15 dimes in the bank. [7.5B]

Cumulative Review Exercises *page 378*

1. $-17x + 28$ [5.1D] **2.** $\frac{1}{15}$ [4.3B] **3.** $\frac{3}{2}$ [6.3A] **4.** -12 [2.4A] **5.** $-\frac{5}{8}$ [2.4A] **6.** $x < -20$ [6.6A]
7. $\angle a = 43°$; $\angle b = 137°$ [6.4B] **8.** $\angle x = 29°$ [6.4A] **9.** $6y^4 + 8y^3 - 16y^2$ [5.1C] **10.** $2\frac{3}{11}$ [3.3C]
11. -6.8 [2.1C] **12.** $\frac{5}{9} > 0.5$ [4.1B] **13.** 7 [2.2B] **14.** $\frac{7}{4}$ [6.3B] **15.** $\frac{2}{3}$ [6.1C] **16.** $15x^2$ [3.2C]
17.

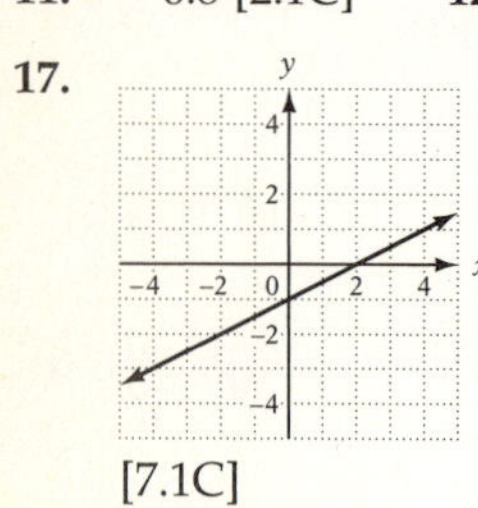

[7.1C]

18.

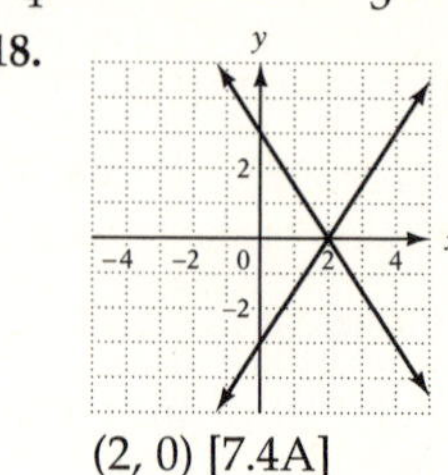

(2, 0) [7.4A]

19. $8x + 4$ [5.4B] **20.** The length of the rectangle is 20 cm. [5.2A] **21.** The distance to the resort is 168 mi. [6.5D]

Answers to Chapter 8 Selected Exercises

Section 8.1 *pages 383–384*

1. $3x^3 + 8x^2 - 2x - 6$ **3.** $5a^3 - 3a^2 + 2a + 1$ **5.** $r^5 + r^3 - 6r^2 + 5$ **7.** $-y^4 + 1$ **9.** $-9a^2 + a + 3$
11. 4 **13.** 6 **15.** 1 **17.** $8y^2 + 4y$ **19.** $3x^2 + 15x - 24$ **21.** $4x^2 - 9x + 9$ **23.** $-y^3 + y^2 - 6y - 2$
25. $5r^3 - 5r^2 + r - 3$ **27.** $y^2 + 7y$ **29.** $8x^2 - 2xy$ **31.** $-3x^2 + 8x + 6$ **33.** $3y^3 + 2y^2 + 8y - 12$
35. $3y^3 - 4y^2 + 4y + 35$ **37.** $7x^3 + 2x^2 - 2x + 1$ **39.** $-6y$ **41.** $-7a^2 - 2a + 4$ **43.** $-2x^2 + 7x - 8$
45. $2y^3 + 5y^2 - 4y + 5$ **47.** $x^3 - 2x^2 - 4x + 6$ **49.** $3x^2 - 4xy$ **51.** $8y^2 - y + 3$
53. $-3x^3 + 2x^2 + 3x + 2$ **55.** $-4b^3 + b^2 - b + 15$ **57.** $-2x^3 + 5x^2 - 2x - 7$ **59.** trinomial **61.** No
63. Yes **65.** No

Section 8.2 *pages 391–394*

1. $2x^2$ **3.** $12x^2$ **5.** $6a^7$ **7.** x^3y^5 **9.** $-10x^9y$ **11.** x^7y^8 **13.** $-6x^3y^5$ **15.** x^4y^5z **17.** $a^3b^5c^4$
19. $-a^5b^8$ **21.** $-6a^5b$ **23.** $40y^{10}z^6$ **25.** $x^3y^3z^2$ **27.** $30x^6y^8$ **29.** $-24a^3b^3c^3$ **31.** 64 **33.** 4
35. -64 **37.** x^9 **39.** x^{14} **41.** x^4 **43.** $4x^2$ **45.** $-8x^6$ **47.** x^4y^6 **49.** $9x^4y^2$ **51.** x^8y^4 **53.** a^4b^6
55. $16x^{10}y^3$ **57.** $-8a^7b^5$ **59.** $-54a^9b^3$ **61.** $-72a^5b^5$ **63.** $x^2 - 2x$ **65.** $-x^2 - 7x$ **67.** $3a^3 - 6a^2$
69. $-5x^4 + 5x^3$ **71.** $12x^3 - 6x^2$ **73.** $6x^2 - 12x$ **75.** $3x^2 + 4x$ **77.** $-x^3y + xy^3$ **79.** $2x^4 - 3x^2 + 2x$
81. $2a^3 + 3a^2 + 2a$ **83.** $3x^6 - 3x^4 - 2x^2$ **85.** $-6y^4 - 12y^3 + 14y^2$ **87.** $-2a^3 - 6a^2 + 8a$ **89.** $6y^4 - 3y^3 + 6y^2$
91. $x^3y - 3x^2y^2 + xy^3$ **93.** $x^3 + 4x^2 + 5x + 2$ **95.** $a^3 - 6a^2 + 13a - 12$ **97.** $-2b^3 + 7b^2 + 19b - 20$
99. $-6x^3 + 31x^2 - 41x + 10$ **101.** $x^3 - 3x^2 + 5x - 15$ **103.** $x^4 - 4x^3 - 3x^2 + 14x - 8$
105. $15y^3 - 16y^2 - 70y + 16$ **107.** $5a^4 - 20a^3 - 5a^2 + 22a - 8$ **109.** $x^2 + 4x + 3$ **111.** $a^2 + a - 12$
113. $y^2 - 5y - 24$ **115.** $y^2 - 10y + 21$ **117.** $2x^2 + 15x + 7$ **119.** $3x^2 + 11x - 4$ **121.** $4x^2 - 31x + 21$
123. $3y^2 - 2y - 16$ **125.** $6a^2 + ab - 2b^2$ **127.** $2x^2 - 3xy - 2y^2$ **129.** $10x^2 + 29xy + 21y^2$
131. $6a^2 - 25ab + 14b^2$ **133.** $2a^2 - 11ab - 63b^2$ **135.** $100a^2 - 100ab + 21b^2$ **137.** $15x^2 + 56xy + 48y^2$
139. $14x^2 - 97xy - 60y^2$ **141.** $y^2 - 36$ **143.** $16x^2 - 49$ **145.** $y^2 - 6y + 9$ **147.** $36x^2 - 60x + 25$
149. $81x^2 - 4$ **151.** $x^2 + 6xy + 9y^2$ **153.** $4x^2 - 12xy + 9y^2$ **155.** $4 - 25x^2$ **157.** $4a^2 - 36ab + 81b^2$
159. $4 + 28x + 49x^2$ **161.** $-16a^9$ **163.** $-7x^4y^4$ **165.** a^{2n} **167.** a^{n+2} **169.** $2x^2 + 6xy$
171. $x^3 + 12x^2 + 48x + 64$ **173.** $b^2 - 14b$ **175.** $x^{2n} - 1$ **177.** $x^{2n} - 2x^n + 1$ **179.** True
181. False; $x^3 + x^3 = 2x^3$ **183.** $x^{(m^n)}$ **185.** $7x^2 - 11x - 8$ **187.** 5

Section 8.3 *pages 399–400*

1. $3x$ **3.** $-x$ **5.** $4x^3$ **7.** $-\frac{4}{x^2}$ **9.** $\frac{a}{b^4}$ **11.** y^3 **13.** $\frac{3}{5}$ **15.** $\frac{24}{b}$ **17.** $-\frac{3}{5ab^2}$ **19.** $-\frac{4b}{9}$
21. $-\frac{2x^2y^2}{11z^5}$ **23.** $-8a^3b^4$ **25.** $\frac{4a^2}{9b^3}$ **27.** $\frac{x^2y^2}{z^3}$ **29.** $\frac{a^2}{b}$ **31.** $x + 1$ **33.** $2a - 5$ **35.** $3a + 2$
37. $4b^2 - 3$ **39.** $x - 2$ **41.** $-x + 2$ **43.** $x^2 + 3x - 5$ **45.** $x^4 - 3x^2 - 1$ **47.** $xy + 2$ **49.** $-3y^3 + 5$
51. $3x - 2 + \frac{1}{x}$ **53.** $-3x + 7 - \frac{6}{x}$ **55.** $4a - 5 + 6b$ **57.** $9x + 6 - 3y$ **59.** $x + 5$ **61.** $b - 7$
63. $y - 5$ **65.** $2y - 7$ **67.** $2y + 6 + \frac{25}{y - 3}$ **69.** $x - 2 + \frac{8}{x + 2}$ **71.** $3y - 5 + \frac{10}{y + 2}$
73. $6x - 12 + \frac{19}{x + 2}$ **75.** $b - 5 - \frac{24}{b - 3}$ **77.** $x^2 - 3$ **79.** xy^2 **81.** $\frac{x^6z^6}{4y^5}$ **83.** $3a^2$
85. $4x^2 - 6x + 3$ **87.** 1

Section 8.4 *pages 405–406*

1. $\frac{1}{5^2} = \frac{1}{25}$ **3.** $\frac{1}{3^3} = \frac{1}{27}$ **5.** $\frac{1}{2^6} = \frac{1}{64}$ **7.** $\frac{1}{x^2}$ **9.** $\frac{1}{a^6}$ **11.** $\frac{x^2}{y^3}$ **13.** $\frac{1}{xy^2}$ **15.** x **17.** $\frac{1}{a^7}$ **19.** $\frac{1}{x^4}$
21. $\frac{1}{a^8}$ **23.** $\frac{y}{x^3}$ **25.** $\frac{b}{a^2}$ **27.** $\frac{1}{a^4}$ **29.** $\frac{1}{a^6}$ **31.** x^6 **33.** a^{18} **35.** 1 **37.** $\frac{y^4}{x^4}$ **39.** $\frac{x}{y^5}$ **41.** $\frac{y^4}{x^4}$
43. $\frac{y^2}{x^4}$ **45.** $-\frac{8x^3}{y^6}$ **47.** $-\frac{5}{a^8}$ **49.** $\frac{3a^4}{8b^5}$ **51.** $\frac{1}{a^5b^6}$ **53.** $\frac{y^4}{4x^3}$ **55.** $\frac{1}{x^3}$ **57.** $\frac{1}{x^{12}y^{12}}$
59. $4.67 \cdot 10^{-6}$ **61.** $4.3 \cdot 10^6$ **63.** 0.000000123 **65.** 634,000 **67.** 20,800,000
69. 150,000 **71.** 0.000000015 **73.** 20,000,000,000 **75.** Light travels $2.592 \cdot 10^{10}$ km in one day.
77. The sun is $3.38983 \cdot 10^5$ times heavier than the earth. **79.** One light year is a distance of $5.785344 \cdot 10^{12}$ mi.
81. $\frac{3}{64}$ **83.** 0.0625
85. $2^{-2} = \frac{1}{4}, 2^{-1} = \frac{1}{2}, 2^0 = 1, 2^1 = 2, 2^2 = 4; 2^{-(-2)} = 4; 2^{-(-1)} = 2, 2^{-0} = 1, 2^{-1} = \frac{1}{2}, 2^{-2} = \frac{1}{4}$

Chapter Review Exercises *page 409*

1. $-8x^3 - 2x^2 + 5x + 3$ [8.1A] **2.** 4 [8.1A] **3.** $3x^3 + 6x^2 - 8x + 3$ [8.1A] **4.** $-5a^3 + 3a^2 - 4a + 3$ [8.1B]
5. $-6x^3y^6$ [8.2A] **6.** x^8y^{12} [8.2B] **7.** $4x^3 - 6x^2$ [8.2C] **8.** $6y^4 - 9y^3 + 18y^2$ [8.2C] **9.** $x^3 - 7x^2 + 17x - 15$ [8.2C]
10. $-4x^4 + 8x^3 - 3x^2 - 14x + 21$ [8.2C] **11.** $a^2 + 3ab - 10b^2$ [8.2D] **12.** $10x^2 - 43xy + 28y^2$ [8.2D]
13. $16y^2 - 9$ [8.2E] **14.** $4x^2 - 20x + 25$ [8.2E] **15.** $-\frac{4}{x^6}$ [8.3A] **16.** $\frac{9y^6}{x}$ [8.3A] **17.** $4x^4 - 2x^2 + 5$ [8.3B]
18. $x + 7$ [8.3C] **19.** $2x + 3 + \frac{2}{2x - 3}$ [8.3C] **20.** $\frac{a^4}{b^6}$ [8.4A] **21.** $-\frac{6b}{a}$ [8.4A] **22.** $5 \cdot 10^{-8}$ [8.4B]
23. 0.039 [8.4B] **24.** It takes the space vehicle 12 h to reach the moon. [8.4C]

Cumulative Review Exercises *page 410*

1. -48 [6.1C] **2.** 22% [6.2A] **3.** $-9x$ [5.1B] **4.** 12 [5.1D] **5.** $(-2, -5)$ [7.1B] **6.** Yes [7.4A]
7. 0 [7.2A] **8.** $x \le -\frac{9}{2}$ [6.6B] **9.** $\frac{1}{13}$ [6.3C] **10.** $10a^3 - 39a^2 + 20a - 21$ [8.2C] **11.** $\frac{1}{2b^2}$ [8.3A]
12. $a - 7$ [8.3C] **13.** 6.8% [4.3C] **14.** $\frac{1}{x^5y^5}$ [8.4A] **15.** -27 [2.4A] **16.** $4 \cdot 10^3$ [8.4B]

17.

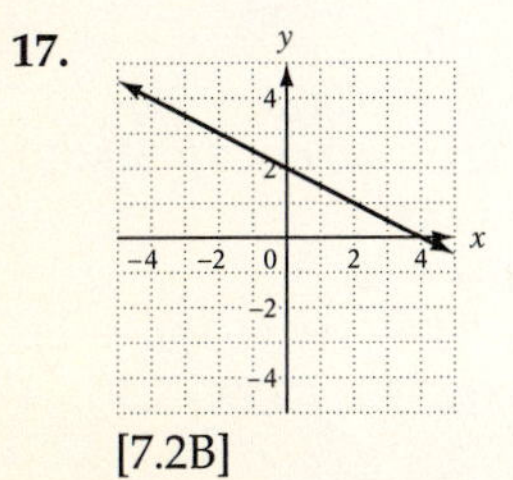

[7.2B]

18.

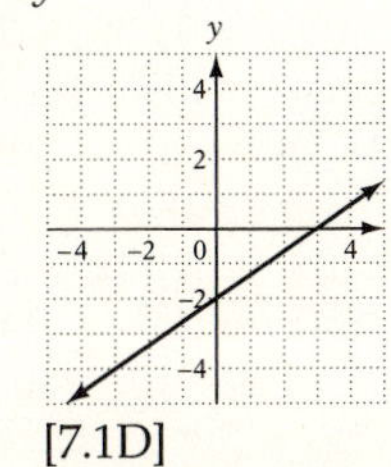

[7.1D]

19. The equation of the line is $y = \frac{1}{2}x - 2$. [7.3A] **20.** The measures of the other two angles are 37° and 90°. [6.4C]

21. The surface area of the sphere is 196π in^2. [5.3B] **22.** The surface area of the pyramid is 400 cm^2. [5.3B]
23. The new dividend is $1.62. [6.2B]

Answers to Chapter 9 Selected Exercises

Section 9.1 *pages 415–416*

1. x^3 **3.** xy^4 **5.** xy^4z^2 **7.** ab^2c^3 **9.** $3x^2$ **11.** $2a$ **13.** $7a^3$ **15.** 1 **17.** $3a^2b^2$ **19.** ab **21.** ab
23. $4x$ **25.** $4a$ **27.** $3m^2n^2$ **29.** $5(a + 1)$ **31.** $6(1 - 3x)$ **33.** $2(x^2 - 10)$ **35.** $8(2 - a^2)$
37. $x(7x - 3)$ **39.** $a^2(3 + 5a^3)$ **41.** $x(9 - 5x)$ **43.** $b^2(6b - 5)$ **45.** $3y(y^3 - 3)$ **47.** $4a^2(3a^3 - 8)$
49. $m^2n^2(m^2 - n^2)$ **51.** $3ab(4ab^4 - 3)$ **53.** $a^2b(1 + a^2b)$ **55.** $5x^2y - 7ab^3$ **57.** $4x^2(2y^3 - 1)$
59. $b(b^2 - 5b - 7)$ **61.** $4(2y^2 - 3y + 8)$ **63.** $5y(y^2 - 4y + 2)$ **65.** $3y^2(y^2 - 3y - 2)$ **67.** $3y(y^2 - 3y + 8)$
69. $a^2(6a^3 - 3a - 2)$ **71.** $ab(2a - 5ab + 7b)$ **73.** $2b(2b^4 + 3b^2 - 6)$ **75.** $x^2(8y^2 - 4y + 1)$ **77.** 496
79. **a.** $2r^2(4 - \pi)$ **b.** $r^2(4 - \pi)$

Section 9.2 *pages 421–422*

1. $(x + 1)(x + 2)$ **3.** $(x + 1)(x - 2)$ **5.** $(a + 4)(a - 3)$ **7.** $(a - 2)(a - 1)$ **9.** $(a + 2)(a - 1)$
11. $(b - 3)(b - 3)$ **13.** $(b + 8)(b - 1)$ **15.** $(y + 11)(y - 5)$ **17.** $(y - 2)(y - 3)$ **19.** $(z - 5)(z - 9)$
21. $(z + 8)(z - 20)$ **23.** $(p + 3)(p + 9)$ **25.** $(x + 10)(x + 10)$ **27.** $(b + 4)(b + 5)$ **29.** $(x + 3)(x - 14)$
31. $(b + 4)(b - 5)$ **33.** $(y + 3)(y - 17)$ **35.** $(p + 3)(p - 7)$ **37.** Nonfactorable over the integers
39. $(x - 5)(x - 15)$ **41.** $(x + 1)(2x + 1)$ **43.** $(y + 3)(2y + 1)$ **45.** $(a - 1)(2a - 1)$ **47.** $(t + 2)(2t - 5)$
49. $(p - 5)(3p - 1)$ **51.** $(3y - 1)(4y - 1)$ **53.** Nonfactorable over the integers **55.** $(2t - 1)(3t - 4)$
57. $(x + 4)(8x + 1)$ **59.** Nonfactorable over the integers **61.** $(3y + 1)(4y + 5)$ **63.** $(2t + 5)(6t - 1)$
65. $(b + 8)(8b + 1)$ **67.** $(5x - 3)(5x - 3)$ **69.** $(4b + 5)(5b + 3)$ **71.** $(3b - 2)(5b - 11)$ **73.** $(3y + 4)(8y + 3)$
75. 6, 10, 12 **77.** 6, 10, 12 **79.** 4, 6

Section 9.3 *pages 429–432*

1. $(x + 2)(x - 2)$ **3.** $(a + 9)(a - 9)$ **5.** $(2x + 1)(2x - 1)$ **7.** $(x^3 + 3)(x^3 - 3)$ **9.** $(5x + 1)(5x - 1)$
11. $(1 + 7x)(1 - 7x)$ **13.** Nonfactorable over the integers **15.** $(x^2 + y)(x^2 - y)$ **17.** $(3x + 4y)(3x - 4y)$
19. $(xy + 2)(xy - 2)$ **21.** $(y + 1)^2$ **23.** $(a - 1)^2$ **25.** Nonfactorable over the integers **27.** $(x + y)^2$
29. $(2a + 1)^2$ **31.** $(8a - 1)^2$ **33.** $(4b + 1)^2$ **35.** $(2b + 7)^2$ **37.** $(5a + 3b)^2$ **39.** $(7x + 2y)^2$
41. $(a + b)(x + 2)$ **43.** $(b + 2)(x - y)$ **45.** $(x - 3)(z - 1)$ **47.** $(b - 2c)(x + y)$ **49.** $(x - 2)(a - 5)$
51. $(y - 2)(b - 2a)$ **53.** $(y - 3)(b - 3)$ **55.** $(x - y)(a + 2)$ **57.** $3(x + 2)(x + 3)$ **59.** $4(x + 1)(x - 2)$
61. $a(b + 8)(b - 1)$ **63.** $x(y + 3)(y + 5)$ **65.** $2a(a + 1)(a + 2)$ **67.** $4y(y + 6)(y - 3)$ **69.** $2x(x + 1)(x - 2)$
71. $6(z + 5)(z - 3)$ **73.** $3a(a + 3)(a - 6)$ **75.** $(x + 7y)(x - 3y)$ **77.** $(a - 5b)(a - 10b)$ **79.** $(s + 8t)(s - 6t)$
81. $2(x + 3)(x - 3)$ **83.** $y(y - 5)^2$ **85.** $a^2(a - 3)(a - 8)$ **87.** $6(y - 2)(y - 6)$
89. Nonfactorable over the integers **91.** $3b(a + 9)(a - 2)$ **93.** $a^2(b + 11)(b - 8)$ **95.** $b^2(a + 3)^2$
97. $xy(x - 2y)(2x - 3y)$ **99.** $3(x + 3)(4x - 1)$ **101.** $10(y + 1)(3y - 2)$ **103.** $x(x + 1)(2x - 5)$
105. $b(a + 3)(2a - 7)$ **107.** Nonfactorable over the integers **109.** $2(x + y)(3x + 2y)$ **111.** $(a - 3b)(2a - 3b)$
113. $(2y + 5z)(y + z)$ **115.** $(1 + x)(2 - x)$ **117.** $(3 - z)(5 + z)$ **119.** $(1 + x)(12 - x)$ **121.** $4(2x + 1)(2x - 3)$
123. $6(2y + 1)(2y - 3)$ **125.** $z(2z - 5)(3z - 4)$ **127.** $y(x - 3)(8x - 3)$ **129.** $5(3x + 4)(4x + 1)$
131. $a^2(3a + 1)(5a + 7)$ **133.** $5(b + 2)(5b - 3)$ **135.** $(2x + 3y)(2x + 5y)$ **137.** $4(9y + 1)(10y - 1)$
139. $(1 + x)(18 - x)$ **141.** $(3a - 5b)(5a - 2b)$ **143.** $3a(2a - 1)^2$ **145.** $x^2(x + 5)(x - 5)$ **147.** $2(x - 1)(a + b)$
149. $(x - 2)(x + 1)(x - 1)$ **151.** $(x + 2)(x - 2)(a + b)$ **153.** 10, −10 **155.** 16, −16 **157.** 4 **159.** 25
161. 19
163. $(2n + 1)^2 - 1 = (2n + 1 - 1)(2n + 1 + 1) = 2n(2n + 2) = 4n(n + 1)$
Since n or $n + 1$ is an even number, $4n(n + 1)$ is divisible by 8.

Section 9.4 *pages 437–440*

1. $-3, -2$ **3.** $7, 3$ **5.** $0, 5$ **7.** $0, 9$ **9.** $0, -\frac{3}{2}$ **11.** $0, \frac{2}{3}$ **13.** $-2, 5$ **15.** $9, -9$ **17.** $\frac{7}{2}, -\frac{7}{2}$
19. $\frac{1}{3}, -\frac{1}{3}$ **21.** $-4, -2$ **23.** $2, -7$ **25.** $2, 3$ **27.** $3, -7$ **29.** $-\frac{1}{2}, 5$ **31.** $-\frac{1}{3}, -\frac{1}{2}$ **33.** $0, 3$
35. $0, 7$ **37.** $-1, -4$ **39.** $2, 3$ **41.** $\frac{1}{2}, -4$ **43.** $\frac{1}{3}, 4$ **45.** $3, 9$ **47.** $9, -2$ **49.** $-1, -2$
51. $5, -9$ **53.** $4, -7$ **55.** $-2, -3$ **57.** $-8, 9$ **59.** $1, 4$ **61.** $-5, 2$ **63.** $3, 4$ **65.** $-\frac{1}{3}, -4$
67. The number is -3. **69.** The consecutive positive integers are 4 and 6.
71. The two consecutive positive even integers are 12 and 14. **73.** The height of the triangle is 14 m.
75. The length of the rectangle is 18 ft. The width is 8 ft. **77.** The length of a side of the original square is 4 m.
79. The radius of the original circle is 3.8 ft. **81.** The width of the border is 2 ft.
83. The object will hit the ground in 4 s. **85.** 15 consecutive natural numbers sum to 120.
87. There are 10 teams in the league. **89.** In 6 s the ball will return to the ground. **91.** $\frac{3}{2}, -4$ **93.** $-1, -9$
95. $0, 7$ **97.** $18, 1$ **99.** 2 or -128 **101.** The length is 20 in. The width is 10 in.

Chapter Review Exercises *page 443*

1. $4ab^3$ [9.1A] **2.** $2x(3x^2 - 4x + 5)$ [9.1B] **3.** $(p + 2)(p + 3)$ [9.2A] **4.** $(a - 3)(a - 16)$ [9.2A]
5. $(x + 5)(x - 3)$ [9.2A] **6.** $(x - 2)(a + b)$ [9.3B] **7.** $5(x^2 - 9x - 3)$ [9.1B] **8.** $2y^2(y + 1)(y - 8)$ [9.3C]
9. Nonfactorable over the integers [9.2B] **10.** $(2x + 1)(3x + 8)$ [9.2B] **11.** $(p + 1)(x - 1)$[9.3B]
12. $4(x + 4)(2x - 3)$ [9.3C] **13.** $3y^2(2x^2 + 3x + 4)$ [9.1B] **14.** $y(y - 3)(y + 3)$ [9.3C]
15. $(2x + 7y)(2x - 7y)$ [9.3A] **16.** $(a + b)(a - b)(x - y)$ [9.3C] **17.** $(2a - 3b)^2$ [9.3A] **18.** $\frac{3}{2}, -7$ [9.4A]
19. $3, 5$ [9.4A] **20.** $\frac{1}{2}, -\frac{1}{2}$ [9.4A] **21.** The length is 15 cm. The width is 6 cm. [9.4B]
22. The measure of the base is 12 in. [9.4B] **23.** The two negative integers are -12 and -13. [9.4B]

Cumulative Review Exercises *page 444*

1. $-32x^8y^7$ [8.2B] **2.** $-3x^2$ [8.3A] **3.** $(-6, 1)$ [7.4B] **4.** $15b^2 - 31b + 14$ [8.2D] **5.** $4\frac{19}{32}$ [3.2C]
6. 0.00081 [8.4B] **7.** -16 [6.3B] **8.** 45 [6.2A] **9.** $-13x + 33$ [5.1D] **10.** $\frac{y^6}{x^8}$ [8.4A]
11. $(p - 10)(p + 1)$ [9.2A] **12.** 3 [6.3C] **13.** $6x^5y^5$ [8.2A] **14.** $x \geq \frac{1}{9}$ [6.6B] **15.** $x(3x - 4)(x + 2)$ [9.3C]
16. $(2, -5)$ [7.4B] **17.** $x^3 - 3x^2 - 6x + 8$ [8.2C] **18.** $x^2 + 2x + 4$ [8.3C]
19.

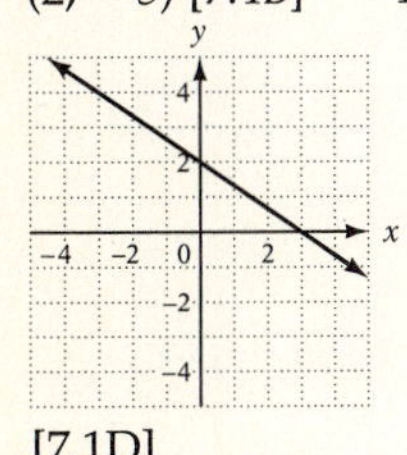

[7.1D]

20. $x = 52°$ [6.4B]

21. The alloy is 70% silver. [6.5B] **22.** The rate of the current is 3 mph. [7.5A]
23. The measures of the two angles are 37° and 53°. [6.4A] **24.** The radius is 1.91 in. [5.2A]

Answers to Chapter 10 Selected Exercises

Section 10.1 pages 451–454

1. $\frac{3}{4x}$ **3.** $\frac{1}{x+3}$ **5.** -1 **7.** $\frac{2}{3y}$ **9.** $-\frac{3}{4x}$ **11.** -5 **13.** $\frac{a}{b}$ **15.** $-\frac{2}{x}$ **17.** $\frac{y-2}{y-3}$ **19.** $\frac{x+5}{x+4}$
21. $\frac{x+4}{x-3}$ **23.** $-\frac{x+2}{x+5}$ **25.** $\frac{2(x+2)}{x+3}$ **27.** $\frac{2x-1}{2x+3}$ **29.** $-\frac{x+7}{x+6}$ **31.** $-\frac{y+2}{y+5}$ **33.** $\frac{2}{3xy}$ **35.** $\frac{8xy^2ab}{3}$
37. $\frac{2}{9}$ **39.** $\frac{y(x+4)}{x(x+1)}$ **41.** $\frac{x^3(x-7)}{y^2(x-4)}$ **43.** $\frac{x+3}{x+1}$ **45.** $\frac{x-5}{x+3}$ **47.** $\frac{x+2}{x+4}$ **49.** $\frac{2x-5}{2x-1}$ **51.** $\frac{3x-4}{2x+3}$
53. $\frac{2xy^2ab^2}{9}$ **55.** $\frac{5}{12}$ **57.** $3x$ **59.** $\frac{y(x+3)}{x(x+1)}$ **61.** $\frac{x+7}{x-7}$ **63.** $-\frac{4ac}{y}$ **65.** $\frac{x-5}{x-6}$ **67.** 1 **69.** $-\frac{x+6}{x+5}$
71. $\frac{2x+3}{x-6}$ **73.** $\frac{4x+3}{2x-1}$ **75.** $\frac{(2x+5)(4x-1)}{(2x-1)(4x+5)}$ **77.** $-6, 1$ **79.** $1, -1$ **81.** $3, -2$ **83.** -3
85. $\frac{4}{3}, -\frac{1}{2}$ **87.** $\frac{4a}{3b}$ **89.** $\frac{8}{9}$ **91.** $\frac{x-4}{y^4}$ **93.** $\frac{(x-2)(x-2)}{(x-4)(x+4)}$

Section 10.2 pages 461–466

1. $24x^3y^2$ **3.** $30x^4y^2$ **5.** $8x^2(x+2)$ **7.** $6x^2y(x+4)$ **9.** $36x(x+2)^2$ **11.** $6(x+1)^2$
13. $(x-1)(x+2)(x+3)$ **15.** $(2x+3)^2(x-5)$ **17.** $(x-1)(x-2)$ **19.** $(x-3)(x+2)(x+4)$
21. $(x+4)(x+1)(x-7)$ **23.** $(x+6)(x-6)(x+4)$ **25.** $(x-10)(x-8)(x+3)$ **27.** $(3x-2)(x-3)(x+2)$
29. $(x+2)(x-3)$ **31.** $(x+6)(x-3)$ **33.** $\frac{4x}{x^2}, \frac{3}{x^2}$ **35.** $\frac{4x}{12y^2}, \frac{3yz}{12y^2}$ **37.** $\frac{xy}{x^2(x-3)}, \frac{6x-18}{x^2(x-3)}$
39. $\frac{9x}{x(x-1)^2}, \frac{6x-6}{x(x-1)^2}$ **41.** $\frac{3x}{x(x-3)}, -\frac{5}{x(x-3)}$ **43.** $\frac{3}{(x-5)^2}, -\frac{2x-10}{(x-5)^2}$ **45.** $\frac{3x}{x^2(x+2)}, \frac{4x+8}{x^2(x+2)}$
47. $\frac{x^2-6x+8}{(x+3)(x-4)}, \frac{x^2+3x}{(x+3)(x-4)}$ **49.** $\frac{3}{(x+2)(x-1)}, \frac{x^2-x}{(x+2)(x-1)}$ **51.** $\frac{5}{(2x-5)(x-2)}, \frac{x^2-3x+2}{(2x-5)(x-2)}$
53. $\frac{x^2-3x}{(x+3)(x-3)(x-2)}, \frac{2x^2-4x}{(x+3)(x-3)(x-2)}$ **55.** $-\frac{x^2-3x}{(x-3)^2(x+3)}, \frac{x^2+2x-3}{(x-3)^2(x+3)}$
57. $\frac{3x^2+12x}{(x-5)(x+4)}, \frac{x^2-5x}{(x-5)(x+4)}, -\frac{3}{(x-5)(x+4)}$ **59.** $\frac{11}{y^2}$ **61.** $-\frac{7}{x+4}$ **63.** $\frac{8x}{2x+3}$ **65.** $\frac{5x+7}{x-3}$
67. $\frac{2x-5}{x+9}$ **69.** $\frac{-3x-4}{2x+7}$ **71.** $\frac{1}{x+5}$ **73.** $\frac{1}{x-6}$ **75.** $\frac{3}{2y-1}$ **77.** $\frac{1}{x-5}$ **79.** $\frac{4y+5x}{xy}$ **81.** $\frac{19}{2x}$
83. $\frac{5}{12x}$ **85.** $\frac{52y-35x}{20xy}$ **87.** $\frac{13x+2}{15x}$ **89.** $\frac{7x}{24}$ **91.** $\frac{x^2+2x+2}{2x^2}$ **93.** $\frac{2x^2+3x-10}{4x^2}$
95. $\frac{16xy-12y+6x^2+3x}{12x^2y^2}$ **97.** $\frac{3xy-6y-2x^2-14x}{24x^2y}$ **99.** $\frac{9x+2}{(x-2)(x+3)}$ **101.** $\frac{2x^2-5x+1}{(x+1)(x-3)}$
103. $\frac{4x^2-34x+5}{(2x-1)(x-6)}$ **105.** $\frac{4x+9}{(x+3)(x-3)}$ **107.** $\frac{-x+9}{(x-3)(x+2)}$ **109.** $\frac{14}{(x-5)(x-5)}$ **111.** $-\frac{2(x+7)}{(x+6)(x-7)}$
113. $\frac{x-4}{x-6}$ **115.** $\frac{2x+1}{x-1}$ **117.** $-\frac{3(x^2+8x+25)}{(x-3)(x+7)}$ **119.** 2 **121.** $\frac{b^2+b-7}{b+4}$ **123.** $\frac{4n}{(n-1)^2}$
125. 1 **127.** $\frac{x^2-x+2}{(x+5)(x+1)}$ **129.** $\frac{2}{3}, \frac{3}{4}, \frac{4}{5}, \frac{50}{51}, \frac{100}{101}, \frac{1000}{1001}$ **131.** $\frac{6}{y}+\frac{7}{x}$ **133.** $\frac{2}{m^2n}+\frac{8}{mn^2}$

Section 10.3 pages 473–478

1. 3 **3.** 12 **5.** 10 **7.** 1 **9.** 9 **11.** -3 **13.** 2 **15.** $\frac{3}{5}$ **17.** 1 **19.** $\frac{1}{4}$ **21.** 3 **23.** $-\frac{3}{2}$
25. 1 **27.** -3 **29.** $\frac{1}{2}$ **31.** 8 **33.** No solution **35.** 5 **37.** -1 **39.** 5 **41.** No solution
43. 0 **45.** No solution **47.** 24 **49.** 10 **51.** 5 **53.** 8 **55.** 2 **57.** $\frac{16}{3}$ **59.** 12.56 **61.** $\frac{2}{9}$
63. The required force is 8 lb. **65.** The yield of a 30-acre farm is 675 bushels.
67. The stopping distance for a car traveling at 60 mph is 244.8 ft. **69.** An object will fall 400 ft in 5 s.
71. The gear which has 36 teeth will make 30 revolutions per minute.
73. When the cost is \$.20 per item, 75 items can be purchased. **75.** The intensity is 48 lumens when the distance is 5 ft.

77. $x = -3y + 6$ **79.** $x = 4y + 12$ **81.** $x = \frac{1}{3}y - \frac{7}{3}$ **83.** $x = -3y + 6$ **85.** $x = -\frac{1}{4a}$

87. $x = -\frac{2}{6a + 1}$ **89.** $h = \frac{2A}{b}$ **91.** $t = \frac{d}{r}$ **93.** $T = \frac{PV}{nR}$ **95.** $L = \frac{P - 2W}{2}$ **97.** $b_1 = \frac{2A - hb_2}{h}$

99. $h = \frac{3V}{A}$ **101.** $S = C - Rt$ **103.** $P = \frac{A}{1 + rt}$ **105.** $w = \frac{A}{S + 1}$ **107.** $h = \frac{V}{\pi r^2}$ **109.** $x = \frac{F_2 d}{F_1 - F_2}$

111. 1 **113.** $0, -\frac{2}{3}$ **115. a.** $S = \frac{F + BV}{B}$
b. The required selling price per desk to break even is $180.
c. The required selling price per camera to break even is $75.

Section 10.4 *pages 485–490*

1. 2.5 ft/s **3.** $325/week **5.** 250 words/page **7.** 52.4 mi/h **9.** 28 mi/gal **11.** $1.65/lb
13. The chef's wage is $12.50 per hour. **15.** The cost of oil is $.96 per quart. **17.** The cost per share is $36.
19. The company's profit is $.20 per pen. **21.** 9 **23.** 12 **25.** 5.7 **27.** 2.2 **29.** 88 **31.** 3.3 **33.** 23.1
35. 21.3 **37.** 7 **39.** 6 **41.** 1 **43.** −6 **45.** 4 **47.** 0.8 **49.** 2.7
51. 10 lb of salt are required for 25 gal of water. **53.** 10,000 people are expected to vote.
55. 20,000 people voted in favor of the amendment.
57. 150 ft^2 of decking can be made from 36 pieces of lumber. **59.** The license fee is $90.
61. 160 ml of syrup are in 280 ml of soft drink. **63.** There are 700 fish in the lake. **65.** 50 lb of fertilizer are required.
67. The distance is 42 mi. **69.** 50 additional shares are needed to earn a dividend of $186.
71. $1\frac{1}{2}$ additional gallons of fruit punch are necessary. **73.** The recommended area is 40 ft^2. **75.** 30° **77.** 13.7 in.
79. 4.9 ft **81.** 22.5 ft **83.** 20.8 ft **85.** 38 cm **87.** 45 cm^2 **89.** 49 m^2 **91.** 15 **93.** 8
95. The player made 210 foul shots. **97.** The math club receives $75. **99.** 16.5

Section 10.5 *pages 494–495*

1. It will take 2 h to fill the fountain with both sprinklers working.
3. With both tractors working, it would take 3 h to plow the field.
5. It would take the new machine 12 h to complete the task. **7.** The assistant working alone would take 90 min.
9. It would take the second welder 2 h to complete the welds.
11. The freight train travels at 30 mph. The express train travels at 50 mph.
13. The rate of the twin-engine plane is 200 mph.
15. The rate of the prop plane was 150 mph. The rate of the jet was 600 mph. **17.** The rate of the plane is 210 mph.
19. The rate of the wind is 20 mph. **21.** It will take $1\frac{1}{19}$ h to fill the tank with all three pipes operating.
23. 2 h was spent traveling by canoe. **25.** The bus usually travels at a rate of 60 mph.

Chapter Review Exercises *page 499*

1. $\frac{2x^3}{3y^3}$ [10.1A] **2.** $-\frac{x + 5}{x + 1}$ [10.1A] **3.** $2250/month [10.4A] **4.** $\frac{3x + 6}{x(x + 2)(x - 2)}, \frac{x^2}{x(x + 2)(x - 2)}$ [10.2B]
5. $\frac{2}{x + 5}$ [10.2C] **6.** $\frac{3}{x + 8}$ [10.2C] **7.** $-\frac{3y}{4x}$ [10.2D] **8.** $\frac{3}{(2x - 1)(x + 1)}$ [10.2D] **9.** $\frac{4}{3x}$ [10.1B]
10. $\frac{x^3(x + 3)}{y(x + 2)}$ [10.1B] **11.** $-\frac{bc}{2}$ [10.1C] **12.** 1 [10.1C] **13.** The plane would use $3\frac{1}{3}$ qt on a 2000-mile trip. [10.4C]
14. 24 in. [10.4D] **15.** 2 [10.3A] **16.** $\frac{1}{3}$ [10.3A] **17.** 35 [10.4B] **18.** 4 [10.4B] **19.** $x = \frac{8}{3}y + \frac{16}{3}$ [10.3C]
20. $t = \frac{d - s}{r}$ [10.3C] **21.** With both pipes working together, it would take 6 min to fill the tank. [10.5A]
22. The rate of the current is 2 mph. [10.5B] **23.** The cold water faucet working alone would fill the tub in 10 min. [10.5A]
24. The rate of the freight train is 30 mph. The rate of the passenger train is 60 mph. [10.5B]

Cumulative Review Exercises *page 500*

1. a^3b^7 [8.2A] **2.** $a^2 + ab - 12b^2$ [8.2D] **3.** $-\frac{9}{2}$ [6.3A] **4.** $a(2a - 3)(a + 5)$ [9.3C] **5.** 10 [6.2A]
6. $\frac{1}{4}$ [10.1C] **7.** $6xy^2$ [9.1A] **8.** $(3x - 2)(3x + 7)$ [9.2B] **9.** (1, 2) [7.4A] **10.** a^9b^{15} [8.2B] **11.** $-\frac{7}{3}$ [7.2A]
12. $\frac{3}{2}$ [6.1C] **13.** $(2x + 7y)^2$ [9.3A] **14.** $y = \frac{1}{2}x - 2$ [7.3A] **15.**

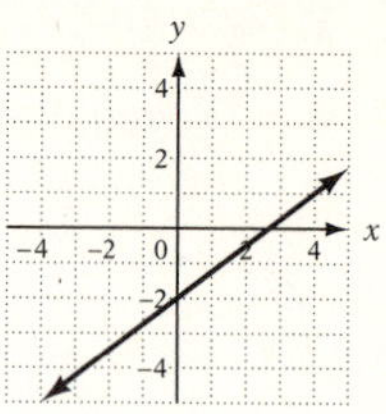

[7.1C]

16. $\angle b = 65°$, $\angle c = 80°$, $\angle y = 100°$ [6.4C]
17. It takes 1.25 s for light to travel to the earth from the moon. [8.4C]
18. The perimeter of the rectangle is $15\frac{1}{6}$ ft. [5.2A]
19. 56 oz of pure water must be added to the salt solution. [6.5B]
20. The rate of the wind is 30 mph. [7.5A]
21. The mixture will cost $2.25/lb. [6.5A]

Answers to Chapter 11 Selected Exercises

Section 11.1 *pages 509–512*

1. 4 **3.** 7 **5.** $4\sqrt{2}$ **7.** $2\sqrt{2}$ **9.** $18\sqrt{2}$ **11.** $10\sqrt{10}$ **13.** $\sqrt{15}$ **15.** $\sqrt{29}$ **17.** $-54\sqrt{2}$
19. $3\sqrt{5}$ **21.** 0 **23.** $48\sqrt{2}$ **25.** $\sqrt{105}$ **27.** 30 **29.** 15.492 **31.** 16.971 **33.** 16 **35.** 16.583
37. 15.652 **39.** 18.762 **41.** x^3 **43.** $y^7\sqrt{y}$ **45.** a^{10} **47.** x^2y^2 **49.** $2x^2$ **51.** $2x\sqrt{6}$ **53.** $xy^3\sqrt{xy}$
55. $ab^5\sqrt{ab}$ **57.** $2x^2\sqrt{15x}$ **59.** $7a^2b^4$ **61.** x^3y **63.** $8a^2b^3\sqrt{5b}$ **65.** $6x^2y^3\sqrt{3y}$ **67.** $4x^3y\sqrt{2y}$
69. $5a + 20$ **71.** $2x^2 + 8x + 8$ **73.** $x + 2$ **75.** $y + 1$ **77.** $3\sqrt{2}$ **79.** $-\sqrt{7}$ **81.** $-11\sqrt{11}$
83. $10\sqrt{x}$ **85.** $-2\sqrt{y}$ **87.** $-11y\sqrt{3b}$ **89.** $2x\sqrt{2}$ **91.** $-3a\sqrt{3a}$ **93.** $-5\sqrt{xy}$ **95.** $8\sqrt{5}$ **97.** $8\sqrt{2}$
99. $15\sqrt{2} - 10\sqrt{3}$ **101.** $\sqrt{x}$ **103.** $-12x\sqrt{3}$ **105.** $2xy\sqrt{x} - 3xy\sqrt{y}$ **107.** $-9x\sqrt{3x}$ **109.** $-13y^2\sqrt{2y}$
111. $4a^2b^2\sqrt{ab}$ **113.** $7\sqrt{2}$ **115.** $6\sqrt{x}$ **117.** $-3\sqrt{y}$ **119.** $-45\sqrt{2}$ **121.** $13\sqrt{3} - 12\sqrt{5}$
123. $32\sqrt{3} - 3\sqrt{11}$ **125.** $6\sqrt{x}$ **127.** $10a\sqrt{3b} + 10a\sqrt{5b}$ **129.** $-2xy\sqrt{3}$ **131.** $-7b\sqrt{ab} + 4a\sqrt{ab}$
133. $3ab\sqrt{2a} - ab + 4ab\sqrt{3b}$ **135.** $0.05ab^2\sqrt{ab}$ **137.** $xy\sqrt{y + x}$
139. No, $\sqrt{9 + 16} \neq \sqrt{9} + \sqrt{16}$; $5 \neq 3 + 4$ **141.** 2 **143.** $-11\sqrt{2x + y}$ **145.** $13a\sqrt{2 + 1}$

Section 11.2 *pages 517–518*

1. 5 **3.** 6 **5.** x **7.** x^3y^2 **9.** $3ab^6\sqrt{2a}$ **11.** $12a^4b\sqrt{b}$ **13.** $2 - \sqrt{6}$ **15.** $x - \sqrt{xy}$
17. $5\sqrt{2} - \sqrt{5x}$ **19.** $4 - 2\sqrt{10}$ **21.** $x - 6\sqrt{x} + 9$ **23.** $3a - 3\sqrt{ab}$ **25.** $10abc$ **27.** $15x - 22y\sqrt{x} + 8y^2$
29. $x - y$ **31.** 4 **33.** 7 **35.** $2x^2\sqrt{2y}$ **37.** $4x\sqrt{y}$ **39.** $\frac{\sqrt{2}}{4}$ **41.** $\frac{2\sqrt{2x}}{x}$ **43.** $\frac{y\sqrt{3}}{3}$ **45.** $\sqrt{2y}$
47. $-\frac{5\sqrt{7} + 15}{2}$ **49.** $-\frac{7\sqrt{2} + 49}{47}$ **51.** $\frac{x + \sqrt{xy}}{x - y}$ **53.** $\frac{3}{2}$ **55.** -1.3 **57.** $-\frac{4}{9}$ **59.** $\frac{3}{2}$
61. **a.** True **b.** True **c.** False: $x + 2\sqrt{x} + 1$ **d.** True

Section 11.3 *pages 525–528*

1. 25 **3.** 144 **5.** No solution **7.** 5 **9.** 16 **11.** 8 **13.** No solution **15.** 12 **17.** $\frac{1}{2}$ **19.** 1
21. -2 **23.** -1 **25.** 15 **27.** 2 **29.** $\frac{7}{3}$ **31.** $\frac{4}{3}$ **33.** 45 **35.** 18 **37.** 15 **39.** 4 **41.** 2
43. -4 **45.** -1 **47.** 0 **49.** 3 **51.** 2 **53.** 13 in. **55.** 11.4 cm **57.** 8.7 ft **59.** 7.9 m
61. 10 cm and 17.3 cm **63.** 8 cm and 13.9 cm **65.** 8.5 in. **67.** 8.6 cm **69.** 16 ft **71.** The distance is 8.5 cm.
73. The perimeter is 24 in. **75.** The perimeter is 14.2 ft. **77.** The perimeter is 17.1 in. **79.** The area is 35.7 cm^2.
81. The area is 12.5 ft^2. **83.** The length of the diagonal is 8.5 in. **85.** The perimeter is 10.9 cm. **87.** 3
89. **a.** $12\sqrt{2}$ cm **b.** 12 cm^2 **91.** 244.78 ft^2

Chapter Review Exercises *page 531*

1. $3\sqrt{5}$ [11.1A] **2.** $5\sqrt{3}$ [11.1A] **3.** $-8\sqrt{5}$ [11.1A] **4.** 13.229 [11.1A] **5.** $11x^4y$ [11.1B]
6. The perimeter is approximately 29.2 cm. [11.3B] **7.** $4a^2b^5\sqrt{2ab}$ [11.B] **8.** $x + 2$ [11.1B] **9.** $5\sqrt{y}$ [11.1C]
10. $-5\sqrt{2}$ [11.1C] **11.** $21\sqrt{2y} - 12\sqrt{2x}$ [11.1C] **12.** $-2xy\sqrt{3xy} - 3xy\sqrt{xy}$ [11.1C] **13.** $4x^2y^2\sqrt{5y}$ [11.2A]
14. $6x^2y\sqrt{y}$ [11.2A] **15.** $a - \sqrt{ab}$ [11.2A] **16.** $y + 2\sqrt{y} - 15$ [11.2A] **17.** 9 [11.2B] **18.** $7ab\sqrt{a}$ [11.2B]
19. $\sqrt{3} + 1$ [11.2B] **20.** $x - 4$ [11.2B] **21.** 11 [11.3A] **22.** 25 [11.3A] **23.** 8 ft and $4\sqrt{3}$ ft [11.3B]
24. 2 [11.3A]

Cumulative Review Exercises *page 532*

1. 750 [6.2A] **2.** $2a(a - 5)(a - 3)$ [9.3C] **3.** $-\frac{x-2}{x+5}$ [10.1A] **4.** -4 [10.4B] **5.** $(4, -3)$ [7.4C]

6. $-8x^3y^6$ [8.2A] **7.** $4b^3 - 4b^2 - 8b - 4$ [8.1A] **8.** $6a - 4 + \frac{2}{a^2}$ [8.3B] **9.** $5xy^2(1 - 2y)(1 + 2y)$ [9.3C]

10. $\frac{5}{3x+1}$ [10.1B] **11.** 4 [10.3A] **12.** $\frac{3}{(2x-1)(x+1)}$ [10.2D] **13.** $\frac{x-1}{x-4}$ [10.1C] **14.** 8 [11.2B] **15.** 6 [11.3A]

16. $\frac{x+3}{x-3}$ [10.2D] **17.** $x \geq 2$ [6.6A] **18.** $3\sqrt{2} - x\sqrt{3}$ [11.2A] **19.** $3x^2y^2(4x - 3y)$ [9.3C] **20.** $14a^5b^2\sqrt{2a}$ [11.2A]

−5 −4 −3 −2 −1 0 1 2 3 4 5

21. The area is 40 m^2. [11.3B] **22.** Side *DE* is 10.5 m. [10.4D]
23. The second panel would take 180 min to heat the water. [10.4B] **24.** 5000 fish are in the lake. [10.4C]
25. Approximately 2309.07 ft^3 of the silo are being used for storage. [5.3A]

Answers to Chapter 12 Selected Exercises

Section 12.1 *pages 541–542*

1. 5 and -3 **3.** -7 and 8 **5.** $-\frac{5}{3}$ and 4 **7.** -5 and 3 **9.** 1 and 3 **11.** -1 and -2 **13.** 3

15. 0 and $-\frac{2}{3}$ **17.** -2 and 5 **19.** $\frac{2}{3}$ and 1 **21.** $\frac{1}{3}$ and -3 **23.** $\frac{2}{3}$ **25.** $-\frac{1}{2}$ and $\frac{3}{2}$ **27.** $\frac{1}{2}$

29. -3 and 3 **31.** $-\frac{1}{2}$ and $\frac{1}{2}$ **33.** -3 and 5 **35.** 1 and 5 **37.** -1 and $\frac{13}{2}$ **39.** -7 and 7

41. -8 and 8 **43.** $-\frac{8}{3}$ and $\frac{8}{3}$ **45.** $-\frac{5}{2}$ and $\frac{5}{2}$ **47.** $-\frac{8}{5}$ and $\frac{8}{5}$ **49.** No real number solution **51.** -2 and 8

53. $-\frac{15}{2}$ and $\frac{3}{2}$ **55.** $\frac{26}{9}$ and $\frac{10}{9}$ **57.** $-5 + 5\sqrt{2}$ and $-5 - 5\sqrt{2}$ **59.** No real number solution

61. $\frac{-3 + 8\sqrt{3}}{4}$ and $\frac{-3 - 8\sqrt{3}}{4}$ **63.** $-\sqrt{2}$ and $\sqrt{2}$ **65.** 0 and $\frac{b}{a}$ **67.** 0 and 1

69. The annual percentage rate is 9%.

Section 12.2 *pages 547–548*

1. 1 and -3 **3.** 8 and -2 **5.** 2 **7.** No real number solution **9.** -1 and -4 **11.** -8 and 1

13. $-2 + \sqrt{3}$ and $-2 - \sqrt{3}$ **15.** $-3 + \sqrt{14}$ and $-3 - \sqrt{14}$ **17.** $1 + \sqrt{2}$ and $1 - \sqrt{2}$

19. $\frac{-3 + \sqrt{13}}{2}$ and $\frac{-3 - \sqrt{13}}{2}$ **21.** 2 and 1 **23.** $\frac{-1 + \sqrt{13}}{2}$ and $\frac{-1 - \sqrt{13}}{2}$ **25.** $-5 + 4\sqrt{2}$ and $-5 - 4\sqrt{2}$

27. $\frac{3 + \sqrt{29}}{2}$ and $\frac{3 - \sqrt{29}}{2}$ **29.** $\frac{1 + \sqrt{17}}{2}$ and $\frac{1 - \sqrt{17}}{2}$ **31.** No real number solution **33.** 1 and $\frac{1}{2}$

35. -3 and $\frac{1}{2}$ **37.** 2 and $\frac{3}{2}$ **39.** 1 and $-\frac{1}{2}$ **41.** -2 and $\frac{1}{3}$ **43.** -2 and $-\frac{2}{3}$ **45.** $\frac{1}{2}$ and $-\frac{3}{2}$ **47.** $\frac{1}{3}$ and $-\frac{3}{2}$

49. $-\frac{1}{2}$ and $\frac{4}{3}$ **51.** $\frac{1+\sqrt{2}}{2}$ and $\frac{1-\sqrt{2}}{2}$ **53.** $\frac{2+\sqrt{5}}{2}$ and $\frac{2-\sqrt{5}}{2}$ **55.** 2 and 4 **57.** $1+\sqrt{6}$ and $1-\sqrt{6}$
59. $1+\sqrt{7}$ and $1-\sqrt{7}$ **61.** 0 and -1 **63.** $4+\sqrt{3}$ and $4-\sqrt{3}$
65. The ball will hit the basket 1.88 s after the player shoots at the basket.

Section 12.3 *pages 551–552*

1. 5 and -1 **3.** -3 and 5 **5.** -7 and 1 **7.** 2 and -3 **9.** 3 and -1 **11.** -5 and 1 **13.** $-\frac{1}{2}$ and 1
15. No real number solution **17.** 0 and 1 **19.** $\frac{3}{2}$ and $-\frac{3}{2}$ **21.** $\frac{3}{2}$ and $-\frac{5}{2}$ **23.** 3 and $-\frac{2}{3}$ **25.** -3 and $\frac{4}{5}$
27. $-\frac{1}{2}$ and $\frac{2}{3}$ **29.** No real number solution **31.** $1+\sqrt{6}$ and $1-\sqrt{6}$ **33.** $-3+\sqrt{10}$ and $-3-\sqrt{10}$
35. $2+\sqrt{13}$ and $2-\sqrt{13}$ **37.** $\frac{-1+\sqrt{2}}{3}$ and $\frac{-1-\sqrt{2}}{3}$ **39.** $-\frac{1}{2}$ **41.** No real number solution
43. $\frac{-4+\sqrt{5}}{2}$ and $\frac{-4-\sqrt{5}}{2}$ **45.** $\frac{1+2\sqrt{3}}{2}$ and $\frac{1-2\sqrt{3}}{2}$ **47.** $\frac{-5+\sqrt{2}}{3}$ and $\frac{-5-\sqrt{2}}{3}$
49. $\frac{-a+\sqrt{a^2-4b}}{2}$ and $\frac{-a-\sqrt{a^2-4b}}{2}$ **51.** 6 **53.** 3 **55.** $5+\sqrt{13}$ and $5-\sqrt{13}$
57. The distance from the corner to the memorial is 250 ft.

Section 12.4 *pages 555–556*

1. up; lowest **3.** up; lowest **5.** up; lowest

7.
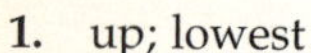

9.
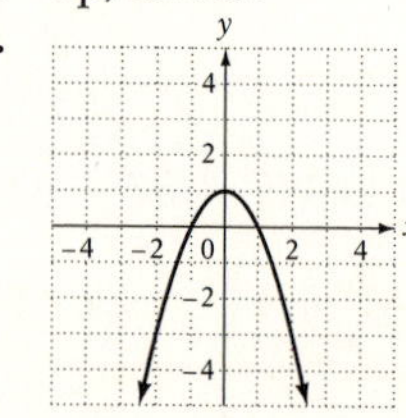

11.
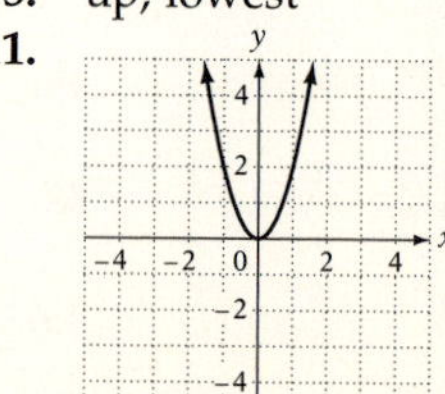

13.
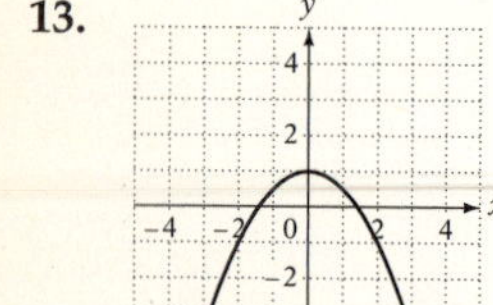

15.
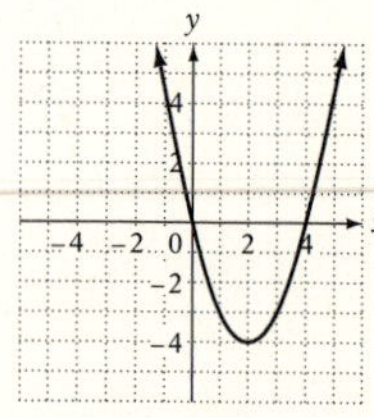

17.
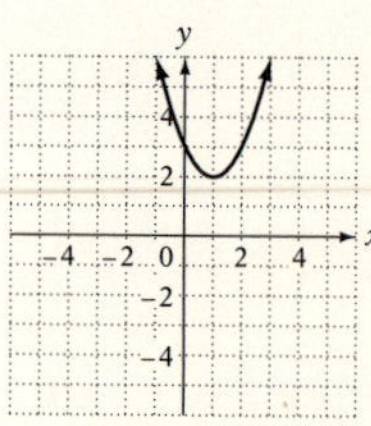

19.
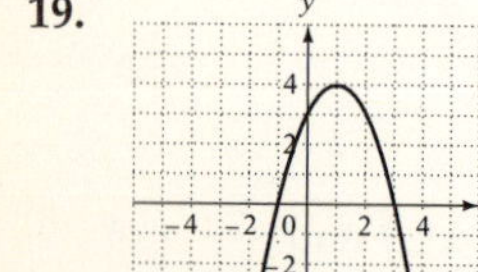

21.
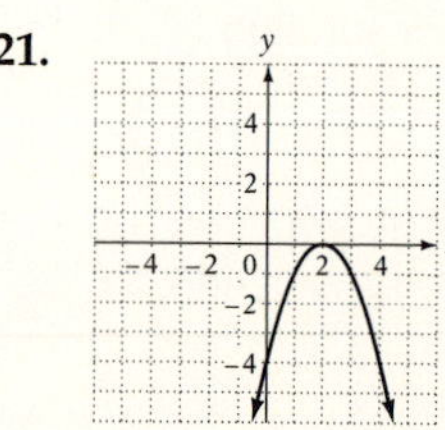

23.
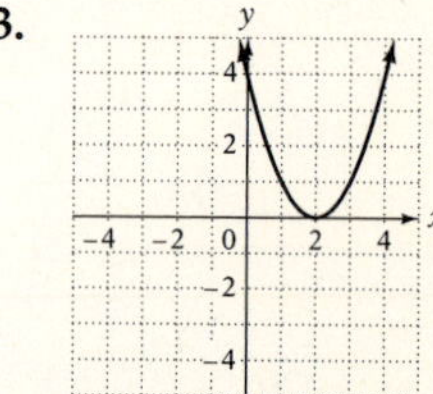

25. $y = x^2 - 8x + 15$ **27.** $y = 2x^2 - 12x + 22$ **29.** (2, 0) and (-2, 0) **31.** (0, 0) and (4, 0)
33. $(-\frac{1}{2}, 0)$ and (1, 0)

Chapter Review Exercises *page 559*

1. 5 and -6 [12.1A] **2.** $-\frac{1}{3}$ and 8 [12.1A] **3.**

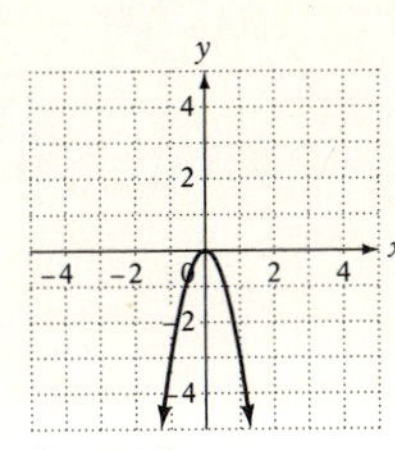

[12.4A]

4. -4 and $\frac{5}{3}$ [12.1A]

5. $\frac{5}{2}$ and $\frac{1}{3}$ [12.1A] **6.** -4 and 5 [12.1A] **7.** -9 and 9 [12.1B] **8.** $-\frac{5}{4}$ and $\frac{5}{4}$ [12.1B] **9.** -2 and 2 [12.1B]

10. [12.4A]

11. $5 + 3\sqrt{2}$ and $5 - 3\sqrt{2}$ [12.1B] **12.** $-4 + 2\sqrt{5}$ and $-4 - 2\sqrt{5}$ [12.1B]

13. $-2 + 2\sqrt{5}$ and $-2 - 2\sqrt{5}$ [12.2A] **14.** [12.4A] **15.** $\frac{5 + \sqrt{33}}{2}$ and $\frac{5 - \sqrt{33}}{2}$ [12.2A]

16. $\frac{-3 + \sqrt{41}}{2}$ and $\frac{-3 - \sqrt{41}}{2}$ [12.2A] **17.** $\frac{3 + \sqrt{7}}{2}$ and $\frac{3 - \sqrt{7}}{2}$ [12.2A]

18. $\frac{-4 + \sqrt{22}}{2}$ and $\frac{-4 - \sqrt{22}}{2}$ [12.2A] **19.** $-2 + \sqrt{2}$ and $-2 - \sqrt{2}$ [12.3A]

20. $\frac{-3 + \sqrt{37}}{2}$ and $\frac{-3 - \sqrt{37}}{2}$ [12.3A] **21.** $\frac{3 + \sqrt{33}}{2}$ and $\frac{3 - \sqrt{33}}{2}$ [12.3A]

22. [12.4A] **23.** $-\frac{1}{2}$ and 3 [12.3A] **24.** $\frac{1 + \sqrt{13}}{6}$ and $\frac{1 - \sqrt{13}}{6}$ [12.3A]

Cumulative Review Exercises *pages 560–561*

1. 14 [11.3A] **2.** $6ab\sqrt{a}$ [11.2B] **3.** $-28x + 27$ [6.3C] **4.** $y = -\frac{4}{3}x - 2$ [7.3A] **5.** $(2, -2)$ [7.4C]

6. $\frac{1}{4(x + 1)}$ [10.1B] **7.** $-4\sqrt{2}$ [11.1C] **8.** $\frac{5}{3}$ [10.3A] **9.** $6 + 3\sqrt{3}$ [11.2B] **10.** $4ab\sqrt{2ab} - 5ab\sqrt{ab}$ [11.1C]

11. 11 and -1 [12.1B] **12.** $2x - 3$ [11.2B] **13.** $3a(3a + 2)(2a + 5)$ [9.3C] **14.** 4 [2.4A]

15. $9b^2 + 12b + 4$ [8.2D] **16.** $\dfrac{2 + \sqrt{14}}{2}$ and $\dfrac{2 - \sqrt{14}}{2}$ [12.3A] **17.** $x + 2 - \dfrac{4}{x - 2}$ [8.3C]

18. $C = Rt + S$ [10.3C] **19.** $-\dfrac{5}{9}$ [10.3A] **20.** $-\dfrac{1}{b^4}$ [8.3A] **21.** $x = 41°$ [6.4B] **22.**

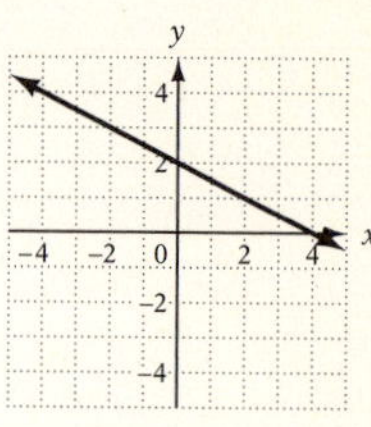

[7.1D]

23. It would take the small pipe 48 h to fill the tank. [10.5A] **24.** 250 additional shares of stock are required. [10.4C]
25. The perimeter is approximately 20.6 in. [11.3B] **26.** The volume of the cube is $(x^3 + 12x^2 + 48x + 64)$ cm^3. [5.3A]

Final Exam *pages 562–565*

1. 144 [1.2B] **2.** 25 [1.4D] **3.** $2\frac{5}{8}$ [3.3B] **4.** 60.205 [4.2A] **5.** 0.69 [4.1C] **6.** −3.8 [2.1C] **7.** $-\dfrac{1}{2}$ [3.2C]

8. $-\dfrac{1}{5}$ [2.4A] **9.** $-3x + 11$ [5.1D] **10.** 5 [6.3A] **11.** $\dfrac{1}{3}$ [6.3C] **12.** $x > \dfrac{2}{3}$ [6.6A] **13.** 144 [1.2B]

14. $1\frac{1}{18}$ [3.2A] **15.** $y < \dfrac{1}{3}$ [6.6B] **16.** 24.5 mi/gal [10.4A] **17.** $\dfrac{45}{2}$ [10.4B] **18.** 37.5% [4.3C] **19.** 45 [6.2A]

20. $-2x^2 + 2x + 12$ [8.1B] **21.** $-32x^{10}y^5$ [8.2B] **22.** $-2x^3 + 8x^2 - 7x + 3$ [8.2C] **23.** $\dfrac{3y^5}{4x}$ [8.3A]

24. $x^2 + 3x - 4$ [8.3C] **25.** $-18y^2$ [8.4A] **26.** $4.73 \cdot 10^{-6}$ [8.4B] **27.** $-3y^2(y^2 - 2y + 7)$ [9.1B]

28. $(x - 9)(x + 1)$ [9.2A] **29.** $(3x + 4)(2x - 3)$ [9.2B] **30.** $-3x(2x + 3)(x + 2)$ [7.3B] **31.** $(7x + 1)(7x - 1)$ [9.3A]

32. $(y - 1)(x - 2)$ [9.3B] **33.** $3x^2(2 - 3y)(2 + 3y)$ [9.3C] **34.** $-\dfrac{3}{x}$ [10.1A] **35.** 1 [10.2C] **36.** 3 [10.2C]

37. $\dfrac{10x + 32}{(x + 5)(x - 1)}$ [10.2D] **38.** 42° [6.4C] **39.** 2 [10.3A] **40.** 4 [10.3B] **41.** $\angle a = 55°$ and $\angle b = 145°$ [6.4C]

42.

[7.1D]

43. $t = \dfrac{L - a}{ac}$ [10.3C] **44.** 5 [6.3D] **45.** −3 [7.3B] **46.** $y = -\dfrac{3}{4}x - 1$ [7.3A]

47. (2, −1) [7.4B] [7.4C] **48.** $16\sqrt{2}$ [11.1A] **49.** $9a^3$ [11.1B] **50.** $38\sqrt{3y}$ [11.1C] **51.** $3x\sqrt{2} - x\sqrt{3}$ [11.2A]

52. $3\sqrt{2} + 3$ [11.2B] **53.** 3 [11.3C] **54.** $\dfrac{3}{2}$ and −1 [12.1A] **55.** $1 + \sqrt{3}$ and $1 - \sqrt{3}$ [12.1B]

56. $\dfrac{1 + \sqrt{10}}{3}$ and $\dfrac{1 - \sqrt{10}}{3}$ [12.2A/12.3A] **57.** 30° [6.4B] **58.** 15 [10.4D]

59. The area of the triangle is 15.36 cm^2. [5.2B] **60.** The perimeter of the triangle is 22.4 in. [11.3C]
61. $\angle b = 75°$, $\angle c = 65°$, and $\angle y = 65°$ [6.4C] **62.** The car can travel 129 mi. [3.2E]
63. The volume of the cylinder is 954.26 in^3. [5.3A] **64.** The surface area of the solid is 184 ft^2. [5.3B]
65. 76,000 people would be expected to vote. [10.4C] **66.** Five years ago, the city's population was 40,000. [6.2B]
67. \$3000 of additional money must be invested. [6.5C] **68.** The cost of the mixture is \$4.50 per pound. [6.5A]
69. The resulting mixture is 25% sugar. [6.5B] **70.** The bicycle travels at a rate of 17 mph. [6.5D]

GLOSSARY

abscissa The first number in an ordered pair. It measures a horizontal distance and is also called the first coordinate. (Sec. 7.1)

absolute value of a number The distance of the number from zero on the number line. (Sec. 2.1)

acute angle An angle whose measure is between 0° and 90°. (Sec. 5.2)

acute triangle A triangle that has three acute angles. (Sec. 5.2)

addend In addition, one of the numbers added. (Sec. 1.1)

addition The process of finding the total of two numbers. (Sec. 1.1)

addition method Method of finding an exact solution of a system of linear equations wherein we use the Addition Property of Equations. (Sec. 7.4)

Addition Property of Zero Zero added to a number does not change the number. (Sec. 1.1)

additive inverses Numbers that are the same distance from zero on the number line, but on opposite sides; also called opposites. (Sec. 2.2)

adjacent angles Two angles that share a common side. (Sec. 6.4)

algebraic fraction A fraction in which the numerator or denominator is a variable expression. (Sec. 10.1)

alternate exterior angles Two angles that are on opposite sides of the transversal and outside the parallel lines. (Sec. 6.4)

alternate interior angles Two angles that are on opposite sides of the transversal and between the parallel lines. (Sec. 6.4)

angle An angle is formed when two rays start at the same point; it is measured in degrees. (Sec. 5.2)

area A measure of the amount of surface in a region. (Sec. 5.2)

Associative Property of Addition Numbers to be added can be grouped (with parentheses, for example) in any order; the sum will be the same. (Sec. 1.1)

Associative Property of Multiplication Numbers to be multiplied can be grouped (with parentheses, for example) in any order; the product will be the same. (Sec. 1.2)

axis of symmetry of a parabola A line of symmetry that passes through the vertex of the parabola. (Sec. 12.4)

axes The two number lines that form a rectangular coordinate system; also called coordinate axes. (Sec. 7.1)

base In exponential notation, the factor that is taken the number of times shown by the exponent. (Sec. 1.2)

basic percent equation Percent times base equals amount. (Sec. 6.2)

binomial A polynomial of two terms. (Sec. 8.1)

circle A plane figure in which all points are the same distance from point O, which is called the center of the circle. (Sec. 5.2)

circumference The distance around a circle. (Sec. 5.2)

clearing denominators Removing denominators from an equation that contains fractions by multiplying each side of the equation by the LCM of the denominators. (Sec. 10.3)

combining like terms Using the Distributive Property to add the coefficients of like variable terms. (Sec. 5.2)

common factor A number that is a factor of two or more numbers is a common factor of those numbers. (Sec. 1.4)

common multiple A number that is a multiple of two or more numbers is a common multiple of those numbers. (Sec. 1.4)

Commutative Property of Addition Two numbers can be added in either order; the sum will be the same. (Sec. 1.1)

Commutative Property of Multiplication Two numbers can be multiplied in either order; the product will be the same. (Sec. 1.2)

complementary angles Two angles whose measures have the sum 90°. (Sec. 5.2)

completing the square Adding to a binomial the constant term that makes it a perfect-square trinomial. (Sec. 12.2)

complex fraction A fraction whose numerator or denominator contains one or more fractions. (Sec. 3.3)

composite number A number that has whole-number factors besides 1 and itself. For instance, 18 is a composite number. (Sec. 1.4)

congruent objects Objects that have the same shape and the same size. (Sec. 10.4)

conjugates Binomial expressions that differ only in the sign of a term. The expressions $a + b$ and $a - b$ are conjugates. (Sec. 11.2)

constant of proportionality k in a variation equation; also called the constant of variation. (Sec. 10.3)

constant of variation k in a variation equation; also called the constant of proportionality. (Sec. 10.3)

constant term A term that includes no variable part; also called a constant. (Sec. 5.1)

coordinate axes The two number lines that form a rectangular coordinate system; also simply called axes. (Sec. 7.1)

coordinates of a point The numbers in the ordered pair that is associated with the point. (Sec. 7.1)

corresponding angles Two angles that are on the same side of the transversal and are both acute angles or are both obtuse angles. (Sec. 6.4)

cube A rectangular solid in which all six faces are squares. (Sec. 5.3)

decimal A number written in decimal notation. (Sec. 4.1)

decimal notation Notation in which a number consists of a whole-number part, a decimal point, and a decimal part. (Sec. 4.1)

decimal part In decimal notation, that part of the number that follows the decimal point. (Sec. 4.1)

decimal point In decimal notation, the point that separates the whole-number part from the decimal part. (Sec. 4.1)

degree Unit used to measure angles; one complete revolution is 360°. (Sec. 5.2)

degree of a polynomial in one variable The largest exponent that appears on the variable. (Sec. 8.1)
denominator The part of a fraction that appears below the fraction bar. (Sec. 3.1)
dependent system of equations A system of equations that has an infinite number of solutions. (Sec. 7.4)
descending order The terms of a polynomial in one variable are arranged in descending order when the exponents of the variable decrease from left to right. (Sec. 8.1)
diameter of a circle A line segment with endpoints on the circle and going through the center. (Sec. 5.2)
diameter of a sphere A line segment with endpoints on the sphere and going through the center. (Sec. 5.3)
difference In subtraction, the result of subtracting two numbers. (Sec. 1.1)
direct variation An equation of the form $y = kx$, where k is a constant value called the constant of variation or the constant of proportionality. (Sec. 10.3)
dividend In division, the number into which the divisor is divided to yield the quotient. (Sec. 1.2)
division The process of finding the quotient of two numbers. (Sec. 1.2)
divisor In division, the number that is divided into the dividend to yield the quotient. (Sec. 1.2)
equation A statement of the equality of two mathematical expressions. (Sec. 6.1)
equilateral triangle A triangle that has three sides of equal length; the three angles are of equal measure. (Sec. 5.2)
equivalent fractions Equal fractions with different denominators. (Sec. 3.1)
estimate An approximation. (Sec. 1.3)
evaluating a variable expression Replacing each variable by its value and then simplifying the resulting numerical expression. (Sec. 1.1)
expanded form The number 46,208 can be written in expanded form as 40,000 + 6000 + 200 + 8. (Sec. 1.1)
exponent In exponential notation, the elevated number that indicates how many times the factor occurs in the multiplication. (Sec. 1.2)
exponential form The expression 2^5 is in exponential form. Compare *factored form.* (Sec. 1.2)
exponential notation The expression of a number to some power, indicated by an exponent. (Sec. 1.2)
exterior angle of a triangle An angle adjacent to an interior angle of the triangle. (Sec. 6.4)
factor A number that divides another number evenly. (Sec. 1.4)
factor a polynomial To write the polynomial as a product of other polynomials. (Sec. 9.1)
factor a trinomial of the form $ax^2 + bx + c$ To express the trinomial as the product of two binomials. (Sec. 9.1)
factored form The expression $2 \cdot 2 \cdot 2 \cdot 2 \cdot 2$ is in factored form. Compare *exponential form.* (Sec. 1.2)
factors In multiplication, the numbers that are multiplied. (Sec. 1.2)
first coordinate The first number in an ordered pair. It measures a horizontal distance and is also called the abscissa. (Sec. 7.1)
FOIL A method of finding the product of two binomials; the letters stand for First, Outer, Inner, and Last. (Sec. 8.2)
fraction Notation used to represent the number of equal parts of a whole. (Sec. 3.1)
fraction bar Bar that separates the numerator of a fraction from the denominator. (Sec. 3.1)
fraction in simplest form A fraction in which there are no common factors in the numerator and the denominator. (Sec. 3.1)
geometric solid A figure in space. (Sec. 5.3)
graph of an equation in two variables A graph of the ordered-pair solutions of an equation. (Sec. 7.1)
graph of an integer A heavy dot directly above that number on the number line. (Sec. 2.1)
graph of an ordered pair The dot drawn at the coordinates of the point in the plane. (Sec. 7.1)
graphing a point in the plane Placing a dot at the location given by the ordered pair; also called plotting a point in the plane. (Sec. 7.1)
greater than A number a is greater than another number b, written $a > b$, if a is to the right of b on the number line. (Sec. 2.1)
greater than or equal to The symbol $\geq$ means "is greater than or equal to." (Sec. 4.4)
greatest common factor The greatest common factor (GCF) of two or more integers is the greatest integer that is a factor of all the integers. The greatest common factor of two or more monomials is the product of the GCF of the coefficients and the common variable factors. (Sec. 1.4, 9.1)
hypotenuse In a right triangle, the side opposite the 90° angle. (Sec. 11.3)
improper fraction A fraction greater than or equal to 1. (Sec. 3.1)
inconsistent system of equations A system of equations that has no solution. (Sec. 7.4)
independent system of equations A system of equations that has one solution. (Sec. 7.4)
inequality An expression that contains the symbol $>$, $<$, $\geq$ (is greater than or equal to), or $\leq$ (is less than or equal to). (Sec. 4.4)
integers The numbers $\ldots, -3, -2, -1, 0, 1, 2, 3, \ldots$. (Sec. 2.1)
interest The amount of money paid for the privilege of using someone else's money. (Sec. 6.5)
interest rate The percent used to determine the amount of interest. (Sec. 6.5)
interior angle of a triangle One of the angles within the region enclosed by the triangle. (Sec. 6.4)
intersecting lines Lines that cross at a point in the plane. (Sec. 5.2)
inverse variation An equation of the form $y = k/x$, where k is a constant value. (Sec. 10.3)
inverting a fraction Interchanging the numerator and denominator. (Sec. 3.2)
irrational number The decimal representation of an irrational number never repeats or terminates and can only be approximated. (Sec. 4.4)
isosceles triangle A triangle that has two sides of equal length; the angles opposite the equal sides are of equal measure. (Sec. 5.2)
least common denominator The least common multiple of denominators. (Sec. 3.2)

least common multiple The smallest common multiple of two or more numbers. (Sec. 1.4)
legs In a right triangle, the sides opposite the acute angles. (Sec. 11.3)
less than A number a is less than another number b, written $a < b$, if a is to the left of b on the number line. (Sec. 2.1)
less than or equal to The symbol $\leq$ means "is less than or equal to." (Sec. 4.4)
like terms Terms of a variable expression that have the same variable part. (Sec. 5.1)
line A line extends indefinitely in two directions in a plane; it has no width. (Sec. 5.2)
linear equation in two variables An equation of the form $y = mx + b$, where m is the coefficient of x and b is a constant; also called a linear function. (Sec. 7.1)
linear model A first-degree equation that is used to describe a relationship between quantities. (Sec. 7.3)
line of best fit A line drawn to approximate data that are graphed as points in a coordinate system. (Sec. 7.3)
line segment Part of a line; it has two endpoints. (Sec. 5.2)
literal equation An equation that contains more than one variable. (Sec. 10.3)
minuend In subtraction, the number from which another number (the subtrahend) is subtracted. (Sec. 1.1)
mixed number A number greater than 1 that has a whole-number part and a fractional part. (Sec. 3.1)
monomial A number, a variable, or a product of numbers and variables; a polynomial of one term. (Sec. 8.1)
multiples of a number The products of that number and the numbers 1, 2, 3, ... (Sec. 1.4)
multiplication The process of finding the product of two numbers. (Sec. 1.2)
Multiplication Property of One The product of a number and one is the number. (Sec. 1.2)
Multiplication Property of Zero The product of a number and zero is zero. (Sec. 1.2)
multiplicative inverse of a number The reciprocal of a number. (Sec. 3.2)
natural numbers The numbers 1, 2, 3, ... (Sec. 1.1)
negative integers The integers to the left of zero on the number line. (Sec. 2.1)
negative numbers Numbers less than zero. (Sec. 2.1)
negative slope A property of a line that slants downward to the right. (Sec. 7.2)
nonfactorable over the integers A polynomial that does not factor using only integers. (Sec. 9.2)
number line A line on which a number can be graphed. (Sec. 2.1)
numerator The part of a fraction that appears above the fraction bar. (Sec. 3.1)
numerical coefficient The number part of a variable term. When the numerical coefficient is 1 or -1, the 1 is usually not written. (Sec. 5.1)
obtuse angle An angle whose measure is between 90° and 180°. (Sec. 5.2)
obtuse triangle A triangle that has one obtuse angle. (Sec. 5.2)
opposite of a polynomial The polynomial created when the sign of each term of the original polynomial is changed. (Sec. 8.1)
opposites Two numbers that are the same distance from zero on the number line, but on opposite sides; also called additive inverses. (Sec. 2.1)
Order of Operations Agreement A set of rules that tell us in what order to perform the operations that occur in a numerical expression. (Sec. 1.4)
ordered pair Pair of numbers, such as (a, b) that can be used to identify a point in the plane determined by the axes of a rectangular coordinate system. (Sec. 7.1)
ordinate The second number in an ordered pair. It measures a vertical distance and is also called the second coordinate. (Sec. 7.1)
origin The point of intersection of the two coordinate axes that form a rectangular coordinate system. (Sec. 7.1)
parabola The graph of a quadratic equation in two variables. (Sec. 12.4)
parallel lines Lines that never meet; the distance between them is always the same. (Sec. 5.2)
parallelogram A quadrilateral that has opposite sides equal and parallel. (Sec. 5.2)
percent Parts of 100. (Sec. 4.3)
percent decrease A decrease of a quantity expressed as a portion of its original value. (Sec. 6.2)
percent increase An increase of a quantity expressed as a portion of its original value. (Sec. 6.2)
perfect square The product of a term and itself. (Sec. 9.3)
perfect-square trinomial A trinomial that is a product of a binomial and itself. (Sec. 9.3)
perimeter The distance around a plane figure. (Sec. 5.2)
period In a number written in standard form, each group of digits separated by a comma. (Sec. 1.1)
perpendicular lines Intersecting lines that form right angles. (Sec. 5.2)
place value The position of each digit in a number in standard form determines that digit's place value. (Sec. 1.1)
plane A flat surface that extends in all directions. (Sec. 5.2)
plane figure A figure that lies totally in a plane. (Sec. 5.2)
point–slope formula If (x_1, y_1) is a point on a line with slope m, then $y - y_1 = m(x - x_1)$. (Sec. 7.3)
polygon A closed figure determined by three or more line segments that lie in a plane. (Sec. 5.2)
polynomial A variable expression in which the terms are monomials. (Sec. 8.1)
positive integers The integers to the right of zero on the number line; also called natural numbers. (Sec. 2.1)
positive numbers Numbers greater than zero. (Sec. 2.1)
positive slope A property of a line that slants upward to the right. (Sec. 7.2)
prime factorization The expression of a number as the product of its prime factors. (Sec. 1.4)
prime number A number whose only whole-number factors are 1 and itself. For instance, 13 is a prime number. (Sec. 1.4)
principal square root The positive square root of a number. (Sec. 11.1)
product In multiplication, the result of multiplying two numbers. (Sec. 1.2)
proper fraction A fraction less than 1. (Sec. 3.1)
proportion An equation that states the equality of two ratios or rates. (Sec. 10.4)

Pythagorean Theorem The square of the hypotenuse of a right triangle is equal to the sum of the squares of the two legs. (Sec. 11.3)

quadrant One of the four regions into which the two axes of a rectangular coordinate system divide the plane. (Sec. 7.1)

quadratic equation An equation of the form $ax^2 + bx + c = 0$, where a is not equal to zero; also called a second-degree equation. (Sec. 12.1)

quadratic equation in two variables An equation of the form $y = ax^2 + bx + c$, where a is not equal to zero. (Sec. 12.4)

quadrilateral A four-sided polygon. (Sec. 5.2)

quotient In division, the result of dividing two numbers. (Sec. 1.2)

radical sign The symbol $\sqrt{\ }$, which is used to indicate the positive, or principal, square root of a number. (Sec. 11.1)

radical equation An equation that contains a variable expression in a radicand. (Sec. 11.3)

radicand In a radical expression, the expression under the radical sign. (Sec. 11.1)

radius of a circle A line segment going from the center to a point on the circle. (Sec. 5.2)

radius of a sphere A line segment going from the center to a point on the sphere. (Sec. 5.3)

rate The quotient of two quantities that have different units. (Sec. 10.4)

rate of work That part of a task that is completed in one unit of time. (Sec. 10.5)

ratio The quotient of two quantities that have the same unit. (Sec. 10.4)

rational number A number that can be written in the form a/b, where a and b are integers and b is not equal to zero. (Sec. 4.4)

rationalizing the denominator The procedure used to remove a radical from the denominator of a fraction. (Sec. 11.2)

ray A ray starts at a point and extends indefinitely in one direction. (Sec. 5.2)

real numbers The rational numbers and the irrational numbers. (Sec. 4.4)

reciprocal of a fraction The fraction with the numerator and denominator interchanged. (Sec. 3.2)

rectangle A paralellogram that has four right angles. (Sec. 5.2)

rectangular coordinate system System formed by two number lines, one horizontal and one vertical, that intersect at the zero point of each line. (Sec. 7.1)

rectangular solid A solid in which all six faces are rectangles. (Sec. 5.3)

regular polygon A polygon in which each side has the same length and each angle has the same measure. (Sec. 5.2)

remainder In division, the quantity left over when it is not possible to separate objects or numbers into a whole number of even groups. (Sec. 1.2)

repeating decimal Decimal that is formed when dividing the numerator of its fractional counterpart by the denominator results in a decimal part wherein a block of digits repeat infinitely. (Sec. 4.3)

right angle A 90° angle. (Sec. 5.2)

right triangle A triangle that contains one right angle. (Sec. 5.2)

right triangle 30°-60°-90° A special right triangle in which the length of the leg opposite the 30° angle is one-half the length of the hypotenuse. (Sec. 11.3)

right triangle 45°-45°-90° A special right triangle in which the sides opposite the 45° angles are equal. (Sec. 11.3)

rounding Giving an approximate value of an exact number. (Sec. 1.3)

scalene triangle A triangle that has no sides of equal length; no two of its angles are of equal measure. (Sec. 5.2)

scatter diagram A graph of collected data as points in a coordinate system. (Sec. 7.3)

scientific notation Notation in which a number is expressed as the product of two factors, one a number between 1 and 10, and the other a power of 10. (Sec. 8.4)

second coordinate The second number in an ordered pair. It measures a vertical distance and is also called the ordinate. (Sec. 7.1)

second-degree equation An equation of the form $ax^2 + bx + c = 0$, where a is not equal to zero; also called a quadratic equation. (Sec. 12.1)

similar objects Objects that have the same shape but not necessarily the same size. (Sec. 10.4)

simplifying a variable expression Combining like terms by adding their numerical coefficients. (Sec. 5.1)

slope of a line A measure of the slant of a line. The symbol for slope is m. The formula for the slope is

$$m = \frac{y_2 - y_1}{x_2 - x_1}$$

where (x_1, y_1) and (x_2, y_2) are the coordinates of two points on the line and $x_1 \neq x_2$. (Sec. 7.2)

slope–intercept form The slope–intercept form of the equation of a straight line is $y = mx + b$. (Sec. 7.2)

solid An object in space. (Sec. 5.3)

solution of an equation A number that, when substituted for the variable, results in a true equation. (Sec. 6.1)

solution of an equation in two variables An ordered pair whose coordinates make the equation a true statement. (Sec. 7.1)

solution of a system of equations in two variables An ordered pair that is a solution of each equation of the system. (Sec. 7.4)

solution set of an inequality A set of numbers, each element of which, when substituted for the variable, results in a true inequality. (Sec. 6.6)

solving an equation Finding a solution of the equation. (Sec. 6.1)

sphere A solid in which all points are the same distance from point O, which is called the center of the sphere. (Sec. 5.3)

square A rectangle that has four equal sides. (Sec. 5.2)

square root A square root of a positive number x is a number a for which $a^2 = x$. (Sec. 11.1)

standard form of a quadratic equation A quadratic equation is in standard form when the polynomial is in descending order and equal to zero. $ax^2 + bx + c = 0$ is in standard form. (Sec. 12.1)

standard form of a whole number A whole number is in standard form when it is written using the digits 0, 1, 2, . . . , 9. An example is 46,208. (Sec. 1.1)

straight angle A 180° angle. (Sec. 5.2)

substitution method Method of finding an exact solution of a system of linear equations wherein we use the Substitution Property of Equality. (Sec. 7.4)

subtraction The process of finding the difference between two numbers. (Sec. 1.1)

subtrahend In subtraction, the number that is subtracted from another number (the minuend). (Sec. 1.1)

sum In addition, the total of the numbers added. (Sec. 1.1)

supplementary angles Two angles whose measures have the sum 180°. (Sec. 5.2)

surface area The total area on the surface of a solid. (Sec. 5.3)

system of equations Equations that are considered together. (Sec. 7.4)

terminating decimal Decimal that is formed when dividing the numerator of its fractional counterpart by the denominator results in a remainder of zero. (Sec. 4.3)

terms of a variable expression The addends of the expression. (Sec. 5.1)

transversal A line intersecting two other lines at two different points. (Sec. 6.4)

triangle A three-sided polygon. (Sec. 5.2)

trinomial A polynomial of three terms. (Sec. 8.1)

undefined slope A property of a vertical line. (Sec. 7.2)

uniform motion The motion of a moving object whose speed and direction do not change. (Sec. 6.5)

unit rate A rate in which the number in the denominator is 1. (Sec. 10.4)

units In the quantity 3 feet, feet are the units in which the measurement is made. (Sec. 10.4)

variable A letter of the alphabet used to stand for a quantity that is unknown or can change. (Sec. 1.1)

variable expression An expression that contains one or more variables. (Sec. 1.1, 5.1)

variable part In a variable term, the variable or variables and their exponents. (Sec. 5.1)

variable term A term composed of a numerical coefficient and a variable part. (Sec. 5.1)

vertex of an angle The common endpoint of the two rays that form the angle. (Sec. 5.2)

vertex of a parabola The lowest point on a parabola that opens up; the highest point on a parabola that opens down. (Sec. 11.4)

vertical angles Two angles that are on opposite sides of the intersection of two lines. (Sec. 6.4)

volume A measure of the amount of space inside a closed surface. (Sec. 5.3)

whole numbers The whole numbers are 0, 1, 2, 3, ... (Sec. 1.1)

whole-number part In decimal notation, that part of the number that appears before the decimal point. (Sec. 4.1)

x-intercept The point at which a graph crosses the x-axis. (Sec. 7.1)

y-intercept The point at which a graph crosses the y-axis. (Sec. 7.1)

zero slope A property of a horizontal line. (Sec. 7.2)

INDEX